ENGINEERING MECHANICS

ENGINEERING MECHANICS

(In S.I. Units)

For

B.E./B. Tech. 1st YEAR

[ANNA UNIVERSITY, TAMIL NADU]

By

Dr. R.K. BANSAL

B.Sc. Engg. (Mech.), M. Tech., Hons. (I.I.T., Delhi),
Ph.D., M.I.E. (India)

Formerly,
Professor in Mechanical Engineering
Department of Mechanical Engineering
Delhi College of Engineering
Delhi

Presently,
Dean (Academics)
Northern India
Engineering College,
New Delhi

AND

SANJAY BANSAL

B.E. (Computer)

LAXMI PUBLICATIONS (P) LTD

(An ISO 9001:2008 Company)

BENGALURU • CHENNAI • GUWAHATI • HYDERABAD • JALANDHAR
KOCHI • KOLKATA • LUCKNOW • MUMBAI • RANCHI • NEW DELHI
BOSTON (USA) • NAIROBI (KENYA)

ENGINEERING MECHANICS

Printed and bound in India
Typeset at Excellent Graphics, Delhi
First Edition : 2005; Second Edition : 2007; Reprint : 2008, 2009, 2010, 2011, Third Edition : 2013
Fourth Edition : 2014, New Edition : 2016
ISBN 978-81-318-0078-2

Branches

✆	Bengaluru	080-26 75 69 30	
✆	Chennai	044-24 34 47 26,	24 35 95 07
✆	Cochin	0484-237 70 04,	405 13 03
✆	Guwahati	0361-254 36 69,	251 38 81
✆	Hyderabad	040-27 55 53 83,	27 55 53 93
✆	Jalandhar	0181-222 12 72	
✆	Kolkata	033-22 27 43 84	
✆	Lucknow	0522-220 99 16	
✆	Mumbai	022-24 93 12 61	
✆	Ranchi	0651-220 44 64	

PUBLISHED IN INDIA BY

Laxmi Publications (P) Ltd.
(An ISO 9001:2008 Company)
113, GOLDEN HOUSE, DARYAGANJ,
NEW DELHI - 110002, INDIA
Telephone : 91-11-4353 2500, 4353 2501
Fax : 91-11-2325 2572, 4353 2528
www.laxmipublications.com info@laxmipublications.com

C—
Printed at: Repro Knowledgecast Limited, Thane

Contents

Chapters *Pages*

Preface to the Fourth Edition

Authors are glad to present the Fourth Edition of the book entitled, 'Engineering Mechanics' to the engineering students of all disciples of Anna University, Tamil Nadu. The course contents are planned in such a way that the book covers the complete course of first year students of Anna University, Tamil Nadu according to the revised syllabus.

The Fourth Edition contains the following five units :

Unit I : Basics and Statics of Particles.

Unit II : Equilibrium of Rigid Bodies.

Unit III : Properties of Surfaces and Solids.

Unit IV : Dynamics of Particles.

Unit V : Friction and Elements of Rigid Body Dynamics.

The Additions or more explanations are given in the following units :

1. Theorems of Pappus have been added in Unit III.
2. Work Energy Equation, Impulse and Momentum have been explained in more details in Unit IV.
3. Wedge Friction has been added in Unit V. Also more details are given for general plane motion in Unit V.

The book is written in a simple and easy-to-follow language, so that even an average student can grasp the subject by self-study. At the end of each chapter highlights, theoretical questions and many unsolved numerical problems with answers are given for the students to solve them.

Mrs. Nirmal Bansal deserves special credit as she not only provided an ideal atmosphere at home for book writing but also gave inspiration and valuable suggestions.

Though every care has been taken in checking the manuscript and proofreading, yet claiming perfection is very difficult. We shall be very grateful to the readers and users of this book for pointing any mistake that might have crept in. Suggestions for improvement are most welcome and would be incorporated in the next edition with a view to make the book more useful.

—Authors

SYLLABUS

(Anna University, Tamil Nadu)

GE6253 **ENGINEERING MECHANICS**

L	T	P	C
3	1	0	4

OBJECTIVES

• To develop capacity to predict the effect of force and motion in the course of carrying out the design functions of engineering.

UNIT I : BASICS AND STATICS OF PARTICLES 12

Introduction—Units and Dimensions—Laws of Mechanics—Lame's theorem, Parallelogram and triangular Law of forces—Vectorial representation of forces—Vector operations of forces-additions, subtractions, dot product, cross product—Coplanar Forces—Rectangular components—Equilibrium of a particle—Forces in space—Equilibrium of a particle in space—Equivalent systems of forces—Principle of transmissibility.

UNIT II : EQUILIBRIUM OF RIGID BODIES 12

Free body diagram—Types of supports—Action and reaction forces—Stable equilibrium—Moments and Couples—Moment of a force about a point and about an axis—Vectorial representation of moments and couples—Scalar components of a moment—Varignon's theorem—Single equivalent force—Equilibrium of Rigid bodies in two dimensions—Equilibrium of Rigid bodies in three dimensions.

UNIT III : PROPERTIES OF SURFACES AND SOLIDS 12

Centroids and centre of mass—Centroids of lines and areas—Rectangular, circular, triangular areas by integration—T-section, I-section, Angle-section, Hollow-section by using standard formula—Theorems of Pappus—Area moments of inertia of plane areas—Rectangular, circular, triangular areas by integration—T-section, I-section, Angle-section, Hollow-section by using standard formula—Parallel axis theorem and perpendicular axis theorem—Principal moments of inertia of plane areas—Principal axes of inertia—Mass moment of inertia—Mass moment of inertia for prismatic, cylindrical and spherical solids from first principle—Relation to area moments of inertia.

UNIT IV : DYNAMICS OF PARTICLES 12

Displacements, Velocity and acceleration, their relationship—Relative motion—Curvilinear motion—Newton's law of motion—Work energy equation—Impulse and Momentum—Impact of elastic bodies.

UNIT V : FRICTION AND ELEMENTS OF RIGID BODY DYNAMICS 12

Friction force—Laws of sliding friction—Equilibrium analysis of simple systems with sliding friction—Wedge friction—Rolling resistance—Translation and Rotation of Rigid Bodies—Velocity and acceleration—General Plane motion of simple rigid bodies such as cylinder, disc/wheel and sphere.

Total : 60 Periods

1

Basics and Statics of Particles

1.1. INTRODUCTION

Engineering mechanics is that branch of science which deals with the behaviour of a body when the body is at rest or in motion. The engineering mechanics may be divided into statics and dynamics. The branch of science, which deals with the study of a body when the body is at rest, is known as *statics* while the branch of science which deals with the study of a body when the body is in motion, is known as *dynamics*. Dynamics is further divided into kinematics and kinetics. The study of a body in motion, when the forces which cause the motion are not considered, is called *kinematics* and if the forces are also considered for the body in motion, that branch of science is called *kinetics*. The classification of Engineering Mechanics are shown in Fig. 1.1.

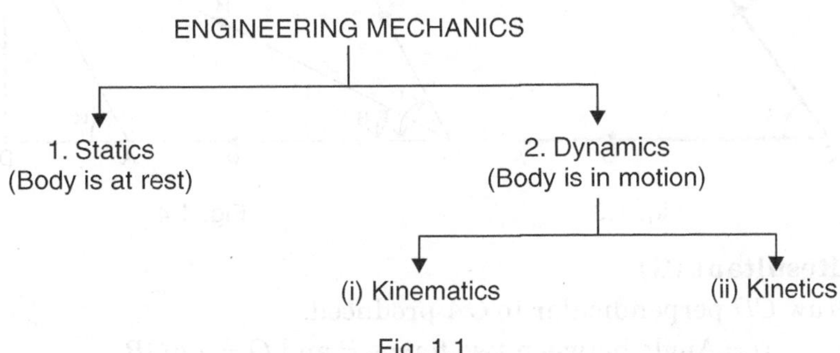

Fig. 1.1

Note. Statics deals with equilibrium of bodies at rest, whereas dynamics deals with the motion of bodies and the forces that cause them.

The following terms are generally used in Mechanics :

1. Vector quantity, 2. Scalar quantity, 3. Particle, 4. Law of parallelogram of forces, 5. Triangle law of forces, and 6. Lame's theorem.

1.1.1. Vector Quantity. A quantity which is completely specified by magnitude and direction, is known as a vector quantity. Some examples of vector quantities are : velocity, acceleration, force and momentum. A vector quantity is represented by means of a straight line with an arrow as shown in Fig. 1.2. The length of the straight line (*i.e.*, *AB*) represents the magnitude and arrow represents the direction of the vector. The symbol $\overrightarrow{AB}$ also represents this vector, which means it is acting from *A* to *B*.

A ——→ B

Fig. 1.2. Vector quantity

1.1.2. Scalar Quantity. A quantity, which is completely specified by magnitude only, is known as a scalar quantity. Some examples of scalar quantity are : mass, length, time and temperature.

1.1.3. A Particle. A particle is a body of infinitely small volume (or a particle is a body of negligible dimensions) and the mass of the particle is considered to be concentrated at a point. Hence a particle is assumed to a point and the mass of the particle is concentrated at this point.

1.1.4. Law of Parallelogram of Forces. The law of parallelogram of forces is used to determine the resultant* of two forces acting at a point in a plane. It states, "If two forces, acting at a point be represented in magnitude and direction by the two adjacent sides of a parallelogram, then their resultant is represented in magnitude and direction by the diagonal of the parallelogram passing through that point."

Let two forces P and Q act at a point O as shown in Fig. 1.3. The force P is represented in magnitude and direction by OA whereas the force Q is presented in magnitude and direction by OB. Let the angle between the two forces be 'α'. The resultant of these two forces will be obtained in magnitude and direction by the diagonal (passing through O) of the parallelogram of which OA and OB are two adjacent sides. Hence draw the parallelogram with OA and OB as adjacent sides as shown in Fig. 1.4. The resultant R is represented by OC in magnitude and direction.

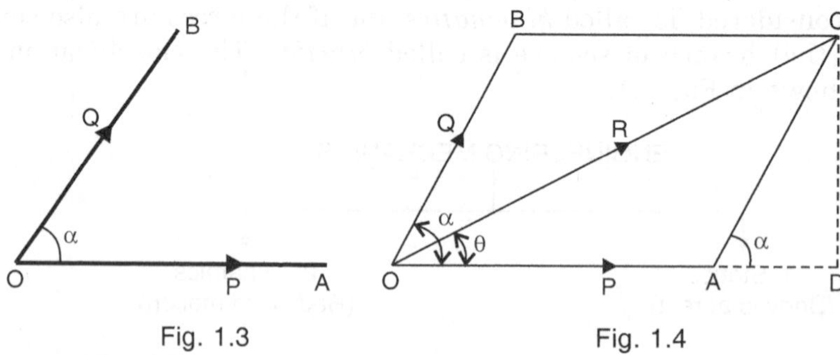

Fig. 1.3 Fig. 1.4

Magnitude of Resultant (R)

From C draw CD perpendicular to OA produced.

Let α = Angle between two forces P and $Q = \angle AOB$

Now $\angle DAC = \angle AOB$ (Corresponding angles)

$= \alpha$

In parallelogram $OACB$, AC is parallel and equal to OB.

$\therefore \quad AC = Q.$

In triangle ACD,

$$AD = AC \cos \alpha = Q \cos \alpha$$

and $$CD = AC \sin \alpha = Q \sin \alpha.$$

In triangle OCD,

$$OC^2 = OD^2 + DC^2.$$

*The resultant of a system of forces may be defined as a single force which has the same effect as system of forces acting on the body.

But $OC = R, OD = OA + AD = P + Q \cos \alpha$

and $DC = Q \sin \alpha.$

$$\therefore \quad R^2 = (P + Q \cos \alpha)^2 + (Q \sin \alpha)^2 = P^2 + Q^2 \cos^2 \alpha + 2PQ \cos \alpha + Q^2 \sin^2 \alpha$$

$$= P^2 + Q^2 (\cos^2 \alpha + \sin^2 \alpha) + 2PQ \cos \alpha$$

$$= P^2 + Q^2 + 2PQ \cos \alpha \qquad (\because \cos^2 \alpha + \sin^2 \alpha = 1)$$

$$\therefore \quad R = \sqrt{P^2 + Q^2 + 2PQ \cos \alpha} \qquad ...(1.1)$$

Equation (1.1) gives the magnitude of resultant force R.

Direction of Resultant

Let θ = Angle made by resultant with OA.

Then from triangle OCD,

$$\tan \theta = \frac{CD}{OD} = \frac{Q \sin \alpha}{P + Q \cos \alpha}$$

$$\therefore \quad \theta = \tan^{-1}\left(\frac{Q \sin \alpha}{P + Q \cos \alpha}\right) \qquad ...(1.2)$$

Equation (1.2) gives the direction of resultant (R).

The direction of resultant can also be obtained by using sine rule [In triangle OAC, $OA = P$, $AC = Q$, $OC = R$, angle $OAC = (180 - \alpha)$, angle $ACO = 180 - [\theta + 180 - \alpha] = (\alpha - \theta)$]

$$\frac{\sin \theta}{AC} = \frac{\sin (180 - \alpha)}{OC} = \frac{\sin (\alpha - \theta)}{OA}$$

$$\frac{\sin \theta}{Q} = \frac{\sin (180 - \alpha)}{R} = \frac{\sin (\alpha - \theta)}{P}$$

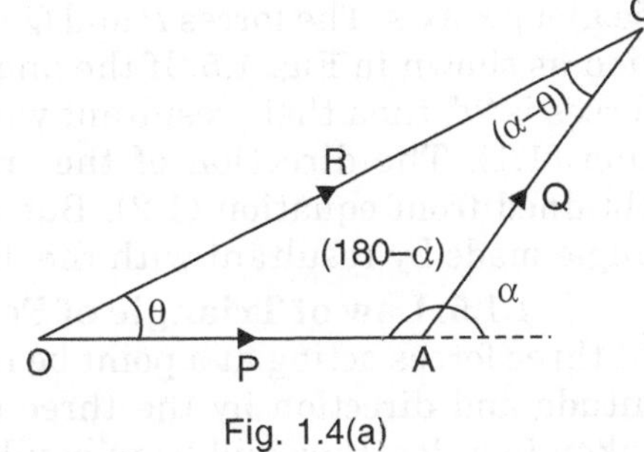

Fig. 1.4(a)

Two cases are important.

1st Case. If the two forces P and Q act at right angles, then

$\alpha = 90°$

From equation (1.1), we get the magnitude of resultant as

$$R = \sqrt{P^2 + Q^2 + 2PQ \cos \alpha} = \sqrt{P^2 + Q^2 + 2PQ \cos 90°}$$

$$= \sqrt{P^2 + Q^2} \qquad (\because \cos 90° = 0) \quad ...(1.2A)$$

From equation (1.2), the direction of resultant is obtained as

$$\theta = \tan^{-1}\left(\frac{Q \sin \alpha}{P + Q \cos \alpha}\right)$$

$$= \tan^{-1}\left(\frac{Q \sin 90°}{P + Q \cos 90°}\right) = \tan^{-1}\frac{Q}{P} \qquad (\because \sin 90° = 1 \text{ and } \cos 90° = 0)$$

2nd Case. The two forces P and Q are equal and are acting at an angle α between them. Then the magnitude and direction of resultant is given as

$$R = \sqrt{P^2 + Q^2 + 2PQ \cos \alpha} = \sqrt{P^2 + P^2 + 2P \times P \times \cos \alpha} \qquad (\because P = Q)$$

$$= \sqrt{2P^2 + 2P^2 \cos \alpha} = \sqrt{2P^2 (1 + \cos \alpha)}$$

$$= \sqrt{2P^2 \times 2\cos^2 \frac{\alpha}{2}} \qquad \left(\because \; 1 + \cos\alpha = 2\cos^2 \frac{\alpha}{2}\right)$$

$$= \sqrt{4P^2 \cos^2 \frac{\alpha}{2}} = 2P \cos \frac{\alpha}{2} \qquad \text{...(1.3)}$$

and

$$\theta = \tan^{-1}\left(\frac{Q\sin\alpha}{P + Q\cos\alpha}\right) = \tan^{-1}\frac{P\sin\alpha}{P + P\cos\alpha} \qquad (\because \; P = Q)$$

$$= \tan^{-1}\frac{P\sin\alpha}{P(1+\cos\alpha)} = \tan^{-1}\frac{\sin\alpha}{1+\cos\alpha}$$

$$= \tan^{-1}\frac{2\sin\frac{\alpha}{2}\cos\frac{\alpha}{2}}{2\cos^2\frac{\alpha}{2}} \qquad \left(\because \; \sin\alpha = 2\sin\frac{\alpha}{2}\cos\frac{\alpha}{2}\right)$$

$$= \tan^{-1}\frac{\sin\frac{\alpha}{2}}{\cos\frac{\alpha}{2}} = \tan^{-1}\left(\tan\frac{\alpha}{2}\right) = \frac{\alpha}{2} \qquad \text{...(1.4)}$$

It is not necessary that one of two forces, should be along x-axis. The forces P and Q may be in any direction as shown in Fig. 1.5. If the angle between the two forces is 'α', then their resultant will be given by equation (1.1). The direction of the resultant would be obtained from equation (1.2). But angle θ will be the angle made by resultant with the direction of P.

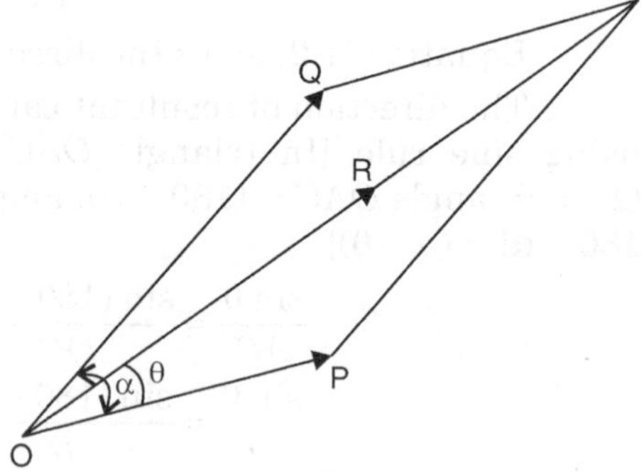

Fig. 1.5

1.1.5. Law of Triangle of Forces. It states that, "if three forces acting at a point be represented in magnitude and direction by the three sides of a triangle, taken in order, they will be in equilibrium."

1.1.6. Lame's Theorem. It states that, "If there forces acting at a point are in equilibrium, each force will be proportional to the sine of the angle between the other two forces."

Suppose the three forces P, Q and R are acting at a point O and they are in equilibrium as shown in Fig. 1.6.

Let α = Angle between force P and Q.

β = Angle between force Q and R.

γ = Angle between force R and P.

Then according to Lame's theorem,

P is proportional sine of angle between Q and R α $\sin\beta$.

Fig. 1.6

$$\therefore \qquad \frac{P}{\sin\beta} = \text{constant}$$

Similarly

$$\frac{Q}{\sin\gamma} = \text{constant and } \frac{R}{\sin\alpha} = \text{constant}$$

or

$$\frac{P}{\sin\beta} = \frac{Q}{\sin\gamma} = \frac{R}{\sin\alpha} \qquad \text{...(1.5)}$$

Proof of Lame's Theorem. The three forces acting on a point, are in equilibrium and hence they can be represented by the three sides of the triangle taken in the same order. Now draw the force triangle as shown in Fig. 1.6 (*a*).

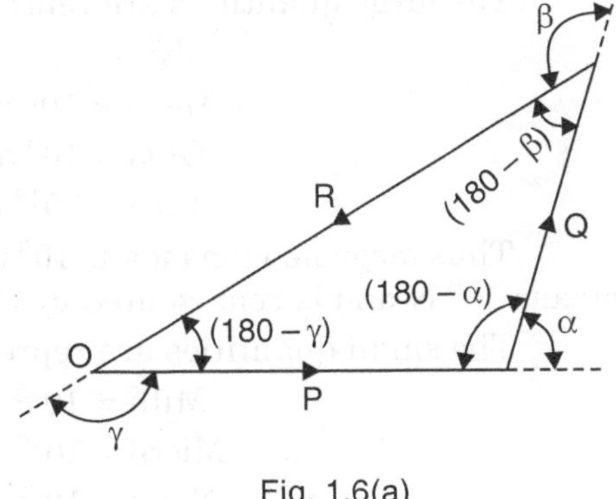

Fig. 1.6(a)

Now applying sine rule, we get

$$\frac{P}{\sin(180-\beta)}=\frac{Q}{\sin(180-\gamma)}=\frac{R}{\sin(180-\alpha)}$$

This can also be written

$$\frac{P}{\sin\beta}=\frac{Q}{\sin\gamma}=\frac{R}{\sin\alpha}$$

This is same equation as equation (1.5).

Note. All the three forces should be acting either towards the point or away from the point.

1.2. UNITS AND DIMENSIONS

The following units of different systems are mostly used :

1. C.G.S. (*i.e.*, Centimetre-Gram Second) system of units.
2. M.K.S. (*i.e.*, Metre-Kilogram-Second) system of units.
3. S.I. (*i.e.*, International) system of units.

1.2.1. C.G.S. System of Units. In this system, length is expressed in centimetre, mass in gram and time in second. The unit of force in this system is *dyne*, which is defined as the force acting on a mass of one gram and producing an acceleration of one centimetre per second square.

1.2.2. M.K.S. System of Units. In this system, length is expressed in metre, mass in kilogram and time in second. The unit of force in this system is expressed as kilogram force and is represented as kgf.

1.2.3. S.I. System of Units. S.I. is abbreviation for 'The System International Units'. It is also called the International System of Units. In this system length is expressed in metre mass in kilogram and time in second. The unit of force in this system is Newton and is represented N. Newton is the force acting on a mass of one kilogram and producing an acceleration of one metre per second square. The relation between newton (N) and dyne is obtained as

$$\text{One Newton} = \text{One kilogram mass} \times \frac{\text{One metre}}{\text{second}^2}$$

$$= 1000 \text{ g} \times \frac{100 \text{ cm}}{\text{s}^2} \qquad (\because \text{ one kg} = 1000 \text{ g})$$

$$= 1000 \times 100 \times \frac{\text{g} \times \text{cm}}{\text{s}^2}$$

$$= 10^5 \text{ dyne} \qquad \left\{\because \text{ dyne} = \frac{\text{g}\times\text{cm}}{\text{s}^2}\right\}$$

When the magnitude of forces is very large, then the unit of force like kilo-newton and mega-newton is used. Kilo-newton is represented by kN.

One kilo-newton = 10^3 newton

or 1 kN = 10^3 N

and One mega newton = 10^6 Newton

The large quantities are represented by kilo, mega, giga and tera. They stand for :

Kilo = 10^3 and represented by ....k

Mega = 10^6 and represented by ....M

Giga = 10^9 and represented by ....G

Tera = 10^{12} and represented by ....T

Thus mega newton means 10^6 newton and is represented by MN. Similarly, giga newton means 10^9 N and is represented by GN. The symbol TN stands for 10^{12} N.

The small quantities are represented by milli, micro, nano and pico. They are equal to

Milli = 10^{-3} and represented by ....m

Micro = 10^{-6} and represented by ....μ

Nano = 10^{-9} and represented by ....n

Pico = 10^{-12} and represented by ....p.

Thus milli newton means 10^{-3} newton and is represented by mN. Micro newton means 10^{-6} N and is represented by μN.

The relation between kilogram force (kgf) and newton (N) is given by One kgf = 9.81 N

Weight of a body is the force with which the body is attracted towards earth. If W = weight of a body, m = mass in kg, then $W = m \times g$ Newtons

If mass, m of the body is 1 kg, then its weight will be,

$$W = 1 \text{ (kg)} \times 9.81 \frac{\text{m}}{\text{s}^2} = 9.81 \text{ N.} \qquad \left(\because \text{ N} = \text{kg} \frac{\text{m}}{\text{s}^2}\right)$$

Table 1.1 shows the multiples and sub-multiples of the S.I. units prefixes.

Table 1.1. S.I. Prefixes

Prefix	*Symbol*	*Multiplying factor*
Tera	T	10^{12} = 1 000 000 000 000
Giga	G	10^9 = 1 000 000 000
Mega	M	10^6 = 1 000 000
Kilo	k	10^3 = 1 000
Milli	m	10^{-3} = 0.001
Micro	μ	10^{-6} = 0.000 001
Nano	n	10^{-9} = 0.000 000 001
Pico	p	10^{-12} = 0.000 000 000 001

The basic, supplementary and some derived units with dimensions are given as :

(A) Basic Units

Physical quantity	*Notation or Unit*	*Dimension or Symbol*
Length	metre	m
Mass	kilogram	kg
Time	Second	s
Electric Current	Ampere	A
Temperature	Kelvin	K
Luminous Intensity	Candella	cd

(B) Supplementary Units

Plane angle	Radian	rad
Solid angle	Steridian	sr

(C) Derived Units

Acceleration	metre/second^2	m/s^2
Angualr velocity	radian/second	rad/s
Angular acceleration	radian/second^2	rad/s^2
Force	Newton	N = kg m/s^2
Moment of force	Newton metre	Nm
Work, Energy	Joule	J = Nm = kg m^2/s^2
Torque	Newton metre	Nm
Power	Watt	W = J/s
Pressure	Pascal	Pa = N/m^2
Frequency	Hertz	Hz = s^{-1}

Problem 1.1. *Two forces of magnitude 10 N and 8 N are acting at a point. If the angle between the two forces is 60°, determine the magnitude of the resultant force.*

Sol. Given :

Force $P = 10$ N

Force $Q = 8$ N

Angle between the two forces, $\alpha = 60°$

The magnitude of the resultant force (R) is given by equation (1.1)

$$R = \sqrt{P^2 + Q^2 + 2PQ\cos\alpha} = \sqrt{10^2 + 8^2 + 2\times 10\times 8\times\cos 60°}$$

$$= \sqrt{100 + 64 + 2\times 10\times 8\times \tfrac{1}{2}} \qquad (\because \cos 60° = \tfrac{1}{2})$$

$$= \sqrt{100 + 64 + 80} = \sqrt{244} = \mathbf{15.62\ N.}\quad \textbf{Ans.}$$

Problem 1.2. *Two equal forces are acting at a point with an angle of 60° between them. If the resultant force is equal to 20 × $\sqrt{3}$ N, find magnitude of each force.*

Sol. Given : Angle between the force, $\alpha = 60°$

Resultant, $R = 20\times\sqrt{3}$

The forces are equal. Let P is the magnitude of each force.

Using equation (1.3), we have

$$R = 2P\cos\frac{\alpha}{2} \quad\text{or}\quad 20\times\sqrt{3} = 2P\times\cos\left(\frac{60°}{2}\right) = 2P\cos 30°$$

$$= 2P\times\frac{\sqrt{3}}{2} = P\times\sqrt{3} \qquad \left(\because \cos 30° = \frac{\sqrt{3}}{2}\right)$$

$$\therefore \quad P = \frac{20\times\sqrt{3}}{\sqrt{3}} = 20\text{ N}$$

∴ Magnitude of each force = **20 N. Ans.**

Problem 1.3. *The resultant of the two forces, when they act at an angle of 60° is 14 N. If the same forces are acting at right angles, their resultant is $\sqrt{136}$ N. Determine the magnitude of the two forces.*

Sol. Given :

Case I

Resultant, $R_1 = 14$ N

Angle, $\alpha = 60°$

Case II

Resultant, $R_2 = \sqrt{136}$ N

Angle, $\alpha = 90°$

Let the magnitude of the two forces are P and Q.

Using equation (1.1) for case I.

$$R = \sqrt{P^2 + Q^2 + 2PQ \cos \alpha}$$

or $$14 = \sqrt{P^2 + Q^2 + 2PQ \times \cos 60°} = \sqrt{P^2 + Q^2 + 2PQ \times \tfrac{1}{2}}$$

or $$14 = \sqrt{P^2 + Q^2 + PQ}$$

Squaring, $196 = P^2 + Q^2 + PQ$ ...(*i*)

Using equation (1.2A) for case II,

$$R = \sqrt{P^2 + Q^2} \quad \text{or} \quad \sqrt{136} = \sqrt{P^2 + Q^2}$$

or $136 = P^2 + Q^2$ (Squaring both sides) ...(*ii*)

Subtracting equation (*ii*) from equation (*i*), we get

$$196 - 136 = P^2 + Q^2 + PQ - (P^2 + Q^2)$$

or $60 = PQ$ ...(*iii*)

Multiplying the above equation by two, we get $120 = 2PQ$ ...(*iv*)

Adding equation (*iv*) to equation (*ii*), we get $136 + 120 = P^2 + Q^2 + 2PQ$

or $256 = P^2 + Q^2 + 2PQ$ or $(16)^2 = (P + Q)^2$

or $16 = P + Q$

$\therefore$ $P = (16 - Q)$ ...(*v*)

Substituting the value of P in equation (*iii*), we get

$$60 = (16 - Q) \times Q = 16Q - Q^2 \quad \text{or} \quad Q^2 - 16Q + 60 = 0$$

$\therefore$ This is a quadratic equation

$$\therefore \quad Q = \frac{16 \pm \sqrt{(-16)^2 - 4 \times 60}}{2} = \frac{16 \pm \sqrt{256 - 240}}{2} = \frac{16 \pm 4}{4}$$

$$= \frac{16 + 4}{2} \text{ and } \frac{16 - 4}{2} = 10 \text{ and } 6$$

Substituting the value of Q in equation (*v*), we get

$$P = (16 - 10) \text{ or } (16 - 6) = 6 \text{ or } 10$$

$\therefore$ Hence the two forces are **10 N** and **6 N**. **Ans.**

Problem 1.4. *Two forces are acting at a point O as shown in Fig. 1.7. Determine the resultant in magnitude and direction.*

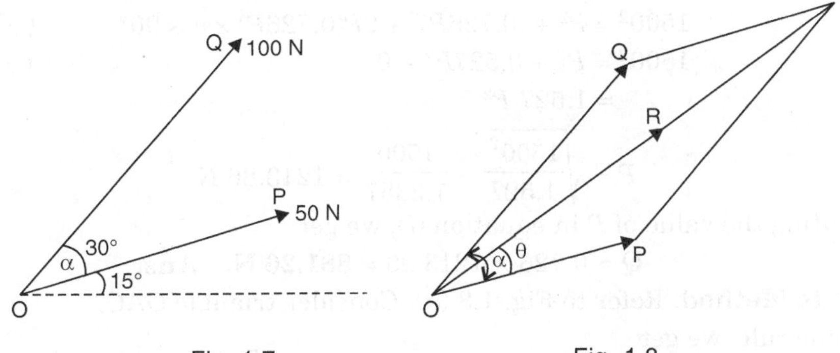

Fig. 1.7 Fig. 1.8

Sol. Given :

Force $P = 50$ N, Force $Q = 100$ N

Angle between the two forces, $\alpha = 30°$

The magnitude of the resultant R is given by equation (1.1) as

$$R = \sqrt{P^2 + Q^2 + 2PQ\cos\alpha} = \sqrt{50^2 + 100^2 + 2 \times 50 \times 100 \times \cos 30°}$$

$$= \sqrt{2500 + 10000 + 8660} = \sqrt{21160} = \textbf{145.46 N.} \quad \textbf{Ans.}$$

The resultant R is shown in Fig. 1.8.

The angle made by the resultant with the direction of P is given by equation (1.2) as

$$\tan\theta = \frac{Q\sin\alpha}{P + Q\cos\alpha}$$

or $$\theta = \tan^{-1}\left(\frac{Q\sin\alpha}{P + Q\cos\alpha}\right) = \tan^{-1}\left(\frac{100 \times \sin 30°}{50 + 100\cos 30°}\right)$$

$$= \tan^{-1} 0.366 = 20.10°$$

$\therefore$ Angle made by resultant with x-axis $= \theta + 15° = 20.10 + 15 =$ **35.10°. Ans.**

Problem 1.5. *The resultant of two concurrent forces is 1500 N and the angle between the forces is 90°. The resultant makes an angle of 36° with one of the force. Find the magnitude of each force.*

Sol. Given :

Resultant, $R = 1500$ N

Angle between the forces, $\alpha = 90°$

Angle made by resultant with one force, $\theta = 36°$

Let P and Q are two forces.

Using equation (1.2), $\tan\theta = \dfrac{Q\sin\alpha}{P + Q\cos\alpha}$

or $$\tan 36° = \frac{Q\sin 90°}{P + Q\cos 90°} = \frac{Q \times 1}{P + Q \times 0} = \frac{Q}{P} \quad \text{or} \quad 0.726 = \frac{Q}{P}$$

or $$Q = 0.726\,P \qquad ...(i)$$

Using equation (1.1), $R = \sqrt{P^2 + Q^2 + 2PQ \cos \alpha}$

or $$R^2 = P^2 + Q^2 + 2PQ \cos \alpha$$

or $$1500^2 = P^2 + (0.726P)^2 + 2P(0.726P) \times \cos 90° \qquad (\because \; Q = 0.726P)$$

or $$1500^2 = P^2 + 0.527P^2 + 0 \qquad (\because \; \cos 90° = 0)$$

$$= 1.527\, P^2$$

$\therefore$ $$P = \sqrt{\frac{1500^2}{1.527}} = \frac{1500}{1.2357} = 1213.86 \text{ N}$$

Substituting the value of P in equation (i), we get

$$Q = 0.726 \times 1213.86 = \mathbf{881.26\ N.}\quad \textbf{Ans.}$$

Alternate Method. Refer to Fig. 1.8 (a). Consider triangle OAC.

Using sine rule, we get

$$\frac{\sin 90°}{R} = \frac{\sin 36°}{Q} = \frac{\sin 54°}{P}$$

or $$\frac{\sin 90°}{R} = \frac{\sin 36°}{Q}$$

or $$Q = \frac{R \sin 36°}{\sin 90°} \quad \text{(where } R = 1500 \text{ N)}$$

$$= \frac{1500 \times 0.5877}{1} = \mathbf{881.67\ N.}\quad \textbf{Ans.}$$

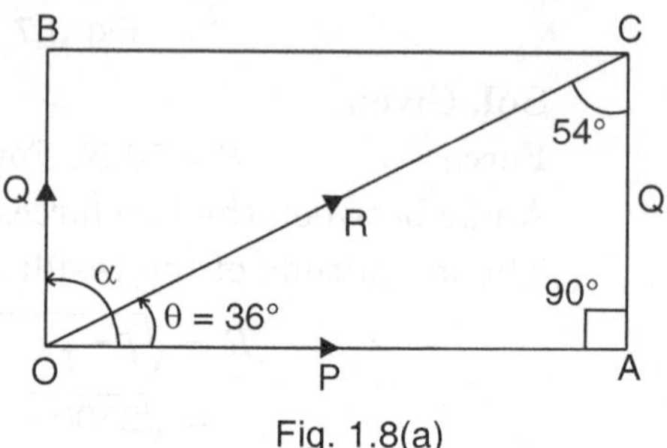

Fig. 1.8(a)

Also, we have $$\frac{\sin 90°}{R} = \frac{\sin 54°}{P}$$

$\therefore$ $$P = \frac{R \sin 54°}{\sin 90°} = \frac{1500 \times 0.8090}{1}$$

$$= \mathbf{1213.52\ N.}\quad \textbf{Ans.}$$

Problem 1.6. *The sum of two concurrent forces P and Q is 270 N and their resultant is 180 N. The angle between the force P and resultant R is 90°. Find the magnitude of each force and angle between them.*

Sol. Given :

Sum of two concurrent forces = 270 N or $P + Q = 270$ N

Resultant, $R = 180$ N

Angle between force P and resultant $R = 90°$

This means $\theta = 90°$

Find : (i) Magnitude of P and Q

(ii) Angle between P and Q ($i.e.$, angle α)

Using equation (1.2), $$\tan \theta = \frac{Q \sin \alpha}{P + Q \cos \alpha}$$

or $$\tan 90° = \frac{Q \sin \alpha}{P + Q \cos \alpha}$$

But $\tan 90° = \infty$ ($i.e.$, infinity). This is only possible when $P + Q \cos \alpha = 0$

$\therefore$ $$P = -\,Q \cos \alpha \qquad \text{...}(i)$$

The above result can also be obtained by using alternate method.

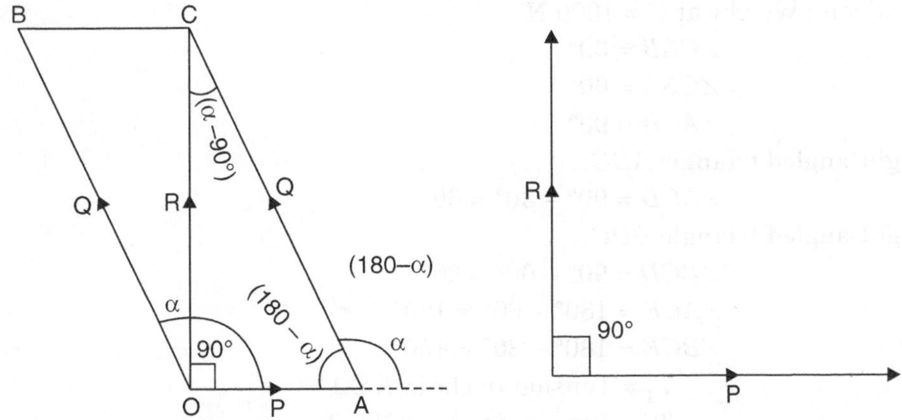

Fig. 1.8(b)

Alternate Method. Refer to Fig. 1.8 (*b*). Consider triangle OAC in which $\theta = 90°$, $\angle OAC = 180 - \alpha$, $\angle ACO = \alpha - \theta = \alpha - 90°$

Using sine rule, we get $\dfrac{\sin 90°}{Q} = \dfrac{\sin (180 - \alpha)}{R} = \dfrac{\sin (\alpha - 90)}{P}$

From first and last terms, we get $\dfrac{\sin 90°}{Q} = \dfrac{\sin (\alpha - 90)}{P}$

or $\dfrac{1}{Q} = \dfrac{-\cos \alpha}{P}$ $\quad [\because \sin (\alpha - 90) = \sin [-(90 - \alpha)] = -\sin (90 - \alpha) = -\cos \alpha]$

$\therefore \quad P = -Q \cos \alpha$

This is the same result as given by equation (*i*) above.

Using equation (1.1), $R = \sqrt{P^2 + Q^2 + 2PQ \cos \alpha}$

Squaring to both sides, we get $R^2 = P^2 + Q^2 + 2PQ \cos \alpha$

$= P^2 + Q^2 + 2P(-P)$

($\because$ From equation (*i*), $Q \cos \alpha = -P$)

$= P^2 + Q^2 - 2P^2 = Q^2 - P^2 = (Q + P)(Q - P)$

or $180^2 = 270\ (Q - P)$ $\quad (\because R = 180, Q + P = 270)$

or $32400 = 270(Q - P)$

$\therefore \quad Q - P = \dfrac{32400}{270} = 120$

But $P + Q = 270$ (given)

Adding the above two equations, we get $2Q = 270 + 120 = 390$

$\therefore$ Q **= 195 N. Ans.**

and $P = 270 - Q = 270 - 195 =$ **75 N. Ans.**

Value of angle α

Substituting the values of P and Q in equation (*i*),

$P = -Q \cos \alpha$ or $75 = -195 \cos \alpha$ or $\cos \alpha = \dfrac{-75}{195} = -0.3846$

$\therefore \quad \alpha = \cos^{-1} (-0.3846) =$ **112.618°. Ans.**

Problem 1.7. *A weight of 1000 N is supported by two chains as shown in Fig. 1.9. Determine the tension in each chain.*

Sol. Given : Weight at C = 1000 N

$$\angle CAB = 30°$$
$$\angle CBA = 60°$$
$$\angle ACB = 90°$$

In right-angled triangle ADC,

$$\angle ACD = 90° - 30° = 60°$$

In right-angled triangle BDC,

$$\angle BCD = 90° - 60° = 30°$$

$\therefore$ $$\angle ACE = 180° - 60° = 120°$$
$$\angle BCE = 180° - 30° = 150°$$

Let T_1 = Tension in chain No. 1

T_2 = Tension in chain No. 2.

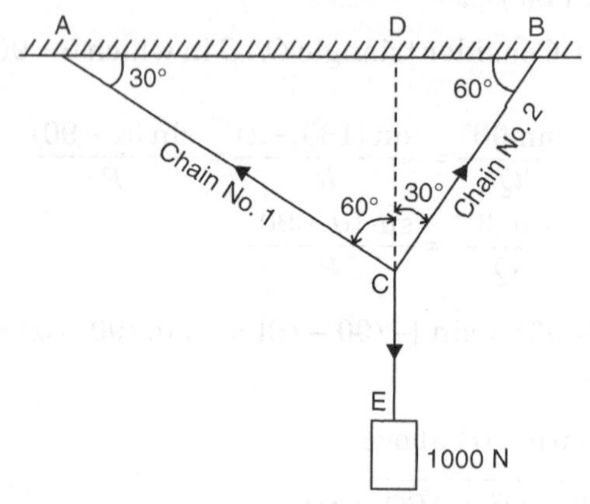

Fig. 1.9

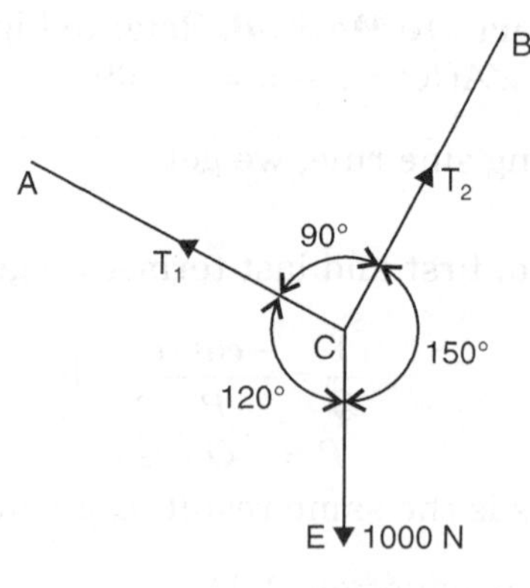

Fig. 1.10

Applying Lame's theorem at point C (Refer to Fig. 1.10).

$$\frac{T_1}{\sin 150°} = \frac{T_2}{\sin 120°} = \frac{1000}{\sin 90°}$$

or
$$\frac{T_1}{\sin 150°} = \frac{T_2}{\sin 120°} = 1000 \qquad (\because \sin 90° = 1)$$

$\therefore$ $T_1 = 1000 \sin 150° = 1000 \times 0.5 =$ **500 N. Ans.**

and $T_2 = 1000 \sin 120° = 1000 \times 0.866 =$ **866 N. Ans.**

Problem 1.8. *A weight of 900 N is supported by two chains of lengths 4 m and 3 m as shown in Fig. 1.11. Determine the tension in each chain.*

Sol. Given : Weight at C = 900 N

Length, AC = 4 m

Length, BC = 3 m

Length, AB = 5 m

In triangle ABC,

$$AC^2 + BC^2 = 4^2 + 3^2 = 16 + 9 = 25$$

$$AB^2 = 5^2 = 25$$

$$\therefore \quad AB^2 = AC^2 + BC^2$$

$\therefore$ Triangle ABC is a right-angled triangle in which

$$\angle ACB = 90°$$

$$\sin \alpha = \frac{BC}{AB} = \frac{3}{5} = 0.6$$

$$\therefore \quad \alpha = 36° \, 52' \quad \text{and} \quad \alpha + \beta = 90°$$

$$\therefore \quad \beta = 90° - \alpha = 90° - (36° \, 52') = 53° \, 8'.$$

Fig. 1.11

Let T_1 = Tension in chain AC

T_2 = Tension in chain BC

In right-angled triangle ADC,

$$\theta_1 = 90° - \alpha = 90° - 36° \, 52' = 53° \, 8'$$

In right-angled triangle BDC,

$$\theta_1 = 90° - \beta = 90° - 53° \, 8' = 36° \, 52'$$

$$\therefore \quad \angle ACE = 180° - \theta_1 = 180° - 53° \, 8' = 126° \, 52'$$

$$\angle BCE = 180° - \theta_2 = 180° - 36° \, 52' = 143° \, 8' \quad \text{and} \quad \angle ACB = 90°.$$

Applying Lame's theorem at C

$$\frac{T_1}{\sin \text{ of } \angle BCE} = \frac{T_2}{\sin \text{ of } \angle ACE} = \frac{900}{\sin 90°}$$

or

$$\frac{T_1}{\sin 143° \, 8'} = \frac{T_2}{\sin 126° \, 52'} = \frac{900}{1} \qquad (\because \ \sin 90° = 1)$$

$$\therefore \quad T_1 = 900 \times \sin 143° \, 8' = 900 \times 0.597 = \mathbf{537.44 \ N.} \quad \textbf{Ans.}$$

$$T_2 = 900 \times \sin 126° \, 52' = \mathbf{720 \ N.} \quad \textbf{Ans.}$$

Problem 1.9. *An electric light fixture weighing 15 N hangs from a point C, by two strings AC and BC. AC is inclined at 60° to the horizontal and BC at 45° to the vertical as shown in Fig. 1.12. Using Lame's theorem or otherwise determine the forces in the strings AC and BC.* (AMIE Summer, 1975)

Sol. Given : Weight at C = 15 N

$$\angle OAC = 60°$$

$$\angle CBD = 45°$$

Let T_1 = Force in string BC

T_2 = Force in string AC

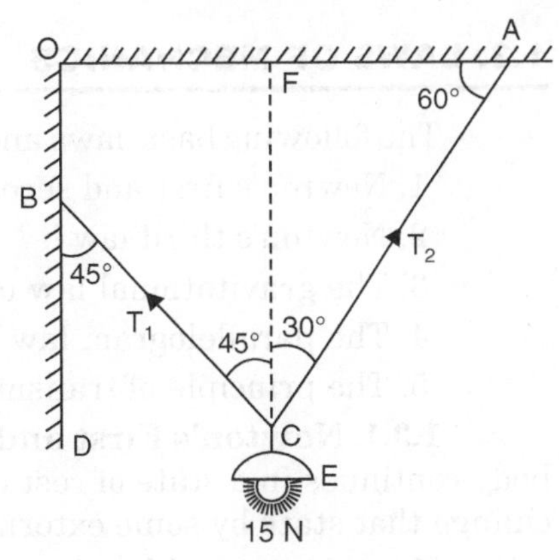

Fig. 1.12

1st Method

Using Lame's theorem at C

$$\frac{15}{\sin \text{ of } \angle BCA} = \frac{T_1}{\sin \text{ of } \angle ACE} = \frac{T_2}{\sin \text{ of } \angle ACE}$$

But

$$\angle BCA = 45° + 30° = 75°$$

$$\angle ACE = 180° - 30° = 150°$$

$$\angle BCE = 180° - 45° = 135°$$

$$\therefore \quad \frac{15}{\sin 75^\circ} = \frac{T_1}{\sin 150^\circ} = \frac{T_2}{\sin 135^\circ}$$

$$\therefore \quad T_1 = \frac{15 \times \sin 150^\circ}{\sin 75^\circ} = \textbf{7.76 N.} \quad \textbf{Ans.}$$

and $$T_2 = \frac{15 \times \sin 135^\circ}{\sin 75^\circ} = \textbf{10.98 N.} \quad \textbf{Ans.}$$

2nd Method

The point C is in the equilibrium. The forces acting at C are 15 N, T_1 and T_2.

Resolving all forces at C in the horizontal direction

$$T_1 \sin 45^\circ = T_2 \sin 30^\circ \quad \text{or} \quad T_1 \times \frac{1}{\sqrt{2}} = T_2 \times \frac{1}{2}$$

$$\therefore \quad T_2 = T_1 \times \frac{2}{\sqrt{2}} = \sqrt{2} \times T_1 \qquad ...(i)$$

Resolving all forces at C in the vertical direction,

$$T_1 \cos 45^\circ + T_2 \cos 30^\circ = 15$$

or $$T_1 \times \frac{1}{\sqrt{2}} + T_2 \times \frac{\sqrt{3}}{2} = 15 \qquad ...(ii)$$

Substituting the value of T_2 from equation (*i*) into equation (*ii*),

$$T_1 \times \frac{1}{\sqrt{2}} + \sqrt{2} \times T_1 \times \frac{\sqrt{3}}{2} = 15$$

or $$\frac{T_1}{\sqrt{2}} + \frac{\sqrt{3}T_1}{\sqrt{2}} = 15 \qquad \left(\because \frac{\sqrt{2}}{2} = \frac{\sqrt{2}}{\sqrt{2} \times \sqrt{2}} = \frac{1}{\sqrt{2}} \right)$$

or $$T_1 + \sqrt{3}\, T_1 = 15 \times \sqrt{2} \quad \text{or} \quad T_1(1 + \sqrt{3}) = 15 \times \sqrt{2}$$

$$\therefore \quad T_1 = \frac{15 \times \sqrt{2}}{1 + \sqrt{3}} = \textbf{7.76 N.} \quad \textbf{Ans.}$$

Substituting this value of T_1 in equation (*i*), we get

$$T_2 = \sqrt{2} \times T_1 = \sqrt{2} \times 7.76 = \textbf{10.98 N.} \quad \textbf{Ans.}$$

1.3. LAWS OF MECHANICS

The following basic laws and principles are considered to be the foundation of mechanics :

1. Newton's first and second laws of motion
2. Newton's third law
3. The gravitational law of attraction
4. The parallelogram law
5. The principle of transmissibility of forces.

1.3.1. Newton's First and Second Laws of Motion. Newton's first law states, "Every body continues in a state of rest or uniform motion in a straight line unless it is compelled to change that state by some external force acting on it."

Newton's second law states, "The net external force acting on a body in a direction is directly proportional to the rate of change of momentum in that direction."

1.3.2. Newton's Third Law. Newton's third law states, "To every action there is always equal and opposite reaction."

Fig. 1.13 shows two bodies A and B which are placed one above the other on a horizontal surface.

Here F_1 = Force exerted by horizontal surface on body A (action)

$-F_1$ = Force exerted by body A on horizontal surface (reaction)

F_2 = Force exerted by body A on body B (action)

$-F_2$ = Force exerted by body B on body A (reaction)

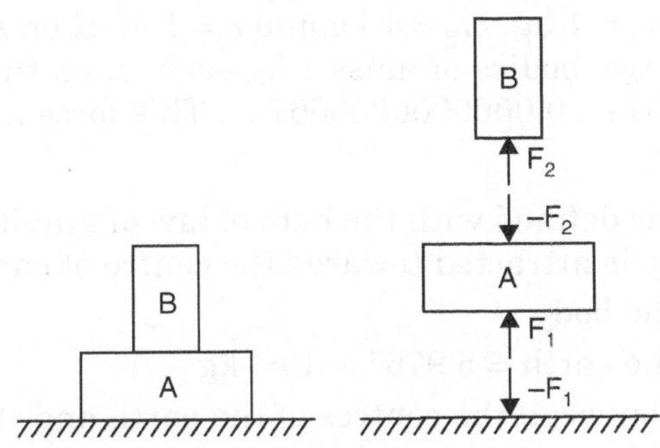

Fig. 1.13

1.3.3. The Gravitational Law of Attraction. It states that two bodies will be attracted towards each other along their connecting line with a force which is directly proportional to the product of their masses and inversely proportional to the square of the distance between their centres.

Refer to Fig. 1.13 (*a*).

Let m_1 = Mass of first body

m_2 = Mass of second body

r = Distance between the centre of bodies

F = Force of attraction between the bodies.

Then according to the law of gravitational attraction.

$$F \propto m_1 \, . \, m_2$$

$$\propto \frac{1}{r^2}$$

or $$F \propto \frac{m_1 m_2}{r^2}$$

or $$F = G \, \frac{m_1 m_2}{r^2} \qquad \text{...(1.6)}$$

Fig. 1.13(a)

where G = Universal gravitational constant of proportionality.

In equation (1.6), F is expressed in N, m_1 and m_2 in kg and r in m. Hence dimensionally, equation (1.6) becomes as

$$N = G \, \frac{\text{kg} \times \text{kg}}{\text{m}^2} \quad \text{or} \quad G = \frac{\text{Nm}^2}{\text{kg}^2} \qquad \text{...}(i)$$

But $$1\text{ N} = 1\text{ kg} \times \frac{1\text{ m}}{\text{s}^2} \quad \text{or} \quad \text{N} = \text{kg} \times \frac{\text{m}}{\text{s}^2}$$

Substituting the value of N is equation (*i*),

$$G = \left(\text{kg} \times \frac{\text{m}}{\text{s}^2}\right) \times \frac{\text{m}^2}{\text{kg}^2} = \frac{\text{m}^3}{\text{kg s}^2} \quad ...(ii)$$

Hence from equations (*i*) and (*ii*), dimension of Universal Gravitational Constant G is N m^2/kg^2 or $\text{m}^3/\text{kg s}^2$.

The value of G is 6.67×10^{-11} N m^2/kg^2 or $\text{m}^3/\text{kg s}^2$.

In equation (1.6), if $m_1 = 1$ kg, $m_2 = 1$ kg and $r = 1$ m, then $F = G$. This means that the force of attraction between two bodies of mass 1 kg each when they are at a distance of 1 m apart, will be 6.67×10^{-11} N *i.e.*, 0.0000000000667 N. This force is very very small.

Weight

The weight of a body is defined with the help of law of gravitation. Weight is defined as the **force** with which a body is attracted towards the centre of earth.

Let M = Mass of the body

M_E = Mass of the earth = 5.9761×10^{24} kg

r = Distance between the centres of the earth and the body

= 6.371×10^6 m (*i.e.*, radius of earth)

G = Universal gravitational constant = 6.67×10^{-11} N m^2/kg^2

F = Force of attraction which is equal to weight (W)

Substituting these values in equation (1.6), we get

$$W = G\,\frac{M_E \times M}{r^2} \qquad (\because \; F = W,\ m_1 = M,\ m_2 = M_E)$$

$$= \frac{6.67 \times 10^{-11}\left(\frac{\text{N m}^2}{\text{kg}^2}\right) \times 5.9761 \times 10^{24}\ (\text{kg}) \times M(\text{kg})}{(6.371 \times 10^6\ \text{m})^2}$$

$$= \frac{6.67 \times 10^{-11} \times 5.9761 \times 10^{24}\, M}{6.371^2 \times 10^{12} \times \text{m}^2}\,\frac{\text{N m}^2}{\text{kg}^2} \times \text{kg} \times \text{kg} = (9.81 \times M)\ \text{N}$$

where 9.81 is acceleration due to gravity and is denoted by 'g'.

$\therefore \quad W = g \times M$ or $M \times g$

Actually the term $\frac{GM_E}{r^2}$ is equal to 9.81 $\frac{\text{m}}{\text{s}^2}$, which is represented by 'g'.

1.3.4. The Parallelogram Law. This law has been already defined. It states that if two forces acting at a point be represented in magnitude and direction by the two adjacent sides of a parallelogram, then their resultant is represented in magnitude and direction by the diagonal of the parallelogram passing through that point.

1.3.5. The Principle of Transmissibility of Forces. It states that if a force, acting at a point on a rigid* body, is shifted to any other point which is on the line of action of the force, the external effect of the force on the body remains unchanged.

*A body which does not deforms under the action of loads or external forces is known as rigid body. Hence in case of a rigid body, the relative movement between the various points of the body are negligible or the distance between any two points remains the same for all the times.

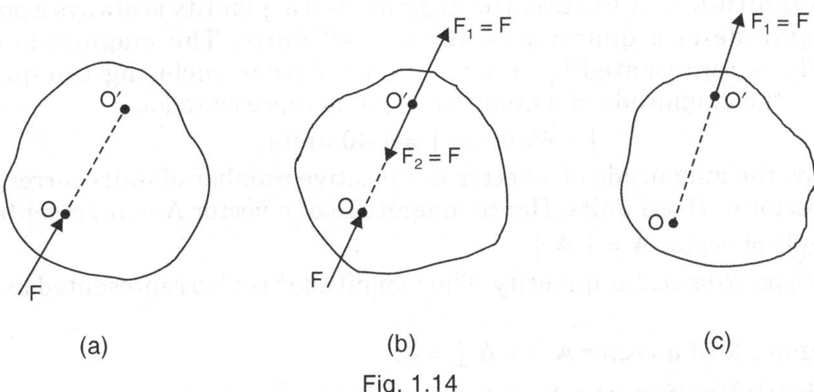

Fig. 1.14

For example, consider a force F acting at point O on a rigid body as shown in Fig. 1.14 (*a*). On this rigid body, "there is another point O' in the line of action of the force F. Suppose at this point O', two equal and opposite forces F_1 and F_2 (each equal to F and collinear with F) are applied as shown in Fig. 1.14 (*b*). The force F and F_2, being equal and opposite, will cancell each other, leaving a force F_1 at point O' as shown in Fig. 1.14 (*c*). But force F_1 is equal to force F.

The original force F acting at point O, has been transferred to point O' which is along the line of action of F without changing the effect of the force on the rigid body. Hence any force acting at a point on a rigid body can be transmitted to act at any other point along its line of action without changing its effect on the rigid body. This proves the principle of transmissibility of a force.

1.4. LAME'S THEOREM

This theorem has been already defined in Art. 1.1.6 (see page 4). It states that "If three forces acting at a point are in equilibrium, each force will be proportional to the sine of the angle between the other two forces."

1.5. PARALLELOGRAM AND TRIANGULAR LAW OF FORCES

These laws have been already defined in articles 1.1.4 and 1.1.5 respectively (see page 2 and 4). *Parallelogram law of forces* states that, "If two forces acting at a point be represented in magnitude and direction by the two adjacent sides of a parallelogram, then their resultant is represented in magnitude and direction by the diagonal of the parallelogram passing through that point.

Triangle law of forces states that, "if three forces acting at a point be represented in magnitude and direction be the three sides of the triangle taken in order, they will be in equilibrium ?

1.6. VECTORS

All quantities that have magnitude and direction are known as vectors. A vector will be represented by a bold face letter (*i.e.*, with thick capital letter). The force F which is a vector will be represented by **F**. The alternate way of representing the force vector is by putting an arrow as $\vec{F}$.

1.6.1. Magnitude of a Vector. The magnitude of a quantity is always a *positive* number. Thus the magnitude of a quantity (– 40) is + 40 units. The magnitude of a quantity mathematically is represented by a set of *vertical lines* enclosing the quantity. Hence mathematically the magnitude of a quantity (– 40) is represented as :

$$| -40 \text{ units} | = +40 \text{ units}$$

Similarly, the magnitude of a vector is a positive number of units corresponding to the *length* of the vector in those units. Hence magnitude of a vector **A** is represented as :

Magnitude of vector $\mathbf{A} = |\ \mathbf{A}\ |$

where $|\ A\ |$ is a *positive* scalar quantity. The magnitude* is also represented by A (*i.e.*, capital letter A).

$\therefore$ Magnitude of a vector $\mathbf{A} = |\ \mathbf{A}\ | = A$.

1.6.2. Multiplication of a Vector by a Scalar

Let $\mathbf{A}$ = a given vector

m = a given scalar

$\therefore$ Vector **A** multiplied by scalar $m = m\mathbf{A}$

where $m\mathbf{A}$ = a vector having same direction as **A** and a magnitude equal to the ordinary scalar product between magnitudes of m and **A**.

1.6.3. Vectorial Representation of Forces and Moments

(A) Vectorial representation of a force. Fig. 1.15 shows a force **F** acting at the origin O. Let the magnitude of this force is equal to length OA. Then this force vector **F** is represented by vector **OA.** Through point A, draw planes parallel to co-ordinate planes.

These planes along with co-ordinate planes make a rectangular box. The force **F** is then represented by the diagonal of the box and its three components F_x, F_y and F_z by its edges.

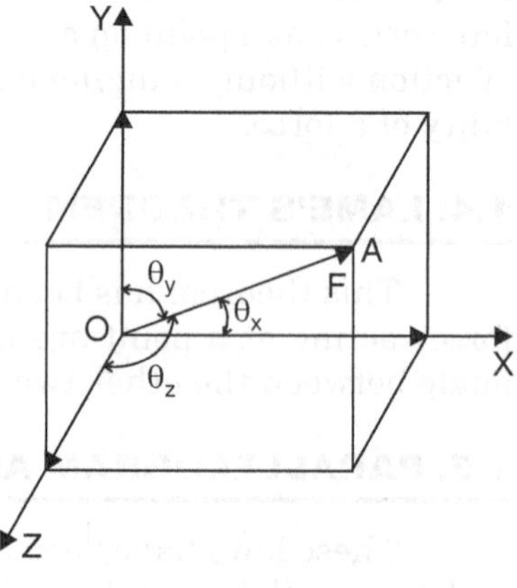

Fig. 1.15

Let θ_x = Angle made by force **F** with x-axis

θ_y = Angle made by force **F** with y-axis

θ_z = Angle made by force **F** with z-axis

Now
$$\left.\begin{aligned} F_x &= F\cos\theta_x \\ F_y &= F\cos\theta_y \\ F_z &= F\cos\theta_z \end{aligned}\right\} \qquad \ldots(1.7)$$

Also $$F = \sqrt{F_x^{\,2} + F_y^{\,2} + F_z^{\,2}} \qquad \ldots(1.7A)$$

The cosines of θ_x, θ_y and θ_z are known as the direction cosines of the force **F** and denoted by

$$l = \cos\theta_x, \quad m = \cos\theta_y \quad \text{and} \quad n = \cos\theta_z \qquad \ldots(1.7B)$$

The three angles** are related by

$$\cos^2\theta_x + \cos^2\theta_y + \cos^2\theta_z = 1 \qquad \ldots(1.8)$$

or $$l^2 + m^2 + n^2 = 1$$

---

*Thick bold face capital letter represents vector whereas capital letter represents magnitude of that vector. Hence **A** is a vector whereas A is the magnitude of vector **A**.

**For x, y co-ordinates $\cos^2\theta_x + \cos^2\theta_y = 1$. This is proved as, $\theta_y = 90 - \theta_x$. Hence $\cos\theta_y = \cos(90 - \theta_x) = \sin\theta_x$. Or $\cos^2\theta_y = \sin^2\theta_x$. Therefore, $\cos^2\theta_x + \cos^2\theta_y = \cos^2\theta_x + \sin^2\theta_x$ which is equal to 1.

Let **i** = vector of unit length in the positive x-direction,

j = vector of unit length in the positive y-direction,

and **k** = vector of unit length in the positive z-direction.

Then force vector **F** is represented by

$$\mathbf{F} = F_x\,\mathbf{i} + F_y\,\mathbf{j} + F_z\,\mathbf{k} \qquad \text{...(1.9)}$$

Magnitude of the vector **F** is given by

$$F = |\,\mathbf{F}\,| = \sqrt{F_x^{\,2} + F_y^{\,2} + F_z^{\,2}}$$

But $F_x = F\cos\theta_x$, $F_y = F\cos\theta_y$ and $F_z = F\cos\theta_z$. Substituting these values in equation (1.9), we get

$$F = (F\cos\theta_x)\,\mathbf{i} + (F\cos\theta_y)\,\mathbf{j} + (F\cos\theta_z)\,\mathbf{k} \qquad \text{...(1.10)}$$

Unit Vector

The vector having a unit length (or unit magnitude) is known as unit vector. Hence unit vector corresponding to force vector **F** is equal to **F** ÷ Magnitude of vector **F**.

$$\therefore \quad \text{Unit vector} = \frac{\text{Force vector } \mathbf{F}}{\text{Magnitude of vector } F} = \frac{\mathbf{F}}{F}$$

$$= \frac{\mathbf{F}}{\sqrt{F_x^{\,2} + F_y^{\,2} + F_z^{\,2}}} \qquad \text{...(1.11)}$$

$$= \frac{F_x\mathbf{i} + F_y\mathbf{j} + F_z\mathbf{k}}{\sqrt{F_x^{\,2} + F_y^{\,2} + F_z^{\,2}}} \qquad (\because \ \mathbf{F} = F_x\mathbf{i} + F_y\mathbf{j} + F_z\mathbf{k})$$

Note. Force vector is represented by **THICK, BOLD FACE** and magnitude is represented by capital letters only.

(B) Vectorial representation of moments. The moment of a force is a vector which is the product of distance and force. Hence in case of moment* of a force the cross-product of distance and force would be taken. Consider the Fig. 1.16.

Let **F** = Force vector $(F_x i + F_y j + F_z k)$

r = Distance (or position) vector with respect to O

$= xi + yj + zk$

M = Moment of force about point O

Then $\mathbf{M} = \mathbf{r} \times \mathbf{F}$

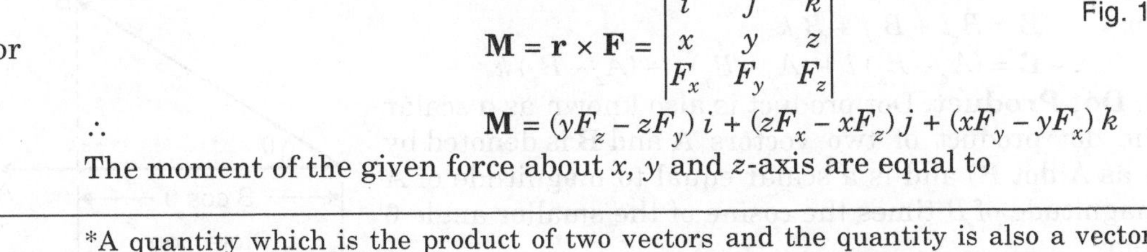

or
$$\mathbf{M} = \mathbf{r} \times \mathbf{F} = \begin{vmatrix} i & j & k \\ x & y & z \\ F_x & F_y & F_z \end{vmatrix}$$

$$\therefore \quad \mathbf{M} = (yF_z - zF_y)\,i + (zF_x - xF_z)\,j + (xF_y - yF_x)\,k$$

Fig. 1.16

The moment of the given force about x, y and z-axis are equal to

*A quantity which is the product of two vectors and the quantity is also a vector, then cross product of the two vectors will be taken. But if the quantity is scalar, then dot product is taken.

or $$M_x = yF_z - zF_y, \quad M_y = zF_x - xF_z, \quad M_z = xF_y - yF_x$$

where M_x = Moment of **F** about x-axis

M_y = Moment of **F** about y-axis, and

M_z = Moment of **F** about z-axis.

Problem 1.10. *A force* **F** $= 2i + 4j - 3k$ *is applied at a point* $P(1, 1, -2)$. *Find the moment of the force F about the point* $(2, -1, 2)$.

Sol. Given :

Force $\mathbf{F} = 2i + 4j - 3k$

The position vector **r** of the point P w.r.t. O.

= Position vector of point P − Position vector of point O.

$= (i + j - 2k) - (2i - j + 2k)$

$\therefore \quad \mathbf{r} = (1 - 2)i + [1 + (1)]j + [-2 - 2]k = -i + 2j - 4k$

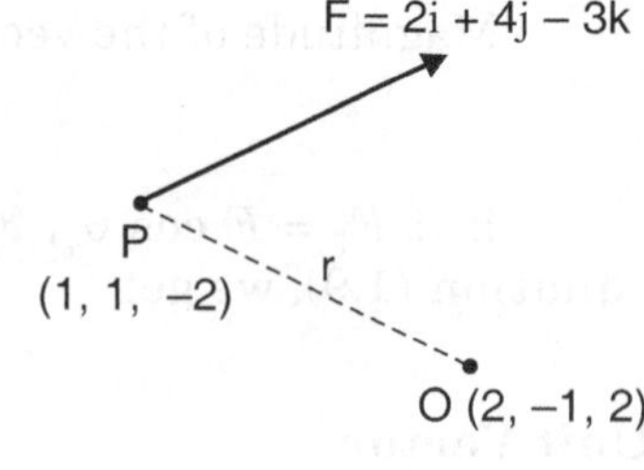

Fig. 1.17

The moment **M** is given by

$$\mathbf{M} = \mathbf{r} \times \mathbf{F} = \begin{vmatrix} i & j & k \\ -1 & 2 & -4 \\ 2 & 4 & -3 \end{vmatrix}$$

$= [(2)(-3) - (-4)(4)]\, i + [(-4)(2) - (-1)(-3)]\, j + [(-1)(4) - (2)(2)]k$

$= (-6 + 16)\, i + (-8 - 3)\, j + (-4 - 4)\, k =$ **10i – 11j – 8k. Ans.**

1.7. VECTOR OPERATIONS [ADDITIONS, SUBTRACTIONS, DOT PRODUCT AND CROSS-PRODUCT]

The followings are the important vector operations :

(*i*) Vector addition, (*ii*) Vector subtraction

(*iii*) Dot product, and (*iv*) Cross product or Vector product.

1.7.1. Vector Addition. The sum of two or more vectors can be obtained by adding the respective components of the vectors. The two vectors **A** and **B** will be added as given below :

Vector $\mathbf{A} = A_x i + A_y j + A_z k$

Vector $\mathbf{B} = B_x i + B_y j + B_z k$

$\therefore \quad \mathbf{A} + \mathbf{B} = (A_x + B_x)\, i + (A_y + B_y)\, j + (A_z + B_z)\, k$

Also $\mathbf{A} + \mathbf{B} = \mathbf{B} + \mathbf{A}$ (Commutative law)

1.7.2. Vector Subtraction. The subtraction of a vector from the other vector is obtained by subtracting the respective components. The vector B can be subtracted from vector A as given below :

Vector $\mathbf{A} = A_x i + A_y j + A_z k$

Vector $\mathbf{B} = B_x i + B_y j + B_z k$

$\therefore \quad \mathbf{A} - \mathbf{B} = (A_x - B_x)\, i + (A_y - B_y)\, j + (A_z - B_z)\, k.$

1.7.3. Dot Product. Dot product is also known as a scalar product. The dot product of two vectors **A** and **B** is denoted by **A . B** (read as **A** dot **B**) and is a scalar equal to magnitude of A times the magnitude of B times the cosine of the smaller angle θ between them.

$\therefore \quad \mathbf{A} \cdot \mathbf{B} = |\mathbf{A}| \cdot |\mathbf{B}| \cos\theta = AB\cos\theta$

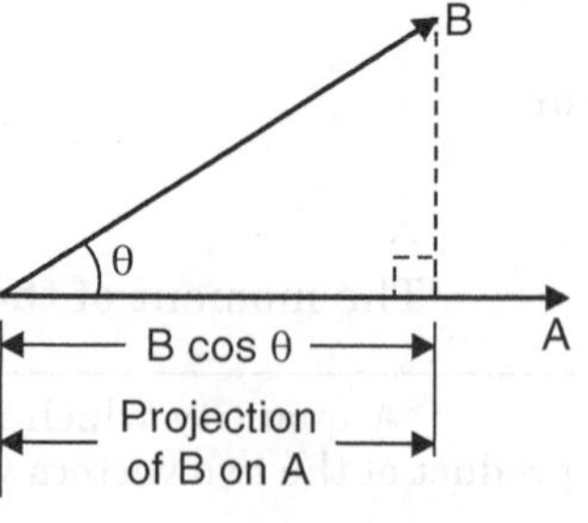

Fig. 1.18

But from Fig. 1.18,

$B \cos \theta$ = Projection of B on A

$\therefore$ $\mathbf{A} \cdot \mathbf{B} = |\mathbf{A}|$ times [Projection of B on A]

Let $\mathbf{A} = A_x i + A_y j + A_z k$

$$\mathbf{B} = B_x i + B_y j + B_z k$$

$$\begin{aligned}\mathbf{A} \cdot \mathbf{B} &= (A_x i + A_y j + A_z k) \cdot (B_x i + B_y j + B_z k) \\ &= A_x B_x i \cdot i + A_x B_y i \cdot j + A_x B_z i \cdot k + A_y B_x j \cdot i + A_y B_y j \cdot j \\ &\quad + A_y B_z j \cdot k + A_z B_x k \cdot i + A_z B_y k \cdot j + A_z B_z k \cdot k \\ &= A_x B_x + 0 + 0 + 0 + A_y B_y + 0 + 0 + 0 + A_z B_z\end{aligned}$$

($\because$ $i \cdot i = 1, j \cdot j = 1$ and $k \cdot k = 1$, as the angle between unit vector and itself is zero and cosine of angle zero is equal to 1. Also the magnitude of unit vector is unity.

And $i \cdot j = j \cdot k = k \cdot i = i \cdot k = j \cdot i = k \cdot j = 0$ as the angle between them is 90° and cosine of 90° is zero.)

$$\therefore \quad \mathbf{A} \cdot \mathbf{B} = A_x B_x + A_y B_y + A_z B_z$$

Hence the dot product of two vectors is equal to the sum of the products of their respective scalar components.

Then scalar product in terms of components

$$\mathbf{AB} = A_x B_x + A_y B_y + A_z B_z \quad \text{...(1.12)}$$

The angle between vectors **A** and **B** is given by

$$\cos \theta = \frac{\mathbf{A} \cdot \mathbf{B}}{|\mathbf{A}||\mathbf{B}|} = \frac{\mathbf{A} \cdot \mathbf{B}}{AB} \quad \text{..(1.13)}$$

Note. (*i*) If $\theta = 0$, the vectors **A** and **B** are acting in the same direction and hence

$$\mathbf{A} \cdot \mathbf{B} = AB$$

(*ii*) If $\theta = 90°$ ($\cos \theta = 0$), the vectors **A** and **B** are perpendicular and hence

$$\mathbf{A} \cdot \mathbf{B} = \mathbf{0}.$$

(*iii*) $\mathbf{A} \cdot \mathbf{B} = \mathbf{B} \cdot \mathbf{A}$ (Commutative law)

(*iv*) The dot product of orthogonal unit vectors *i, j, k* is given as : $i \cdot i = 1$, $j \cdot j = 1$, $k \cdot k = 1$ because the dot product of a unit vector with itself is unity as it is the product of unit magnitude by cosine of zero angle between their directions and $i \cdot j = j \cdot k = k \cdot i = 0$ as now the angle between their direction is 90° and cosine of 90° is zero.

1.7.4. Cross Product or Vector Product. Cross product is also known as vector product. The cross product of two vectors **A** and **B** is denoted by **A** × **B** (read as **A** cross **B**) and is a vector :

(*i*) whose magnitude is equal to the magnitude of *A* times the magnitude of *B* times the sine of the smaller angle between the two vectors ;

(*ii*) whose direction is perpendicular to the plane containing the vectors **A** and **B** ; and

(*iii*) whose sense is given by right-hand screw rule*.

* The sense corresponds to the advance of a right-hand screw rotated about an axis perpendicular to the plane containing vectors *A* and *B* while turning from vector *A* to vector *B* through angle θ.

Hence cross product of vector **A** and vector **B** inclined to each other at an angle θ is a vector whose magnitude is $AB \sin \theta$ and direction is perpendicular to the plane containing vector **A** and vector **B** as shown in Fig. 1.19.

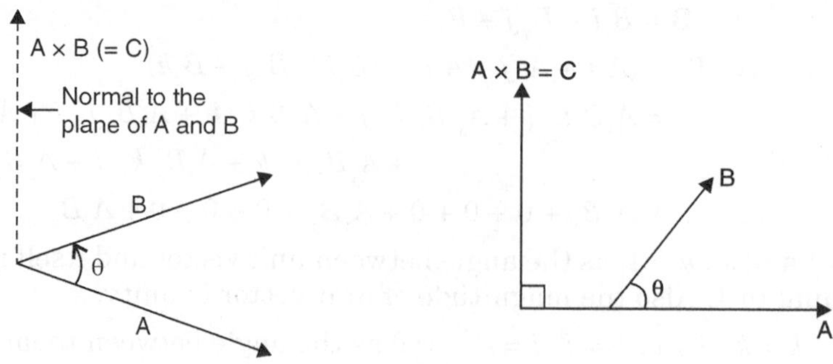

Fig. 1.19

The two vectors **A** and **B** should be directed either away from the point or towards the point. Then the direction of the cross-product is determined either by :

(*i*) **Right hand thumb rule.** Curl the fingers of right hand from vector **A** to vector **B**. Then the direction of the erect thumb will point in the direction of **A × B.** [Refer to Fig. 1.20(*a*)].

(*ii*) **Right hand screw rule.** Hold a right hand screw with its axis perpendicular to the plane containing vector **A** and **B.** Now turn the screw from vector **A** to vector **B**. The direction of advance of screw gives the direction of **A × B**. [Refer to Fig. 1.20(*b*)].

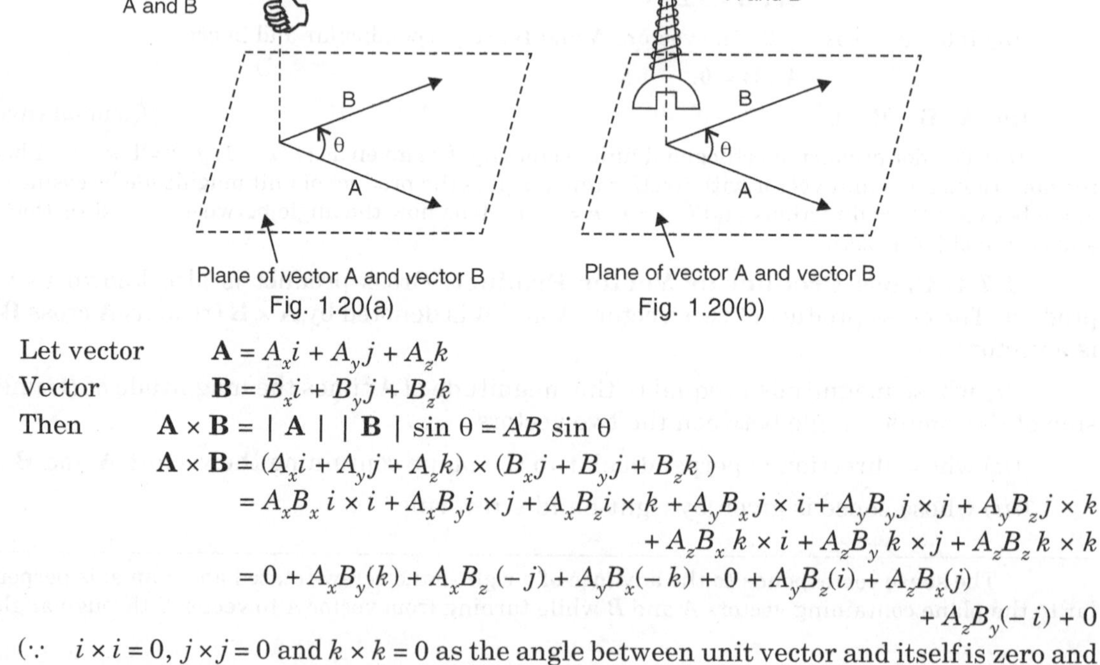

Fig. 1.20(a) Fig. 1.20(b)

Let vector $\mathbf{A} = A_x i + A_y j + A_z k$

Vector $\mathbf{B} = B_x i + B_y j + B_z k$

Then $\mathbf{A} \times \mathbf{B} = |\ \mathbf{A}\ |\ |\ \mathbf{B}\ | \sin \theta = AB \sin \theta$

$$\mathbf{A} \times \mathbf{B} = (A_x i + A_y j + A_z k) \times (B_x j + B_y j + B_z k)$$

$$= A_x B_x\, i \times i + A_x B_y i \times j + A_x B_z i \times k + A_y B_x j \times i + A_y B_y j \times j + A_y B_z j \times k + A_z B_x\, k \times i + A_z B_y\, k \times j + A_z B_z\, k \times k$$

$$= 0 + A_x B_y (k) + A_x B_z\, (-j) + A_y B_x (-k) + 0 + A_y B_z (i) + A_z B_x (j) + A_z B_y (-i) + 0$$

($\because$ $i \times i = 0$, $j \times j = 0$ and $k \times k = 0$ as the angle between unit vector and itself is zero and sine of zero angle is zero).

And $\quad i \times j = k, j \times k = i, k \times i = j$ and (cyclic order)

$j \times i = -k, k \times j = -i, i \times k = -j)$ (cyclic order)

$$\therefore \quad \mathbf{A} \times \mathbf{B} = k(A_xB_y - A_yB_x) + j(A_zB_x - A_xB_z) + i(A_yB_z - A_zB_y)$$

$$= i(A_yB_z - A_zB_y) + j(A_zB_x - A_xB_z) + k(A_xB_y - A_yB_x)$$

The same result can be obtained as given below

$$\mathbf{A} \times \mathbf{B} = \begin{vmatrix} i & j & k \\ A_x & A_y & A_z \\ B_x & B_y & B_z \end{vmatrix}$$

$$= i(A_yB_z - A_zB_y) + j(A_zB_x - A_xB_z) + k(A_xB_y - A_yB_x)$$

In terms of rectangular components

$$\mathbf{A} \times \mathbf{B} = \begin{vmatrix} i & j & k \\ A_x & A_y & A_z \\ B_x & B_y & B_z \end{vmatrix}$$

or $$\mathbf{A} \times \mathbf{B} = (A_yB_z - A_zB_y)\,\mathbf{i} + (A_zB_x - A_xB_z)\,\mathbf{j} + (A_xB_y - A_yB_z)\,\mathbf{k}.$$

Note. (*i*) The cross product of orthogonal unit vectors i, j, k is given by

(*i*) $i \times i = 0, \quad j \times j = 0, k \times k = 0$ ... Included angle is zero and sine of zero angle is zero

$i \times j = k, \quad j \times i = -k, j \times k = i,$

$k \times j = -i, \quad k \times i = j, i \times k = -j.$

(*ii*) If $\theta = 0$ or $\pi, A \times B = 0$

$\therefore \quad \mathbf{A} \times \mathbf{A} = 0$

$\mathbf{A} \times \mathbf{B} \neq \mathbf{B} \times \mathbf{A}$ (Not commutative)

$\mathbf{A} \times \mathbf{B} = -\mathbf{B} \times \mathbf{A}.$

Problem 1.11. *Three vectors* **A, B** *and* **C** *are given as :*

$$\mathbf{A} = \mathbf{i} + \mathbf{j} + 2\mathbf{k}, \quad \mathbf{B} = 4\mathbf{i} - 3\mathbf{j} + 2\mathbf{k} \quad and \quad \mathbf{C} = 2\mathbf{i} - \mathbf{j} - \mathbf{k}.$$

Find the resultant vector and a unit vector in the direction of the resultant.

Sol. Given :

$$\mathbf{A} = i + j + 2k \qquad \mathbf{B} = 4i - 3j + 2k \qquad \mathbf{C} = 2i - j - k$$

Resultant vector, $\quad \mathbf{R} = \mathbf{A} + \mathbf{B} + \mathbf{C}$

$$= (i + j + 2k) + (4i - 3j + 2k) + (2i - j - k)$$

$$= (1 + 4 + 2)\, i + (1 - 3 - 1)\, j + (2 + 2 - 1)\, k$$

$$= \mathbf{7i - 3j + 3k}. \quad \textbf{Ans.}$$

Unit vector in the direction of the resultant R is

$$= \frac{7i - 3j + 3k}{\sqrt{7^2 + (-3)^2 + (3)^2}} = \frac{7i - 3j + 3k}{\sqrt{49 + 9 + 9}} = \frac{7i - 3j + 3k}{\sqrt{67}} = \frac{7i - 3j + 3k}{8.185}$$

$$= \frac{7}{8.185}\, i - \frac{3}{8.185}\, j + \frac{3}{8.185}\, k = \mathbf{0.855\, i - 0.366\, j + 0.366\, k}. \quad \textbf{Ans.}$$

Problem 1.12. *Two vectors* **A** *and* **B** *are given as :*

$$\mathbf{A} = 4i - 3j - 2k \quad and \quad \mathbf{B} = 3i + 2j - k.$$

Determine : (i) their dot product and (ii) angle between them.

Sol. Given :

Vector $\mathbf{A} = 4i - 3j - 2k$

Vector $\mathbf{B} = 3i + 2j - k$

(i) Dot product

The dot product of two vectors is given by equation (1.12) as

$$\mathbf{A.B} = A_xB_x + A_yB_y + A_zB_z$$

$\therefore$ $\mathbf{A.B} = 4 \times 3 + [(-3) \times (2)] + [(-2) \times (-1)] = 12 + (-6) + 2 = \mathbf{8}.$ **Ans.**

(ii) Angle between the vectors

Let θ = Angle between the two vectors.

Using equation (1.13),

$$\cos\theta = \frac{\mathbf{A.B}}{|\mathbf{A}||\mathbf{B}|} = \frac{\mathbf{A.B}}{AB} \qquad ...(i)$$

But $|\mathbf{A}| = A = \sqrt{4^2 + (-3)^2 + (-2)^2} = \sqrt{16 + 9 + 4} = \sqrt{29} = 5.385$

and $|\mathbf{B}| = B = \sqrt{3^2 + 2^2 + (-1)^2} = \sqrt{9 + 4 + 1} = \sqrt{14} = 3.74$

Substituting these values in equation *(i)*,

$$\cos\theta = \frac{8}{5.385 \times 3.74} = 0.3972 \qquad (\because\ A.B = 8)$$

$\therefore$ $\theta = \cos^{-1} 0.3972 = \mathbf{66.59°}.$ **Ans.**

Problem 1.13. *A force vector* **F** *is equal to 10i + 5j – 8k. The point of application of this force moves from the point 2i + k to the point 4i – j – 4k. Find the work done by the force.*

Sol. Given :

Force vector, $\mathbf{F} = 10i + 5j - 8k$

Initial point $= 2i + k$

Final point $= 4i - j - 4k$.

$\therefore$ Distance vector **d** = Final point – Initial point

$= (4i - j - 4k) - (2i + k) = (4 - 2)i + (-1)j + (-4 - 1)k = 2i - j - 5k$

Work is a product of force and distance. Work is a scalar quantity. As work is scalar, hence *dot product* of force and distance will give work done.

$\therefore$ Work done = Force vector . Distance vector

$= (10i + 5j - 8k) . (2i - j - 5k)$ [See equation (1.12)]

$= 10 \times 2 + [(5) \times (-1) + (-8)(-5)] = 20 + (-5) + 40 = \mathbf{55}.$ **Ans.**

Problem 1.14. *A vector* **A** *is equal to 3i – 2j + 2k. Find the projection of this vector on the line joining the points P(1, 2, – 3) and Q(– 1, – 2, 2).*

Sol. Given :

Vector, $\mathbf{A} = 3i - 2j + 2k$

Initial point, $P = i + 2j - 3k$

Final point, $Q = -i - 2j - 2k$

The line joining points P and Q is given by vector = Find point – Initial point

$= (-i - 2j - 2k) - (i + 2j - 3k) = (-1 - 1)\,i + (-2 - 2)\,j + (-2 + 3)\,k$

$= -2i - 4j + k.$

Let the above vector is represented by **B**

$\therefore$ $\mathbf{B} = -2i - 4j + k.$

To find the projection of vector **A** on vector **B** (*i.e.,* on line *PQ*)

We know **A.B** = | **A** | times projection of *B* on *A* = | **B** | times projection of *A* on *B*

$$\therefore \quad \text{Projection of } A \text{ on } B = \frac{\mathbf{AB}}{|\mathbf{B}|} \quad \text{...(i)}$$

Now $\mathbf{A.B} = A_xB_x + A_yB_y + A_zB_z$

$= (3)(-2) + (-2)(-4) + (2)(1) = -6 + 8 + 2 = 4$

$|\mathbf{B}| = \sqrt{(-2)^2 + (-4)^2 + (1)^2} = \sqrt{4 + 16 + 1} = \sqrt{21} = 4.582$

Substituting these values in equation (*i*), we get

$$\text{Projection of } A \text{ on } B = \frac{4}{4.582} = \mathbf{0.8729.} \quad \textbf{Ans.}$$

Problem 1.15. *Determine the cross product of vectors* **A** = $3i - 5j - 4k$ *and* **B** = $3i + 4j - 2k$ *and the angle between them.*

Sol. Given :

Vector $\mathbf{A} = 3i - 5j - 4k$

Vector $\mathbf{B} = 3i + 4j - 2k$

The cross product of vectors **A** and **B** is given by

$$A \times B = \begin{vmatrix} i & j & k \\ 3 & -5 & -4 \\ 3 & 4 & -2 \end{vmatrix}$$

$= [(-5)(-2) - (-4)(4)]i + [(-4)(3) - (3)(-2)]j + [(3)(4) - (-5)(3)]k$

or $\mathbf{A} \times \mathbf{B} = [10 - (-16)]\,i + [-12 - (-6)]\,j + [12 - (-15)]\,k$

$= \mathbf{26\,i - 6\,j + 27\,k.}$ **Ans.**

The angle between the vector *A* and *B* is given by

$$\sin\theta = \frac{|\mathbf{A} \times \mathbf{B}|}{|\mathbf{A}||\mathbf{B}|} \quad \text{...(i)}$$

Now $|\mathbf{A} \times \mathbf{B}| = \sqrt{26^2 + (-6)^2 + (27)^2} = \sqrt{676 + 36 + 729} = \sqrt{1441} = 37.96$

$|\mathbf{A}| = \sqrt{3^2 + (-5)^2 + (-4)^2} = \sqrt{9 + 25 + 16} = 7.071$

$|\mathbf{B}| = \sqrt{3^2 + 4^2 + (-2)^2} = \sqrt{9 + 16 + 4} = 5.385$

Substituting these values in equation (*i*),

$$\sin\theta = \frac{37.96}{7.071 \times 5.385} = 0.9969$$

$\therefore$ $\theta = \sin^{-1} 0.9969 = \mathbf{85.5°}.$ **Ans.**

1.8. RESOLUTION AND COMPOSITION OF A FORCE [Rectangular Components]

Resolution of a force means "finding the components of a given force in two given directions."

Let a given force be R which makes an angle θ with X-axis as shown in Fig. 1.21. It is required to find the components of the force R along X-axis and Y-axis.

Components of R along X-axis $= R \cos \theta$.

Components of R along Y-axis $= R \sin \theta$.

Hence, the resolution of forces is the process of finding components of forces in specified directions.

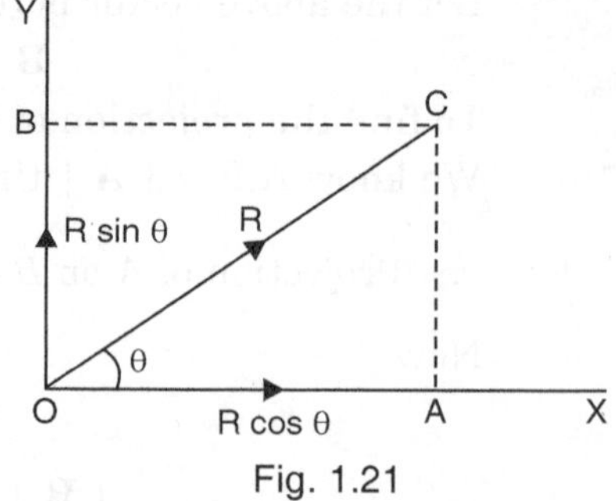

Fig. 1.21

1.8.1. Resolution of a Number of Coplanar Forces. Let a number of coplanar forces (forces acting in one plane are called co-planar forces) R_1, R_2, R_3, ... are acting at a point as shown in Fig. 1.22.

Let θ_1 = Angle made by R_1 with X-axis

θ_2 = Angle made by R_2 with X-axis

θ_3 = Angle made by R_3 with X-axis

H = Resultant component of all forces along X-axis

V = Resultant component of all forces along Y-axis

R = Resultant of all forces

θ = Angle made by resultant with X-axis.

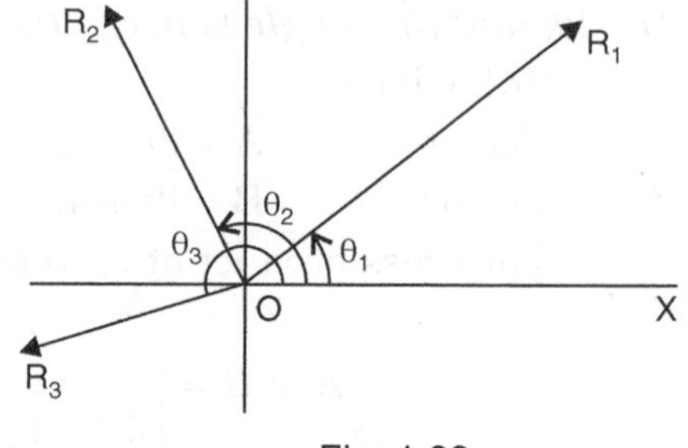

Fig. 1.22

Each force can be resolved into two components, one along X-axis and other along Y-axis.

Component of R_1 along X-axis $= R_1 \cos \theta_1$

Component of R_1 along Y-axis $= R_1 \sin \theta_1$.

Similarly, the components of R_2 and R_3 along X-axis and Y-axis are $(R_2 \cos \theta_2, R_2 \sin \theta_2)$ and $(R_3 \cos \theta_3, R_3 \sin \theta_3)$ respectively.

Resultant components along X-axis

= Sum of components of all forces along X-axis.

$$\therefore \quad H = R_1 \cos \theta_1 + R_2 \cos \theta_2 + R_3 \cos \theta_3 + \ldots \qquad \ldots(1.14)$$

Resultant component along Y-axis.

= Sum of components of all forces along Y-axis.

$$\therefore \quad V = R_1 \sin \theta_1 + R_2 \sin \theta_2 + R_3 \sin \theta_3 + \ldots \qquad \ldots(1.15)$$

Then resultant of all the forces, $R = \sqrt{H^2 + V^2}$ ...(1.16)

The angle made by R with X-axis is given by, $\tan \theta = \dfrac{V}{H}$ ...(1.17)

Problem 1.16. *Two forces are acting at a point O as shown in Fig. 1.23. Determine the resultant in magnitude and direction.*

Sol. The above problem has been solved earlier. Hence it will be solved by resolution of forces.

Force $P = 50$ N and force $Q = 100$ N.

Let us first find the angles made by each force with X-axis.

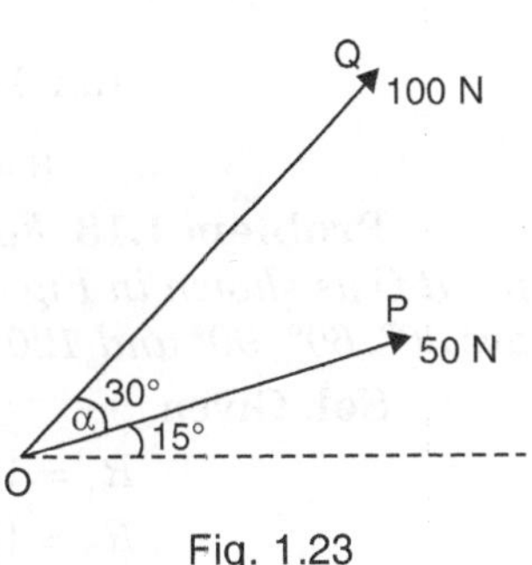

Fig. 1.23

Angle made by P with x-axis = 15°

Angle made by Q with x-axis = 15 + 30 = 45°

Let H = Sum of components of all forces along X-axis.

V = Sum of components of all forces along Y-axis.

The sum of components of all forces along X-axis is given by,

$$H = P \cos 15° + Q \cos 45°$$
$$= 50 \times \cos 15° + 100 \cos 45° = 119 \text{ N}$$

The sum of components of all forces along Y-axis is given by,

$$V = P \sin 15° + Q \sin 45°$$
$$= 50 \sin 15° + 100 \sin 45° = 83.64 \text{ N}$$

The magnitude of the resultant force is given by equation (1.16),

$$R = \sqrt{H^2 + V^2} = \sqrt{119^2 + 83.64^2} = \mathbf{145.46\ N.}\quad \textbf{Ans.}$$

The direction of the resultant force is given by equation (1.17), $\tan \theta = \frac{V}{H} = \frac{83.64}{119}$

$$\therefore \quad \theta = \tan^{-1} \frac{83.64}{119} = \mathbf{35.10°.}\quad \textbf{Ans.}$$

Here θ is the angle made by resultant R with x-axis.

Problem 1.17. *Three forces of magnitude 40 kN, 15 kN and 20 kN are acting at a point O as shown in Fig. 1.24. The angles made by 40 kN, 15 kN and 20 kN forces with X-axis are 60°, 120° and 240° respectively. Determine the magnitude and direction of the resultant force.*

Sol. Given :

$$R_1 = 10 \text{ kN}, \theta_1 = 60°$$
$$R_2 = 15 \text{ kN}, \theta_2 = 120°$$
$$R_3 = 20 \text{ kN}, \theta_3 = 240°$$

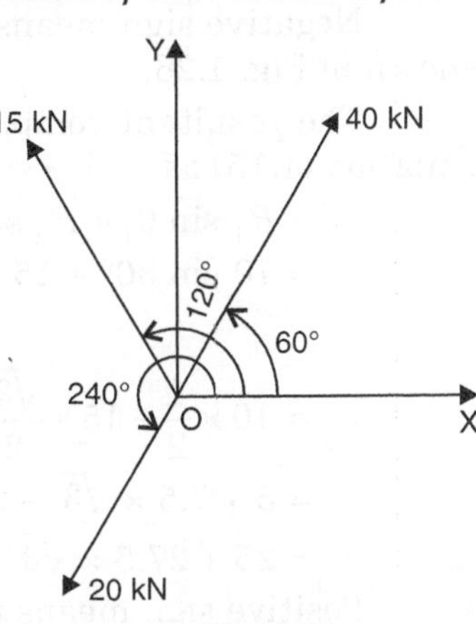

Fig. 1.24

The sum of components of all forces along X-axis is given by equation (1.14) as

$$H = R_1 \cos \theta_1 + R_2 \cos \theta_2 + R_3 \cos \theta_3$$
$$= 40 \times \cos 60° + 15 \times \cos 120° + 20 \times \cos 240°$$
$$= 40 \times \tfrac{1}{2} + 15 \times (-\tfrac{1}{2}) + 20 \times (-\tfrac{1}{2})$$
$$= 20 - 7.5 - 10 = 2.5 \text{ kN.}$$

The resultant component along Y-axis is given by equation (1.15) as

$$V = R_1 \sin \theta_1 + R_2 \sin \theta_2 + R_3 \sin \theta_3$$
$$= 40 \times \sin (60°) + 15 \times \sin (120°) + 20 \times \sin (240°)$$

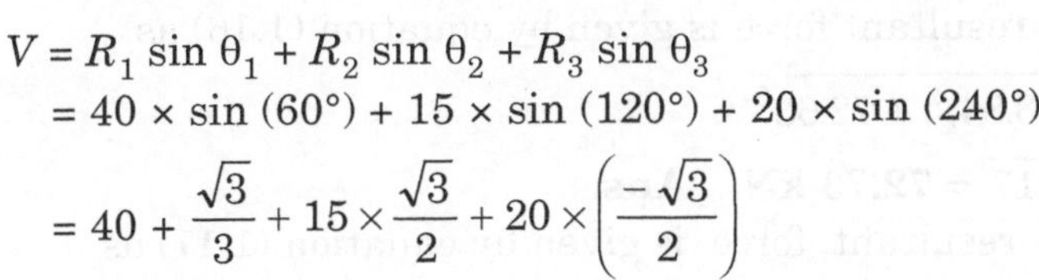

$$= 40 + \frac{\sqrt{3}}{3} + 15 \times \frac{\sqrt{3}}{2} + 20 \times \left(\frac{-\sqrt{3}}{2}\right)$$

$$= 20 \times \sqrt{3} + 7.5 \times \sqrt{3} - 10 \times \sqrt{3} = 17.5 \times \sqrt{3} \text{ kN} = 30.31 \text{ kN.}$$

The magnitude of the resultant force is given by equation (1.16)

$$R = \sqrt{H^2 + V^2} = \sqrt{2.5^2 + 30.31^2} = \mathbf{30.41\ kN.}\quad \textbf{Ans.}$$

The direction of the resultant force is given by equation (1.17)

$$\tan \theta = \frac{V}{H} = \frac{30.31}{2.5} = 12.124 = \tan 85.28°$$

$\therefore$ $\theta = \mathbf{85.28°}$ **or** **85° 16.8′. Ans.**

Problem 1.18. *Four forces of magnitude 10 kN, 15 kN, 20 kN and 40 kN are acting at a point O as shown in Fig. 1.25. The angles made by 10 kN, 15 kN, 20 kN and 40 kN with X-axis are 30°, 60°, 90° and 120° respectively. Find the magnitude and direction of the resultant force.*

Sol. Given :

$R_1 = 10$ kN and $\theta_1 = 30°$

$R_2 = 15$ kN and $\theta_2 = 60°$

$R_3 = 20$ kN and $\theta_3 = 90°$

$R_4 = 40$ kN and $\theta_4 = 120°$

Fig. 1.25

The resultant components along X-axis is given by eqn. (1.14) as

$$H = R_1 \cos \theta_1 + R_2 \cos \theta_2 + R_3 \cos \theta_3 + R_4 \cos \theta_4$$

$$= 10 \cos 30° + 15 \cos 60° + 20 \cos 90° + 40 \cos 120°$$

$$= 10 \times \frac{\sqrt{3}}{2} + 15 \times \frac{1}{2} + 20 \times 0 + 40 \times \left(-\frac{1}{2}\right)$$

$$\{\because \cos 90° = 0 \text{ and } \cos 120° = -\tfrac{1}{2}\}$$

$$= 5 \times \sqrt{3} + 7.5 - 20 = 8.66 + 7.5 - 20 = -3.84 \text{ kN.}$$

Negative sign means that H is acting along OX' as shown in Fig. 1.26.

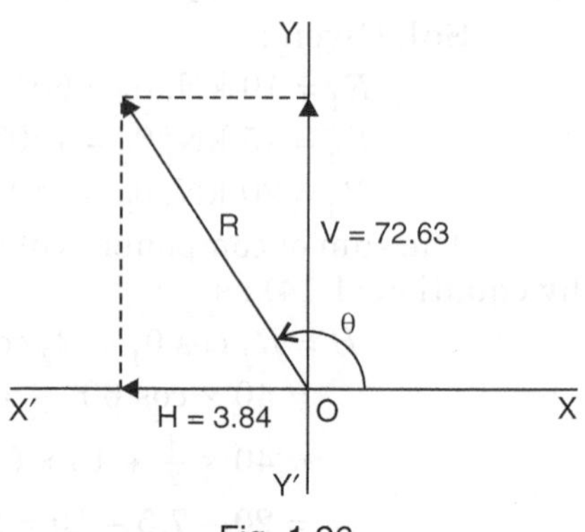

Fig. 1.26

The resultant component along Y-axis is given by equation (1.15) as

$$V = R_1 \sin \theta_1 + R_2 \sin \theta_2 + R_3 \sin \theta_3 + R_4 \sin \theta_4$$

$$= 10 \sin 30° + 15 \sin 60° + 20 \sin 90° + 40 \sin 120°$$

$$= 10 \times \frac{1}{2} + 15 \times \frac{\sqrt{3}}{2} + 20 \times 1 + 40 \times \frac{\sqrt{3}}{2}$$

$$= 5 + 7.5 \times \sqrt{3} + 20 + 20 \times \sqrt{3}$$

$$= 25 + 27.5 \times \sqrt{3} = 72.63 \text{ kN.}$$

Positive sign means that V is acting along OY as shown in Fig. 1.26.

The magnitude of the resultant force is given by equation (1.16) as

$$R = \sqrt{H^2 + V^2} = \sqrt{(-3.84)^2 + 72.63^2}$$

$$= \sqrt{14.745 + 5275.117} = \mathbf{72.73 \text{ kN.}} \quad \textbf{Ans.}$$

The direction of the resultant force is given by equation (1.17) as

$$\tan \theta = \frac{V}{H} = \frac{72.63}{-3.84} = -18.91.$$

From Fig. 1.26 it is clear that θ lies between 90° and 180°.

The angle whose tangent is 18.91° is 86.97.

$\therefore$ $\theta = (180° - 86.97°) = \mathbf{93.03°}$. **Ans.**

1.8.2. Moment of a Force. The product of a force and the perpendicular distance of the line of action of the force from a point is known as moment of the force about that point.

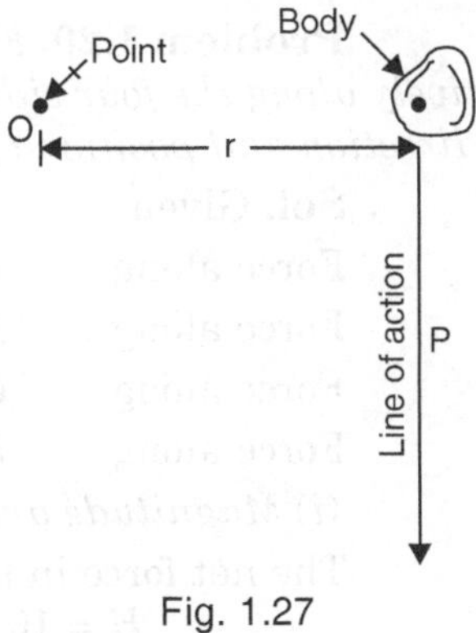

Fig. 1.27

P = A force acting on a body as shown in Fig. 1.27.

r = Perpendicular distance between the point O and line of action of the force P.

The moment of the force P about $O = P \times r$

The tendency of the moment $P \times r$ is to rotate the body in the clockwise direction about O.

Hence this moment is called clockwise moment. If the tendency of rotation is anti-clockwise, the moment is called anti-clockwise moments.

1.8.3. Units of Moment. In M.K.S. system the moment is expressed as kgf m whereas in S.I. system, moment is expressed as newton metre (N m).

1.8.4. Effect of Force and Moment on a Body. The force acting on a body causes linear displacement while moment causes an angular displacement. Hence a body when acted by a number of coplanar forces will be in equilibrium if :

(i) Resultant component of forces along any direction is zero *i.e.*, resultant component of forces in the direction of x, in the direction of y and in the direction of z are zero.

(ii) Resultant moments of the forces about any point in the plane of the forces is zero or clockwise moment is equal to anti-clockwise moments.

Note. If the resultant component of forces acting on a body along any direction is zero but the net moment of the forces about any point is not zero the body will not be in equilibrium. The body will have the tendency to rotate about the point.

Problem 1.19. *A beam of span 10 m is carrying a point load of 200 N at a distance 4 m from A. Determine the beam reactions.*

Sol. Given :

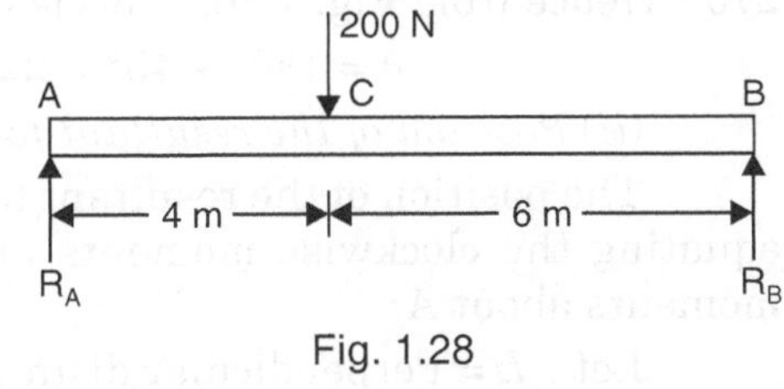

Fig. 1.28

Span $AB = 10$ m

Load at C, $W = 200$ N

Distance, $AC = 4$ m

Distance, $BC = 10 - 4 = 6$ m

Let R_A = Reaction at A and

R_B = Reaction at B

As the beam is in equilibrium, the clockwise moments of all forces about any point must be equal to anti-clockwise moments about that point. Also the resultant force in any direction must be zero.

Taking moments about A,

Clockwise moment = Anti-clockwise moments

$$200 \times 4 = R_B \times 10$$

$$\therefore \quad R_B = \frac{200 \times 4}{10} = \textbf{80 N. Ans.}$$

Also $R_A + R_B = 200$ N

$$\therefore \quad R_A = 200 - R_B = 200 - 80 = \textbf{120 N. Ans.}$$

Problem 1.20. *Four forces of magnitudes 10 N, 20 N, 30 N and 40 N are acting respectively along the four sides of a square ABCD as shown in Fig. 1.29. Determine the magnitude, direction and position of the resultant force.*

Sol. Given :

Force along $AB = 10$ N

Force along $BC = 20$ N

Force along $CD = 30$ N

Force along $DA = 40$ N

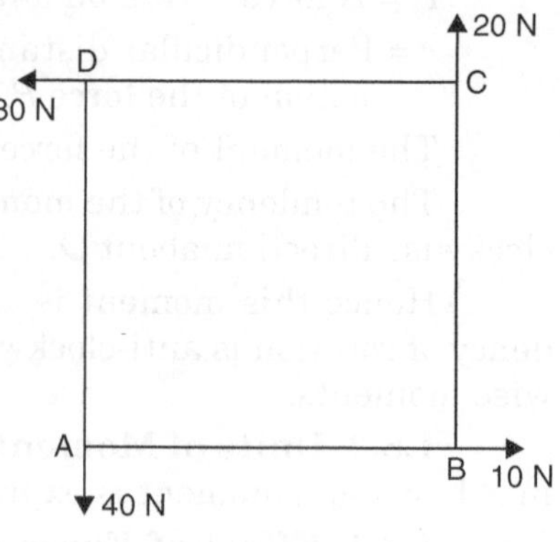

Fig. 1.29

(*i*) *Magnitude and direction of the resultant force*

The net force in the horizontal direction is given as,

$$H = 10 - 30 = -20 \text{ N}$$

The net force in the vertical direction given as,

$$V = 20 - 40 = -20 \text{ N}$$

The resultant force is given by equation (1.16) as

$$R = \sqrt{H^2 + V^2} = \sqrt{(-20)^2 + (-20)^2}$$

$$= \sqrt{400 + 400} = \sqrt{2 \times 400}$$

$$= \mathbf{20 \times \sqrt{2} \text{ N}.} \quad \textbf{Ans.}$$

The direction of the resultant force is given by equation (1.17) as

$$\tan\theta = \frac{V}{H} = \frac{-20}{-20} = 1$$

$$\therefore \qquad \theta = 45°.$$

Since H and V are –ve, hence θ lies between 180° and 270°. Hence from Fig. 1.30, it is clear that actual

$$\theta = 180° + 45° = \mathbf{225°}. \quad \textbf{Ans.}$$

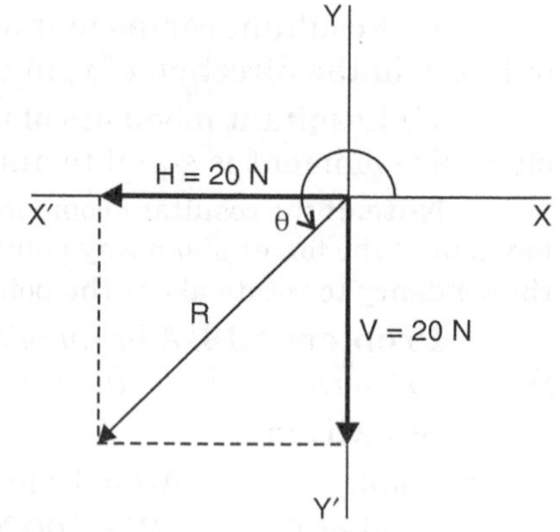

Fig. 1.30

(*ii*) *Position of the resultant force*

The position of the resultant force is obtained by equating the clockwise moments and anti-clockwise moments about A.

Let L = Perpendicular distance between A and line of action of the resultant force and

a = Side of the square $ABCD$.

Taking moments of all forces about A,

$$40 \times 0 + 10 \times 0 + 20 \times a + 30 \times a$$

$$= R \times \text{perpendicular distance of } R \text{ from } A$$

$$= 20 \times \sqrt{2} \times L \qquad (\because \quad R = 20 \times \sqrt{2})$$

or $$20a + 30a = 20 \times \sqrt{2} \times L$$

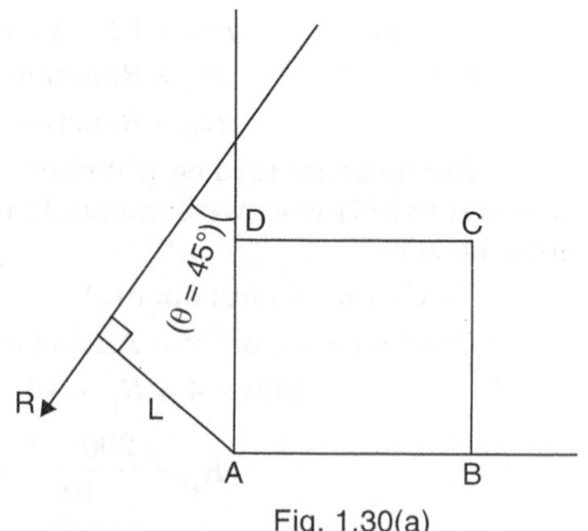

Fig. 1.30(a)

or

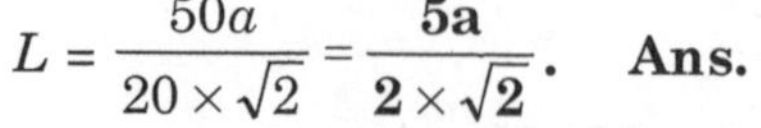

$$L = \frac{50a}{20 \times \sqrt{2}} = \frac{\mathbf{5a}}{\mathbf{2 \times \sqrt{2}}}. \quad \textbf{Ans.}$$

The position of the resultant force is shown in Fig. 1.30 (*a*).

1.9. COPLANAR FORCES

Coplanar forces means the forces in a plane. When several forces act on a body, then they are called *a force system* or *a system of forces*. In a system in which all the forces lie in the same plane, it is known as *coplanar force system*. Hence this article deals with a system of forces which are acting in the same plane and the forces are either having a common line of action or intersecting at a common point. If the forces are having common line of action, then they are known as collinear whereas if the forces intersect at a common point, then they are known as concurrent.

A force system may be coplanar or non-coplanar. If in a system all the forces lie in the same plane then the force system is known as coplanar. But if in a system all the forces lie in different planes, then the force system is known as non-coplanar. Hence a force system is classified as shown in Fig. 1.31.

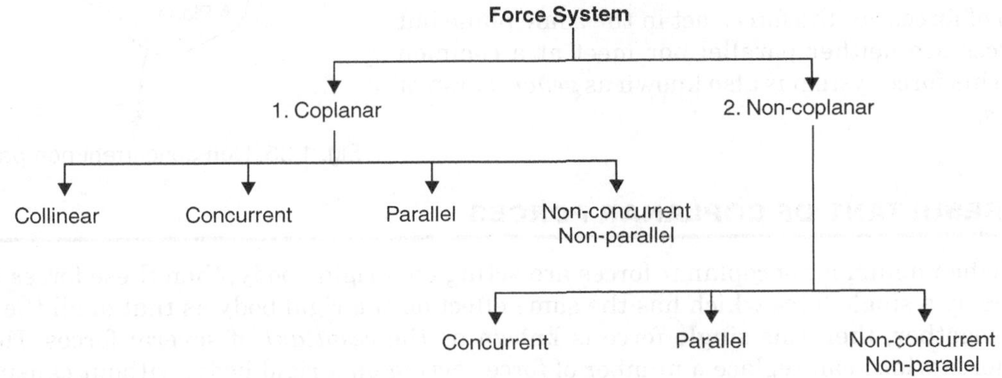

Fig. 1.31

Here we shall discuss only coplanar force system, in which the forces may be :

(*i*) Collinear

(*ii*) Concurrent

(*iii*) Parallel

(*iv*) Non-concurrent, non-parallel (or General system of forces).

1.9.1. Coplanar Collinear. Fig. 1.32 shows three forces F_1, F_2 and F_3 acting in a plane. These three forces are in the same line *i.e.*, these three forces are having common line of action. This system of forces is known as coplanar collinear force system. Hence in coplanar collinear system of forces, all the forces act in the same plane and have a common line of action.

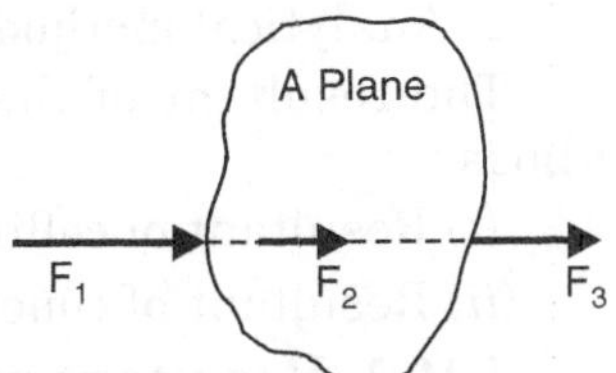

Fig. 1.32. Coplanar collinear forces

1.9.2. Coplanar Concurrent. Fig. 1.33 shows three forces F_1, F_2 and F_3 acting in a plane and these forces intersect or meet at a common point *O*. This system of forces is known as coplanar concurrent force system. Hence in coplanar concurrent system of forces, all the forces act in the same plane and they intersect at a common point.

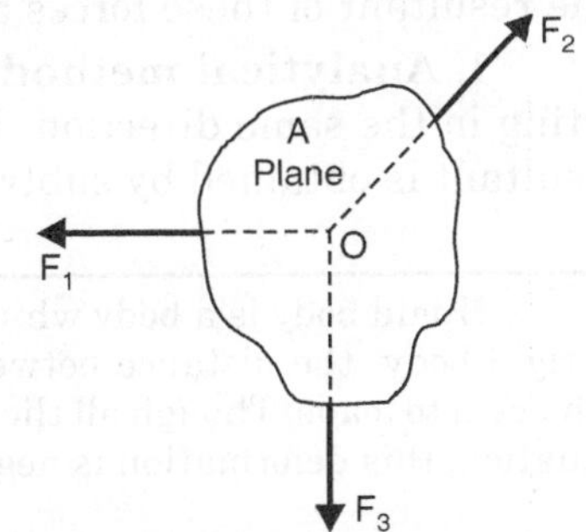

Fig. 1.33. Concurrent coplanar forces

1.9.3. Coplanar Parallel. Fig. 1.34 shows three forces F_1, F_2 and F_3 acting in a plane and these forces are parallel. This system of forces is known as coplanar parallel force system. Hence in coplanar parallel system of forces, all the forces act in the same plane and are parallel.

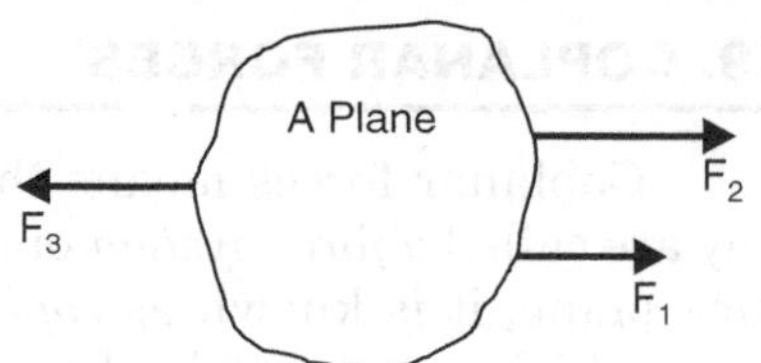

Fig. 1.34. Coplanar parallel forces.

1.9.4. Coplanar Non-concurrent Non-parallel. Fig. 1.35 shows four forces F_1, F_2, F_3 and F_4 acting in a plane. The lines of action of these forces lie in the same plane but they are neither parallel nor meet or intersect at a common point. This system of forces is known as coplanar non-concurrent non-parallel force system. Hence in coplanar non-concurrent non-parallel system of forces, all the forces act in the same plane but the forces are neither parallel nor meet at a common point. This force system is also known as *general system of forces*.

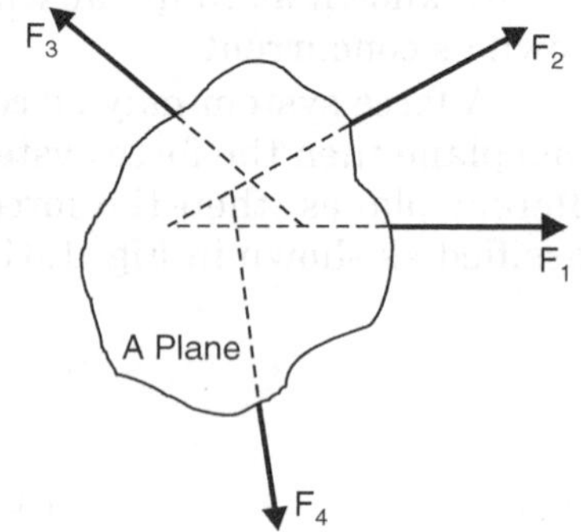

Fig. 1.35. Non-concurrent non-parallel

1.10. RESULTANT OF COPLANAR FORCES

When a number of coplanar forces are acting on a rigid* body, then these forces can be replaced by a single force which has the same effect on the rigid body as that of all the forces acting together, then this single force is known as the *resultant* of several forces. Hence a single force which can replace a number of forces acting on a rigid body, without causing any change in the external effects on the body, is known as the *resultant force*.

The resultant of coplanar forces may be determined by the following two methods :

1. Graphical method
2. Analytical method.

The resultant of the following coplanar forces will be determined by the above two methods :

(*i*) Resultant of collinear coplanar forces

(*ii*) Resultant of concurrent coplanar forces.

1.10.1. Resultant of Collinear Coplanar Forces. As defined in Art. 1.9.1, collinear coplanar forces are those forces which act in the same plane and have a common line of action. The resultant of these forces are obtained by analytical method or graphical method.

1. **Analytical method.** The resultant is obtained by adding all the forces if they are acting in the same direction. If any one of the forces is acting in the opposite direction, then resultant is obtained by subtracting that force.

---

*Rigid body is a body which does not deform under the action of loads or external forces. In case of rigid body, the distance between any two points of the body remains constant, when this body is subjected to loads. Though all the bodies do deform to same extent under the action of loads, but in many situation, this deformation is negligible small.

Fig. 1.36 shows three collinear coplanar forces F_1, F_2 and F_3 acting on a rigid body in the same direction. Their resultant R will be sum of these forces.

$$\therefore \quad R = F_1 + F_2 + F_3 \qquad \text{...(1.18)}$$

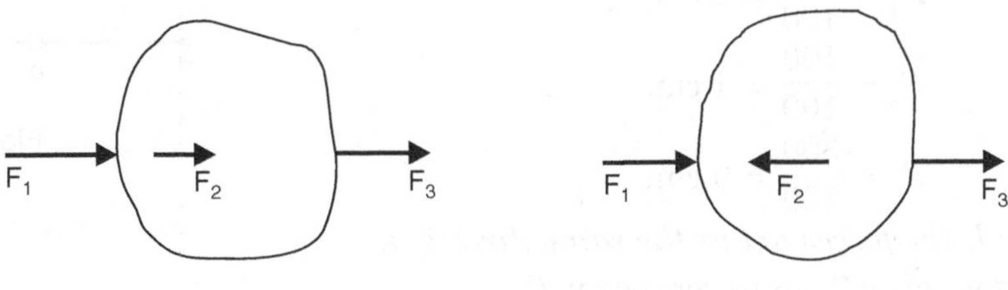

Fig. 1.36 Fig. 1.37

If any one of these forces (say force F_2) is acting in the opposite direction, as shown in Fig. 1.37, then their resultant will be given by

$$R = F_1 - F_2 + F_3 \qquad \text{...(1.19)}$$

2. **Graphical method.** Some suitable scale is chosen and vectors are drawn to the chosen scale. These vectors are added/ or subtracted to find the resultant. The resultant of the three collinear forces F_1, F_2 and F_3 acting in the same direction will be obtained by adding all the vectors. In Fig. 1.38, the force $F_1 = ab$ to some scale, force $F_2 = bc$ and force $F_3 = cd$. Then the length ad represents the magnitude of the resultant on the scale chosen.

F1 F2 F3
a b c d
$R = F_1 + F_2 + F_3$

Fig. 1.38

The resultant of the forces F_1, F_2 and F_3 acting on a body shown in Fig. 1.37 will be obtained by subtracting the vector F_2. This resultant is shown in Fig. 1.39, in which the force $F_1 = ab$ to some suitable scale. This force is acting from a to b. The force F_2 is taken equal to bc on the same scale in opposite direction. This force is acting from b to c. The force F_3 is taken equal to cd. This force is acting from c to d. The resultant force is represented in magnitude by ad on the chosen scale.

F2 F1 F3
a c b d
$R = F_1 - F_2 + F_3$

Fig. 1.39

Problem 1.21. *Three collinear horizontal forces of magnitude 200 N, 100 N and 300 N are acting on a rigid body. Determine the resultant of the forces analytically and graphically when*

(*i*) *all the forces are acting in the same direction,*

(*ii*) *the force 100 N acts in the opposite direction.*

Sol. Given : $F_1 = 200$ N, $F_2 = 100$ N and $F_3 = 300$ N

(*a*) **Analytical method**

(*i*) When all the forces are acting in the same direction, then resultant is given by equation (1.18) as

$$R = F_1 + F_2 + F_3 = 200 + 100 + 300 = \mathbf{600\ N.} \quad \textbf{Ans.}$$

(*ii*) When the force 100 N acts in the opposite direction, then resultant is given by equation (1.19) as

$$R = F_1 - F_2 + F_3 = 200 - 100 + 300 = \mathbf{400\ N.} \quad \textbf{Ans.}$$

(*b*) **Graphical method**

Select a suitable scale. Suppose 100 N = 1 cm. Then to this scale, we have

$$F_1 = \frac{200}{100} = 2 \text{ cm},$$

$$F_2 = \frac{100}{100} = 1 \text{ cm},$$

and $$F_3 = \frac{300}{100} = 3 \text{ cm}.$$

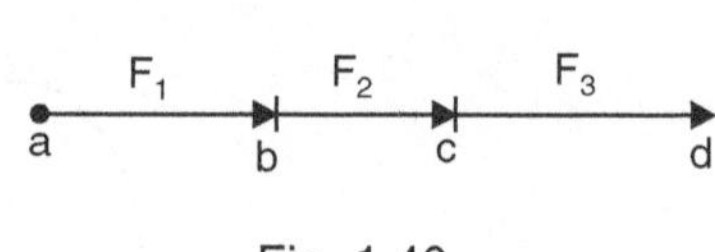

Fig. 1.40

(*i*) *When all the forces act in the same direction.*

Draw vector ab = 2 cm to represent F_1,

vector bc = 1 cm to represent F_2 and

vector cd = 3 cm to represent F_3 as shown in Fig. 1.40.

Measure vector ad which represents the resultant.

By measurement length ad = 6 cm

∴ Resultant = Length ad × chosen scale (∵ Chosen scale is 1 cm = 100 N)

= 6 × 100 = **600 N. Ans.**

(*ii*) *When force 100 N = F_2, acts in the opposite direction*

Draw length ab = 2 cm to represent force F_1.

From b, draw bc = 1 cm in the opposite direction to represent F_2. From c draw cd = 3 cm to represent F_3 as shown in Fig. 1.40(*a*).

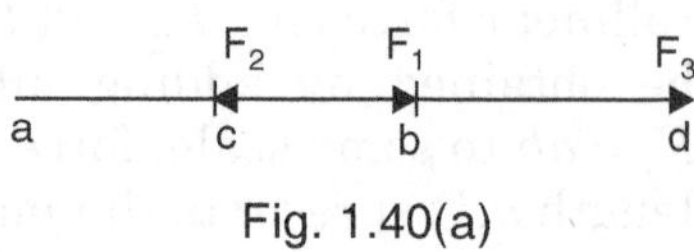

Fig. 1.40(a)

Measure length ad. This gives the resultant.

By measurement, length ad = 4 cm

∴ Resultant = Length ad × chosen scale

= 4 × 100 = **400 N. Ans.**

1.10.2. Resultant of Concurrent Coplanar Forces. As defined in Art. 1.9.2, concurrent coplanar forces are those forces which act in the same plane and they intersect or meet at a common point. We will consider the following two cases :

(*i*) When two forces act at a point

(*ii*) When more than two forces act at a point.

1. **When two forces act at a point**

(*a*) **Analytical method**

In Art. 1.1.4, we have mentioned that when two forces act at a point, their resultant is found by the law of parallelogram of forces. The magnitude of resultant is obtained from equation (1.1) and the direction of resultant with one of the forces is obtained from equation (1.2).

Suppose two forces P and Q act at point O as shown in Fig. 1.41 and α is the angle between them. Let θ is the angle made by the resultant R with the direction of force P.

Forces P and Q form two sides of a parallelogram and according to the law, the diagonal through the point O gives the resultant R as shown.

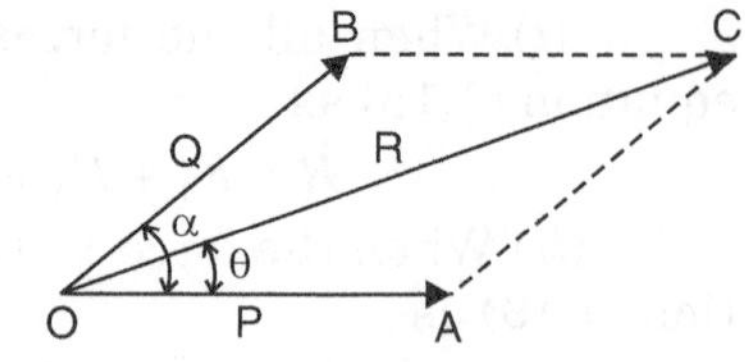

Fig. 1.41

The magnitude* of resultant is given by

$$R = \sqrt{P^2 + Q^2 + 2PQ \cos \alpha}$$

The above method of determining the resultant is also known as the *cosine law method.*

The direction* of the resultant with the force P is given by

$$\theta = \tan^{-1}\left(\frac{Q \sin \alpha}{P + Q \cos \alpha}\right)$$

(*b*) **Graphical method**

(*i*) Choose a convenient scale to represent the forces P and Q.

(*ii*) From point O, draw a vector $Oa = P$.

(*iii*) Now from point O, draw another vector $Ob = Q$ and at an angle of α as shown in Fig. 1.42.

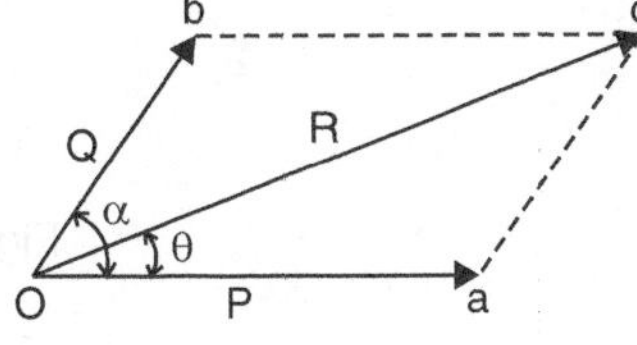

Fig. 1.42

(*iv*) Complete the parallelogram by drawing lines $ac \parallel$ to Ob and $bc \parallel$ to Oa.

(*v*) Measure the length Oc.

Then resultant R will be equal to length $Oc \times$ chosen scale.

(*vi*) Also measure the angle θ, which will give the direction of resultant.

The resultant can also be determine graphically by drawing a triangle *oac* as explained below and shown in Fig. 1.43.

(*i*) Draw a line *oa* parallel to P and equal to P.

(*ii*) From a, draw a vector ac at an angle α with the horizontal and cut ac equal to Q.

(*iii*) Join oc. Then oc represents the magnitude and direction of resultant R.

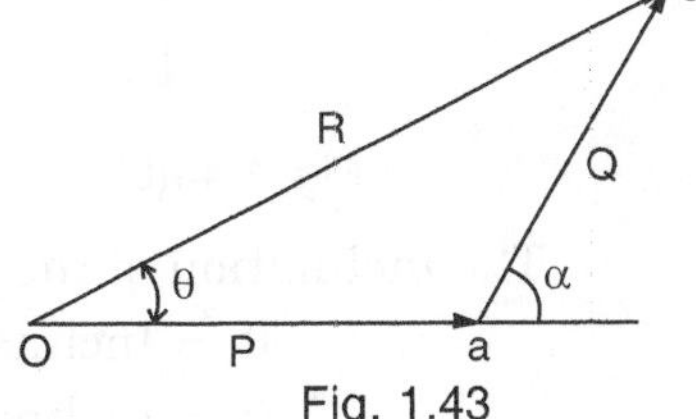

Fig. 1.43

Magnitude of resultant R = Length $Oc \times$ chosen scale.

The direction of resultant is given by angle θ. Hence measure the angle θ.

2. **When more than two forces act at a point**

(*a*) **Analytical method.** The resultant of three or more forces acting at a point is found analytically by a method which is known as rectangular components methods (Refer to Art. 1.8.1). According to this method all the forces acting at a point are resolved into horizontal and vertical components and then *algebraic summation*** of horizontal and vertical components is done separately. The summation of horizontal component is written as ΣH and that of vertical as ΣV. Then resultant R is given by

$$R = \sqrt{(\Sigma H)^2 + (\Sigma V)^2}\,.$$

The angle made by the resultant with horizontal is given by

$$\tan \theta = \frac{(\Sigma V)}{(\Sigma H)}$$

---

*Refer Art. 1.1.4, for the derivation of magnitude and direction of resultant on page 2.

**Summation means addition. Algebraic summation of horizontal components means that if all the horizontal components are in the same direction then they are added. But if one horizontal component is in opposite direction then it is subtracted.

∴ Let four forces F_1, F_2, F_3 and F_4 act at a point O as shown in Fig. 1.44.

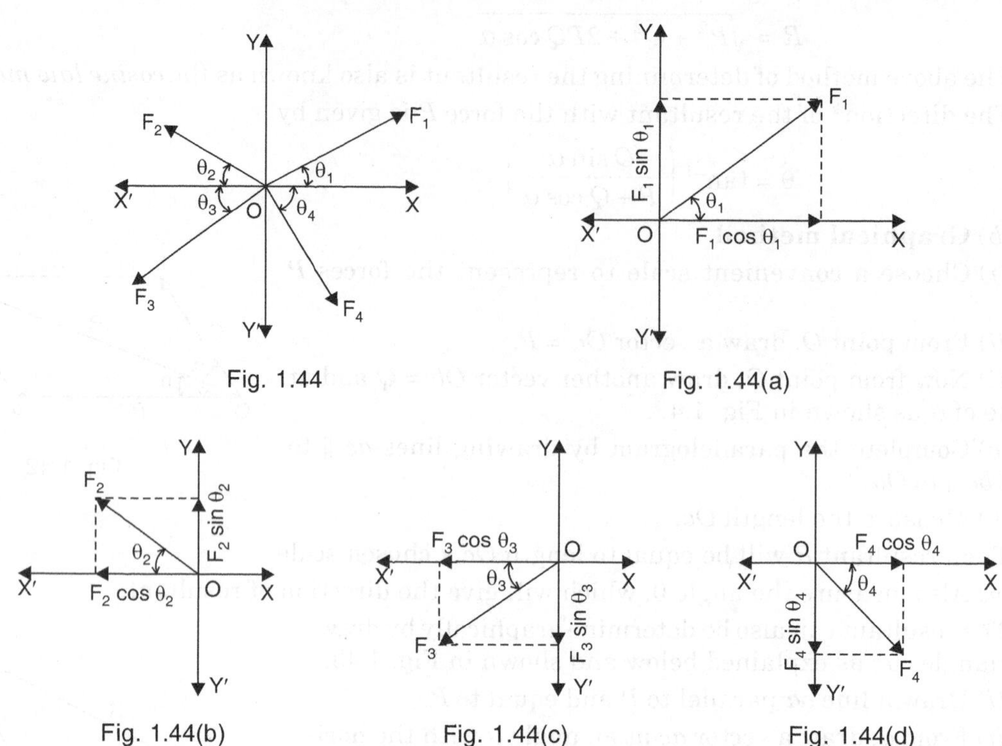

Fig. 1.44 Fig. 1.44(a)

Fig. 1.44(b) Fig. 1.44(c) Fig. 1.44(d)

The inclination of the forces is indicated with respect to horizontal direction. Let

θ_1 = Inclination of force F_1 with OX

θ_2 = Inclination of force F_2 with OX'

θ_3 = Inclination of force F_3 with OX'

θ_4 = Inclination of force F_4 with OX.

The force F_1 is resolved into horizontal and vertical components and these components are shown in Fig. 1.44(*a*). Similarly, Figs. 1.44(*b*), (*c*) and (*d*) show the horizontal and vertical components of forces F_2, F_3 and F_4 respectively. The various horizontal components are :

$F_1 \cos \theta_1 \rightarrow (+)$

$F_2 \cos \theta_2 \leftarrow (-)$

$F_3 \cos \theta_3 \leftarrow (-)$

$F_4 \cos \theta_4 \rightarrow (+)$

∴ Summation or algebraic sum of horizontal components :

$$\Sigma H = F_1 \cos \theta_1 - F_2 \cos \theta_2 - F_3 \cos \theta_3 + F_4 \cos \theta_4$$

Similarly, various vertical components of all forces are :

$F_1 \sin \theta_1 \uparrow (+)$

$F_2 \sin \theta_2 \uparrow (+)$

$F_3 \sin \theta_3 \downarrow (-)$

$F_4 \sin \theta_4 \downarrow (-)$

∴ Summation or algebraic sum of vertical components :

$$\Sigma V = F_1 \sin \theta_1 + F_2 \sin \theta_2 - F_3 \sin \theta_3 - F_4 \sin \theta_4$$

Then the resultant will be given by $R = \sqrt{(\Sigma H)^2 + (\Sigma V)^2}$ ...(1.20)

And the angle (θ) made by resultant with x-axis is given by $\tan \theta = \dfrac{(\Sigma V)}{(\Sigma H)}$ ...(1.21)

(*b*) **Graphical method.** The resultant of several forces acting at a point is found graphically with the help of the *polygon law of forces*, which may be stated as

"If a number of coplanar forces are acting at a point such that they can be represented in magnitude and direction by the sides of a polygon taken in the same order, then their resultant is represented in magnitude and direction by the closing side of the polygon taken in the opposite order.

Let the four forces F_1, F_2, F_3 and F_4 act at a point O as shown in Fig. 1.45. The resultant is obtained graphically by drawing polygon of forces as explained below and shown in Fig. 1.45(*a*).

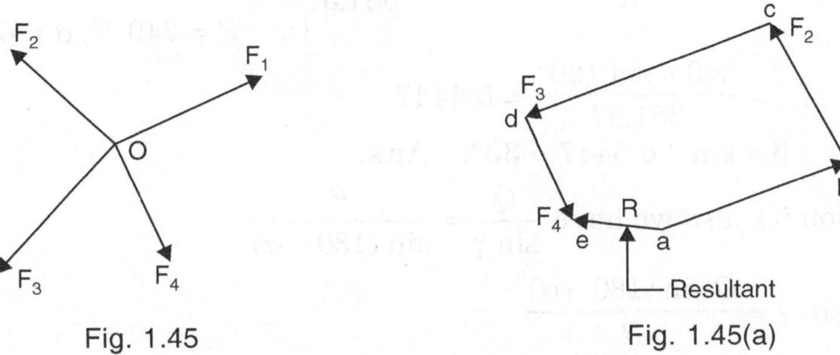

Fig. 1.45 Fig. 1.45(a)

(*i*) Choose a suitable scale to represent the given forces.

(*ii*) Take any point a. From a, draw vector ab parallel to force OF_1. Cut ab = force F_1 to the scale.

(*iii*) From point b, draw bc parallel to OF_2. Cut bc = force F_2.

(*iv*) From point C, draw cd parallel to OF_3. Cut cd = force F_3.

(*v*) From point d, draw de parallel to OF_4. Cut de = force F_4.

(*vi*) Join point a to e. This is the closing side of the polygon. Hence ae represents the resultant in magnitude and direction.

Magnitude of resultant R = Length ae × scale.

The resultant is acting from a to e.

Problem 1.22. *Two forces of magnitude 240 N and 200 N are acting at a point O as shown in Fig. 1.46. If the angle between the forces is 60°, determine the magnitude of the resultant force. Also determine the angle β and γ as shown in the figure.*

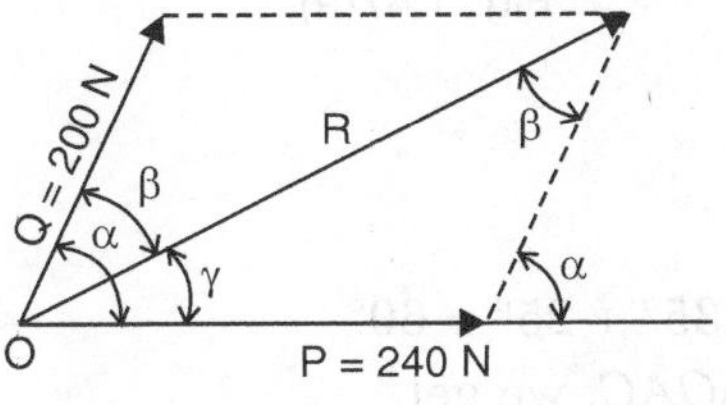

Fig. 1.46

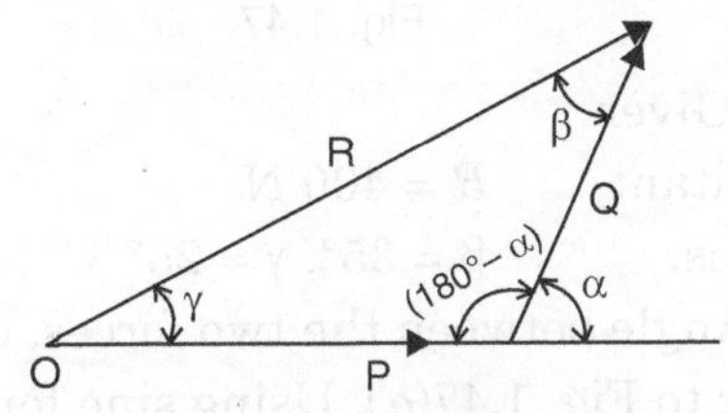

Fig. 1.46(a)

Sol. Given :

Force $P = 240$ N, $Q = 200$ N

Angle between the forces, $\alpha = 60°$

The magnitude of resultant R is given by,

$$R = \sqrt{P^2 + Q^2 + 2PQ\cos\alpha} = \sqrt{240^2 + 200^2 + 2 \times 240 \times 200 \times \cos 60°}$$

$$= \sqrt{57600 + 40000 + 48000} = \mathbf{381.57\ N.}\quad \textbf{Ans.}$$

Now refer to Fig. 1.46(*a*). Using sine formula, we get

$$\frac{P}{\sin\beta} = \frac{Q}{\sin\gamma} = \frac{R}{\sin(180° - \alpha)} \qquad ...(i)$$

or

$$\frac{P}{\sin\beta} = \frac{R}{\sin(180° - \alpha)}$$

$$\therefore \quad \sin\beta = \frac{P\sin(180° - \alpha)}{R} = \frac{240\sin(180 - 60)}{381.57}$$

$$(\because \quad P = 240\text{ N}, \alpha = 60°, R = 381.57\text{ N})$$

$$= \frac{240 \times \sin 120°}{381.57} = 0.5447$$

$$\therefore \quad \beta = \sin^{-1} 0.5447 = \mathbf{33°.}\quad \textbf{Ans.}$$

From equation (*i*), also we have $\dfrac{Q}{\sin\gamma} = \dfrac{R}{\sin(180 - \alpha)}$

$$\therefore \quad \sin\gamma = \frac{Q\sin(180 - \alpha)}{R}$$

$$= \frac{200\sin(180 - 60)}{381.57} = \frac{200\sin 120°}{381.57} = 0.4539$$

$$\gamma = \sin^{-1} 0.4539 = \mathbf{26.966°.}\quad \textbf{Ans.}$$

Problem 1.23. *Two forces P and Q are acting at a point Q as shown in Fig. 1.47. The resultant force is 400 N and angles β and γ are 35° and 25° respectively. Find the two forces P and Q.*

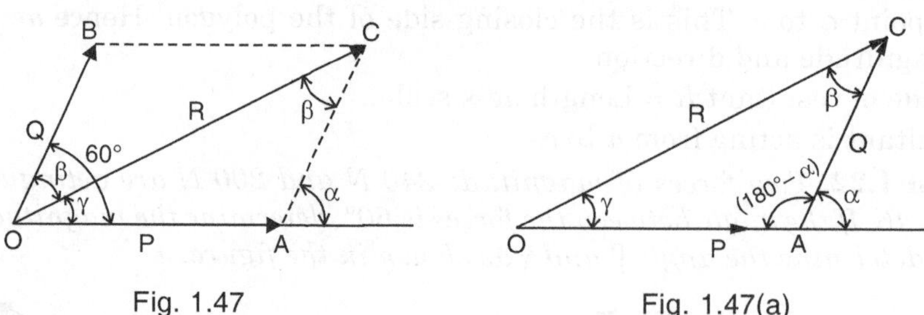

Fig. 1.47 Fig. 1.47(a)

Sol. Given :

Resultant, $R = 400$ N

Angles, $\beta = 35°, \gamma = 25°$

$\therefore$ Angle between the two forces, $\alpha = \beta + \gamma = 35° + 25° = 60°$

Refer to Fig. 1.47(*a*). Using sine formula for ΔOAC, we get

$$\frac{P}{\sin\beta} = \frac{Q}{\sin\gamma} = \frac{R}{\sin(180-\alpha)} \quad ...(i)$$

$$\frac{P}{\sin\beta} = \frac{R}{\sin(180-\alpha)}$$

$\therefore$ $$P = \frac{R\sin\beta}{\sin(180-\alpha)} = \frac{400\times\sin 35°}{\sin(180-60)} \quad (\because R = 400, \beta = 35, \alpha = 60°)$$

$$= \frac{400\times 0.5736}{0.866} = \mathbf{264.93\ N.} \quad \textbf{Ans.}$$

Also from equation (*i*), we have $\dfrac{Q}{\sin\gamma} = \dfrac{R}{\sin(180-\alpha)}$

$\therefore$ $$Q = \frac{R\sin\gamma}{\sin(180-\alpha)} = \frac{400\times\sin 25°}{\sin(180°-60°)} = \frac{400\times 0.4226}{0.866} = \mathbf{195.19\ N.} \quad \textbf{Ans.}$$

Problem 1.24. *Two forces P and Q are acting at a point O as shown in Fig. 1.48. The force P = 240 N and force Q = 200 N. If the resultant of the forces is equal to 400 N, then find the values of angles β, γ and α.*

Sol. Given :

Forces, $P = 240$ N, $Q = 200$ N

Resultant, $R = 400$ N

Let β = Angle between R and Q,

γ = Angle between R and P.

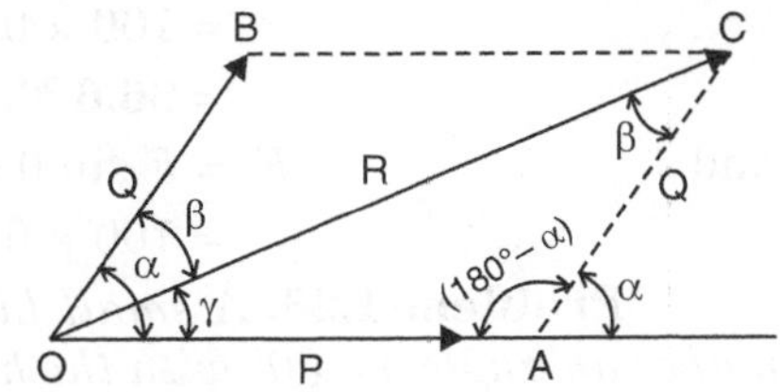

Fig. 1.48

From Fig. 1.48, it is clear that, $\alpha = \beta + \gamma$.

Let us first calculate the angle α (*i.e.*, angle between the two forces).

Using the relation,

$$R = \sqrt{P^2 + Q^2 + 2PQ\cos\alpha} \quad \text{or} \quad R^2 = P^2 + Q^2 + 2PQ\cos\alpha$$

or $$400^2 = 240^2 + 200^2 + 2\times 240\times 200\times\cos\alpha$$

or $$16000 = 57600 + 40000 + 96000\times\cos\alpha$$

$\therefore$ $$\cos\alpha = \frac{16000-57600-40000}{96000} = 0.65$$

$\therefore$ $$\alpha = \cos^{-1} 0.65 = 49.458° = 49° (0.458\times 60') = 49° 27.5'$$

Now using sine formula for ΔOAC of Fig. 1.48, we get

$$\frac{P}{\sin\beta} = \frac{Q}{\sin\gamma} = \frac{R}{\sin(180-\alpha)} \quad ...(i)$$

or $$\frac{P}{\sin\beta} = \frac{R}{\sin(180-\alpha)}$$

$\therefore$ $$\sin\beta = \frac{P\sin(180-\alpha)}{R} = \frac{240\sin(180-49.458)}{400} \quad (\because P = 240, \alpha = 49.458°)$$

$$= \frac{240\sin(130.542°)}{400} = 0.4559$$

$\therefore$ $$\beta = \sin^{-1} 0.4559 = \mathbf{27.12°.} \quad \textbf{Ans.}$$

Also from equation (*i*), we have $\dfrac{Q}{\sin \gamma} = \dfrac{R}{\sin (180 - \alpha)}$

$$\therefore \quad \sin \gamma = \frac{Q \sin (180 - \alpha)}{R} = \frac{200 \times \sin (180 - 49.458)}{400}$$

$$= \frac{200 \times \sin (130.542^\circ)}{400} = 0.3799$$

$$\therefore \quad \gamma = \sin^{-1} 0.3799 = \mathbf{22.33^\circ}. \quad \textbf{Ans.}$$

Problem 1.25. *A force of 100 N is acting at a point making an angle of 30° with the horizontal. Determine the components of this force along X and Y directions.*

Sol. Given :

Force, $F = 100$ N

Angle made by F with horizontal, $\theta = 30°$

Let F_x = Component along x-axis

F_y = Component along y-axis

Then $F_x = F \cos \theta = 100 \cos 30°$

$= 100 \times 0.866$

$=$ **86.6 N. Ans.**

and $F_y = F \sin \theta = 100 \sin 30°$

$= 100 \times 0.5 =$ **50 N. Ans.**

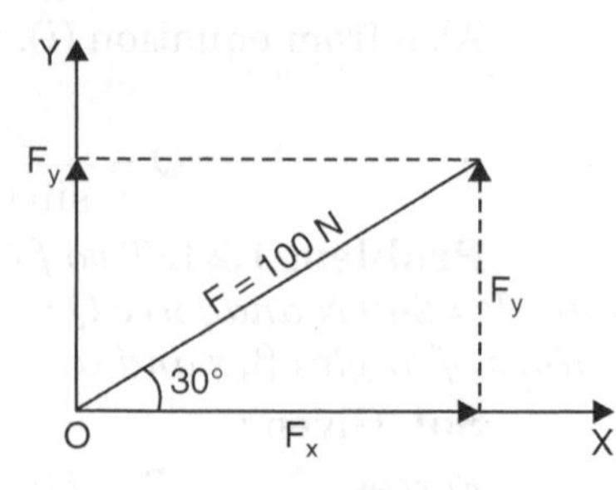

Fig. 1.49

Problem 1.26. *A small block of weight 100 N is placed on an inclined plane which makes an angle θ = 30° with the horizontal. What is the component of this weight ; (ii) parallel to the inclined plane and (iii) perpendicular to the inclined plane ?*

Sol. Given :

Weight of block, $W = 100$ N

Inclination of plane, $\theta = 30°$

The weight of block $W = 100$ N is acting vertically downwards through the C.G. of the block. Resolve this weight into two components *i.e.*, one perpendicular to the inclined plane and other parallel to the inclined plane as shown in Fig. 1.50. The perpendicular (normal) component makes an angle of 30° with the direction of W.

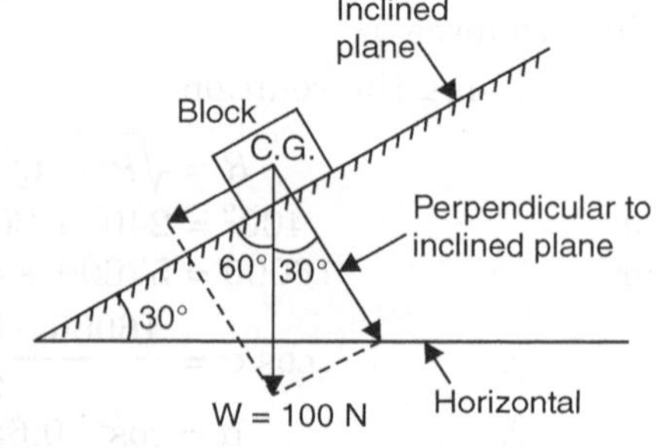

Fig. 1.50

Hence component of the weight perpendicular to the inclined plane

$= W \cos 30° = 100 \times 0.866 =$ **8.66 N. Ans.**

Component of the weight (W) parallel to the inclined plane

$= W \sin 30° = 100 \times 0.5 =$ **50 N. Ans.**

Problem 1.27. *Fig. 1.51 shows a particular position of the connecting rod BA and crank AO. At this position, the connecting rod of the engine exerts a force 2500 N on the crank pin at A. Resolve this force into horizontal and vertical components at A. Also resolve the given force at A along AO and along a direction perpendicular to AO.*

Sol. Given :

Length $BA = 50$ cm

Length $AO = 25$ cm.

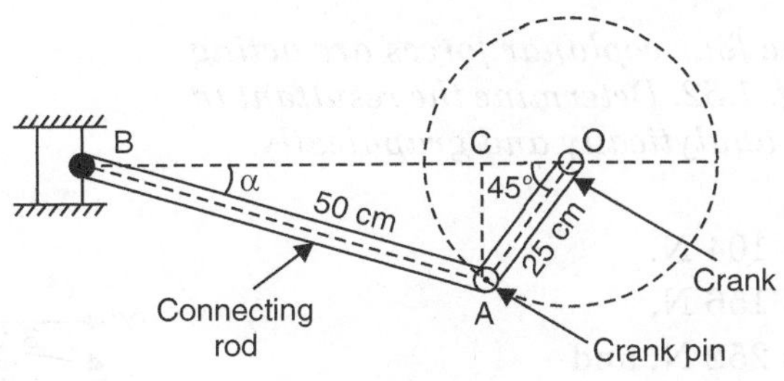

Fig. 1.51

Ist Part

Force exerted by connecting rod BA at A = 2500 N. This force is acting along BA at point A.

Let α = Angle made by BA with horizontal.

This angle can be calculated by drawing a perpendicular AC from point A on the horizontal axis. Now the side AC is common in triangles ABC and AOC.

In triangle ABC, $\sin \alpha = \dfrac{AC}{AB}$ or $AC = AB \sin \alpha = 50 \sin \alpha$ $\quad (\because AB = 50)$

In triangle AOC, $\sin 45° = \dfrac{AC}{AO}$ or $AC = AO \sin 45° = 25 \sin 45°$ $\quad (\because AO = 25)$

Equating the two values of AC, we get $50 \sin \alpha = 25 \sin 45°$

$$\therefore \quad \sin \alpha = \frac{25 \sin 45°}{50} = 0.3535 \quad \text{or} \quad \alpha = \sin^{-1} 0.3535 = 20.7°$$

Now the force 2500 N is acting along BA at point A as shown in Fig. 1.51(a).

$\therefore$ Horizontal component of this force at

$A = 2500 \cos \alpha$

$= 2500 \cos 20.7°$

$=$ **2338.61 N. Ans.**

Vertical component $= 2500 \sin \alpha$

$= 2500 \times \sin 20.7$

$=$ **883.75 N. Ans.**

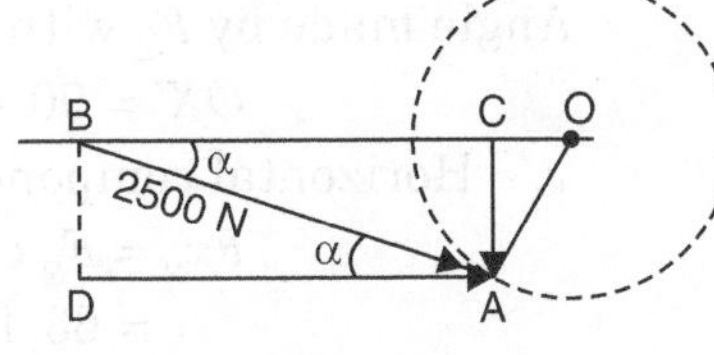

Fig. 1.51(a)

IInd Part

The force of 2500 N is acting along BA at point A as shown in Fig. 1.51(b). This force at point A is represented by AD. Hence BAD is a straight line. Resolve the force AD along AO and perpendicular to AO *i.e.*, in the direction AE where AE is perpendicular to AO at point A.

Now angle OAD $= 45 + 20.7 = 65.7°$

Component of force AD along

$AO = AD \cos 65.7°$

$= 2500 \cos 65.7°$

$=$ **1028.78 N. Ans.**

Component of force AD along $AE = AD \sin 65.7°$

$= 2500 \times \sin 65.7$

$=$ **2278.5 N. Ans.**

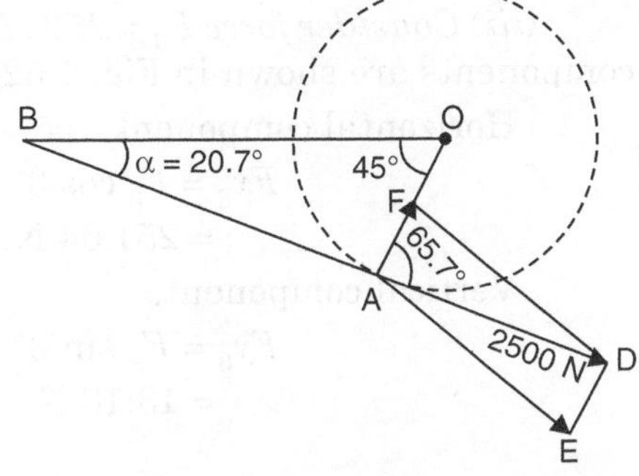

Fig. 1.51(b)

Problem 1.28. *The four coplanar forces are acting at a point as shown in Fig. 1.52. Determine the resultant in magnitude and direction analytically and graphically.*

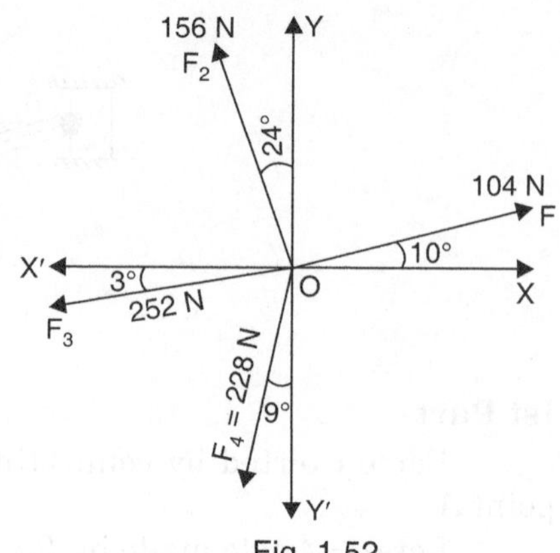

Fig. 1.52

Sol. Given :

Forces, $F_1 = 104$ N,

$F_2 = 156$ N,

$F_3 = 252$ N, and

$F_4 = 228$ N.

(*a*) **Analytical method.** Resolve each force along horizontal and vertical axes. The horizontal components along *OX* will be considered as +ve whereas along *OX′* as –ve. Similarly, vertical components in upward direction will be +ve whereas in downward direction as –ve.

(*i*) *Consider force F_1 = 104 N.* Horizontal and vertical components are shown in Fig. 1.52(*a*).

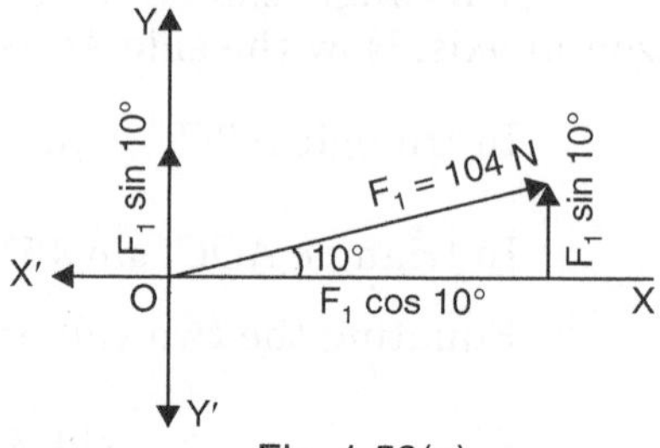

Fig. 1.52(a)

Horizontal component,

$$Fx_1 = F_1 \cos 10° = 104 \times 0.9848$$
$$= 102.42 \text{ N}$$

Vertical component,

$$Fy_1 = F_1 \sin 10° = 104 \times 0.1736$$
$$= 18.06 \text{ N}.$$

(*ii*) *Consider force F_2 = 156 N.* Horizontal and vertical components are shown in Fig. 1.52(*b*).

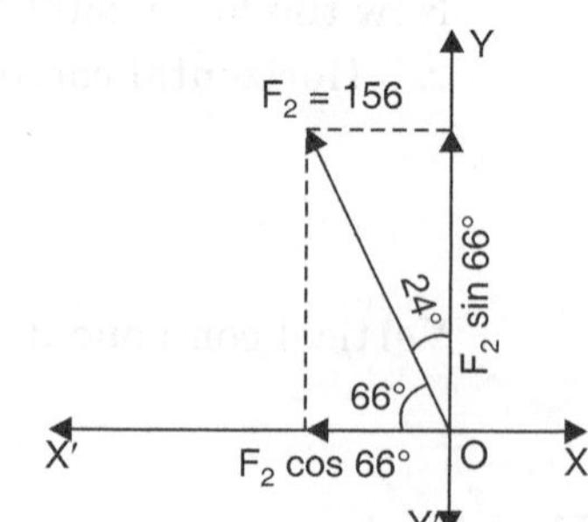

Fig. 1.52(b)

Angle made by F_2 with horizontal axis

$$OX' = 90 - 24 = 66°$$

∴ Horizontal components,

$$Fx_2 = F_2 \cos 66° = 156 \times 0.4067$$
$$= 63.44 \text{ N}.$$

It is negative as it is acting along *OX′*.

Vertical component,

$$Fy_2 = F_2 \sin 66° = 156 \times 0.9135$$
$$= 142.50 \text{ N}. \quad (+\text{ve})$$

(*iii*) *Consider force F_3 = 252 N.* Horizontal and vertical components are shown in Fig. 1.52(*c*).

Horizontal component,

$$Fx_3 = F_3 \cos 3° = 252 \times 0.9986$$
$$= 251.64 \text{ N}. \quad (-\text{ve})$$

Vertical component,

$$Fy_3 = F_3 \sin 3° = 252 \times 0.0523$$
$$= 13.18 \text{ N}. \quad (-\text{ve})$$

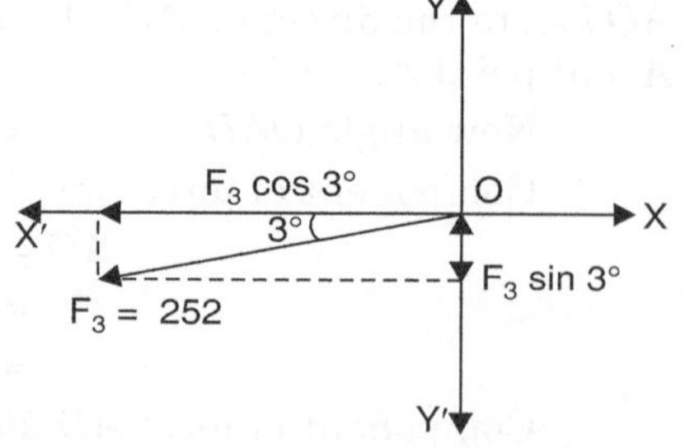

Fig. 1.52(c)

(*iv*) *Consider force* $F_4 = 228$ *N*. Horizontal and vertical components are shown in Fig. 1.52(*d*).

Angle made by F_4 with horizontal axis

$$OX' = 90 - 9 = 81°.$$

∴ Horizontal component,

$$Fx_4 = F_4 \cos 81° = 228 \times 0.1564$$
$$= 35.66 \text{ N} \qquad (-\text{ve})$$

Vertical component,

$$Fy_4 = F_4 \sin 81° = 228 \times 0.9877$$
$$= 225.2 \text{ N}. \qquad (-\text{ve})$$

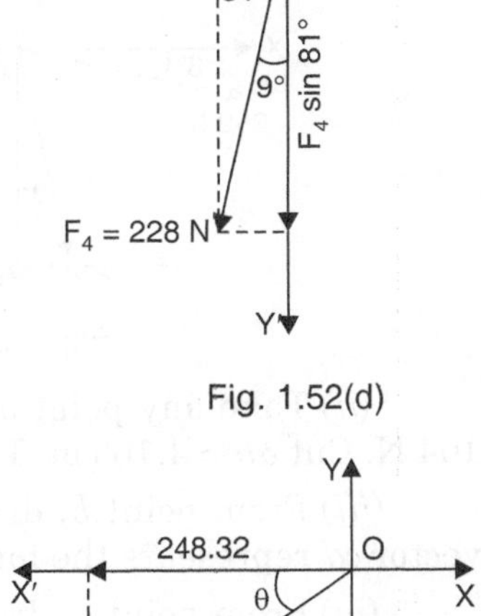

Fig. 1.52(d)

Now algebraic sum of horizontal components is given by,

$$\Sigma H = Fx_1 - Fx_2 - Fx_3 - Fx_4$$
$$= 102.4 - 63.44 - 251.64 - 35.66$$
$$= -\ 248.32 \text{ N}.$$

–ve sign means that ΣH is acting along OX' as shown in Fig. 1.52(*e*).

Similarly, the algebraic sum of vertical components is given by,

$$\Sigma V = 18.06 = 142.50 + 13.18 - 225.2$$
$$= -\ 77.82 \text{ N}.$$

– ve sign means that ΣV is acting along OY' as shown in Fig. 1.52(*e*).

Fig. 1.52(e)

The magnitude of resultant (*i.e.*, *R*) is obtained by using equation (1.20).

$$\therefore \qquad R = \sqrt{(\Sigma H)^2 + (\Sigma V)^2}$$
$$= \sqrt{(248.32)^2 + (77.82)^2}$$
$$= \mathbf{260.2\ N.} \quad \textbf{Ans.}$$

The direction of resultant is given by equation (1.21).

$$\therefore \qquad \tan\theta = \frac{\Sigma V}{\Sigma H} = \frac{77.82}{248.32} = 0.3134$$

$$\therefore \qquad \theta = \tan^{-1} 0.3134 = 17.4°. \quad \textbf{Ans.}$$

(*b*) **Graphical method.** Fig. 1.53(*a*), shows the point at which four forces 104 N, 156 N, 252 N and 228 N are acting. The resultant force is obtained graphically by drawing polygon of forces as explained below and shown in Fig. 1.53(*b*) :

(*i*) Choose a suitable scale to represent the given forces. Let the scale is 25 N = 1 cm. Hence the force 104 N will be represented by $\frac{104}{25} = 4.16$ cm, force 156 N will be represented by $\frac{156}{25} = 6.24$ cm force 252 N will be represented by $\frac{252}{25} = 10.08$ cm and the force 228 N will be represented by $\frac{228}{25} = 9.12$ cm.

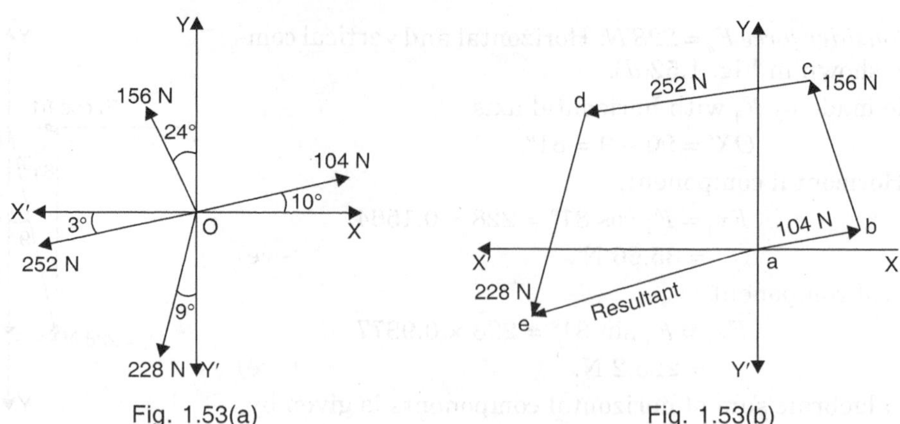

Fig. 1.53(a) Fig. 1.53(b)

(*ii*) Take any point *a*. From point *a*, draw vector *ab* parallel to line of action of force 104 N. Cut *ab* = 4.16 cm. Then *ab* represents the force 104 N in magnitude and direction.

(*iii*) From point *b*, draw vector *bc* parallel to force 156 N and cut *bc* = 6.24 cm. Then vector *cd* represents the force 156 N in magnitude and direction.

(*iv*) From point *c*, draw a vector *cd* parallel 252 N force and cut *cd* = 10.08 cm. Then vector *cd* represents the force 252 N in magnitude and direction.

(*v*) Now from point *d*, draw the vector *de* parallel to 228 N force and cut *de* = 9.12 cm. Then vector *de* represents the force 228 N in magnitude and direction.

(*vi*) Join point *a* to *e*. The line *ae* is the closing side of the polygon. Hence the side *ae* represents the resultant in magnitude and direction. Measure the length of *ae*.

By measurement, length *ae* = 10.4 cm

$\therefore$ Resultant, R = Length *ae* × Scale = 10.4 × 25 ($\because$ 1 cm = 25 N)

= **260 N. Ans.**

Now measure angle made by *ae* with horizontal. This angle is 17.4° with axis *OX*′. **Ans.**

Problem 1.29. *The resultant of four forces which are acting at a point O as shown in Fig. 1.54, is along Y-axis. The magnitude of forces F_1, F_3 and F_4 are 10 kN, 20 kN and 40 kN respectively. The angles made by 10 kN, 20 kN and 40 kN with X-axis are 30°, 90° and 120° respectively. Find the magnitude and direction of force F_2 if resultant is 72 kN.*

Sol. Given :

$$F_1 = 10 \text{ kN}, \theta_1 = 30°$$
$$F_2 = ?, \quad \theta_2 = \theta$$
$$F_3 = 20 \text{ kN}, \theta_3 = 90°$$
$$F_4 = 40 \text{ kN}, \theta_4 = 120°$$

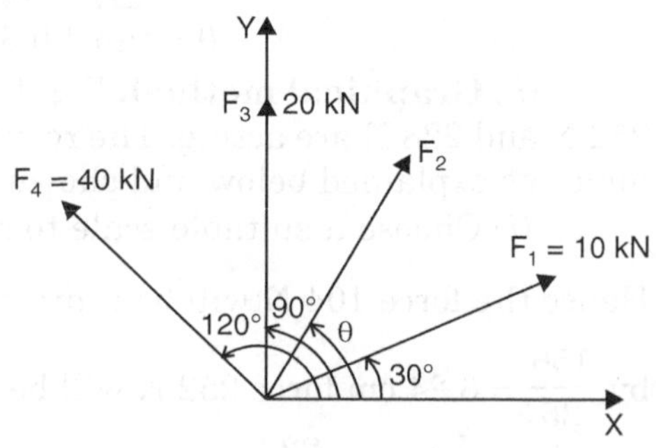

Fig. 1.54

Resultant, R = 72 kN

Resultant is along *Y*-axis.

Hence the algebraic sum of horizontal component should be zero and algebraic sum of vertical components should be equal to the resultant.

$\therefore$ $\Sigma H = 0$ and $\Sigma V = R = 72$ kN

But $\Sigma H = F_1 \cos 30° + F_2 \cos \theta + F_3 \cos 90° + F_4 \cos 120°$

$= 10 \times 0.866 + F_2 \cos \theta + 20 \times 0 + 40 \times (-\frac{1}{2})$

$= 8.66 + F_2 \cos \theta + 0 - 20 = F_2 \cos \theta - 11.34$

$\therefore$ $\Sigma H = 0$ or $F_2 \cos \theta - 11.34 = 0$

or $F_2 \cos \theta = 11.34$ ...(*i*)

Now $\Sigma V = F_1 \sin 30° + F_2 \sin \theta + F_3 \sin 90° + F_4 \sin 120°$

$= 10 \times \frac{1}{2} + F_2 \sin \theta + 20 \times 1 + 40 \times 0.866$

$= 5 + F_2 \sin \theta + 20 + 34.64 = F_2 \sin \theta + 59.64$

But $\Sigma V = R$

$\therefore$ $F_2 \sin \theta + 59.64 = 72$

$\therefore$ $F_2 \sin \theta = 72 - 59.64 = 12.36$ ...(*ii*)

Dividing equation (*ii*) and (*i*),

$$\frac{F_2 \sin \theta}{F_2 \cos \theta} = \frac{12.36}{11.34} \quad \text{or} \quad \tan \theta = 1.0899$$

$\therefore$ $\theta = \tan^{-1} 1.0899 = \mathbf{47.46°}$. **Ans.**

Substituting the value of θ in equation (*ii*), we get $F_2 \sin (47.46°) = 12.36$

or $$F_2 = \frac{12.36}{\sin (47.46°)} = \frac{12.36}{0.7368} = \mathbf{16.77\ kN.}\ \textbf{Ans.}$$

Problem 1.30. *Determine the magnitude, direction and position of a single force P, which keeps in equilibrium the system of forces acting at the corners of a rectangular block as shown in Fig. 1.55. The position of the force P may be stated by reference to axes with origin O and coinciding with the edges of the block.*

Sol. Given :

Length $OC = 4$ m,

Length $BC = 3$ m

Force at $O = 20$ N (←)

Force at $C = 35$ N (↓)

Force at $B = 25$ N (→)

Force at $A = 50$ N (↓)

Let O be the origin and OX and OY be the reference axes as shown in Fig. 1.56.

Forces 50 N and 20 N form a concurrent system and their line of action intersect at O.

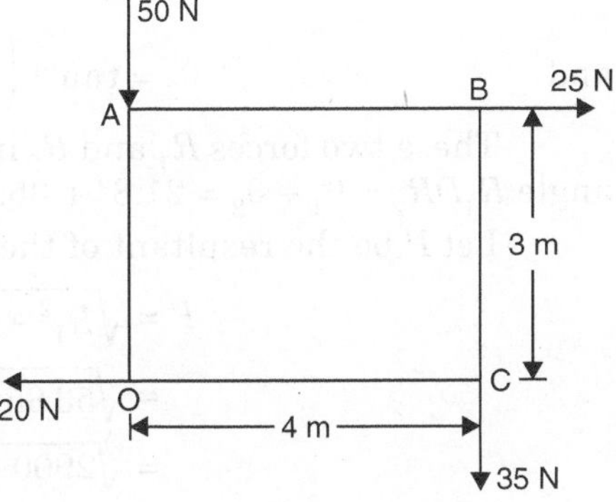

Fig. 1.55

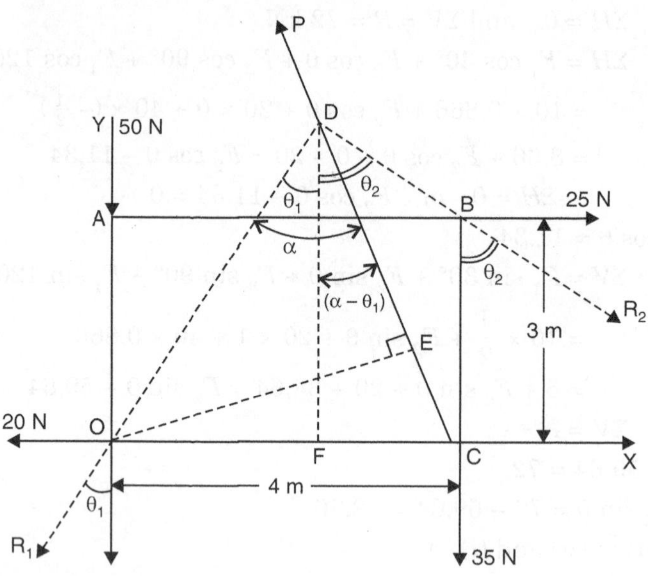

Fig. 1.56

The resultant of these forces

$$R_1 = \sqrt{50^2 + 20^2} = \sqrt{2900} = 53.85 \text{ N}$$

and $$\theta_1 = \tan^{-1}\left(\frac{20}{50}\right) = 21.8° \text{ with vertical axis.}$$

Similarly the forces 35 N and 25 N form a concurrent system and their line of action intersect at B.

The resultant of these forces

$$R_2 = \sqrt{25^2 + 35^2} = \sqrt{1850} = 43.01 \text{ N}$$

and $$\theta_2 = \tan^{-1}\left(\frac{25}{35}\right) = 35.53° \text{ with } BC \text{ i.e., with vertical line.}$$

These two forces R_1 and R_2 intersect at D. The angle between these forces is $\theta_1 + \theta_2$ *i.e.*, angle $R_1DR_2 = \theta_1 + \theta_2 = 21.8° + 35.53° = 57.33°$.

Let P be the resultant of the forces R_1 and R_2.

$$\therefore \quad P = \sqrt{R_1^2 + R_2^2 + 2R_1 \times R_2 \times \cos(57.33°)}$$

$$= \sqrt{53.85^2 + 43.01^2 + 2 \times 53.85 \times 43.01 \times \cos 57.33°}$$

$$= \sqrt{2900 + 1850 + 4632.17 \times 0.5398} = \mathbf{85.15\ N.} \quad \textbf{Ans.}$$

The angle made by the resultant P with R_1 is given by

$$\tan\alpha = \frac{R_2 \sin 57.33°}{R_1 + R_2 \cos 57.33°} = \frac{43.01 \sin 57.33°}{53.85 + 43.01 \times \cos 57.33°}$$

$$= \frac{43.01 \times 0.8418}{53.85 + 23.21} = \frac{36.2058}{77.06} = 0.4698$$

$$\therefore \quad \alpha = \tan^{-1} 0.4698 = 25.16°$$

Hence the resultant P makes $(\alpha - \theta_1)$ angle with vertical in anti-clockwise direction *i.e.*, P makes (25.16 – 21.8 = 3.36°). **Ans.**

Position of the Force P

The position of the force P is obtained by equating the clockwise moments and anti-clockwise moments about O (Refer to Fig. 1.56).

Let OE = Perpendicular distance between O and line of action of the force P.

Taking moments of all forces about O,

$$20 \times 0 + 50 \times 0 + 35 \times 4 + 25 \times 3 = P \times OE$$

or

$$0 + 0 + 140 + 75 = 85.147 \times OE \quad \text{or} \quad OE = \frac{215}{85.147} = 2.525 \text{ m}$$

From right angled triangle OED, $\sin \alpha = \dfrac{OE}{OD}$

$$\therefore \qquad OD = \frac{OE}{\sin \alpha} = \frac{OE}{\sin 25.16°} = \frac{2.525}{0.4241} = 5.939$$

Let x and y are the co-ordinates of the force P with reference to the axes with origin O.

Then $\qquad x = OF$ and $y = DF$

In right angled triangle OFD,

$$OF = OD \times \sin \theta_1 = 5.939 \times \sin 21.8° \qquad (\because \ \theta_1 = 21.8°)$$

$$= 2.20 \text{ m}$$

Also $\qquad FD = OD \times \cos \theta_1 = 5.939 \times \cos 21.8° = 5.514$ m

$\therefore \qquad x = OF =$ **2.20 m. Ans.**

and $\qquad y = FD =$ **5.514 m. Ans.**

Graphical method [Refer to Fig. 1.57(*a*)]

(*i*) To a suitable scale, take OG = 50 N and GH = 20 N. Join OH. Then OH represents the resultant R_1 in magnitude and direction. Produce the line HO backward.

(*ii*) From point B, Take BJ = 35 N and JK = 25 N. Join BK, which represents the resultant R_2 in magnitude and direction. Produce KB in the backward direction to interest the line of action of R_1 at point D.

(*iii*) To find the resultant of R_1 and R_2 (*i.e.*, force P) refer to Fig. 1.57(*b*).

(*iv*) Take any point '*a*'. From this point draw line *ab* parallel to R_1 and equal to R_1. From point '*b*', draw line *bc* parallel to R_2 and equal to R_2. Join the point *c* to *a*.

(*v*) Then *ca* represents in magnitude and direction the force P. Hence measure *ca*. Then

$P = ca =$ **85.15 N. Ans.**

(*vi*) From point D, draw the line DL parallel to *ca*. Hence DL represents the direction of the force P.

(*vii*) To find the position of the force P which is acting at point D, draw DF parallel to axes OY. Then OF represents the x-coordinate and FD represents the y-coordinate of the force P. Measure OF and FD. Then by measurement,

$OF = x =$ **2.20 m. Ans.**

and $\qquad FD = y =$ **5.514. Ans.**

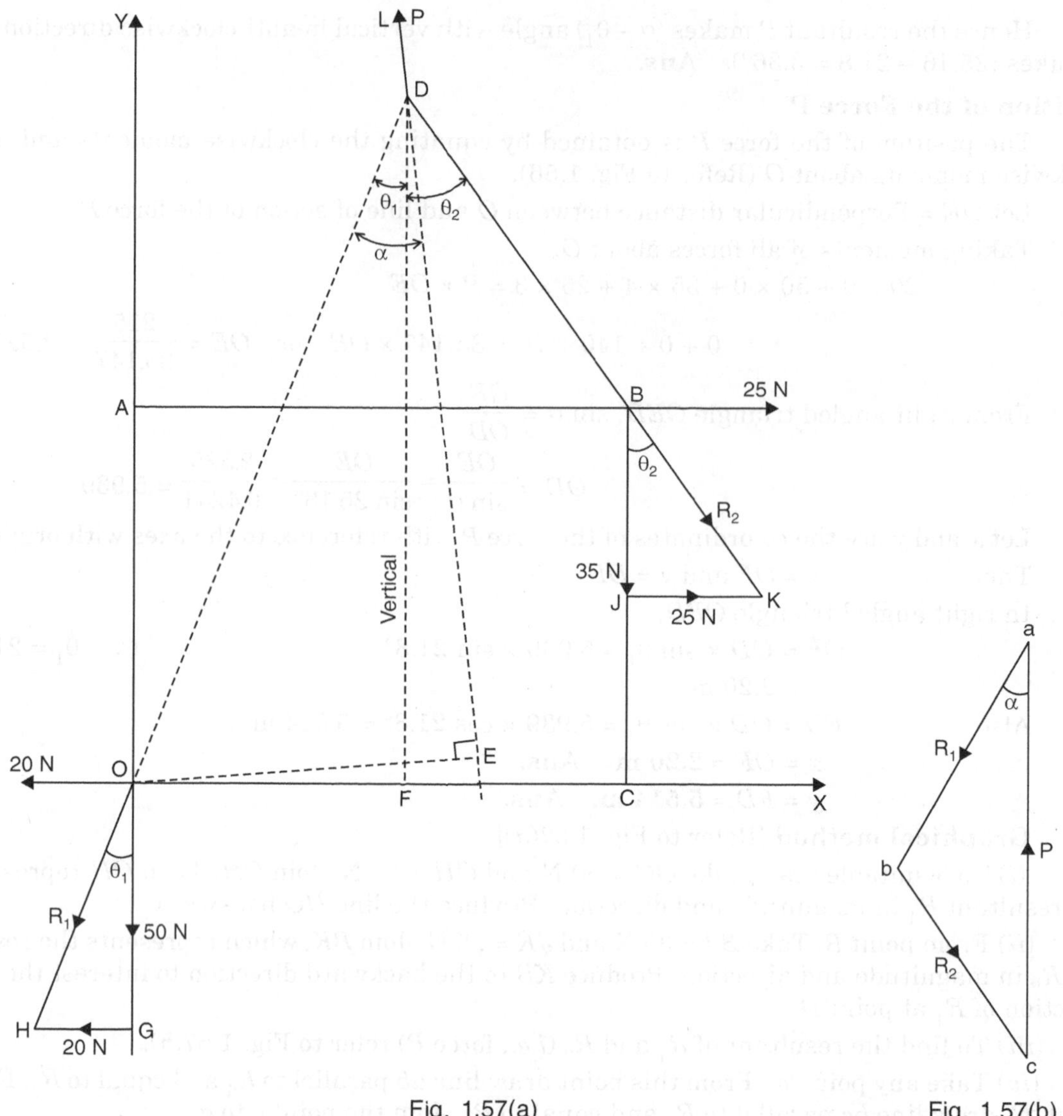

Fig. 1.57(a) Fig. 1.57(b)

1.10.3. Resultant of Parallel Coplaner Forces. The forces, which are having their line of action parallel to each other, are known parallel forces. The two parallel forces will not intersect at a point.

The following are the important types of parallel forces :

1. Like parallel forces,
2. Unlike parallel forces.

1. **Like parallel forces.** The parallel forces which are acting in the same direction, are known as like parallel forces. In Fig. 1.58, two parallel forces F_1 and F_2 are shown. They are acting in the same direction. Hence they are called as like parallel forces. These forces may be equal or unequal in magnitude.

F_1 F_2

Fig. 1.58

2. **Unlike parallel forces.** The parallel forces which are acting in the opposite direction, are known as unlike parallel forces. In Fig. 1.59, two parallel forces F_1, F_2 are acting in opposite direction. Hence they are called as unlike parallel forces. These forces may be equal or unequal in magnitude.

The unlike parallel forces may be divided into : (*i*) unlike equal parallel forces, and (*ii*) unlike unequal parallel forces.

Unlike equal parallel forces are those which are acting in opposite direction and are equal in magnitude.

Unlike unequal parallel forces are those which are acting in opposite direction and are unequal in magnitude.

The resultant of following two parallel forces will be considered :

Fig. 1.59

1. Two parallel forces are like.
2. Two parallel forces are unlike and are unequal in magnitude.
3. Two parallel forces are unlike but equal in magnitude.

1. **Resultant of two like parallel forces.** Fig. 1.60 shows a body on which two like parallel forces F_1 and F_2 are acting. It is required to determine the resultant (R) and also the point at which the resultant R is acting. For the two parallel forces which are acting in the same direction, obviously the resultant R is given by,

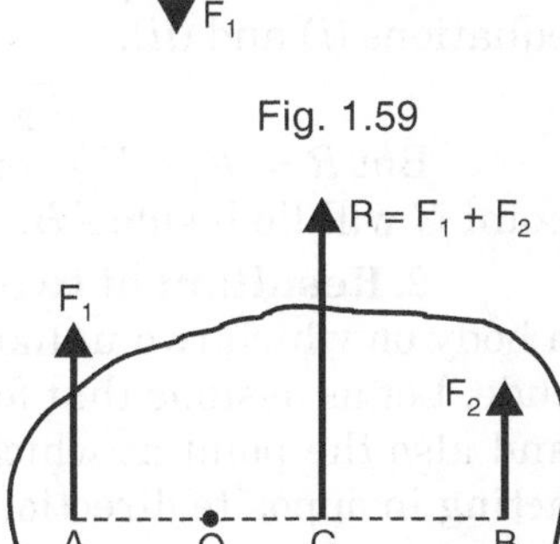

Fig. 1.60

$$R = F_1 + F_2$$

In order to find the point at which the resultant is acting, Varignon's principle (or method of moments) is used. According to this, the algebraic sum of moments of F_1 and F_2 about any point should be equal to the moment of the resultant (R) about that point. Now arbitrarily choose any point O along line AB and take moments of all forces about this point.

Moment of F_1 about $O = F_1 \times AO$ (clockwise) (–)

Moment of F_2 about $O = F_2 \times BO$ (anti-clockwise) (+ve)

$\therefore$ Algebraic sum of moments of F_1 and F_2 about O

$$= -F_1 \times AO + F_2 \times BO$$

Moment of resultant about $O = R \times OC$ (anti-clockwise)(+)

But according to principle of moments the algebraic sum of moments of F_1 and F_2 about O should be equal to the moment of resultant about the same point O.

$$\therefore \quad -F_1 \times AO + F_2 \times BO = +R \times CO = (F_1 + F_2) \times CO \qquad (\because \; R = F_1 + F_2)$$

or $$F_1(AO + CO) = F_2(BO - CO)$$

or $$F_1 \times AC = F_2 \times BC \qquad (\because \; AO + CO = AC \text{ and } BO - CO = BC)$$

or $$\frac{F_1}{F_2} = \frac{BC}{AC}$$

The above relation shows that the resultant R acts at the point C, parallel to the lines of action of the given forces F_1 and F_2 in such a way that the resultant divides the distance AB in the ratio inversely proportional to the magnitudes of F_1 and F_2. Also the point C lies in line AB *i.e.*, point C is not outside AB.

The location of the point C, at which the resultant R is acting, can also be determined by taking moments about points A of Fig. 1.60. As the force F_1 is passing through A, the moment of F_1 about A will be zero.

The moment of F_2 about $A = F_2 \times AB$ (anti-clockwise) (+)

Algebraic sum of moments of F_1 and F_2 about O

$$= O + F_2 \times AB = F_2 \times AB \text{ (anti-clockwise) (+)} \quad ...(i)$$

The moment of resultant R about A

$$= R \times AC \text{ (anti-clockwise)(+)} \quad ...(ii)$$

But according to the principle of moments, the algebraic sum of moments of F_1 and F_2 about A should be equal to the moment of resultant about the same point A. Hence equating equations (i) and (ii),

$$F_2 \times AB = R \times AC$$

But $R = (F_1 + F_2)$ hence the distance AC should be less than AB. Or in other words, the point C will lie inside AB.

2. **Resultant of two unlike parallel forces (unequal in magnitude).** Fig. 1.61 shows a body on which two unlike parallel forces F_1 and F_2 are acting which are unequal in magnitude. Let us assume that force F_1 is more than F_2. It is required to determined the resultant R and also the point at which the resultant R is acting. For the two parallel forces, which are acting in opposite direction, obviously the resultant is given by,

$$R = F_1 - F_2$$

Let the resultant R is acting at C as shown in Fig. 1.61.

In order to find the point C, at which the resultant is acting, principle of moments is used.

Choose arbitrarily any point O in line AB. Take the moments of all forces (*i.e.*, F_1, F_2 and R) about this point.

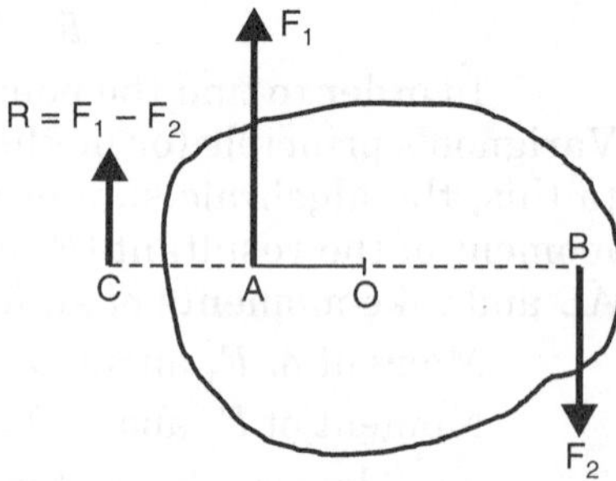

Fig. 1.61

Moment of F_1 about $O = F_1 \times AO$ (clockwise)

Moment of F_2 about $O = F_2 \times BO$ (clockwise)

Algebraic sum of moments of F_1 and F_2 about O

$$= F_1 \times AO + F_2 \times BO \quad ...(i)$$

Moment of resultant force R about O

$$= R \times CO \text{ (clockwise)}$$

$$= (F_1 - F_2) \times CO \quad (\because R = F_1 - F_2)$$

$$= F_1 \times CO - F_2 \times CO \quad ...(ii)$$

But according to the principle of moments, the algebraic sum of moments of all forces about any point should be equal to the moment of resultant about that point. Hence equating equations (i) and (ii), we get

$$F_1 \times AO + F_2 \times BO = F_1 \times CO - F_2 \times CO$$

or
$$F_2(BO + CO) = F_1(CO - AO)$$

$$F_2 \times BC = F_1 \times AC \quad (\because BO + CO = BC \text{ and } CO - AO = AC)$$

or
$$\frac{BC}{AC} = \frac{F_1}{F_2} \quad \text{or} \quad \frac{F_1}{F_2} = \frac{BC}{AC}$$

But $F_1 > F_2$, hence BC will be more than AC. Hence point O lies outside of AB and on the same side as the larger force F_1. *Thus in case of two unlike parallel forces the resultant lies outside the line joining the points of action of the two forces and on the same side as the larger force.*

The location of the point C, at which the resultant R is acting, can also be determined by taking moments about point A, of Fig. 1.61. As the force F_1 is passing through A, the moment of F_1 about A will be zero.

The moment of F_2 about $A = F_2 \times AB$ (clockwise) (–)

Algebraic sum of moments of F_1 and F_2 about A

$$= O + F_2 \times AB = F_2 \times AB \text{ (clockwise) (–)} \quad ...(i)$$

The moment of resultant R about A should be equal to the algebraic sum of moments of F_1 and F_2 (*i.e.*, $= F_2 \times AB$) according to the principle of moments. Also the moment of resultant R about A should be clockwise. As R is acting upwards [$\because$ $F_1 > F_2$ and $R = (F_1 - F_2)$ so R is acting in the direction of F_1], the moment of resultant R about A would be clockwise only if the points C is towards the left of point A. Hence the point C will be outside the line AB and on the side of F_1 (*i.e.*, larger force).

Now the moment of resultant R about A

$$= R \times AC \text{ (clockwise) (–)} \quad ...(ii)$$

Equating equations (*i*) and (*ii*),

$$F_2 \times AB = R \times AC$$

$$= (F_1 - F_2) \times AC \quad (\because \ R = F_1 - F_2)$$

As F_1, F_2 and AB are known, hence AC can be calculated. Or in other words, the location of point C is known.

3. **Resultant of two unlike parallel forces which are equal in magnitude.** When two equal and opposite parallel forces act on a body, at some distance apart, the two forces from a *couple* which has a tendency to rotate the body. The perpendicular distance between the parallel forces is known as *arm of the couple.*

Fig. 1.62 shows a body on which two parallel forces, which are acting in opposite direction but equal in magnitude are acting. These two forces will form a couple which will have a tendency to rotate the body in clockwise direction. The moment of the couple is the product of either one of the forces and perpendicular distance between the forces.

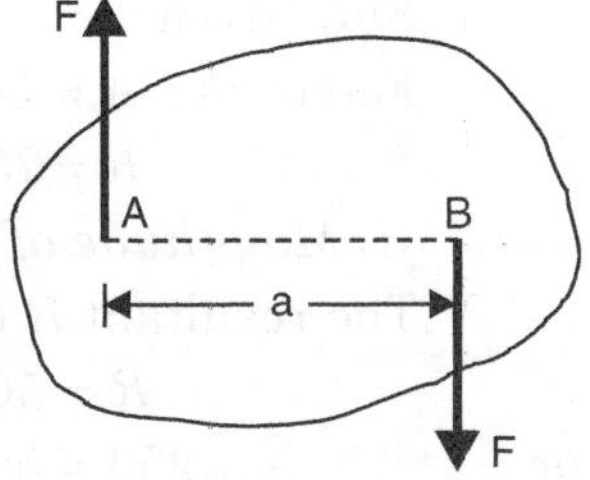

Fig. 1.62

Let F = Force at A or at B

a = Perpendicular distance (or arm of the couple)

The moment (M) of the couple is given by, $M = F \times a$.

The units of moment will be Nm.

Problem 1.31. *Three like parallel forces 100 N, 200 N and 300 N are acting at points A, B and C respectively on a straight line ABC as shown in Fig. 1.63. The distances are AB = 30 cm and BC = 40 cm. Find the resultant and also the distance of the resultant from point A on line ABC.*

Sol. Given :

Force at A = 100 N

Force at B = 200 N

Force at C = 300 N

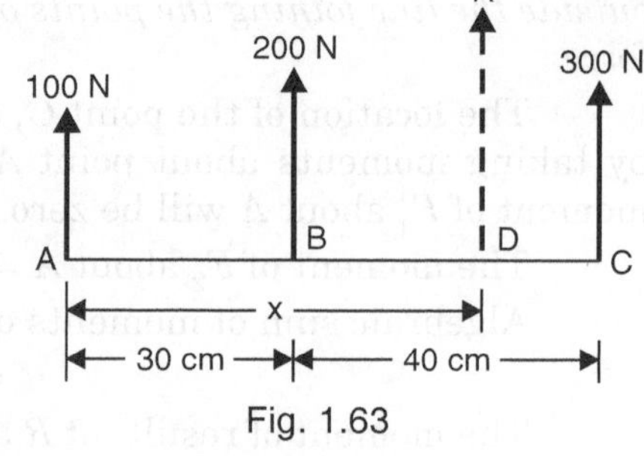

Fig. 1.63

Distance $AB = 30$ cm, $BC = 40$ cm. As all the forces are parallel and acting in the same direction, their resultant R is given by

$$R = 100 + 200 + 300 = 600 \text{ N}$$

Let the resultant is acting at a distance of x cm from the point A as shown in Fig. 1.63.

Now take the moments of all forces about point A. The force 100 N is passing A, hence its moment about A will be zero.

$\therefore$ Moment of 100 N force about $A = 0$

Moment of 200 N force about $A = 200 \times 30 = 6000$ N cm (anti-clockwise)

Moment of 300 N force about $A = 300 \times AC$

$= 300 \times 70 = 21000$ N cm (anti-clockwise)

Algebraic sum of moments of all forces about A

$= 0 + 6000 + 21000 = 27000$ N cm (anti-clockwise)

Moment of resultant R about $A = R \times x$

$= 600 \times x$ N cm ($\because$ $R = 600$)

But algebraic sum of moments of all forces about A

= Moment of resultant about A

or $27000 = 600 \times x$ or $x = \dfrac{27000}{600} =$ **45 cm. Ans.**

Problem 1.32. *The three like parallel forces of magnitude 50 N, F and 100 N are shown in Fig. 1.64. If the resultant R = 250 N and is acting at a distance of 4 m from A, then find*

(i) Magnitude of force F.

(ii) Distance of F from A.

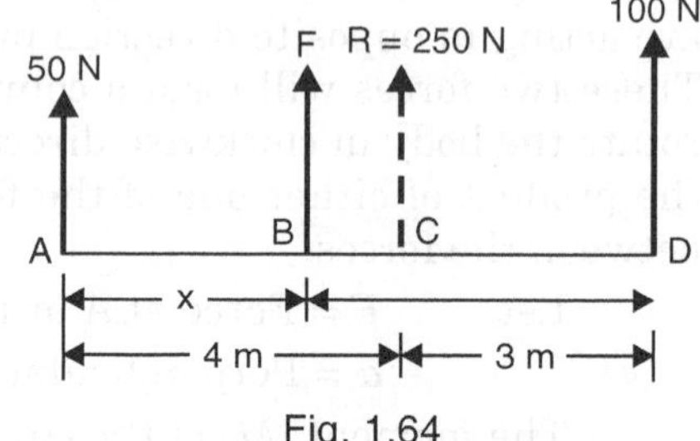

Fig. 1.64

Sol. Given :

Forces at $A = 50$ N, at $B = F$ and $D = 100$ N

$R = 250$ N, Distance $AC = 4$ m, $CD = 3$ m.

(i) Magnitude of force F

The resultant R of three like forces is given by,

$$R = 50 + F + 100$$

or $250 = 50 + F + 100$ ($\because$ $R = 250$)

$\therefore$ $F = 250 - 50 - 100 =$ **100 N. Ans.**

(ii) Distance of F from A

Take the moments of all forces about point A.

Moment of force 50 N about $A = 0$ ($\because$ force 50 N is passing through)

Moment of force F about $A = F \times x$ (anti-clockwise)

Moment of force 100 N about $A = 100 \times AD = 100 \times 7 = 700$ N m (anti-clockwise)

$\therefore$ Algebraic sum of moments of all forces about A

$= 0 + F \times x + 700$ N m

$= F \times x + 700$ N m (anti-clockwise)

Moment of resultant R about $A = R \times 4 = 250 \times 4 = 1000$ N m (anti-clockwise)

But algebraic sum of moments of all forces about A must be equal to the moment of resultant R about A.

$\therefore \quad F \times x + 700 = 1000 \quad \text{or} \quad F \times x = 1000 - 700 = 300$

or $\quad x = \frac{300}{F} = \frac{300}{100} \quad (\because \; F = 100 \text{ N})$

$= 3$ **m. Ans.**

Problem 1.33. *Four parallel forces of magnitudes 100 N, 150 N, 25 N and 200 N are shown in Fig. 1.65. Determine the magnitude of the resultant and also the distance of the resultant from point A.*

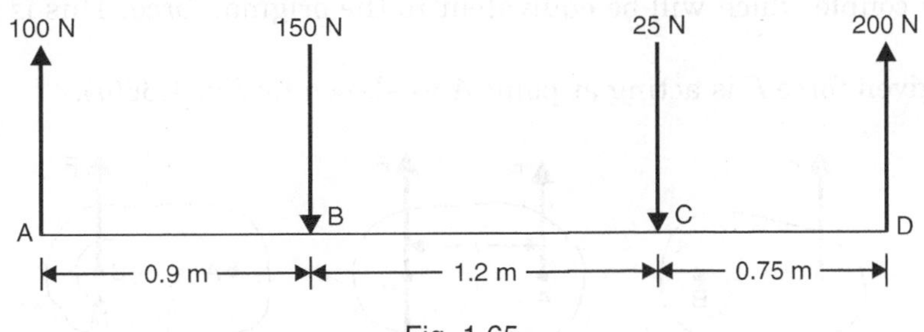

Fig. 1.65

Sol. Given :

Forces are 100 N, 150 N, 25 N and 200 N.

Distances $\quad AB = 0.9$ m, $BC = 1.2$ m, $CD = 0.75$ m.

As all the forces are acting vertically, hence their resultant R is given by

$$R = 100 - 150 - 25 + 200$$

(Taking upward force +ve and downward as –ve)

$$= 300 - 175 = 125 \text{ N}$$

+ve sign shows that R is acting vertically upwards. To find the distance of R from point A, take the moments of all forces about point A.

Let $\quad x$ = Distance of R from A in metre.

As the force 100 N is passing through A, its moment about A will be zero.

Moment of 150 N force about $A = 150 \times AB$

$= 150 \times 0.9$ (clockwise) (–) $= -135$ Nm

Moment of 25 N force about $A = 25 \times AC = 25 \times (0.9 + 1.2)$

$= 25 \times 2.1$ (clockwise) (–) $= -52.5$ Nm.

Moment of 200 N force about $A = 200 \times AD$

$= 200 \times (0.9 + 1.2 + 0.75)$

$= 200 \times 2.85$ (anti-clockwise) (+) $= 570$ Nm

Algebraic sum of moments of all forces about A

$= -135 - 52.5 + 570 = 382.5$ Nm ...(*i*)

+ve sign shows that this moment is anti-clockwise. Hence the moment of resultant R about A must be 382.5 Nm, *i.e.*, moment of R should be anti-clockwise about A. The moment of R about A will be anti-clockwise if R is acting upwards and towards the right of A.

Now moment of R about $A = R \times x$. But $R = 125$

$$= 125 \times x \quad \text{(anti-clockwise) (+)}$$

$$= +\ 125 \times x \quad ...(ii)$$

Equating (*i*) and (*ii*), $382.5 = 125 \times x$ or $x = \frac{382.5}{125} = \mathbf{3.06\ m.}$ **Ans.**

$\therefore$ Resultant (R = 125 N) will be 125 N upwards and is acting at a distance of 3.06 m to the right of point A.

1.10.4. Resolution of a Force into a Force and a Couple. A given force F applied to a body at any point A can always be replaced by an equal force applied at another point B together with a couple which will be equivalent to the original force. This is proved as given below :

Let the given force F is acting at point A as shown in Fig. 1.66(*a*).

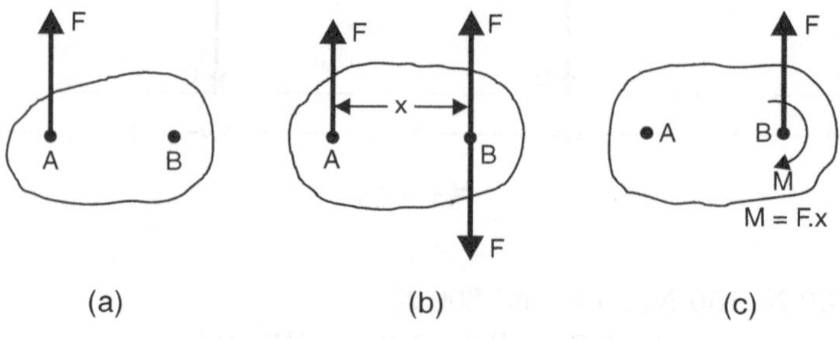

Fig: 1.66

This force is to be replaced at the point B. Introduce two equal and opposite forces at B, each of magnitude F and acting parallel to the force at A as shown in Fig. 1.66(*b*). The force system of Fig. 1.66(*b*) is equivalent to the single force acting at A of Fig. 1.66(*a*). In Fig. 1.66(*b*) three equal forces are acting. The two forces *i.e.*, force F at A and the oppositely directed force F at B (*i.e.*, vertically downward force at B) from a couple. The moment of this couple is $F \times x$ clockwise where x is the perpendicular distance between the lines of action of forces at A and B. The third force is acting at B in the same direction in which the force at A is acting. In Fig. 1.66(*c*), the couple is shown by curved arrow with symbol M. The force system of Fig. 1.66(*c*) is equivalent to Fig. 1.66(*b*). Or in other words the Fig. 1.66(*c*) is equivalent to Fig. 1.66(*a*). Hence the given force F acting at A has been replaced by an equal and parallel force applied at point B in the same direction together with a couple of moment $F \times x$.

Thus a force acting at a point in a rigid body can be replaced by an equal and parallel force at any other point in the body, and a couple.

Problem 1.34. *A system of parallel forces are acting on a rigid bar as shown in Fig. 1.67. Reduce this system to :*

(*i*) *a single force*

(*ii*) *a single force and a couple at A*

(*iii*) *a single force and a couple at B.*

Sol. Given :

Forces at A, C, D and B are 32.5 N, 150 N, 67.5 N and 10 N respectively.

Distances AC = 1 m, CD = 1 m and BD = 1.5 m.

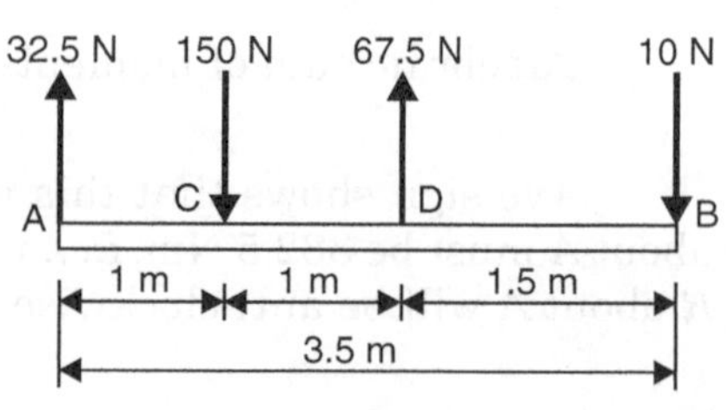

Fig. 1.67

(*i*) *Single force system.* The single force system will consist only resultant force in magnitude and location. All the forces are acting in the vertical direction and hence their resultant (R) in magnitude is given by

$$R = 32.5 - 150 + 67.5 - 10 = \mathbf{-60\ N.}\quad \textbf{Ans.}$$

Negative sign shows that resultant is acting vertically downwards.

Let x = Distance of resultant from A towards right. To find the location of the resultant take the moments of all forces about A, we get moment of resultant about A.

= Algebraic sum of moments of all forces about A

or $$R \times x = -50 \times AC + 67.5 \times AD - 10 \times AB$$

(Taking clockwise moment –ve and anticlockwise moment +ve)

or $$(-60) \times x = 50 \times 1 + 67.5 \times 2 - 10 \times 3.5 \quad (\because\ R = -60)$$

or $$-60x = -150 + 135 - 35 = -50$$

$\therefore$ $$x = \frac{-50}{-60} = \mathbf{0.833\ m.}\quad \textbf{Ans.}$$

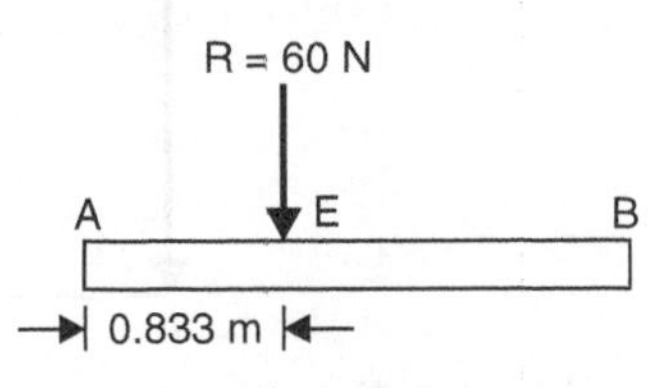

Fig. 1.67(a)

Hence the given system of parallel forces is equivalent to a single force 60 N acting vertically downwards at point E at a distance of 0.833 m from A shown in Fig. 1.67(*a*).

(*ii*) *A single force and a couple at A.* The resultant force R acting at point E as shown in Fig. 1.67(*a*) can be replaced by an equal force applied at point A in the same direction together with a couple. This is shown in Fig. 1.67(*c*).

The moment of the couple = 60 × 0.833 Nm (clockwise)

= **– 49.98 Nm. Ans.** (–ve sign is due to clockwise)

(*iii*) *A single force and a couple at B.* First find distance BE. But from Fig. 1.67(*b*), the distance

$$BE = AB - AE = 3.5 - 0.833 = 2.667\ \text{m}.$$

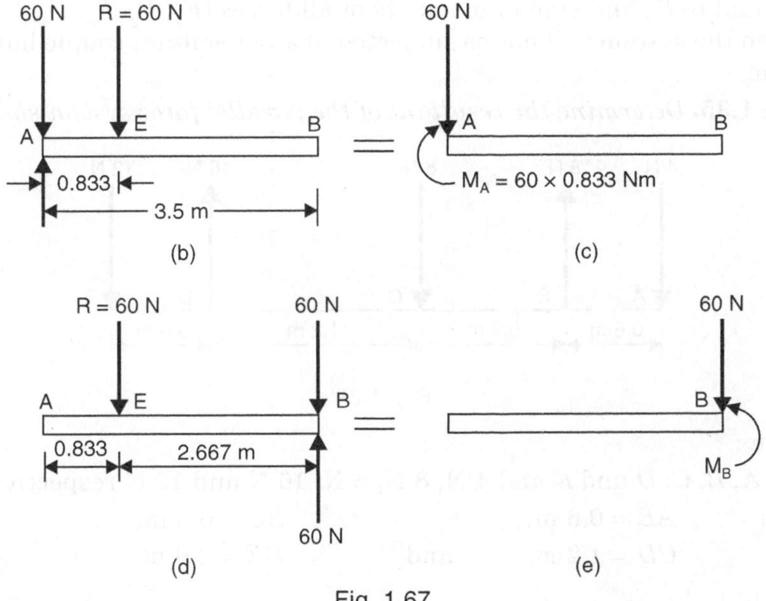

Fig. 1.67

Now if the force $R = 60$ N is moved to the point B, it will be accompanied by a couple of moment $60 \times BE$ or 60×2.667 Nm. This is shown in Fig. 1.67(*e*).

The moment of the couple $= 60 \times 2.667$ Nm (anti-clockwise)

$= \mathbf{160\ Nm.}$ **Ans.**

1.10.5. General Case of Parallel Forces in a Plane. Fig. 1.68 shows a number of parallel forces acting on a body in one plane. The forces F_1, F_2 and F_4 are acting in one direction, whereas the forces F_3 and F_5 are acting in the opposite direction. Let R_1 = Resultant of forces F_1, F_2 and F_5 and R_2 = Resultant of forces F_3 and F_5. The resultants R_1 and R_2 are acting in opposite direction and are parallel to each other. Now three important cases are possible.

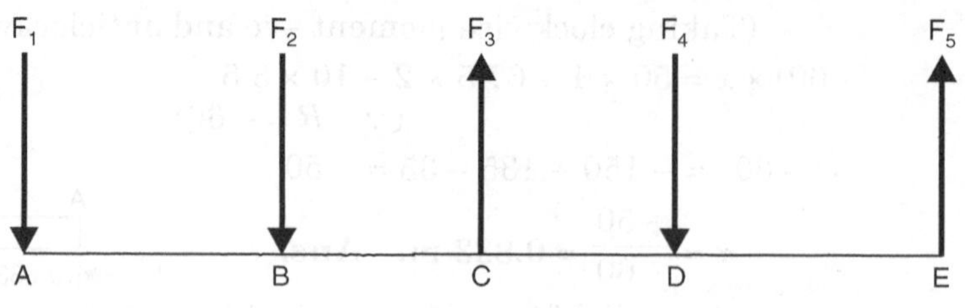

Fig. 1.68

1. R_1 may not be equal to R_2. Then we shall have two unequal parallel forces (R_1 and R_2) acting in the opposite direction. The resultant R of these two forces (R_1 and R_2) can be easily obtained. The point of application of resultant R can be obtained by equating the moment of R about any point to the algebraic sum of the moments of individual forces about the same point.

2. R_1 is equal to R_2. Then we shall have two equal parallel forces (R_1 and R_2) acting in the opposite direction. The resultant R of these two forces will be zero. Now the system may reduce to a couple or the system is in equilibrium. To distinguish between these two cases, the algebraic sum of moments of all forces (F_1, F_2, ...F_5) about any point is taken. If the sum of moments is not zero, the system reduces a resultant couple. The calculated moment gives the moment of this couple.

3. R_1 is equal to R_2 and sum of moments of all forces (F_1, F_2, F_3, F_4, F_5, ...) about any point is zero, then the system will not be subjected to any resultant couple but the system will be in equilibrium.

Problem 1.35. *Determine the resultant of the parallel force system shown in Fig. 1.69.*

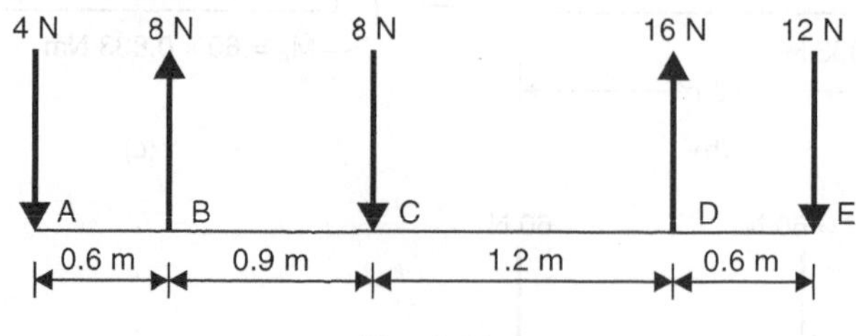

Fig. 1.69

Sol. Given :

Forces at A, B, C, D and E and 4 N, 8 N, 8 N, 16 N and 12 N respectively.

Distances $AB = 0.6$ m, $BC = 0.9$ m,

$CD = 1.2$ m, and $DE = 0.6$ m.

Since all the forces are vertical and parallel, hence their resultant is given by

$$R = -4 + 8 - 8 + 16 - 12 = 0$$

As the resultant force on the system is zero, there will be two possibilities. The system has a resultant couple or the system is in equilibrium. To distinguish between these two possibilities, take the sum of moments of all forces about any point. Let us take the moments about point A.

$\therefore$ Algebraic sum of moments of all forces about A

$$= 4 \times 0 + 8 \times AB - 8 \times AC + 16 \times AD - 12 \times AE$$
$$= 0 + 8 \times 0.6 - 8 \times (0.6 + 0.9) + 16 \times (0.6 + 0.9 + 1.2) - 12 \times (0.6 + 0.9 + 1.2 + 0.6)$$
$$= 0 + 4.8 - 12 + 16 \times 2.7 - 12 \times 3.3 \text{ Nm}$$
$$= 4.8 - 12 + 43.2 - 39.6 = 48 - 51.6$$
$$= -3.6 \text{ Nm}$$

As the algebraic sum of moments of all forces about any point is not zero, the system will have a resultant couple of magnitude -3.6 Nm *i.e.*, a clockwise couple. **Ans.**

Problem 1.36. *Determine the resultant of the parallel forces acting on a body as shown in Fig. 1.70.*

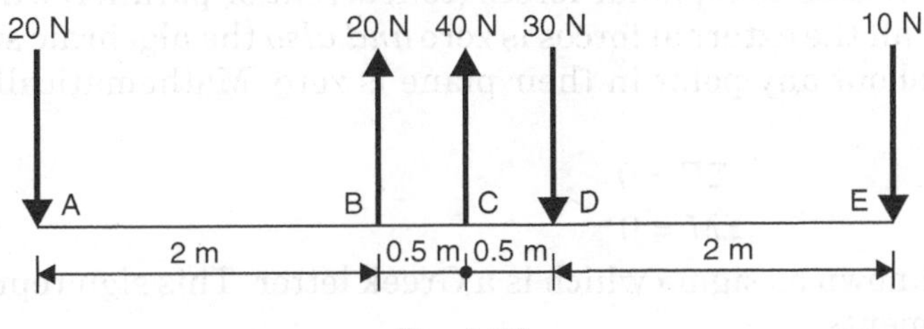

Fig. 1.70

Sol. Since all the forces are vertical and parallel, hence their resultant is given by

$$R = -20 + 20 + 40 - 30 - 10 = 0$$

Taking moments of all forces about the point A, we get

Resultant moment $= 20 \times 0 + 20 \times 2 + 40 \times 2.5 - 30 \times 3 - 10 \times 5$

$$= 0 + 40 + 100 - 90 - 50 = 140 - 140 = 0$$

As the resultant moment is zero and also the resultant force on the body is zero, the body will be in equilibrium. **Ans.**

1.11. EQUILIBRIUM OF A PARTICLE

If a single force is acting on a particle, the particle cannot be in equilibrium. But if two forces or three forces or more than three forces are acting on the particle, the particle can be in equilibrium under certain conditions.

When two forces are acting. If two forces are acting on a particle, the particle will be in equilibrium when the two forces are *equal, opposite* and *collinear.*

When three forces are acting. If three forces are acting on a particle, the particle will be in equilibrium when the three forces are concurrent and

(*i*) Resultant force in x and y direction is zero *i.e.*,

$$\Sigma F_x = 0 \quad \text{and} \quad \Sigma F_y = 0$$

(*ii*) The three forces obey Lame's Theorem* which states that "If a particle (or a body) is in equilibrium under the action of three forces, each force is proportional to the sine of the angle between other two."

Or

(*iii*) The three forces form a closed triangle.

When more than three forces are acting. If more than three forces are acting on a particle, the particle will be in equilibrium when the resultant force in x and y directions is zero *i.e.,*

$$\Sigma F_x = 0 \quad \text{and} \quad \Sigma F_y = 0.$$

1.12. EQUILIBRIUM OF A RIGID BODY

When some external forces (which may be concurrent or parallel) are acting on a stationary body, the body may start moving or may start rotating about any point. But if the body does not start moving *and also* does not start rotating about any point, then the body** is said to be in equilibrium.

1.12.1. Principle of Equilibrium. The principle of equilibrium states that, a stationary body which is subjected to coplanar forces (concurrent or parallel) will be in equilibrium if the algebraic sum of all the external forces is zero *and also* the algebraic sum of moments of all the external forces about any point in their plane is zero. Mathematically, it is expressed by the equations :

$$\Sigma F = 0 \qquad ...(1.22)$$

$$\Sigma M = 0 \qquad ...(1.23)$$

The sign Σ is known as sigma which is a Greek letter. This sign represents the *algebraic sum* of forces or moments.

The equation (1.22) is also known as force law of equilibrium whereas the equation (1.23) is known as moment law of equilibrium.

The forces are generally resolved into horizontal and vertical components. Hence equation (1.22) is written as

$$\Sigma F_x = 0 \qquad ...(1.24)$$

and

$$\Sigma F_y = 0 \qquad ...(1.25)$$

where ΣF_x = Algebraic sum of all horizontal components

and ΣF_y = Algebraic sum of all vertical components.

1.12.2. Equations of Equilibrium for Non-concurrent Force Systems. A non-concurrent force systems will be in equilibrium if the resultant of all forces and moment is zero. Hence the equations of equilibrium are

$$\Sigma F_x = 0, \Sigma F_y = 0 \quad \text{and} \quad \Sigma M = 0.$$

1.12.3. Equations of Equilibrium for Concurrent Force System. For the concurrent forces, the lines of action of all forces meet at a point, and hence the moment of those force about that very point will be zero or $\Sigma M = 0$ automatically.

Thus for concurrent force system, the condition $\Sigma M = 0$ becomes redundant and only two conditions, *i.e.*, $\Sigma F_x = 0$ and $\Sigma F_y = 0$ are required.

*Lame's Theorem has been explained in more details in Art 1.1.6 on page 4.

**A body will be in equilibrium if both the resultant of the forces and resultant moment are zero, whereas a particle will be in equilibrium if the resultant force acting on it is zero.

1.12.4. Force Law of Equilibrium. Force law of equilibrium is given by equation (1.22) or by equations (1.24) and (1.25). Let us apply this law to the following important force system :

(*i*) Two force system (*ii*) Three force system

(*iii*) Four force system.

1.12.5. Two Force System. When a body is subjected to two forces, then the body will be in equilibrium if the two forces are *collinear, equal and opposite* as shown in Fig. 1.71.

Fig. 1.71

If the two forces acting on a body are equal and opposite but are parallel, as shown in Fig. 1.71(*a*), then the body will not be in equilibrium. This is due to the fact that the three conditions of equilibrium will not be satisfied. This is proved as given below :

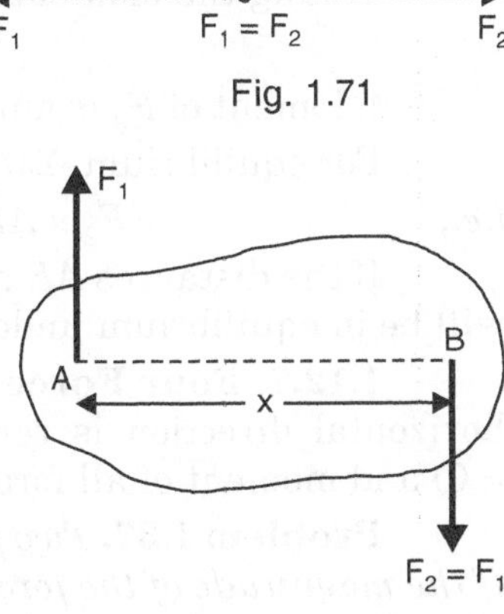

Fig. 1.71(a)

(*i*) Here $\Sigma F_x = 0$ as there is no horizontal force acting on the body. Hence first condition of equilibrium is satisfied.

(*ii*) Also here $\Sigma F_y = 0$ as $F_1 = F_2$.

Hence second condition of equilibrium is also satisfied.

(*iii*) ΣM about any point should be zero. The resultant moment about point A is given by

$$M_A = -F_2 \times AB \qquad \text{(–ve sign is due to clockwise moment)}$$

But M_A is not equal to zero. Hence the third condition is not satisfied.

Hence a body will not be in equilibrium under the action of two equal and opposite parallel forces.

Two equal and opposite parallel forces produce a couple and moment of the couple is $-F_1 \times AB$ [Fig. 1.71(*a*)].

1.12.6. Three Force System. The three forces acting on a body which is in equilibrium may be either concurrent or parallel. Let us first consider that the body is in equilibrium when three forces, acting on the body, are concurrent. This is shown in Fig. 1.72.

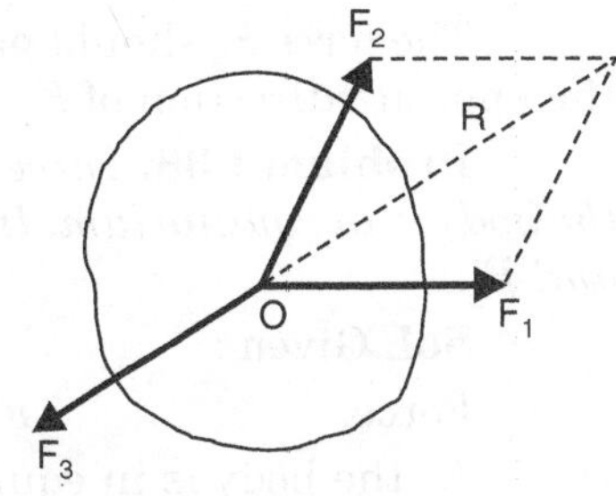

Fig. 1.72

(*a*) *When three forces are concurrent.* The three concurrent forces F_1, F_2 and F_3 are acting on a body at point O and the body is in equilibrium. The resultant of F_1 and F_2 is given by R. If the force F_3 is collinear, equal and opposite to the resultant R, then the body will be in equilibrium. The force F_3 which is equal and opposite to the resultant R is known as *equilibrant.* Hence for three concurrent forces acting on a body when the body is in equilibrium, the resultant of the two forces should be equal and opposite to the third force.

(*b*) *When three forces are parallel.* Fig. 1.73 shows a body on which three parallel forces F_1, F_2 and F_3 are acting and the body is in equilibrium. If three forces F_1, F_2 and F_3 are acting in the same direction, then there will be a resultant $R = F_1 + F_2 + F_3$ and body will not be in equilibrium. The three forces are acting in opposite direction and their magnitude is so adjusted that there is no resultant force and body is in equilibrium. Let us suppose that F_2 is acting in opposite direction as shown in Fig. 1.73.

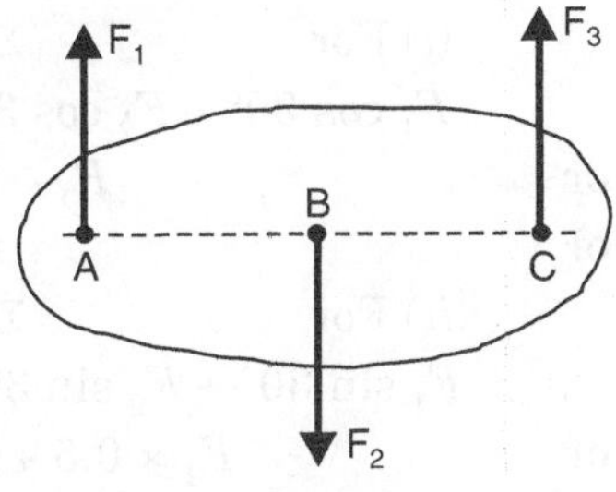

Fig. 1.73

Now let us apply the three conditions of equilibrium :

(*i*) $\Sigma F_x = 0$ as there is no horizontal force acting on the body

(*ii*) $\Sigma F_y = 0$ *i.e.*, $F_1 + F_3 = F_2$

(*iii*) $\Sigma M = 0$ about any point.

Taking the moments of F_1, F_2 and F_3 about point A,

$$\Sigma M_A = -F_2 \times AB + F_3 \times AC$$

(Moment of F_3 is anti-clockwise whereas moment of F_2 is clockwise)

For equilibrium, ΣM_A should be zero

i.e., $$-F_2 \times AB + F_3 \times AC = 0$$

If the distances AB and AC are such that the above equation is satisfied, then the body will be in equilibrium under the action of three parallel forces.

1.12.7. Four Force System. The body will be in equilibrium if the resultant force in horizontal direction is zero (*i.e.*, $\Sigma F_x = 0$), resultant force in vertical direction is zero (*i.e.*, $\Sigma F_y = 0$) and moment of all forces about any point in the plane of forces is zero (*i.e.*, $\Sigma M = 0$).

Problem 1.37. *Two forces F_1 and F_2 are acting on a body and the body is in equilibrium. If the magnitude of the force F_1 is 100 N and its acting at O long x-axis as shown in Fig. 1.74, then determine the magnitude and direction of force F_2.*

Sol. Given :

Force, $F_1 = 100$ N

The body is in equilibrium under the action of two forces F_1 and F_2.

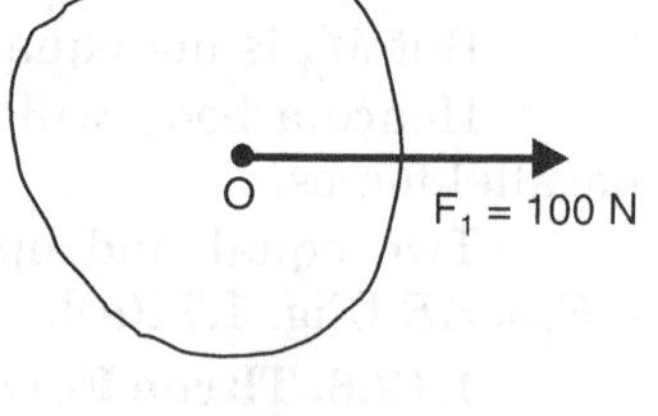

Fig. 1.74

When two forces are acting on a body and the body is in equilibrium, then the two forces should be collinear, equal and opposite.

$\therefore$ $$F_2 = F_1 = 100 \text{ N}$$

The force F_2 should pass through O, and would be acting in the opposite direction of F_1.

Problem 1.38. *Three forces F_1, F_2 and F_3 are acting on a body as shown in Fig. 1.75 and the body is in equilibrium. If the magnitude of force F_3 is 400 N, find the magnitudes of force F_1 and F_2.*

Sol. Given :

Force, $F_3 = 400$ N

As the body is in equilibrium, the resultant force in x-direction should be zero and also the resultant force in y-direction should be zero.

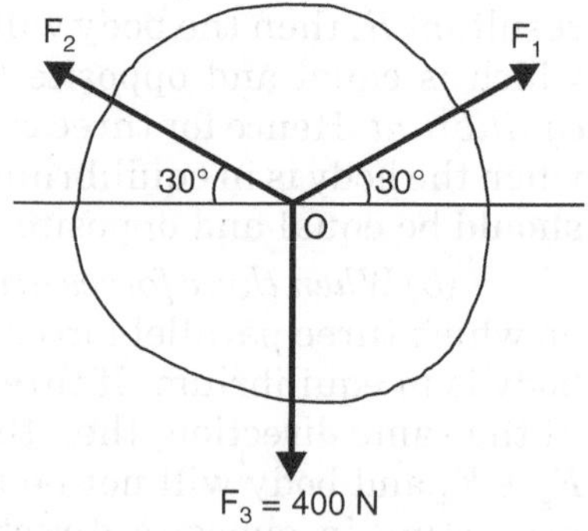

Fig. 1.75

(*i*) For $\Sigma F_x = 0$, we get

$$F_1 \cos 30° - F_2 \cos 30° = 0$$

or $$F_1 - F_2 = 0$$

or $$F_1 = F_2 \quad \ldots(i)$$

(*ii*) For $\Sigma F_y = 0$, we get

$$F_1 \sin 30° + F_2 \sin 30° - 400 = 0$$

or $$F_1 \times 0.5 + F_2 \times 0.5 = 400$$

or $\quad F_1 \times 0.5 + F_2 \times 0.5 = 400 \qquad (\because \; F_2 = F_1)$

or $\quad F_1 = \mathbf{400\ N.}$ **Ans.**

Also $\quad F_2 = F_1 = \mathbf{400\ N.}$ **Ans.**

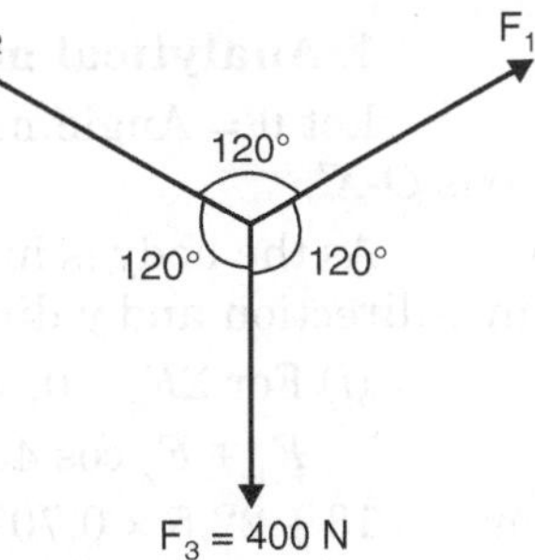

Fig. 1.75(a)

2nd Method

If three forces are acting on a body at a point and the body is in equilibrium, Lame's Theorem can be applied.

Using Lame's theorem,

$$\frac{F_1}{\sin 120°} = \frac{F_2}{\sin 120°} = \frac{400}{\sin 120°}$$

or $\quad F_1 = F_2 = \mathbf{400\ N.}$ **Ans.**

Problem 1.39. *Three parallel forces F_1, F_2 and F_3 are acting on a body as shown in Fig. 1.76 and the body is in equilibrium. If force F_1 = 250 N and F_3 = 1000 N and the distance between F_1 and F_2 = 1.0 m, then determine the magnitude of force F_2 and the distance of F_2 from force F_3.*

Sol. Given :

Force, $\quad F_1 = 250$ N

Force, $\quad F_3 = 1000$ N

Distance $\quad AB = 1.0$ m

The body is in equilibrium.

Find F_2 and distance BC.

Fig. 1.76

For the equilibrium of the body, the resultant force in the vertical direction should be zero (here there is no force in horizontal direction).

$\therefore$ For $\quad \Sigma F_y = 0$, we get

$$F_1 + F_3 - F_2 = 0$$

or $\quad 250 + 1000 - F_2 = 0$

or $\quad F_2 = 250 + 1000 = \mathbf{1250\ N.}$ **Ans.**

For the equilibrium of the body, the moment of all forces about any point must be zero.

Taking moments of all forces about point A and considering distance $BC = x$, we get

$$F_2 \times AB - AC \times F_3 = 0$$

or $\quad 1250 \times 1.0 - (1 + x) \times 1000 = 0 \qquad (\because \; AC = AB + BC = 1 + x)$

or $\quad 1250 - 1000 - 1000x = 0$

or $\quad 250 = 1000x$

or $\quad x = \dfrac{250}{1000} = \mathbf{0.25\ m.}$ **Ans.**

Problem 1.40. *The five forces F_1, F_2, F_3, F_4 and F_5 are acting at a point on a body as shown in Fig. 1.77 and the body is in equilibrium. If F_1 = 18 N, F_2 = 22.5 N, F_3 = 15 N and F_4 = 30 N, find the force F_5 in magnitude and direction.*

Sol. Given :

Forces, $\quad F_1 = 18$ N, $\quad F_2 = 2.25$ N,

$F_3 = 15$ N and $F_4 = 30$ N.

The body is in equilibrium. Find force F_5 in magnitude and direction. This problem can be solved analytically and graphically.

1. Analytical method

Let θ = Angle made by force F_5 with horizontal axis $O\text{-}X'$.

As the body is in equilibrium, the resultant force in x-direction and y-direction should be zero.

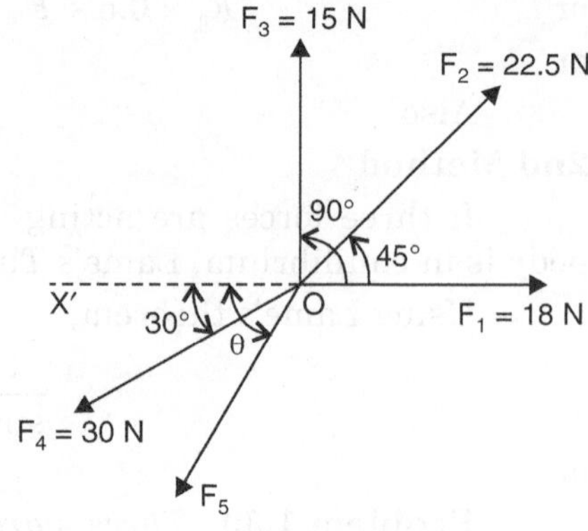

Fig. 1.77

(*i*) For $\Sigma F_x = 0$, we get

$$F_1 + F_2 \cos 45° - F_4 \cos 30° - F_5 \cos \theta = 0$$

or $$18 + 22.5 \times 0.707 - 30 \times 0.866 - F_5 \cos \theta = 0$$

or $$18 + 15.9 - 25.98 - F_5 \cos \theta = 0$$

or $$F_5 \cos \theta = 18 + 15.9 - 25.98$$

or $$F_5 \cos \theta = 7.92 \quad ...(i)$$

(*ii*) For $\Sigma F_y = 0$, we get

$$F_2 \sin 45° + F_3 - F_4 \sin 30° - F_5 \sin \theta = 0$$

or $$22.5 \times 0.707 + 15 - 30 \times 0.5 - F_5 \sin \theta = 0$$

or $$15.9 + 15 - 15 - F_5 \sin \theta = 0$$

or $$F_5 \sin \theta = 15.9 \quad ...(ii)$$

Dividing equation (*ii*) by equation (*i*), we get $\dfrac{F_5 \sin \theta}{F_5 \cos \theta} = \dfrac{15.9}{7.92}$ or $\tan \theta = 2.0075$

$$\therefore \quad \theta = \tan^{-1} 2.0075 = \mathbf{63.52°}. \quad \textbf{Ans.}$$

Substituting the value of θ in equation (*i*), we get

$$F_5 \cos 63.52° = 7.92$$

$$\therefore \quad F_5 = \frac{7.92}{\cos 63.52} = \mathbf{17.76\ N}. \quad \textbf{Ans.}$$

2. Graphical method

(*i*) First draw a space diagram with given four forces F_1, F_2, F_3 and F_4 at correct angles as shown in Fig. 1.78(*a*).

(*ii*) Now choose a suitable scale, say 1 cm = 5 N for drawing a force diagram. Take any point O in the force diagram as shown in Fig. 1.78(*b*).

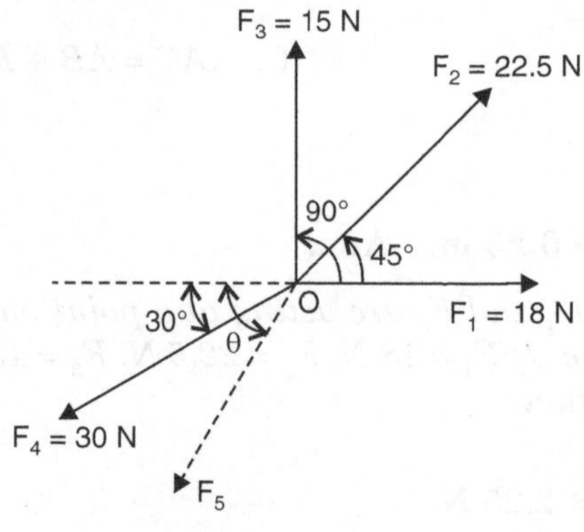

(a) Space diagram

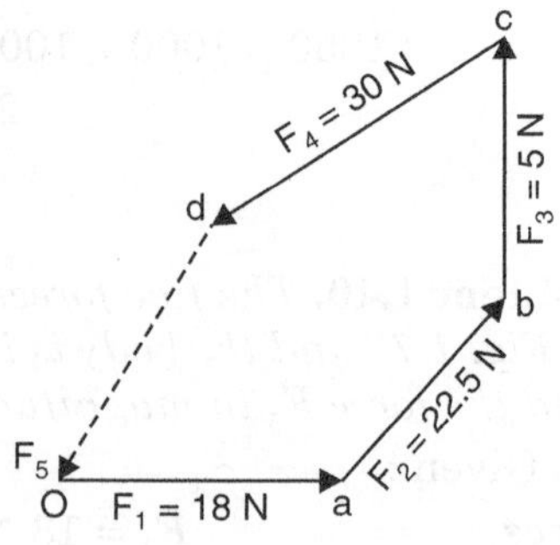

(b) Force diagram

Fig. 1.78

(*iii*) Draw line Oa parallel to force F_1 and cut $Oa = F_1 = 18$ N to the same scale.

(*iv*) From a, draw the line ab parallel to F_2 and cut $ab = F_2 = 22.5$ N

(*v*) From b, draw the line bc parallel to F_3 and cut $bc = F_3 = 15$ N

(*vi*) From c, draw the line cd parallel to F_4 and cut $cd = F_4 = 30$ N

(*vii*) Now join d to O. Then the closing side dO represents the force F_5 in magnitude and direction. Now measure the length dO.

By measurement, length $dO = 3.55$ cm.

$\therefore$ Force F_5 = Length dO × Scale = 3.55 × 5 = **17.75 N. Ans.**

The direction is obtained in the space diagram by drawing the force F_5 parallel to line dO.

Measure the angle θ, which is equal to 63.5°. Or the force F_5 is making an angle of 180 + 63.5 = 243.5° with the force F_1.

Problem 1.41. *Fig. 1.78(c) shows the coplanar system of forces acting on a flat plate. Determine : (i) the resultant and (ii) x and y intercepts of the resultant.*

Sol. Given :

Force at $A = 2240$ N

Angle with x-axis = 63.43°

Force at $B = 1805$ N

Angle with x-axis = 33.67°

Force at $C = 1500$ N

Angle with x-axis = 60°

Lengths $OA = 4$ m

$DB = 3$ m

$DC = 2$ m

and $OD = 3$ m.

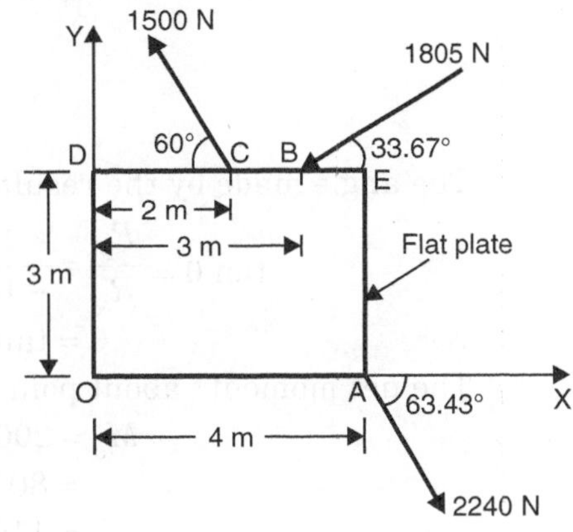

Fig. 1.78(c)

Each force is resolved into X and Y components as shown in Fig. 1.78(*d*).

(*i*) Force at $A = 2240$ N

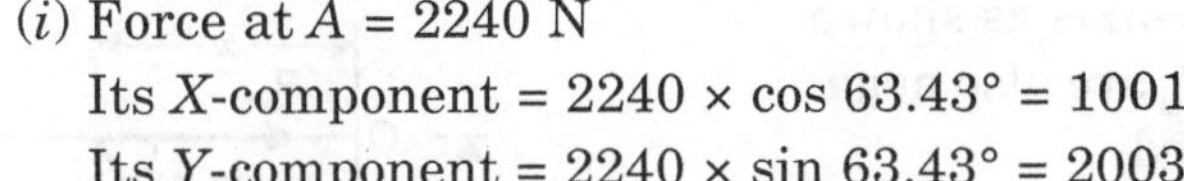

Its X-component = 2240 × cos 63.43° = 1001.9 N

Its Y-component = 2240 × sin 63.43° = 2003.4 N

(*ii*) Force at $B = 1805$ N

Its X-component = 1805 × cos 33.67° = 1502.2 N

Its Y-component = 1805 × sin 33.67° = 1000.7 N

(*iii*) Force at $C = 1500$ N.

Its X-component = 1500 × cos 60° = 750 N

Its Y-component = 1500 × sin 60° = 1299 N

The net force along X-axis,

$$R_x = \Sigma F_x = 1001.9 - 1502.2 - 750 = -1250.3 \text{ N}$$

The net force along Y-axis,

$$R_y = \Sigma F_y = -2003.4 - 1000.7 + 1299 = -1705.1 \text{ N}$$

(i) The resultant force is given by,

$$R = \sqrt{R_x^2 + R_y^2} = \sqrt{(-1250.3)^2 + (-1705.1)^2}$$

$$= \sqrt{1563250 + 2907366} = \mathbf{2114.4\ N.\quad Ans.}$$

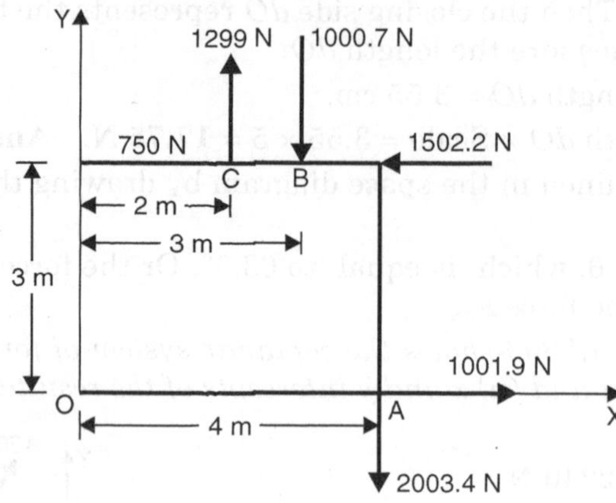

Fig. 1.78(d)

The angle made by the resultant with x-axis is given by

$$\tan\theta = \frac{R_y}{R_x} = \frac{-1705.1}{-1250.3} = 1.363$$

$$\therefore \qquad \theta = \tan^{-1} 1.363 = 53.70$$

The net moment* about point O,

$$M_o = 2003.4 \times 4 + 1000.7 \times 3 - 1299 \times 2 - 1502.2 \times 3 - 750 \times 3$$

$$= 8012.16 + 3002.1 - 2598 - 4506.6 - 2250$$

$$= 11014.26 - 9354.6 = 1659.55 \text{ Nm}$$

As the net moment about O is clockwise, hence the resultant must act towards right of origin O, making an angle = 53.7° with x-axis as shown in Fig. 1.78(*e*). The components R_x and R_y are also negative. Hence this condition is also satisfied.

(ii) Intercepts of resultant on x-axis and y-axis [Refer to Fig. 1.78(*e*)].

Let x = Intercept of resultant along x-axis.

y = Intercept of resultant along y-axis.

The moment of a force about a point is equal to the sum of the moments of the components of the force about the same point. Resolving the resultant (R) into its component R_x and R_y at F.

Moment of R about O = Sum of moments of R_x and R_y at O

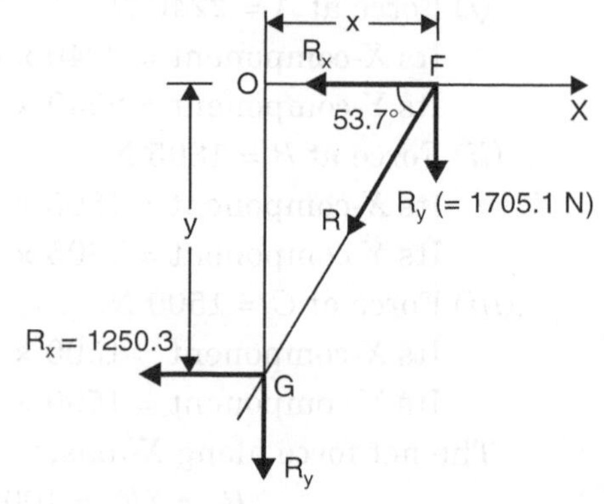

Fig. 1.78(e)

*Considering clockwise moment positive and anti-clockwise moment as negative. At A, the X component of 1001.9 N passes through O and hence has no moment.

But moment of R about $O = 1659.66$ $(M_o = 1659.66)$

$\therefore \quad 1659.66 = R_x \times O + R_y \times x$

(as R_x at F passes through O hence it has no moment)

$\therefore \quad 1659.66 = 1705.1 \times x$ $(\because \; R_y = 1705.1)$

$$\therefore \quad x = \frac{1659.66}{1705.10} = \textbf{0.97 m right of O.} \quad \textbf{Ans.}$$

To find y-intercept, resolve the resultant R at G into its component R_x and R_y.

$\therefore$ Moment of R about O = Sum of moments of R_x and R_y at O

or $\quad 1659.66 = R_x \times y + R_y \times O.$

(At G, R_y passes through O and hence has no moment)

$\therefore \quad 1659.66 = 1250.3 \times y$

$$\therefore \quad y = \frac{1659.66}{1250.30} = \textbf{1.32 m below O.} \quad \textbf{Ans.}$$

Problem 1.42. *A lamp weighing 5 N is suspended from the ceiling by a chain. It is pulled aside by a horizontal cord until the chain makes an angle of 60° with the ceiling as shown in Fig. 1.79. Find the tensions in the chain and the cord by applying Lame's theorem and also by graphical method.*

Sol. Given :

Weight of lamp = 5 N

Angle made by chain with ceiling = 60°

Cord is horizontal as shown in Fig. 1.79.

(*i*) **By Lame's theorem**

Let $\quad T_1$ = Tension (or pull) in the cord

$\quad T_2$ = Tension (or pull) in the chain.

Now from the geometry, it is obvious that angles between T_1 and lamp will be 90°, between lamp and T_2 150° and between T_2 and T_1 120°. [Refer to Fig. 1.79(*b*)].

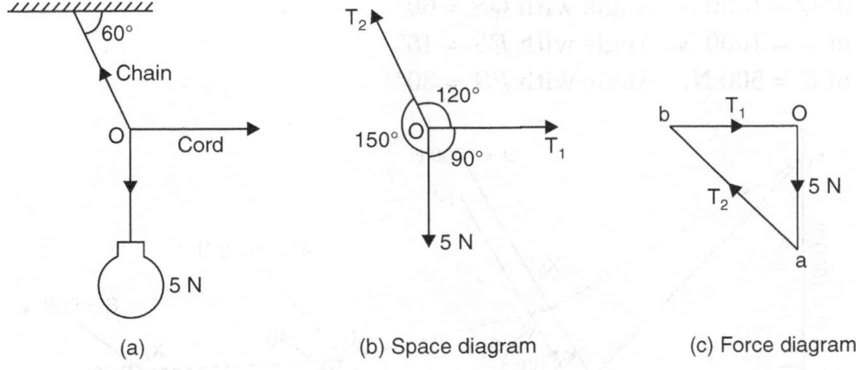

(a) (b) Space diagram (c) Force diagram

Fig. 1.79

Applying Lame's theorem, we get

$$\frac{T_1}{\sin 150°} = \frac{T_2}{\sin 90°} = \frac{5}{\sin 120°}$$

$\therefore \qquad T_1 = 5 \times \frac{\sin 150°}{\sin 120°} = \textbf{2.887 N.} \quad \textbf{Ans.}$

and $\qquad T_2 = 5 \times \frac{\sin 90°}{\sin 120°} = \textbf{5.774 N.} \quad \textbf{Ans.}$

(*ii*) **By Graphical method**

(1) First draw the space diagram at correct angles as shown in Fig. 1.79(*b*). Now choose a suitable scale say 1 cm = 1 N for drawing a force diagram as shown in Fig. 1.79(*c*). Take any point *O* in the force diagram.

(2) From *O*, draw the line *Oa* vertically downward to represent the weight of the lamp. Cut *Oa* = 5 N.

(3) From *a*, draw the line *ab* parallel to T_2. The magnitude of T_2 is unknown. Now from *O*, draw the line *Ob* horizontally (*i.e.*, parallel to T_1) cutting the line *ab* at point *b*.

(4) Now measure the lengths *ab* and *bO*.

Then *ab* represents T_2 and *bO* represents T_1. By measurements, *ab* = 5.77 cm and *bO* = 2.9 cm.

$\therefore$ Pull in the cord = *bO* = 2.9 cm × scale = 2.9 × 1

= **2.9 N. Ans.**

Pull in the chain = *ab* = 5.77 cm × scale = 5.77 × 1

= **5.77 N. Ans.**

Problem 1.43. *On a horizontal line PQRS 12 cm long, where PQ = QR = RS = 4 cm, forces of 1000 N, 1500 N, 1000 N and 500 N are acting at P, Q, R and S respectively, all downwards, their lines of action making angles of 90°, 60°, 45° and 30° respectively with PS. Obtain the resultant of the system completely in magnitude, direction and position graphically and check the answer analytically.*

Sol. Given :

PQ = *QR* = *RS* = 4 cm

Force at *P* = 1000 N. Angle with *PS* = 90°

Force at *Q* = 1500 N. Angle with *QS* = 60°

Force at *R* = 1000 N. Angle with *RS* = 45°

Force at *S* = 500 N. Angle with *PS* = 30°

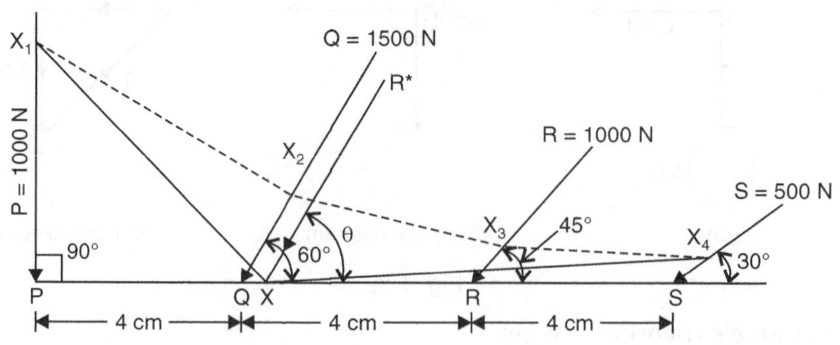

Fig. 1.79(d). Space diagram.

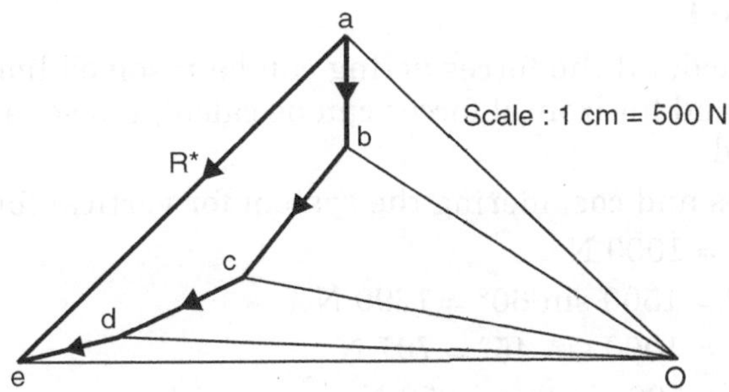

Fig. 1.79(e). Force diagram.

Graphical method

Draw the space diagram of the forces as shown in Fig. 1.79(*d*). The procedure is as follows :

(*i*) Draw a horizontal line *PQRS* = 12 cm in which take *PQ* = *QR* = *RS* = 4 cm.

(*ii*) Draw the line of action of forces *P, Q, R, S* of magnitude 1000 N, 1500 N, 1000 N and 500 N respectively at an angle of 90°, 60°, 45° and 30° respectively with line *PS* as shown in Fig. 1.79.

Magnitude and Direction of Resultant Force (R*)

To find the magnitude and direction of the resultant force, the force diagram is drawn as shown in Fig. 1.79(*e*) as given above :

(*i*) Draw the vector *ab* to represent the force 1000 N to a scale of 1 cm = 500 N. The vector *ab* is parallel to the line of action of force *P*.

(*ii*) From point *b*, draw vector *bc* = 1500 N and parallel to the line of action of force *Q*. Similarly the vectors, *cd* = 1000 N and parallel to line of action of force *R* and *de* = 500 N and parallel to the line of action of force *S*, are drawn.

(*iii*) Join *ae* which gives the magnitude of the resultant. Measuring *ae*, the resultant force is equal to 3770 N.

(*iv*) To get the line of action of the resultant, choose any point *O* on force diagram (called the pole) and join *Oa, Ob, Oc, Od* and *Oe*.

(*v*) Now choose any point X_1 on the line of action of force *P* and draw a line parallel to *Oa*.

(*vi*) Also from the point X_1, draw another line parallel to *Ob*, which cuts the line of action of force *Q* at X_2. Similarly from point X_2, draw a line parallel to *Oc* to cut the line of action of force *R* at X_3. From point X_3, draw a line parallel to *Od* to cut the line of action of force *S* at X_4.

(*vii*) From point X_4, draw a line parallel *Oe*.

(*viii*) Produce the first line (*i.e.*, the line from X_1 and parallel to *Oa*) and the last line (*i.e.*, the line from X_4 and parallel to *Oe*) to interest at *X*. Then the resultant must pass through this point.

(*ix*) From point *X*, draw a line parallel to *ae* which determines the line of action of resultant force. Measure *PX*. By measurements :

Resultant force, $R^* = 3770$ N

Point of action, $PX = 4.20$ cm

Direction, $\theta = 60° \, 30'$ with *PS*.

Analytical method

In analytical method, all the forces acting can be resolved horizontally and vertically. Resultant of all vertical and horizontal forces can be calculated separately and then the final resultant can be obtained.

Resolving all forces and considering the system for vertical forces only.

Vertical force at P = 1000 N

Vertical force at Q = 1500 sin 60° = 1299 N

Vertical force at R = 1000 sin 45° = 707 N

Vertical force at S = 500 sin 30° = 250 N

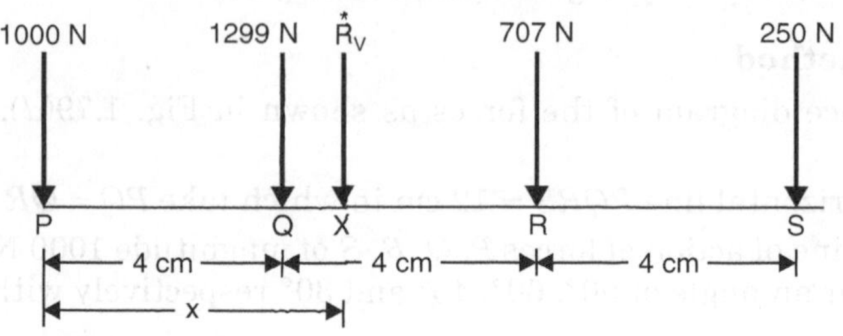

Fig. 1.79(f)

Let R_V^* = the resultant of all vertical forces and acting at a distance x cm from P.

$$= 1000 + 1299 + 707 + 250 = 3256 \text{ N}$$

Taking moments of all vertical forces about point P,

$$R_V^* \times x = 1299 \times 4 + 707 \times 8 + 250 \times 12 = 13852$$

$$\therefore \quad x = \frac{13852}{R_V^*} = \frac{13852}{3256} = 4.25 \text{ cm}$$

Now consider the system for horizontal forces only, horizontal force at P = 0

Horizontal force at Q = 1500 × cos 60° = 750 N

Horizontal force at R = 1000 × cos 45° = 707 N

Horizontal force at S = 500 × cos 30° = 433 N

Resultant of all horizontal forces will be,

$$R_H^* = 0 + 750 + 707 + 433 = 1890 \text{ N}$$

The resultant R^* of R_V^* and R_H^* will also pass through point X which is at a distance of 4.25 cm from P.

$$\therefore \quad R^* = \sqrt{R_V^{*2} + R_H^{*2}} = \sqrt{3256^2 + 1890^2} = \mathbf{3764\ N.} \quad \textbf{Ans.}$$

The resultant will make an angle θ with PS and is given by

$$\tan \theta = \frac{R_V^*}{R_H^*} = \frac{3256}{1890} = 1.723$$

$$\therefore \quad \theta = \tan^{-1} 1.723 = 59.9°$$

Thus the resultant of 3764 N makes an angle 59.9° with *PS* and passing through point *X* which is at a distance of 4.25 cm from point *P*.

This result confirms closely with the values obtained by graphical method.

1.13. FORCES IN SPACE

When a force is represented in three dimensions, then the force is known as 'Force in space'. Forces in space are generally represented in vector form. In Fig. 1.80, the force **F** is represented by vector **OA**. Through point *A*, draw planes parallel to co-ordinate planes (*i.e.*, Co-ordinate planes are *xy*, *yz* and *zx* as shown in Fig. 1.80).

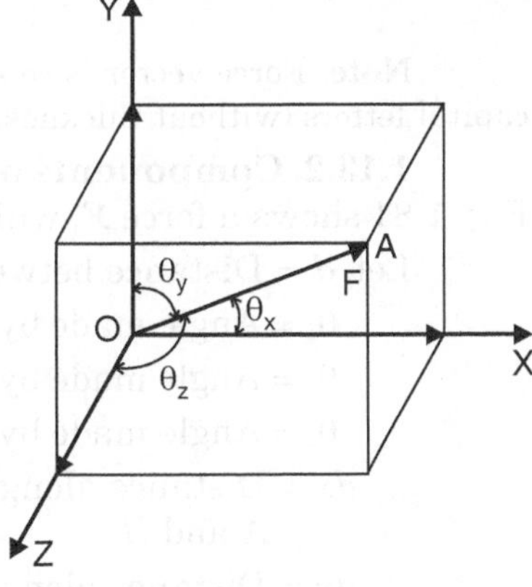

Fig. 1.80

These planes along with co-ordinate planes make a rectangular box. The force **F** is then represented by the diagonal of the box and its three components F_x, F_y and F_z by its edges.

Let θ_x = Angle made by force **F** with *x*-axis

θ_y = Angle made by force **F** with *y*-axis

θ_z = Angle made by force **F** with *z*-axis

Now
$$\left.\begin{aligned} F_x &= F\cos\theta_x \\ F_y &= F\cos\theta_y \\ F_z &= F\cos\theta_z \end{aligned}\right\} \qquad \text{...(1.26)}$$

Also
$$F = \sqrt{F_x^{\,2} + F_y^{\,2} + F_z^{\,2}} \qquad \text{...(1.27)}$$

The cosines of θ_x, θ_y and θ_z are known as the direction cosines of the force **F** and denoted by

$$l = \cos\theta_x, \quad m = \cos\theta_y \quad \text{and} \quad n = \cos\theta_z \qquad \text{...(1.27A)}$$

The three angles* are related by

$$\cos^2\theta_x + \cos^2\theta_y + \cos^2\theta_z = 1 \qquad \text{...(1.28)}$$

or
$$l^2 + m^2 + n^2 = 1$$

Let **i** = vector of unit length in the positive *x*-direction,

j = vector of unit length in the positive *y*-direction,

and **k** = vector of unit length in the positive *z*-direction then force vector **F** is represented by

$$\mathbf{F} = F_x\,\mathbf{i} + F_y\,\mathbf{j} + F_z\,\mathbf{k} \qquad \text{...(1.29)}$$

Magnitude of the vector **F** is given by

$$F = |\,\mathbf{F}\,| = \sqrt{F_x^{\,2} + F_y^{\,2} + F_z^{\,2}}$$

But $F_x = F\cos\theta_x$, $F_y = F\cos\theta_y$ and $F_z = F\cos\theta_z$. Substituting these values in equation (1.29), we get

$$F = (F\cos\theta_x)\,\mathbf{i} + (F\cos\theta_y)\,\mathbf{j} + (F\cos\theta_z)\,\mathbf{k} \qquad \text{...(1.30)}$$

1.13.1. Unit Vector. The vector having a unit length (or unit magnitude) is known as unit vector. Hence unit vector corresponding to force vector **F** is equal to **F** ÷ Magnitude of vector **F**.

---

*For *x, y* co-ordinates $\cos^2\theta_x + \cos^2\theta_y = 1$. This is proved as, $\theta_y = 90 - \theta_x$. Hence $\cos\theta_y = \cos(90 - \theta_x) = \sin\theta_x$. Or $\cos^2\theta_y = \sin^2\theta_x$. Therefore, $\cos^2\theta_x + \cos^2\theta_y = \cos^2\theta_x + \sin^2\theta_x$ which is equal to 1.

$$\therefore \quad \text{Unit vector} = \frac{\text{Force vector } \mathbf{F}}{\text{Magnitude of vector } F} = \frac{\mathbf{F}}{F}$$

$$= \frac{\mathbf{F}}{\sqrt{F_x^2 + F_y^2 + F_z^2}} \quad ...(1.31)$$

$$= \frac{F_x\mathbf{i} + F_y\mathbf{j} + F_z\mathbf{k}}{\sqrt{F_x^2 + F_y^2 + F_z^2}} \quad (\because \; \mathbf{F} = F_x\mathbf{i} + F_y\mathbf{j} + F_z\mathbf{k})$$

Note. Force vector is represented by **THICK, BOLD FACE** and magnitude is represented by capital letters (without thickness) only.

1.13.2. Components of a Force when Two Points on its Line of Action are Given. Fig. 1.81 shows a force **F**, which is defined by two points $A(x_1, y_1, z_1)$ and $B(x_2, y_2, z_2)$

Let d = Distance between A and B

θ_x = Angle made by force **F** with x-axis

θ_y = Angle made by force **F** with y-axis

θ_z = Angle made by force **F** with z-axis

dx = Distance along x-axis between A and B

dy = Distance along y-axis between A and B

dz = Distance along z-axis between A and B.

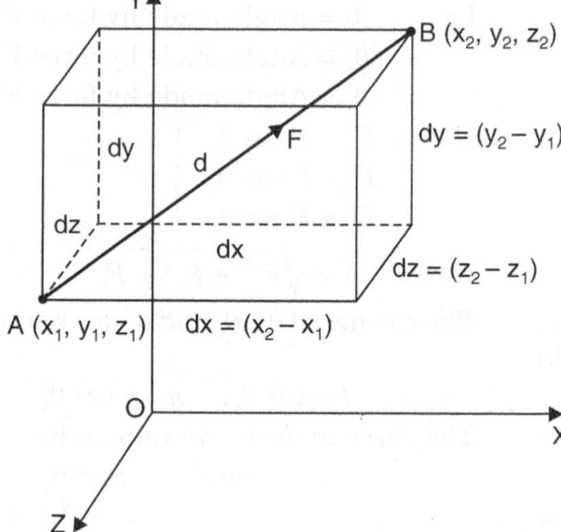

Fig. 1.81

Then $\quad dx = (x_2 - x_1), dy = (y_2 - y_1)$

and $\quad dz = (z_2 - z_1) \quad ...(1.32)$

Also $\quad dx = d \cos \theta_x, dy = d \cos \theta_y$

and $\quad dz = d \cos \theta_z$

and $\quad d = \sqrt{dx^2 + dy^2 + dz^2} \quad ...(1.33)$

Also we know that

$$F_x = F \cos \theta_x, F_y = F \cos \theta_y$$

and $\quad F_z = F \cos \theta_z$

Now $\quad \dfrac{F_x}{dx} = \dfrac{F \cos \theta_x}{d \cos \theta_x} = \dfrac{F}{d}$

Similarly $\quad \dfrac{F_y}{dy} = \dfrac{F}{d}$ and $\dfrac{F_z}{dz} = \dfrac{F}{d} \quad \therefore \quad \dfrac{F_x}{dx} = \dfrac{F_y}{dy} = \dfrac{F_z}{dz} = \dfrac{F}{d} \quad ...(1.34)$

or

$$\frac{F_x}{(x_2 - x_1)} = \frac{F_y}{(y_2 - y_1)} = \frac{F_z}{(z_2 - z_1)} = \frac{F}{d}.$$

$$\therefore \quad F_x = (x_2 - x_1)\frac{F}{d}, F_y = (y_2 - y_1)\frac{F}{d} \text{ and } F_2 = (z_2 - z_1)\frac{F}{d} \quad ...(1.34A)$$

From equation (1.34*A*) the components of the force can be obtained, provided we know '*F*' and '*d*'.

1.13.3. Position Vector of a Given Point. The position vector of a point can be given with respect to origin or with respect to another point.

(*i*) The position vector of a point *A* with respect to origin *O* is the vector **OA.** It is represented by **r.** It is given by (see Fig. 1.82)

$$\mathbf{r} = x\mathbf{i} + y\mathbf{j} + z\mathbf{k}$$

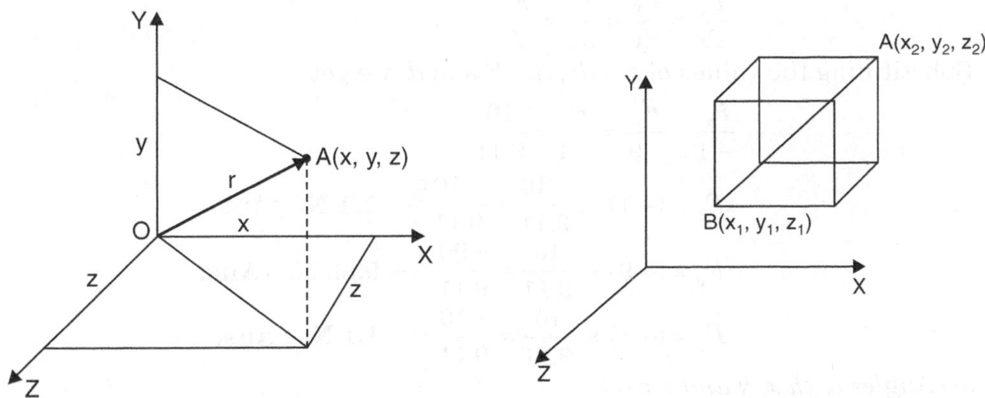

Fig. 1.82 Fig. 1.82(a)

(*ii*) The position vector of a point *A* with co-ordinates (x_2, y_2, z_2) with respect to point $B(x_1, y_1, z_1)$ is given by (see Fig. 1.82*a*)

$$\mathbf{r} = (x_2 - x_1)\,\mathbf{i} + (y_2 - y_1)\,\mathbf{j} + (z_2 - z_1)\,\mathbf{k}.$$

Problem 1.44. *A force vector is represented by a line AB. The co-ordinates of point A are (2, 4, 3) and of point B(1, – 5, 2) respectively. If the magnitude of force = 10 N, then determine :*

(*i*) *the components of the force along x, y and z-axis,*

(*ii*) *angles with the x, y and z-axis, and*

(*iii*) *specify the force vector.*

Sol. Given :

Co-ordinates of point *A* are (2, 4, 3)

$\therefore \quad x_1 = 2, y_1 = 4 \quad$ and $\quad z_1 = 3$

Co-ordinates of point *B* are (1, – 5, 2)

$\therefore \quad x_2 = 1, y_2 = -5 \quad$ and $\quad z_2 = 2.$

Magnitude of force, $\quad F = 10$ N

Here the magnitude of the force and co-ordinates of two points through which force passes, are given. The force components can be obtained if we know '*d*' (*i.e.*, distance between two points). But $\quad d = \sqrt{dx^2 + dy^2 + dz^2}$.

From equation (1.32),

$$dx = x_2 - x_1 = 1 - 2 = -1$$
$$dy = y_2 - y_1 = -5 - 4 = -9$$
$$dz = z_2 - z_1 = 2 - 3 = -1$$

and from equation (1.33),

$$d = \sqrt{dx^2 + dy^2 + dz^2} = \sqrt{(-1)^2 + (-9)^2 + (-1)^2} = \sqrt{1 + 81 + 1} = \sqrt{83} = 9.11.$$

(i) Components of the force along the axes

Let F_x = Component of the force **F** along x-axis,

F_y = Component of the force **F** along y-axis, and

F_z = Component of the force **F** along z-axis.

Now using equation (1.34), we get

$$\frac{F_x}{dx} = \frac{F_y}{dy} = \frac{F_z}{dz} = \frac{F}{d}$$

Substituting the values of dx, dy, dz, F and d, we get

$$\frac{F_x}{-1} = \frac{F_y}{-9} = \frac{F_z}{-1} = \frac{10}{9.11}$$

$$\therefore \quad F_x = (-1) \times \frac{-10}{9.11} = \frac{-10}{9.11} = -\ \textbf{1.1 N.} \quad \textbf{Ans.}$$

$$F_y = (-9) \times \frac{10}{9.11} = \frac{-90}{9.11} = -\ \textbf{9.88 N.} \quad \textbf{Ans.}$$

$$F_z = (-1) \times \frac{10}{9.11} = \frac{-10}{9.11} = -\ \textbf{1.1 N.} \quad \textbf{Ans.}$$

(ii) Angles with x, y and z-axes

Let θ_x = The angle that the force F makes with x-axis,

θ_y = The angle made by F with y-axis, and

θ_z = The angle made by F with z-axis.

Using equation (1.26),

$$F_x = F \cos \theta_x$$

$$\therefore \quad \cos \theta_x = \frac{F_x}{F} = \frac{-1.1}{10} = -0.11$$

$$\therefore \quad \theta_x = \cos^{-1}(-0.11) = \textbf{96.3°.} \quad \textbf{Ans.}$$

Similarly, $$F_y = F \cos \theta_y$$

$$\therefore \quad \cos \theta_y = \frac{F_y}{F} = \frac{-9.88}{10} = -0.988$$

$$\therefore \quad \theta_y = \cos^{-1}(-0.988) = \textbf{171.09°.} \quad \textbf{Ans.}$$

Also $$F_z = F \cos \theta_z$$

$$\therefore \quad \cos \theta_z = \frac{F_z}{F} = \frac{-1.097}{10} = -0.11 \quad \text{or} \quad \theta_z = \cos^{-1}(-0.11) = \textbf{96.3°.} \quad \textbf{Ans.}$$

(iii) Specify the force vector

The force vector **F** is represented by equation (1.29) as

$$\mathbf{F} = F_x\mathbf{i} + F_y\mathbf{j} + F_z\mathbf{k} = -1.1\,\mathbf{i} + (-9.88)\,\mathbf{j} + (-1.1)\,\mathbf{k}$$

$$(\because \quad F_x = -1.1, F_y = -9.88 \text{ and } F_z = -1.1)$$

$$= -1.1\,\mathbf{i} - 9.88\,\mathbf{j} - 1.1\,\mathbf{k}. \quad \textbf{Ans.}$$

IInd Method

The force vector **F** can also be obtained as given below :

The vector joining the points (2, 4, 3) and (1, – 5, 2) is given by

$$= dx\,\mathbf{i} + dx\,\mathbf{j} + dz\,\mathbf{k} = (x_2 - x_1)\,\mathbf{i} + (y_2 - y_1)\,\mathbf{j} + (z_2 - z_1)\,\mathbf{k}$$

$$= (1 - 2)\,\mathbf{i} + (-5 - 4)\,\mathbf{j} + (2 - 3)\,\mathbf{k} = -1\,\mathbf{i} - 9\,\mathbf{j} - 1\,\mathbf{k}.$$

Now find the unit vector in the direction of above vector *i.e.*, in the direction of vector ($-1\,\mathbf{i} - 9\,\mathbf{j} - 1\,\mathbf{k}$).

This unit vector is obtained from equation (1.31) and is equal to the given vector divided by the square root of the sum of squares of its components.

$\therefore$ Unit vector in the direction of the vector ($-1\,\mathbf{i} - 9\,\mathbf{j} - 1\,\mathbf{k}$) is

$$\frac{(-1\mathbf{i} - 9\,\mathbf{j} - 1\mathbf{k})}{\sqrt{(-1)^2 + (-9)^2 + (-1)^2}} = \frac{(-1\mathbf{i} - 9\,\mathbf{j} - 1\mathbf{k})}{\sqrt{1 + 81 + 1}} = \frac{(-1\mathbf{i} - 9\,\mathbf{j} - 1\mathbf{k})}{9.11}$$

The force vector $\mathbf{F}$ = Magnitude of force times unit vector in the direction of force

$$= \frac{10\,(-1\mathbf{i} - 9\,\mathbf{j} - 1\mathbf{k})}{9.11} = -\frac{10 \times 1}{9.11}\,\mathbf{i} - \frac{10 \times 9}{9.11}\,\mathbf{j} - \frac{10 \times 1}{9.11}\,\mathbf{k}$$

$$= -1.1\,\mathbf{i} - 9.88\,\mathbf{j} - 1.1\,\mathbf{k}.\quad \textbf{Ans.}$$

Problem 1.45. *With respect to origin (0, 0, 0), a point A is located at (– 5, 2, 14). Specify the position of the point :*

(*i*) *in terms of orthogonal components,* (*ii*) *in terms of directional cosine, and*

(*iii*) *in terms of unit vector.*

Sol. Given :

Co-ordinates of origins are (0, 0, 0)

Co-ordinates of point A are (– 5, 2, 14).

(*i*) *Position of the point in terms of orthogonal components*

The components of the vector along x-axis, y-axis and z-axis are given as :

Along x-axis = (– 5 – 0) = – 5

Along z-axis = (2 – 0) = 2

Along z-axis = (14 – 0) = 14

$\therefore$ Position vector, $r = -5\,\mathbf{i} + 2\,\mathbf{j} + 14\,\mathbf{k}$. **Ans.**

(*ii*) *Position of the point in terms of direction cosine*

The magnitude of vector is given by,

$$|\,\mathbf{r}\,| = r = \sqrt{(-5)^2 + (2)^2 + (14)^2} = \sqrt{25 + 4 + 196} = \sqrt{255} = 15.$$

The direction cosines are given by equation (1.27 A) as

$$l = \cos\theta_x, \qquad m = \cos\theta_y, \qquad n = \cos\theta_z$$

Also from eqn. (1.26), we have

$$\cos\theta_x = \frac{\text{Component of vector along } x\text{-axis}}{\text{Magnitude of vector}} = \frac{-5}{15} = -0.333$$

$$\cos\theta_y = \frac{\text{Component of vector along } y\text{-axis}}{\text{Magnitude of vector}} = \frac{2}{15} = 0.0133$$

and
$$\cos\theta_z = \frac{\text{Component of vector along } z\text{-axis}}{\text{Magnitude of vector}} = \frac{14}{15} = 0.933$$

$\therefore$ Direction cosines are

$$\left.\begin{aligned} l &= \cos\theta_x = -0.333 \\ m &= \cos\theta_y = 0.0133 \\ n &= \cos\theta_z = 0.933 \end{aligned}\right] \quad \textbf{Ans.}$$

(iii) In terms of unit vector

The unit vector in the direction of the vector ($r = -5\,\mathbf{i} + 2\,\mathbf{j} + 14\,\mathbf{k}$) is $\dfrac{-5\,\mathbf{i}+2\,\mathbf{j}+14\,\mathbf{k}}{|r|}$

$$= \frac{-5\,\mathbf{i}+2\,\mathbf{j}+14\,\mathbf{k}}{15} = -0.333\,\mathbf{i} + 0.0133\,\mathbf{j} + 0.933\,\mathbf{k}. \quad \textbf{Ans.}$$

Problem 1.46. *A force P is acting as shown in Fig. 1.83. Express the force P as vector. Magnitude of force P = 500 N.*

Sol. Given :

Length $a = 2$ cm.

Magnitude of $P = 500$ N.

The force P is acting from A to G.

Hence point A corresponds to 1 and point G corresponds to 2.

The point G is lying on x-axis.

Hence co-ordinates of point G are $(a, 0, 0)$ or $(2, 0, 0)$.

The point A is in y-z plane

$\therefore \quad x_2 = 2, y_2 = 0 \quad \text{and} \quad z_2 = 0$

The co-ordinates of the point A are $(0, a, a)$ or $(0, 2, 2)$.

$\therefore \quad x_1 = 0, y_1 = 2 \quad \text{and} \quad z_1 = 2$

Fig. 1.83

The vector joining the points A (0, 2, 2) and G (2, 0, 0) is given by

$$= d_x\,\mathbf{i} + d_y\,\mathbf{j} + d_z\,\mathbf{k}$$
$$= (x_2 - x_1)\,\mathbf{i} + (y_2 - y_1)\,\mathbf{j} + (z_2 - z_1)\,\mathbf{k}$$
$$= (2 - 0)\,\mathbf{i} + (0 - 2)\,\mathbf{j} + (0 - 2)\,\mathbf{k}$$
$$= 2\,\mathbf{i} - 2\,\mathbf{j} - 2\,\mathbf{k}.$$

The unit vector in the direction of vector $2\,\mathbf{i} - 2\,\mathbf{j} - 2\,\mathbf{k}$ is

$$= \frac{2\,\mathbf{i}-2\,\mathbf{j}-2\,\mathbf{k}}{\sqrt{(2)^2+(-2)^2+(-2)^2}} = \frac{2\,\mathbf{i}-2\,\mathbf{j}-2\,\mathbf{k}}{\sqrt{12}}$$

The force vector P is given by

= Magnitude of force P times unit vector in the direction of force

$$= 500\left(\frac{2\,\mathbf{i}-2\,\mathbf{j}-2\,\mathbf{k}}{\sqrt{12}}\right) = \frac{1000}{\sqrt{12}}\,\mathbf{i} - \frac{1000}{\sqrt{12}}\,\mathbf{j} - \frac{1000}{\sqrt{12}}\,\mathbf{k}$$

$$= 288.683\,\mathbf{i} - 288.683\,\mathbf{j} - 288.683\,\mathbf{k}. \quad \textbf{Ans.}$$

IInd Method

The force vector **P** can also be obtained as

$$\mathbf{P} = P_x i + P_y j + P_z k \qquad ...(i)$$

where $P_x = \dfrac{P}{d}(x_2 - x_1) = \dfrac{P}{d}(dx) = \dfrac{2P}{d}$ $\qquad (\because \; dx = 2)$

$P = 500$ N $\qquad$ (given)

$d = \sqrt{dx^2 + dy^2 + dz^2} = \sqrt{2^2 + (-2)^2 + (-2)^2}$ $\qquad (\because \; dy = -2, dz = -2)$

$= \sqrt{12}$

$$\therefore \qquad P_x = \frac{P}{d} dx = \frac{2P}{d} = \frac{2 \times 500}{\sqrt{12}} = \frac{1000}{\sqrt{12}}$$

Similarly, $$P_y = \frac{P}{d} dy = \frac{500}{\sqrt{12}} \times (-2) = \frac{-1000}{\sqrt{12}}$$

$$P_z = \frac{P}{d} dz = \frac{500}{\sqrt{12}} \times (-2) = \frac{-1000}{\sqrt{12}}$$

$\therefore$ Force vector $\mathbf{P} = P_x i + P_y j + P_z k$

$$= \frac{1000}{\sqrt{12}} i + \left(\frac{(-1000)}{\sqrt{12}}\right) j + \left(\frac{(-1000)}{\sqrt{12}}\right) k$$

$$= \frac{1000}{\sqrt{12}} i - \frac{1000}{\sqrt{12}} j - \frac{1000}{\sqrt{12}} k$$

$= 288.683\mathbf{i} - 288.683\mathbf{j} - 288.683\mathbf{k}$. **Ans.**

Problem 1.47. *The lines of action of three forces concurrent at the origin O passes respectively through points A, B and C having coordinates :*

$$x_A = -1, y_A = 2, z_A = 4 \; ; x_B = 3, y_B = 0, z_B = -3 \quad \textit{and} \quad x_C = 2, y_C = -2 \quad \textit{and} \quad z_C = 4.$$

The magnitude of the forces are $F_A = 40$ N, $F_B = 10$ N and $F_C = 30$ N. Find the magnitude and direction of their resultant.

Sol. Given :

Co-ordinates of point A are (– 1, 2, 4)

Co-ordinates of point B are (3, 0, – 3)

Co-ordinates of point C are (2, – 2, 4)

The forces passes through the origin and points A, B, C respectively.

Magnitude of forces are

$$F_A = 40 \text{ N}, F_B = 10 \text{ N and } F_C = 30 \text{ N}.$$

The position vector of force F_A

$$= (-1 - 0)\,\mathbf{i} + (2 - 0)\,\mathbf{j} + (4 - 0)\,\mathbf{k} = -1\,\mathbf{i} + 2\,\mathbf{j} + 4\,\mathbf{k}$$

The unit vector in the direction of force F_A

$$= \frac{(-1\mathbf{i} + 2\,\mathbf{j} + 4\,\mathbf{k})}{\sqrt{1^2 + 2^2 - 4^2}} = \frac{(-1\mathbf{i} + 2\,\mathbf{j} + 4\,\mathbf{k})}{\sqrt{21}} = \frac{(-1\mathbf{i} + 2\,\mathbf{j} + 4\,\mathbf{k})}{4.5825}$$

Magnitude of force $F_A = 40$ N

$\therefore$ Force vector F_A = Magnitude of F_A times unit vector

$$= 40\left(\frac{-\mathbf{i} + 2\,\mathbf{j} + 4\,\mathbf{k}}{4.5825}\right) = -\frac{40}{4.5825}\mathbf{i} + \frac{40 \times 2}{4.5825}\mathbf{j} + \frac{40 \times 4}{4.5825}\mathbf{k}$$

$$= -8.728\,\mathbf{i} + 17.456\,\mathbf{j} + 34.912\,\mathbf{k} \qquad ...(i)$$

Similarly, the position vector of force F_B

$$= 3\,\mathbf{i} + 0\,\mathbf{j} - 3\,\mathbf{k} = 3\,\mathbf{i} - 3\,\mathbf{k}$$

The unit vector in the direction of force F_B

$$= \frac{3\,\mathbf{i} - 3\,\mathbf{k}}{\sqrt{3^2 + 3^2}} = \frac{3\,\mathbf{i} - 3\,\mathbf{k}}{4.242} = \frac{3}{4.242}\mathbf{i} - \frac{3}{4.242}\mathbf{k} = 0.7072\,\mathbf{i} - 0.7072\,\mathbf{k}$$

∴ Force vector $\mathbf{F_B}$ = Magnitude of F_B times unit vector in the direction of force F_B

$= 10\ (0.7072\ \mathbf{i} - 0.7072\ \mathbf{k})$

$= 7.072\ \mathbf{i} - 7.072\ \mathbf{k}$ ...(*ii*)

Similarly force vector $\mathbf{F_C}$ = Magnitude of F_C times unit vector in the direction of F_C

$$= 30\ \frac{(2\,\mathbf{i} - 2\,\mathbf{j} + 4\,\mathbf{k})}{\sqrt{2^2 + 2^2 + 4^2}} = 30\ \frac{(2\,\mathbf{i} - 2\,\mathbf{j} + 4\,\mathbf{k})}{\sqrt{24}} = \frac{(60\,\mathbf{i} - 60\,\mathbf{j} + 120\,\mathbf{k})}{4.899}$$

$$= \frac{60}{4.899}\,\mathbf{i} - \frac{60}{4.899}\,\mathbf{j} + \frac{120}{4.899}\,\mathbf{k}$$

$= 12.247\ \mathbf{i} - 12.247\ \mathbf{j} + 24.494\ \mathbf{k}$ ...(*iii*)

The resultant force vector is obtained by adding force vectors $\mathbf{F_A}$, $\mathbf{F_B}$ and $\mathbf{F_C}$ [or by adding equations (*i*), (*ii*) and (*iii*)].

∴ Resultant force vector

= Force vector $\mathbf{F_A}$ + Force vector $\mathbf{F_B}$ + Force vector $\mathbf{F_C}$

$= (-\ 8.728\ \mathbf{i} + 17.456\ \mathbf{j} + 34.912\ \mathbf{k}) + (7.072\ \mathbf{i} - 7.072\ \mathbf{k}) + (12.247\ \mathbf{i} - 12.247\ \mathbf{j} + 24.494\ \mathbf{k})$

$= (-\ 8.728\ \mathbf{i} + 7.072\ \mathbf{i} + 12.247\ \mathbf{i}) + (17.456\ \mathbf{j} - 12.247\ \mathbf{j}) + (34.912\ \mathbf{k} - 7.072\ \mathbf{k} + 24.494\ \mathbf{k})$

$= 10.591\mathbf{i} + 5.209\ \mathbf{j} + 52.334\ \mathbf{k}$

Force Vector $\mathbf{F_A}$

Let us find the force vector $\mathbf{F_A}$ first. It is given as

$\mathbf{F_A} = (F_A)_x\, i + (F_A)_y j + (F_A)_z\, k$ ...(*i*)

The force F_A passes through point $A(-\ 1, 2, 4)$ and origin (0, 0, 0)

∴ $dx = (x_2 - x_1) = (-\ 1 - 0) = -\ 1$, $dy = (y_2 - y_1) = 2 - 0 = 2$

$dz = (z_2 - z_1) = (4 - 0) = 4$

and $d = \sqrt{dx^2 + dy^2 + dz^2} = \sqrt{(-\ 1)^2 + 2^2 + 4^2} = \sqrt{21} = 4.5825$

Now $(F_A)_x = \frac{F_A}{d}\, dx = \frac{40}{4.5825}\,(-\ 1) = \frac{-40}{4.5825} = -\ 8.728$

($\because$ $F_A = 40$ N, $d = 4.5825$, $dx = -\ 1$)

$(F_A)_y = \frac{F_A}{d}\, dy = \frac{40}{4.5825}\,(2) = \frac{80}{4.5825} = 17.456$

and $(F_A)_z = \frac{F_A}{d}\, dz = \frac{40}{4.5825}\,(4) = \frac{160}{4.5825} = 34.912$

Substituting these values in equation (*i*), we get

$\mathbf{F_A} = -\ 8.728\, i + 17.456\, j + 34.912\, k$ ...(*A*)

Now calculate similarly force vectors $\mathbf{F_B}$ and $\mathbf{F_C}$

Force Vector $\mathbf{F_B}$

$\mathbf{F_B} = (F_B)_x\, i + (F_B)_y\, j + (F_B)_z\, k$ ...(*ii*)

The force F_B = 10 N passes through point $B(3, 0, -\ 3)$ and origin (0, 0, 0)

∴ $dx = (3 - 0) = 3$, $dy = (0 - 0) = 0$, $dz = (-\ 3 - 0) = -\ 3$,

and $d = \sqrt{dx^2 + dy^2 + dz^2} = \sqrt{3^2 + 0^2 + (-\ 3)^2} = \sqrt{18} = 4.242$

Now $(F_B)_x = \frac{F_B}{d} dx = \frac{10}{4.242} \times (3) = \frac{30}{4.242} = 7.072$ $(\because \; F_B = 10 \text{ N}, dx = 3)$

$$(F_B)_y = \frac{F_B}{d} . dy = \frac{10}{4.242} (0) = 0$$

$$(F_B)_z = \frac{F_B}{d} . dz = \frac{10}{4.242} (-3) = -7.072$$

Substituting these values in equation (*ii*), we get

$$\mathbf{F_B} = 7.072\, i - 7.072\, k$$

Force Vector $\mathbf{F_C}$

$$\mathbf{F_C} = (F_C)_x\, i + (F_C)_y j + (F_C)_z\, k \qquad ...(iii)$$

The force F_C = 30 N, passes through point $C(2, -2, 4)$ and origin (0, 0, 0)

$\therefore$ $dx = 2 - 0 = 2, dy = -2 - 0 = -2, dz = 4 - 0 = 4$

and $d = \sqrt{dx^2 + dy^2 + dz^2} = \sqrt{2^2 + (-2)^2 + 4^2} = \sqrt{24} = 4.899$

Now $(F_C)_x = \frac{F_C}{d} . dx = \frac{30}{4.899} \times 2 = 12.247$

$$(F_C)_y = \frac{F_C}{d} . dy = \frac{30}{4.899} (-2) = \frac{-60}{4.899} = -12.247$$

$$(F_C)_z = \frac{F_C}{d} . dz = \frac{30}{4.899} (4) = \frac{120}{4.899} = 24.494$$

Substituting these values in equation (*iii*), we get

$$\mathbf{F_C} = (F_C)_x\, i + (F_C)_y j + (F_C)_z\, k$$
$$= 12.247\, i - 12.247\, j + 24.494\, k$$

The resultant force vector is obtained by adding force vectors $\mathbf{F_A}$, $\mathbf{F_B}$ and $\mathbf{F_C}$ [or by adding equations (*i*), (*ii*) and (*iii*)]

$\therefore$ Resultant force vector

$$= \mathbf{F_A} + \mathbf{F_B} + \mathbf{F_C}$$
$$= [(-8.728\, i + 17.456\, j + 34.912\, k)] + [7.072\, i - 7.072\, k] + [12.247\, i - 12.247\, j + 24.494\, k]$$
$$= [-8.728\, i + 7.072\, i + 12.247\, i] + [17.456\, j - 12.247\, j] + [34.912\, k - 7.072\, k + 24.494\, k]$$
$$= 10.591\, \mathbf{i} + 5.209\, \mathbf{j} + 52.334\, \mathbf{k}. \quad \textbf{Ans.}$$

$\therefore$ Magnitude of resultant force

$$= \sqrt{10.591^2 + 5.209^2 + 52.334^2} = \sqrt{112.169 + 27.133 + 2738.847}$$
$$= \mathbf{53.648\ N.} \quad \textbf{Ans.}$$

Direction of the resultant force is obtained by finding unit vector in the direction of resultant force vector.

$\therefore$ Unit vector in the direction of resultant force vector $10.591\, \mathbf{i} + 5.209\, \mathbf{j} + 52.334\, \mathbf{k}$ is given by

$$= \frac{10.591\,\mathbf{i} + 5.209\, \mathbf{j} + 52.0334\, \mathbf{k}}{\sqrt{10.591^2 + 5.209^2 + 52.334^2}} = \frac{10.591\,\mathbf{i} + 5.209\, \mathbf{j} + 52.334\, \mathbf{k}}{53.648}$$

$$= \frac{10.591}{53.648}\,\mathbf{i} + \frac{5.209}{53.648}\,\mathbf{j} + \frac{52.334}{53.648}\,\mathbf{k} = 0.1974\,\mathbf{i} + 0.0971\,\mathbf{j} + 0.9755\,\mathbf{k}$$

∴ Direction of resultant force is 0.1974 **i** + 0.0971 **j** + 0.9755 **k**. **Ans.**

Problem 1.48. *A force P = 911 N is directed from a point A (4, 1, 4) metres towards a point B(– 3, 4, 1) metres. Determine the force vector P.*

Sol. Given :

Magnitude of P = 911 N

Co-ordinates of point A = (4, 1, 4)

∴ Position vector of point A = (4 **i** + 1 **j** + 4 **k**)

Co-ordinates of point B = (– 3, 4, – 1)

∴ Position vector of point B = (– 3 **i** + 4 **j** – **k**)

∴ Position* vector of force P

= Position vector of B – Position vector of A

$= (-3-4)\,\mathbf{i} + (4-1)\,\mathbf{j} + (-1-4)\,\mathbf{k}$

$= -7\,\mathbf{i} + 3\,\mathbf{j} - 5\,\mathbf{k}$

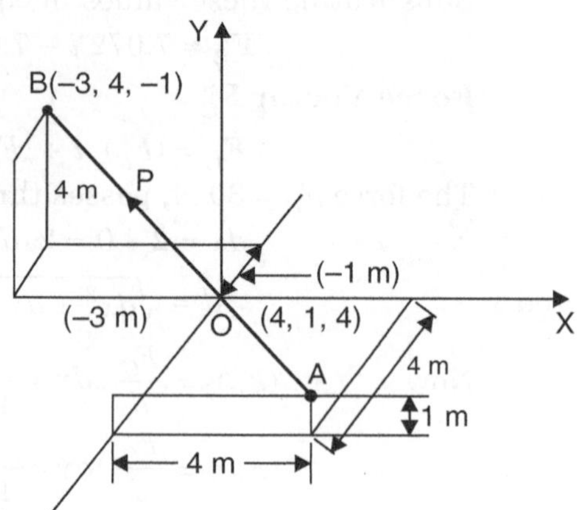

Fig. 1.84

∴ Unit vector in the direction of force P

$$= \frac{-7\,\mathbf{i} + 3\,\mathbf{j} - 5\,\mathbf{k}}{\sqrt{(-7)^2 + 3^2 + (-5)^2}}$$

$$= \frac{-7\,\mathbf{i} + 3\,\mathbf{j} - 5\,\mathbf{k}}{\sqrt{49+9+25}} = \frac{-7\,\mathbf{i} + 3\,\mathbf{j} - 5\,\mathbf{k}}{9.11}$$

∴ Force vector **P** = Magnitude of P times unit vector in the direction of force

$$= 9.11\,\frac{(-7\,\mathbf{i} + 3\,\mathbf{j} - 5\,\mathbf{k})}{9.11}$$

$$= \frac{911(-7)}{9.11}\,\mathbf{i} + \frac{911 \times 3}{9.11}\,\mathbf{j} - \frac{911 \times 5}{9.11}\,\mathbf{k} = -700\,\mathbf{i} + 300\,\mathbf{j} - 500\,\mathbf{k}.$$ **Ans.**

Problem 1.49. *A force P is directed from a point A(4, 1, 4) metres towards a point B(– 3, 4, – 1) metres. If it causes a moment M_z = 1900 Nm, determine the moment of P about x and y-axis.*

Sol. Given :

Co-ordinates of point A = (4, 1, 4)

∴ Position vector of point A = $4i + j + 4k$

Co-ordinates of point B = (– 3, 4, – 1)

∴ Position vector of point B = $-3i + 4j - k$

Moment about z-axis, M_z = 1900 Nm

Let magnitude of force = P

---

*As force is directed from A to B. Hence co-ordinates of A(4, 1, 4) means $x_1 = 4, y_1 = 1, z_1 = 4$. The co-ordinates of B(– 3, 4, – 1) means $x_2 = -3, y_2 = 4$ and $z_2 = -1$.

Therefore position vector is $(x_2 - x_1)\,\mathbf{i} + (y_2 - y_1)\,\mathbf{j} + (z_2 - z_1)\,\mathbf{k}$.

The force P is directed from A to B. Hence position vector of force P is given by

$$\mathbf{r} = \text{Position vector of point } B - \text{position vector of point } A$$
$$= (-3i + 4j - k) - (4i + j + 4k) = -7i + 3j - 5k$$

Now unit vector in the direction of force

$$= \frac{-7i + 3j - 5k}{\sqrt{(-7)^2 + 3^2 + (-5)^2}} = \frac{-7i + 3j - 5k}{\sqrt{49 + 9 + 25}} = \frac{-7i + 3j - 5k}{9.11}$$

Let $\mathbf{F}$ = Force vector

= Magnitude of force times the unit vector in the direction of force.

$$= \frac{P.(-7i + 3j - 5k)}{9.11}$$

$\therefore$ Force vector, $\mathbf{F} = \dfrac{P(-7i + 3j - 5k)}{9.11}$

The position vector of point $A = (4i + j + 4k)$

$\therefore$ Moment of the force about origin O is given by cross product of position* vector of point A with the given force vector.

Let M = Moment about origin O

Then M = Position vector of point A × Force vector $\mathbf{F}$

$$= (4i + j + 4k) \times \frac{P(-7i + 3j - 5k)}{9.11} = \frac{P}{9.11}(4i + j + 4k) \times (-7i + 3j - 5k)$$

$$= \frac{P}{9.11}\begin{vmatrix} i & j & k \\ 4 & 1 & 4 \\ -7 & 3 & 5 \end{vmatrix}$$

$$= \frac{P}{9.11}[(-5 - 12)i + (-28 + 20)j + (12 + 7)k] = \frac{P}{9.11}(-17i - 8j + 19k)$$

But $M = M_x i + M_y j + M_z k = \dfrac{P}{9.11}(-17)i - \dfrac{8P}{9.11}j + \dfrac{19P}{9.11}k$

$\therefore$ Moment of the given force about z-axis** $= \dfrac{P}{9.11} \times 19$

But moment of given force about z-axis = 1900 Nm

$$\therefore \quad \frac{P}{9.11} \times 19 = 1900$$

$$\therefore \quad P = \frac{1900 \times 9.11}{19} = 911 \text{ N}$$

The moment of the given force about x and y-axis is given by

$$M_x = \frac{P}{9.11}(-17) = \frac{911}{9.11}(-17) = -\textbf{1700 Nm.} \quad \textbf{Ans.}$$

$$M_y = \frac{P}{9.11}(-8) = \frac{911}{9.11}(-8) = -\textbf{800 Nm.} \quad \textbf{Ans.}$$

---

*The moment of the given force about origin is also equal to the cross product of position vector of point B with given force vector. The result will be same.

**In the moment equation (*i.e.*, M), the co-efficient of 'k' gives moment about z-axis.

Problem 1.50. *Fig. 1.84(a) shows a rectangular parallel piped subjected to four forces in the directions shown. Reduce them to a resultant force applied at the origin and a moment*

Sol. Given :

Force at C = 400 kN in the direction of CO

Force at F = 1000 kN in the direction of OF

Force at B = 800 kN in the direction of CB

Force at E = 707 kN in the direction of AE.

Co-ordinates of point C = (5, 0, 0)

∴ Position vector of point $C = 5i + 0j + 0k = 5i$

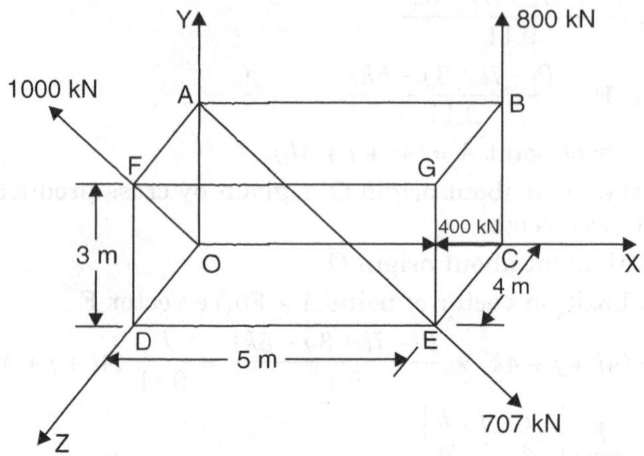

Fig. 1.84(a)

Co-ordinates of point B = (5, 3, 0)

∴ Position vector of point $B = (5i + 3j)$

Co-ordinates of point F = (0, 3, 4)

∴ Position vector of point $F = (0i + 3j + 4k)$

Co-ordinates of point E = (5, 0, 4)

∴ Position vector of point $E = (5i + 0j + 4k)$

Co-ordinates of point A = (0, 3, 0)

∴ Position vector of point $A = (0i + 3j + 0k)$

∴ Position vector of force 400 kN

= Position vector of point O* – Position vector of point C

$= (0i + 0j + 0k) - (5i) = -5i$

Position vector of force 1000 kN

= Position vector of point F – Position vector of O

$= (0i + 3j + 4k) - (0i + 0j + 0k) = 3j + 4k$

Position vector of force 800 kN

= Position vector of point B – Position vector of C

$= (5i + 3j) - (5i) = 3j$

*Co-ordinates of point O is (0, 0, 0). Hence position vector of point $O = (0i + 0j + 0k)$.

Position vector of force 707 kN

$$= \text{Position vector of point } E - \text{Position vector of } A$$
$$= (5i + 0j + 4k) - (0i + 3j + 0k) = (5i - 3j + 4k)$$

Now the unit vectors in the direction of forces can be obtained from the position vector of forces.

Unit vector in the direction of force 400 kN

$$= \frac{-5i}{\sqrt{(-5)^2}} = \frac{-5i}{5} = -i$$

Unit vector in the direction of force 1000 kN

$$= \frac{3j + 4k}{\sqrt{3^2 + 4^2}} = \frac{3j + 4k}{5}.$$

Unit vector in the direction of force 800 kN

$$= \frac{3j}{\sqrt{3^2}} = \frac{3j}{3} = j$$

Unit vector in the direction of force 707 kN

$$= \frac{5i - 3j + 4k}{\sqrt{5^2 + (-3)^2 + 4^2}} = \frac{5i - 3j + 4k}{\sqrt{25 + 9 + 16}} = \frac{(5i - 3j + 4k)}{\sqrt{50}}$$

Now different force vectors can be obtained by multiplying the force with the unit vector in the direction of the force.

∴ Force vector of 400 kN

$$= (400 \text{ kN}) \text{ times the unit vector in the direction of } 400 \text{ kN}$$
$$= 400 \text{ times } (-i) = -400\, i \qquad ...(i)$$

Similarly, force vector of 1000 kN

$$= 1000 \text{ times unit vector in the direction of } 1000 \text{ kN}$$
$$= 1000\left(\frac{3j + 4k}{5}\right) = 200\,(3j + 4k) = 600j + 800k \qquad ...(ii)$$

Force vector of 800 kN

$$= 800 \text{ (unit vector in the direction of 800 kN)}$$
$$= 800\,(j) = 800j \qquad ...(iii)$$

Force vector of 707 kN

$$= 707 \text{ (unit vector in the direction of 707 kN)}$$
$$= 707\,\frac{(5i - 3j + 4k)}{\sqrt{50}} = \frac{707\,(5i + 3j + 4k)}{7.07} = 100\,(5i - 3j + 4k)$$
$$= 500i - 300j + 400k \qquad ...(iv)$$

∴ The resultant force vector (**R**) will be obtained by adding equations (*i*), (*ii*), (*iii*) and (*iv*).

∴ Resultant force vector **R**

$$= (-400i) + (600j + 800k) + (800j) + (500i - 300j + 400k)$$
$$= (-400 + 500)i + (600 + 800 - 300)j + (800 + 400)k$$

$$= \mathbf{100i + 1100j + 1200k}.$$ **Ans.**

The magnitude of resultant force R is given by

$$R = \sqrt{100^2 + 1100^2 + 1200^2} = \sqrt{10000 + 1210000 + 1440000} = \sqrt{2660000}$$
$$= 1630.95 \simeq 1631 \text{ kN.}$$

Now let us find the moment of each force vector about O the origin.

We have to find resultant force applied at the origin and the moment.

The moment of the resultant force about origin will be zero, as resultant force is passing through the origin.

The moments of force 400 kN and 1000 kN about origin will also be zero as these forces passing through origin.

The position vector **r** of 800 kN

$$= \text{Position vector of point } C - \text{Position vector of } O = 5i$$

The position vector **r** of 707 kN with respect to origin

$$= \text{Position vector of point } A - \text{Position vector of } O = 3j$$

$\therefore$ Moment of force 800 kN about origin

$$= \text{Position vector } \mathbf{r} \text{ of force 800 kN} \times \text{Force vector 800 kN}$$
$$= 5i \times 800j \qquad (\because \text{ Force vector of 800 kN} = 800\,j)$$
$$= 4000\,k \qquad (\because\ i \times j = k)$$

Moment of force 707 kN about origin

$$= \text{Position vector } \mathbf{r} \text{ of force 707 kN} \times \text{Force vector 707 kN}$$
$$= 3j \times (500i - 300j + 400k) = 1500j \times i - 900j \times j + 1200j \times k$$
$$= 1500\,(-k) - 900\,(0) + 1200\,(i) \qquad [\because\ j \times i = -k, j \times j = 0, j \times k = i]$$
$$= -1500\,k + 1200\,i$$
$$= 1200\,i - 5000\,k$$

$\therefore$ Sum of moments of all forces at origin

$$= 4000\,k + (1200\,i - 1500\,k) = \mathbf{1200\,i + 2500\,k}. \quad \textbf{Ans.}$$

Problem 1.51. *Three forces P, Q and R are applied along the diagonals of rectangular parallelopiped as shown in Fig. 1.85. If 2P = 2Q = R = 1000 N. Determine the equivalent force and moment at the origin.*

Sol. Given :

$$2P = 2Q = R = 1000 \text{ N}$$

$\therefore$ $$P = \frac{1000}{2} = 500 \text{ N}, Q = 500 \text{ N}$$

and $$R = 1000 \text{ N}$$

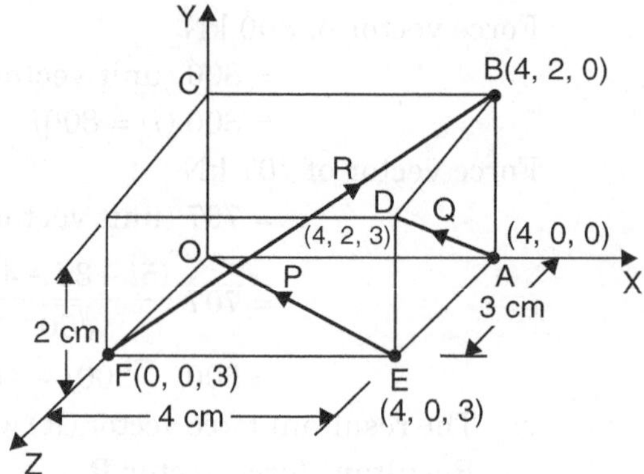

Fig. 1.85

Now position vector of force P

$$= \text{Position vector of point } O$$
$$- \text{Position vector of point } E$$
$$= (0i + 0j + 0k) - (4i + 0j + 3k)$$
$$= (-4i - 3k)$$

Position vector of force Q

$$= \text{Position vector of point } D$$
$$- \text{Position vector of point } A$$
$$= (4i + 2j + 3k) - (4i + 0j + 0k)$$
$$= (2j + 3k)$$

Position vector of force R

$= \text{Position vector of point } B - \text{Position vector of point } F$

$= (4i + 2j + 0k) - (0i + 0j + 3k) = (4i - 2j - 3k)$

Unit vector in the direction of force P

$$= \frac{-4i - 3k}{\sqrt{(-4)^2 + (-3)^2}} = \frac{-4i - 3k}{5}$$

Unit vector in the direction of force Q

$$= \frac{2j + 3k}{\sqrt{2^2 + 3^2}} = \frac{2j + 3k}{\sqrt{13}} = \frac{2j + 3k}{3.6}$$

Unit vector in the direction of force R

$$= \frac{4i + 2j - 3k}{\sqrt{4^2 + 2^2 + (-3)^2}} = \frac{4i + 2j - 3k}{\sqrt{16 + 4 + 9}} = \frac{4i + 2j - 3k}{\sqrt{29}} = \frac{4i + 2j - 3k}{5.385}$$

Now force vector **P**

$= \text{Magnitude of } P \text{ times the unit vector in the direction of } P$

$$= 500\left(\frac{-4i - 3k}{5}\right) \qquad (\because \quad P = 500 \text{ N})$$

$= 100\,(-4i - 3k) = -400i - 300k$

Similarly, force vector **Q**

$= \mathbf{Q}$ times the unit vector in the direction of $\mathbf{Q}$

$$= 500\left(\frac{2j + 3k}{3.6}\right) = \frac{1000j}{3.6} + \frac{1500k}{3.6} = 277.77\,j + 416.66k$$

Also force vector **R**

$= \text{Magnitude of } R \text{ times unit vector in the direction of } R$

$$= 1000\left(\frac{4i + 2j - 3k}{5.385}\right) = \frac{4000}{5.385}\,i + \frac{2000}{5.385}\,j - \frac{3000}{5.385}\,k$$

$= 742.8\,i + 371.4\,j - 557.1\,k$

$\therefore$ Resultant force vector

$= \text{Force vector } \mathbf{P} + \text{Force vector } \mathbf{Q} + \text{Force vector } \mathbf{R}$

$= (-400i - 300k) + (277.77j + 416.66k) + (742.8i + 371.4j - 557.1k)$

$= (-400 + 742.8)i + (277.77 + 371.4)j + (-300 + 416.66 - 557.1)k$

$= \mathbf{342.8i + 649.17j - 440.44k}$. **Ans.**

Magnitude of resultant force

$$= \sqrt{342.8^2 + 649.17^2 + 440.44^2} = \sqrt{117511.84 + 421421.68 + 193987.39}$$

$= \mathbf{856.1}$ **N. Ans.**

Moment of all forces at O is given by

$$M_o = r_1 \times \text{Force vector } \mathbf{P} + r_2 \times \text{Force vector } \mathbf{Q} + r_3 \times \text{Force vector } \mathbf{R}$$

where $\mathbf{r}_1$ = Position vector of a point on force vector $\mathbf{P} = 0$

($\because$ Force P is passing through origin)

$\mathbf{r}_2$ = Position vector of a point on force vector $\mathbf{Q} = 4i$ or $(4i + 2j + 3k)$

$\mathbf{r}_3$ = Position vector of a point on force vector $\mathbf{R} = 3k$ or $(4i + 2j + 0k)$

$\therefore \quad M_o = 0 \times (-400i - 300k) + 4i \times (277.77j + 416.66k)$
$+ 3k \times (742.8i + 371.4j - 557.1k)$
$= 0 + (4)(277.77)i \times j + (4)(416.66)i \times k + (3)(742.8)k \times i$
$+ (3)(371.4)k \times j + (3)(-557.4)k \times k$
$= (4)(277.77)k + (4)(416.66)(-j) + (3)(742.8)(j) + (3)(371.4)(-i) + 0$
$(\because \; i \times j = k, i \times k = -j, k \times i = j, k \times j = -i, k \times k = 0)$
$= 1111.08k - 1666.64j + 2228.4j - 1114.2i$
$= -1114.2i + (2228.4 - 1666.64)j + 1111.08k$
$= \mathbf{-1114.2i + 561.76j + 1111.08k}$. **Ans.**

Problem 1.52. *A bar AB is acted upon by six forces as shown in Fig. 1.86. The forces are :*

$\mathbf{F_1} = 80i, \qquad \mathbf{F_2} = 130k,$
$\mathbf{F_3} = -50j, \qquad \mathbf{F_4} = 210k,$
$\mathbf{F_5} = -100i + 190j - 50k, \qquad \mathbf{F_6} = 315j.$

All forces are in Newtons. Determine the equivalent force and moment acting at A.

Sol. Given :

$\mathbf{F_1} = 80i, \qquad \mathbf{F_2} = 130k$
$\mathbf{F_3} = -50j, \qquad \mathbf{F_4} = -210k$
$\mathbf{F_5} = -100i + 190j - 50k,$
$\mathbf{F_6} = 315j.$

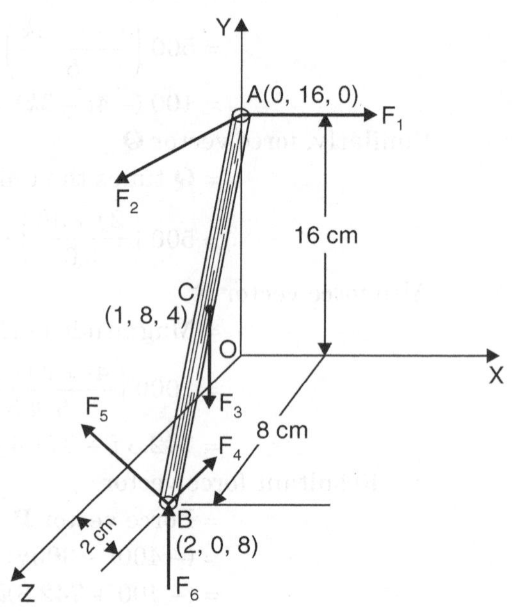

Fig. 1.86

Position vector of point $A = 16j$
Position vector of point $B = (2i + 8k)$
Position vector of point $C = (i + 8j + 4k)$.

(i) To find the equivalent force

Equivalent force is the resultant force.

The resultant force **R** is given by

$\mathbf{R} = \mathbf{F_1} + \mathbf{F_2} + \mathbf{F_3} + \mathbf{F_4} + \mathbf{F_5} + \mathbf{F_6}$
$= 80i + 130k + (-50j) + (-210k)$
$+ (-100i + 190j - 50k) + 315j$
$= (80 - 100)i + (-50 + 190 + 315)j$
$+ (130 - 210 - 50)k$
$= \mathbf{-20i + 455j - 130k}$. **Ans.**

(ii) To find the moment of all forces about point A

Let $\mathbf{r_1}$ = Position vector of force F_1 from point A
= 0 (as force F_1 is passing through A)
$\mathbf{r_2}$ = Position vector of force F_2 from point $A = 0$
$\mathbf{r_3}$ = Position vector of force F_3 from point A
= Position vector of point C – position vector of point A
$= (i + 8j + 4k) - (16j) = (i - 8j + 4k)$

$\mathbf{r}_4$ = Position vector of force F_4 with respect to point A
= Position vector of point B – position vector of point A
$= (2i + 8k) - (16j) = (2i - 16j + 8k)$

$\mathbf{r}_5$ = Position vector of force F_5 from point A
$= 2i - 16j + 8k$

$\mathbf{r}_6$ = Position vector of force F_6 from point A
$= 2i - 16j + 8k.$

(Force F_5 and force F_6 are acting at point B, hence their position vector will be same as of force F_4.)

∴ Moment* of all forces about point A is given by,

$$M_A = \mathbf{r}_1 \times \mathbf{F}_1 + \mathbf{r}_2 \times \mathbf{F}_2 + \mathbf{r}_3 \times \mathbf{F}_3 + \mathbf{r}_4 \times \mathbf{F}_4 + \mathbf{r}_5 \times \mathbf{F}_5 + \mathbf{r}_6 \times \mathbf{F}_6$$

$$= 0 + 0 + (i - 8j + 4k) \times (-50j) + (2i - 16j + 8k) \times (-210k) + (2i - 16j + 8k) \times (-100i + 190j - 50k) + (2i - 16j + 8k) \times (315j)$$

$$= (-50i \times j + 400j \times j - 200k \times j) + (-420i \times k + 3360j \times k - 1680k \times k) + (-200i \times i + 380i \times j - 100i \times k + 1600j \times i - 16 \times 190 \times j \times j + 800j \times k - 800k \times i + 1520k \times j - 400k \times k) + (630i \times j - 16 \times 315 \times j \times j + 2520\, k \times j)$$

But $i \times j = k, j \times j = 0, i \times i = 0, k \times k = 0, k \times j = -i, i \times k = -j,$
$j \times k = i, j \times i = -k, k \times i = j, k \times j = -i.$

$$\therefore \quad M_A = [-50k + 0 - 200(-i)] + [-420(-j) + 3360i - 0] + [0 + 380k - 100(-j) + 1600(-k) - 0 + 800i - 800j + 1520(-i) - 0] + [630k - 0 + 2520(-i)]$$

$$= -50k + 200i + 420j + 3360i + 380k + 100j - 1600k + 800i - 800j - 1520i + 630k - 2520i$$

$$= (200 + 3360 + 800 - 1520 - 2520)i + (420 + 100 - 800)j + (-50 + 380 - 1600 + 630)k$$

$= \mathbf{320i + 290j - 640k}$. **Ans.**

1.14. EQUILIBRIUM OF A PARTICLE IN SPACE

When a particle (or a rigid body) is in space, the forces acting on the rigid body (or on the particle) may be concurrent or non-concurrent. If the forces acting are concurrent, then the equations of equilibrium are

$$\Sigma F_x = 0, \quad \Sigma F_y = 0 \quad \text{and} \quad \Sigma F_z = 0$$

i.e., the resultant force in x, y and z directions are zero.

But if the forces acting are non-concurrent then the resultant force in x, y and z directions should be zero also the resultant moment about x, y and z axis should be zero. *i.e.*,

$$\Sigma F_x = 0, \qquad \Sigma F_y = 0 \qquad \text{and} \qquad \Sigma F_z = 0$$

Also $\Sigma M_{xy} = 0, \quad \Sigma M_{xz} = 0 \quad$ and $\quad \Sigma M_{yz} = 0$

*Moment is the cross product of position vector and force vector

∴ $\mathbf{M} = \mathbf{r} \times \mathbf{F}$.

Thus there will be six equations of equilibrium, three equations of equilibrium will be for force and three equations of equilibrium will be for moment. In vector form, these equations can be written as

$$\Sigma \mathbf{F} = 0 \quad \text{and} \quad \Sigma \mathbf{M} = 0.$$

1.15. EQUIVALENT SYSTEMS OF FORCES

We have learnt that a force can be replaced by a force and a couple. When a number of forces are acting on a body in space then each force is transferred to an arbitrary reference point along with a couple for each force transferred. Thus, we get

(*i*) a system of concurrent forces and

(*ii*) a system of couples.

The system of concurrent forces can be added to get the resultant single force **(R).** Also the system of couples can be added to get the resultant moment **(M).**

Hence the equations become as

$$\mathbf{R} = \mathbf{F}_1 + \mathbf{F}_2 + \mathbf{F}_3 + \ldots\ldots$$

$$\mathbf{M} = \mathbf{M}_1 + \mathbf{M}_2 + \mathbf{M}_3 + \ldots\ldots$$

An equivalent system for a given system of coplanar forces, is a combination of a force passing through a given point and a moment about that point. The force is the resultant of all forces acting on the body. And the moment is the sum of all the moments about that point.

Hence equivalent system consists of :

(*i*) a single force R passing through the given point P and

(*ii*) a single moment M_R

where R = the resultant of all force acting on the body.

M_R = sum of all moments of all the forces about point P.

Problem 1.53. *Three external forces are acting on a L-shaped body as shown in Fig. 1.87. Determine the equivalent system through point O.*

Sol. Given :

Force at A = 2000 N, Angle = 30°

Force at B = 1500 N

Force at C = 1000 N

Distance OA = 200 mm, OB = 100 mm

and BC = 200 mm

Angle COA = 90°

Determine the equivalent system through O. This means find

(*i*) single resultant force, R

(*ii*) single moment through O.

Taking x-axis along OA and y-axis along OC.

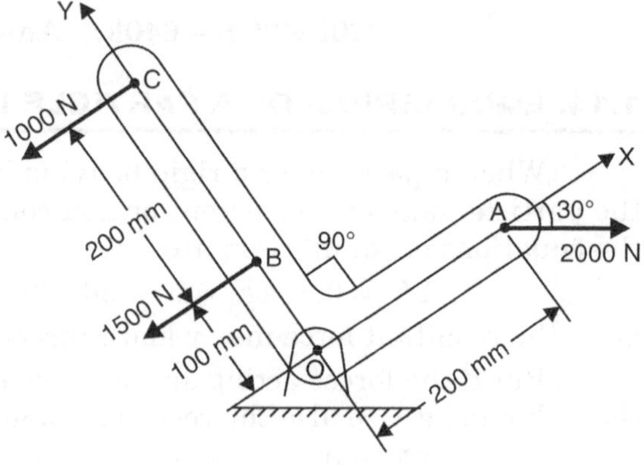

Fig. 1.87

The force at A is resolved into two components.

Component along x-axis = 2000 × cos 30° = 1732 N

Component along y-axis = 2000 × sin 30° = 1000 N

Resolving all forces along X-axis *i.e.*,

$$\Sigma F_x = 2000 \cos 30° - 1500 - 1000 = -768 \text{ N}$$

Similarly $\Sigma F_y = -2000 \times \sin 30° = -1000$ (–ve sign is due to downward)

$$\therefore \text{ Resultant, } R = \sqrt{\Sigma F_x^2 + \Sigma F_y^2} = \sqrt{(-768)^2 + (-1000)^2}$$

$$= \sqrt{589824 + 1000000} = 1260.88 \text{ N}$$

Taking moments of all forces about point O,

$$M_o = (-2000 \sin 30) \times 200 + 1500 \times 100 + 1000 \times 300$$

$$= -200000 + 150000 + 300000$$

$$= 250000 \text{ Nmm} = 250 \text{ Nm}$$

$\therefore$ Equivalent system through point O is

$$R = 1260.88 \text{ N}$$

$$M = 250 \text{ Nm.}$$

Problem 1.54. *Fig. 1.88 shows two vertical forces and a couple of moment 2000 Nm acting on a horizontal rod which is fixed at end A.*

(i) Determine the resultant of the system.

(ii) Determine an equivalent system through A.

Fig. 1.88

Sol. Given :

Force at C = 4000 N, Force at B = 2500 N

Moment at D = 2000 Nm

Distance AC = 1 m, BC = 1.5 m

CD = 0.8 m, BD = 0.7 m

(i) Resultant of the system

This means to find the resultant of all the forces and also the point at which the resultant is acting. There are two vertical forces only.

Hence resultant, $R = 4000 - 2500 = 1500$ N acting downward

The point at which the resultant is acting is obtained by taking moments about point A. For moments there are two forces (4000 N at C and 2000 N at B) and also a moment at D.

Moment of force 4000 N about point $A = 4000 \times 1$

= 4000 Nm (clockwise)

Moment of force 2500 N about point A

$= 2500 \times (1 + 1.5) = 2500 \times 2.5$

= 6250 Nm (anti-clockwise)

Moment at D = 2000 Nm (clockwise)

$\therefore$ Sum of all moments about A

= 4000 (clockwise) – 6250 (anti-clockwise) + 2000 (clockwise)

= – 250 (anti-clockwise)

The resultant is acting vertically downward. If it is acting towards right of A, then it will give clockwise moment. But we want anti-clockwise moment. Hence the resultant must act towards the left of A.

Let x = Distance of resultant force (1500 N) from A

$\therefore$ Moment of resultant force (R) about A

$$= 1500 \times x$$

$$\therefore \quad 1500 \times x = 250$$

$$\therefore \quad x = \frac{250}{1500} = \mathbf{0.166\ m.}\quad \textbf{Ans.}$$

Hence resultant of the system is 1500 N ↓ acting at a distance of 0.166 m left to A. **Ans.**

(ii) Equivalent system through A

This means to find a single resultant force and a single moment through A.

Single resultant force, R = 1500 N

Single moment through, A = **250 Nm. Ans.**

1.16. PRINCIPLE OF TRANSMISSIBILITY

The principle of transmissibility of forces has been already defined in Art. 1.3.5(see page 16). It states that "If a force, acting at a point on a rigid body, is shifted to any other point which is on the line of action of the force, the external effect of the force on the body remains unchanged". For more details and explanation, please refer to Art. 1.3.5.

HIGHLIGHTS

(i) Introduction

1. Engineering mechanics is divided into statics and dynamics. The study of a body at rest is known as statics whereas the study of a body in motion is known as dynamics.
2. A quantity which is completely specified by magnitude and direction is known as vector quantity.
3. A particle is a body of infinitely small volume and is considered to be concentrated at a point.
4. Law of parallelogram of forces states that "If two forces, acting at a point be represented in magnitude and direction by the two adjacent sides of a parallelogram, then their resultant is represented in magnitude and direction by the diagonal of the parallelogram passing through that point.
5. If two forces P and Q act at a point and the angle between the two forces be α, then the resultant is given by

$$R = \sqrt{P^2 + Q^2 + 2PQ\cos\alpha}$$

and the angle made by the resultant with the direction of force P is expressed as

$$\tan\theta = \frac{Q\sin\alpha}{P + Q\cos\alpha}.$$

6. If the two forces P and Q are equal and are acting at an angle α between them, then the resultant is given by

$$R = 2P\cos\frac{\alpha}{2}$$

and angle made by the resultant is expressed as $\theta = \frac{\alpha}{2}$.

7. According to Lame's theorem, "If three forces acting at a point are equilibrium, each force will be proportional to the sine of the angle between the other two forces."

8. The relation between newton and dyne is given by One newton = 10^5 dyne.
9. Moment of a force about a point = Force × perpendicular distance of the line of action of the force from that point.
10. The force causes linear displacement while moment causes angular displacement. A body will be in equilibrium if (*i*) resultant force in any direction is zero, and (*ii*) the net moment of the forces about any point is zero.
11. Gravitational law of attraction is given by,

$$F = G\,\frac{m_1 \times m_2}{r^2}$$

where G = Universal gravitational constant

m_1, m_2 = Mass of bodies

r = Distance between the bodies

F = Force of attraction between the bodies.

(ii) Coplanar Concurrent Forces

12. Coplanar forces means the forces are acting in one plane.
13. Concurrent forces means the forces are intersecting at a common point.
14. Collinear forces means the forces are having same line of action.
15. The resultant of coplanar forces are determined by analytical and graphical methods.
16. The resultant (R) of three collinear forces F_1, F_2 and F_3 acting in the same direction, is given by $R = F_1 + F_2 + F_3$. If the force F_2 is acting in opposite direction then their resultant will be, $R = F_1 - F_2 + F_3$.
17. The resultant of three or more forces acting at a point is given by, $R = \sqrt{(\Sigma H)^2 + (\Sigma V)^2}$, where ΣH = Algebraic sum of horizontal components of all forces, ΣV = Algebraic sum of vertical components of all forces. The angle made by the resultant with horizontal is given by, $\tan\theta = \frac{(\Sigma V)}{(\Sigma H)}$.
18. The resultant of several forces acting at a point is found graphically by using polygon law of forces.
19. Polygon law of forces states that if a number of coplanar forces are acting at a point such that they can be represented in magnitude and direction by the sides of a polygon taken in the same order, then their resultant is represented in magnitude and direction by the closing side of the polygon taken in the opposite order.

(iii) Coplanar Parallel Forces

20. Parallel forces are having their lines of action parallel to each other.
21. The moment of a force about any point is the product of force and perpendicular distance between the point and line of action of force.
22. Anti-clockwise moment is taken +ve whereas clockwise moment is taken –ve.
23. Varignon's principle states that the moment of a force about any point is equal to the algebraic sum of moments of its components about that point.
24. Like parallel forces are parallel to each other and are acting in the same direction, whereas the unlike parallel forces are acting in opposite direction.
25. The resultant of two like parallel forces is the sum of the two forces and acts at a point between the line in such a way that the resultant divides the distance in the ratio inversely proportional to the magnitudes of the forces.
26. When two equal and opposite parallel forces act on a body at some distance apart, the two forces form a couple which has a tendency to rotate the body. The moment of this couple is the product of either one of the forces and perpendicular distance between the forces.
27. A given force F applied to a body at any point A can always be replaced by an equal force applied at another point B in the same direction together with a couple.

28. If the resultant of a number of parallel forces is not zero, the system can be reduced to a single force, whose magnitude is equal to the algebraic sum of all forces. The point of application of this single force is obtained by equating the moment of this single force about any point to the algebraic sum of moments of all forces acting on the system about the same point.

29. If the resultant of a number of parallel forces is zero, then the system may have a resultant couple or may be in equilibrium. If the algebraic sum of moments of all forces about any point is not zero, then system will have a resultant couple. But if the algebraic sum of moments of all forces about any point is zero, the system will be in equilibrium.

EXERCISE 1

(A) Theoretical Questions

(i) Introduction

1. What do you mean by scalar and vector quantities ?
2. Define the law of parallelogram of forces. What is the use of this law ?
3. State triangle law of forces and Lame's theorem.
4. Two forces P and Q are acting at a point in a plane. The angle between the forces is 'α'. Prove that the resultant (R) of the two forces is given by $R = \sqrt{P^2 + Q^2 + 2PQ \cos \alpha}$.
5. Define the following terms : dyne, newton, meganewton and moment of a force.
6. Prove that one newton is equal to 10^5 dyne.
7. Explain the terms : clockwise moments and anti-clockwise moments.
8. What is the effect of force and moment on a body ?
9. Indicate whether the following statement is true or false.
   "The resultant components of the forces acting on a body along any direction is zero but the net moment of the forces about any point is not zero, the body will be in equilibrium'. [**Ans.** False]
10. Write the S.I. units of : Force, moment and velocity.
11. What do you mean by resolution of a force ?
12. A number of coplanar forces are acting at a point making different angles with x-axis. Find an expression for the resultant force. Find also the angle made by the resultant force with x-axis.
13. State and explain the pinciple of transmissibility of forces.
14. State and explain the following laws :
    (*i*) Newton's laws of motion.
    (*ii*) The gravitational law of attraction.
15. Using gravitational law of attraction, prove that $W = m \times g$.
16. Explain fully the following terms :
    (*i*) Resolved part of a given force in a given direction.
    (*ii*) Lame's theorem.

(ii) Coplanar Concurrent Forces

17. Define and explain the following terms :
    (*i*) Coplanar and non-coplanar forces (*ii*) Collinear and concurrent forces
    (*iii*) Parallel and non-parallel forces.
18. What is the difference between collinear and concurrent forces ?
19. State and explain the following laws of forces :
    (*i*) Law of parallelogram of forces (*ii*) Law of triangle of forces
    (*iii*) Law of polygon of forces.

20. Derive an expression for the resultant in magnitude and direction of two coplanar concurrent forces using cosine law method.

21. Explain in detail the method of finding resultant in magnitude and direction of three or more forces acting at a point by analytical and graphical method.

22. Explain the procedure of resolving a given force into two components at right angles to each other.

23. Three collinear forces F_1, F_2 and F_3 are acting on a body. What will be the resultant of these forces, if

(*a*) all are acting in the same direction (*b*) force F_3 is acting in opposite direction.

24. State the law of parallelogram of forces and show that the resultant $R = \sqrt{P^2 + Q^2}$ when the two forces P and Q are acting at right angles to each other. Find the value of R if the angle between the forces is zero.

(iii) Coplanar Parallel Forces

25. Define the terms : Coplanar parallel forces, like parallel forces and unlike parallel forces.

26. Define and explain the moment of a force. Differentiate between clockwise moment and anti-clockwise moment.

27. (*a*) State the Varignon's principle. Also give the proof of Varignon's principle.

(*b*) Differentiate between :

(*i*) Concurrent and non-concurrent forces,

(*ii*) Coplanar and non-coplanar forces,

(*iii*) Moment of a force and couple.

28. Define moment of a force about a point and show that the algebraic sum of the moments of two coplanar forces about a point is equal to the moment of their resultant about that point.

29. What are the different types of parallel forces ? Distinguish between like and unlike parallel forces.

30. Prove that the resultant of two like parallel forces F_1 and F_2 is $F_1 + F_2$. Also prove that the resultant divides the line of joining the points of action of F_1 and F_2 internally in the inverse ratio of the forces.

31. Prove that in case of two unlike parallel forces the resultant lies outside the line joining the points of action of the two forces and on the same side as the larger force.

32. Describe the method of finding the line of action of the resultant of a system of parallel forces.

33. The resultant of a system of parallel forces is zero, what does it signify ?

34. Describe the method of finding the resultant of two unlike parallel forces which are equal in magnitude.

35. Indicate whether the following statements are True or False :

(*i*) Force is an agency which tends to cause motion.

(*ii*) The value of g reduces slightly as we move from poles towards the equator.

(*iii*) Coplanar forces are those which have the same magnitude and direction.

(*iv*) A couple consists of two unequal and parallel forces acting on a body, having the same line of action.

(*v*) A vector diagram of a force represents its magnitude, direction, sense and point of application.

(*vi*) The force of gravitation on a body is called its weight.

(*vii*) The centre of gravity of a body is the point, through which the resultant of parallel forces passes in whatever position may the body be placed.

[**Ans.** (*i*) True (*ii*) True (*iii*) False (*iv*) False (*v*) False (*vi*) True (*vii*) True]

(B) Numerical Problems

(i) Introduction

1. Determine the magnitude of the resultant of the two forces of magnitude 12 N and 9 N acting at a point when the angle between the two forces is 30°. **[Ans.** 20.3 N]
2. Find the magnitude of two equal forces acting at a point with an angle of 60° between them, if the resultant is equal to $30 \times \sqrt{3}$ N. **[Ans.** 30 N]
3. The resultant of two forces when they act at right angles is 10 N, whereas when they act at an angle of 60° the resultants is $\sqrt{148}$. Determine the magnitude of the two forces. **[Ans.** 8 N and 6 N]
4. Three forces of magnitude 30 kN, 10 kN and 15 kN are acting at a point *O*. The angles made by 30 kN force, 10 kN force and 15 kN force with *x*-axis are 60°, 120° and 240° respectively. Determine the magnitude and direction of the resultant force. **[Ans.** 21.79 kN, 83° 24]
5. A weight of 800 N is supported by two chains as shown in Fig. 1.89. Determine the tension in each chain. **[Ans.** 273.5 N, 751.7 N]

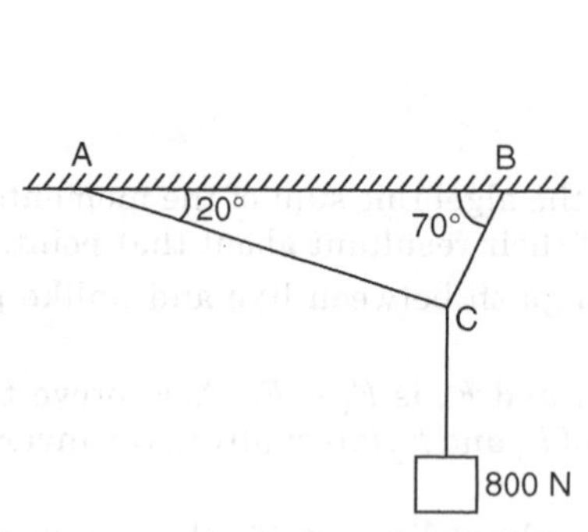

Fig. 1.89

Fig. 1.90

6. An electric light fixture weighing 20 N hangs from a point *C*, by two strings *AC* and *BC*. *AC* is inclined at 60° to the horizontal and *BC* at 30° to the vertical as shown in Fig. 1.90. Using Lame's theorem or otherwise determine the forces in the strings *AC* and *BC*. **[Ans.** 8.929 N, 13.05 N]
7. A beam *AB* of span 6 m carries a point load of 100 N at a distance 2 m from *A*. Determine the beam reaction. **[Ans.** R_A = 66.67 N ; R_B = 33.33 N]
8. Four forces of magnitudes 20 N, 30 N, 40 N and 50 N are acting respectively along the four sides of a square taken in order. Determine the magnitude, direction and position of the resultant force. $\left[\textbf{Ans. } 20 \times \sqrt{2}\text{ N}, 225°, \dfrac{7a}{2 \times \sqrt{2}}\right]$
9. Two forces magnitude 15 N and 12 N are acting at a point. If the angle between the two forces is 60°, determine the resultant of the forces in magnitude and direction. **[Ans.** 23.43 N, 26.3°]

10. Four forces of magnitude P, $2P$, $3 \times \sqrt{3}\,P$ and $4P$ are acting at a point O. The angles made by these forces with x-axis are 0°, 60°, 150° and 300° respectively. Find the magnitude and direction of the resultant force. **[Ans.** P, 120°**]**

(ii) Coplanar Concurrent Forces

11. Three collinear horizontal forces of magnitude 300 N, 100 N and 250 N are acting on rigid body. Determine the resultant of the forces analytically and graphically when : (*i*) all the forces are acting in the same direction ; (*ii*) the force 100 N acts in the opposite direction.

[Ans. (*i*) 650 N, (*ii*) 450 N**]**

12. Two forces of magnitude 15 N and 12 N are acting at a point. The angle between the forces is 60°. Find the resultant in magnitude. **[Ans.** 20.43 N**]**

13. A force of 1000 N is acting at a point, making an angle of 60° with the horizontal. Determine the components of this force along horizontal and vertical directions. **[Ans.** 500 N, 866 N**]**

14. A small block of weight 100 N is placed on an inclined plane which makes an angle of 60° with the horizontal. Find the components of this weight (*i*) perpendicular to the inclined plane and (*ii*) parallel to the inclined plane. **[Ans.** 50 N, 86.6 N**]**

15. Two forces P and Q are acting at a point O as shown in Fig. 1.91. The force P = 264.9 N and force Q = 195.2 N. If the resultant of the forces is equal to 400 N then find the values of angles β, γ, α.

[Ans. $\beta = 35°$, $\gamma = 25°$, $\alpha = 60°$**]**

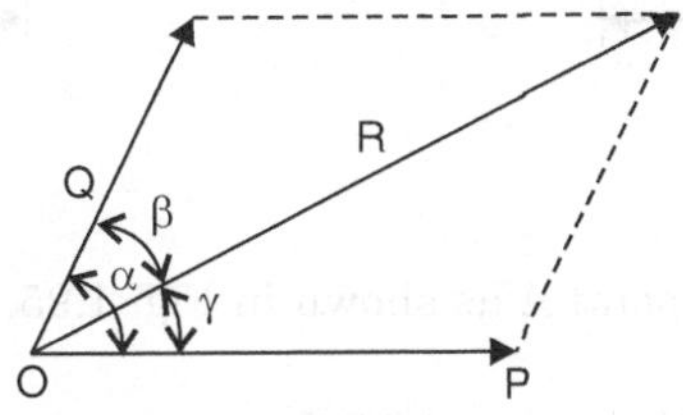

Fig. 1.91

16. A small block of unknown weight is placed on an inclined plane which makes an angle of 30° with horizontal plane. The component of this weight parallel to the inclined plane is 100 N. Find the weight of the block. **[Ans.** 200 N**]**

17. In question 16, find the component of the weight perpendicular to the inclined plane.

[Ans. 173.2 N**]**

18. The four coplanar forces are acting at a point as shown in Fig. 1.92. Determine the resultant in magnitude and direction analytically and graphically. **[Ans.** 1000 N, $\theta = 60°$ with OX**]**

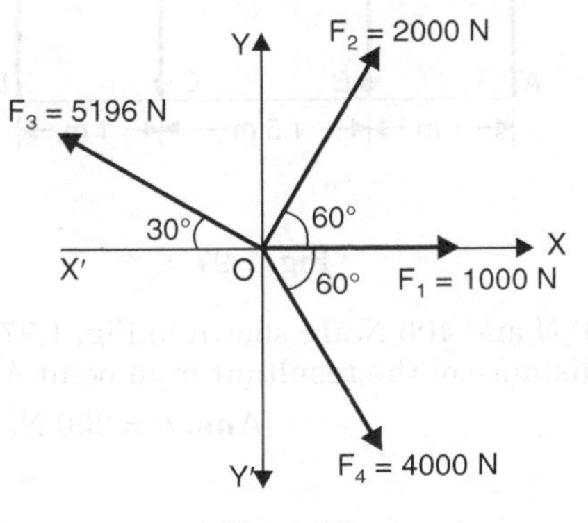

Fig. 1.92

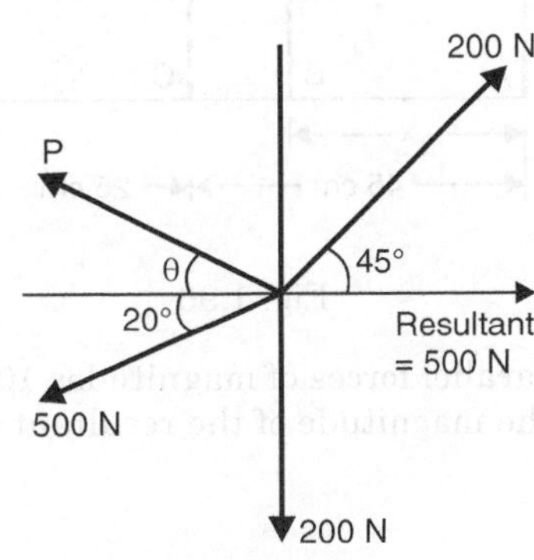

Fig. 1.93

19. The four coplanar forces are acting at a point as shown in Fig. 1.93. One of the forces is unknown and its magnitude is shown by P. The resultant is having a magnitude 500 N and is acting along x-axis. Determine the unknown force P and its inclination with x-axis.

[**Ans.** P = 286.5 N and θ = 53° 15′]

(iii) Coplanar Parallel Forces

20. Four forces of magnitudes 20 N, 40 N, 60 N and 80 N are acting respectively along the four sides of a square $ABCD$ as shown in Fig. 1.94. Determine the resultant moment about point A. Each side of square is 2 m. [**Ans.** 200 Nm anti-clockwise]

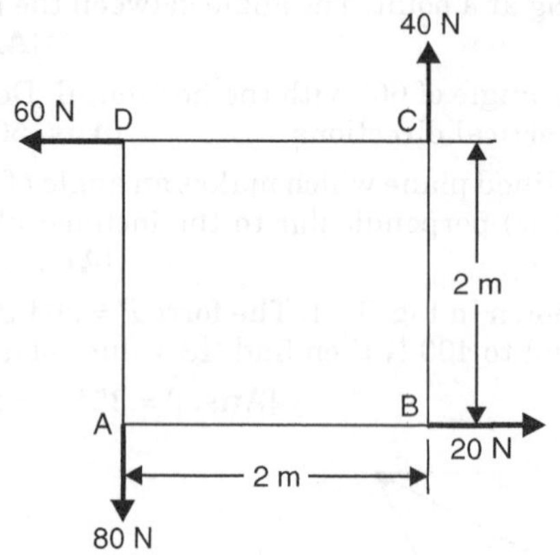

Fig. 1.94

Fig. 1.95

21. A force of 50 N is acting at a point A as shown in Fig. 1.95. Determine the moment of this force about O. [**Ans.** 100 Nm clockwise]

22. Three like parallel forces 20 N, 40 N and 60 N are acting at points, A, B and C respectively on a straight line ABC. The distances are AB = 3 m and BC = 4 m. Find the resultant and also the distance of the resultant from point A on line ABC. [**Ans.** 120 N, 4.5 m]

23. The three like parallel forces 101 N, F and 300 N are acting as shown in Fig. 1.96. If the resultant R = 600 N and is acting at a distance of 45 cm from A, then find the magnitude of force F and distance of F and A. [**Ans.** 200 N, 30 cm]

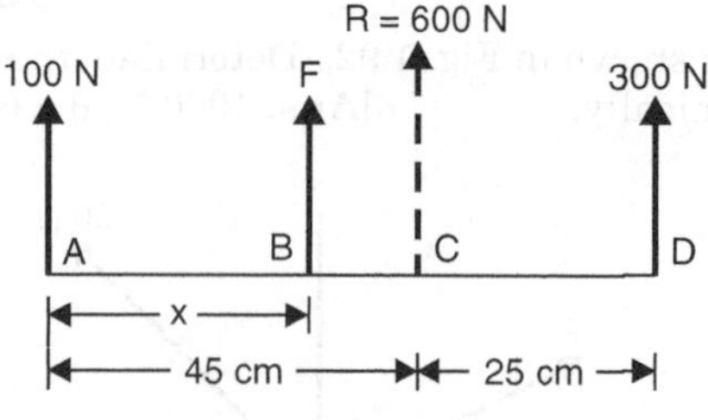

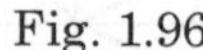

Fig. 1.96

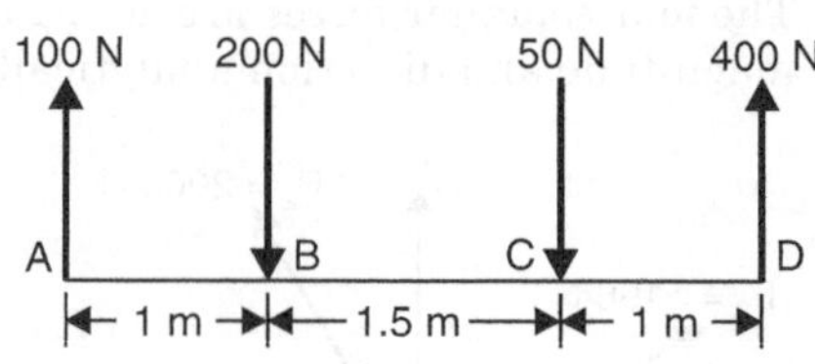

Fig. 1.97

24. Four parallel forces of magnitudes 100 N, 200 N, 50 N and 400 N are shown in Fig. 1.97. Determine the magnitude of the resultant and also the distance of the resultant from point A.

[**Ans.** R = 350 N, 3.07 m]

25. A system of parallel forces are acting on a rigid bar as shown in Fig. 1.98. Reduce this system to :

(*i*) a single force, [**Ans.** (*i*) R = 120 N at 2.83 m from A

(*ii*) a single force and a couple at A (*ii*) R = 120 N and M_A = – 340 Nm

(*iii*) a single force and a couple at B. (*iii*) R = 120 N and M_B = 120 Nm]

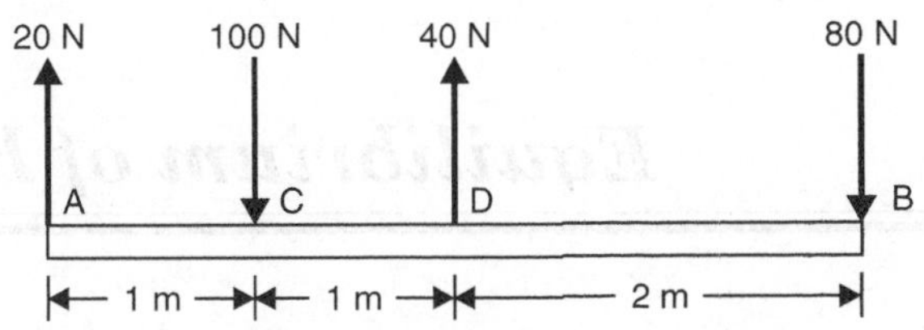

Fig. 1.98

26. Five forces are acting on a body as shown in Fig. 1.99. Determine the resultant.

[**Ans.** R = 0, Resultant couple = 10 Nm]

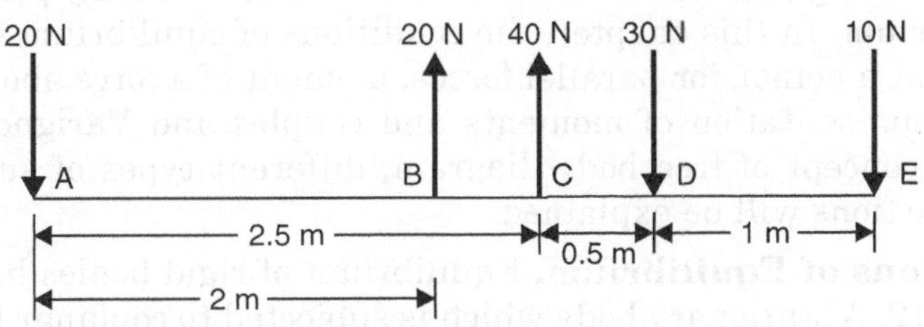

Fig. 1.99

27. Determine the resultant of the parallel forces shown in Fig. 1.100.

[**Ans.** Body is in equilibrium]

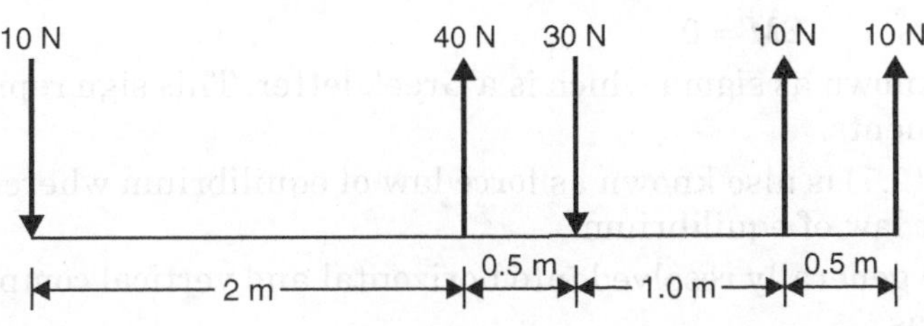

Fig. 1.100

2

Equilibrium of Rigid Bodies

2.1. INTRODUCTION

When some external forces (which may be concurrent or parallel) are acting on a stationary body, the body may start moving or may start rotating about any point. But if the body does not start moving *and also* does not start rotating about any point, then the body* is said to be in equilibrium. In this chapter, the conditions of equilibrium for concurrent forces (*i.e.*, forces meeting at a point), for parallel forces, moment of a force about a point and about an axis, vectorial representation of moments and couples and Varignon's theorem will be described. Also the concept of free body diagram, different types of support reactions and determination of reactions will be explained.

2.1.1. Equations of Equilibrium. Equilibrium of rigid bodies has been described in chapter 1, Article 1.12. A stationary body which is subjected to coplanar forces (concurrent or parallel) will be in equilibrium if the algebraic sum of all the external forces is zero *and also* the algebraic sum of moments of all the external forces about any point in their plane is zero. Mathematically, it is expressed by the equations :

$$\Sigma F = 0 \qquad \text{...(2.1)}$$

$$\Sigma M = 0 \qquad \text{...(2.2)}$$

The sign Σ is known as sigma which is a Greek letter. This sign represents the *algebraic sum* of forces or moments.

The equation (2.1) is also known as force law of equilibrium whereas the equation (2.2) is known as moment law of equilibrium.

The forces are generally resolved into horizontal and vertical components. Hence equation (2.1) is written as

$$\Sigma F_x = 0$$

and $$\Sigma F_y = 0$$

where ΣF_x = Algebraic sum of all horizontal components

and ΣF_y = Algebraic sum of all vertical components.

2.1.2. Equations of Equilibrium for Non-concurrent Force Systems. A non-concurrent force systems will be in equilibrium if the resultant of all forces and moment is zero.

Hence the equations of equilibrium are

$$\Sigma F_x = 0, \Sigma F_y = 0 \quad \text{and} \quad \Sigma M = 0.$$

*A body will be in equilibrium if both the resultant of the forces and resultant moment are zero, whereas a paraticle will be in equilibrium if the resultant force acting on it is zero.

2.1.3. Equations of Equilibrium for Concurrent Force System. For the concurrent forces, the lines of action of all forces meet at a point, and hence the moment of those force about that very point will be zero or $\Sigma M = 0$ automatically.

Thus for concurrent force system, the condition $\Sigma M = 0$ becomes redundant and only two conditions, *i.e.*, $\Sigma F_x = 0$ and $\Sigma F_y = 0$ are required.

2.2. FREE BODY DIAGRAM, ACTION AND REACTION FORCES

From the Newton's third law of motion, we know that to every action there is equal and opposite reaction. Hence reaction is always equal and opposite to the action.

Fig. 2.1(*a*) shows a ball placed on a horizontal surface (or horizontal plane) such that it is free to move along the plane but cannot move vertically downward. Hence the ball will exert a force vertically downwards at the support as shown in Fig. 2.1(*b*). This force is known as *action*. The support will exert an equal force vertically upwards on the ball at the point of contact as shown in Fig. 2.1(*c*).

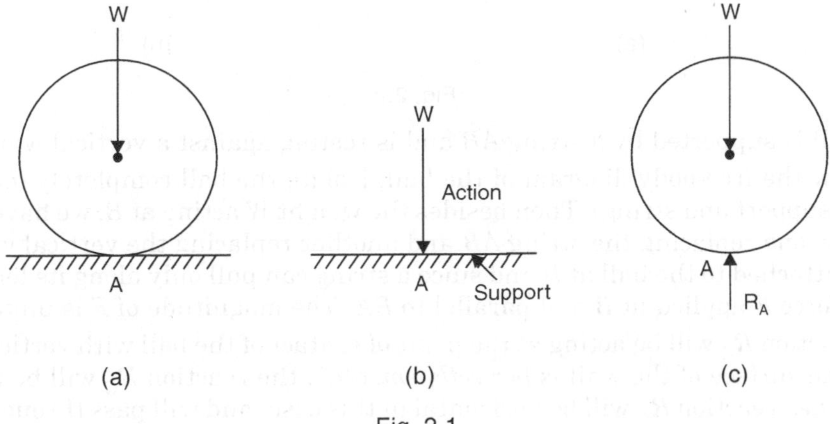

Fig. 2.1

The force, exerted by the support on the ball, is known as *reaction*. Hence '*any force on a support causes an equal and opposite force from the support so that action and reaction are two equal and opposite forces*'.

The equilibrium of the bodies which are placed on the supports can be considered if we remove the supports and replace them by the reactions which they exert on the body. In Fig. 2.1 (*a*), if we remove the supporting surface and replace it by the reaction R_A that the surface exerts on the balls as shown in Fig. 2.1 (*c*), we shall get *free-body diagram*.

The point of application of the reaction R_A will be the point of contact A, and from the law of equilibrium of two forces, we conclude that the reaction R_A must be vertical and equal to the weight W.

Hence Fig. 2.1 (*c*), in which the ball is *completely isolated* from its support and in which all forces acting on the ball are shown by vectors, is known a free-body diagram. Hence to draw the free-body diagram of a body we remove all the supports (like wall, floor, hinge or any other body) and replace them by the reactions which these support exert on the body. Also the body should be completely isolated.

Problem 2.1. *Draw the free body diagram of ball of weight W supported by a string AB and resting against a smooth vertical wall at C as shown in Fig. 2.2(a).*

Sol. Given :

Weight of ball = W

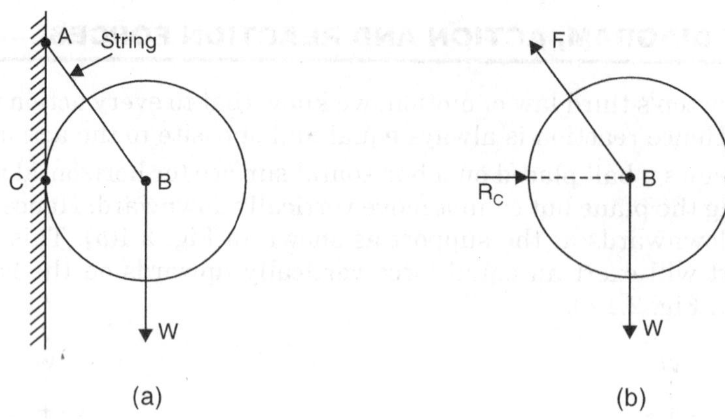

Fig. 2.2

The ball is supported by a string AB and is resting against a vertical wall at C.

To draw the free-body diagram of the ball, isolate the ball completely (*i.e.*, isolate the ball from the support and string). Then besides the weight W acting at B, we have two reactive forces to apply one replacing the string AB and another replacing the vertical wall AC. Since the string is attached to the ball at B and since a string can pull only along its length, we have the reactive force F applied at B and parallel to BA. The magnitude of F is unknown.

The reaction R_C will be acting at the point of contact of the ball with vertical wall *i.e.*, at point C. As the surface of the wall is *perfectly smooth**, the reaction R_C will be normal to the vertical wall (*i.e.*, reaction R_C will be horizontal in this case) and will pass through the point B. The magnitude of R_C is also unknown. The complete free-body diagram is shown in Fig. 2.2(*b*).

Problem 2.2. *A circular roller of weight 100 N and radius 10 cm hangs by a tie rod AB = 20 cm and rests against a smooth vertical wall at C as shown in Fig. 2.3 (a). Determine : (i) the force F in the tie rod, and (ii) the reaction R_C at point C.*

Sol. Given :

Weight of roller, $W = 100$ N

Radius of roller, $BC = 10$ cm

Length of tie rod, $AB = 20$ cm

From ΔABC, we get $\sin\theta = \dfrac{BC}{AB} = \dfrac{10}{20} = 0.5$

$$\theta = \sin^{-1} 0.5 = 30°$$

The free-body diagram of the roller is shown in Fig. 2.3 (*b*) in which

R_C = Reaction at C

F = Force in the tie rod AB

* The reaction at a perfectly smooth surface is always normal to the surface.

Free-body diagram shows the equilibrium of the roller. Hence the resultant force in x-direction and y-direction should be zero.

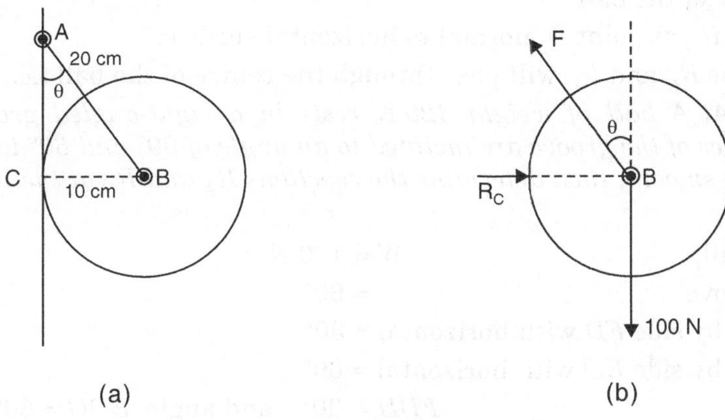

Fig. 2.3

For $\Sigma F_x = 0$, we get $R_C - F \sin \theta = 0$

or $R_C = F \sin \theta$ ...(*i*)

For $\Sigma F_y = 0$, we get $100 - F \cos \theta = 0$

or $100 = F \cos \theta$

or $F = \dfrac{100}{\cos \theta} = \dfrac{100}{\cos 30°}$ ($\because$ θ = 30°)

= **115.47 N. Ans.**

Substituting the value of F in equation (*i*),

$R_C = 115.47 \times \sin 30° =$ **57.73 N. Ans.**

Problem 2.3. *Draw the free-body diagram of a ball of weight W, supported by a string AB and resting against a smooth vertical wall at C and also resting against a smooth horizontal floor at D as shown in Fig. 2.4(a).*

Sol. Given :

To draw the free-body diagram of the ball, the ball should be isolated completely from the vertical support, horizontal support and string AB. Then the forces acting on the isolated ball as shown in Fig. 2.4(b), will be :

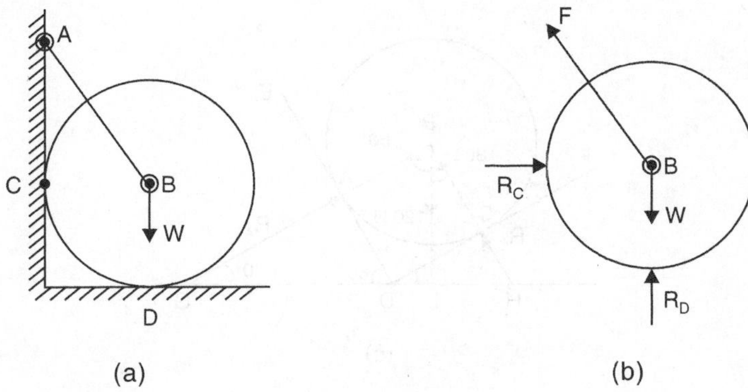

Fig. 2.4

(*i*) Reaction R_C at point *C*, normal to *AC*.

(*ii*) Force *F* in the direction of string.

(*iii*) Weight *W* of the ball.

(*iv*) Reaction R_D at point *D*, normal to horizontal surface.

The reactions R_C and R_D will pass through the centre of the ball *i.e.*, through point *B*.

Problem 2.4. *A ball of weight 120 N rests in a right-angled groove, as shown in Fig. 2.5(a). The sides of the groove are inclined to an angle of 30° and 60° to the horizontal. If all the surfaces are smooth, then determine the reactions R_A and R_C at the points of contact.*

Sol. Given :

Weight of ball, $W = 120$ N

Angle of groove $= 90°$

Angle made by side *FD* with horizontal = 30°

Angle made by side *ED* with horizontal = 60°

∴ Angle $FDH = 30°$ and angle $EDG = 60°$

Consider the equilibrium of the ball. For this draw the free body diagram of the ball as shown in Fig. 2.5(*b*).

The forces acting on the isolated ball will be :

(*i*) Weight of the ball = 120 N and acting vertically downwards.

(*ii*) Reaction R_C acting at *C* and normal to *FD*.

(*iii*) Reaction R_A acting at *A* and normal to *DE*.

The reactions R_A and R_C will pass through *B*, *i.e.*, centre of the ball. The angles made by R_A and R_C at point *B* will be obtained as shown in Fig. 2.5(*c*).

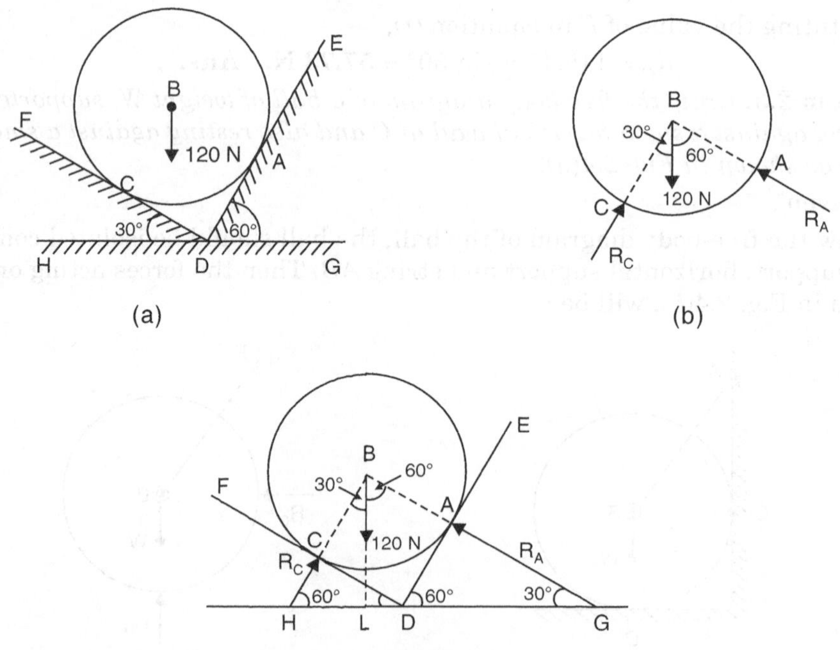

Fig. 2.5

In ΔHDC, $\angle CDH = 30°$ and $\angle DCH = 90°$. Hence $\angle DHC$ will be 60°. Now in ΔHBL, $\angle BLH = 90°$ and angle $LHB = 60°$. Hence $\angle HBL$ will be 30°.

Similarly, $\angle GBL$ may be calculated. This will be equal to 60°.

For the equilibrium of the ball,

$$\Sigma F_x = 0 \text{ and } \Sigma F_y = 0$$

For $\Sigma F_x = 0$, we have $R_C \sin 30° - R_A \sin 60° = 0$

or $$R_C \sin 30° = R_A \sin 60°$$

or $$R_C = R_A \times \frac{0.866}{\sin 30°} = 1.732\, R_A \quad ...(i)$$

For $\Sigma F_y = 0$, we have $120 - R_A \cos 60° - R_C \cos 30° = 0$

or $$120 = R_A \cos 60° + R_C \cos 30°$$
$$= R_A \times 0.5 + (1.732\, R_A) \times 0.866 \quad (\because \; R_C = 1.732\, R_A)$$
$$= 0.5\, R_A + 1.5\, R_A = 2R_A$$

$\therefore$ $$R_A = \frac{120}{2} = \mathbf{60\ N.}\ \ \textbf{Ans.}$$

Substituting this value in equation (*i*), we get

$$R_C = 1.732 \times 60 = \mathbf{103.92\ N.}\ \ \textbf{Ans.}$$

Problem 2.5. *A circular roller of radius 5 cm and of weight 100 N rests on a smooth horizontal surface and is held in position by an inclined bar AB of length 10 cm as shown in Fig. 2.6. A horizontal force of 200 N is acting at B. Find the tension (or Force) in the bar AB and the vertical reaction at C.*

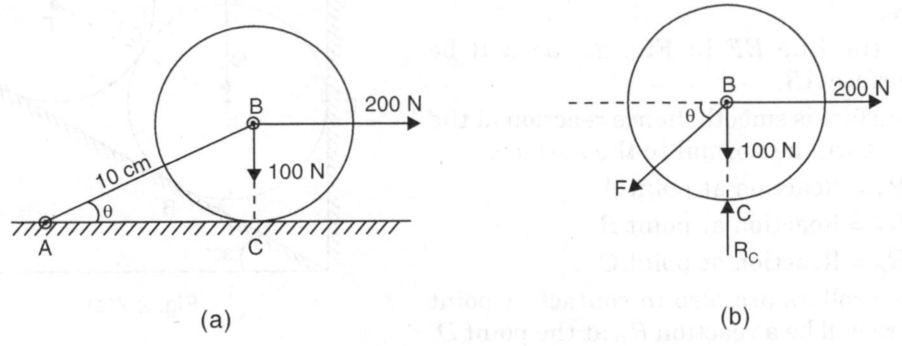

Fig. 2.6

Sol. Given :

Weight, $W = 100$ N

Radius *i.e.*, $BC = 5$ cm

Length of bar, $AB = 10$ cm

Horizontal force at $B = 200$ N

In ΔABC, $$\sin \theta = \frac{BC}{AB} = \frac{5}{10} = 0.5$$

$\therefore$ $$\theta = \sin^{-1} 0.5 = 30°$$

Let F = Tension in the string AB.

Consider the equilibrium of the roller. For this draw the free body diagram of the roller as shown in Fig. 2.6(*b*).

The reaction R_C at point *C* will pass through point *B*.

The tension (or force *F*) will be acting along the length of the string.

As the roller is in equilibrium in Fig. 2.6(*b*), the resultant force in *x*-direction and *y*-direction should be zero.

For $\Sigma F_x = 0$, we have $F \cos \theta - 200 = 0$

$$\therefore \quad F = \frac{200}{\cos \theta} = \frac{200}{\cos 30°} \qquad (\because \quad \theta = 30°)$$

$$= \mathbf{230.94\ N.} \quad \textbf{Ans.}$$

For $\Sigma F_y = 0$, we have $R_C - W - F \sin \theta = 0$

or $R_C = W + F \sin \theta = 100 + 230.94 \times \sin 30°$

$$= \mathbf{215.47\ N.} \quad \textbf{Ans.}$$

Problem 2.6. *Two identical rollers P and Q, each of weight W, are supported by an inclined plane and a vertical wall as shown in Fig. 2.7 (a). Assume all the surfaces to be smooth. Draw the free body diagrams of :*

(a) roller Q, (b) roller P and (c) rollers P and Q taken together.

Sol. Given :

Weight of each roller = *W*

Radius of each roller = *R*

Identical rollers means the radius of each roller is same.

Hence the line *EF* in Fig. 2.7(*a*) will be parallel to surface *AB*.

Each surface is smooth, hence reaction at the point of contact will be normal to the surface.

Let R_A = Reaction at point *A*

R_B = Reaction at point *B*

R_C = Reaction at point *C*

The two rollers are also in contact at point *D*. Hence there will be a reaction R_D at the point *D*.

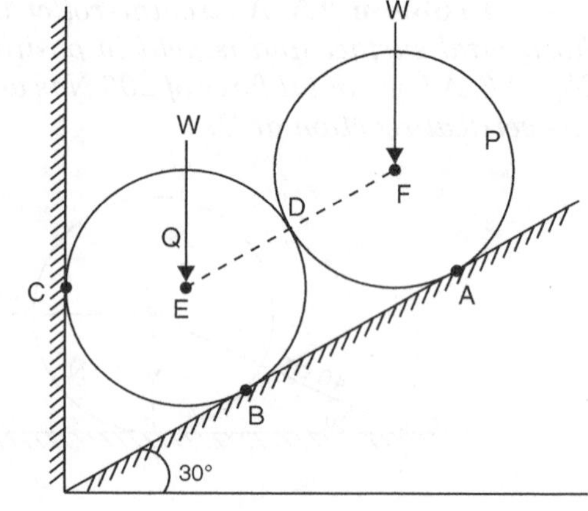

Fig. 2.7(a)

(*a*) **Free-body diagram of roller Q.** To draw the free-body diagram of roller *Q*, isolate the roller *Q* completely and find the forces acting on the roller *Q*. The roller *Q* has points of contact at *B, C* and *D*. The forces acting on the roller *Q* will be [Refer to Fig. 2.7(*b*)] :

(*i*) Weight of roller *W*.

(*ii*) Reaction R_B at point *B*.

This will be normal to the surface *BA* at point *B*.

(*iii*) Reaction R_C at point *C*. This will be normal to the vertical surface at point *C*.

(*iv*) Reaction R_D at point *D*. This will be normal to the tangent at point *D*.

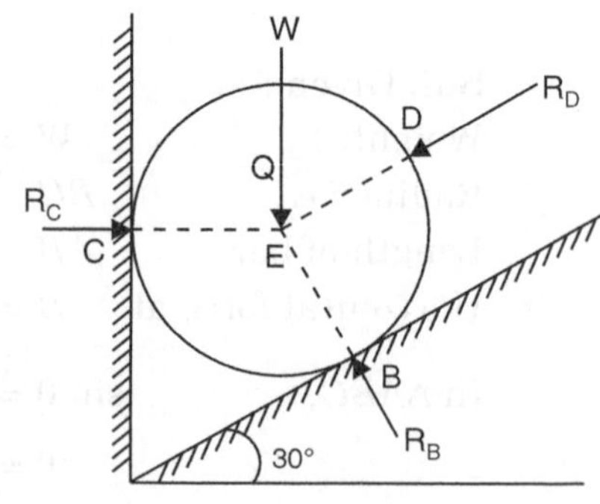

Fig. 2.7(b)

The reactions R_B, R_C and R_D will pass through the centre E of the roller Q. These three reactions are unknown.

(*b*) **Free-body diagram of roller P.** Free-body diagram of roller P is shown in Fig. 2.7(*c*). The roller P has points of contact at A and D. The forces acting on the roller P are :

(*i*) Weight W

(*ii*) Reaction R_A at point A

(*iii*) Reaction R_D at point D.

The reactions R_A and R_D will pass through point F, *i.e.*, centre of roller P. These two reactions are unknown. If W is given, then these reactions can be calculated.

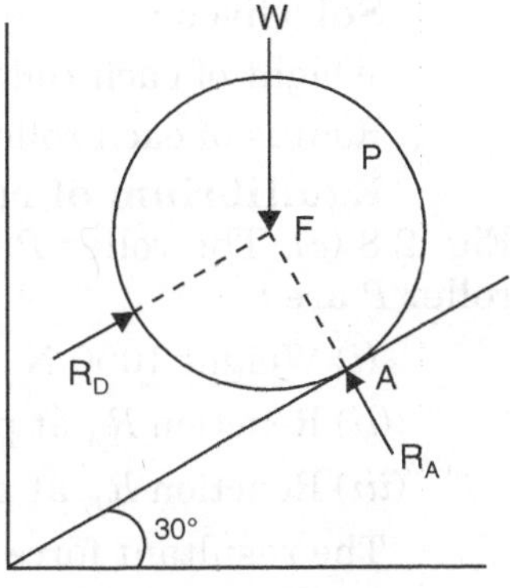

Fig. 2.7(c)

(*c*) **Free-body diagram of rollers P and Q taken together.** When the rollers P and Q are taken together, then points of contacts are A, B and C. The free-body diagram of this case is shown in Fig. 2.7(*d*). The forces acting are :

(*i*) Weight W on each roller

(*ii*) Reaction R_A at point A

(*iii*) Reaction R_B at point B

(*iv*) Reaction R_C at point C.

In this case there will be no reaction at point D.

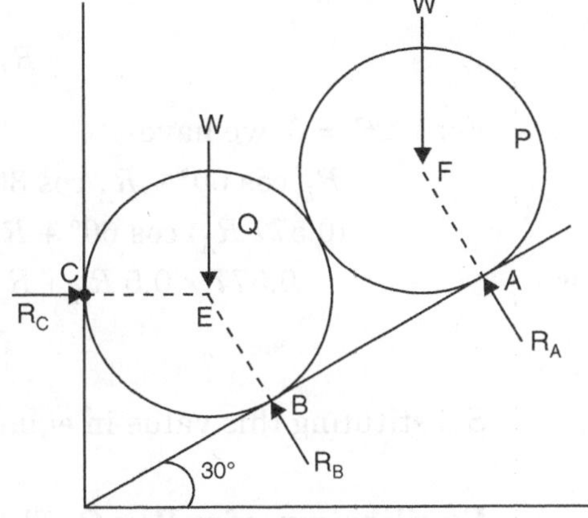

Fig. 2.7(d)

Problem 2.7. *Two identical rollers, each of weight W = 1000 N, are supported by an inclined plane and a vertical wall as shown in Fig. 2.8(a). Find the reactions at the points of supports A, B and C. Assume all the surfaces to be smooth.*

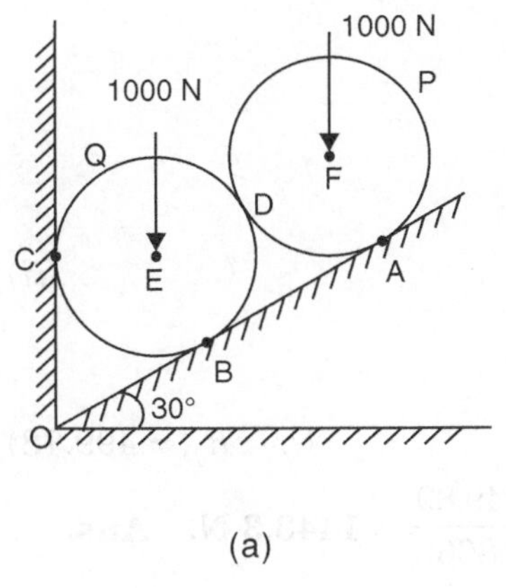

(a)

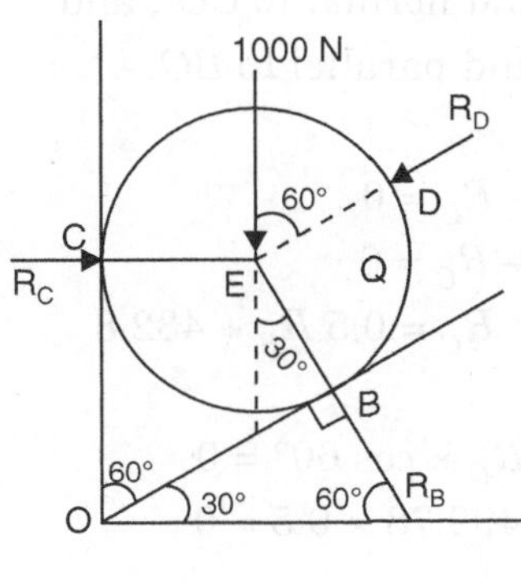

(b)

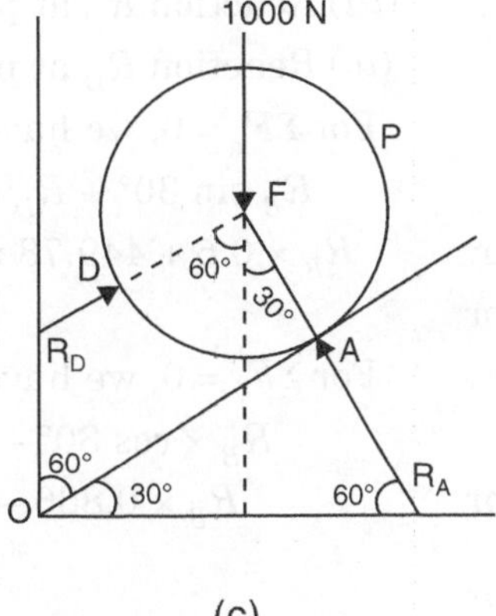

(c)

Fig. 2.8

Sol. Given :

Weight of each roller = 1000 N

Radius of each roller is same. Hence line *EF* will be parallel to *AB*.

Equilibrium of roller P. First draw the free-body diagram of roller *P* as shown in Fig. 2.8 (*c*). The roller *P* has points of contact at *A* and *D*. Hence the forces acting on the roller *P* are :

(*i*) Weight 1000 N acting vertically downward.

(*ii*) Reaction R_A at point *A*. This is normal to *OA*.

(*iii*) Reaction R_D at point *D*. This is parallel to line *OA*.

The resultant force in *x* and *y* directions on roller *P* should be zero.

For $\Sigma F_x = 0$, we have

$$R_D \sin 60° - R_A \sin 30° = 0 \quad \text{or} \quad R_D \sin 60° = R_A \sin 30°$$

$$\therefore \quad R_D = R_A \frac{\sin 30°}{\sin 60°} = 0.577\, R_A \qquad \text{...}(i)$$

For $\Sigma F_y = 0$, we have

$$R_D \cos 60° + R_A \cos 30° - 1000 = 0$$

$$(0.577\, R_A) \cos 60° + R_A \cos 30° = 1000 \qquad (\because\ R_D = 0.577\, R_A)$$

or $$0.577 \times 0.5\, R_A + R_A \times 0.866 = 1000$$

$$1.1545\, R_A = 1000 \quad \text{or} \quad R_A = \frac{1000}{1.1545} = \mathbf{866.17\ N.\ \ Ans.}$$

Substituting this value in equation (*i*), we get

$$R_D = 0.577 \times 866.17 = 499.78$$

Equilibrium of roller Q. The free-body diagram of roller *Q* is shown in Fig. 2.8 (*b*). The roller *Q* has points of contact at *B, C* and *D*.

The forces acting on the roller *Q* are :

(*i*) Weight *W* = 1000 N ;

(*ii*) Reaction R_B at point *B* and normal to *BO* ;

(*iii*) Reaction R_C at point *C* and normal to *CO* ; and

(*iv*) Reaction R_D at point *D* and parallel to *BO*.

For $\Sigma F_x = 0$, we have

$$R_B \sin 30° + R_D \sin 60° - R_C = 0$$

or $$R_B \times 0.5 + 449.78 \times 0.866 - R_C = 0$$

or $$R_C = 0.5\, R_B + 432.8 \qquad \text{...}(ii)$$

For $\Sigma F_y = 0$, we have

$$R_B \times \cos 30° - 1000 - R_D \times \cos 60° = 0$$

or $$R_B \times 0.866 - 1000 - 499.78 \times 0.5 = 0 \qquad (\because\ R_D = 499.78)$$

or $$0.866\, R_B - 1249.89 = 0 \quad \text{or} \quad R_B = \frac{1249.89}{0.866} = \mathbf{1443.3\ N.\ \ Ans.}$$

Substituting this value in equation (*ii*), we get

$$R_C = 0.5 \times 1443.3 + 432.8 = \mathbf{1154.45\ N.\ \ Ans.}$$

Problem 2.8. *Two spheres, each of weight 1000 N and of radius 25 cm rest in a horizontal channel of width 90 cm as shown in Fig. 2.9. Find the reactions on the points of contact A, B and C.*

Sol. Given :

Weight of each sphere, $W = 1000$ N

Radius of each sphere, $R = 25$ cm

$\therefore$ $AF = BF = FD = DE = CE = 25$ cm

Width of horizontal channel = 90 cm

Join the centre E to centre F as shown in Fig. 2.9(*b*).

Now $EF = 25 + 25 = 50$ cm, $FG = 40$ cm

In $\Delta EFG, EG = \sqrt{EF^2 - FG^2} = \sqrt{50^2 - 40^2} = \sqrt{2500 - 1600} = 30$

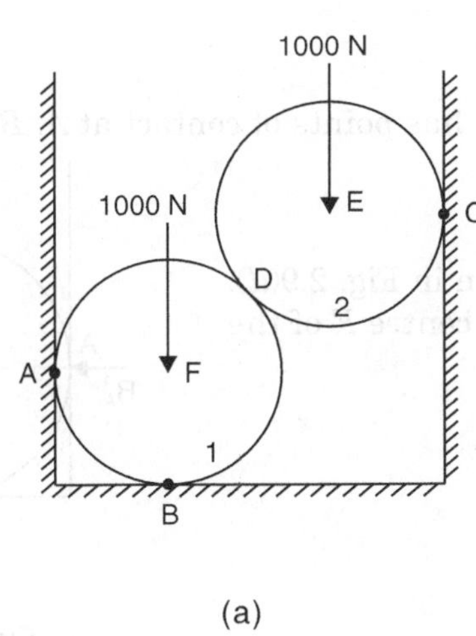

(a)

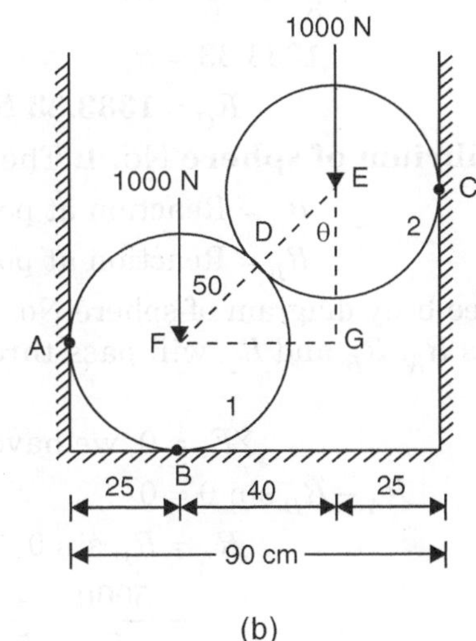

(b)

Fig. 2.9

$$\therefore \quad \cos\theta = \frac{EG}{EF} = \frac{30}{50} = \frac{3}{5} \quad \text{and} \quad \sin\theta = \frac{FG}{EF} = \frac{40}{50} = \frac{4}{5}.$$

Equilibrium of Sphere No. 2. The sphere 2 has points of contact at C and D.

Let R_C = Reaction at C

and R_D = Reaction at D

The free-body diagram of sphere No. 2 is shown in Fig. 2.9 (*c*).

The reaction R_D at point D, will pass through the centre E of the sphere No. 2, as any line normal to any point on the circumference of the circle will pass through the centre of circle. For the equilibrium of the sphere No. 2, the resultant force in x and y directions should be zero.

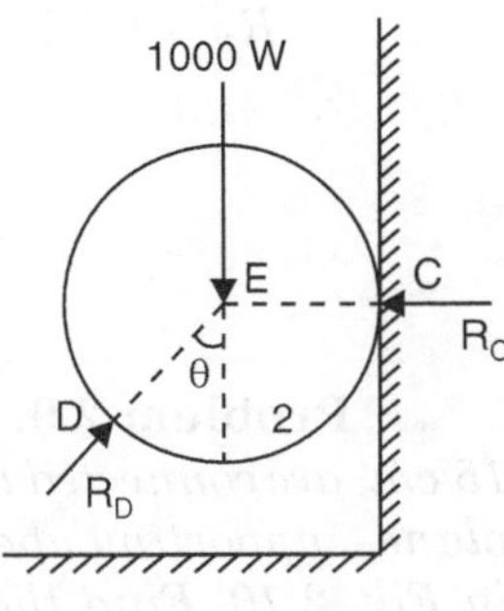

Fig. 2.9(c)

For $\Sigma F_x = 0$, we have $R_D \sin \theta = R_C$ ...(*i*)

For $\Sigma F_y = 0$, we have $R_D \cos \theta = 1000$

or $$R_D = \frac{1000}{\cos \theta} = \frac{1000}{\left(\frac{3}{5}\right)} \qquad \left(\because \cos \theta = \frac{3}{5}\right)$$

$$= 1000 \times \frac{5}{3} = \frac{5000}{3} \text{ N} \qquad \ldots(ii)$$

Substituting the value of R_D in equation (*i*),

$$\frac{5000}{3} \times \sin \theta = R_C$$

or $$\frac{5000}{3} \times \frac{4}{5} = R_C \qquad \left(\because \sin \theta = \frac{4}{5}\right)$$

or $$1333.33 = R_C$$

$\therefore$ R_C = **1333.33 N. Ans.**

Equilibrium of sphere No. 1. The sphere 1 has points of contact at *A*, *B* and *D*.

Let R_A = Reaction at point *A*

R_B = Reaction at point *B*

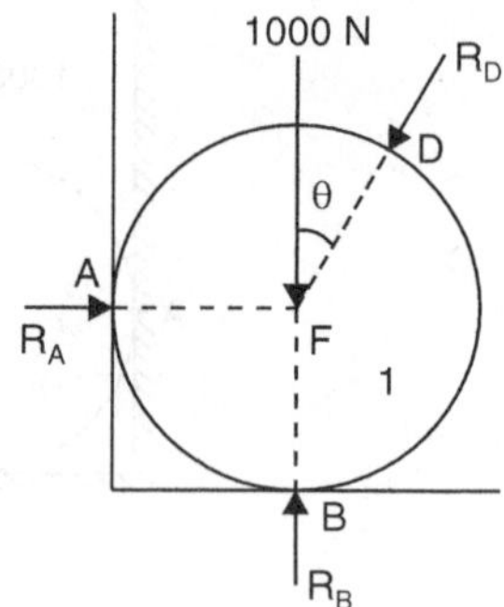

Fig. 2.9(d)

The free-body diagram of sphere No. 1 is shown in Fig. 2.9(*d*). The reactions R_A, R_B and R_D will pass through the centre *F* of the sphere No. 1.

For $\Sigma F_x = 0$, we have

$$R_A - R_D \sin \theta = 0$$

or $$R_A = R_D \sin \theta$$

$$= \frac{5000}{3} \times \frac{4}{5}$$

$$\left(\because R_D = \frac{5000}{3} \text{ and } \sin \theta = \frac{4}{5}\right)$$

= **1333.33 N. Ans.**

For $\Sigma F_y = 0$, we have

$$R_B - 1000 - R_D \cos \theta = 0$$

$\therefore$ $$R_B = 1000 + R_D \cos \theta$$

$$= 1000 + \frac{5000}{3} \times \frac{3}{5} \qquad \left(\because \cos \theta = \frac{3}{5}\right)$$

= **2000 N. Ans.**

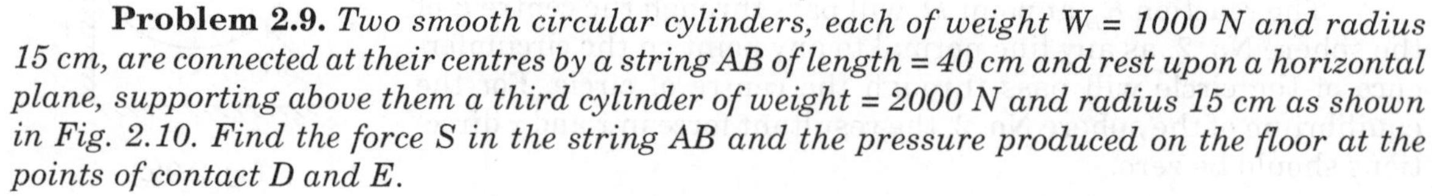

Problem 2.9. *Two smooth circular cylinders, each of weight W = 1000 N and radius 15 cm, are connected at their centres by a string AB of length = 40 cm and rest upon a horizontal plane, supporting above them a third cylinder of weight = 2000 N and radius 15 cm as shown in Fig. 2.10. Find the force S in the string AB and the pressure produced on the floor at the points of contact D and E.*

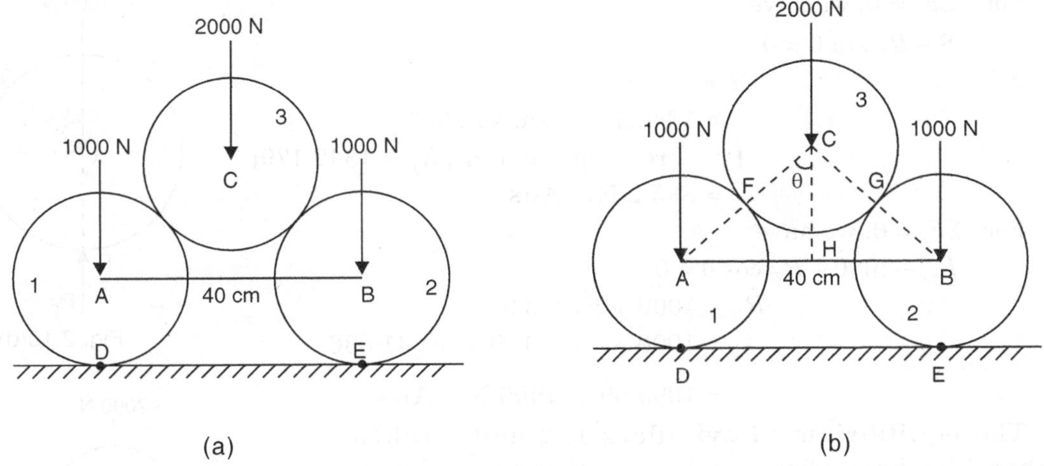

Fig. 2.10

Sol. Given :

Weight of cylinders 1 and 2 = 1000 N

Weight of cylinder 3 = 2000 N

Radius of each cylinder = 15 cm

Length of string AB = 40 cm

From Fig. 2.10 (*b*), $AC = AF + FC = 15 + 15 = 30$ cm

$$AH = \frac{1}{2} \times AB = \frac{1}{2} \times 40 = 20 \text{ cm.}$$

From ΔACH,
$$\sin \theta = \frac{AH}{AC} = \frac{20}{30} = 0.667$$

$\therefore$
$$\theta = \sin^{-1} 0.667 = 41.836°.$$

Equilibrium of cylinder 3. The cylinder 3 has points of contact at F and G. The reactions R_F and R_G will pass through the centre of sphere 3. The free-body diagram is shown in Fig. 2.10(*c*). Resolving forces horizontally,

$$R_F \sin \theta - R_G \sin \theta = 0$$

or
$$R_F = R_G \qquad \ldots(i)$$

Resolving forces vertically,

$$R_F \cos \theta + R_G \cos \theta = 2000$$

or
$$R_F \cos \theta + R_F \cos \theta = 2000 \qquad (\because\ R_F = R_G)$$

$\therefore$
$$R_F = \frac{2000}{2 \times \cos \theta} = \frac{1000}{\cos 41.836°} = 1342.179 \text{ N} \qquad \ldots(ii)$$

2000 N
3
C
θ θ
F
G
R_F
R_G

Fig. 2.10(c)

Equilibrium of cylinder 1. The cylinder 1 has points of contact at D and F. Also the cylinder 1 is connected to cylinder 2 by a string AB. To draw the free-body diagram of cylinder 1, there will be reactions R_F and R_D at points F and D as shown in Fig. 2.9 (*d*). Also there will be a force S in the direction of the string AB.

For $\Sigma F_x = 0$, we have

$S - R_F \sin\theta = 0$

$\therefore \quad S = R_F \sin\theta$

$= 1342.179 \times \sin 41.836°$

$[\because$ From equation (*ii*), $R_F = 1342.179]$

$=$ **895.2 N. Ans.**

1000 N
R_F
F
θ
A
S
1
D
R_D

Fig. 2.10(d)

For $\Sigma F_y = 0$, we have

$R_D - 1000 - R_F \cos\theta = 0$

or $\quad R_D = 1000 + R_F \cos\theta$

$= 1000 + 1342.179 \times \cos 41.836$

$= 1999.99 \simeq$ **2000 N. Ans.**

The equilibrium of cylinders 1, 2 and 3 taken together. The three cylinders taken together have points of contact at *D* and *E*. The free-body diagram is shown in Fig. 2.9 (*d*). In this case only vertical forces exist. Hence resultant force in *y*-direction should be zero.

$\therefore \quad R_D + R_E - 1000 - 2000 - 1000 = 0$

or $\quad R_E = 1000 + 2000 + 1000 - R_D$

or $\quad R_E = 4000 - R_D$

$= 4000 - 2000 \qquad (\because R_D = 2000)$

$=$ **2000 N. Ans.**

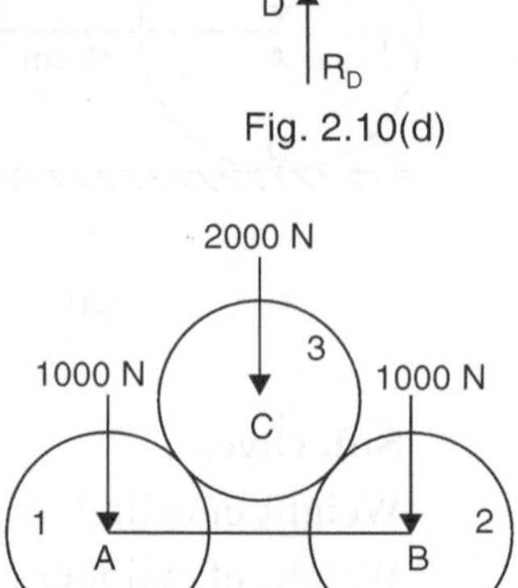

Fig. 2.10(e)

Problem 2.10. *A roller of radius 40 cm, weighing 3000 N is to be pulled over a rectangular block of height 20 cm as shown in Fig. 2.11, by a horizontal force applied at the end of a string wound round the circumference of the roller. Find the magnitude of the horizontal force which will just turn the roller over the corner of the rectangular block. Also determine the magnitude and direction of reactions at A and B. All surfaces may be taken as smooth.*

Sol. Given :

Radius of roller = 40 cm

Weight, $\quad W = 3000$ N

Height of block = 20 cm

Find horizontal force *P*, reaction R_A and reaction R_B when the roller just turns over the block.

When the roller is about to turn over the corner of the rectangular block, the roller lifts at the point *A* and then there will be no contact between the roller and the point *A*. Hence reaction R_A at point *A* will become zero.

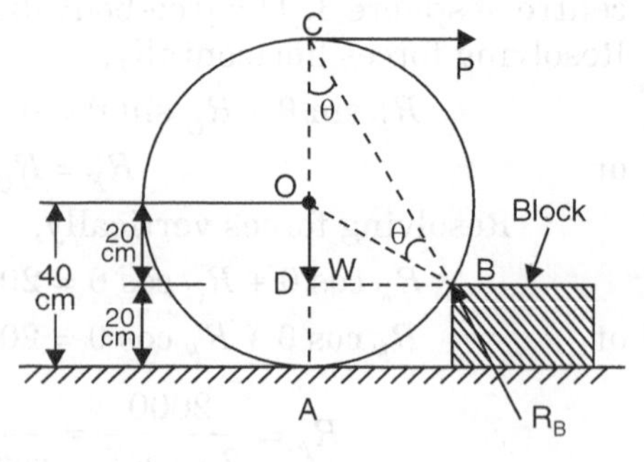

Fig. 2.11

Now the roller will be in equilibrium under the action of the following three forces :

(*i*) its weight *W* acting vertically downward

(*ii*) horizontal force *P*

(*iii*) reaction R_B at point *B*. The direction of R_B is unknown.

For the equilibrium, these three forces should pass through a common point. As the force P and weight W is passing through point C, hence the reaction R_B must also pass through the point C. Therefore, the line BC gives the direction of the reaction R_B.

In ΔBOD, BO = Radius = 40 cm,

$$OD = OA - AD = 40 - 20 = 20 \text{ cm}$$

$\therefore$ $$BD = \sqrt{BO^2 - OD^2} = \sqrt{40^2 - 20^2} = \sqrt{1200} = 34.64$$

Now in ΔBCD, $$\tan\theta = \frac{BD}{CD} = \frac{34.64}{CO + OD} = \frac{34.64}{(40+20)} = 0.5773$$

$\therefore$ $$\theta = \tan^{-1} 0.5773 = 29.999° \simeq 30°$$

Resolving forces horizontally, we get $P - R_B \sin\theta = 0$

or $$P = R_B \sin\theta = R_B \times \sin 30° = 0.5\, R_B \quad ...(i)$$

Resolving forces vertically, we get

$$W - R_B \cos\theta = 0 \quad \text{or} \quad 3000 - R_B \times \cos 30° = 0$$

or $$R_B = \frac{3000}{\cos 30°} = \mathbf{3464.2\ N.}\quad \textbf{Ans.}$$

Substituting this value of R_B in equation (*i*), we get

$$P = 0.5 \times 3464.2 = \mathbf{1732.1\ N.}\quad \textbf{Ans.}$$

2nd Method

This problem can also be solved by taking moments of all the three forces about the point B (*i.e.*, corner of the rectangular block) as shown below :

$$P \times CD = W \times BD$$

or $$P \times 60 = 3000 \times 34.64 \qquad (\because\ BD = 34.64)$$

$\therefore$ $$P = \frac{3000 \times 34.64}{60} = \mathbf{1732.0\ N.}\quad \textbf{Ans.}$$

Problem 2.11. *If in the problem 2.10, the force P is applied horizontally at the centre of the roller, what would be the magnitude of this force ? Also determine the least force and its line of action at the roller centre, for turning the roller over the rectangular block.*

Sol. Given :

Radius of roller = 40 cm

Weight, W = 3000 N

Height of block = 20 cm

When the roller is about to turn over the corner of the rectangular block, the roller lifts at the point A, and then there will be no contact between the roller and the point A. Hence reaction R_A at point A will be zero.

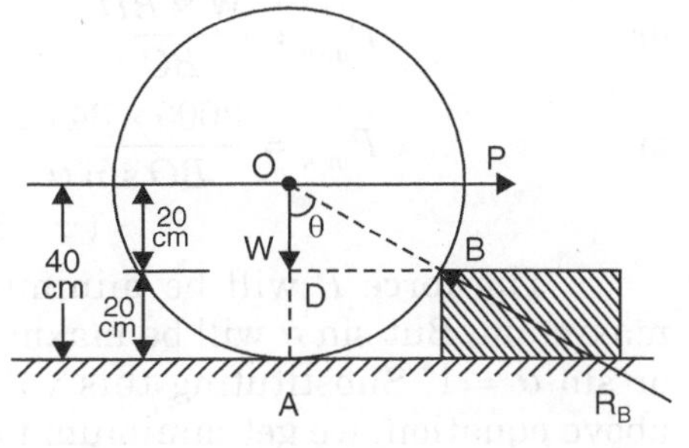

Fig. 2.12

Now the roller will be in equilibrium under the action of the following three forces :

(*i*) Weight of the roller W acting vertically downward through point O.

(*ii*) Horizontal force P applied at the centre of the roller.

(*iii*) Reaction R_B at point B. The direction of reaction R_B is unknown.

The above three forces should pass through a common point, as roller is in equilibrium. But the weight W and force P is passing through point O, hence R_B should also pass through point O. Hence line joining B to O, gives the direction of R_B as shown in Fig. 2.12.

From ΔBOD, we have $\cos\theta = \frac{OD}{BO} = \frac{20}{40} = 0.5$

$\therefore \quad \theta = \cos^{-1} 0.5 = 60°$

Now resolving forces vertically,

$$R_B \cos\theta - W = 0$$

or $$R_B = \frac{W}{\cos\theta} = \frac{3000}{\cos 60°} = 6000 \text{ N} \quad ...(i)$$

Resolving forces horizontally, we get $P - R_B \sin\theta = 0$

or $$P = R_B \sin\theta = 6000 \times \sin 60° = \mathbf{5196.\ N.} \quad \textbf{Ans.}$$

Method of moments

The force P can also be calculated by the method of moments. Taking the moments of all the forces about the point B, we get

$$P \times OD - W \times BD = 0$$

or $$P = \frac{W \times BD}{OD} \quad ...(ii)$$

$$(\because \quad BD = \sqrt{BO^2 - OD^2} = \sqrt{40^2 - 20^2} = 34.64)$$

$$= \frac{3000 \times 34.64}{20} = \mathbf{5196\ N.} \quad \textbf{Ans.}$$

Least force and its line of action

Let P_{min} = Least force applied as shown in Fig. 2.13.

α = Angle of the least force

From ΔOBC, $\quad BC = BO \sin\alpha$

Taking moments of all forces about point B, we get $P_{min} \times BC - W \times BD = 0$

or $$P_{min} = \frac{W \times BD}{BC}$$

or $$P_{min} = \frac{3000 \times 34.64}{BO \sin\alpha} = \frac{3000 \times 34.64}{40 \times \sin\alpha}$$

$$(\because \quad BO = \text{Radius} = 40)$$

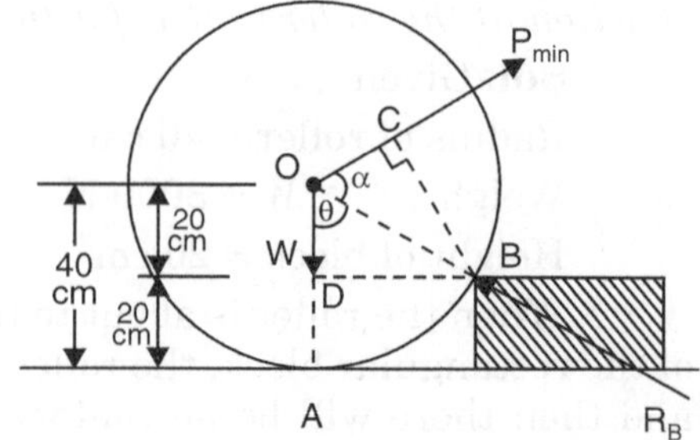

Fig. 2.13

The force P will be minimum when $\sin\alpha$ is maximum. But $\sin\alpha$ will be maximum, when $\alpha = 90°$ or $\sin\alpha = 1$. Substituting this value of $\sin\alpha$ in the above equation, we get minimum force.

$$\therefore \quad P_{min} = \frac{3000 \times 34.64}{40 \times 1} = \mathbf{2598\ N.} \quad \textbf{Ans.}$$

The direction of least force (*i.e.*, P_{min}) is at right angles to the line BO.

Problem 2.12. *A L-shaped body ABC is hinged at A with a force F acting at its end C. Determine the angle θ which this force should make with the horizontal to keep the edge AB of the body vertical.*

Sol. Given :

Length $BC = 25$ cm

Length $AB = 40$ cm.

Let us assume that force F is making an angle θ with the horizontal as shown in Fig. 2.14.

The forces acting on the given body ABC are :

(*i*) Force F acting at point C.

(*ii*) Reaction R_A at a point A.

Only two forces are acting on the body and the body is in equilibrium. But the two forces can be in equilibrium only if they are equal in magnitude, opposite in direction and have the same line of action. As the force F is acting at point C, hence reaction R_A should also pass through point C. Therefore, the line of action of reaction R_A is along line AC. The line of action of F should also be along AC, *i.e.*, the force F will be making an angle α with the horizontal.

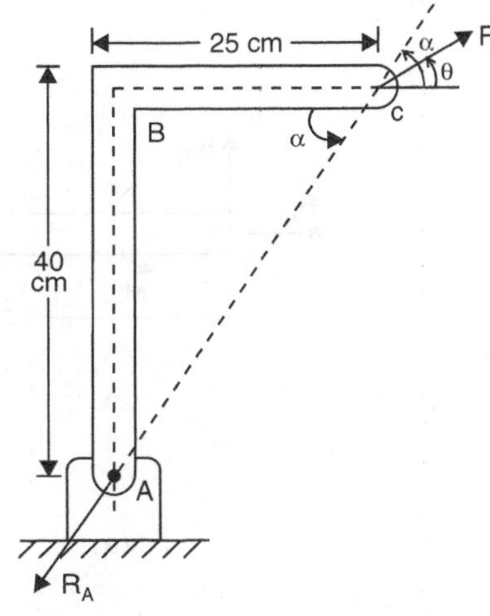

Fig. 2.14

Now from ΔABC, $\tan \alpha = \dfrac{AB}{BC} = \dfrac{40}{25} = 1.6$

$\therefore \quad \alpha = \tan^{-1} 1.6 = 57.99°$

$\therefore \quad \theta = \alpha =$ **57.99°. Ans.**

Problem 2.13. *A horizontal force 200 N is applied to the sloping bar BCD whose bottom rests on a horizontal plane, as shown in Fig. 2.15. Its upper end is pinned at B to the horizontal bar AB which has a pinned support at A. What couple M must be applied to AB to hold the system in equilibrium ? What is the magnitude of the pin reaction at B ? Assume the bars to be weightless and pins at A and B to be smooth.*

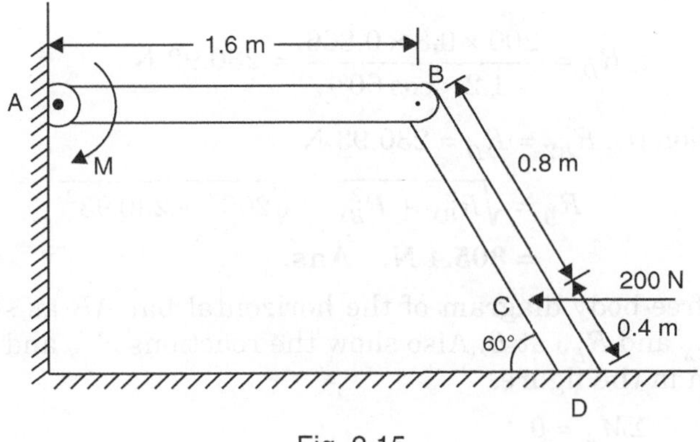

Fig. 2.15

Sol. Given :

Length $AB = 1.6$ m, Length $BD = 1.2$ m, $BC = 0.8$ m and $CD = 0.4$ m

Horizontal force at $C = 200$ N

Let $\quad M =$ Couple applied A to bar AB to keep the system in equilibrium.

$R_B =$ Reaction at B

R_{BX} and R_{BY} are the horizontal and vertical components of reaction R_B.

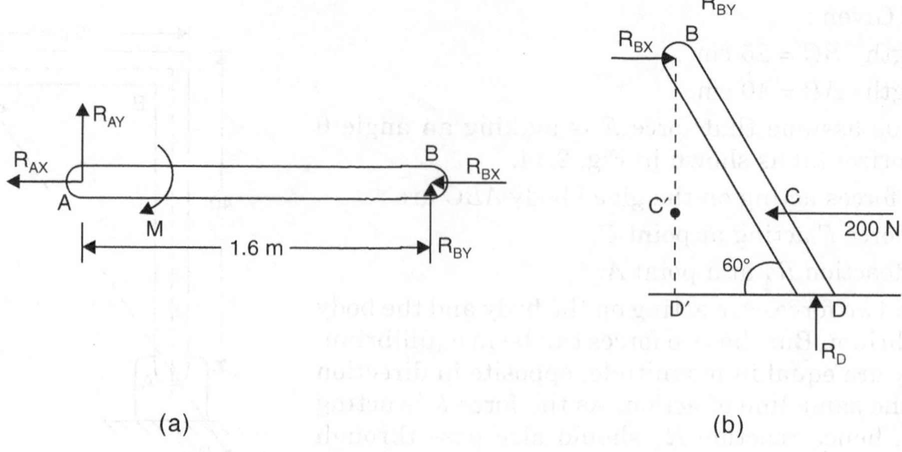

Fig. 2.16

The free-body diagram of the sloping bar *BCD* is shown in Fig. 2.16 (*b*). The reaction R_D at point *D* is vertical. Also show R_{BX} and R_{BY} at *B*.

Since the bar is in equilibrium, first apply $\Sigma F_x = 0$, which gives

$$R_{BX} = 200 \text{ N}$$

Now apply $\Sigma F_y = 0$, then $R_{BY} = R_D$ ...(*i*)

The magnitude of R_{BY} and R_D is unknown. To find their values apply ΣM at the pin *B*.

Taking moments of all forces at point *B*, we get

$$R_D \times DD' = 200 \times BC'$$

or $R_D \times BD \cos 60° = 200 \times BC \sin 60$ ($\because$ $DD' = BD \cos 60°$ and $BC' = BC \sin 60°$)

or $R_D \times 1.2 \cos 60° = 200 \times 0.8 \sin 60°$

or $$R_D = \frac{200 \times 0.8 \times 0.866}{1.2 \times \cos 60°} = 230.93 \text{ N}$$

$\therefore$ From equation (*i*), $R_{BY} = R_D = 230.93$ N

Reaction at *B*, $$R_B = \sqrt{R_{BX}^2 + R_{BY}^2} = \sqrt{200^2 + 230.93^2}$$

$$= \textbf{305.4 N.} \quad \textbf{Ans.}$$

Now draw the free-body diagram of the horizontal bar *AB* as shown in Fig. 2.16 (*a*). Show the reactions R_{BX} and R_{BY} at *B*. Also show the reactions R_{AY} and R_{AX} at *A*. The applied couple *M* is also shown in the figure.

Apply $\Sigma M_A = 0$

$$M = R_{BY} \times 1.6 = 230.93 \times 1.6$$

$$= \textbf{369.44 Nm.} \quad \textbf{Ans.}$$

Problem 2.14. *A body weighing 2000 N is suspended with a chain AB 2 m long. It is pulled by a horizontal force of 320 N as shown in Fig. 2.17. Find the force in the chain and the lateral displacement (i.e., x) of the body.*

Sol. Given :

Weight suspended at B = 2000 N

Length AB = 2 m

Horizontal force at B = 320 N

Find : Force in AB and value of x

Let F = Force in chain AB

θ = Angle made by AB with horizontal.

The free-body diagram of the point B is shown in Fig. 2.17(b).

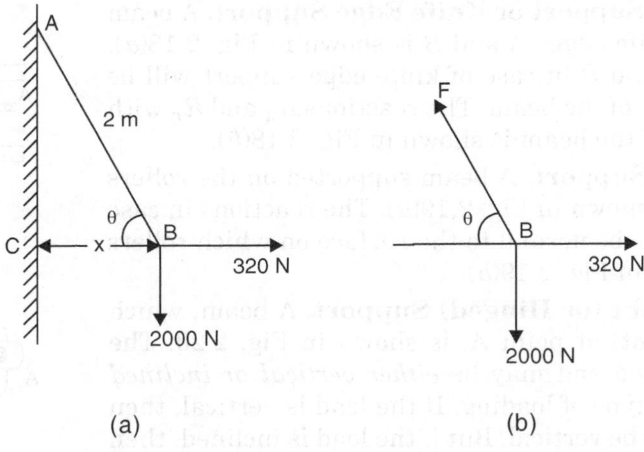

Fig. 2.17

The point B is in equilibrium under the action of three forces. Hence using Lami's theorem, we get

$$\frac{F}{\sin 90^\circ} = \frac{2000}{\sin (180^\circ - \theta)} = \frac{320}{\sin (90^\circ + \theta)}$$

or $$\frac{F}{1} = \frac{2000}{\sin \theta} = \frac{320}{\cos \theta} \quad [\because \sin (180 - \theta) = \sin \theta, \sin (90 + \theta) = \cos \theta]$$

or $$F \sin \theta = 2000 \qquad ...(i)$$

and $$F \cos \theta = 320 \qquad ...(ii)$$

Dividing equation (i) by equation (ii), we get

$$\tan \theta = \frac{2000}{320} = 6.25$$

$\therefore$ $$\theta = \tan^{-1} 6.25 = 80.9^\circ$$

Substituting this value of θ in equation (i), we get

$$F \sin 80.9^\circ = 2000 \quad \text{or} \quad F = \frac{2000}{\sin 80.9} = \mathbf{2025.5\ N.}$$ **Ans.**

Now from Fig. 2.17 (a), $\cos \theta = \frac{x}{2}$ or $x = 2 \times \cos \theta = 2 \times \cos 80.9^\circ$

$= \mathbf{0.3163\ m.}$ **Ans.**

2.3. TYPES OF SUPPORTS AND THEIR REACTIONS

Though there are many types of supports, yet the following are important from the subject point of view:

1. Simple supports or knife edge supports
2. Roller support
3. Pin-joint (or hinged) support
4. Smooth surface support
5. Fixed or built-in support.

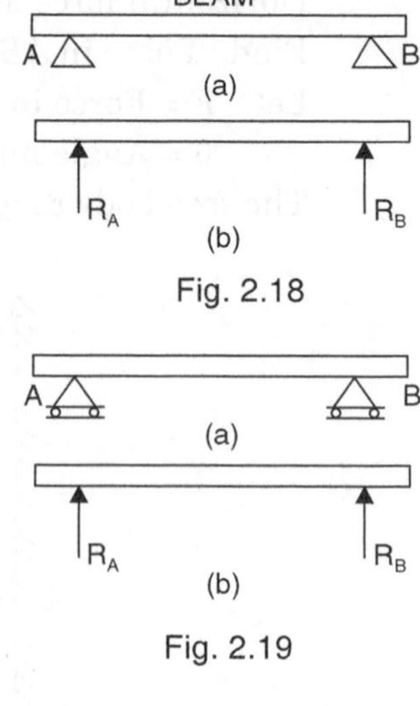

Fig. 2.18

Fig. 2.19

Fig. 2.20

2.3.1. Simple Support or Knife Edge Support. A beam supported on the knife edges A and B is shown in Fig. 2.18(*a*). The reactions at A and B in case of knife edge support will be normal to the surface of the beam. The reactions R_A and R_B with free-body diagram of the beam is shown in Fig. 2.18(*b*).

2.3.2. Roller Support. A beam supported on the rollers at points A and B is shown in Fig. 2.19(*a*). The reactions in case of roller supports will be normal to the surface on which rollers are placed as shown in Fig. 2.19(*b*).

2.3.3. Pin Joint (or Hinged) Support. A beam, which is hinged (or pin-joint) at point A, is shown in Fig. 2.20. The reaction at the hinged end may be *either vertical or inclined* depending upon the type of loading. If the load is vertical, then the reaction will also be vertical. But if the load is inclined, then the reaction at the hinged end will also be inclined.

2.3.4. Smooth Surface Support. Fig. 2.21 shows a body in contact with a smooth surface. The reaction will always act normal to the support as shown in Fig. 2.21(*a*) and 2.21(*b*).

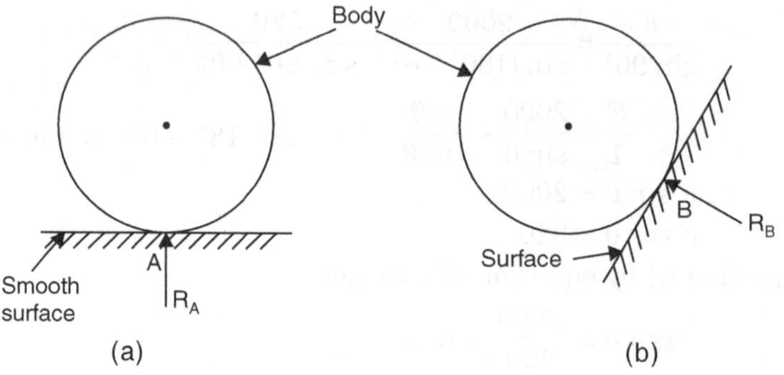

Fig. 2.21

Fig. 2.22 shows a rod AB resting inside a sphere, whose surface are smooth. Here the rod becomes body and sphere becomes surface. The reactions on the ends of the rod (*i.e.*, at point A and B) will be normal to the sphere surface at A and B. The normal at any point on the surface of the sphere will always pass through the centre of the sphere. Hence reactions R_A and R_B will have directions AO and BO respectively as shown in Fig. 2.22.

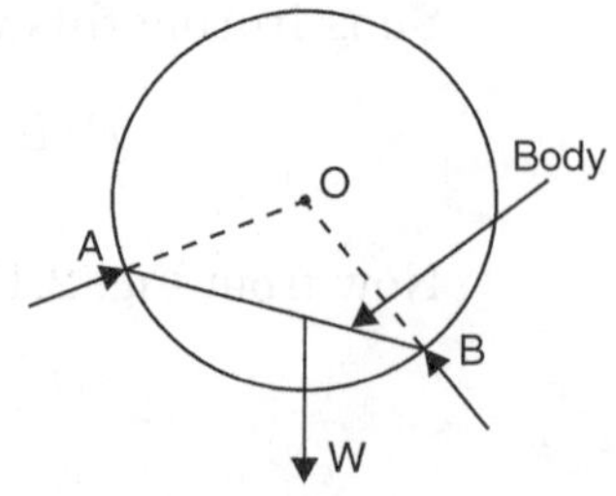

Fig. 2.22

2.3.5. Fixed or Built-in Support. Fig. 2.23 shows the end *A* of a beam, which is fixed. Hence the support at *A* is known as a fixed support. In case of fixed support, the reaction will be inclined. Also the fixed support will provide a couple.

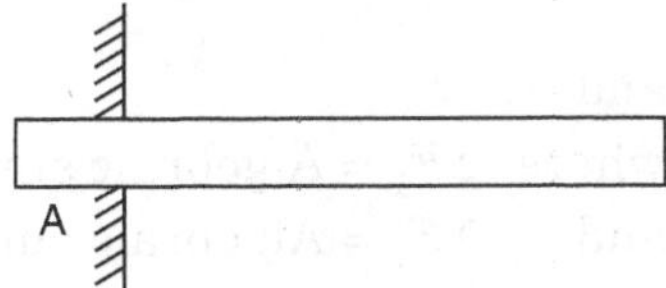

Fig. 2.23

2.3.6. Types of Loading. The following are the important types of loading :

(*a*) Concentrated or point load,

(*b*) Uniformly distributed load, and

(*c*) Uniformly varying load.

(*a*) **Concentrated or point load.** Fig. 2.24 shows a beam *AB*, which is simply supported at the ends *A* and *B*. A load *W* is acting at the point *C*. This load is known as point load (or concentrated load). Hence any load acting at a point on a beam, is known as point load.

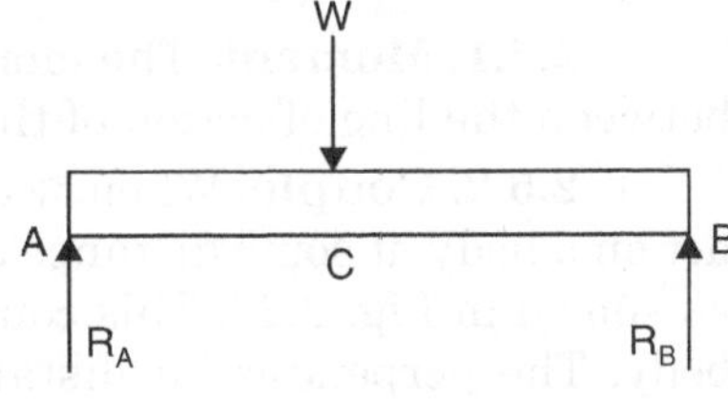

Fig. 2.24

In actual practice, it is not possible to apply a load at a point (*i.e.*, at a mathematical point) as it must have some contact area. But this area in comparison to the length of the beam is very very small (or area is negligible).

(*b*) **Uniformly distributed load.** If a beam is loaded in such a way, that each unit length of the beam carries same intensity of the load, then that type of load is known as uniformly distributed load (which is written as U.D.L.). Fig. 2.25 shows a beam *AB*, which carries a uniformly distributed load.

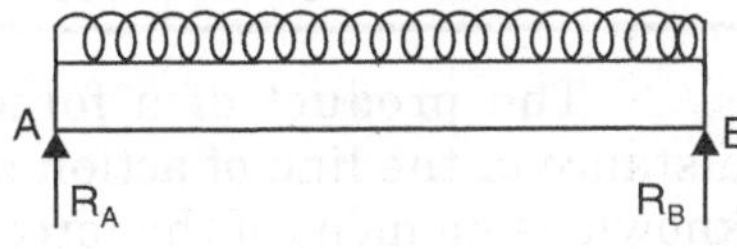

Fig. 2.25

For finding the reactions the total uniformly distributed load is assumed to act at the C.G. of the load.

(*c*) **Uniformly varying load.** Fig. 2.26 shows a beam *AB*, which carries load in such a way that the rate of loading on each unit length of the beam varies uniformly. This type of load is known as uniformly varying load. The total load on the beam is equal to the area of the load diagram. The total load acts at the C.G. of the load diagram.

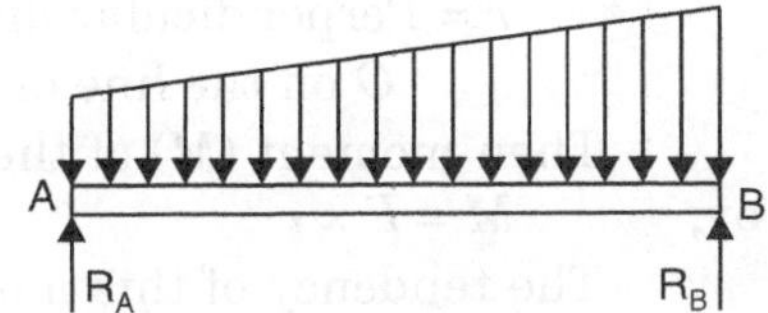

Fig. 2.26

2.4. STABLE EQUILIBRIUM

For stable equilibrium of a body, the algebraic sum of all the external forces should be zero *and also* the algebraic sum of moments of all the external forces about any point in their plane is zero. Mathematically, it is expressed by the equations :

$$\Sigma F = 0$$

$$\Sigma M = 0$$

The sign Σ is known as sigma which is a Greek letter. This sign represents the *algebraic sum* of forces or moments.

For two-dimensional bodies, the forces are generally resolved into horizontal and vertical components. Hence we have

$$\Sigma F_x = 0$$

and $$\Sigma F_y = 0$$

where ΣF_x = Algebraic sum of all horizontal components

and ΣF_y = Algebraic sum of all vertical components.

Similarly for moments, we have

$\Sigma M_x = 0$ and $\Sigma M_y = 0$

2.5. MOMENTS AND COUPLES

2.5.1. Moment. The moment is the product of the force and the perpendicular distance between the line of action of the force and the point about which moment is to be taken.

2.5.2. Couple. When two equal and opposite parallel forces act on a body at some distance apart, the two forces form a couple as shown in Fig. 2.27. This couple has the tendency to rotate the body. The perpendicular distance between the parallel forces is known as *arm of the couple*. The moment of the couple is the porduct of either one of the forces and perpendicular distance between the forces.

∴ Moment of the couple = $F \times a$.

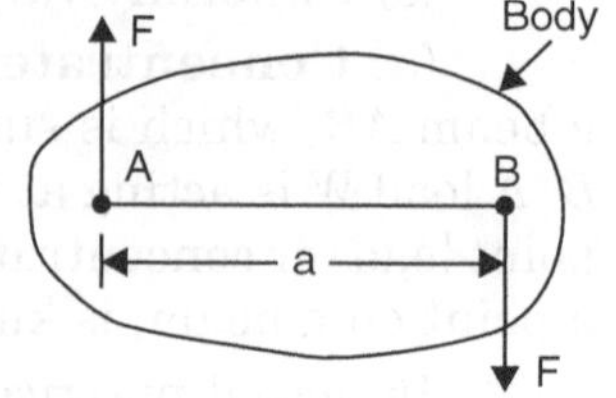

Fig. 2.27

2.6. MOMENT OF A FORCE ABOUT A POINT AND ABOUT AN AXIS

The product of a force and the perpendicular distance of the line of action of the force from a point is known as moment of the force about that point.

Let F = A force acting on a body as shown in Fig. 2.28.

r = Perpendicular distance from the point O on the line of action of force F.

Then moment (M) of the force F about O is given by, $M = F \times r$

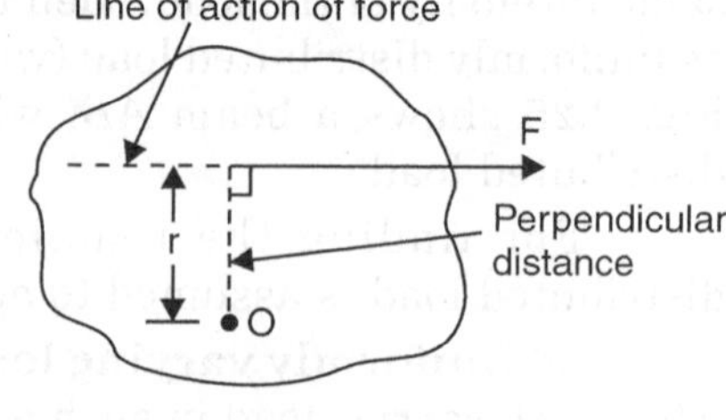

Fig. 2.28

The tendency of this moment is to rotate the body in the clockwise direction about O. Hence this moment is called *clockwise moment.* If the tendency of a moment is to rotate the body in anti-clockwise direction, then that moment is known as *anti-clockwise moment.* If clockwise moment is taken – ve then anti-clockwise moment will be +ve.

In S.I. system, moment is expressed in Nm (Newton metre).

Fig. 2.29 shows a body on which three forces F_1, F_2 and F_3 are acting. Suppose it is required to find the resultant moments if these forces about point O.

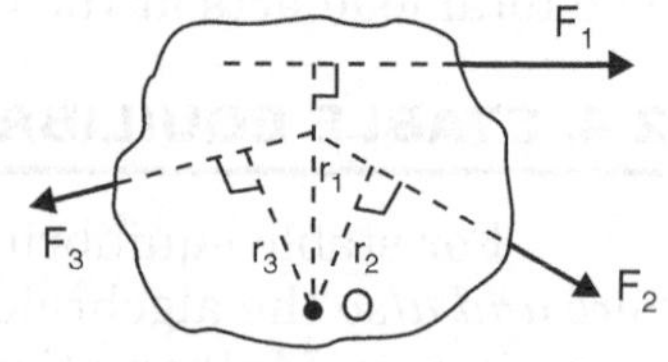

Fig. 2.29

Let r_1 = Perpendicular distance from O on the line of action of force F_1.

r_2 and r_3 = Perpendicular distances from O on the lines of action of force F_2 and F_3 respectively.

Moment of F_1 about $O = F_1 \times r_1$ (clockwise) (–)

Moment of F_2 about $O = F_2 \times r_2$ (clockwise) (–)

Moment of F_3 about $O = F_3 \times r_3$ (anti-clockwise) (+)

The resultant moment will be *algebraic sum* of all the moments.

∴ The resultant moment of F_1, F_2 and F_3 about $O = -F_1 \times r_1 - F_2 \times r_2 + F_3 \times r_3$.

Problem 2.15. *Four forces of magnitude 10 N, 20 N, 30 N and 40 N are acting respectively along the four sides of a square ABCD as shown in Fig. 2.30. Determine the resultant moment about the point A. Each side of the square is given 2 m.*

Sol. Given :

Length $AB = BC = CD = DA = 2$ m

Force at $B = 10$ N,

Force at $C = 20$ N,

Force at $D = 30$ N,

Force at $A = 40$ N,

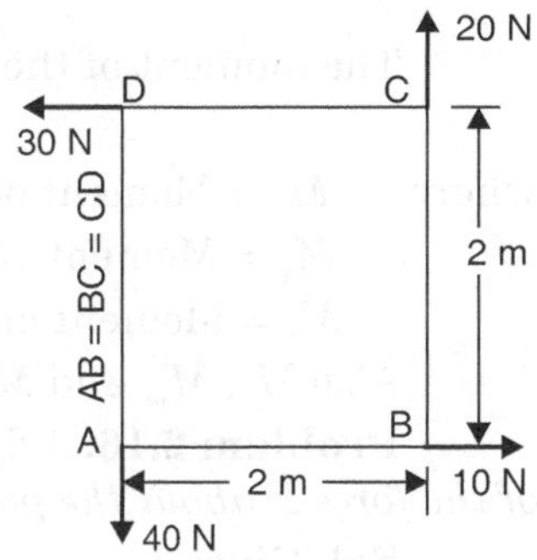

Fig. 2.30

The resultant moment about point A is to be determined.

The forces at A and B passes through point A. Hence perpendicular distance from A on the lines of action of these forces will be zero.

Hence their moments about A will be zero. The moment of the force at C about point A

= Force at C × ⊥ distance from A on the line of action of force at C

= (20 N) × (Length AB)

= 20 × 2 N m = 40 N m (anti-clockwise)

The moment of force at D about point A

= Force at D × ⊥ distance from A on the line of action of force at D

= (30 N) × (Length AD)

= 30 × 2 N m = 60 N m (anti-clockwise)

∴ Resultant moment of all forces about A

= 40 + 60 = **100 N m** (anti-clockwise). **Ans.**

2.7. VECTORIAL REPRESENTATION OF MOMENTS AND COUPLES

2.7.1. Vectorial Representation of Moments. The moment of a force is a vector which is the product of distance and force. Hence in case of moment* of a force the cross-product of distance and force would be taken. Consider the Fig. 2.31.

Let **F** = Force vector $(F_x i + F_y j + F_z k)$

r = Distance (or position) vector with respect to O

$= xi + yj + zk$

M = Moment of force about point O

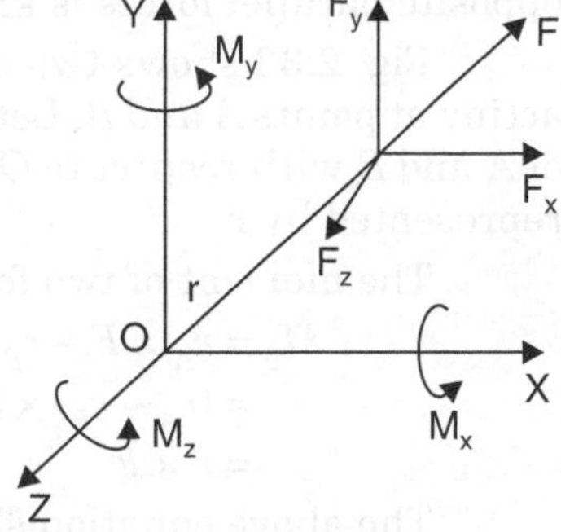

Fig. 2.31

*A quantity which is the product of two vectors and the quantity is also a vector, then cross product of the two vectors will be taken. But if the quantity is scalar, then dot product is taken.

Then $$\mathbf{M} = \mathbf{r} \times \mathbf{F}$$

or $$\mathbf{M} = \mathbf{r} \times \mathbf{F} = \begin{vmatrix} i & j & k \\ x & y & z \\ F_x & F_y & F_z \end{vmatrix}$$

$\therefore$ $$\mathbf{M} = (yF_z - zF_y)\, i + (zF_x - xF_z)\, j + (xF_y - yF_x)\, k$$

The moment of the given force about x, y and z-axis are equal to

$$M_x = yF_z - zF_y, \quad M_y = zF_x - xF_z, \quad M_z = xF_y - yF_x$$

where M_x = Moment of **F** about x-axis

M_y = Moment of **F** about y-axis, and

M_z = Moment of **F** about z-axis.

Also M_x, M_y, and M_z are known as scalar components of moment.

Problem 2.16. *A force $\mathbf{F} = 2i + 4j - 3k$ is applied at a point $P(1, 1, -2)$. Find the moment of the force F about the point $(2, -1, 2)$.*

Sol. Given :

Force $\mathbf{F} = 2i + 4j - 3k$

The position vector **r** of the point P w.r.t. O.

= Position vector of point P − Position vector of point O.

$= (i + j - 2k) - (2i - j + 2k)$

$\therefore$ $\mathbf{r} = (1 - 2)i + [1 + (1)]\, j + [-2 - 2]k$

$= -i + 2j - 4k$

Fig. 2.32

The moment **M** is given by

$$\mathbf{M} = \mathbf{r} \times \mathbf{F} = \begin{vmatrix} i & j & k \\ -1 & 2 & -4 \\ 2 & 4 & -3 \end{vmatrix}$$

$= [(2)(-3) - (-4)(4)]\, i + [(-4)(2) - (-1)(-3)]\, j + [(-1)(4) - (2)(2)]k$

$= (-6 + 16)\, i + (-8 - 3)\, j + (-4 - 4)\, k =$ **10i − 11j − 8k. Ans.**

2.7.2. Vectorial Representation of Couples. The moment produced by two equal, opposite parallel forces is known as couple.

Fig. 2.33 shows two equal opposite and parallel forces acting at points A and B. Let $\mathbf{r_A}$ and $\mathbf{r_B}$ are the position vectors of A and B with respect to O. The vector which joints B to A is represented by **r**.

The moment of two forces about point O is given by

$$M_o = r_A \times F - r_B \times F$$
$$= (r_A - r_B) \times F$$
$$= r \times F \qquad (\because\ r_A - r_B = r)$$

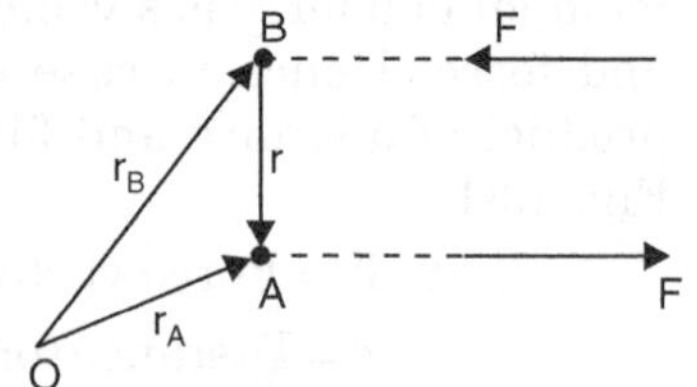

Fig. 2.33

The above equation shows that moment vector is independent of moment centre O.

$\therefore$ $M = r \times F$

This moment is known as couple.

The effect of couple is to produce pure rotation about an axis normal of the plane of force which constitute couple.

2.7.3. Scalar Components of a Moment. The moment of a force about any point O, is given by,

$$\mathbf{M} = \mathbf{r} \times \mathbf{F}$$

$$= \begin{vmatrix} i & j & k \\ x & y & z \\ F_x & F_y & F_z \end{vmatrix}$$

$$= (yF_z - zF_y)i + (zF_x - xF_z)j + (xF_y - yF_x)k$$

$$= M_x i + M_y j + M_z k$$

where M_x = Moment of $\mathbf{F}$ about x-axis = $(yF_z - zF_y)$,

M_y = Moment of $\mathbf{F}$ about y-axis = $(zF_x - xF_z)$ and

M_z = Moment of $\mathbf{F}$ about z-axis = $(xF_y - yF_x)$.

Also M_x, M_y and M_z are known as *scalar components of Moment M.*

2.8. VARIGNON'S THEOREM (OR PRINCIPLE OF MOMENTS)

Varignon's Theorem states that the moment of a force about any point is equal to the *algebraic sum* of the moments of its components about *that point.*

Principle of moments states that the moment of the resultant of a number of forces about any point is equal to the *algebraic sum* of the moments of all the forces of the system about the same point.

Proof of Varignon's Theorem

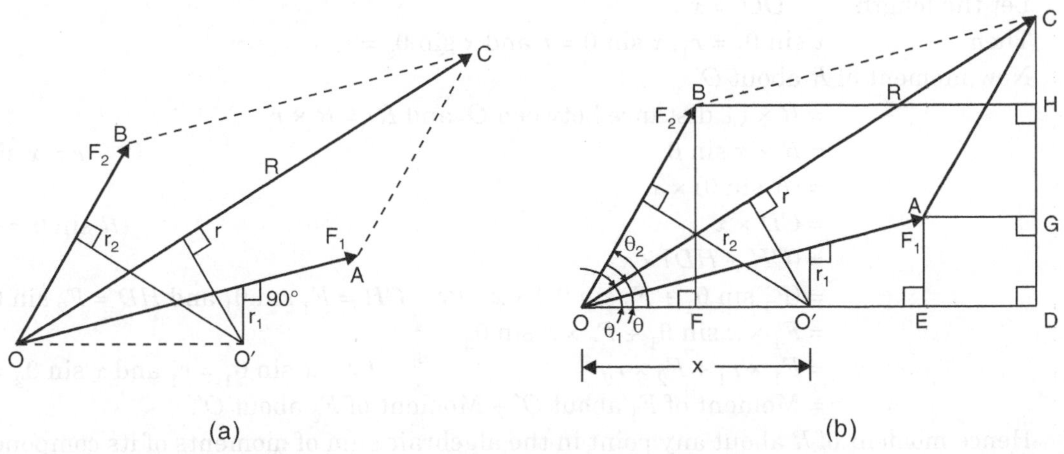

Fig. 2.34

Fig. 2.34 (*a*) shows two forces F_1 and F_2 acting at point O. These forces are represented in magnitude and direction by OA and OB. Their resultant R is represented in magnitude and direction by OC which is the diagonal of parallelogram $OACB$. Let O' is the point in the plane about which moments of F_1, F_2 and R are to be determined. From point O', draw perpendiculars on OA, OC and OB.

Let r_1 = Perpendicular distance between F_1 and O'.

r = Perpendicular distance between R and O'.

r_2 = Perpendicular distance between F_2 and O'.

Then according to Varignon's principle ;

Moment of R about O' must be equal to algebraic sum of moments of F_1 and F_2 about O'.

or $$R \times r = F_1 \times r_1 + F_2 \times r_2$$

Now refer to Fig. 2.34 (*b*). Join OO' and produce it to D. From points C, A and B draw perpendiculars on OD meeting at D, E and F respectively. From A and B also draw perpendiculars on CD meeting the line CD at G and H respectively.

Let θ_1 = Angle made by F_1 with OD,

θ = Angle made by R with OD, and

θ_2 = Angle made by F_2 with OD.

In Fig. 2.34 (*b*), $OA = BC$ and also OA parallel to BC, hence the projection of OA and BC on the same vertical line CD will be equal *i.e.*, $GD = CH$ as GD is the projection of OA on CD and CH is the projection of BC on CD.

Then from Fig. 2.34 (*b*), we have

$$P_1 \sin \theta_1 = AE = GD = CH$$

$$F_1 \cos \theta_1 = OE$$

$$F_2 \sin \theta_1 = BF = HD$$

$$F_2 \cos \theta_2 = OF = ED$$ $\because$ ($OB = AC$ and also $OB \parallel AC$. Hence projections of OB and AC on the same horizontal line OD will be equal *i.e.*, $OF = ED$)

$$R \sin \theta = CD$$

$$R \cos \theta = OD$$

Let the length $OO' = x$.

Then $x \sin \theta_1 = r_1$, $x \sin \theta = r$ and $x \sin \theta_2 = r_2$

Now moment of R about O'

$= R \times (\perp$ distance between O' and $R) = R \times r$

$= R \times x \sin \theta$ $\quad (\because \; r = x \sin \theta)$

$= (R \sin \theta) \times x$

$= CD \times x$ $\quad (\because \; R \sin \theta = CD)$

$= (CH + HD) \times x$

$= (F_1 \sin \theta_1 + F_2 \sin \theta_2) \times x$ $\quad (\because \; CH = F_1 \sin \theta_1$ and $HD = F_2 \sin \theta_2)$

$= F_1 \times x \sin \theta_1 + F_2 \times x \sin \theta_2$

$= F_1 \times r_1 + F_2 \times r_2$ $\quad (\because \; x \sin \theta_1 = r_1$ and $x \sin \theta_2 = r_2)$

= Moment of F_1 about O' + Moment of F_2 about O'.

Hence moment of R about any point in the algebraic sum of moments of its components (*i.e.*, F_1 and F_2) about the same point. Hence Varignon's principle is proved.

The principle of moments (or Varignon's principle) is not restricted to only two concurrent forces but is also applicable to any coplanar force system, *i.e.*, concurrent or non-concurrent or parallel force system.

Problem 2.17 *A force of 100 N is acting at a point A as shown in Fig. 2.35. Determine the moments of this force about O.*

Sol. Given :

Force at A = 100 N

Draw a perpendicular from O on the line of action of force 100 N. Hence OB is the perpendicular on the line of action of 100 N as shown in Fig. 2.35.

1st Method

Triangle OBC is a right-angled triangle. And angle

$$OCB = 60°.$$

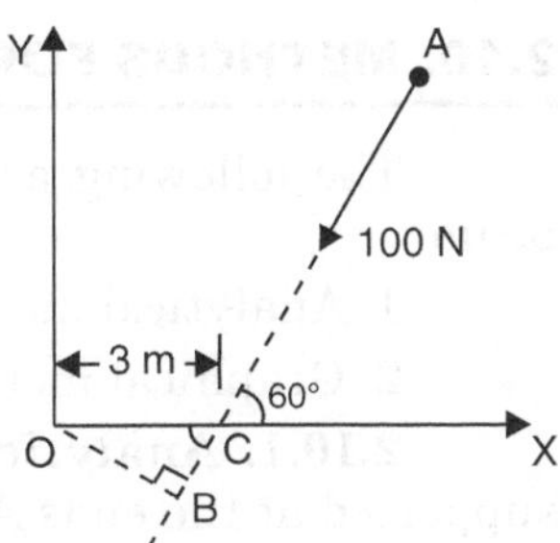

Fig. 2.35

$$\therefore \quad \sin 60° = \frac{OB}{OC}$$

$$\therefore \quad OB = OC \sin 60°$$

$$= 3 \times 0.866 = 2.598 \text{ m}$$

Moment of the force 100 N about O

$$= 100 \times OB = 100 \times 2.598$$

$= \mathbf{259.8\ N\ m}$ (clockwise). **Ans.**

2nd Method

The moment of force 100 N about O, can also be determined by using Varignon's principle. The force 100 N is replaced by its two rectangular components at any convenient point. Here the convenient point is chosen as C. The horizontal and vertical components of force 100 N acting at C are shown in Fig. 2.36.

(*i*) The horizontal component

$$= 100 \times \cos 60° = 50 \text{ N}$$

But this force is passing through O and hence has no moment about O.

Fig. 2.36

(*ii*) The vertical component

$$= 100 \times \sin 60° = 100 \times 0.866 = 86.6 \text{ N}$$

This force is acting vertically downwards at C. Moment of this force about O

$= 86.6 \times OC = 86.6 \times 3$ $\quad (\because \ OC = 3 \text{ m})$

$= \mathbf{259.8\ N}$ (clockwise). **Ans.**

2.9. EQUILIBRIUM OF RIGID BODIES IN TWO DIMENSIONS AND IN THREE DIMENSIONS

For the equilibrium of rigid bodies in two-dimensions or in three dimensions the algebraic sum of all the external forces should be zero and also the algebraic sum of moments of all the external forces about any point in their plane should be zero. Mathematically it is expressed by the equations :

$$\Sigma F = 0 \quad \text{and} \quad \Sigma M = 0$$

The sign Σ is known as sigma which represents the algebraic sum of forces or moments

(*i*) For two-dimensional bodies, the forces are written as

$$\Sigma F_x = 0 \quad \text{and} \quad \Sigma F_y = 0$$

(*ii*) For three-dimensional bodies, the forces are expressed as

$$\Sigma F_x = 0, \Sigma F_y = 0 \quad \text{and} \quad \Sigma F_z = 0$$

where ΣF_x = Algebraic sum of all components of forces in x-direction,

ΣF_y = Algebraic sum of all components of forces in y-direction, and

ΣF_z = Algebraic sum of all components of forces in z-direction.

2.10. METHODS FOR FINDING OUT THE REACTIONS OF A BEAM

The following are the methods of finding out the reactions at the two supports of a beam:

1. Analytical method, and
2. Graphical method.

2.10.1. Analytical Method. Fig. 2.37 shows a beam AB of length L and is simply supported at the ends A and B. The beam carries two point loads W_1 and W_2 at a distance L_1 and L_2 from the end A.

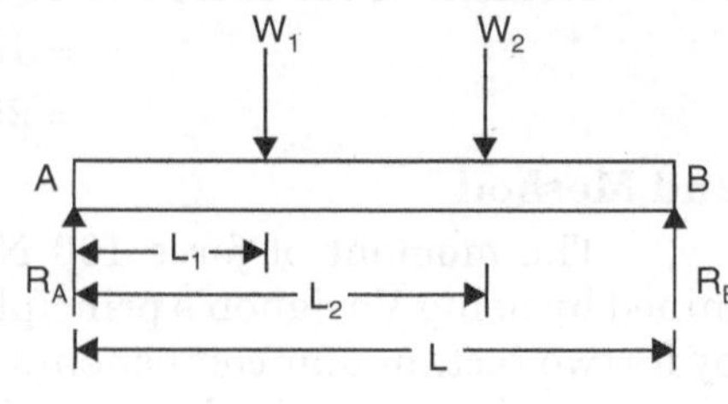

Fig. 2.37

Let R_A = Reaction at A

and R_B = Reaction at B

As the beam is in equilibrium, the equations of the equilibrium *i.e.*, $\Sigma F_x = 0$, $\Sigma F_y = 0$ and $\Sigma M = 0$ should be satisfied. In this case there is no horizontal force, hence the equations of equilibrium are $\Sigma F_y = 0$ and $\Sigma M = 0$.

For $\Sigma y = 0$, we have

$$R_A + R_B = W_1 + W_2 \quad \text{...(i)}$$

For $\Sigma M = 0$, the moments about any point of all the forces should be zero.

Taking the moments about point A, we get

$$W_1 \times L_1 + W_2 \times L_2 - R_B \times L = 0$$

or
$$W_1 \times L_1 + W_2 \times L_2 = R_B \times L \quad \text{or} \quad R_B = \frac{W_1 \times L_1 + W_2 \times L_2}{L}$$

As W_1, W_2, L_1, L_2 and L are given, hence value of R_B can be calculated.

Now from equation (*i*), we have $R_A = (W_1 + W_2) - R_B$.

2.10.2. Graphical Method for Finding Out the Reactions of a Beam. The graphical method consists of the following steps :

(*a*) Construction of space diagram ;

(*b*) Use of Bow's notations ; and

(*c*) Construction of vector diagram.

The given beam is drawn to a suitable scale along with the loads and the reactions R_A and R_B. This step is known as construction of space diagram.

The different loads and forces (*i.e.*, reactions R_A and R_B) are named by two capital letters, placed on their either side of the space diagram as shown in Fig. 2.38. This step is known as *Bow's notation*. The load W_1 is named by *PQ*, W_2 by *QR*, reaction R_B by *SR* and reaction R_A by *SP*.

Now the vector diagram is drawn according to the following steps :

(*i*) Choose a suitable scale to represent the various loads. Now take any point p and draw pq parallel and equal to the load *PQ* (*i.e.*, W_1) vertically downward to the same scale.

(*ii*) Now through q, draw qr parallel and equal to *QR* vertically downward to the same scale.

(*iii*) Select any suitable point O. Now join the point O to points p, q and r as shown in Fig. 2.38 (*b*).

(*iv*) Now in Fig. 2.38 (*a*), extend the lines of action of the loads and the two reactions. Take any point 1, on the line of action of the reaction R_A. Through 1, draw the line 1-2 parallel to pO, intersecting the line of action of load W_1 at 2.

(*v*) Now from point 2, draw line 2-3 parallel to qO, intersecting the line of action of load W_2 at 3. Similarly, from point 3, draw the line 3-4 parallel to rO, intersecting the line of action of reaction R_B at point 4.

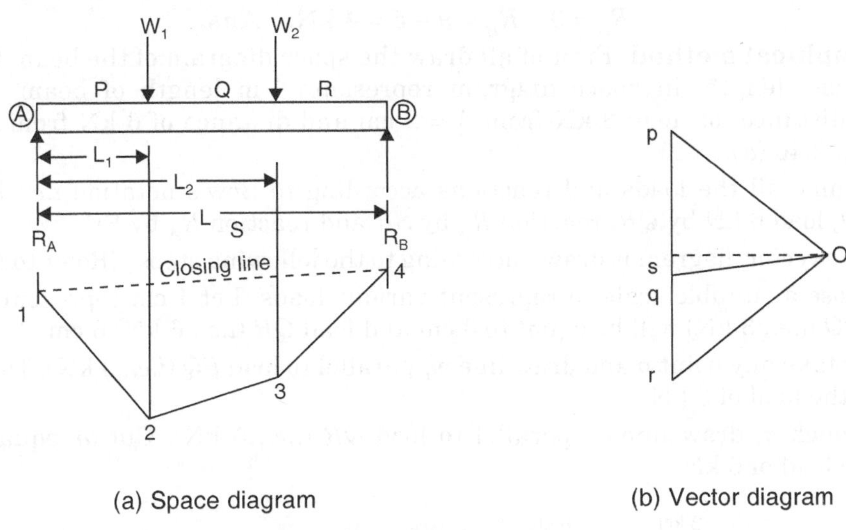

(a) Space diagram (b) Vector diagram

Fig. 2.38

(*vi*) Now join the point 1 to point 4. The line 1-4 is known as closing line. Now from point O (*i.e.*, from vector diagram) draw line Os parallel to line 1-4.

(*vii*) Now in the vector diagram the length sp represents the magnitude of reaction R_A to the same scale. Similarly, the length rs represents the magnitude of reaction R_B to the same scale.

2.11. PROBLEMS FOR EQUILIBRIUM OF RIGID BODIES IN TWO DIMENSIONS AND THREE DIMENSIONS

2.11.1. Problems on simple supported beam

Problem 2.18. *A simply supported beam AB of span 6 m carries point loads of 3 kN and 6 kN at a distance of 2 m and 4 m from the left end A as shown in Fig. 2.39. Find the reactions at A and B analytically and graphically.*

Sol. Given :

Span of beam = 6 m

Let R_A = Reaction at A

R_B = Reaction at B

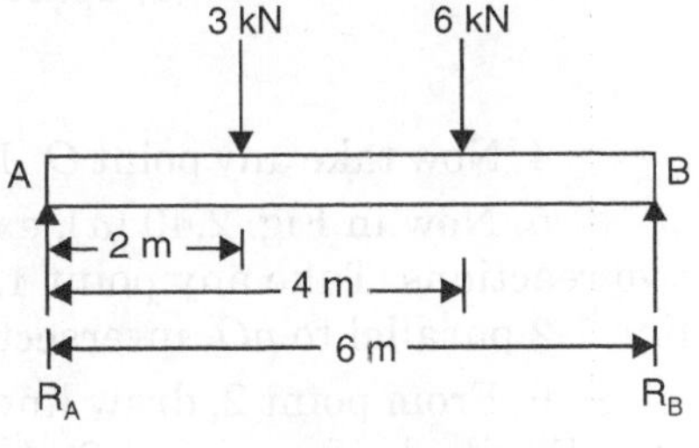

Fig. 2.39

(*a*) **Analytical method.** As the beam is in equilibrium, the moments of all the forces about any point should be zero.

Now taking the moment of all forces about A, and equating the resultant moment to zero, we get

$$R_B \times 6 - 3 \times 2 - 6 \times 4 = 0$$

or $$6R_B = 6 + 24 = 30$$

or $$R_B = \frac{30}{6} = \textbf{5 kN. Ans.}$$

Also for equilibrium, $\Sigma F_y = 0$

$\therefore$ $$R_A + R_B = 3 + 6 = 9$$

$\therefore$ $$R_A = 9 - R_B = 9 - 5 = \textbf{4 kN. Ans.}$$

(*b*) **Graphical method.** First of all draw the space diagram of the beam to a suitable scale. Let 1 cm length in space diagram represents 1 m length of beam. Hence take AB = 6 cm, distance of load 3 kN from A = 2 cm and distance of 6 kN from A = 4 cm as shown in Fig. 2.40 (*a*).

Now name all the loads and reactions according to Bow's notation *i.e.*, load 3 kN is named by PQ, load 6 kN by QR, reaction R_B by SR and reaction R_A by SP.

Now the vector diagram is drawn according to the following steps : [Refer to Fig. 2.40 (*b*)].

1. Choose a suitable scale to represent various loads. Let 1 cm represents 1 kN load. Hence load PQ (*i.e.*, 3 kN) will be equal to 3 cm and load QR (*i.e.*, 6 kN) 6 cm.

2. Now take any point p and draw line pq parallel to load PQ (*i.e.*, 3 kN). Take pq = 3 cm to represent the load of 3 kN.

3. Through q, draw line qr parallel to load QR (*i.e.*, 6 kN). Cut qr equal to 6 cm to represent the load of 6 kN.

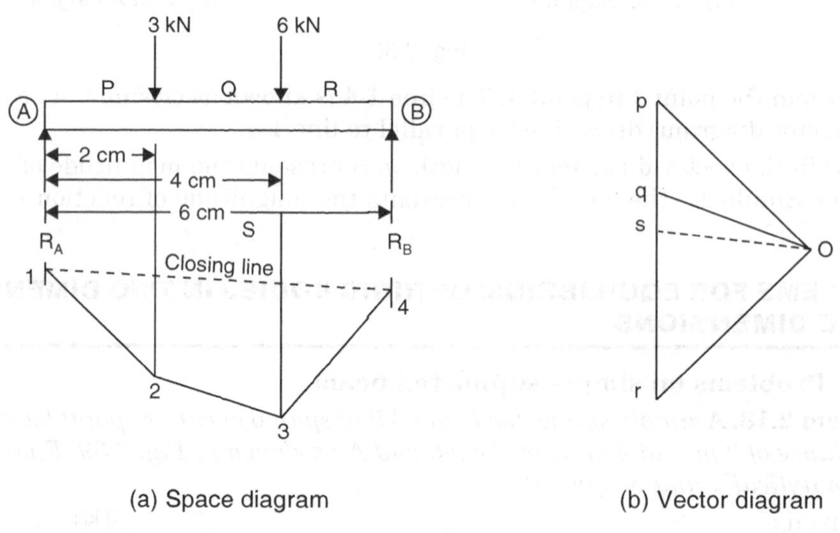

(a) Space diagram (b) Vector diagram

Fig. 2.40

4. Now take any point O. Join the point O to the points p, q and r as shown in Fig. 2.40 (*b*).

5. Now in Fig. 2.40 (*a*), extend the lines of action of the loads (3 kN and 6 kN), and the two reactions. Take any point 1, on the line of action of the reaction R_A. Through 1, draw the line 1-2 parallel to pO, intersecting the line of action of load 3 kN at point 2.

6. From point 2, draw line 2-3 parallel to qO, intersecting the line of action of load 6 kN at 3. Similarly, from point 3, draw a line 3-4 parallel to rO, intersecting the line of action of reaction R_B at point 4.

7. Join 1 to 4. The line 1-4 is known as closing line. From the vector diagram, from point O, draw line Os parallel to line 1-4.

8. Measure the length sp and rs. The length sp represents the reaction R_A and length rs represents the reaction R_B.

By measurement, $sp = 4$ cm and $rs = 5$ cm

$\therefore$ R_A = Length sp × scale = 4 × 1 kN = **4 kN. Ans.**

R_B = Length rs × scale = 5 × 1 kN = **5 kN. Ans.**

Problem 2.19. *A simply supported beam AB of length 9 m, carries a uniformly distributed load of 10 kN/m for a distance of 6 m from the left end. Calculate the reactions at A and B.*

Sol. Given :

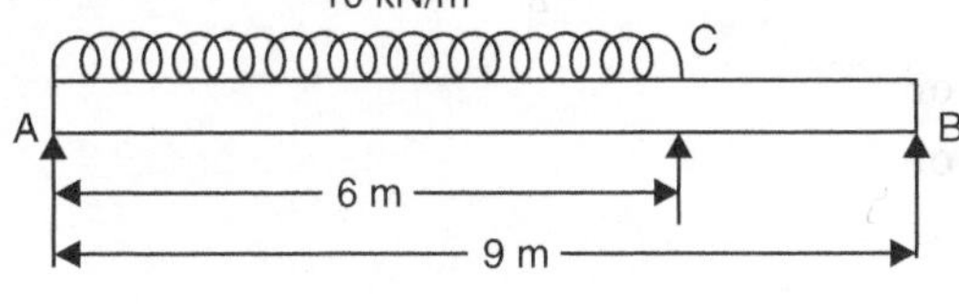

Fig. 2.41

Length of beam = 9 m

Rate of U.D.L. = 10 kN/m

Length of U.D.L. = 6 m

Total load due to U.D.L.

= (Length of U.D.L.) × Rate of U.D.L.

= 6 × 10 = 60 kN

This load of 60 kN will be acting at the middle point of AC *i.e.*, at a distance of $\frac{6}{2} = 3$ m from A.

Let R_A = Reaction at A and R_B = Reaction at B

Taking the moments of all forces about point A, and equating the resultant moment to zero, we get

$$R_B \times 9 - (6 \times 10) \times 3 = 0 \quad \text{or} \quad 9R_B - 180 = 0$$

$$\therefore \quad R_B = \frac{180}{9} = \textbf{20 kN. Ans.}$$

Also for equilibrium, $\Sigma F_y = 0$

or $$R_A + R_B = 6 \times 10 = 60$$

$$\therefore \quad R_A = 60 - R_B = 60 - 20 = \textbf{40 kN. Ans.}$$

Problem 2.20. *A simply supported beam of length 10 m, carries the uniformly distributed load and two point loads as shown in Fig. 2.42. Calculate the reactions R_A and R_B.*

Sol. Given :

Length of beam = 10 m

Length of U.D.L. = 4 m

Rate of U.D.L. = 10 kN/m

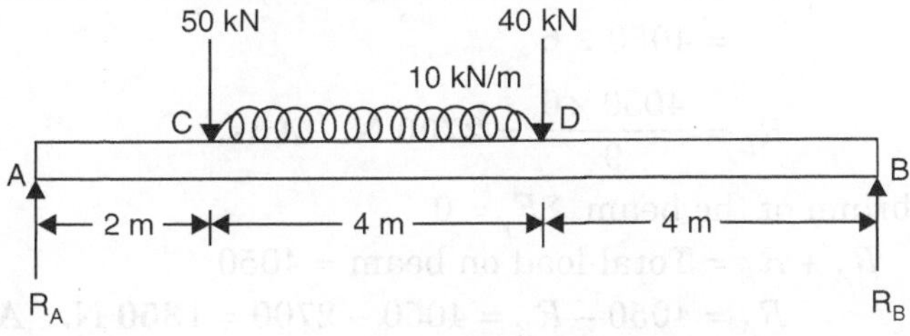

Fig. 2.42

$\therefore$ Total load due to U.D.L. = 4 × 10 = 40 kN

This load of 40 kN due to U.D.L. will be acting at the middle point of *CD, i.e.,* at a distance of $\frac{4}{2}$ = 2 m from *C* (or at a distance of 2 + 2 = 4 m from point *A*).

Let R_A = Reaction at *A*

and R_B = Reaction at *B*

Taking the moments of all forces about point *A* and equating the resultant moment to zero, we get

$$R_B \times 10 - 50 \times 2 - 40 \times (2 + 4) - (10 \times 4)\left(2 + \frac{4}{2}\right) = 0$$

or $$10R_B - 100 - 240 - 160 = 0$$

or $$10R_B = 100 + 240 + 160 = 500$$

$$\therefore \quad R_B = \frac{500}{10} = \mathbf{50\ kN.}\quad \textbf{Ans.}$$

Also for equilibrium of the beam, $\Sigma F_y = 0$

$\therefore$ $R_A + R_B$ = Total load on the beam = 50 + 10 × 4 + 40 = 130

$\therefore$ $R_A = 130 - R_B = 130 - 50 =$ **80 kN. Ans.**

Problem 2.21. *A simply supported beam of span 9 m carries a uniformly varying load from zero at end A to 900 N/m at end B. Calculate the reactions at the two ends of the support.*

Sol. Given :

Span of beam = 9 m

Load at end *A* = 0

Load at end *B* = 900 N/m

Total load on the beam = Area of $ABC = \frac{AB \times BC}{2} = \frac{9 \times 900}{2}$

= 4050 N

This load will be acting at the C.G. of the ΔABC,

i.e., at a distance of $\frac{2}{3} \times AB = \frac{2}{3} \times 9 = 6$ m from end *A*.

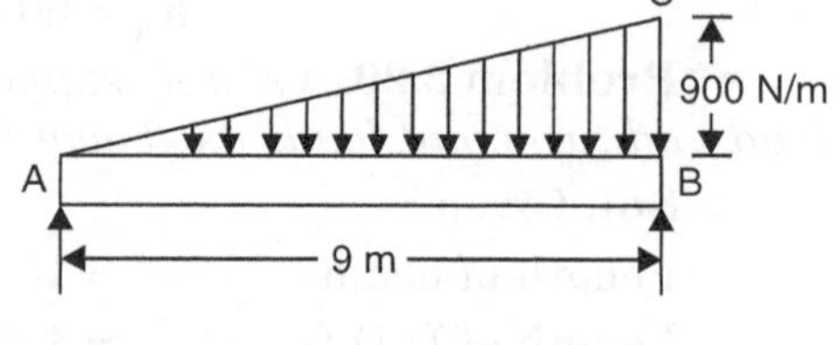

Fig. 2.43

Let R_A = Reaction at *A*

and R_B = Reaction at *B*.

Taking the moments of all forces about point *A* and equating the resultant moment to zero, we get

$R_B \times 9$ = (Total load on beam) × Distance of total load from *A*

= 4050 × 6

$$\therefore \quad R_B = \frac{4050 \times 6}{9} = \mathbf{2700\ N.}\quad \textbf{Ans.}$$

Also for equilibrium of the beam, $\Sigma F_y = 0$

or $R_A + R_B$ = Total load on beam = 4050

$\therefore$ $R_A = 4050 - R_B = 4050 - 2700 =$ **1350 N. Ans.**

Problem 2.22. *A simply supported beam of length 5 m carries a uniformly increasing load of 800 N/m at one end to 1600 N/m at the other end. Calculate the reactions at both ends.*

Sol. Given :

Length of beam = 5 m

Load at A = 800 N/m

Load at B = 1600 N/m

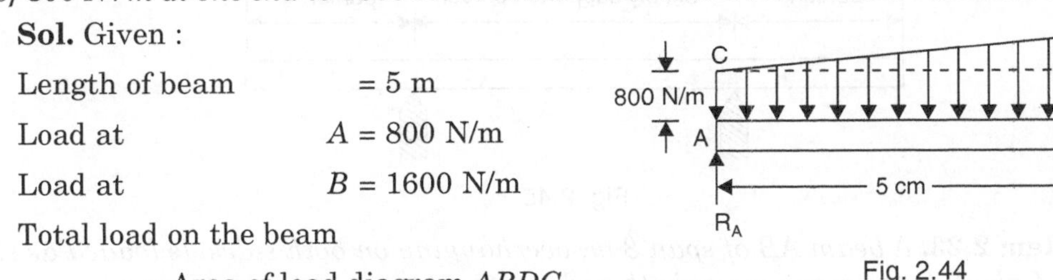

Fig. 2.44

Total load on the beam

= Area of load diagram $ABDC$

= Area of rectangle $ABEC$ + Area of ΔCED

$$= AB \times AC + \frac{CE \times ED}{2}$$

$$= 5 \times 800 + \tfrac{1}{2} \times 5 \times 800 \qquad (\because \; CE = AB = 5 \text{ m}, ED = 1600 - 800 = 800)$$

$$= 4000 + 2000 = 6000 \text{ N}$$

The C.G. of the rectangle $ABEC$ will be at a distance of $\frac{5}{2} = 2.5$ m from A, whereas the C.G. of the triangle CED will be at a distance of $\frac{2}{3} \times 5 = 3.33$ m from A.

Let R_A = Reaction at A

and R_B = Reaction at B.

Taking the moments of all forces about point A and equating the resultant moment to zero, we get

$R_B \times 5$ – (Load due to rectangle) × Distance of C.G. of rectangle from A – (Load due to triangle) × Distance of C.G. of triangle from A = 0

or
$$5R_B - (5 \times 800) \times 2.5 - \left(\frac{1}{2} \times 5 \times 800\right) \times \left(\frac{2}{3} \times 5\right) = 0$$

or
$$5R_B - 1000 - 6666.66 = 0$$

or
$$5R_B = 1000 + 6666.66 = 16666.66$$

or
$$R_B = \frac{16666.66}{5} = \mathbf{3333.33 \text{ N.}} \quad \textbf{Ans.}$$

Also for the equilibrium of the beam, $\Sigma F_y = 0$

$\therefore$ $R_A + R_B$ = Total load on the beam

= 6000 $\qquad (\because$ Total load on beam = 6000 N)

$\therefore$ $R_A = 6000 - R_B = 6000 - 3333.33 =$ **2666.67 N. Ans.**

2.11.2. Problems on Overhanging Beams. If the end portion of a beam is extended beyond the support, then the beam is known as overhanging beam. Overhanging portion may be at one end of the beam or at both ends of the beam as shown in Fig. 2.45.

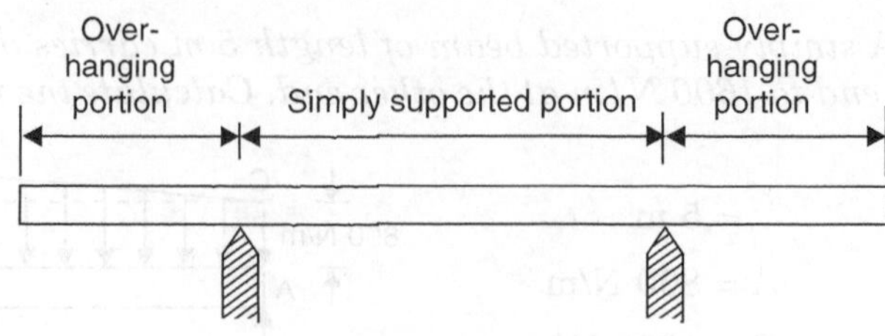

Fig. 2.45

Problem 2.23. *A beam AB of span 8 m, overhanging on both sides, is loaded as shown in Fig. 2.46. Calculate the reactions at both ends.*

Sol. Given :

Span of beam = 8 m

Let R_A = Reaction at A

and R_B = Reaction at B.

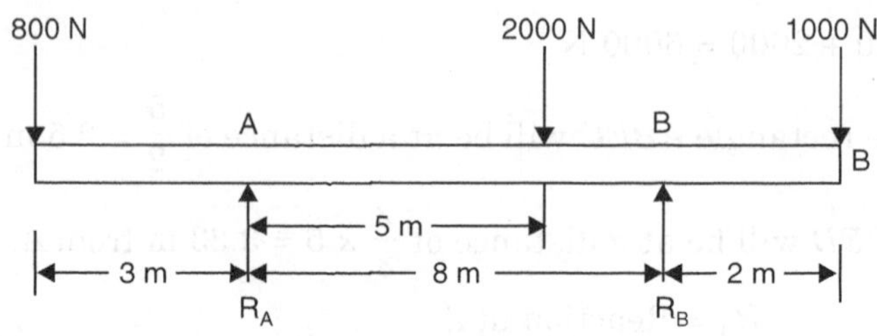

Fig. 2.46

Taking the moments of all forces about point A and equating the resultant moment to zero, we get

$$R_B \times 8 + 800 \times 3 - 2000 \times 5 - 1000 \times (8 + 2) = 0$$

or $$8R_B + 2400 - 10000 - 10000 = 0$$

or $$8R_B = 20000 - 2400 = 17600$$

$$\therefore \quad R_B = \frac{17600}{8} = \mathbf{2200\ N.} \quad \textbf{Ans.}$$

Also for the equilibrium of the beam, we have

$$R_A + R_B = 800 + 2000 + 1000 = 3800$$

$$\therefore \quad R_A = 3800 - R_B = 3800 - 2200 = \mathbf{1600\ N.} \quad \textbf{Ans.}$$

Problem 2.24. *A beam AB of span 4 m, overhanging on one side upto a length of 2 m, carries a uniformly distributed load of 2 kN/m over the entire length of 6 m and a point load of 2 kN/m as shown in Fig. 2.47. Calculate the reactions at A and B.*

Sol. Given :

Span of beam = 4 m

Total length = 6 m

Rate of U.D.L. = 2 kN/m

Total load due to U.D.L. = 2 × 6 = 12 kN

The load of 12 kN (*i.e.*, due to U.D.L.) will act at the middle point of *AC*, *i.e.*, at a distance of 3 m from *A*.

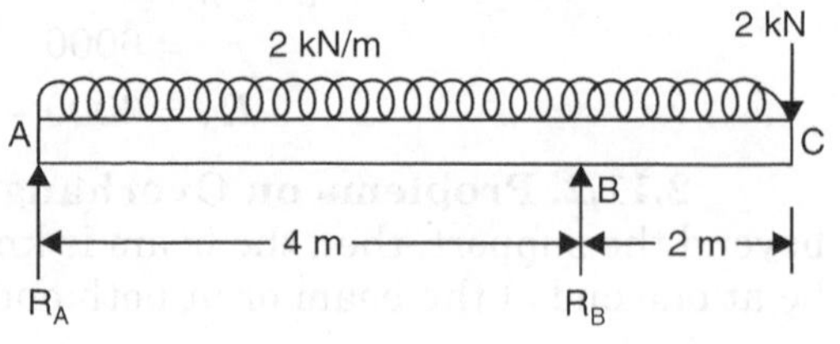

Fig. 2.47

Let R_A = Reaction at A

and R_B = Reaction at B.

Taking the moments of all forces about point A and equating the resultant moment to zero, we get

$$R_B \times 4 - (2 \times 6) \times 3 - 2 \times (4 + 2) = 0$$

or $$4R_B - 36 - 12 = 0$$

or $$4R_B = 36 + 12 = 48$$

$\therefore$ $$R_B = \frac{48}{4} = \textbf{12 kN.} \quad \textbf{Ans.}$$

Also for equilibrium, $\Sigma F_y = 0$ or $R_A + R_B = 12 + 2 = 14$

$\therefore$ $$R_A = 14 - R_B = 14 - 12 = \textbf{2 kN.} \quad \textbf{Ans.}$$

2.11.3. Problems on Roller and Hinged Supported Beams. In case of roller supported beams, the reaction on the roller end is always *normal* to the support. All the steel trusses of the bridges is generally having one of their ends supported on rollers. The main advantage of such a support is that beam, due to change in temperature, can move easily towards left or right, on account of expansion or contraction.

In case of a hinged supported beam, the reaction on the hinged end may be either vertical or inclined, depending upon the type of loading. The main advantage of a hinged end is that the beam remains stable. Hence all the steel trusses of the bridges, have one of their end on rollers and the other end as hinged.

Problem 2.25. *A beam AB 1.7 m long is loaded as shown in Fig. 2.48. Determine the reactions at A and B.*

Sol. Given :

Length of beam = 1.7 m

Let R_A = Reaction at A

and R_B = Reaction at B.

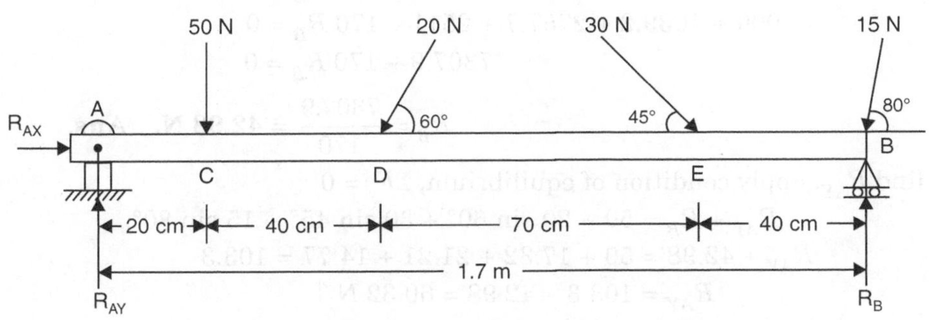

Fig. 2.48

Since the beam is supported on rollers at B, therefore the reaction R_B will be vertical.

The beam is hinged at A, and is carrying inclined load, therefore the reaction R_A will be inclined. This means reaction R_A will have two components, *i.e.*, vertical component and horizontal component.

Let R_{AX} = Horizontal component of reaction R_A

R_{AY} = Vertical component of reaction R_A.

First resolve all the inclined loads into their vertical and horizontal components.

(*i*) Vertical component of load at D

$$= 20 \sin 60° = 20 \times 0.866 = 17.32 \text{ N}$$

and its horizontal component

$$= 20 \cos 60° = 10 \text{ N} \leftarrow$$

(*ii*) Vertical component of load at E

$$= 30 \sin 45° = 21.21 \text{ N}$$

and its horizontal component

$$= 30 \cos 45° = 21.21 \text{ N} \rightarrow$$

(*iii*) Vertical component of load at B

$$= 15 \sin 80° = 14.77 \text{ N}$$

and its horizontal component

$$= 15 \cos 80° = 2.6 \text{ N} \leftarrow$$

From condition of equilibrium, $\Sigma F_x = 0$

or $R_{AX} - 10 + 21.21 - 2.6 = 0$

or $R_{AX} = 10 - 21.21 + 2.6 = -8.61$ N

– ve sign shows that the assumed direction of R_{AX} (*i.e.*, horizontal component of R_A) is wrong. Correct direction will be opposite to the assumed direction. Assumed direction of R_{AX} is towards right. Hence correct direction of R_{AX} will be towards left at A.

$\therefore$ $R_{AX} = 8.61 \text{ N} \leftarrow$

To find R_B, take moments* of all forces about A.

For equilibrium, $\Sigma M_A = 0$

$$\therefore \quad 50 \times 20 + (20 \sin 60°) \times (20 + 40) + (30 \times \sin 45°) \times (20 + 40 + 70) + (15 \sin 80°) \times (170) - 170\, R_B = 0$$

or $1000 + 1039.2 + 2757.7 + 2511 - 170\, R_B = 0$

or $7307.9 - 170\, R_B = 0$

$$\therefore \quad R_B = \frac{7307.9}{170} = \mathbf{42.98\ N.\quad Ans.}$$

To find R_{AY}, apply condition of equilibrium, $\Sigma F_y = 0$

or $R_{AY} + R_B = 50 + 20 \sin 60° + 30 \sin 45° + 15 \sin 80°$

or $R_{AY} + 42.98 = 50 + 17.32 + 21.21 + 14.77 = 103.3$

$\therefore$ $R_{AY} = 103.3 - 42.98 = 60.32 \text{ N} \uparrow$

$\therefore$ Reaction at A, $R_A = \sqrt{R_{AX}^2 + R_{AY}^2}$

$$= \sqrt{8.61^2 + 60.32^2} = 60.32 \text{ N}$$

---

*The moment of all horizontal components about point A, will be zero.

The angle made by R_A with x-direction is given by

$$\tan\theta = \frac{R_{AY}}{R_{AX}} = \frac{60.32}{8.61} = 7.006$$

$\therefore$ $\theta = \tan^{-1} 7.006 = \mathbf{81.87°}$. **Ans.**

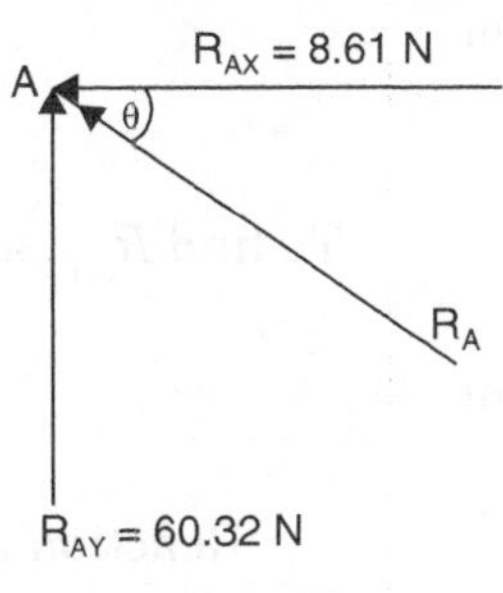

Fig. 2.48 (a)

Problem 2.26. *A beam AB 6 m long is loaded as shown in Fig. 2.49. Determine the reactions at A and B by (a) analytical method, and (b) graphical method.*

Sol. Given :

Length of beam = 6 m

Let R_A = Reaction at A

R_B = Reaction at B.

The reaction R_B will be vertical as the beam is supported on rollers at end B.

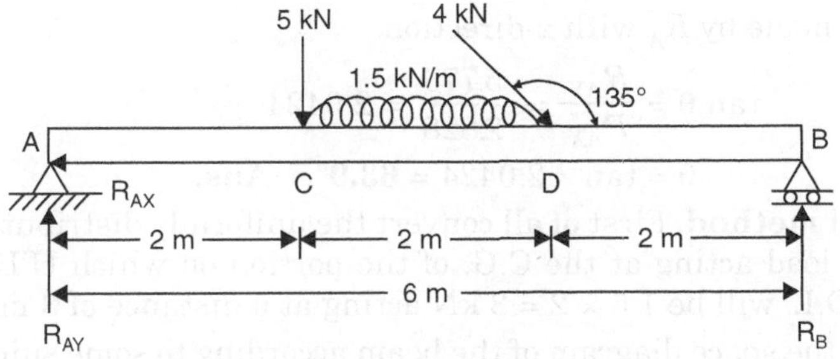

Fig. 2.49

The reaction R_A will be inclined, as the beam is hinged at A and carries inclined load.

Let R_{AX} = Horizontal component of reaction R_A

R_{AY} = Vertical component of reaction R_A.

(*a*) **Analytical method.** First resolve the inclined load of 4 kN into horizontal and vertical components.

Horizontal component of 4 kN at D

$$= 4\cos 45° = 2.828 \text{ kN} \rightarrow$$

and its vertical component

$$= 4\sin 45° = 2.828 \text{ kN} \downarrow$$

For equilibrium, $\Sigma F_x = 0$

$\therefore$ $-R_{AX} + 2.828 = 0$

or $R_{AX} = 2.828$ N

To find R_B, take the moments* of all forms about point A.

$\therefore$ For equilibrium, $\Sigma M_A = 0$

$$R_B \times 6 - 5 \times 2 - (2 \times 1.5)\left(2 + \frac{2}{2}\right) - (4\sin 45°)(2 + 2) = 0$$

or $$6R_B - 10 - 9 - 11.312 = 0$$

*The moment of horizontal component about A, will be zero.

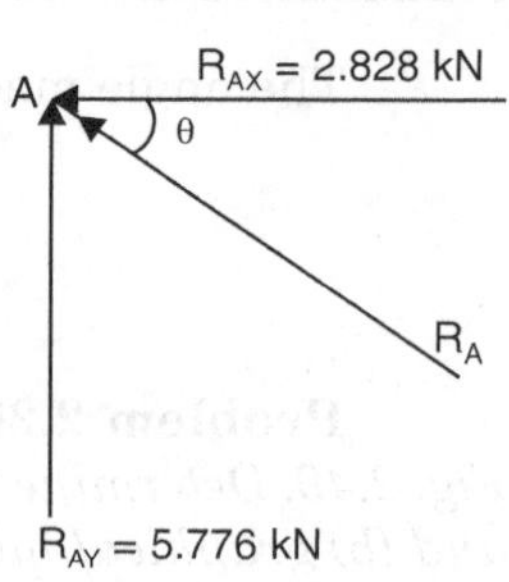

Fig. 2.49(a)

or $$6R_B = 10 + 9 + 11.312 = 30.312$$

$\therefore$ $$R_B = \frac{30.312}{6} = \mathbf{5.052\ kN.}\quad \textbf{Ans.}$$

To find R_{AY}, apply the condition of equilibrium, $\Sigma F_y = 0$

$\therefore$ $$R_{AY} + R_B - 5 - (1.5 \times 2) - 4 \sin 45° = 0$$

or $$R_{AY} + 5.052 - 5 - 3 - 2.828 = 0$$

$\therefore$ $$R_{AY} = -5.052 + 5 + 3 + 2.828 = 5.776 \text{ kN}$$

$\therefore$ Reaction at A is given by

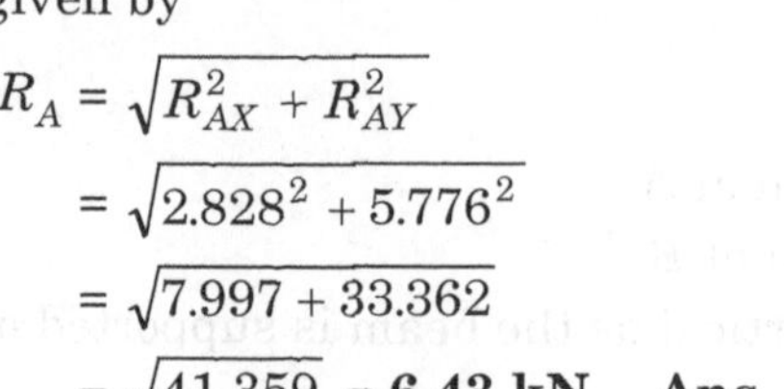

$$R_A = \sqrt{R_{AX}^2 + R_{AY}^2}$$
$$= \sqrt{2.828^2 + 5.776^2}$$
$$= \sqrt{7.997 + 33.362}$$
$$= \sqrt{41.359} = \mathbf{6.43\ kN.}\quad \textbf{Ans.}$$

Let θ = Angle made by R_A with x-direction.

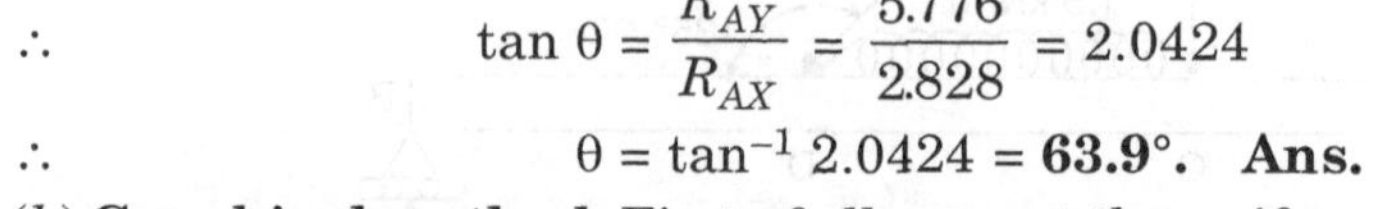

$\therefore$ $$\tan\theta = \frac{R_{AY}}{R_{AX}} = \frac{5.776}{2.828} = 2.0424$$

$\therefore$ $$\theta = \tan^{-1} 2.0424 = \mathbf{63.9°.}\quad \textbf{Ans.}$$

(*b*) **Graphical method.** First of all convert the uniformly distributed load (U.D.L.) into its equivalent point load acting at the C.G. of the portion on which U.D.L. is acting. Hence total load due to U.D.L. will be 1.5 × 2 = 3 kN acting at a distance of 3 m from point A :

(*i*) Now draw the space diagram of the beam according to some suitable scale, as shown in Fig. 2.50(*a*).

(*ii*) Name all the loads and reactions according to Bow's notation. Now draw the vector diagram as shown in Fig. 2.50(*b*). Choose any suitable scale for vector diagram.

(*iii*) Take any point *p* for drawing vector diagram. From *p*, draw line *pq* parallel and equal to load 5 kN (*i.e.*, load *PQ*). From *q*, draw *qr* parallel and equal to 3 kN. From *r*, draw *rs* parallel and equal to 4 kN load.

(*iv*) Now take any point *O*, and join *Op, Oq, Or* and *Os*.

(*v*) Now in space diagram [*i.e.*, Fig. 2.50(*a*)], extend the lines of actions of loads *PQ, QR, RS* and reaction R_B.

(*vi*) Take any point 1, vertically below the point *A* as shown in Fig. 2.50(*a*). From point 1, draw line 1-2 parallel to line *pO*, intersecting the line of action of 5 kN at point 2.

(*vii*) Similarly, draw lines 2-3, 3-4 and 4-5 parallel to *qO, rO* and *sO* respectively. Join point 1 to 5. Line 1-5 is the closing line in space diagram.

(*viii*) From *O* in vector diagram, draw a line parallel to closing line 1-5. Now through *s*, draw a line *st* vertical (as the reaction R_B is vertical), intersecting the line through *O* at *t*. Join *t* to *p*.

(*ix*) The length *st* represents the reaction R_B in magnitude and direction whereas the length *tp* gives the magnitude and direction of reaction R_A. At point *A*, draw a line parallel to *tp* as shown in Fig. 2.50(*a*). By measurement, we get

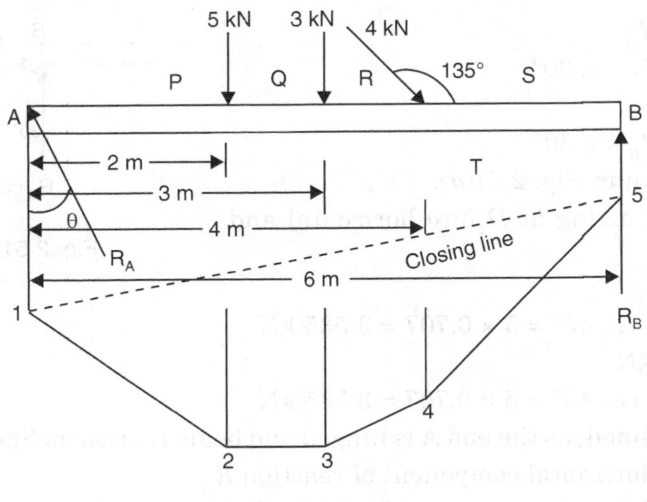

(a) Space diagram

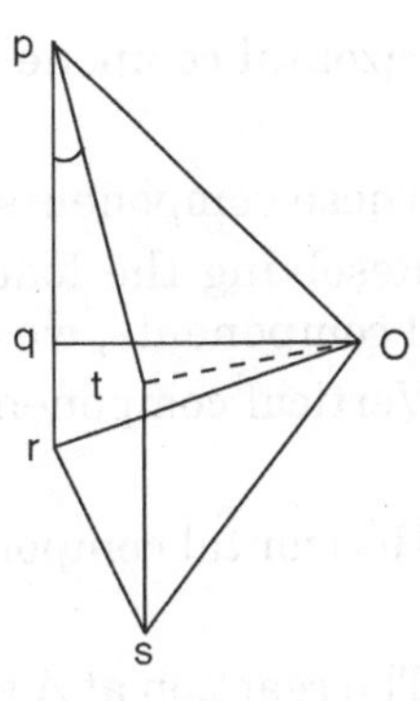

(b) Vector diagram

Fig. 2.50

R_A = length tp = 6.43 kN

R_B = length st = 5.052 kN

and $\theta = 26.1°$.

Problem 2.27. *A beam AB 10 m long is hinged at A and supported on rollers over a smooth surface inclined at 30° to the horizontal at B. The beam is loaded as shown in Fig. 2.51. Determine reactions at A and B.*

Sol. Given :

Length of beam = 10 m

Let R_A = Reaction at A

and R_B = Reaction at B

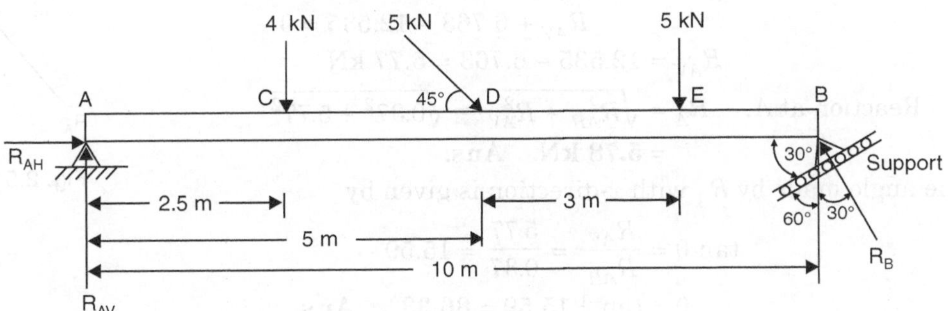

Fig. 2.51

The reaction R_B will be normal to the support as the beam at B is supported on the rollers. But the support at B is making an angle 30° with the horizontal or 60° with the vertical as shown in Fig. 2.51. Hence the reaction R_B is making an angle of 30° with the vertical.

The vertical component of R_B

$$= R_B \cos 30°$$

and horizontal component of R_B

$$= R_B \sin 30°$$

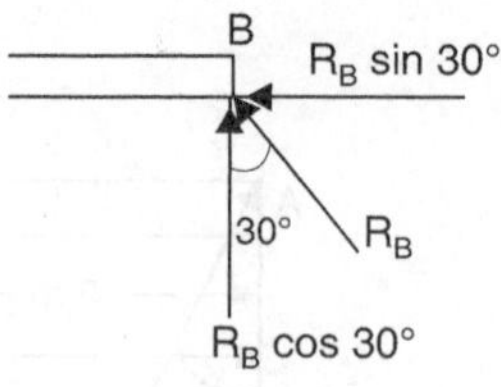

Fig. 2.51(a)

These components are shown in Fig. 2.51(*a*).

Resolving the load of 5 kN acting at *D* into horizontal and vertical components, we get

Vertical component of 5 kN

$$= 5 \sin 45° = 5 \times 0.707 = 3.535 \text{ kN}$$

Horizontal component of 5 kN

$$= 5 \cos 45° = 5 \times 0.707 = 3.535 \text{ kN}$$

The reaction at *A* will be inclined, as the end *A* is hinged and beam carries inclined load.

Let R_{AX} = Horizontal component of reaction R_A

R_{AY} = Vertical component of reaction R_A

For equilibrium of the beam, the moments of all forces about any point should be zero.

Taking the moments* about point *A*,

$$(R_B \cos 30°) \times 10 - 4 \times 2.5 - (5 \sin 45°) \times 5 - 5 \times 8 = 0.$$

$$8.66\, R_B - 10 - 17.675 - 40 = 0$$

or $$R_B = \frac{10 + 17.675 + 40}{8.66} = \mathbf{7.81\ kN} \quad \textbf{Ans.}$$

For equilibrium, $\Sigma F_x = 0$

or $$R_{AH} + 5 \cos 45° - R_B \sin 30° = 0$$

or $$R_{AH} + 3.535 - 7.81 \times 0.5 = 0$$

$\therefore$ $$R_{AH} = 7.81 \times 0.5 - 3.535 = 0.37 \text{ kN}$$

For equilibrium, $\Sigma F_y = 0$

$\therefore$ $$R_{AV} + R_B \cos 30° - 4 - 5 \sin 45° - 5 = 0$$

or $$R_{AV} + 7.81 \times 0.866 - 4 - 3.535 - 5 = 0$$

or $$R_{AV} + 6.763 - 12.535 = 0$$

or $$R_{AV} = 12.535 - 6.763 = 5.77 \text{ kN}$$

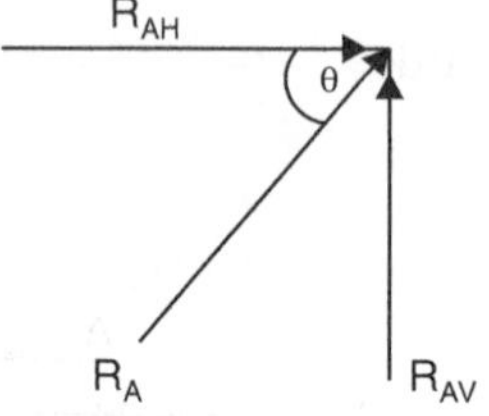

Fig. 2.51(b)

$\therefore$ Reaction at *A*, $$R_A = \sqrt{R_{AH}^2 + R_{AV}^2} = \sqrt{0.37^2 + 5.77^2}$$

$$= \mathbf{5.78\ kN.} \quad \textbf{Ans.}$$

The angle made by R_A with *x*-direction is given by

$$\tan \theta = \frac{R_{AV}}{R_{AH}} = \frac{5.77}{0.37} = 15.59$$

$\therefore$ $$\theta = \tan^{-1} 15.59 = \mathbf{86.33°.} \quad \textbf{Ans.}$$

Problem 2.28. *Determine the reactions at the hinged support A and the roller support B as shown in Fig. 2.52(a).*

Sol. Given :

The support at *A* is hinged whereas the support at *B* is placed on the roller. Hence the reaction at the roller support will be perpendicular to the inclined surface.

---

*The moments of horizontal components of 5 kN at *D* and of reaction R_B will be zero about the point *A*.

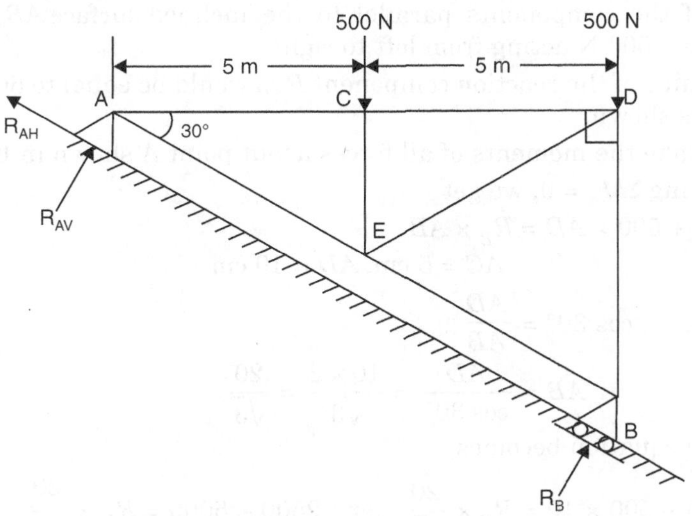

Fig. 2.52(a)

The reaction at the hinged support A (*i.e.*, reaction R_A) will be inclined at some angle to the inclined surface AB.

Let R_{AV} = Component of reaction R_A normal to inclined surface AB

R_{AH} = Component of reaction R_A along the inclined surface AB.

The given vertical force of 500 N at C is resolved parallel and perpendicular to the inclined surface AB. The value of its component parallel to the inclined surface is equal to 500 sin 30° or 250 N and its value perpendicular to the inclined surface is equal to 500 cos 30° or $250 \times \sqrt{3}$ N. The directions of these components are shown in Fig. 2.52(*b*).

Similarly, the vertical force of 500 N at D is resolved as shown in Fig. 2.52(*b*).

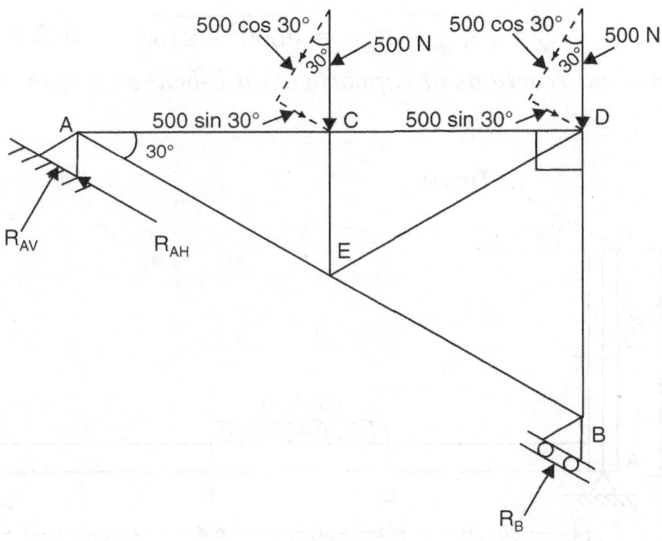

Fig. 2.52(b)

Now sum of the components parallel to the inclined surface AB = 500 sin 30° + 500 sin 30° = 250 + 250 = 500 N acting from left to right.

Hence the value of the reaction component R_{AH} would be equal to 500 N and should act from right to left as shown.

To find R_B, take the moments of all forces about point A shown in Fig. 2.52 (*a*).

Hence applying $\Sigma M_A = 0$, we get

$$500 \times AC + 500 \times AD = R_B \times AB$$

But $AC = 5 \text{ cm}, AD = 10 \text{ cm}$

From ΔABD, $\cos 30° = \dfrac{AD}{AB}$.

Hence $AB = \dfrac{AD}{\cos 30°} = \dfrac{10 \times 2}{\sqrt{3}} = \dfrac{20}{\sqrt{3}}$

Hence above equation becomes,

$$500 \times 5 + 500 \times 10 = R_B \times \frac{20}{\sqrt{3}} \quad \text{or} \quad 2500 + 5000 = R_B \times \frac{20}{\sqrt{3}}$$

$$\therefore \quad R_B = \frac{(2500 + 5000) \times \sqrt{3}}{20} = \mathbf{649.5\ N.}\quad \textbf{Ans.}$$

Now from Fig. 2.52 (*b*), equating all the forces perpendicular to the inclined surface AB, we get

$$R_{AV} + R_B = 500 \times \cos 30° + 500 \cos 30°$$

or $$R_{AV} + 649.5 = 1000 \cos 30° = 1000 \times \frac{\sqrt{3}}{2}$$

$$\therefore \quad R_{AV} = 1000 \times \frac{\sqrt{3}}{2} - 649.5 = 216.5 \text{ N}$$

But $$R_{AH} = 500 \text{ N}$$

$$\therefore \quad R_A = \sqrt{R_{AH}^2 + R_{AV}^2} = \sqrt{500^2 + 216.5^2} = \mathbf{545\ N.}\quad \textbf{Ans.}$$

Problem 2.29. *Find reactions at supports of an L-bent shown in Fig. 2.53.*

Sol. Given :

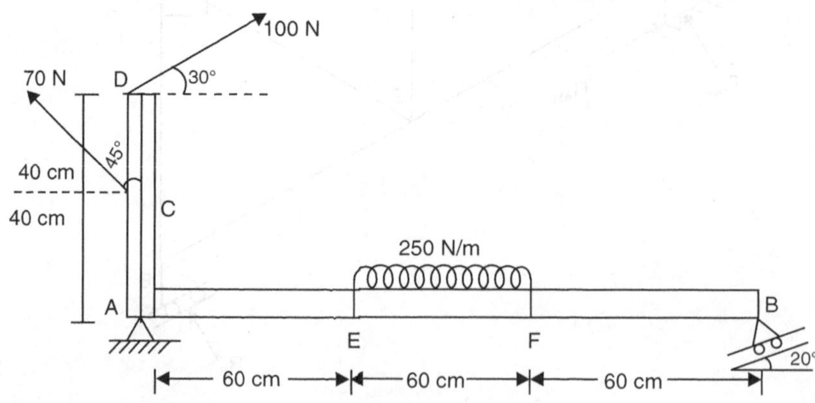

Fig. 2.53

Force at point D = 100 N at an angle of 30° with horizontal
Force at point C = 70 N at an angle of 45° with vertical
Load on EF = 250 N/m = 250 × Length EF in metre
= 250 × 0.6 = 150 N

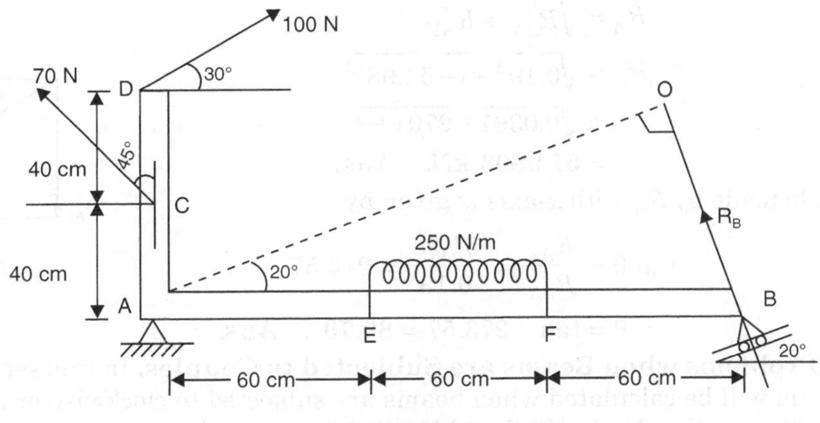

Fig. 2.53(a)

The load on EF will be acting at the middle point of EF *i.e.*, at a distance of 0.6/2 = 0.3 m from E or at a distance of 0.6 + 0.3 = 0.9 m from A. The point B is placed on roller at an angle of 20° with the horizontal. Hence reaction at B will be normal to the surface of the roller.

The perpendicular distance from A on the line of action of $R_B = AO = AB \cos 20° = 180 \times \cos 20°$ cm = 1.8 cos 20° m as shown in Fig. 2.53(*a*). For equilibrium of the beam, the moments of all forces about any point should be zero. Taking moments of all forces about point A, we get

[Horizontal component at D] × AD – [Horizontal component at C] × AC + Load on EF × 90 – R_B × AO = 0

or $$(100 \cos 30) \times 80 - (70 \times \sin 45) \times 40 + 150 \times 90 - R_B \times 180 \cos 20 = 0$$

(**Note.** The vertical components at D and C, pass through the point A. Hence moments of these vertical components about A are zero).

or $$6928 - 1979.6 + 13500 - 169.14\, R_B = 0$$

or $$18448.4 = 169.14\, R_B$$

∴ $$R_B = \frac{18448.4}{169.14} = 109.07$$

Let R_A = Reaction at the point A

The reaction at A can be resolved in two components *i.e.*, R_{AX} and R_{AY}.

Fig. 2.53(b)

For equilibrium, $\Sigma F_x = 0$

or $$R_{AX} + 100 \cos 30° - 70 \sin 45° - R_B \sin 20° = 0$$

or $$R_{AX} = R_B \sin 20° + 70 \sin 45° - 100 \cos 30°$$
$$= 109.07 \times 0.342 + 70 \times 0.707 - 100 \times 0.866$$
$$= 37.3 + 49.49 - 86.6 = 0.19 \text{ N}$$

For equilibrium, $\Sigma F_y = 0$

or $$R_{AX} + 100 \sin 30° + 70 \cos 45° + R_B \cos 20° = 150$$

or $$R_{AY} = 150 - 100 \sin 30° - 70 \cos 45° - R_B \cos 20°$$
$$= 150 - 50 - 49.49 - 109.07 \times 0.9396 = -51.98 \text{ N}$$

(–ve sign means, R_{Ay} will be acting vertically downward)

Refer to Fig. 2.53(*c*)

$\therefore$ $$R_A = \sqrt{R_{AX}^2 + R_{AY}^2}$$

or $$R_A = \sqrt{0.19^2 + (-51.98)^2}$$
$$= \sqrt{0.0361 + 2701.92}$$
$$= \mathbf{51.9803 \text{ kN.}} \quad \textbf{Ans.}$$

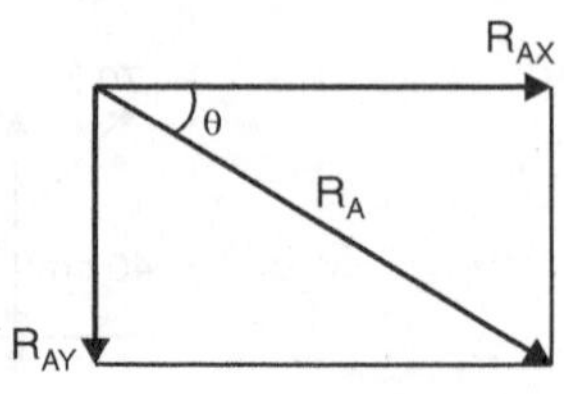

Fig. 2.53(c)

The angle made by R_A with *x*-axis is given by

$$\tan \theta = \frac{R_{AY}}{R_{AX}} = \frac{51.98}{0.19} = 273.57$$

$\therefore$ $$\theta = \tan^{-1} 273.57 = \mathbf{89.79°.} \quad \textbf{Ans.}$$

2.11.4. Problems when Beams are Subjected to Couples. In this section, the reactions of the beam will be calculated when beams are subjected to clockwise or anti-clockwise couple along with the other loads. While taking the moments about any point, the magnitude and sense of the couple is taken into consideration. But when the total load on the beam is calculated the magnitude and sense of the couple is not considered.

Problem 2.30. *A simply supported beam AB of 7 m span is subjected to : (i) 4 kN m clockwise couple at 2 m from A, (ii) 8 kN m anti-clockwise couple at 5 m from A, and (iii) a triangular load with zero intensity at 2 m from A increasing to 4 kN per m at a point 5 m from A. Determine reactions at A and B.*

Sol. Given :

Span of beam = 7 m

Couple at *C* (*i.e.*, at 2 m from A) = 4 kN m (clockwise)

Couple at *D* (*i.e.*, at 5 m from *A*) = 8 kN m (anti-clockwise)

Triangular load from *C* to *D* with :

Vertical load at *C* = 0

Vertical load at *D* = 4 kN/m

$\therefore$ Total load on beam = Area of triangle $CDE = \frac{CD \times DE}{2} = \frac{3 \times 4}{2} = 6$ kN

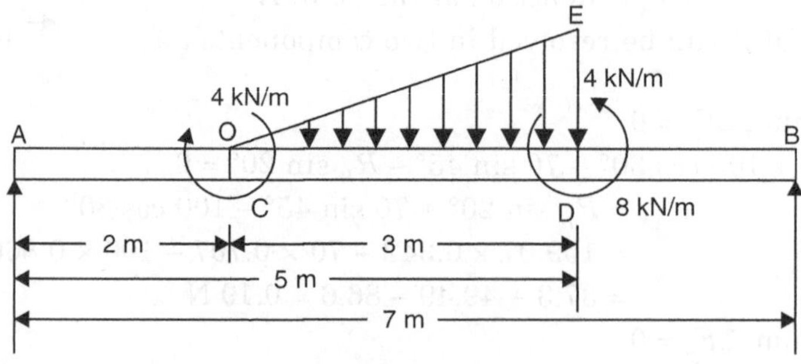

Fig. 2.54

This load will be acting at the C.G. of the ΔCDE *i.e.*, at a distance of $\frac{2}{3} \times CD = \frac{2}{3} \times 3$ = 2 m from C or 2 + 2 + 4 m from end A.

Let R_A = Reaction at A.

R_B = Reaction at B.

Taking the moments of all forces about point A and equating the resultant moment to zero (*i.e.*, $\Sigma M_A = 0$ and considering clockwise moment positive and anti-clockwise moment negative), we get

$$-R_B \times 7 + 4^* - 8^{**} + (\text{Total load on beam}) \times \text{Distance of total load from } A) = 0$$

or $$-7R_B + 4 - 8 + 6 \times 4 = 0$$

$$-7R_B + 4 - 8 + 24 = 0$$

or $$20 = 7R_B$$

or $$R_B = \frac{20}{7} \text{ kN. Ans.}$$

Also for the equilibrium of the beam $\Sigma F_y = 0$

or $$R_A + R_B = \text{Total load on the beam} = 6 \text{ kN}$$

$\therefore$ $$R_A = 6 - R_B = 6 - \frac{20}{7} = \frac{22}{7} \text{ kN. Ans.}$$

2.12. SINGLE EQUIVALENT FORCE

When a number of forces are acting on a rigid body, then these forces can be represented by a single force which has the same effect on the rigid body as that of all the forces acting together, then this single force is known as **'Single Equivalent Force'.** This single equivalent force is also known as resultant of several forces. Hence a single force which can replace a number of forces acting on a rigid body, without causing any change in the external effects on the body, is known as single equivalent force (or resultant force).

HIGHLIGHTS

1. The principle of equilibrium states that a stationary body will be in equilibrium if the algebraic sum of all the forces is zero and also the algebraic sum of moments of all the external forces is zero.
2. The conditions of equilibrium are written mathematically as $\Sigma F_x = 0$, $\Sigma F_y = 0$, and $\Sigma M = 0$. The sign Σ is known as sigma and this sign represents the algebraic sum.
3. When a body is subjected to two forces, the body will be in equilibrium if the two forces are collinear, equal and opposite.
4. Two equal and opposite parallel forces produce a couple whose moment is equal to either force multiplied by their perpendicular distance.
5. If three concurrent forces are acting on a body and the body is in equilibrium, then the resultant of two forces should be equal and opposite to the third force.
6. Free body diagram of a body is a diagram in which the body is completely isolated from its support and the supports are replaced by the reactions which these supports exert on the body.

---

*The couple at C is 4 kN m clockwise. Hence its sense is positive.

**The couple at D is 8 kN m anti-clockwise. Hence its sense is negative. The couple is also moment.

7. The reaction at the knife edge support will be normal to the surface of the beam.
8. The reaction in case of roller support will be normal to the surface of roller base.
9. The reaction at the hinged end (or pinned end) will be either vertical or inclined depending upon the type of loading. If the load is vertical, then reaction will be vertical. But if the load is inclined, then the reaction will also be inclined.
10. For a smooth surface, the reaction is always normal to the support.
11. A load, acting at a point on a beam, is known as point load or concentrated load.
12. If each unit length of the beam carries same intensity of load, then that type of load is known as uniformly distributed load which is written as U.D.L.
13. The reactions of a beam can be determined by analytical method and graphical method.
14. The reactions by analytical method are obtained by using equations of equilibrium, *i.e.*, $\Sigma F_x = 0$, $\Sigma F_y = 0$ and $\Sigma M = 0$.
15. The reactions by graphical method are obtained by drawing a space diagram and a vector diagram.
16. If a beam is loaded with inclined loads, then the inclined loads are resolved normal to the beam and along the beam. Now the equations of equilibrium are used for finding reactions.
17. The rection on a roller support is at right angles to the roller base.

EXERCISE 2

(A) Theoretical Questions

1. Define and explain the terms. Principle of equilibrium, force law of equilibrium and moment law of equilibrium.
2. A number of forces are acting on a body. What are conditions of equilibrium, so that the body is in equilibrium?
3. Two forces are acting on a body and the body is in equilibrium. What conditions should be fulfilled by these two forces?
4. How will you prove that a body will not be in equilibrium when the body is subjected to two forces which are equal and opposite but are parallel?
5. Explain the statement "Two equal and opposite parallel forces produces a couple."
6. (*a*) What conditions must be fulfilled by a set of three parallel forces which are acting on a body and body is in equilibrium?

   (*b*) State the graphical conditions that must be satisfied for the equilibrium of a system of coplanar forces.

   (*c*) What are the conditions that must be satisfied in order that a body may have plane motion?

   (*d*) Discuss the various laws governing the equilibrium of coplanar forces.
7. Three concurrent forces are acting on a body which is in equilibrium, then the resultant of the two forces should be equal and opposite to the third force. Prove this statement.
8. State and explain the Lami's theorem.
9. What do you mean by action and reaction? Give examples.
10. Explain and define the term : 'Free-body Diagram'. Draw the free-body diagram of a ball of weight W, placed on a horizontal surface.
11. State the conditions of equilibrium of a system of forces acting on a body as applicable to (*i*) graphical method, and (*ii*) analytical method.
12. Explain the term 'support reactions.' What are the different types of support?
13. What is the difference between a roller support and a hinged support?

14. What are the important types of loading on a beam? Differentiate between uniformly distributed load and uniformly varying load on a beam.

15. Name the different methods of finding the reactions at the two supports of a beam.

16. A beam AB of length L is simply supported at the ends A and B. It carries two point loads W_1 and W_2 at a distance L_1 and L_2 from the end A respectively. How will you find the reactions R_A and R_B by analytical method?

17. Describe in details the different steps involved in finding the reactions of a beam by graphical method.

18. Define and explain an overhanging beam.

19. What is the main advantage of roller support in case of the steel trusses of the bridges?

(B) Numerical Problems

1. Three forces F_1, F_2 and F_3 are acting on a body as shown in Fig. 2.55 and the body is in equilibrium. If the magnitude of force F_3 is 250 N, find the magnitudes of forces F_1 and F_2. [**Ans.** F_1 = 125 N and F_2 = 215.6 N]

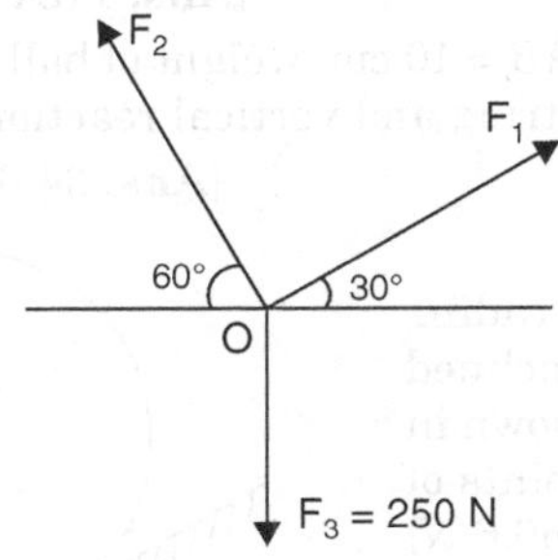

Fig. 2.55

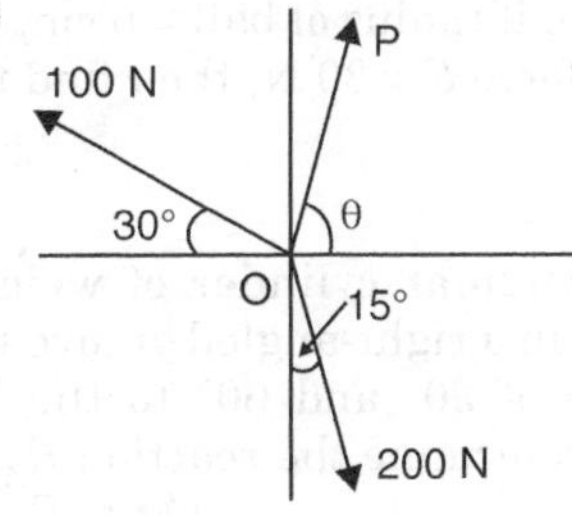

Fig. 2.56

2. Three forces of magnitudes P, 100 N and 200 N are acting at a point O as shown in Fig. 2.56. Determine the magnitude and direction of the force P. [**Ans.** P = 147 N and θ = 76.8°]

3. Three parallel forces F_1, F_2 and F_3 are acting on a body as shown in Fig. 2.57 and the body is in equilibrium. If force F_1 = 300 N and F_3 = 1000 N and the distance between F_1 and F_2 = 2.0 m, then determine the magnitude of force F_2 and distance of F_3 from force F_2. [**Ans.** 1300 N, 0.6 m]

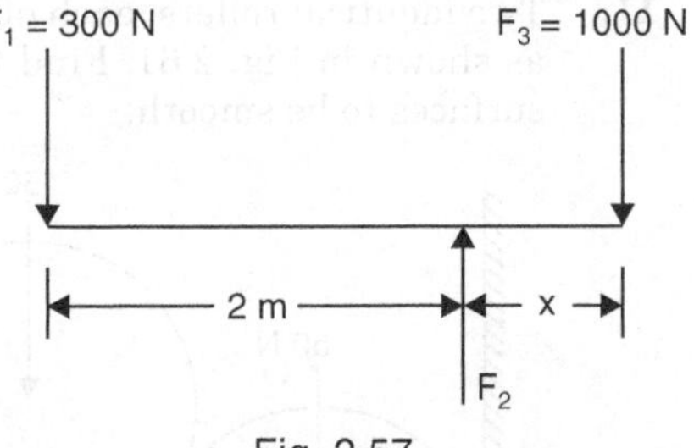

Fig. 2.57

4. Three forces of magnitude 40 kN, 15 kN and 20 kN are acting at a point O. The angles made by 40 kN, 15 kN and 20 kN forces with x-axis are 60°, 120° and 240° respectively. Determine the magnitude and direction of the resultant force. [**Ans.** 30.41 kN and 85.28° with x-axis]

5. A lamp weighing 10 N is suspended from the ceiling by a chain. It is pulled aside by a horizontal cord until the chain makes an angle of 60° with the ceiling. Find the tensions in the chain and the cord by applying Lami's theorem and also by graphical method. [**Ans.** 11.54 N and 5.77 N]

6. Draw the free-body diagram of a ball of weight W supported by a string AB and resting on a smooth horizontal surface at C when a horizontal force is applied to the ball as shown in Fig. 2.58.

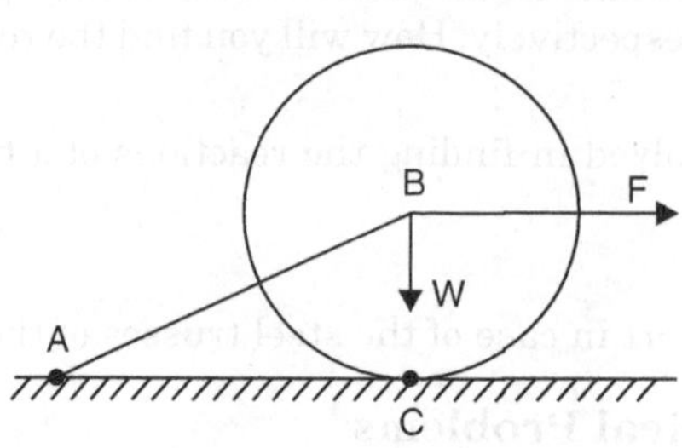

Fig. 2.58

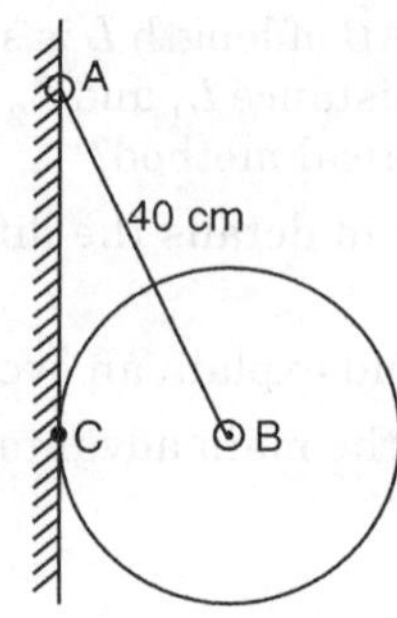

Fig. 2.59

7. A circular roller of weight 1000 N and radius 20 cm hangs by a tie rod AB = 40 cm and rests against a smooth vertical wall at C as shown in Fig. 2.59. Determine the tension in the tie rod and reaction R_C at point C. [**Ans.** 1154.7 N and 577.3 N]

8. In problem 6 if radius of ball = 5 cm, length of string AB = 10 cm, weight of ball W = 40 N and the horizontal force F = 30 N, then find the tension the string and vertical reaction R_C at point C. [**Ans.** 34.64 N and 57.32 N]

9. A smooth circular cylinder of weight 1000 N and radius 10 cm rests in a right-angled groove whose sides are inclined at an angle of 30° and 60° to the horizontal as shown in Fig. 2.60. Determine the reaction R_A and R_C at the points of contact. [**Ans.** R_A = 500 N, R_C = 866.6 N]

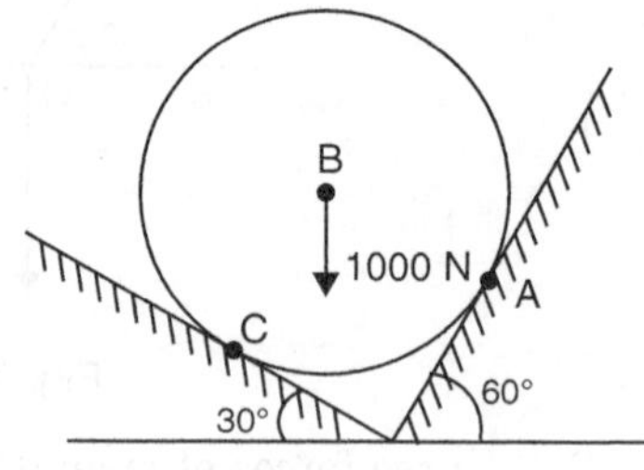

Fig. 2.60

10. If in the above problem, the sides of the groove makes an angle of 45° with the horizontal, then find the reactions R_A and R_C. [**Ans.** $R_A = R_C$ = 707 N]

11. Two identical rollers, each of weight 50 N, are supported by an inclined plane and a vertical wall as shown in Fig. 2.61. Find the reactions at the points of supports A, B and C. Assume all the surfaces to be smooth. [**Ans.** R_A = 43.3 N, R_B = 72 N, R_C = 57.5 N]

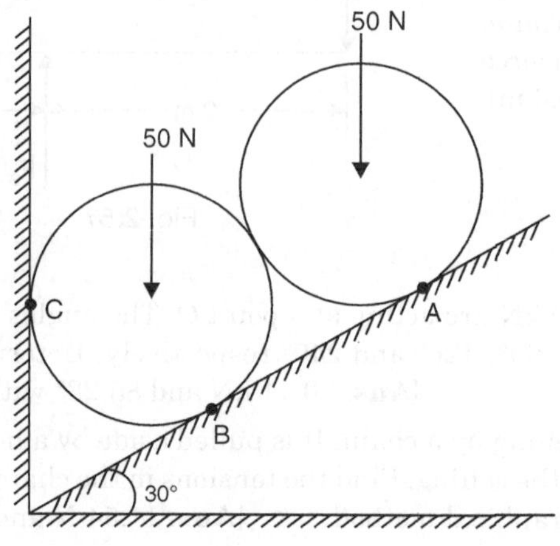

Fig. 2.61

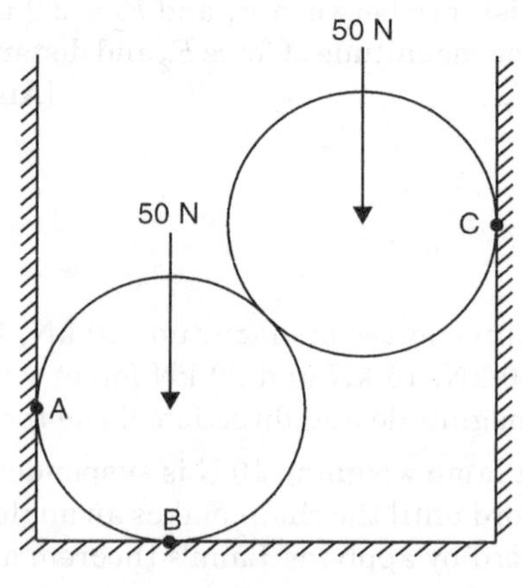

Fig. 2.62

12. Two spheres, each of weight 50 N and of radius 10 cm rest in a horizontal channel of width 36 cm as shown in Fig. 2.62. Find the reactions on the points of contact *A*, *B* and *C*.

[**Ans.** $R_A = R_C = 66.67$ N, $R_B = 100$ N]

13. A simply supported beam of length 8 m carries point loads of 4 kN and 6 kN at a distance of 2 m and 4 m from the left end. Find the reactions at both ends analytically and graphically.

[**Ans.** 6 kN, 4 kN]

14. A simply supported beam of length 8 m carries a uniformly distributed load of 10 kN/m for a distance of 4 m, starting from a point which is at a distance of 1 m from the left end. Calculate the reactions at both ends.

[**Ans.** 25 kN, 15 kN]

15. A beam 6 m long is simply supported at the ends and carries a uniformly distributed load of 1.5 kN/m and three concentrated loads 1 kN, 2 kN and 3 kN acting respectively at a distance of 1.5 m, 3 m and 4.5 m from the left end. Calculate the reactions at both ends. [**Ans.** 7 kN, 8 kN]

16. A simply supported beam of span 10 m carries a uniformly varying load from zero at the left end to 1200 N/m at the right end. Calculate the reactions at both ends of the beam.

[**Ans.** 2000 N and 4000 N]

17. A simply supported beam *AB* is subjected to a distributed load increasing from 1500 N/m to 4500 N/m from end *A* to end *B* respectively. The span *AB* = 6 m. Determine the reactions at the supports.

[**Ans.** $R_A = 7500$ N, $R_B = 10500$ N]

18. An overhanging beam carries the loads as shown in Fig. 2.63. Calculate the reactions at both ends.

[**Ans.** $R_A = 1$ kN, $R_B = 6$ kN]

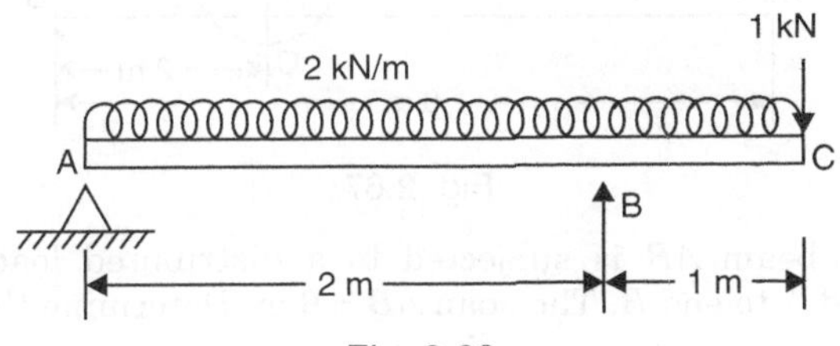

Fig. 2.63

19. An overhanging beam carries the loads as shown in Fig. 2.64. Calculate the reactions at both ends.

[**Ans.** $R_A = 10$ kN, $R_B = 11$ kN]

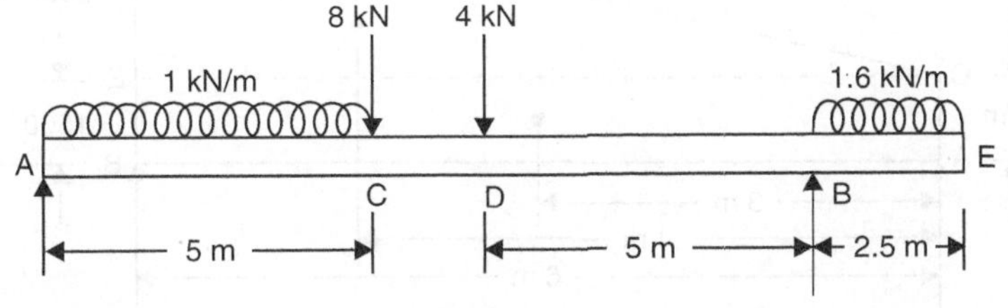

Fig. 2.64

20. A beam is loaded as shown in Fig. 2.65. Determine the reactions at both ends.

[**Ans.** $R_{AV} = 2.875$ kN, $R_{AH} = 5.196$ kN →, $R_B = 7.125$ kN]

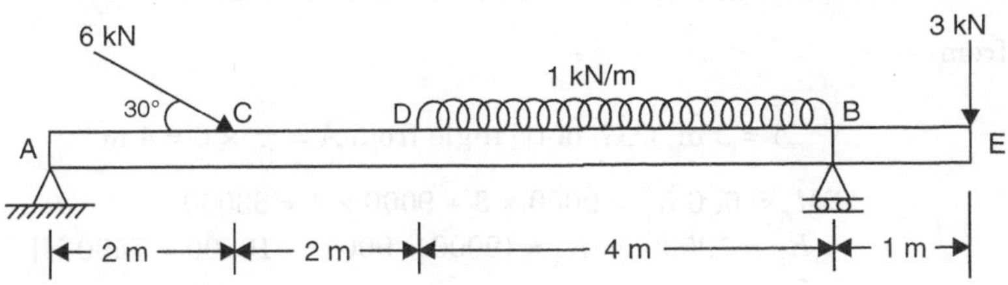

Fig. 2.65

21. A beam *AB* of span 6 m is hinged at *A* and supported on rollers at end *B* and carries load as shown in Fig. 2.66. Determine the reactions at *A* and *B*.

[**Ans.** R_{AV} = 5.87 kN, R_{AH} = 3.222 kN →, R_B = 7.3 kN]

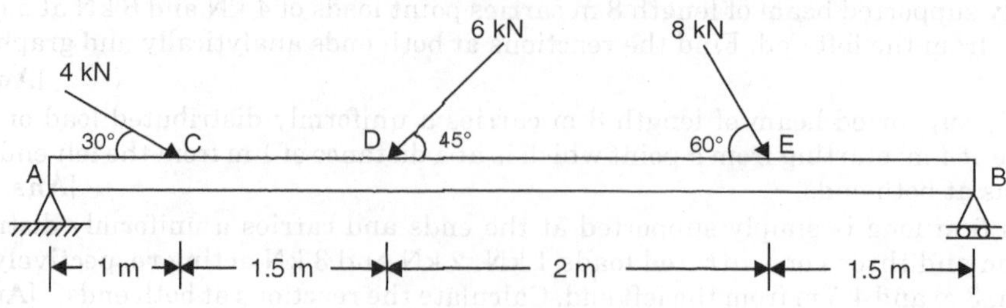

Fig. 2.66

22. A beam *AB* of span 8 m is subjected to the uniformly distributed load of 1 kN/m over the entire length and the moment 32 kN/m at *C* as shown in Fig. 2.67. Determine the reactions at the both ends.

[**Ans.** R_A = 0, R_B = 8 kN]

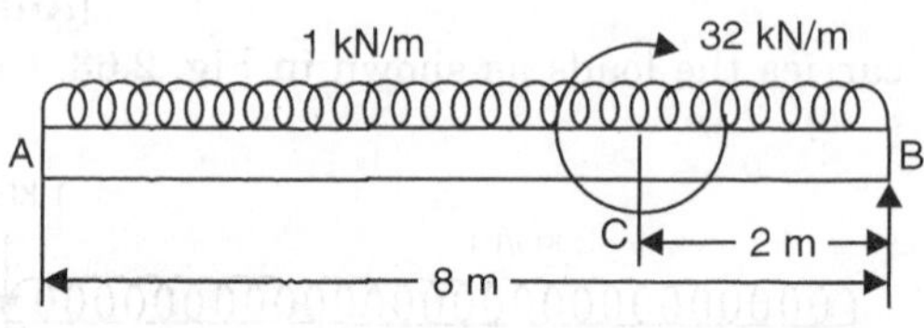

Fig. 2.67

23. A simply supported beam *AB* is subjected to a distributed load increasing from 1500 N/m to 4500 N/m from end *A* to end *B*. The span *AB* = 6 m. Determine the reactions at the supports.

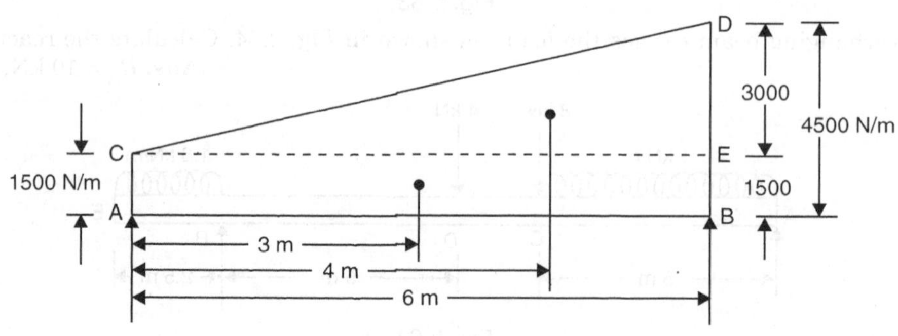

Fig. 2.68

[**Hint.** Area of rectangle = 1500 × 6 = 9000 N, Area of triangle = $\frac{3000 \times 6}{2}$ = 9000 N. C.G. of rectangle from

$$A = 3 \text{ m, C.G. of triangle from } A = \frac{2}{3} \times 6 = 4 \text{ m.}$$

$$\Sigma M_A = 0,\ 6\,R_B = 9000 \times 3 + 9000 \times 4 = 63000$$

$$\therefore \quad R_B = 10500 \text{ N},\ R_A = (9000 + 9000) - 10500 = 7500 \text{ N}]$$

3

Properties of Surfaces and Solids

3.1. CENTROIDS AND CENTRE OF MASS

The point at which the total area of a plane figure (like rectangle, square, triangle, quadrilateral, circle etc.) is assumed to be concentrated, is known as the centroid of that area. The centroid is also represented by C.G. or simply G.

Centre of mass is the point at which the total mass of a body, is assumed to be concentrated. A body is having only one centre of mass for all positions of the body.

3.1.1. Centroid of Simple Plane Figures. (*i*) The centre of gravity (C.G.) of a uniform rod lies at its middle point.

(*ii*) The centre of gravity of a triangle lies at the point where the three medians* of the triangle meet.

(*iii*) The centre of gravity of a rectangle or of a parallelogram is at the point, where its diagonals meet each other. It is also the point of intersection of the lines joining the middle points of the opposite sides.

(*iv*) The centre of gravity of a circle is at its centre.

3.1.2. Centroid of Areas of Plane Figures by the Method of Moments. Fig. 3.1 shows a plane figure of total area A whose centre of gravity is to be determined. Let this area A is composed of a number of small areas a_1, a_2, a_3, a_4, ... etc.

$\therefore \quad A = a_1 + a_2 + a_3 + a_4 + ...$

Let x_1 = The distance of the C.G. of the area a_1 from axis OY

x_2 = The distance of the C.G. of the area a_2 from axis OY

x_3 = The distance of the C.G. of the area a_3 from axis OY

x_4 = The distance of the C.G. of the area a_4 from axis OY and so on.

Fig. 3.1

The moment of small area a_1 about the axis OY

= area a_1 × distance of C.G. of area a_1 from axis OY

$= a_1 \times x_1$

The above equation is known as first moment of area about the axis OY.

This first moment of area is used to determine the centroid of the area.

The moments of all small areas about the axis OY

$$= a_1x_1 + a_2x_2 + a_3x_3 + a_4x_4 + ... \qquad ...(i)$$

*The line connecting the vertex and the middle point of the opposite side of a triangle is known as median of the triangle.

Let G is the centre of gravity of the total area A whose distance from the axis OY is $\bar{x}$.

Then moment of total area about $OY = A\bar{x}$ ...(ii)

The moments of all small areas about the axis OY must be equal to the moment of total area about the same axis. Hence equating equations (i) and (ii), we get

$$a_1x_1 + a_2x_2 + a_3x_3 + a_4x_4 + \ldots = A\bar{x}$$

or
$$\bar{x} = \frac{a_1x_1 + a_2x_2 + a_3x_3 + a_4x_4 + \ldots}{A} \quad \ldots(3.1)$$

where $A = a_1 + a_2 + a_3 + a_4 \ldots$

If we take the moments of the small areas about the axis OX and also the moment of total area about the axis OX, we will get

$$\bar{y} = \frac{a_1y_1 + a_2y_2 + a_3y_3 + a_4y_4 + \ldots}{A} \quad \ldots(3.2)$$

where $\bar{y}$ = The distance of G from axis OX

y_1 = The distance of C.G. of the area a_1 from axis OX

y_2, y_3, y_4 = The distance of C.G. of area a_2, a_3, a_4 from axis OX respectively.

3.1.3. Centroid of Areas of Plane Figures by Integration Method. The equations (3.1) and (3.2) can be written as

$$\bar{x} = \frac{\Sigma\, a_ix_i}{\Sigma a_i} \quad \text{and} \quad \bar{y} = \frac{\Sigma\, a_iy_i}{\Sigma a_i}$$

where $i = 1, 2, 3, 4, \ldots..$

x_i = Distance of C.G. of area a_i from axis OY and

y_i = Distance of C.G. of area a_i from axis OX.

The value of i depends upon the number of small areas. If the small areas are large in number (mathematically speaking infinite in number), then the summations in the above equations can be replaced by integration. Let the small areas are represented by dA instead of 'a', then the above equations are written as :

$$\bar{x} = \frac{\int x^* \, dA}{\int dA} \quad \ldots(3.2A)$$

and
$$\bar{y} = \frac{\int y^* \, dA}{\int dA} \quad \ldots(3.2B)$$

where $\int x^* \, dA = \Sigma x_i a_i$

$\int dA = \Sigma a_i$

$\int y^* dA = \Sigma y_i a_i$

Also x^* = Distance of C.G. of area dA from axis OY,

y^* = Distance of C.G. of area dA from axis OX.

x^*dA = First moment of area dA about the axis OY, and

y^*dA = First moment of area dA about the axis OX.

3.2. CENTROIDS OF LINES AND AREAS [RECTANGULAR, CIRCULAR, TRIANGULAR AREAS BY INTEGRATION]

3.2.1. Centroid of a Line. The centroid of a line which may be straight or curved, is obtained by dividing the given line, into a large number of small lengths as shown in Fig. 3.1 (a).

The centroid is obtained by replacing dA by dL in equations (3.2 A) and (3.2 B).

Then these equations become

$$\bar{x} = \frac{\int x^* dL}{\int dL} \qquad ...(3.2C)$$

and

$$\bar{y} = \frac{\int y^* dL}{\int dL} \qquad ...(3.2D)$$

where x^* = Distance of C.G. of length dL from y-axis, and

y^* = Distance of C.G. of length dL from x-axis.

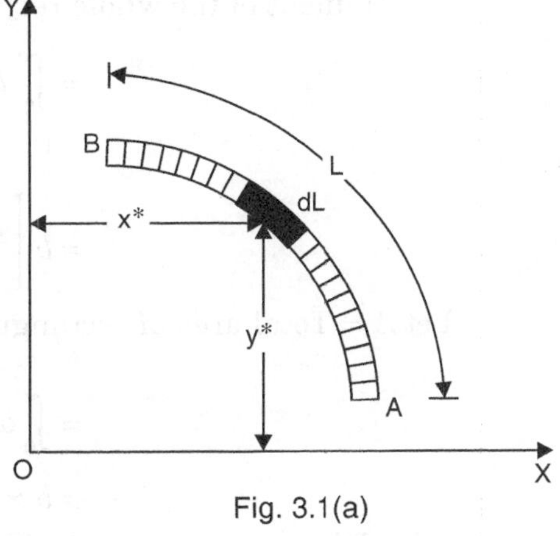

Fig. 3.1(a)

If the lines are straight, then the above equations are written as :

$$\bar{x} = \frac{L_1x_1 + L_2x_2 + L_3x_3 + ...}{L_1 + L_2 + L_3 + ...} \qquad ...(3.2E)$$

and

$$\bar{y} = \frac{L_1y_1 + L_2y_2 + L_3y_3 + ...}{L_1 + L_2 + L_3 + ...} \qquad ...(3.2F)$$

Important Points. (*i*) The axis, about which moments of areas are taken, is known as axis of reference. In the above article, axis OX and OY are called axes of reference.

(*ii*) The axis of reference of plane figures, is generally taken as the lowest line of the figure for determining $\bar{y}$, and left line of the figure for calculating $\bar{x}$.

(*iii*) If the given section is symmetrical about X axis or Y axis, then the C.G. of the section will lie on the axis of symmetry.

3.2.2. Centroid of Sections—Rectangle, Circle, Triangle by Integration

The first moment of area as defined in Art. 3.1 will be used to determine the centroids of the following sections by the method of integration :

1. Rectangular section, 2. Circular section, and
3. Triangular section.

3.2.3. Centroid of a Rectangular Section by Integration. Fig 3.2 shows a rectangular section $ABCD$ having width = b and depth = d. Consider a rectangular elementary strip of thickness 'dy' at a distance y from the axis OX.

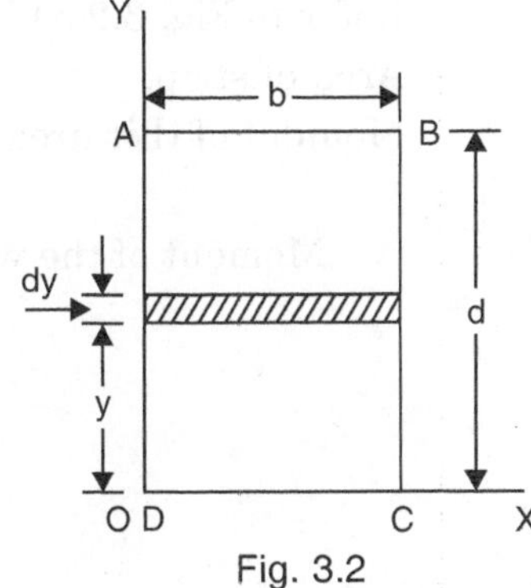

Fig. 3.2

Let $\qquad dA$ = Area of strip = $b \,.\, dy$

Moment* of the area dA about axis OX

$$= dA \times y = (b \,.\, dy) \times y \qquad (\because dA = b \,.\, dy)$$

$$= by \times dy$$

The moment of the whole area about axis OX (or first moment of the whole area about axis OX) will be obtained by integrating the above equation between the limits 0 to d.

*This moment of area dA about axis OX is also known as first moment of area dA about axis OX.

$\therefore$ Moment of the whole rectangular section about axis OX

$$= \int_0^d by \times dy = b \int_0^d y \, dy$$

($\because$ b is constant and can be taken outside the integral sign)

$$= b\left[\frac{y^2}{2}\right]_0^d = \frac{bd^2}{2} \qquad ...(i)$$

Let A = Total area of rectangular section

$$= \int_0^d dA = \int_0^d b \, . \, dy = b\left[y\right]_0^d \qquad [\because \; dA = b \, . \, dy]$$

$$= b \times d$$

$\bar{y}$ = Distance of the centroid of the rectangular section from axis OX.

$\therefore$ Moment of total area of rectangular section about axis OX

$$= A \times \bar{y} = (b \times d) \times \bar{y} \qquad ...(ii)$$

Equating the equations (i) and (ii), we get

$$(b \times d) \times \bar{y} = \frac{bd^2}{2}$$

$$\therefore \qquad \bar{y} = \frac{bd^2}{2} \times \frac{1}{bd} = \frac{d}{2}$$

Similarly, the distance of the centroid of the rectangular section from the axis OY is given by

$$\bar{x} = \frac{b}{2}$$

Refer to Fig. 3.2(a)

Area of strip, $\qquad dA = d \, . \, dx$

Moment of this area dA about axis OY

$$= dA \times x$$

$\therefore$ Moment of the whole area about axis OY

$$= \int_0^b dA \times x = \int_0^b (d \, . \, dx)x \qquad [\because \; dA = d \, . \, dx]$$

$$= d \int_0^b x \, dx = d\left[\frac{x^2}{2}\right]_0^b = \frac{d \times b^2}{2} \qquad ...(iii)$$

Fig. 3.2(a)

Let $\bar{x}$ = Distance of centroid of whole area from axis OY.

A = Total area = $b \times d$

$\therefore$ Moment of total area A about axis OY

$$= A \times \bar{x} \qquad ...(iv)$$

Equating equations (*iii*) and (*iv*), we get

$$A \times \bar{x} = \frac{d \times b^2}{2}$$

or
$$\bar{x} = \frac{d \times b^2}{2} \times \frac{1}{A} = \frac{d \times b^2}{2} \times \frac{1}{b \times d} = \frac{b}{2}.$$

3.2.4. Centroid of Circular Section by Integration. Fig. 3.3 shows a circular section of radius R with O as centre. The equation of the circle is

$$x^2 + y^2 = R^2$$

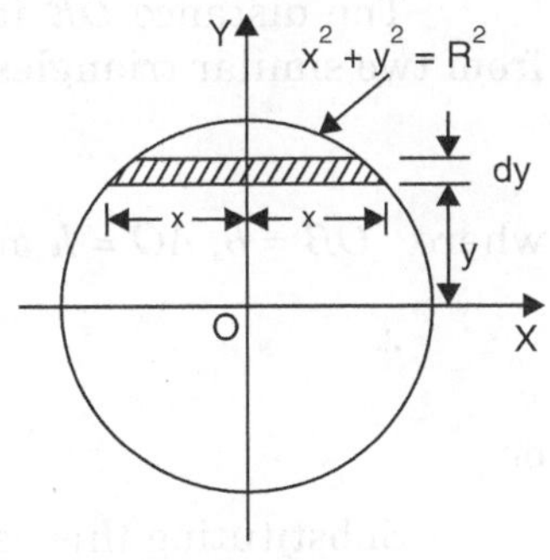

Fig. 3.3

Consider a rectangular elementary strip of thickness 'dy' at a distance of y from the axis OX.

The area of strip, $dA = 2x \,.\, dy$.

Moment of this area dA about x-axis

$$= dA \,.\, y$$
$$= (2x \,.\, dy) \,.\, y \quad (\because \quad dA = 2\,x \,.\, dy)$$
$$= 2xy \,.\, dy \qquad ...(i)$$

But the equation of the circle is

$$x^2 + y^2 = R^2$$

or
$$x = \sqrt{R^2 - y^2}$$

Substituting the above value of x in equation (*i*), we get

Moment of area dA about x-axis

$$= 2\sqrt{R^2 - y^2} \,.\, y \,.\, dy$$

Moment of total area A about x-axis will be obtained by integrating the above equation from $-R$ to R.

∴ Moment of area A about x-axis

$$= \int_{-R}^{+R} 2\sqrt{R^2 - y^2} \,.\, y \,.\, dy \qquad (\because \quad y \text{ varies from } -R \text{ to } +R)$$

$$= -\int_{-R}^{+R} \sqrt{R^2 - y^2}\,(-2y) \,.\, dy = -\left[\frac{(R^2 - y^2)^{3/2}}{3/2}\right]_{-R}^{R}$$

$$= -\frac{2}{3}\,[(R^2 - R^2)^{3/2} - \{R^2 - (-R)^2\}^{3/2}]$$

$$= \frac{-2}{3}\,[0 - 0] = 0 \qquad ...(ii)$$

Also the moment of total area A about x-axis $= A \times \bar{y}$

$$...(iii)$$

where $\bar{y}$ = Distance of centroid of total area A from x-axis.

Equating the two values given by equations (*ii*) and (*iii*), we get

$$A \times \bar{y} = 0 \quad \text{or} \quad \bar{y} = 0$$

This means that the centroid of the circle is at the centre of the circle.

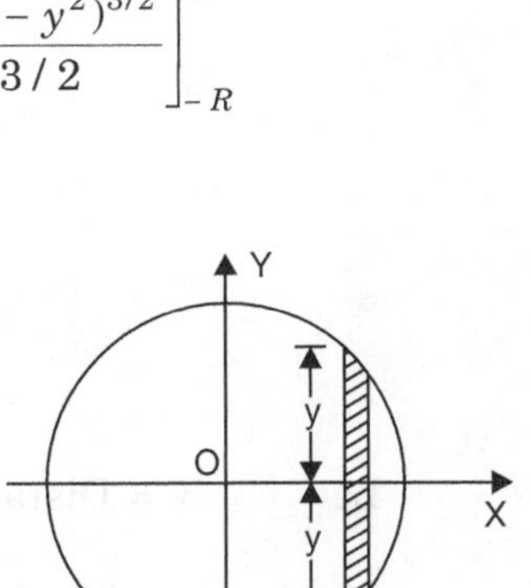

Fig. 3.3(a)

Similarly, taking a strip of thickness dx parallel to y-axis at a distance x from y-axis as shown in Fig. 3.3 (a), it can be proved that

$$\bar{x} = 0$$

3.2.5. Centroid of a Triangular Section by Integration. Fig. 3.4 shows a triangular section AOB of base width $= b$ and height $= h$. Consider a small strip of thickness dy at a distance y from the axis OX.

Area of strip, dA = Length $DE \times dy$ ...(i)

The distance DE in terms of y, b and h is obtained from two similar triangles ADE and AOB as

$$\frac{DE}{OB} = \frac{AD}{AO}$$

where $OB = b, AO = h$ and $AD = (h - y)$

$$\therefore \quad \frac{DE}{b} = \frac{(h-y)}{h}$$

or

$$DE = \frac{b(h-y)}{h}$$

Fig. 3.4

Substituting this value of DE in equation (i), we get

Area of strip, $$dA = \frac{b(h-y)}{h} \cdot dy$$

Moment of this area dA about axis OX

$$= dA \cdot y = \frac{b(h-y)}{h} \cdot dy \cdot y$$

$$= \frac{b}{h}(h-y) \cdot y \cdot dy$$

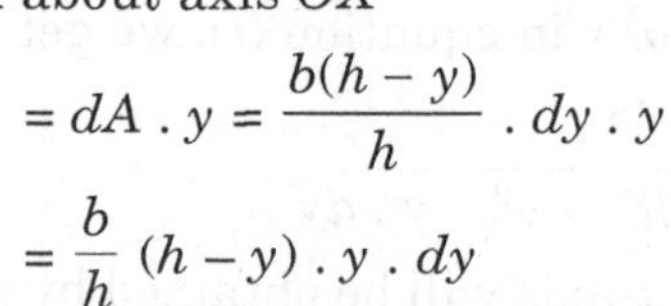

The moment of the total area A of the triangular section is obtained by integrating the above equation between the limits 0 to h.

$\therefore$ Moment of total area A about axis OX

$$= \int_0^h \frac{b}{h}(h-y) \cdot y \cdot dy = \frac{b}{h}\int_0^h (h-y) \cdot y \cdot dy$$

[$\because$ b and h are constants and can be taken outside the integral sign]

$$= \frac{b}{h}\int_0^h (hy - y^2)\, dy = \frac{b}{h}\left[\frac{hy^2}{2} - \frac{y^3}{3}\right]_0^h$$

$$= \frac{b}{h}\left[\left(\frac{h \cdot h^2}{2} - \frac{h^3}{3}\right) - 0\right] = \frac{b}{h}\left(\frac{h^3}{2} - \frac{h^3}{3}\right) = \frac{b}{h} \times \frac{h^3}{6} \quad \text{...(ii)}$$

Let $\bar{y}$ = Distance of centroid of total area A from axis OX

$$A = \text{Total area} = \int_0^h dA = \int_0^h \frac{b}{h}(h-y) \cdot dy \qquad \left[\because \; dA = \frac{b}{h}(h-y) \cdot dy\right]$$

$$= \frac{b}{h}\left[hy - \frac{y^2}{2}\right]_0^h = \frac{b}{h}\left[h^2 - \frac{h^2}{2}\right] = \frac{b}{h} \cdot \frac{h^2}{2} = \frac{bh}{2}$$

Then moment of total are A about axis OX

$$= A \times \bar{y}$$

$$= \left(\frac{b \times h}{2}\right) \times \bar{y} \quad \ldots(iii)$$

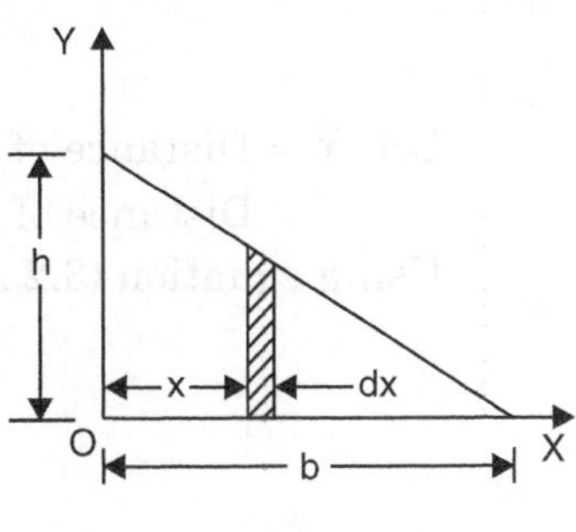

Fig. 3.4(a)

Equating the two values given by equations (*ii*) and (*iii*), we get

$$\left(\frac{b \times h}{2}\right) \times \bar{y} = \frac{b}{h} \cdot \frac{h^3}{6}$$

or
$$\bar{y} = \left(\frac{b}{h} \cdot \frac{h^3}{6}\right) \times \frac{1}{\left(\frac{b \times h}{2}\right)} = \frac{h}{3}$$

Similarly, taking a strip of thickness dx parallel to y-axis at a distance of x from the y-axis [Refer to Fig. 3.4 (*a*)], it can be proved that

$$\bar{x} = \frac{b}{3}.$$

3.2.6. Problems of Finding Centre of Gravity of Areas by Integration Method

Problem 3.1. *Determine the co-ordinates of the C.G. of the area OAB shown in Fig. 3.5, if the curve OB represents the equation of a parabola, given by*

$$y = kx^2$$

in which $OA = 6$ *units*

and $AB = 4$ *units.*

Sol. The equation of parabola is $y = kx^2$ ...(*i*)

First determine the value of constant k. The point B is lying on the curve and having co-ordinates

$$x = 6 \quad \text{and} \quad y = 4$$

Substituting these values of equation (*i*), we get

$$4 = k \times 6^2 = 36\,k$$

$$\therefore \quad k = \frac{4}{36} = \frac{1}{9}$$

Substituting the value of k in equation (*i*), we get

$$y = \frac{1}{9} x^2 \quad \ldots(ii)$$

or
$$x^2 = 9y$$

or
$$x = 3\sqrt{y} \quad \ldots(iii)$$

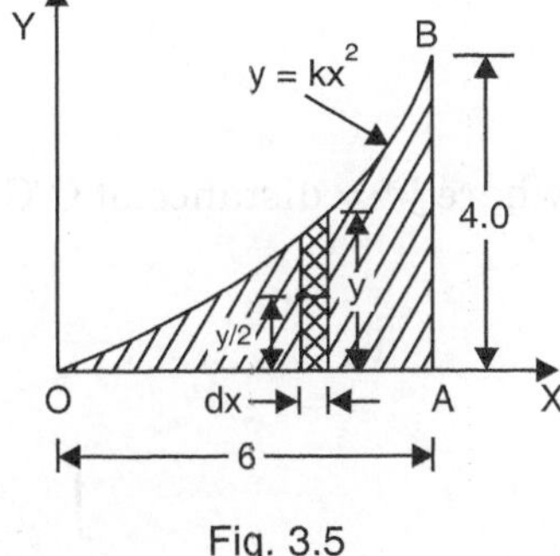

Fig. 3.5

Consider a strip of height y and width dx as shown in Fig. 3.5. The area dA of the strip is given by

$$dA = y \times dx$$

The co-ordinates of the C.G. of this area dA are x and $\frac{y}{2}$

$\therefore$ Distance of C.G. of area dA from y-axis $= x$

and distance of C.G. of area dA from x-axis $= \frac{y}{2}$

$\therefore \qquad x^* = x \quad \text{and} \quad y^* = \dfrac{y}{2}$

Let $\bar{x}$ = Distance of C.G. of total area OAB from axis OY

$\bar{y}$ = Distance of C.G. of total area OAB from axis OX.

Using equation (3.2 A), we get

$$\bar{x} = \frac{\int x^* \, dA}{\int dA} = \frac{\int_0^6 x \times y dx}{\int_0^6 y dx} \qquad (\because \; dA = ydx, x^* = x)$$

But $y = \dfrac{x^2}{9}$ from equation (ii).

$$\therefore \qquad \bar{x} = \frac{\int_0^6 x \times \frac{x^2}{9} \times dx}{\int_0^6 \frac{x^2}{9} dx} = \frac{\frac{1}{9}\int_0^6 x^3 \, dx}{\frac{1}{9}\int_0^6 x^2 dx}$$

$$= \frac{\int_0^6 x^3 \, dx}{\int_0^6 x^2 \, dx} = \frac{\left[\frac{x^4}{4}\right]_0^6}{\left[\frac{x^3}{3}\right]_0^6} = \frac{\frac{1}{4} \times 6^4}{\frac{1}{3} \times 6^3}$$

$$= \frac{1}{4} \times \frac{3}{1} \times 6 = 4.5. \quad \textbf{Ans.}$$

Using equation (3.2 B), we get

$$\bar{y} = \frac{\int y^* \, dA}{\int dA}$$

where y^* = distance of C.G. of area dA from x-axis

$$= \frac{y}{2} \text{ (here)}$$

$$dA = ydx$$

$$\therefore \qquad \int y^* dA = \int \frac{y}{2} \times dA = \int_0^6 \frac{y}{2} \times ydx = \int_0^6 \frac{y^2}{2} dx$$

$$= \frac{1}{2}\int_0^6 y^2 dx = \frac{1}{2}\int_0^6 \left(\frac{x^2}{9}\right)^2 dx \qquad \left(\because \; y = \frac{x^2}{9}\right)$$

$$= \frac{1}{2}\int_0^6 \frac{x^4}{81} dx = \frac{1}{2} \times \frac{1}{81}\int_0^6 x^4 \, dx = \frac{1}{2} \times \frac{1}{81}\left[\frac{x^5}{5}\right]_0^6$$

$$= \frac{1}{2} \times \frac{1}{81} \times \frac{6^5}{5} = \frac{6^5}{810}$$

Also $$\int dA = \int_0^6 ydx = \int_0^6 \frac{x^2}{9}\,dx = \frac{1}{9}\left[\frac{x^3}{3}\right]_0^6 = \frac{1}{9}\times\frac{6^3}{3} = \frac{6^3}{27}$$

$$\therefore \quad \bar{y} = \frac{\int y^* \, dA}{\int dA} = \frac{\frac{6^5}{810}}{\frac{6^3}{27}} = \frac{27}{810}\times\frac{6^5}{6^3}$$

$$= \frac{1}{30}\times 6^2 = \frac{36}{30} = \mathbf{\frac{6}{5}}. \quad \textbf{Ans.}$$

Problem 3.2. *Determine the co-ordinates of the C.G. of the shaded area between the parabola* $y = \frac{x^2}{4}$ *and the straight line* $y = x$ *as shown in Fig. 3.6.*

Sol. The equations of parabola and straight line are

$$y = \frac{x^2}{4} \qquad ...(i)$$

$$y = x \qquad ...(ii)$$

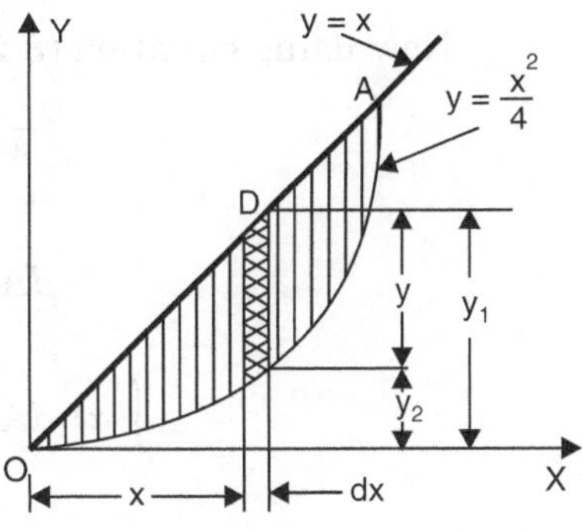

Fig. 3.6

The point A is lying on the straight line as well as on the given parabola. Hence both the above equations hold good for point A. Let the co-ordinates of point A are x, y.

Substituting the value of y from equation (ii) in equation (i), we get

$$x = \frac{x^2}{4} \quad \text{or} \quad 4 = \frac{x^2}{x} = x$$

Substituting the value of $x = 4$, in equation (ii),

$$y = 4$$

Hence the co-ordinates of point A are 4, 4.

Now divide the shaded area into large small areas each of height y and width dx as shown in Fig. 3.6. Then area dA of the strip is given by

$$dA = ydx = (y_1 - y_2)\,dx \qquad ...(iii)$$

where y_1 = Co-ordinate of point D which lies on the straight line OA

y_2 = Co-ordinate of the point E which lies on the parabola OA.

The horizontal co-ordinates of the points D and E are same.

The values of y_1 and y_2 can be obtained in terms of x from equations (ii) and (i),

$$y_1 = x \quad \text{and} \quad y_2 = \frac{x^2}{4}$$

Substituting these values in equation (iii),

$$dA = \left(x - \frac{x^2}{4}\right)dx \qquad ...(iv)$$

The distance of the C.G. for the area dA from y-axis is given by,

$$x^* = x$$

And the distance of the C.G. of the area dA from x-axis is given by,

$$y^* = y_2 + \frac{y}{2} = y_2 + \frac{y_1 - y_2}{2} \qquad (\because \quad y = y_1 - y_2)$$

$$= \frac{2y_2 + y_1 - y_2}{2} = \frac{y_1 + y_2}{2}$$

$$= \frac{x + \frac{x^2}{4}}{2} \qquad \left(\because \quad y_1 = x \text{ and } y_2 \frac{x^2}{4}\right)$$

$$= \frac{1}{2}\left(x + \frac{x^2}{4}\right) \qquad \ldots(v)$$

Now let $\bar{x}$ = Distance of C.G. of shaded area of Fig. 3.6 from y-axis

$\bar{y}$ = Distance of C.G. of shaded area of Fig. 3.6 from x-axis.

Now using equation (3.2 A),

$$\bar{x} = \frac{\int x^* \, dA}{\int dA} \text{ where } x^* = x$$

$$dA = \left(x - \frac{x^2}{4}\right) dx \qquad \text{[See equation } (iv)]$$

$$\therefore \qquad \int x^* \, dA = \int_0^4 x\left(x - \frac{x^2}{4}\right) dx \qquad (\because \quad x \text{ varies from 0 to 4})$$

$$= \int_0^4 \left(x^2 - \frac{x^3}{4}\right) dx = \left[\frac{x^3}{3} - \frac{x^4}{4 \times 4}\right]_0^4$$

$$= \frac{4^3}{3} - \frac{4^4}{4 \times 4} = \frac{64}{3} - 16 = \frac{64 - 48}{3} = \frac{16}{3}$$

and

$$\int dA = \int_0^4 \left(x - \frac{x^2}{4}\right) dx = \left[\frac{x^2}{2} - \frac{x^3}{3 \times 4}\right]_0^4 = \frac{4^2}{2} - \frac{4^3}{3 \times 4}$$

$$= \frac{16}{2} - \frac{16}{3} = \frac{48 - 32}{6} = \frac{16}{6} \qquad \ldots(vi)$$

$$\therefore \qquad \bar{x} = \frac{\int x^* \, dA}{\int dA} = \frac{\frac{16}{3}}{\frac{16}{6}} = \frac{16}{3} \times \frac{6}{16} = 2. \textbf{ Ans.}$$

Now using equation (3.2 B), $\bar{y} = \dfrac{\int y^* \, dA}{\int dA}$

where $y^* = \dfrac{1}{2}\left[x + \dfrac{x^2}{4}\right]$ [From equation (v)]

$$dA = \left(x - \frac{x^2}{4}\right) dx$$

$$\therefore \quad \int y^* \, dA = \int_0^4 \frac{1}{2}\left(x + \frac{x^2}{4}\right)\left(x - \frac{x^2}{4}\right) dx$$

$$= \frac{1}{2}\int_0^4 \left(x^2 - \frac{x^4}{16}\right) dx = \frac{1}{2}\left[\frac{x^3}{3} - \frac{x^5}{5 \times 16}\right]_0^4$$

$$= \frac{1}{2}\left[\frac{4^3}{3} - \frac{x^5}{5 \times 16}\right] = \frac{1}{2}\left[\frac{64}{3} - \frac{64}{5}\right]$$

$$= \frac{64}{2}\left[\frac{1}{3} - \frac{1}{5}\right] = 32\left(\frac{5-3}{15}\right)$$

$$= 32 \times \frac{2}{15} = \frac{64}{15}$$

and $$\int dA = \frac{16}{6} \qquad \text{[From equation } (vi)]$$

$$\therefore \quad \bar{y} = \frac{\int y^* \, dA}{\int dA} = \frac{\frac{64}{15}}{\frac{16}{6}} = \frac{64}{15} \times \frac{6}{16} = \frac{8}{5}. \quad \textbf{Ans.}$$

3.2.7. Problems of Finding Centroid or Centre of Gravity of a Line-segment by Integration Method

Problem 3.3. *Determine the centre of gravity of a quadrant AB of the arc of a circle of radius R as shown in Fig. 3.7.*

Sol. The centre of gravity of the line AB, which is an arc of a circle radius R, is obtained by dividing the curved line AB into a large number of elements of length dL as shown in Fig. 3.7.

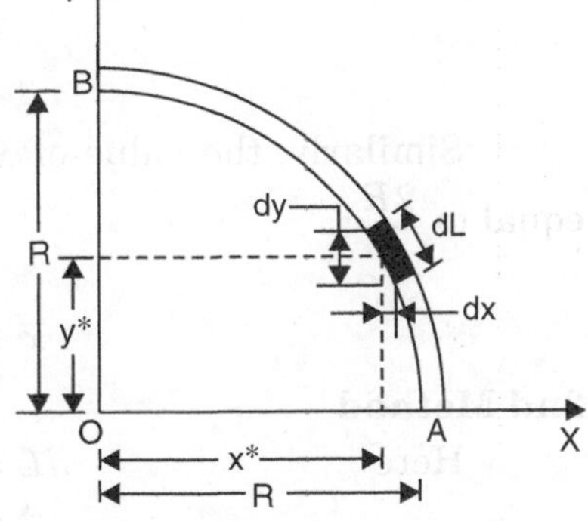

Fig. 3.7

The equation of curve AB is the equation of circle of radius R.

$\therefore$ The equation of curve AB is given by

$$x^2 + y^2 = R^2$$

Differentiating the above equation,

$$2x\,dx + 2y\,dy = 0 \qquad [\because \; R \text{ is constant}]$$

or $$2y\,dy = -\,2x\,dx$$

or $$dy = \frac{-\,2x\,dx}{2y} = \frac{-\,x\,dx}{y} \qquad \ldots(i)$$

1st Method

Consider an element of length dL as shown in Fig. 3.7. The C.G. of the length dL is at a distance x^* from y-axis and y^* from x-axis.

Now using equation (3.2 D) for $\bar{y}$, we get

$$\bar{y} = \frac{\int y^* \, dL}{\int dL} \qquad \ldots(ii)$$

Let us express dL in terms of dx and dy.

But $$dL = \sqrt{dx^2 + dy^2}$$

$$= \sqrt{dx^2 + \left(\frac{-x\,dx}{y}\right)^2} \qquad \left(\because \text{ From } (i),\ dy = \frac{-x\,dx}{y}\right)$$

$$= \sqrt{dx^2 + \frac{x^2}{y^2}dx^2} = dx\sqrt{1 + \frac{x^2}{y^2}} = dx\sqrt{\frac{y^2 + x^2}{y^2}}$$

$$= dx\sqrt{\frac{R^2}{y^2}} \qquad (\because x^2 + y^2 = R^2)$$

$$= \frac{R}{y}\,.\,dx.$$

Substituting the value of dL in equation (ii),

$$\bar{y} = \frac{\int y^* \times \frac{R}{y}\,.\,dx}{\int dL} = \frac{\int y \times \frac{R}{y}\,.\,dx}{\int dL} \qquad (\because \quad y^* = y)$$

$$= \frac{\int R\,dx}{\int dL} = \frac{R\int_0^R dx}{\int dL} = \frac{R\big[x\big]_0^R}{\left(\frac{2\pi R}{4}\right)}$$

($\because \quad \int dL$ is total length of arc of one quadrant of a circle)

$$= \frac{R \times R}{\frac{2\pi R}{4}} = \frac{\mathbf{2R}}{\pi}. \quad \textbf{Ans.}$$

Similarly, the value of $\bar{x}$ can be calculated. Due to symmetry this value will also be equal to $\frac{2R}{\pi}$.

$$\therefore \qquad \bar{x} = \bar{y} = \frac{\mathbf{2R}}{\pi}. \quad \textbf{Ans.}$$

2nd Method

Here
$$dL = R\,d\theta$$
$$y^* = R\sin\theta$$
$$x^* = R\cos\theta$$

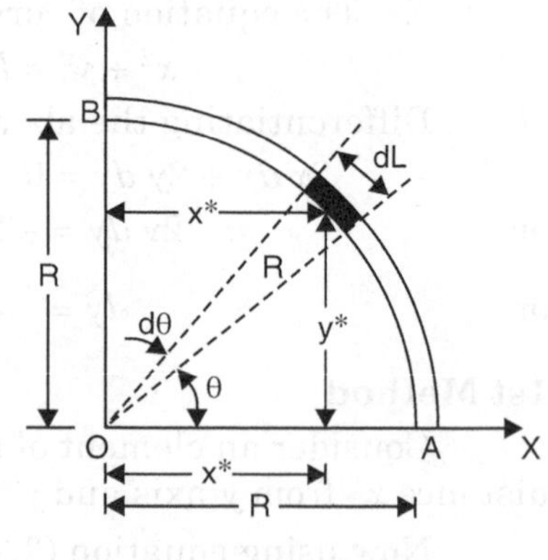

Fig. 3.7(a)

Now $$\bar{y} = \frac{\int y^*\,dL}{\int dL} = \frac{\int_0^{\pi/2}(R\sin\theta)\times(R\,d\theta)}{\int_0^{\pi/2}R\,d\theta}$$

$$= \frac{\int_0^{\pi/2}R^2\sin\theta\,d\theta}{\int_0^{\pi/2}R\,d\theta} = \frac{R^2\int_0^{\pi/2}\sin\theta\,d\theta}{R\int_0^{\pi/2}d\theta}$$

$$= \frac{R^2\left[-\cos\theta\right]_0^{\pi/2}}{R\left[\theta\right]_0^{\pi/2}} = \frac{-R\left[\cos\left(\frac{\pi}{2}\right)-\cos 0\right]}{\left[\frac{\pi}{2}-0\right]}$$

$$= \frac{-R\,[0-1]}{\frac{\pi}{2}} = \frac{\mathbf{2R}}{\pi}.\ \ \textbf{Ans.}$$

Similarly,
$$\bar{x} = \frac{\int x^*\,dL}{\int dL} = \frac{\int_0^{\pi/2}(R\sin\theta)\times(R\,d\theta)}{\int_0^{\pi/2} R\,d\theta} = \frac{R^2\int_0^{\pi/2}\cos\theta\,d\theta}{R\int_0^{\pi/2} d\theta}$$

$$= \frac{R\left[\sin\theta\right]_0^{\pi/2}}{R\left[\theta\right]_0^{\pi/2}} = \frac{R\,[\sin 90^\circ - \sin 0^\circ]}{\left(\frac{\pi}{2}-0\right)}$$

$$= \frac{R}{\left(\frac{\pi}{2}\right)} = \frac{\mathbf{2R}}{\pi}.\ \ \textbf{Ans.}$$

Problem 3.4. *Determine the centre of gravity of the area of the circular sector OAB of radius R and central angle α as shown in Fig. 3.8.*

Sol. The given area is symmetrical about x-axis. Hence the C.G. of the area will lie on x-axis. This means $\bar{y} = 0$. To find $\bar{x}$, the moment of small areas are to be taken along y-axis. Divide the area *OAB* into a large number of triangular elements each of altitude R and base $Rd\theta$ as shown in Fig. 3.8. Such triangular element is shown by *OCD* in which altitude $OC = R$ and base $CD = Rd\theta$. The area dA of this triangular element is given by,

$$dA = \frac{OC\times CD}{2} = \frac{R\times Rd\theta}{2} = \frac{R^2 d\theta}{2}$$

The C.G. of this triangular element is at G

where $OG = \frac{2}{3}\times OC = \frac{2}{3}\times R$

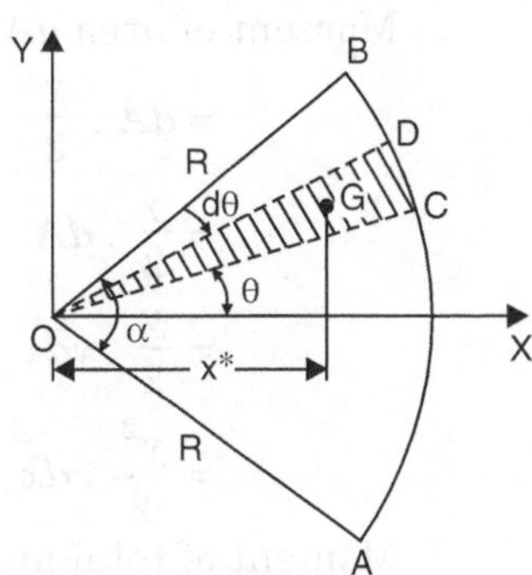

Fig. 3.8

The distance of C.G. of area dA from y-axis is given by,

$$x^* = OG\times\cos\theta = \frac{2}{3}R\times\cos\theta$$

Now using equation (3.2*A*),

$$\bar{x} = \frac{\int x^*\,dA}{\int dA} = \frac{2\int_0^{\alpha/2}\left(\frac{2}{3}R\cos\theta\right)\left(\frac{R^2 d\theta}{2}\right)}{2\int_0^{\alpha/2}\frac{R^2}{2}\,d\theta}$$

$$= \frac{\frac{R^3}{3}\int_0^{\alpha/2} \cos d\theta}{\frac{R^2}{2}\int_0^{\alpha/2} d\theta} = \frac{2R}{3}\frac{\left[\sin\theta\right]_0^{\alpha/2}}{\left[\theta\right]_0^{\alpha/2}}$$

$$= \frac{2R}{3}\frac{\sin\left(\frac{\alpha}{2}\right)}{\left(\frac{\alpha}{2}\right)} = \frac{\mathbf{4R}}{\mathbf{3a}}\ \mathbf{sin}\left(\frac{\alpha}{\mathbf{2}}\right). \quad \textbf{Ans.}$$

The area *OAB* is symmetrical about the *x*-axis, hence

$$\bar{y} = \mathbf{0}. \quad \textbf{Ans.}$$

For a semi-circle, $\alpha = \pi = 180°$, hence

$$\bar{x} = \frac{4R}{3\alpha}\sin\left(\frac{\pi}{2}\right)$$

$$= \frac{4R}{3\times\pi}\sin\left(\frac{180}{2}\right) = \frac{\mathbf{4R}}{\mathbf{3\pi}}. \quad \textbf{Ans.}$$

Problem 3.5. *Determine the centre of gravity of a semicircle of radius R as shown in Fig. 3.9.*

Sol. This problem can also be solved by the method given in problem 3.4. The following other methods can also be used. Due to symmetry, $\bar{x} = 0$, the area *AOB* is symmetrical about the *Y*-axis, hence $\bar{x} = 0$. The value of $\bar{y}$ is obtained by taking the moments of small areas and total area about *x*-axis.

1. Considering the strip parallel to y-axis

Area of strip, $dA = y \,.\, dx$

The distance of the C.G. of the area *dA* from *x*-axis is equal to $\frac{y}{2}$.

Moment of area *dA* about *x*-axis

$$= dA \,.\, \frac{y}{2}$$

$$= \frac{y}{2} \,.\, dA$$

$$= \frac{y}{2} \,.\, ydx \qquad (\because \quad dA = y.dx)$$

$$= \frac{y^2}{2} \,.\, dx$$

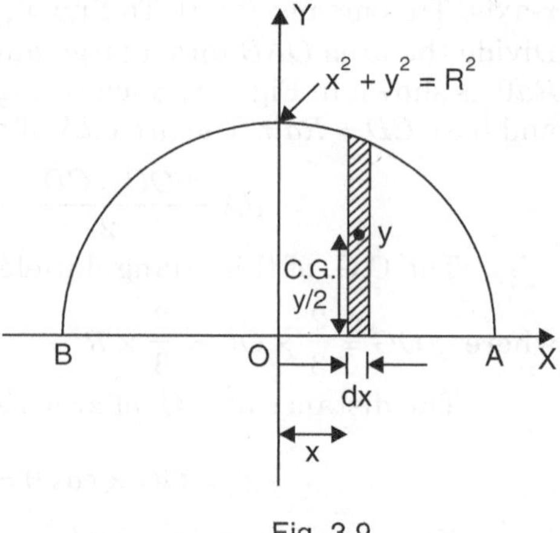

Fig. 3.9

Moment of total area *A* about *x*-axis is obtained by integrating the above equation.

$\therefore$ Moment of total area *A* about *x*-axis

$$= \int \frac{y^2}{2} \,.\, dx$$

$$= \int_{-R}^{R} \frac{y^2}{2}\, dx \qquad (\because \ x \text{ varies from } -R \text{ to } R)$$

But equation of semicircle is

$$x^2 + y^2 = R^2 \text{ or } y^2 = R^2 - x^2$$

Substituting this value of y^2 in the above equation, we get

Moment of total area A about x-axis

$$= \int_{-R}^{R} \frac{(R^2 - x^2)}{2}\, dx = \frac{1}{2}\left[R^2 \,.\, x - \frac{x^3}{3}\right]_{-R}^{R}$$

$$= \frac{1}{2}\left[\left(R^2 \,.\, R - \frac{R^3}{3}\right) - \left\{R^2\,(-R) - \frac{(-R)^2}{3}\right\}\right]$$

$$= \frac{1}{2}\left[\left(R^3 - \frac{R^3}{3}\right) - \left\{-R^3 - \frac{(-R^3)}{3}\right\}\right]$$

$$= \frac{1}{2}\left[\frac{2R^3}{3} - \left(-R^3 + \frac{R^3}{3}\right)\right] = \frac{1}{2}\left[\frac{2R^3}{3} - \left(-\frac{2R^3}{3}\right)\right]$$

$$= \frac{1}{2}\left[\frac{2R^3}{3} + \frac{2R^3}{3}\right] = \frac{1}{2} \times \frac{4R^3}{3} = \frac{2R^3}{3} \qquad ...(i)$$

Let $\bar{y}$ = Distance of C.G. of the total area of semicircle from x-axis.

The total area of semicircle is also equal to $\frac{\pi R^2}{2}$

$\therefore$ Moment of this total area about x-axis

$$= \bar{y} \times \frac{\pi R^2}{2} \qquad ...(ii)$$

Equating the two values given by equations (i) and (ii), we get

$$\bar{y} \times \frac{\pi R^2}{3} = \frac{2R^3}{3}$$

$$\therefore \quad \bar{y} = \frac{2R^3}{3} \times \frac{2}{\pi R^2} = \frac{\mathbf{4R}}{\mathbf{3\pi}}. \quad \textbf{Ans.}$$

Hence the location of C.G. of semicircle is $\left(\mathbf{0}, \frac{\mathbf{4R}}{\mathbf{3\pi}}\right)$. **Ans.**

2. Considering the strip parallel to x-axis

Area of strip, $dA = 2x \,.\, dy$

The distance of the C.G. of this area from x-axis is y

$\therefore$ Moment of this area about x-axis

$$= y \,.\, dA$$

$$= y \,.\, 2x dy$$

$$= 2xy\, dy \qquad ...(i)$$

But, we know $x^2 + y^2 = R^2$

$$\therefore \quad x^2 = R^2 - y^2$$

or

$$x = \sqrt{R^2 - y^2}$$

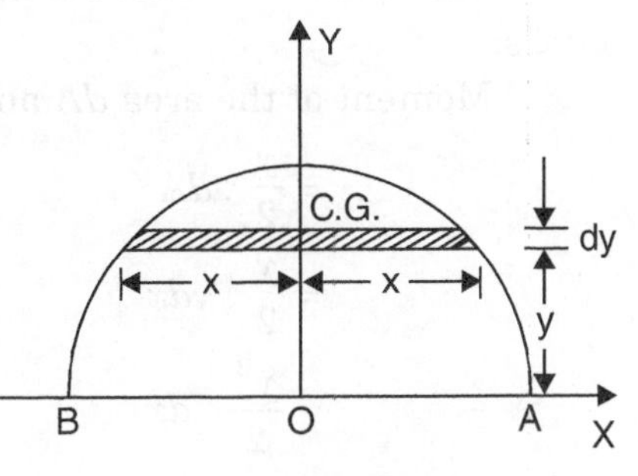

Fig. 3.9(a)

Substituting the above value of x in equation (*i*), we get

Moment of area dA about x-axis

$$= 2\sqrt{R^2 - y^2} \,.\, y \,.\, dy$$

Moment of total area A about x-axis will be obtained by integrating the above equation from O to R.

$\therefore$ Moment of area A about x-axis

$$= \int_0^R 2\sqrt{R^2 - y^2} \,.\, y \, dy \qquad (\because \; y \text{ varies from } O \text{ to } R)$$

$$= - \int_0^R \sqrt{R^2 - y^2} \,.\, (-2y)\, dy = - \left[\frac{(R^2 - y^2)^{3/2}}{3/2} \right]_0^R$$

$$= - \frac{2}{3} [0 - R^2] = \frac{2R^3}{3} \qquad \text{...(i)}$$

Also the moment of total area A about x-axis $= A \times \bar{y}$

where A = total area of semi-circle $= \dfrac{\pi R^2}{2}$

$\bar{y}$ = Distance of C.G. of area A from x-axis

$\therefore$ Moment of total area A about x-axis $= \dfrac{\pi R^2}{2} \times \bar{y}$ ...(*ii*)

Equating the two values given by equations (*i*) and (*ii*),

$$\frac{\pi R^2}{2} \times \bar{y} = \frac{2R^3}{3}$$

or

$$\bar{y} = \frac{2R^3}{3} \times \frac{2}{\pi R^2} = \mathbf{\frac{4R}{3\pi}}. \quad \textbf{Ans.}$$

Problem 3.6. *To determine the centre of gravity of the area shown in Fig. 3.9(b) given by*

$$\frac{x^2}{a^2} + \frac{y^2}{b^2} = 1.$$

Sol. Consider a small strip of thickness dx parallel to y-axis at a distance of x from the y-axis.

Area of the strip, $dA = y.dx$

The C.G. of area dA is at a distance $\dfrac{y}{2}$ from x-axis.

Moment of the area dA about x-axis

$$= \frac{y}{2} \,.\, dA$$

$$= \frac{y}{2} \,.\, ydx \qquad (\because \; dA = y.dx)$$

$$= \frac{y^2}{2} \,.\, dx$$

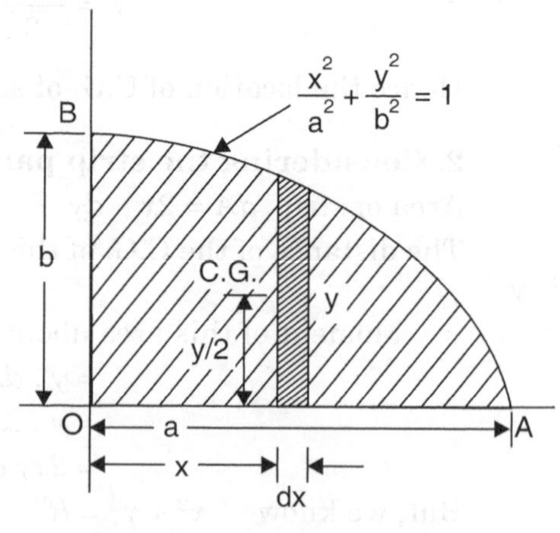

Fig. 3.9(b)

∴ Moment of the total area about x-axis

$$= \int_0^a \frac{y^2}{2} \, . \, dx \qquad (\because \; x \text{ varies from } O \text{ to } a) \quad ...(i)$$

Let us substitute the value of y^2 in terms of x.

The given equation is $\dfrac{x^2}{a^2} + \dfrac{y^2}{b^2} = 1$

or
$$\frac{y^2}{b^2} = 1 - \frac{x^2}{a^2} = \frac{a^2 - x^2}{a^2}$$

or
$$y^2 = \frac{b^2}{a^2}(a^2 - x^2) \quad ...(ii)$$

Substituting the value of y^2 in equation (i), we get

Moment of total area about x-axis

$$= \frac{1}{2}\int_0^a \frac{b^2}{a^2}(a^2 - x^2)dx = \frac{b^2}{2a^2}\left[a^2 x - \frac{x^3}{3}\right]_0^a$$

$$= \frac{b^2}{2a^2}\left[a^3 - \frac{a^3}{3}\right] = \frac{b^2}{2a^2} \times \frac{2a^3}{3} = \frac{ab^2}{2} \quad ...(iii)$$

The total area A of the given figure is given by

$$A = \int dA = \int y \, . \, dx \quad ...(iv)$$

From equation (ii),
$$y = \left[\frac{b^2}{a^2}(a^2 - x^2)\right]^{1/2} = \frac{b}{a}(a^2 - x^2)^{1/2} \quad ...(v)$$

Now equation (iv) is, $A = \displaystyle\int_0^a \frac{b}{a}(a^2 - x^2)^{1/2} \, . \, dx$

$$= \frac{b}{a}\left[\int_0^a (a^2 - x^2)^{1/2} \, . \, dx\right]^* = \frac{b}{a}\left[\frac{\pi a^2}{4}\right]^*$$

$$= \frac{\pi \, . \, ab}{4} \qquad \left(\because \; \int_0^a \sqrt{a^2 - x^2} \, dx = \frac{\pi a^4}{4}\right) \quad ...(vi)$$

Let $\bar{y}$ = the distance of C.G. of the total area A from x-axis.

Then moment of total area A about x-axis

$$= A \times \bar{y}$$

$$= \frac{\pi ab}{4} \, . \, \bar{y} \quad ...(vii)$$

The equations (iii) and (vii) give the moment of total area about x-axis. Hence equating these equations, we get

---

*Please refer some standard Textbook of Mathematics.

$$\int_0^a \sqrt{a^2 - x^2} \, dx = \left[\frac{1}{2} x \sqrt{a^2 - x^2} + \frac{1}{2} a^2 \sin^{-1}\frac{x}{a}\right]_0^a = \left[0 + \frac{1}{2} a^2 \sin^{-1}(1)\right]$$

$$= \frac{1}{2} a^2 \left(\frac{\pi}{2}\right) = \frac{\pi a^4}{4}.$$

$$\frac{\pi ab}{4} \cdot \bar{y} = \frac{ab^2}{3}$$

$$\therefore \quad \bar{y} = \frac{ab^2}{3} \cdot \frac{4}{\pi ab} = \mathbf{\frac{4b}{3\pi}}. \quad \textbf{Ans.}$$

To find $\bar{x}$, take the moment of small area dA about y-axis.

The C.G. of area dA is at a distance of x from y-axis.

$\therefore$ Moment of area dA about y-axis $= x.dA$

$$= x.y.dx \qquad (\because \; dA = ydx)$$

Moment of total area A about y-axis is obtained by integration

Now moment of total area A about y-axis

$$= \int_0^a x.y.dx \qquad (\because \; x \text{ varies from } O \text{ to } a)$$

$$= \int_0^a x \cdot \frac{b}{a}(a^2 - x^2)^{1/2} \cdot dx \qquad \left[\because \; y = \frac{b}{a}(a^2 - x^2)^{1/2} \text{ from equation } (v)\right]$$

$$= \frac{b}{a}\int_0^a x \cdot (a^2 - x^2)^{1/2}\, dx = \frac{b}{a}\int_0^a \frac{(-2)}{(-2)} \cdot x\,(a^2 - x^2)^{1/2} \cdot dx$$

$$= \frac{b}{-2a}\left[\frac{(a^2 - x^2)^{3/2}}{3/2}\right]_0^a = \frac{-b}{3a}[0 - a^3] = \frac{ba^2}{3} \qquad ...(viii)$$

Also the moment of total area A about y-axis

$$= A \times \bar{x} \qquad ...(ix)$$

where $\bar{x}$ = Distance of C.G. of total area A from y-axis.

Equating the two values given by equations $(viii)$ and (ix),

$$A \times \bar{x} = \frac{ba^2}{3}$$

$$\therefore \quad \bar{x} = \frac{ba^2}{3A} = \frac{ba^2}{3 \times \frac{\pi ab}{4}} \qquad \left[\because \; A = \frac{\pi ab}{4} \text{ see equation } (vi)\right]$$

$$= \mathbf{\frac{4a}{3\pi}}. \quad \textbf{Ans.}$$

The co-ordinates of the C.G. of given area are

$$\bar{x} = \frac{4a}{3\pi} \text{ and } \bar{y} = \frac{4b}{3\pi}.$$

3.3. CENTROID OF T-SECTION, I-SECTION, ANGLE-SECTION, HOLLOW-SECTION ETC. BY USING STANDARD FORMULA

The centroid of structural sections like T-section, I-section, L-section etc. are obtained by splitting them into rectangular components. Then equations (3.1) and (3.2) are used.

Problem 3.7. *Find the centre of gravity of the T-section shown in Fig. 3.10.*

Sol. The given *T*-section is split up into two rectangles *ABCD* and *EFGH* as shown in Fig. 3.10 (*a*). The given *T*-section is symmetrical about *Y-Y* axis. Hence the C.G. of the section will lie on this axis. The lowest line of the figure is line *GF*. Hence the moments of the areas are taken about this line *GF*, which is the axis of reference in this case.

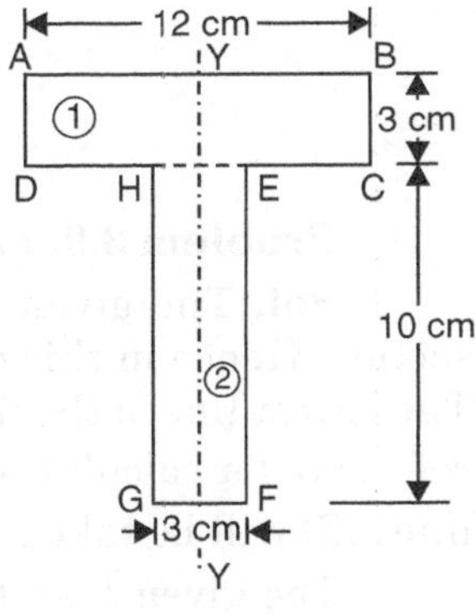

Fig. 3.10

Fig. 3.10(a)

Let $\bar{y}$ = The distance of the C.G. of the *T*-section from the bottom line *GF* (which is axis of reference)

a_1 = Area of rectangle $ABCD = 12 \times 3 = 36 \text{ cm}^2$

y_1 = Distance of C.G. of area a_1 from bottom line

$$GF = 10 + \frac{3}{2} = 11.5 \text{ cm}$$

a_2 = Area of rectangle $EFGH = 10 \times 3 = 30 \text{ cm}^2$

y_2 = Distance of C.G. of area a_2 from bottom line

$$GF = \frac{10}{2} = 5 \text{ cm.}$$

Using equation (3.2), we have

$$\bar{y} = \frac{a_1y_1 + a_2y_2}{A} = \frac{a_1y_1 + a_2y_2}{a_1 + a_2} \qquad (\because \; A = a_1 + a_2)$$

$$= \frac{36 \times 11.5 + 30 \times 5}{36 + 30} = \frac{414 + 150}{66}$$

$= \mathbf{8.545 \text{ cm.}}$ **Ans.**

Problem 3.8. *Find the centre of gravity of the I-section shown in Fig. 3.11(a).*

Sol. The *I*-section is split up into three rectangles *ABCD, EFGH* and *JKLM* as shown in Fig. 3.11 (*b*). The given *I*-section is symmetrical about *Y* axis. Hence the C.G. of the section will lie on this axis. The lowest line of the figure is *ML*. Hence the moment of areas are taken about this line, which is the axis of reference.

Let $\bar{y}$ = Distance of the C.G. of the *I*-section from the bottom line *ML*

a_1 = Area of rectangle $ABCD = 10 \times 2 = 20 \text{ cm}^2$

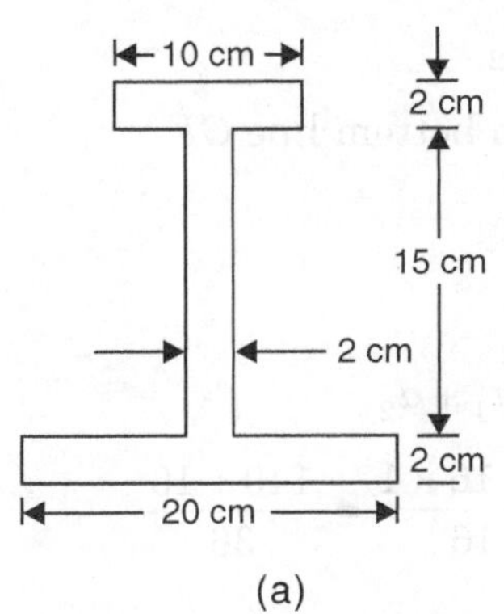

(a)

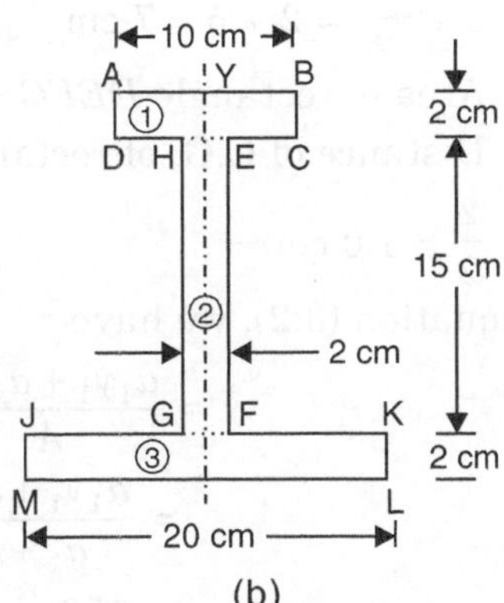

(b)

Fig. 3.11

y_1 = Distance of C.G. of rectangle *ABCD* from bottom line $ML = 2 + 15 + \frac{2}{2} = 18$ cm

a_2 = Area of rectangle *EFGH* = 15 × 2 = 30 cm²

y_2 = Distance of C.G. of rectangle *EFGH* from bottom line $ML = 2 + \frac{15}{2} = 2 + 7.5 = 9.5$ cm

a_3 = Area of rectangle *JKLM* = 20 × 2 = 40 cm²

y_3 = Distance of C.G. of rectangle *JKLM* from bottom line $ML = \frac{2}{2} = 1.0$ cm

Now using equation (3.2), we have $\bar{y} = \frac{a_1y_1 + a_2y_2 + a_3y_3}{A}$

$$= \frac{a_1y_1 + a_2y_2 + a_3y_3}{a_1 + a_2 + a_3} \qquad (\because \quad A = a_1 + a_2 + a_3)$$

$$= \frac{20 \times 18 + 30 \times 9.5 + 40 \times 1}{20 + 30 + 40}$$

$$= \frac{360 + 285 + 40}{90} = \frac{685}{90} = \mathbf{7.611\ cm.} \quad \textbf{Ans.}$$

Problem 3.9. *Find the centre of gravity of the L-section shown in Fig. 3.12.*

Sol. The given *L*-section is not symmetrical about any section. Hence in this case, there will be two axes of references. The lowest line of the figure (*i.e.*, line *GF*) will be taken as axis of reference for calculating $\bar{y}$. And the left line of the *L*-section (*i.e.*, line *AG*) will be taken as axis of reference for calculating $\bar{x}$.

The given *L*-section is split up into two rectangles *ABCD* and *DEFG*, as shown in Fig. 3.12.

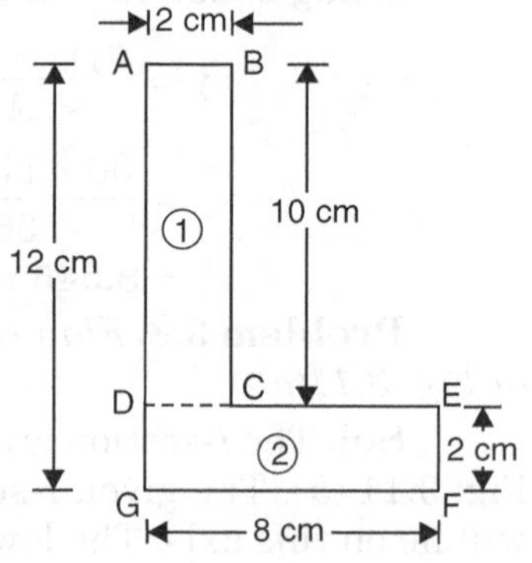

Fig. 3.12

To Find $\bar{y}$

Let $\bar{y}$ = Distance of the C.G. of the *L*-section from bottom line *GF*

a_1 = Area of rectangle *ABCD* = 10 × 2 = 20 cm²

y_1 = Distance of C.G. of rectangle *ABCD* from bottom line *GF*

$$= 2 + \frac{10}{2} = 2 + 5 = 7 \text{ cm}$$

a_2 = Area of rectangle *DEFG* = 8 × 2 = 16 cm²

y_2 = Distance of C.G. of rectangle *DEFG* from bottom line *GF*

$$= \frac{2}{2} = 1.0 \text{ cm.}$$

Using equation (3.2), we have

$$\bar{y} = \frac{a_1y_1 + a_2y_2}{A} \text{ where } A = a_1 + a_2$$

$$= \frac{a_1y_1 + a_2y_2}{a_1 + a_2} = \frac{20 \times 7 + 16 \times 1}{20 + 16} = \frac{140 + 16}{36}$$

$$= \frac{156}{36} = \frac{13}{3} = 4.33 \text{ cm.}$$

To Find $\bar{x}$

Let $\bar{x}$ = Distance of the C.G. of the *L*-section from left line *AG*

x_1 = Distance of the rectangle *ABCD* from left line *AG*

$$= \frac{2}{2} = 1.0 \text{ cm}$$

x_2 = Distance of the rectangle *DEFG* from left line *AG*

$$= \frac{8}{2} = 4.0 \text{ cm.}$$

Using equation (3.1), we get

$$\bar{x} = \frac{a_1x_1 + a_2x_2}{A}, \quad \text{where } A = a_1 + a_2$$

$$= \frac{a_1x_1 + x_2y_2}{a_1 + a_2} = \frac{20 \times 1 + 16 \times 4}{20 + 16} \quad (\because \quad a_1 = 20 \text{ and } a_2 = 16)$$

$$= \frac{20 + 64}{36} = \frac{84}{36} = \frac{7}{3} = 2.33 \text{ cm.}$$

Hence the C.G. of the *L*-section is at a distance of 4.33 cm from the bottom line *GF* and 2.33 cm from the left line *AG*. **Ans.**

Problem 3.10. *Using the analytical method, determine the centre of gravity of the plane uniform lamina shown in Fig. 3.13.* (AMIE Summer, 1975)

Sol. Let $\bar{y}$ be the distance between C.G. of the lamina and the bottom line *AB*.

Area 1

$$a_1 = 10 \times 5 = 50 \text{ cm}^2$$

$$y_1 = \frac{5}{2} = 2.5 \text{ cm}$$

Area 2

$$a_2 = \frac{\pi}{2} \times r^2 = \frac{\pi}{2} \times 2.5^2 = 9.82 \text{ cm}^2$$

$$y_2 = \frac{5}{2} = 2.5 \text{ cm}$$

Area 3

$$a_2 = \frac{5 \times 5}{2} = 12.5 \text{ cm}^2$$

$$y_2 = 5 + \frac{5}{3} = 6.67 \text{ cm.}$$

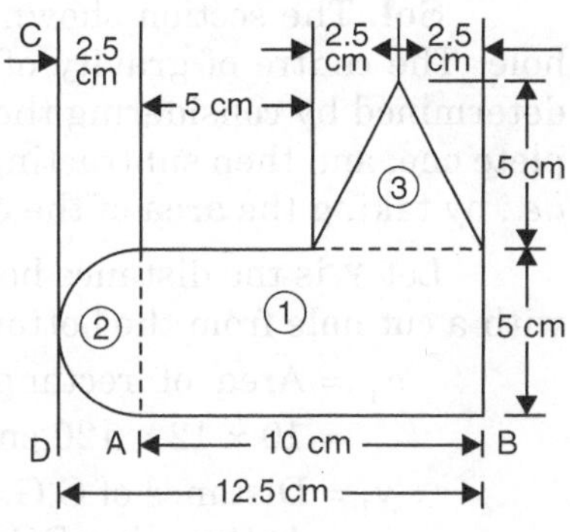

Fig. 3.13

Using the relation,

$$\bar{y} = \frac{a_1y_1 + a_2y_2 + a_3y_3}{a_1 + a_2 + a_3}$$

$$= \frac{50 \times 2.5 + 9.82 + 2.5 + 12.5 \times 6.67}{50 + 9.82 + 12.5} \text{ cm} = \frac{232.9}{72.32} = 3.22 \text{ cm.}$$

Similarly, let $\bar{x}$ be the distance between C.G. of the lamina and the left line *CD*.

Area 1

$$a_1 = 50 \text{ cm}^2$$

$$x_1 = 2.5 + \frac{10}{2} = 7.5 \text{ cm}$$

Area 2

$$a_2 = 9.82 \text{ cm}^2$$

$$x_2 = 2.5 - \frac{4r}{3\pi} = 2.5 - \frac{4.25}{3\pi} \text{ cm} = 1.44 \text{ cm}$$

Area 3

$$a_3 = 12.5 \text{ cm}^2$$

$$x_3 = 2.5 + 5 + 2.5 = 10 \text{ cm.}$$

Now using the relation,

$$\bar{x} = \frac{a_1x_1 + a_2x_2 + a_3x_3}{a_1 + a_2 + a_3} = \frac{50 \times 7.5 + 9.82 \times 1.44 + 12.5 \times 10}{50 + 9.82 + 12.5} \text{ cm}$$

$$= \frac{514.14}{72.32} = 7.11 \text{ cm.}$$

Hence the C.G. of the uniform lamina is at a distance of 3.22 cm from the bottom line *AB* and 7.11 cm from the left line *CD*. **Ans.**

Problem 3.11. *From a rectangular lamina ABCD 10 cm × 12 cm a rectangular hole of 3 cm × 4 cm is cut as shown in Fig. 3.14.*

Find the C.G. of the remainder lamina.

Sol. The section shown in Fig. 3.14, is having a cut hole. The centre of gravity of a section with a cut hole is determined by considering the main section first as a complete one, and then subtracting the area of the cut-out hole, *i.e.*, by taking the area of the cut-out hole as negative.

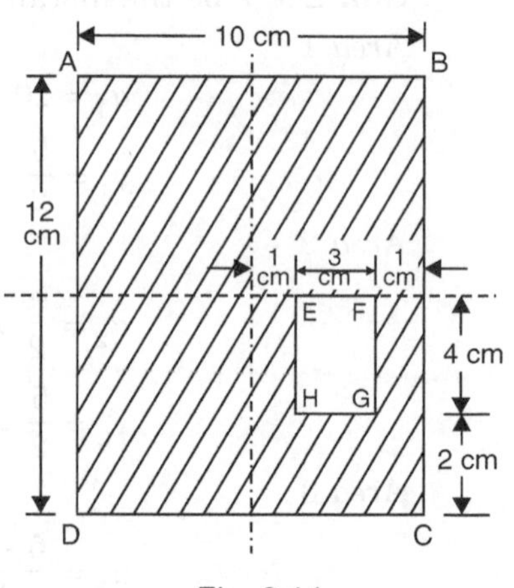

Fig. 3.14

Let $\bar{y}$ is the distance between the C.G. of the section with a cut hole from the bottom line *DC*.

a_1 = Area of rectangle *ABCD*

$= 10 \times 12 = 120 \text{ cm}^2$

y_1 = Distance of C.G. of the rectangle *ABCD* from bottom line *DC*

$$= \frac{12}{2} = 6 \text{ cm}$$

a_2 = Area of cut-out hole, *i.e.*, rectangle *EFGH*,

$= 4 \times 3 = 12 \text{ cm}^2$

y_2 = Distance of C.G. of cut-out hole from bottom line *DC*

$$= 2 + \frac{4}{2} = 2 + 2 = 4 \text{ cm.}$$

Now using equation (3.2) and taking the area (a_2) of the cut-out hole as negative, we get

$$\bar{y} = \left(\frac{a_1y_1 - a_2y_2}{A}\right)^*, \text{ where } A = a_1 - a_2$$

$$= \frac{a_1y_1 - a_2y_2}{a_1 - a_2} \qquad \text{(–ve sign is taken due to cut-out hole)}$$

$$= \frac{120 \times 6 - 12 \times 4}{120 - 12} = \frac{720 - 48}{108} = 6.22 \text{ cm.}$$

To Find $\bar{x}$

Let $\bar{x}$ = Distance between the C.G. of the section with a cut hole from the left line *AD*

x_1 = Distance of the C.G. of the rectangle *ABCD* from the left line *AD*

$$= \frac{10}{2} = 5 \text{ cm}$$

x_2 = Distance of the C.G. of the cut-out hole from the left line *AD*

$$= 5 + 1 + \frac{3}{2} = 7.5 \text{ cm.}$$

Using equation (3.1) and taking area (a_2) of the cut hole as negative, we get

$$\bar{x} = \frac{a_1x_1 - a_2x_2}{a_1 - a_2} \qquad (\because \; A = a_1 - a_2)$$

$$= \frac{120 \times 5 - 12 \times 7.5}{120 - 12} = \frac{600 - 90}{108} = \frac{510}{108} = 4.72 \text{ cm.}$$

Hence the C.G. of the section with a cut hole will be at a distance of 6.22 cm from bottom line *DC* and 4.72 cm from the line *AD*. **Ans.**

3.4. THEOREMS OF PAPPUS

By rotating a plane curve (say a straight line) about a fixed axis for one complete rotation, a surface is generated which is called surface of revolution.

Similarly by rotating a plane area about a fixed axis for one complete revolution, a body is generated, which is called body of revolution. The Pappus theorems are used for determining the surface area or volumes. There are two theorems.

Theorem 1. This is used for determining the surface area. This theorem states that the surface area generated by rotating a plane curve about a non-intersecting axis in the plane of the curve is equal to the product of the length of the curve and the distance travelled by the centroid of the curve for one complete revolution.

According to this theorem,

Surface area generated = Length of plane curve × Distance travelled by C.G. of the plane curve for one revolution.

or
$$A = L \times (\bar{x}\theta) \qquad \text{...(3.3}A\text{)}$$

where A = Surface area generated,

L = Length of curve,

$\bar{x}$ = Distance of C.G. of plane curve from the axis,

θ = Angle turned by C.G. of curve for one revolution = 2π

$$\therefore \quad A = L \times \bar{x} \times 2\pi$$

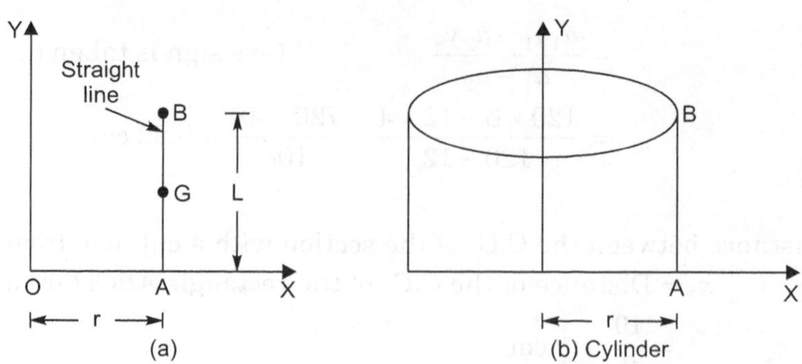

Fig. 3.15

A line *AB* of length *L* and at a distance *r* from axis *OY* as shown in Fig. 3.15 is rotated for one revolution about axis *Y*. This will generate the surface whose area will be equal to

$$A = L \times (\bar{x} \times \theta)$$

where $\bar{x}$ = Distance of C.G. of line *AB* from axis *Y* = *r*,

θ = Angle burned for one revolution = 2π

$\therefore$ Surface area *A* will become as

$$A = L \times r \times 2\pi = 2\pi r \times L$$

Hence the surface of a cylinder is generated when straight line is rotated about *Y*-axis.

Theorem 2. This is used for determining the volume. This theorem states that volume generated by revolving an area about a non-intersecting axis in the plane of area is equal to the product of area and the distance moved by the C.G. of the area for one revolution.

$\therefore$ Volume (*V*) generated

= Area × distance travelled by C.G. of area for one revolution

$= A \times \bar{x} \times \theta$ ...(3.3*B*)

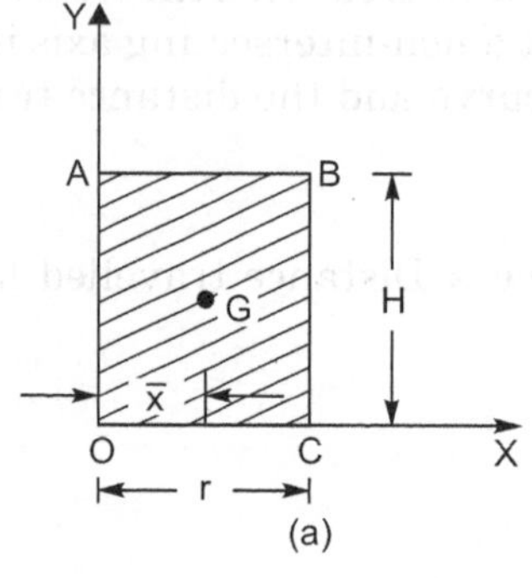

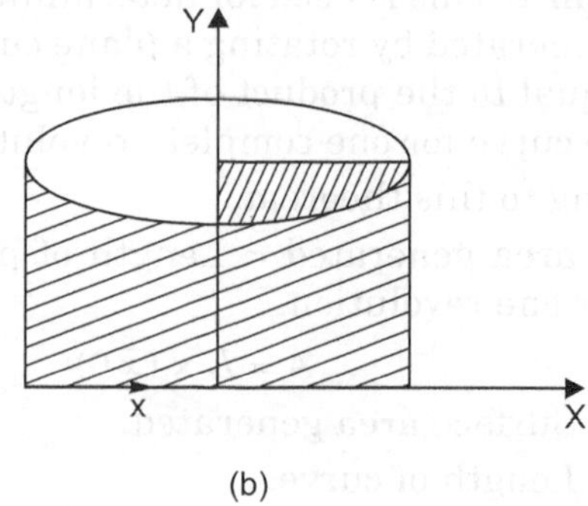

Fig. 3.16

Fig. 3.16 (*a*) shows a rectangular area *OABC* (shaded) of dimensions *r* an *H*. The distance of C.G. of the area from *Y*-axis is $\bar{x} = \frac{OC}{2} = \frac{r}{2}$. It is rotated about *Y*-axis for one revolution. The volume of the body generated will be,

V = Area of rectangle × Distance travelled by C.G. for one revolution

$= (r \times H) \times \bar{x} \times \theta$

$= (r \times H) \times \frac{r}{2} \times 2\pi$ $\left[\because \quad \bar{x} = \frac{r}{2}, \theta = 2\pi\right]$

$= \pi r^2 \times H$ ...(3.3C)

Hence a solid cylinder is generated by rotating a rectangular area.

Other Examples

(*i*) A solid cone is generated by rotating a triangular area [Fig. 3.17 (*a*)]

(*ii*) A solid sphere is generated by rotating a semi-circular area [Fig. 3.17(*b*)]

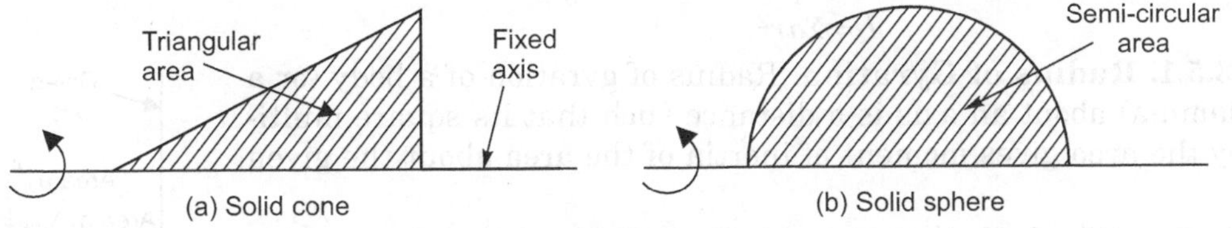

Fig. 3.17

3.5. AREA MOMENT OF INERTIA OF PLANE AREAS

Consider a thick lamina of area A as shown in Fig. 3.18.

Let x = Distance of the C.G. of area A from the axis OY.

y = Distance of the C.G. of area A from the axis OX.

Then moment of area about the axis OY

= Area × perpendicular distance of C.G. of area from axis OY

$= Ax$ ...(3.3D)

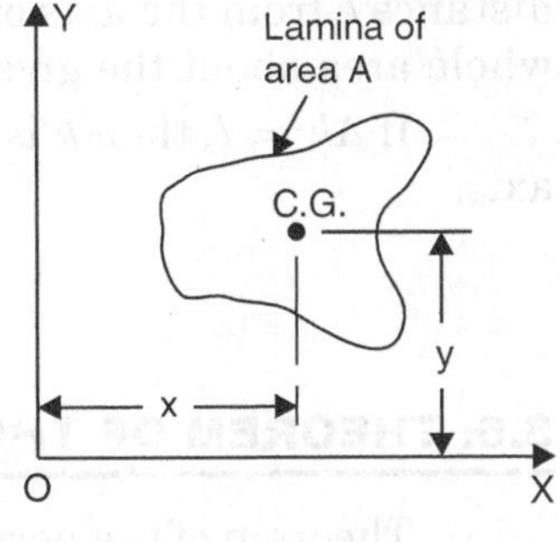

Fig. 3.18

Equation (3.3*D*) is known as *first moment of area* about the axis OY. This first moment of area is used to determine the centroid of the area.

If the moment of area given by equation (3.3*D*) is again multiplied by the perpendicular distance between the C.G. of the area and axis OY (*i.e.*, distance x), then the quantity (Ax). $x = Ax^2$ is known as moment of the moment of area or **second moment of area** or area moment of inertia about the axis OY. This second moment of area is used in the study of mechanics of fluids and mechanics of solids.

Similarly, the moment of area (or first moment of area) about the axis $OX = Ay$.

And second moment of area (or area moment of inertia) about the axis $OX = (Ay) \, . \, y = Ay^2$.

If, instead of area, the mass (m) of the body is taken into consideration then the second moment is known as second moment of mass. This second moment of mass is also known as mass moment of inertia.

Hence moment of inertia when mass is taken into consideration about the axis $OY = mx^2$ and about the axis $OX = my^2$.

Hence the product of the area (or mass) and the square of the distance of the centre of gravity of the area (or mass) from an axis is known as moment of inertia of the area (or mass) about that axis. Moment of inertia is represented by I. Hence moment of intertia about the axis OX is represented by I_{XX} whereas about the axis OY by I_{YY}.

The product of the area (or mass) and the square of the distance of the centre of gravity of the area (or mass) from an axis perpendicular to the plane of the area is known as polar moment of inertia and is represented by J.

Consider a plane area which is split up into small areas $a_1, a_2, a_3, \ldots$ etc. Let the C.G. of the small areas from a given axis be at a distance of $r_1, r_2, r_3, \ldots$ etc. as shown in Fig. 3.19.

Then the moment of inertia of the plane area about the given axis is given by

$$I = a_1r_1^2 + a_2r_2^2 + a_3r_3^2 + \ldots \qquad \ldots(3.4)$$

or

$$I = \Sigma ar^2. \qquad \ldots(3.5)$$

3.5.1. Radius of Gyration. Radius of gyration of a body (or a given lamina) about an axis is a distance such that its square multiplied by the area gives moment of inertia of the area about the given axis.

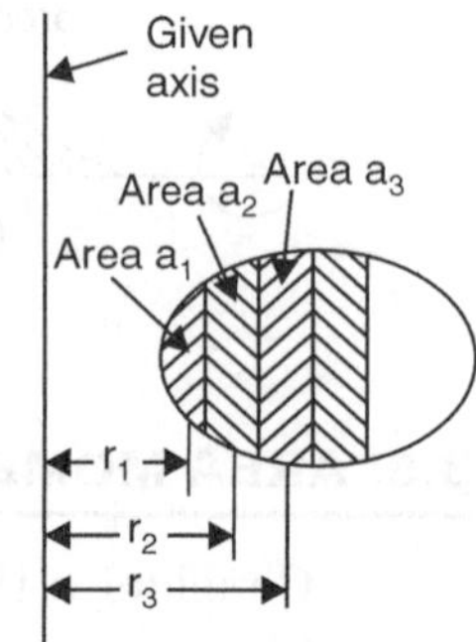

Fig. 3.19

For the Fig. 3.19, the moment of inertia about the given axis is given by equation (3.4) as

$$I = a_1r_1^2 + a_2r_2^2 + a_3r_3^2 + \ldots \qquad \ldots(i)$$

Let the whole mass (or area) of the body is concentrated at a distance k from the axis of reference, then the moment of inertia of the whole area about the given axis will be equal to Ak^2.

If $Ak^2 = I$, then k is known as radius of gyration about the given axis.

$$\therefore \qquad k = \sqrt{\frac{I}{A}}. \qquad \ldots(3.6)$$

3.6. THEOREM OF THE PERPENDICULAR AXIS

Theorem of the perpendicular axis states that if I_{XX} and I_{YY} be the moments of inertia of a plane section about two mutually perpendicular axes X-X and Y-Y in the plane of the section, then the moment of inertia of the section I_{ZZ} about the axis Z-Z, perpendicular to the plane and passing through the intersection of X-X and Y-Y is given by

$$I_{ZZ} = I_{XX} + I_{YY}.$$

The moment of inertia I_{ZZ} is also known as *polar moment of inertia*.

Proof. A plane section of area A and lying in plane x-y is shown in Fig. 3.20. Let OX and OY be the two mutually perpendicular axes, and OZ be the perpendicular axis. Consider a small area dA.

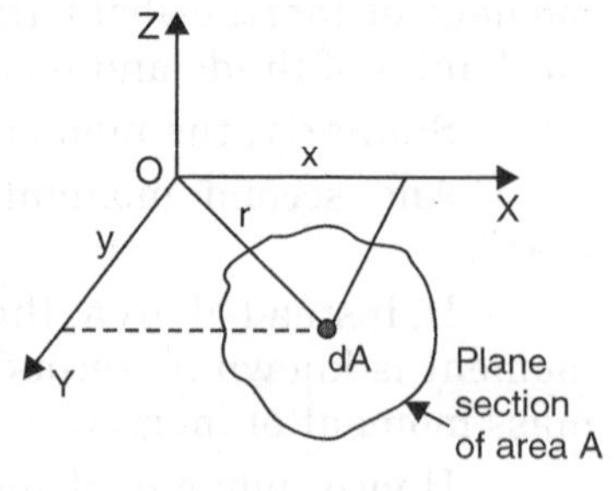

Fig. 3.20

Let x = Distance of dA from the axis OY

y = Distance of dA from axis OX

r = Distance of dA from axis OZ

Then $r^2 = x^2 + y^2$.

Now moment of inertia of dA about x-axis

$= dA \times$ (Distance of dA from x-axis)^2

$= dA \times y^2$.

$\therefore$ Moment of inertia of total area A about x-axis, $I_{XX} = \Sigma dAy^2$.

Similarly, moment of inertia of total area A about y-axis, $I_{YY} = \Sigma dAx^2$

and moment of inertia of total area A about z-axis, $I_{ZZ} = \Sigma dAr^2$

$$= \Sigma dA\ [x^2 + y^2] \qquad (\because\ r^2 = x^2 + y^2)$$

$$= \Sigma dA\ x^2 + \Sigma dA\ y^2$$

$$= I_{YY} + I_{XX}$$

or
$$I_{ZZ} = I_{XX} + I_{YY}. \qquad ...(3.7)$$

The above equation shows that the moment of inertia of an area about an axis at origin normal to x, y plane is the sum of moments of inertia about the corresponding x and y-axes.

3.7. THEOREM OF PARALLEL AXIS

It states that if the moment of inertia of a plane area about an axis in the plane of area through the C.G. of the plane area be represented by I_G, then the moment of the inertia of the given plane area about a parallel axis AB in the plane of area at a distance h from the C.G. of the area is given by

$$I_{AB} = I_G + Ah^2.$$

where I_{AB} = Moment of inertia of the given area about AB

I_G = Moment of inertia of the given area about C.G.

A = Area of the section

h = Distance between the C.G. of the section and the axis AB.

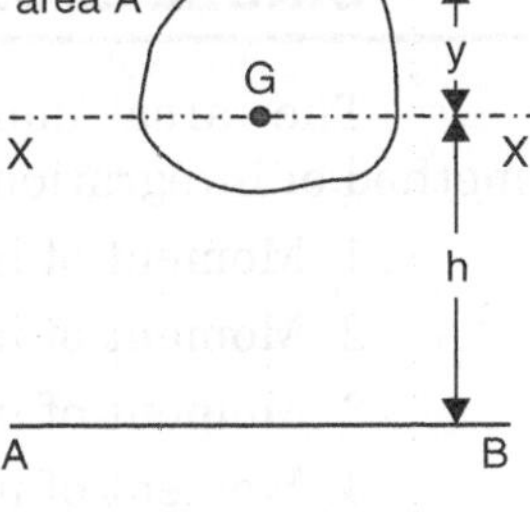

Fig. 3.21

Proof. A lamina of plane area A is shown in Fig. 3.21.

Let X-X = The axis in the plane of area A and passing through the C.G. of the area.

AB = The axis in the plane of area A and parallel to axis X-X.

h = Distance between AB and X-X.

Consider a strip parallel to X-X axis at a distance y from the X-X axis.

Let the area of the strip = dA

Moment of inertia of area dA about X-X axis = dAy^2.

$\therefore$ Moment of inertia of the total area about X-X axis,

$$I_{XX} \text{ or } I_G = \Sigma dAy^2 \qquad ...(i)$$

Moment of inertia of the area dA about AB

$$= dA(h + y)^2$$

$$= dA[h^2 + y^2 + 2hy].$$

$\therefore$ Moment of inertia of the total area A about AB,

$$I_{AB} = \Sigma dA[h^2 + y^2 + 2hy]$$

$$= \Sigma dAh^2 + \Sigma dAy^2 + \Sigma dA^2hy.$$

As h or h^2 is constant and hence they can be taken outside the summation sign. Hence the above equation becomes

$$I_{AB} = h^2\Sigma dA + \Sigma dAy^2 + 2h\Sigma dAy.$$

But $\Sigma dA = A$. Also from equation (*i*), $\Sigma dAy^2 = I_G$. Substituting these values in the above equation, we get

$$I_{AB} = h^2 . A + I_G + 2h\Sigma dAy. \qquad ...(ii)$$

But $dA . y$ represents the moment of area of strip about *X-X* axis. And ΣdAy represents the moments of the total area about *X-X* axis. But the moments of the total area about *X-X* axis is equal to the product of total area (*A*) and the distance of the C.G. of the total area from *X-X* axis. As the distance of the C.G. of the total area from *X-X* axis is zero, hence ΣdAy will be equal to zero.

Substituting this value in equation (*ii*), we get

$$I_{AB} = h^2 . A + I_G + 0$$

or

$$I_{AB} = I_G + Ah^2 \qquad ...(3.8)$$

Thus if the moment of inertia of an area with respect to an axis in the plane of area (and passing through the C.G. of the area) is known, the moment of inertia with respect to any parallel axis in the plane may be determined by using the above equation.

3.8. AREA MOMENT OF INERTIA OF PLANE AREA LIKE RECTANGLE, TRIANGLE, CIRCLE ETC. BY INTEGRATION

The area moment of inertia of the following sections will be determined by the method of integration :

1. Moment of inertia of a rectangular section,
2. Moment of inertia of a circular section,
3. Moment of inertia of a triangular section,
4. Moment of inertia of a uniform thin rod.

3.8.1. Moment of Inertia of a Rectangular Section

1st Case. ***Moment of inertia of the rectangular section about the X-X axis passing through the C.G. of the section.***

Fig. 3.22 shows a rectangular section *ABCD* having width $= b$ and depth $= d$. Let *X-X* is the horizontal axis passing through the C.G. of the rectangular section. We want to determine the moment of inertia of the rectangular section about *X-X* axis. The moment of inertia of the given section about *X-X* axis is represent by I_{XX}.

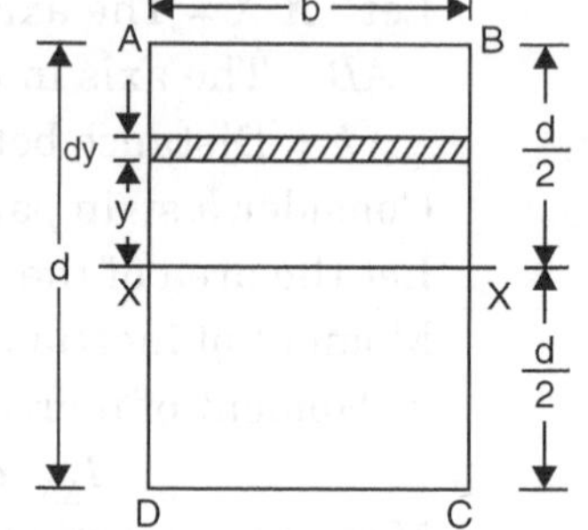

Fig. 3.22

Consider a rectangular elementary strip of thickness dy at a distance y from the *X-X* axis as shown in Fig. 3.22.

Area of the strip $= b \cdot dy$.

Moment of inertia of the area of the strip about *X-X* axis

$$= \text{Area of strip} \times y^2$$

$$= (b \cdot dy) \times y^2 = by^2 dy.$$

Moment of inertia of the whole section will be obtained by integrating the above equation between the limits $-\dfrac{d}{2}$ to $\dfrac{d}{2}$.

$$\therefore \qquad I_{XX} = \int_{-\frac{d}{2}}^{\frac{d}{2}} by^2 dy = b \int_{-\frac{d}{2}}^{\frac{d}{2}} y^2 dy$$

($\because$ **b is constant and can be taken outside the integral sign)**

$$= b\left[\frac{y^3}{3}\right]_{-\frac{d}{2}}^{\frac{d}{2}} = \frac{b}{3}\left[\left(\frac{d}{2}\right)^3 - \left(-\frac{b}{2}\right)^2\right]$$

$$= \frac{b}{3}\left[\frac{d^3}{8} - \left(-\frac{d^3}{8}\right)\right] = \frac{b}{3}\left[\frac{d^3}{8} + \frac{d^3}{8}\right]$$

$$= \frac{b}{3} \cdot \frac{2d^3}{8} = \frac{bd^3}{12}. \qquad ...(3.9)$$

Similarly, the moment of inertia of the rectangular section about Y-Y axis passing through the C.G. of the section is given by

$$I_{YY} = \frac{db^3}{12}. \qquad ...(3.10)$$

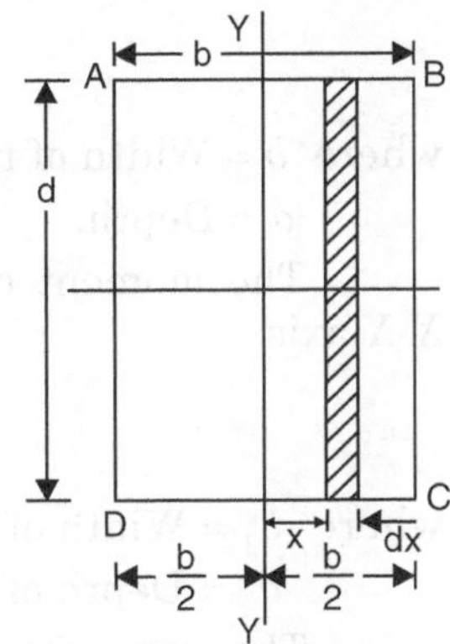

Fig. 3.22(a)

Refer to Fig. 3.22 (*a*)

Area of strip, $dA = d \times dx$

M.O.I. of strip about Y-Y axis $= dA \times x^2$

$$= (d \times dx) \times x^2 \qquad (\because \; dA = d \,.\, dx)$$

$$= d \times x^2 \times dx$$

$$\therefore \qquad I_{YY} = \int_{-\frac{b}{2}}^{\frac{b}{2}} d \times x^2 \times dx = d\left[\frac{x^3}{3}\right]_{-\frac{b}{2}}^{\frac{b}{2}}$$

$$= \frac{d}{3}\left[\left(\frac{b}{2}\right)^3 - \left(-\frac{b}{2}\right)^3\right]$$

$$= \frac{d}{3}\left[\frac{b^3}{8} + \frac{b^3}{8}\right] = \frac{d}{3} \cdot \frac{b^3}{4} = \frac{db^3}{12}$$

2nd Case. *Moment of inertia of the rectangular section about a line passing through the base.*

Fig. 3.23 shows a rectangular section $ABCD$ having width $= b$ and depth $= d$. We want to find the moment of inertia of the rectangular section about the line CD, which is the base of the rectangular section.

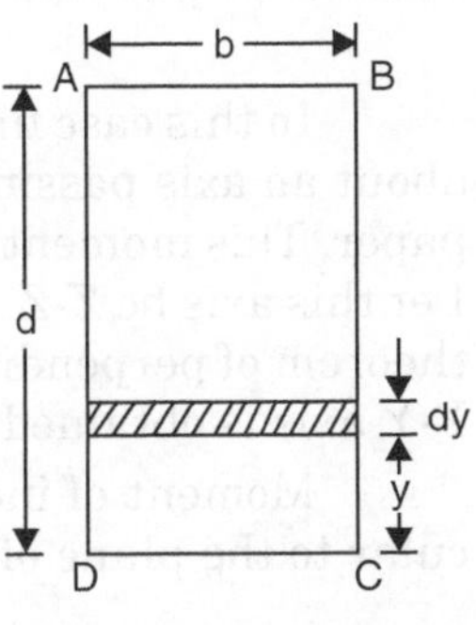

Fig. 3.23

Consider a rectangular elementary strip of thickness dy at a distance y from the line CD as shown in Fig. 3.23.

Area of strip $= b \,.\, dy$.

Moment of inertia of the area of strip about the line CD

$$= \text{Area of strip} \,.\, y^2$$

$$= b \,.\, dy \,.\, y^2 = by^2 \, dy.$$

Moment of inertia of the whole section about the line *CD* is obtained by integrating the above equation between the limits *o* to *d*.

∴ Moment of inertia of the whole section about the line *CD*

$$= \int_0^d by^2 dy = b \int_0^d y^2 dy$$

$$= b \left[\frac{y^3}{3} \right]_0^d = \frac{bd^3}{3}. \qquad ...(3.11)$$

3rd Case. ***Moment of inertia of a hollow rectangular section.***

Fig. 3.24 shows a hollow rectangular section in which *ABCD* is the main section and *EFGH* is the cut-out section.

The moment of inertia of the main section *ABCD* about *X-X* axis is given by equation (3.7),

$$= \frac{bd^3}{12}$$

where b = Width of main section

d = Depth.

The moment of inertia of the cut-out section *EFGH* about *X-X* axis

$$= \frac{b_1 {d_1}^3}{12}$$

Fig. 3.24

where b_1 = Width of the cut-out section and

d_1 = Depth of the cut-out section.

Then moment of inertia of hollow rectangular section about *X-X* axis,

I_{XX} = Moment of inertia of rectangle *ABCD* about *X-X* axis – moment of inertia of rectangle *EFGH* about *X-X* axis

$$= \frac{bd^3}{12} - \frac{b_1 {d_1}^3}{12}.$$

3.8.2. Moment of Inertia of a Circular Section. Fig. 3.25 shows a circular section of radius *R* with *O* as centre. Consider an elementary circular ring of radius '*r*' and thickness '*dr*'. Area of circular ring

$$= 2\pi r.\ dr.$$

In this case first find the moment of inertia of the circular section about an axis passing through *O* and perpendicular to the plane of the paper. This moment of inertia is also known as *polar moment of inertia*. Let this axis be *Z-Z*. (Axis *Z-Z* is not shown in Fig. 3.25). Then from the theorem of perpendicular axis, the moment of inertia about *X-X* axis or *Y-Y* axis is obtained.

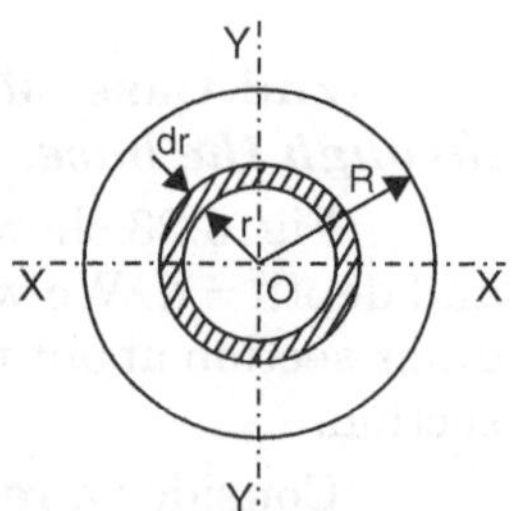

Fig. 3.25

Moment of inertia of the circular ring about an axis passing through *O* and perpendicular to the plane of the paper

= (Area of ring) × (radius of ring from O)^2

$= (2\pi r\ .\ dr)\ .\ r^2$

$= 2\pi r^3 dr \qquad ...(i)$

Moment of inertia of the whole circular section is obtained by integrating equation (*i*) between the limit *O* to *R*.

∴ Moment of inertia of the whole section about an axis passing through *O* and perpendicular to the plane of paper is given as

$$I_{ZZ} = \int_0^R 2\pi r^3\, dr = 2\pi \int_0^R r^3 dr$$

$$= 2\pi \left[\frac{r^4}{4}\right]_0^R = 2\pi \frac{R^4}{4} = \frac{\pi R^4}{2}.$$

But $R = \dfrac{D}{2}$

where D = Diameter of the circular section

$$\therefore \quad I_{ZZ} = \frac{\pi}{2} \times \left(\frac{D}{2}\right)^4 = \frac{\pi D^4}{32} \quad \text{...(3.12)}$$

But from the theorem of perpendicular axis given by equation (3.7), we have

$$I_{ZZ} = I_{XX} + I_{YY}.$$

But due to symmetry, $I_{XX} = I_{YY}$

$$\therefore \quad I_{XX} = I_{YY} = I_{ZZ}/2.0$$

$$= \frac{\pi D^4}{32} \times \frac{1}{2} = \frac{\pi D^4}{64} \quad \text{...(3.13)}$$

Moment of inertia of a hollow circular section

Fig. 3.26 shows a hollow circular section.

Let D = Diameter of outer circle, and

d = Diameter of cut-out circle.

Then from equation (3.13), the moment of inertia of the outer circle about *X-X* axis $= \dfrac{\pi}{64} D^4.$

And moment of inertia of the cut-out circle about *X-X* axis

$$= \frac{\pi}{64} d^4.$$

∴ Moment of inertia of the hollow circular section, about *X-X* axis,

I_{XX} = Moment of inertia of outer circle – moment of inertia of cut-out circle

$$= \frac{\pi}{64} D^4 - \frac{\pi}{64} d^4 = \frac{\pi}{64} [D^4 - d^4]$$

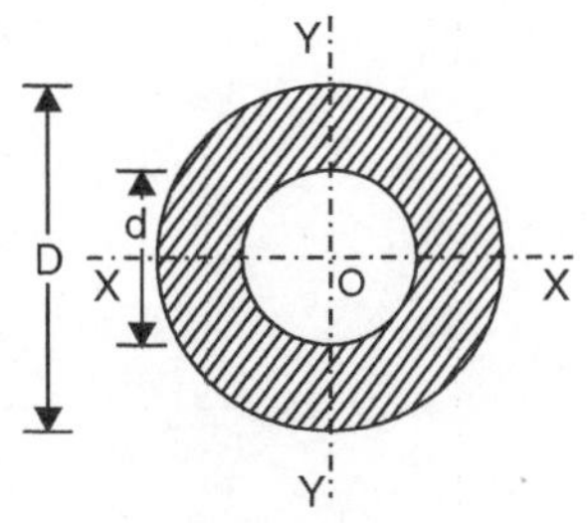

Fig. 3.26

Similarly, $I_{YY} = \dfrac{\pi}{64} [D^4 - d^4].$

3.8.3. Moment of inertia of a triangular section

1st Case. ***Moment of inertia of a triangular section about its base.***

Fig. 3.27 shows a triangular section *ABC* of base width = *b* and height = *h*. Consider a small strip of thickness *dy* at a distance *y* from the vertex *A*.

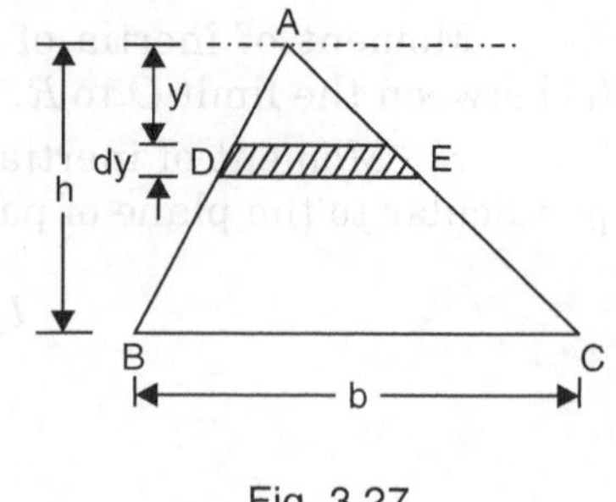

Fig. 3.27

Area of the strip = $DE \,.\, dy$ ...(i)

The distance DE in terms of y, b and h is obtained from two similar triangles ADE and ABC as

$$\frac{DE}{BC} = \frac{y}{h}$$

$$\therefore \quad DE = BC \,.\, \frac{y}{h}$$

$$= \frac{b \,.\, y}{h} \qquad (\because \; BC = b)$$

Substituting this value of DE in equation (i), we get

$$\text{Area of strip} = \frac{by}{h} \,.\, dy.$$

Distance of the strip from the base = $(h - y)$

∴ Moment of inertia of the strip about the base

= Area of strip × (Distance of strip from base)^2

$$= \frac{by}{h} \,.\, dy \,.\, (h - y)^2 = \frac{by}{h} (h - y)^2 \,.\, dy.$$

The moment of inertia of the whole triangular section about the base (I_{BC}) is obtained by integrating the above equation between the limits O to h.

$$\therefore \quad I_{BC} = \int_0^h \frac{by}{h} [h - y]^2 \, dy$$

$$= \frac{b}{h} \int_0^h y(h - y)^2 \, dy$$

($\because$ b and h are constants and can be taken outside the integral sign)

$$= \frac{b}{h} \int_0^h y\,(h^2 + y^2 - 2hy)\, dy = \frac{b}{h} \int_0^h (yh^2 + y^3 - 2hy^2)\, dy$$

$$= \frac{b}{h} \left[\frac{y^2 h^2}{2} + \frac{y^4}{4} - \frac{2hy^3}{2} \right]_0^h$$

$$= \frac{b}{h} \left[\frac{h^2 \,.\, h^2}{2} + \frac{h^4}{4} - \frac{2h \,.\, h^3}{3} \right] = \frac{b}{h} \left[\frac{h^4}{2} + \frac{y^4}{4} - \frac{2h^4}{3} \right]$$

$$= \frac{b}{h} \,.\, h^4 \left[\frac{6 + 3 - 8}{12} \right] = bh^3 . \frac{1}{12}$$

$$= \frac{bh^3}{12} \qquad \text{...(3.14)}$$

2nd Case. ***Moment of inertia of the triangular section about an axis passing through the C.G. and parallel to the base.***

Consider a triangular section of base = b and height = h as shown in Fig. 3.28. Let X-X is the axis passing through the C.G. of the triangular section and parallel to the base.

The distance between the C.G. of the triangular section and base $AB = \dfrac{h}{3}$.

Now from the theorem of parallel axis, given by equation (3.8), we have

Moment of inertia about

BC = Moment of inertia about C.G. + Area × (Distance between X-X and $BC)^2$

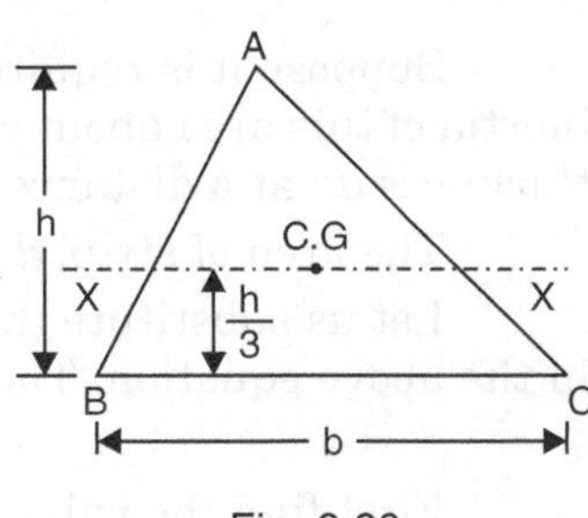

Fig. 3.28

or $$I_{BC} = I_G + A \times \left(\frac{h}{3}\right)^2$$

$$\therefore \quad I_G = I_{BC} - A \times \left(\frac{h}{3}\right)^2$$

$$= \frac{bh^3}{12} - \left(\frac{b \times h}{2}\right) . \left(\frac{h}{3}\right)^2 \qquad \left(\because \; I_{BC} = \frac{bh^3}{12} \text{ and Area} = \frac{b \times h}{2}\right)$$

$$= \frac{bh^3}{12} - \frac{bh^3}{18} = \frac{bh^3(3-2)}{36}$$

$$= \frac{bh^3}{36}. \qquad \text{...(3.15)}$$

3.8.4. Moment of Inertia of a Uniform Thin Rod. Consider a uniform thin rod AB of length L as shown in Fig. 3.29.

Let m = Mass per unit length of rod, and

M = Total mass of the rod

$= m \times L$ ...(*i*)

Suppose it is required to find the moment of inertia of the rod about the axis Y-Y. Consider a strip of length dx at a distance x from the axis Y-Y.

Fig. 3.29

Mass of the strip = Length of strip × Mass per unit length

$= dx \,.\, m$ or $m \,.\, dx$.

Moment of inertia of the strip about Y-Y axis

= Mass of strip × x^2

$= (m \,.\, dx) \,.\, x^2 = mx^2dx$.

Moment of inertia of the whole rod (I_{YY}) will be obtained by integrating the above equation between the limits O to L.

$$\therefore \quad I_{YY} = \int_0^L mx^2dx = m\int_0^L x^2\,dx \qquad (\because \; m \text{ is constant})$$

$$= m\left[\frac{x^3}{3}\right]_0^L = \frac{mL^3}{3}$$

$$= \frac{mL \,.\, L^2}{3} = \frac{mL^2}{3} \qquad [\because \; m \,.\, L = M \text{ from equation } (i)]$$

3.8.5. Moment of Inertia of Area under a Curve of Given Equation. Fig. 3.30 shows an area under a curve whose equation is parabolic and is given by

$x = ky^2$

in which $y = b$ when $x = a$

Suppose it is required to find the moment of inertia of this area about y-axis. Consider a strip of thickness dx at a distance x from y-axis.

The area of strip, $dA = y\,dx$ ...(*i*)

Let us substitute the value of y in terms of x in the above equation. The equation of curve is

$$x = ky^2 \quad ...(ii)$$

First find the value of k.

When $y = b$, $x = a$. Hence above equation becomes

$$a = kb^2$$

or

$$k = \frac{a}{b^2}$$

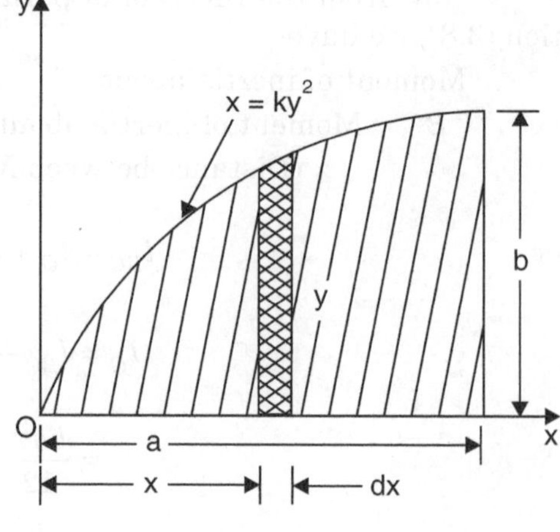

Fig. 3.30

Substituting the value of k in equation (*ii*), we get

$$x = \frac{a}{b^2} \cdot y^2 \quad \text{or} \quad y^2 = \frac{b^2 x}{a}$$

or

$$y = \left(\frac{b^2 x}{a}\right)^{1/2} = \frac{b}{\sqrt{a}} \sqrt{x} \quad ...(iii)$$

Substituting this value of y in equation (*i*), we get

$$dA = \frac{b}{\sqrt{a}} \cdot \sqrt{x} \cdot dx$$

The moment of inertia of elemental area (dA) about y-axis

$$= x^2 . dA = x^2 \cdot \frac{b}{\sqrt{a}} \cdot \sqrt{x}\, dx$$

$\therefore$ Moment of inertia of the total area about y-axis is obtained by integrating the above equation between the limits O to a. ($\because$ x varies from O to a)

$$\therefore \quad I_{YY} = \int_0^a x^2 \cdot \frac{b}{\sqrt{a}} \cdot \sqrt{x} \cdot dx = \frac{b}{\sqrt{a}} \cdot \int_0^a x^{5/2} . dx$$

$$= \frac{b}{\sqrt{a}} \left[\frac{x^{7/2}}{7/2}\right]_0^a = \frac{2}{7} \cdot \frac{b}{\sqrt{a}} \cdot a^{7/2} = \frac{2}{7}\,\mathbf{ba^2}. \quad \textbf{Ans.}$$

To find the moment of inertia of the given area about x-axis, the element shown in Fig. 3.30 can be considered to be a rectangle of thickness dx. The moment of inertia of this element about x-axis is equal to the moment of inertia of the rectangle about its base.

$\therefore$ Moment of inertia of the element about x-axis

$$= \frac{dx \,.\, y^3}{3} \qquad \left[\because \text{ it is } \frac{bd^3}{3} \text{ where } b = dx \text{ and } d = y\right]$$

The moment of inertia of the given area about x-axis is obtained by integrating the above equation between the limits O to a.

$$\therefore \qquad I_{XX} = \int_0^a \frac{dx \cdot y^3}{3} = \int_0^a \frac{y^3}{3} \cdot dx$$

$$= \int_0^a \frac{\left[\frac{b}{\sqrt{a}} \cdot \sqrt{x}\right]^3 dx}{3} \qquad \left[\because \; y \frac{b}{\sqrt{a}} \sqrt{x} \text{ from equation } (iii)\right]$$

$$= \frac{b^3}{3a^{3/2}} \int_0^a \cdot x^{3/2} \, dx = \frac{b^3}{3a^{3/2}} \left[\frac{x^{5/2}}{5/2}\right]_0^a$$

$$= \frac{b^3}{3 \cdot a^{3/2}} \cdot \frac{2}{5} \cdot a^{5/2} = \frac{2}{15} b^3 \cdot a$$

$$= \frac{2}{15} \mathbf{ab^3}. \quad \textbf{Ans.}$$

3.9. MOMENT OF INERTIA OF T-SECTION, I-SECTION, ANGLE-SECTION, HOLLOW-SECTION ETC. BY USING STANDARD FORMULA

First of all, the location of the centroid of the given section is determined. Then the given section is splitted into rectangles or triangles. The moment of inertia of the rectangles is determined about its centroid. Then this moment of inertia is transferred about the axis passing through the centroid of the given section, using theorem of parallel axis.

Problem 3.12. *Fig. 3.31 shows a T-section of dimensions 10 × 10 × 2 cm. Determine the moment of inertia of the section about the horizontal and vertical axes, passing through the centre of gravity of the section.*

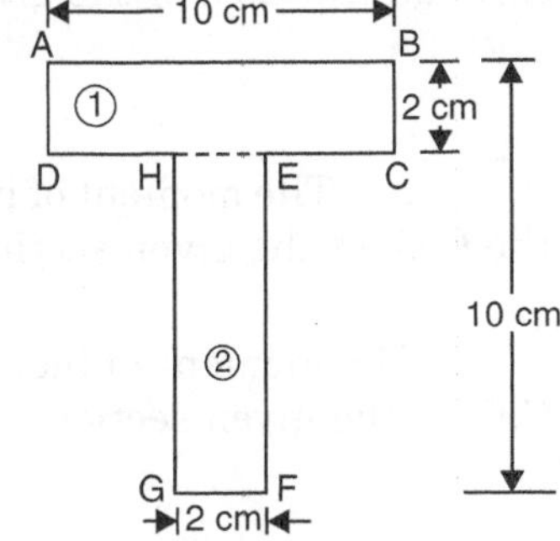

Fig. 3.31

Sol. First of all, find the location of centre of gravity of the given *T*-section. The given section is symmetrical about the axis *Y-Y* and hence the C.G. of the section will lie on *Y-Y* axis. The given section is split up into two rectangles *ABCD* and *EFGH* for calculating the C.G. of the section.

Let $\bar{y}$ = Distance of the C.G. of the section from the bottom line *GF*

a_1 = Area of rectangle *ABCD* = 10 × 2 = 20 cm²

y_2 = Distance of C.G. of the area a_1 from the bottom line *GF* = 8 + 1 = 9 cm

a_2 = Area of rectangle *EFGH* = 8 × 2 = 16 cm²

y_2 = Distance of C.G. of rectangle *EFGH* from the bottom line *GF* = $\frac{8}{2}$ = 4 cm

Using the relation, $$\bar{y} = \frac{a_1y_1 + a_2y_2}{a_1 + a_2} = \frac{20 \times 9 + 16 \times 4}{20 + 16} = \frac{180 + 64}{36}$$

$$= \frac{244}{36} = 6.777 \text{ cm.}$$

Hence the C.G. of the given section lies at a distance of 6.777 cm from *GF*. Now find the moment of inertia of the *T*-section.

Now, Let I_{G_1} = Moment of inertia of rectangle ① about the horizontal axis and passing through its C.G.

I_{G_2} = Moment of inertia of rectangle ② about the horizontal axis and passing through the C.G. of the rectangle ②

h_1 = The distance between the C.G. of the given section and the C.G. of the rectangle ①

$= y_1 - \bar{y} = 9.0 - 6.777 = 2.223$ cm

h_2 = The distance between the C.G. of the given section and the C.G. of the rectangle ②

$= \bar{y} - y_2 = 6.777 - 4.0 = 2.777$ cm.

Now,
$$I_{G_1} = \frac{10 \times 2^3}{2} = 6.667 \text{ cm}^4$$
$$I_{G_2} = \frac{2 \times 8^3}{12} = 85.333 \text{ cm}^4.$$

From the theorem of parallel axes, the moment of inertia of the rectangle ① about the horizontal axis passing through the C.G. of the given section

$$= I_{G_1} + a_1 h_1^2 = 6.667 + 20 \times (2.223)^2$$
$$= 6.667 + 98.834 = 105.501 \text{ cm}^4.$$

Similarly, the moment of inertia of the rectangle ② about the horizontal axis passing through the C.G. of the given section

$$= I_{G_2} + a_2 h_2^2 = 85.333 + 16 \times (2.777)^2$$
$$= 85.333 + 123.387 = 208.72 \text{ cm}^4.$$

∴ The moment of inertia of the given section about the horizontal axis passing through the C.G. of the given section

$$= 105.501 + 208.72 = \mathbf{314.221 \text{ cm}^4}. \quad \textbf{Ans.}$$

The moment of inertia of the given section about the vertical axis passing through the C.G. of the given section

$$= \frac{2 \times 10^3}{12} + \frac{8 \times 2^3}{12}$$
$$= 166.67 + 5.33 = \mathbf{172 \text{ cm}^4}. \quad \textbf{Ans.}$$

Problem 3.13. *Find the moment of inertia of the section shown in Fig. 3.32 about the centroidal axis X-X perpendicular to the web.* (AMIE Summer, 1977)

Sol. First of all find the location of centre of gravity of the given figure. The given section is symmetrical about the axis *Y-Y* and hence the C.G. of the section will lie on *Y-Y* axis. The given section is split up into three rectangles *ABCD, EFGH* and *JKLM*. The centre of gravity of the section is obtained by using

$$\bar{y} = \frac{a_1 y_1 + a_2 y_2 + a_3 y_3}{a_1 + a_2 + a_3} \quad \ldots(i)$$

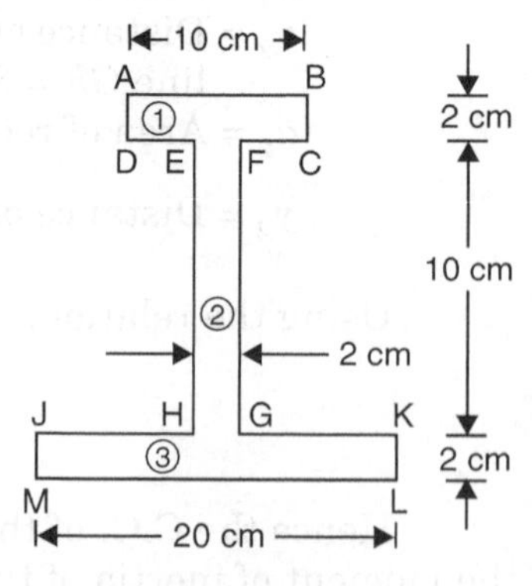

Fig. 3.32

where $\bar{y}$ = Distance of the C.G. of the section from the bottom line *ML*

a_1 = Area of rectangle *ABCD* = 10 × 2 = 20 cm²

y_1 = Distance of the C.G. of the rectangle *ABCD* from the bottom line *ML*

$$= 2 + 10 + \frac{2}{2} = 12 + 1 = 13 \text{ cm}$$

a_2 = Area of rectangle *EFGH* = 10 × 2 = 20 cm²

y_2 = Distance of the C.G. of rectangle *EFGH* from the bottom line *ML*

$$= 2 + \frac{10}{2} = 2 + 5 = 7 \text{ cm}$$

a_3 = Area of rectangle *JKLM* = 20 × 2 = 40 cm²

y_3 = Distance of the C.G. of rectangle *JKLM* from the bottom line *ML*

$$= \frac{2}{2} = 1.0 \text{ cm.}$$

Substituting the above values in equation (*i*), we get

$$\bar{y} = \frac{20 \times 13 + 20 \times 7 + 40 \times 1}{20 + 20 + 40}$$

$$= \frac{260 + 140 + 40}{80} = \frac{440}{80} = 5.50 \text{ cm.}$$

The C.G. of the given section lies at a distance of 5.50 cm from the bottom line *ML*. We want to find the moment of inertia of the given section about a horizontal axis passing through the C.G. of the given section.

Let I_{G_1} = Moment of inertia of rectangle ① about the horizontal axis passing through its C.G.

I_{G_2} = Moment of inertia of rectangle ② about the horizontal axis passing through the C.G. of rectangle ②

I_{G_3} = Moment of inertia of rectangle ③ about the horizontal axis passing through the C.G. of rectangle ③

h_1 = The distance between the C.G. of the rectangle ① and the C.G. of the given section

$= y_1 - \bar{y} = 13.0 - 5.50 = 7.50$ cm

h_2 = The distance between the C.G. of rectangle ② and the C.G. of the given section

$= y_2 - \bar{y} = 7.0 - 5.50 = 1.50$ cm

h_3 = The distance between the C.G. of the rectangle ③ and the C.G. of the given section

$= \bar{y} - y_3 = 5.50 - 1.0 = 4.5$ cm

Now,
$$I_{G_1} = \frac{10 \times 2^3}{12} = 6.667 \text{ cm}^4$$

$$I_{G_2} = \frac{2 \times 10^3}{12} = 166.667 \text{ cm}^4$$

$$I_{G_3} = \frac{20 \times 2^3}{12} = 13.333 \text{ cm}^4.$$

From the theorem of parallel axes, the moment of inertia of the rectangle ① about the horizontal axis passing through the C.G. of the given section

$= I_{G_1} + a_1h_1^2 = 6.667 + 20 \times (7.5)^2$

$= 6.667 + 1125 = 1131.667 \text{ cm}^4.$

Similarly, the moment of inertia of the rectangle ② about the horizontal axis passing through the C.G. of the given section

$= I_{G_2} + a_2h_2^2 = 166.667 + 20 \times 1.5^2$

$= 166.667 + 45 = 211.667 \text{ cm}^4.$

And moment of inertia of the rectangle ③ about the horizontal axis, passing through the C.G. of the given section

$= I_{G_3} + a_3h_3^2 = 13.333 + 40 \times 4.5^2$

$= 13.333 + 810 = 823.333 \text{ cm}^4$

Now moment of inertia of the given section about the horizontal axis, passing through the C.G. of the given section

= Sum of the moment of inertia of the rectangles ①, ② and ③ about the horizontal axis, passing through the C.G. of the given section

$= 1131.667 + 211.667 + 823.333 =$ **2166.667 cm⁴. Ans.**

Problem 3.14. *Determine the polar moment of inertia of I-section shown in Fig. 3.32 (a). (All dimensions are in mm).* (U.P. Tech. University, 2001–2002)

Sol. Let us first find the location of C.G. of the given section. It is symmetrical about the vertical axis, hence C.G. lies on this section.

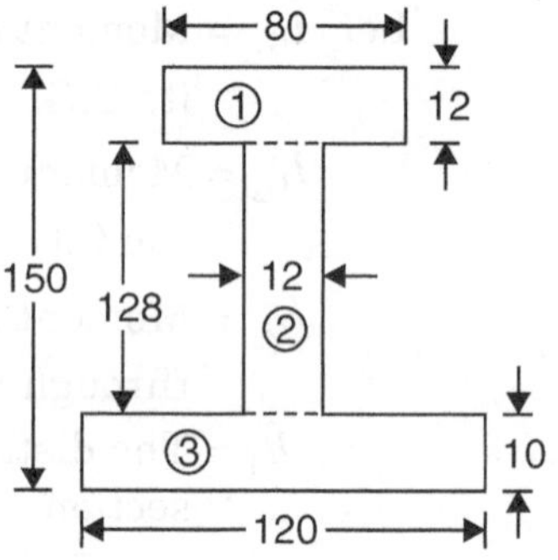

Fig. 3.32(a)

Now, A_1 = Area of first rectangle = $80 \times 12 = 960 \text{ mm}^2$

A_2 = Area of second rectangle $[(150 - 12 - 10) \times 12]$

$= 128 \times 12 = 1536 \text{ mm}^2$

A_3 = Area of third rectangle = $120 \times 10 = 1200 \text{ mm}^2$

y_1 = Distance of C.G. of area A_1 from bottom line

$= 150 - \frac{12}{2} = 144 \text{ mm}$

y_2 = Distance of C.G. of area A_2 from bottom line

$= 10 + \frac{128}{2} = 74 \text{ mm}$

y_3 = Distance of C.G. of area A_3 from bottom line $= \frac{10}{2} = 5$ cm.

$\bar{y}$ = Distance of C.G. of the given section from bottom line.

The C.G. of the section is obtained by using,

$$\bar{y} = \frac{A_1y_1 + A_2y_2 + A_3y_3}{A_1 + A_2 + A_3}$$

$$= \frac{960 \times 144 + 1536 \times 74 + 1200 \times 5}{960 + 1536 + 1200}$$

$$= \frac{138240 + 113664 + 6000}{3696} = \frac{257904}{3696} = 69.779 \simeq 69.78 \text{ cm}$$

Location of centroidal axis is shown in Fig. 3.32 (*b*).

(*i*) **Moment of inertia of the given section about X-X**

M.O.I. of the rectangle ① about centroid axis *X-X* is given by,

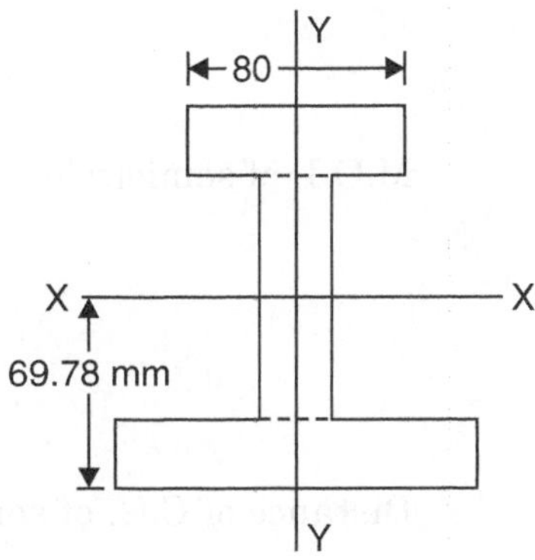

Fig. 3.32(b)

$$I_{XX1} = (I_{G_1})_X + A_1 \times h_1^2 \text{ where } h_1 = (y_1 - \bar{y})$$

$$= \frac{80 \times 12^3}{12} + 960(144 - 69.78)^2 = 5.3 \times 10^6 \text{ mm}^4$$

M.O.I. of rectangle ② about centroid axis *X-X* is given by,

$$I_{XX2} = (I_{G_2})_X + A_2 \times h_2^2 \text{ where } h_2 = (y_2 - \bar{y})$$

$$= \frac{12 \times 128^3}{12} + 1536 \times (74 - 69.78) = 2.12 \times 10^6 \text{ mm}^4$$

and $$I_{XX3} = (I_{G_3})_X + A_3 \times h_3^2 \text{ where } h_3 = (y_3 - \bar{y})$$

$$= \frac{120 \times 10^3}{12} + 1200 \times (5 - 69.78)^2 = 5.04 \times 10^6 \text{ mm}^4$$

$$\therefore \quad I_{XX} = I_{XX1} + I_{XX2} + I_{XX3}$$

$$= 5.3 \times 10^6 + 2.12 \times 10^6 + 5.04 \times 10^6 \text{ mm}^4$$

$$= 12.46 \times 10^6 \text{ mm}^4$$

(*ii*) **Moment of inertia of the given section about Y-Y**

$$I_{YY1} = (I_{G_1})_Y = \frac{12 \times 80^3}{12} = 521 \times 10^3 \text{ mm}^4 = 0.521 \times 10^6 \text{ mm}^4$$

$$I_{YY2} = (I_{G_2})_Y = \frac{128 \times 12^3}{12} = 18.432 \times 10^3 \text{ mm}^4 = 0.018432 \times 10^6 \text{ mm}^4$$

$$I_{YY3} = (I_{G_3})_Y = \frac{10 \times 120^3}{12} = 1.44 \times 10^6 \text{ mm}^4$$

$$\therefore \quad I_{Y\text{-}Y} = I_{YY1} + I_{YY2} + I_{YY3}$$

$$= 0.521 \times 10^6 + 0.018432 \times 10^6 + 1.44 \times 10^6 \text{ mm}^4 = 1.979 \times 10^6 \text{ mm}^4$$

∴ Polar moment of inertia (I_{ZZ}) is given by,

$$I_{ZZ} = I_{XX} + I_{YY}$$

$$= 12.46 \times 10^6 + 1.979 \times 10^6 \text{ mm}^4$$

$$= \mathbf{14.439 \times 10^6 \text{ mm}^4}.$$ **Ans.**

Example 3.15. *Find the moment of inertia of the area shown shaded in Fig. 3.33, about edge AB.*

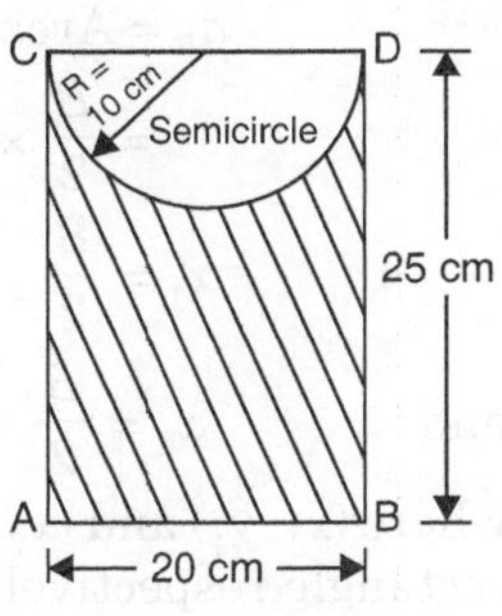

Fig. 3.33

Sol. Given :

Radius of semi-circle, $R = 10$ cm

Width of rectangle, $b = 20$ cm

Depth of rectangle, $d = 25$ cm

Moment of inertia of the shaded portion about *AB*

= M.O.I. of rectangle *ABCD* about *AB*

– M.O.I. of semi-circle on *DC* about *AB*

M.O.I. of rectangle $ABCD$ about AB

$$= \frac{bd^3}{3} \qquad \text{[see equation (3.11)]}$$

$$= \frac{20 \times 25^3}{12} = 104.167 \text{ cm}^4$$

M.O.I. of semicircle about DC

$$= \frac{1}{2} \times \text{[M.O.I. of a circle of radius 10 cm about a diameter]}$$

$$= \frac{1}{2} \times \left[\frac{\pi}{64} d^4\right] = \frac{1}{2} \times \frac{\pi}{64} \times 20^4 = 3.925 \text{ cm}^4$$

Distance of C.G. of semicircle from DC

$$= \frac{4r}{3\pi} = \frac{4 \times 10}{3\pi} = 4.24 \text{ cm}$$

Area of semicircle, $\quad A = \frac{\pi r^2}{2} = \frac{\pi \times 10^2}{2} = 157.1 \text{ cm}^2$

M.O.I. of semicircle about a line through its C.G. parallel to CD

= M.O.I. of semi-circle about CD – Area × [Distance of C.G. of semicircle from DC]^2

$$= 3925 - 157.1 \times 4.24^2$$

$$= 3925 - 2824.28 = 1100.72 \text{ cm}^4$$

Distance of C.G. of semicircle from AB

$$= 25 - 4.24 = 20.76 \text{ cm}$$

M.O.I. of semicircle about AB = $1100.72 + 157.1 \times 20.76^2$

$$= 1100.72 + 67706.58 = 68807.30 \text{ cm}^4$$

$\therefore$ M.O.I. of shaded portion about AB

$$= 104.167 - 68807.30 = \mathbf{35359.7 \text{ cm}^4}. \quad \textbf{Ans.}$$

Problem 3.16. *Find the moments of inertia about the centroidal XX and YY axes of the section shown in Fig. 3.34 (a).*

(U.P. Tech. University, 2002–2003)

Sol. First find the location of the C.G. of the given figure.

Let a_1 = Area of complete rectangle = $B \times D$

a_2 = Area of removed rectangle portion

$$= \frac{B}{2} \times \frac{D}{2} = \frac{BD}{4}$$

$$x_1 = \frac{B}{2}, \; y_1 = \frac{D}{2}$$

and $\quad x_2 = \frac{B}{2} + \frac{1}{2}\left(\frac{B}{2}\right) = \frac{3B}{4}, \; y_2 = \frac{D}{2} + \frac{1}{2}\left(\frac{D}{2}\right) = \frac{3D}{4}$

Fig. 3.34(a)

where (x_1, y_1) and (x_2, y_2) are the co-ordinates of the C.G. of the complete rectangle and cut out rectangle respectively. Area a_2 is negative.

Now $\quad \bar{x} = \dfrac{a_1x_1 - a_2y_2}{a_1 - a_2} = \dfrac{BD \times \dfrac{B}{2} - \dfrac{B \times D}{4} \times \dfrac{3B}{4}}{\dfrac{3}{4}BD}$

$$= \frac{\dfrac{B^2 \times D}{2} - \dfrac{3}{16}B^2D}{\dfrac{3}{4}BD} = \frac{\dfrac{5}{16}BD^2}{\dfrac{3}{4}BD} = \frac{5}{12} \times D$$

Similarly $\quad \bar{y} = \dfrac{a_1y_1 - a_2y_2}{a_1 - a_2} = \dfrac{BD \times \dfrac{D}{2} - \dfrac{BD}{4} \times \dfrac{3D}{4}}{\dfrac{3}{4}BD}$

$$= \frac{\dfrac{BD^2}{2} - \dfrac{3}{16}BD^2}{\dfrac{3}{4}BD} = \frac{\dfrac{5}{16}BD^2}{\dfrac{3}{4}BD} = \frac{5}{12} \times D$$

Now draw the centroidal axes *XX* and *YY* as shown in Fig. 3.34(*b*).

Let I_{XX1} = M.O.I. of complete rectangle ① about centroidal axis *X-X*

= M.O.I. of complete rectangle ① about horizontal axis passing through its C.G.

+ Area of complete rectangle ①

× Distance between *X-X* axis and horizontal axis passing through the C.G. of rectangle ① (By theorem of parallel axis) $[\because I_{XX1} = I_{G_1XX} + A_1h_1^2]$

$$= \frac{BD^3}{12} + (B \times D)\,[y_1 - \bar{y}]^2$$

$$= \frac{BD^3}{12} + BD\left[\frac{D}{2} - \frac{5D}{12}\right]^2 \qquad \left[\because \; y_1 = \frac{D}{2}, \bar{y} = \frac{5D}{2}\right]$$

$$= \frac{BD^3}{12} + BD\left[\frac{D}{12}\right]^2$$

$$= \frac{BD^3}{12} + \frac{BD^3}{144} = \frac{13}{144}BD^3$$

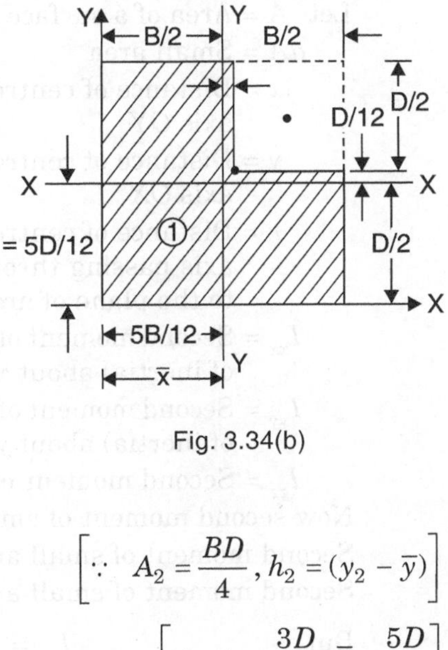

Fig. 3.34(b)

Similarly $\quad I_{XX2} = (I_{G2X}) + A_2 \times h_2^2$

$$= \frac{\dfrac{B}{2} \times \left(\dfrac{D}{2}\right)^3}{12} + \frac{BD}{4} \times [y_2 - \bar{y}]^2 \qquad \left[\because \; A_2 = \frac{BD}{4}, h_2 = (y_2 - \bar{y})\right]$$

$$= \frac{BD^3}{192} + \frac{BD}{4}\left[\frac{3D}{4} - \frac{5D}{12}\right]^2 \qquad \left[\because \; y_2 = \frac{3D}{2}, \bar{y} = \frac{5D}{12}\right]$$

$$= \frac{BD^3}{192} + \frac{BD}{4} \times \left(\frac{4D}{12}\right)^2 = \frac{BD^3}{192} + \frac{16BD^3}{4 \times 144}$$

$$= \frac{BD^3}{192} + \frac{BD^3}{36} = \frac{3BD^3 + 16BD^3}{576} = \frac{19BD^3}{576}$$

Now I_{XX} = M.O.I. of given section about centroidal axis X-X

$$= I_{XX1} - I_{XX2} = \frac{13BD^3}{144} - \frac{19BD^3}{576}$$

$$= \frac{52BD^3 - 19BD^3}{576} = \frac{33BD^3}{576} = \mathbf{0.0573\ BD^3}.\quad \textbf{Ans.}$$

Similarly the M.O.I. of the given section about centroidal axis Y-Y is given by

$$I_{YY} = I_{YY1} - I_{YY2}$$

where I_{YY1} = M.O.I. of rectangle ① about centroidal axis Y-Y

$$= I_{G1y} + A_1 \times [x_1 - \bar{x}]^2$$

$$= \frac{DB^3}{12} + BD \times \left[\frac{B}{2} - \frac{5B}{12}\right]^2 = \frac{DB^2}{12} + \frac{BD \times B^2}{144} = \frac{13}{144} DB^3$$

and $I_{YY2} = I_{G2y} + A_2[x_2 - \bar{x}]^2$

$$= \frac{\frac{D}{2} \times \left(\frac{B}{2}\right)^3}{12} + \frac{BD}{4}\left[\frac{3B}{4} - \frac{5B}{12}\right]^2 = \frac{DB^3}{192} + \frac{DB^3}{36} = \frac{19DB^3}{576}$$

$$\therefore\ I_{YY} = \frac{13}{144} DB^3 - \frac{19DB^3}{576} = \frac{33}{576} DB^3 = \mathbf{0.0573\ DB^3}.\quad \textbf{Ans.}$$

3.10. POLAR MOMENT OF INERTIA

Refer to Fig. 3.35.

Let A = Area of a surface in x-y plane

dA = Small area

x = Distance of centroid of small area dA from axis OY

y = Distance of centroid of small area dA from axis OX

r = Distance of centroid of small area dA from axis passing through O and perpendicular to the plane of area A *i.e.,* from axis OZ.

I_{xx} = Second moment of total area A (*i.e.,* moment of inertia) about x-axis.

I_{yy} = Second moment of total area A (*i.e.,* moment of inertia) about y-axis.

I_{zz} = Second moment of total area A about z-axis.

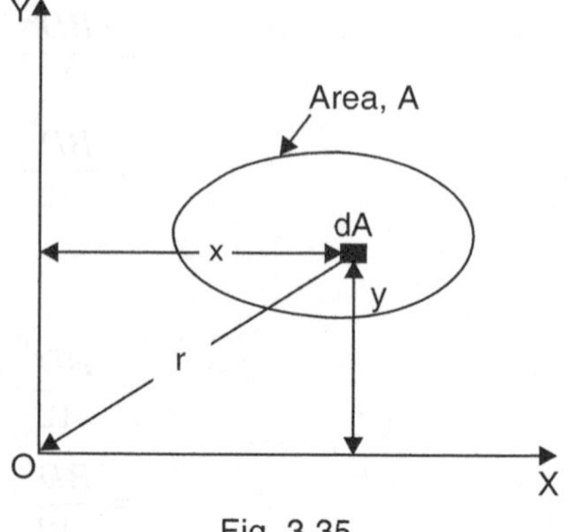

Fig. 3.35

Now second moment of small area dA about x-axis $= dA \cdot y^2$ or $y^2 \cdot dA$

Second moment of small area dA about y-axis. $= dA \cdot x^2$ or $x^2 \cdot dA$

Second moment of small area dA about z-axis $= dA \cdot r^2$ or $r^2 \cdot dA$

But

$$I_{xx} = \int y^2 \cdot dA$$

$$I_{yy} = \int x^2 \cdot dA$$

$$I_{zz} = \int r^2 \cdot dA$$

The second moment of area about z-axis is known as **polar moment of inertia.**

$\therefore$ Polar moment of Inertia $= I_{zz} = \int r^2 \, . \, dA$ or $I_{zz} = \int r^2 \, dA$

$$= \int (x^2 + y^2)\, dA \qquad (\because r^2 = x^2 + y^2)$$

$$= \int x^2 \, . \, dA + \int y^2 \, . \, dA$$

$$= I_{yy} + I_{xx} \qquad \left(\because \int x^2 \, dA = I_{yy} \text{ and } \int y^2 dA = I_{xx}\right)$$

$$= I_{xx} + I_{yy} \qquad \ldots(3.16)$$

Hence the polar moment of inertia for an area with respect to an axis perpendicular to the plane is equal to the sum of the moments of inertia about two mutually perpendicular axes in the plane of the area.

3.11. PRODUCT OF INERTIA

The Fig. 3.36 shows a body of area A. Consider a small area dA. The moment of this area about x-axis is $y \, . \, dA$. Now the moment of $y \, . \, dA$ about y-axis is $xydA$. Then $xydA$ is known as the product of inertia of area dA with respect to x-axis and y-axis. The integral $\int xydA$ is known as the product of inertia of area A with respect to x and y axes. This product of inertia is represented by I_{xy}.

$$\therefore \qquad I_{xy} = \int xydA \qquad \ldots(3.17)$$

Hence the product of inertia of the plane area is obtained if an elemental area is multiplied by the product of its co-ordinates and is integrated for entire area.

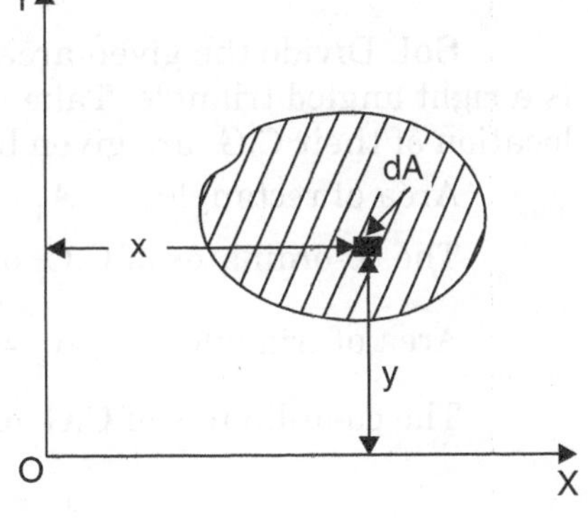

Fig. 3.36

The product of inertia (I_{xy}) can also be written mathematically as

$$I_{xy} = \Sigma x_i y_i A_i = x_1 y_1 A_1 + x_2 y_2 A_2 + \ldots\ldots \qquad \ldots(3.18)$$

where $x_i y_i$ = co-ordinates of the C.G. of area A_i.

Note. (*i*) The product of inertia may be positive, negative or zero depending upon distance x and y which could be positive, negative or zero.

(*ii*) If area is symmetrical with respect to one or both of the axes, the product of inertia will be zero as shown in Fig. 3.37. The total area A is symmetrical about y-axis. The small area dA which is symmetrical about y-axis has co-ordinates (x, y) and $(-x, y)$. The corresponding product of inertia for small area are $xydA$ and $-xydA$ respectively. Hence product of inertia for total area becomes zero.

(*iii*) The product of inertia with respect to centroidal axis will also be zero.

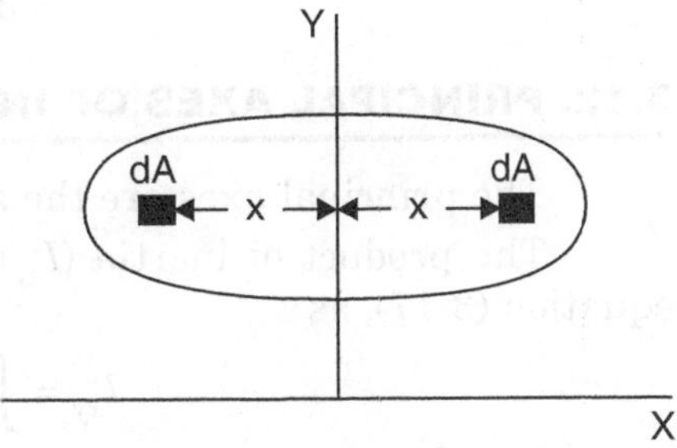

Fig. 3.37

Problem 3.17. *Fig. 3.38 (a) shows a plane area. Determine the product moment of inertia of the given area. All dimensions are in mm.*

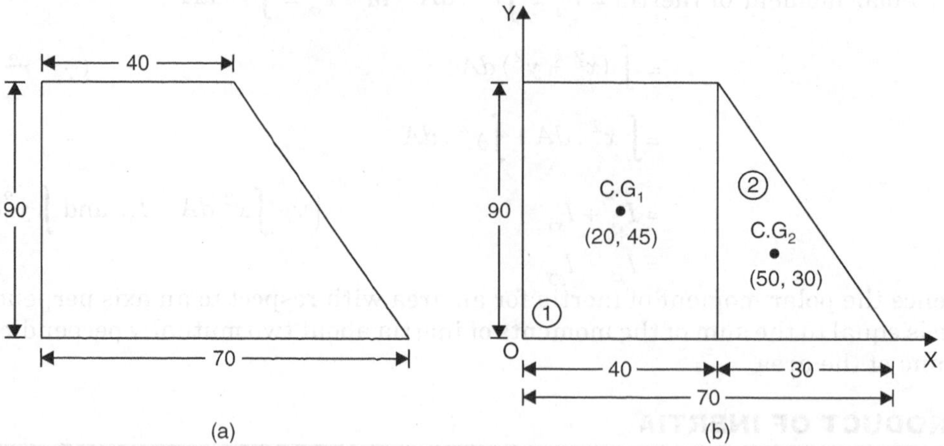

Fig. 3.38

Sol. Divide the given area into two parts. The first part is a rectangle and second part is a right angled triangle. Take x-axis and y-axis as shown in the Fig. 3.38 (*b*). The areas and location of their C.G. are given below :

Area of rectangle, $A_1 = 90 \times 40 = 3600 \text{ mm}^2$.

The co-ordinates of C.G. of rectangle ① are : $x_1 = 20$ mm, $y_1 = 45$ mm.

Area of triangle, $A_2 = \dfrac{90 \times 30}{2} = 1350 \text{ mm}^2$.

The co-ordinates of C.G. of triangle ② are :

$$x_2 = 40 + \frac{1}{3} \times 30 = 40 + 10 = 50 \text{ mm} \; ; \; y_2 = \frac{1}{3} \times 90 = 30 \text{ mm}.$$

The product of inertia of given area is given by equation (3.18), as

$$\begin{aligned} I_{xy} &= x_1 y_1 A_1 + x_2 y_2 A_2 \\ &= A_1 x_1 y_1 + A_2 x_2 y_2 \\ &= 3600 \times 20 \times 45 + 1350 \times 50 \times 30 \\ &= 3240000 + 2025000 = \mathbf{5265000 \text{ mm}^4}. \quad \textbf{Ans.} \end{aligned}$$

3.12. PRINCIPAL AXES OF INERTIA

The principal axes are the axes about which the product of inertia is zero.

The product of inertia (I_{xy}) of plane area A with respect to x and y-axes is given by equation (3.17), as

$$I_{xy} = \int xy \, dA$$

But the moment of inertia of plane area A about x-axis [I_{xx}] or about y-axis [I_{yy}] is given by

$$I_{xx} = \int y^2 \, dA \text{ and } I_{yy} = \int x^2 \, dA$$

The moment of inertia is always positive but product of inertia may be positive (if both x and y are positive), may be negative (if one co-ordinate is positive and other is negative) or may be zero (if any co-ordinate is zero).

Fig. 3.39 (*a*) shows a body of area A. Consider a small area dA. The product of inertia of the total area A with respect to x and y-axes is given as

$$I_{xy} = \int xydA \qquad ...(i)$$

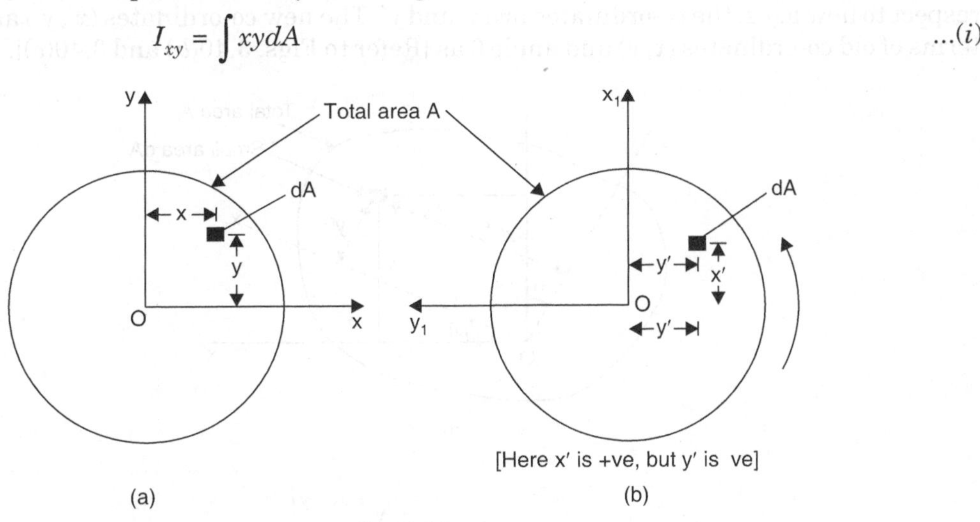

Fig. 3.39

Let now the axes are rotated anticlockwise by 90° as shown in Fig. 3.39 (*b*) keeping the total area A in the same position. Let x_1 and y_1 are the new axes. The co-ordinates of the same small area dA with respect to new axes are x' and y'.

Hence the product of inertia of the total area A with respect to new axes x_1 and y_1 becomes as

$$I_{x_1y_1} = \int x'y'\,dA \qquad ...(ii)$$

Now let us find the relation between old and new co-ordinates. From Figs. 3.39 (*a*) and 3.39 (*b*), we get

$$x = -y' \quad \text{and} \quad y = x'$$

or

$$y' = -x \quad \text{and} \quad x' = y$$

Substituting the values of x' and y' in equation (*ii*), we get

$$I_{x_1y_1} = \int (y)(-x)\,dA = -\int xydA = -I_{xy} \qquad \left(\because \int xydA = I_{xy}\right)$$

The above result shows that by rotating the axes through 90°, the product of inertia has become negative. This means that the product of inertia which was positive previously has now become negative by rotating the axes through 90°. Hence product of inertia has changed its sign. It was positive initially and now it is negative. Between these two positions, it is also possible that by rotating the axes through certain angle, the product of inertia will become zero. The new axes about which product of inertia is zero, are known as **principal axes.**

Note. (*i*) The product of inertia is zero about principal axes.

(*ii*) As the product of inertia is zero about symmetrical axis, hence symmetrical axis is the principal axis of inertia for the area.

(*iii*) The product of inertia depends upon the orientation of the axes.

3.13. PRINCIPAL MOMENTS OF INERTIA OF PLANE AREAS

Fig. 3.40(*a*) shows a body of area *A* with respect to old axes (x, y) and new axes (x_1, y_1). The new axes x_1 and y_1 have been rotated through an angle θ in anticlockwise direction. Consider a small area *dA*. The co-ordinates of the small area with respect to old axes are (x, y) whereas with respect to new axes, the co-ordinates are x' and y'. The new co-ordinates (x', y') are expressed in terms of old co-ordinates (x, y) and angle θ as [Refer to Figs. 3.40(*b*) and 3.40(*c*)].

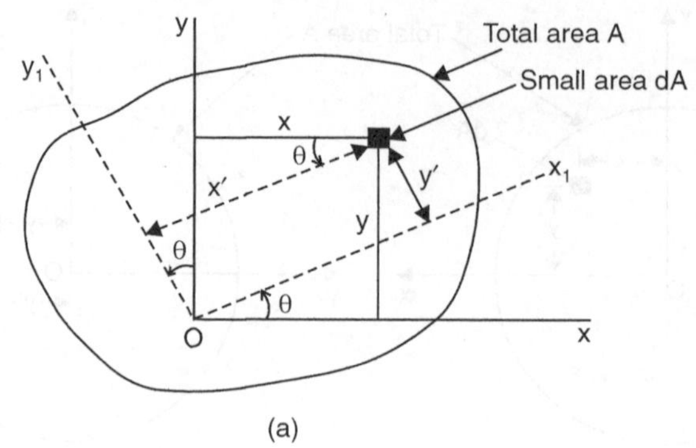

(a)

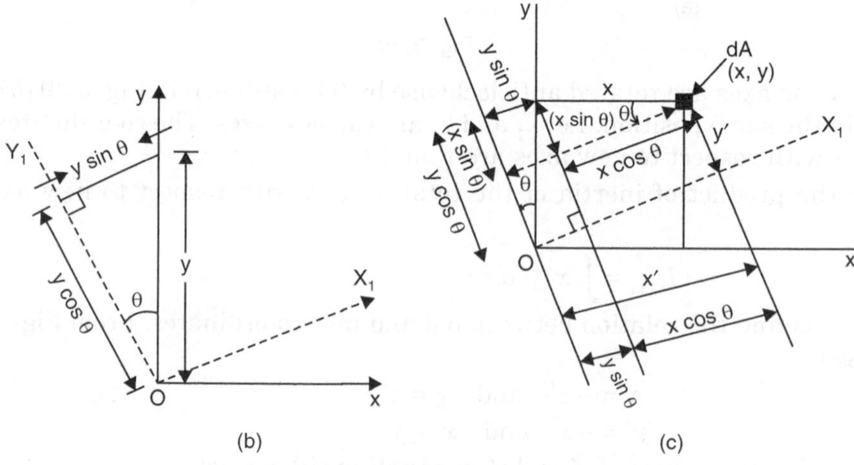

(b) (c)

Fig. 3.40

$$x' = y \sin \theta + x \cos \theta \qquad \text{...}(i)$$

and
$$y' = y \cos \theta - x \sin \theta \qquad \text{...}(ii)$$

The moment of inertia and product of inertia of area *A* with respect to old axes are

$$I_{xx} = \int y^2 dA, \; I_{yy} = \int x^2 dA \quad \text{and} \quad I_{xy} = \int xy dA \qquad \text{...(3.19)}$$

Also the moment of inertia and product of inertia of area *A* with respect to new axes will be

$$I_{x_1x_1} = \int (y')^2 \, dA, \; I_{y_1y_1} = \int (x')^2 \, dA \text{ and } I_{x_1x_1} = \int x'y' \, dA$$

Let us substitute the values of x', y' from equations (*i*) and (*ii*) in the above equations, we get

$$I_{x_1x_1} = \int (y')^2\, dA$$

$$= \int (y\cos\theta - x\sin\theta)^2\, dA \qquad [\because\ y' = y\cos\theta - x\sin\theta]$$

$$= \int (y^2\cos^2\theta + x^2\sin^2\theta - 2xy\cos\theta\sin\theta)\, dA$$

$$= \int y^2\cos^2\theta\, dA + \int x^2\sin^2\theta\, dA - \int 2xy\cos\theta\sin\theta\, dA$$

$$= \cos^2\theta \int y^2\, dA + \sin^2\theta \int x^2\, dA - 2\cos\theta\sin\theta \int xy\, dA$$

($\because$ After rotation, the angle θ is constant and hence $\cos^2\theta$, $\sin^2\theta$ and $2\cos\theta\sin\theta$ are constant)

$$= (\cos^2\theta) I_{xx} + (\sin^2\theta) I_{yy} - (2\cos\theta\sin\theta) I_{xy} \qquad \text{...(3.20)}$$

$$\left(\because\ \int y^2\, dA = I_{xx}, \int x^2\, dA = I_{yy} \text{ and } \int xy dA = I_{xy}\right)$$

Similarly, $$I_{y_1y_1} = \int (x')^2\, dA$$

$$= \int (y\sin\theta + x\cos\theta)^2\, dA \qquad [\because\ x' = y\sin\theta + x\cos\theta]$$

$$= \int (y^2\sin^2\theta + x^2\cos^2\theta + 2xy\sin\theta\cos\theta)\, dA$$

$$= \int y^2\sin^2\theta\, dA + \int x^2\cos^2\theta\, dA + \int 2xy\sin\theta\cos\theta\, dA$$

$$= \sin^2\theta \int y^2\, dA + \cos^2\theta \int x^2\, dA + 2\sin\theta\cos\theta \int xy dA$$

($\because$ θ is constant and hence $\sin\theta$ and $\cos\theta$ are constants)

$$= \sin^2\theta \,.\, I_{xx} + \cos^2\theta\, I_{yy} + 2\sin\theta\cos\theta\, I_{xy} \qquad \text{...(3.21)}$$

$$\left(\because\ \int y^2\, dA = I_{xx}, \int x^2\, dA = I_{yy} \text{ and } \int xy dA = I_{xy}\right)$$

Adding equations (3.20) and (3.21), we get

$$I_{x_1x_1} + I_{y_1y_1} = I_{xx}[\sin^2\theta + \cos^2\theta] + I_{yy}[\sin^2\theta + \cos^2\theta] + 2\sin\theta\cos\theta\, I_{xy} - 2\sin\theta\cos\theta\, I_{xy}$$

$$= I_{xx} + I_{yy} \qquad [\because\ \sin^2\theta + \cos^2\theta = 1] \qquad \text{...(3.22)}$$

The equation (3.22) shows that sum of moments of inertia about old axes (x, y) and new axes (x_1, y_1) are same. Hence the sum of moments of inertia of area A is independent of orientation of axes.

Now let us find the value of $I_{x_1x_1} - I_{y_1y_1}$.

Subtracting equation (3.21) from equation (3.20), we get

$$I_{x_1x_1} - I_{y_1y_1} = \cos^2\theta\, I_{xx} + \sin^2\theta\, I_{yy} - 2\cos\theta\sin\theta\, I_{xy} - [\sin^2\theta\, I_{xx} + \cos^2\theta\, I_{yy} + 2\cos\theta\sin\theta\, I_{xy}]$$

$$= I_{xx}(\cos^2\theta - \sin^2\theta) + I_{yy}(\sin^2\theta - \cos^2\theta) - 4\cos\theta\sin\theta \,.\, I_{xy}$$

$$= I_{xx}(\cos^2\theta - \sin^2\theta) - I_{yy}(\cos^2\theta - \sin^2\theta) - 4\cos\theta\sin\theta\, I_{xy}$$

$$= (I_{xx} - I_{yy})(\cos^2\theta - \sin^2\theta) - 2 \times 2\cos\theta\sin\theta \times I_{xy}$$

$$= (I_{xx} - I_{yy})\cos 2\theta - 2I_{xy}\sin^2\theta \qquad ...(3.23)$$

$$\left(\because \quad \cos^2\theta = \frac{1+\cos 2\theta}{2}, \sin^2\theta = \frac{1-\cos 2\theta}{2}\right.$$

$\therefore$ $\quad \cos^2\theta - \sin^2\theta = \cos 2\theta$ and $2\sin\theta\cos\theta = \sin 2\theta$)

Now let us find the values of $I_{x_1x_1}$ and $I_{y_1y_1}$ in terms of I_{xx}, I_{yy} and θ.

Adding equations (3.22) and (3.23), we get

$$2I_{x_1x_1} = [I_{xx} + I_{yy}] + [(I_{xx} - I_{yy})\cos 2\theta - 2I_{xy}\sin 2\theta]$$

or
$$I_{x_1x_1} = \frac{(I_{xx} + I_{yy})}{2} + \frac{(I_{xx} - I_{yy})}{2}\cos 2\theta - I_{xy}\sin 2\theta \qquad ...(3.24)$$

To find the values of $I_{y_1y_1}$, subtract equation (3.23) from (3.22). Now subtracting equation (3.23) from equation (3.22), we get

$$2I_{y_1y_1} = (I_{xx} + I_{yy}) - [(I_{xx} - I_{yy})\cos 2\theta - 2I_{xy}\sin 2\theta]$$

$\therefore$
$$I_{y_1y_1} = \frac{(I_{xx} + I_{yy})}{2} - \frac{(I_{xx} - I_{yy})}{2}\cos 2\theta + I_{xy}\sin 2\theta \qquad ...(3.25)$$

Product of inertia about new axes

Let us now find the value of $I_{x_1y_1}$ in terms of I_{xy} and angle θ.

We know that $\quad I_{x_1y_1} = \int (x')(y')\, dA$

Substituting the values of x' and y', we get

$$I_{x_1y_1} = \int (y\sin\theta + x\cos\theta)(y\cos\theta - x\sin\theta)\, dA$$

$$(\because \quad x' = y\sin\theta + x\cos\theta \text{ and } y' = y\cos\theta - x\sin\theta)$$

or
$$I_{x_1y_1} = \int (y^2\sin\theta\cos\theta - xy\sin^2\theta + xy\cos^2\theta - x^2\cos\theta\sin\theta)\, dA$$

$$= \int y^2\sin\theta\cos\theta\, dA - \int xy\sin^2\theta\, dA + \int xy\cos^2\theta\, dA - \int x^2\cos\theta - \sin\theta\, dA$$

$$= \sin\theta\cos\theta\int y^2\, dA - \sin^2\int xydA + \cos^2\theta\int xydA\cos\theta\sin\theta\int x^2\, dA$$

($\because$ θ is constant and hence $\sin\theta$, $\cos\theta$ are constants)

$$I_{x_1y_1} = \frac{2\sin\theta\cos\theta}{2}\int y^2\, dA - \sin^2\theta\, I_{xy} + \cos^2\theta\, I_{xy} - \frac{2\cos\theta\sin\theta}{2}\int x^2\, dA$$

$$\left(\because \int xydA = I_{xy}\right)$$

$$= \frac{\sin 2\theta}{2}\,.\, I_{xx} + I_{xy}(\cos^2\theta - \sin^2\theta) - \frac{\sin 2\theta}{2}\, I_{yy}$$

$$\left(\because \int y^2\, dA = I_{xx}, \int x^2\, dA = I_{yy}\right)$$

$$= \frac{I_{xx}}{2}\sin 2\theta + I_{xy}(\cos^2\theta - \sin^2\theta) - \frac{I_y}{2}\,.\,\sin\theta$$

$$= \frac{(I_{xx} - I_{yy})}{2} \sin 2\theta + I_{xy} (\cos^2 \theta - \sin^2 \theta)$$

$$= \frac{(I_{xx} - I_{yy})}{2} \sin 2\theta + I_{xy} \cos 2\theta \quad \text{...(3.26)}$$

$$(\because \quad \cos^2 \theta - \sin^2 \theta = \cos 2\theta)$$

Direction of principal axes

We have already defined the principal axes. Principal axes are the axes about which the product of inertia is zero. Now the new axes (x_1, y_1) will become principal axes if the product of inertia given by equation (3.26) is zero (*i.e.*, $I_{x_1y_1} = 0$).

$\therefore$ For principal axes, $I_{x_1y_1} = 0$

or
$$\frac{(I_{xx} - I_{yy})}{2} \sin 2\theta + I_{xy} \cos 2\theta = 0$$

or
$$\frac{I_{xx} - I_{yy}}{2} \sin 2\theta = - I_{xy} \cos 2\theta$$

or
$$\frac{\sin 2\theta}{\cos 2\theta} = \frac{-2I_{xy}}{I_{xx} - I_{yy}} = \frac{2I_{xy}}{I_{yy} - I_{xx}}$$

or
$$\tan 2\theta = \frac{2I_{xy}}{I_{yy} - I_{xx}} \quad \text{...(3.27)}$$

The above equation will give the two values of 2θ or θ. These two values of θ will differ by 90°. By substituting the values of θ in equations (3.24) and (3.25), the values of **principal moments of inertia** ($I_{x_1x_1}$ and $I_{y_1y_1}$) can be obtained. If from equation (3.27), the values of sin 2θ and cos 2θ in terms of I_{xy}, I_{xx} and I_{yy} are substituted in equation (3.24), we get

$$I_{x_1x_1} = \frac{I_{xx} + I_{yy}}{2} \pm \sqrt{\frac{(I_{xx} - I_{yy})^2}{2} + I_{xy^2}}\ .$$

These are the values of principal moment of inertia.

Problem 3.18. *For the section shown in Fig. 3.41(a) determine :*

(i) Moment of inertia about its centroid along (x, y) axis.

(ii) Moment of inertia about new axes which is turned through an angle of 30° anticlockwise to the old axis.

(iii) Principal moments of inertia about its centroid.

All dimensions are in cm.

Sol. Given :

The Fig. 3.41(*a*) shows the given section. It is symmetrical about *x*-axis. The C.G. of the section lies at O (origin of the axes). To find moment of inertia of the given section, it is divided into three rectangles as shown in Fig. 3.41(*b*). First the moment of inertia of each rectangle about its centroid is calculated. Then by using parallel axis theorem, the moment of inertia of the given section about its centroid is obtained.

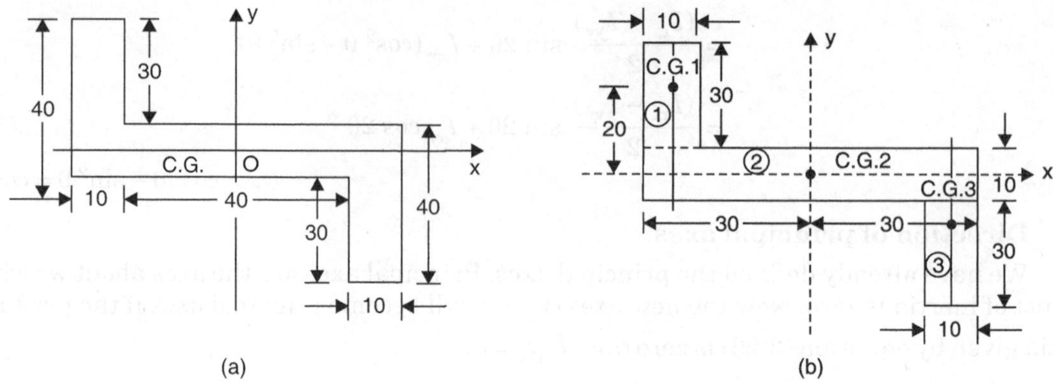

Fig. 3.41

***(a)* Consider rectangle ①**

The C.G. of rectangle ① is at a distance of 20 cm from x-axis and at a distance of 25 cm from y-axis.

$$(I_{xx})_1 = (I_G)_{1x} + A_1(k_{1x})^2 \qquad ...(i)$$

where $(I_{xx})_1$ = M.O.I. of rectangle ① about x-axis passing through the centroid of the given section.

$(I_G)_{1x}$ = M.O.I. of rectangle ① about an axis passing through C.G. of rectangle ① and parallel to x-axis

$$= \frac{bd^3}{12}$$

$$= \frac{10 \times 30^3}{12} \text{ (Here } b = 10 \text{ and } d = 30)$$

$$= 2.25 \times 10^4 \text{ cm}^4$$

A_1 = Area of rectangle ① = 10 × 30 = 300

(k_{1x}) = Distance of C.G. of rectangle ① from x-axis

= 20

Substituting the above values in equation (i), we get

$$(I_{xx})_1 = 2.25 \times 10^4 + 300 \times 20^2$$

$$= 2.25 \times 10^4 + 12 \times 10^4$$

$$= 14.25 \times 10^4 \text{ cm}^4 \qquad ...(A)$$

Similarly the M.O.I. of rectangle ① about y-axis passing through the centroid of the given figure is given by,

$$(I_{yy})_1 = (I_G)_{1y} + A_1(k_{1y})^2 \qquad ...(ii)$$

where $(I_G)_{1y} = \frac{bd^3}{12} = \frac{30 \times 10^3}{12} = 0.25 \times 10^4 \text{ cm}^4$

(k_{1y}) = Distance of C.G. of rectangle ① from y-axis = 25

$$\therefore \quad (I_{yy})_1 = 0.25 \times 10^4 + 300 \times 25 \qquad (\because \; A_1 = 300)$$

$$= 0.25 \times 10^4 + 18.75 \times 10^4$$

$$= 19 \times 10^4 \text{ cm}^4 \qquad ...(B)$$

(*b*) **Consider rectangle ②**

The C.G. of this rectangle coincides with the C.G. of the given section. Hence

$$(I_{xx})_2 = \frac{bd^3}{12} = \frac{60 \times 10^3}{12} = 0.5 \times 10^4 \text{ cm}^4 \quad ...(C)$$

and
$$(I_{yy})_2 = \frac{10 \times 60^3}{12} = 18 \times 10^4 \text{ cm}^4 \quad ...(D)$$

(*c*) **Consider rectangle ③**

The C.G. of rectangle ③ is at a distance of 20 cm from *x*-axis and at a distance of 25 cm from *y*-axis. Hence k_{3x} = 20 cm and k_{3y} = 25 cm.

Now
$$(I_{xx})_3 = (I_G)_{3x} + A_3(k_{3x})^2$$
$$= \frac{10 \times 30^3}{12} + (10 \times 30)(20)^2 = 2.25 \times 10^4 + 12 \times 10^4 = 14.25 \times 10^4 \text{ cm}^4$$

and
$$(I_{yy})_3 = (I_G)_{3y} + A_3(k_{3y})^2$$
$$= \frac{30 \times 10^3}{12} + 300 \times 25^2 = 0.25 \times 10^4 + 18.75 \times 10^4 = 19 \times 10^4 \text{ cm}^4.$$

(*i*) *Moment of inertia of complete section about its centroid*

$$I_{xx} = (I_{xx})_1 + (I_{xx})_2 + (I_{xx})_3$$
$$= 14.25 \times 10^4 + 0.5 \times 10^4 + 14.25 \times 10^4 \text{ cm}^4$$
$$= \mathbf{29 \times 10^4 \text{ cm}^4}. \quad \textbf{Ans.}$$

and
$$I_{yy} = (I_{yy})_1 + (I_{yy})_2 + (I_{yy})_3$$
$$= 19 \times 10^4 + 18 \times 10^4 + 19 \times 10^4$$
$$= \mathbf{56 \times 10^4 \text{ cm}^4}. \quad \textbf{Ans.}$$

(*ii*) *Moment of inertia of complete section about new axes which is turned through an angle of 30° anticlockwise.*

Here θ = 30°.

Let us first calculate the product of inertia of whole area about old axes *x*, *y*.

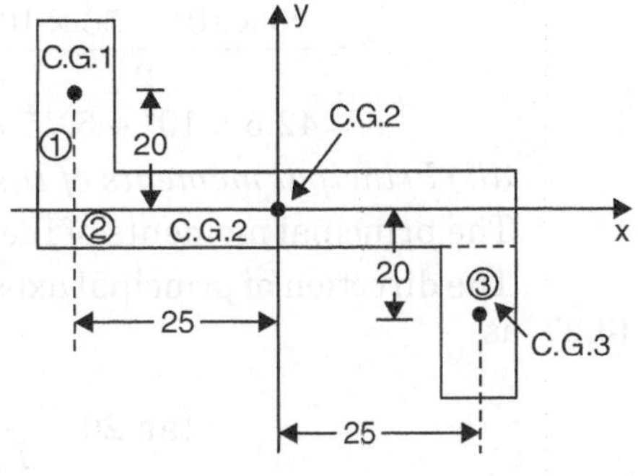

Fig. 3.41(c)

(*a*) *Consider rectangle ①*.

$$A_1 = 10 \times 30 = 300.$$

The C.G. of rectangle ① is at a distance of 20 cm above *x*-axis and at a distance of 25 cm from *y*-axis. Hence co-ordinates of this C.G. are

$$x_1 = -25 \text{ cm and } y_1 = 20 \text{ cm}.$$

(*b*) *For rectangle ②*

$A_2 = 10 \times 60 = 600$ cm². The C.G. of rectangle ② lies on the origin (*O*). Hence $x_2 = 0$ and $y_2 = 0$.

(*c*) *For rectangle ③*, $A_3 = 10 \times 30 = 300$ cm²

The C.G. of rectangle ③ is at a distance of 20 cm below *x*-axis and at a distance of 25 cm from *y*-axis.

Hence co-ordinates of this C.G. are : x_3 = 25 cm and y_3 = (– 20 cm).

The product of inertia (I_{xy}) of the whole figure is given by equation (3.18) as

$$I_{xy} = A_1x_1y_1 + A_2x_2y_2 + A_3x_3y_3$$
$$= 300 \times (-25) \times 20 + 600 \times 0 \times 0 + 300 \times 25 \times (-20)$$
$$= -15 \times 10^4 + 0 + (-15 \times 10^4)$$
$$= -30 \times 10^4 \text{ cm}^4$$

Now the moment of inertia of the complete section about the new axes (x_1, y_1) can be obtained from equations (3.24) and (3.25) as

Fig. 3.41(d)

$$I_{x_1x_1} = \frac{(I_{xx} + I_{yy})}{2} + \frac{(I_{xx} - I_{yy})}{2} \cos 2\theta - I_{xy} \sin 2\theta$$

where $I_{xx} = 29 \times 10^4 \text{ cm}^4$, $I_{yy} = 56 \times 10^4 \text{ cm}^4$,

$I_{xy} = -30 \times 10^4 \text{ cm}^4$ and $\theta = 30°$

$$\therefore \quad I_{x_1x_1} = \frac{29 \times 10^4 + 56 \times 10^4}{2} + \frac{29 \times 10^4 - 56 \times 10^4}{2} \cos 60° - (-30 \times 10^4) \sin 60°$$

$$= 42.5 \times 10^4 - 13.5 \times 10 \times \frac{1}{2} + 30 \times 10^4 \times 0.866$$

$$= 35.75 \times 10^4 + 26 \times 10^4 = \mathbf{61.75 \times 10^4\ cm^4.\ Ans.}$$

and
$$I_{y_1y_1} = \frac{I_{xx} + I_{yy}}{2} - \frac{I_{xx} - I_{yy}}{2} \cos \theta + I_{xy} \sin 2\theta$$

$$= \frac{29 \times 10^4 + 56 \times 10^4}{2} - \frac{29 \times 10^4 - 56 \times 10^4}{2} \cos 60° + (-30 \times 10^4) \sin 60°$$

$$= 42.5 \times 10^4 + 6.75 \times 10^4 - 26 \times 10^4 = \mathbf{23.25 \times 10^4\ cm^4.\ Ans.}$$

(iii) Principal moments of Inertia about the centroid

The principal moments of inertia are the moments of inertia about the principal axes.

The direction of principal axes is given by equation (3.27) as

$$\tan 2\theta = \frac{2I_{xy}}{I_{yy} - I_{xx}}$$

$$= \frac{2 \times (-30 \times 10^4)}{56 \times 10^4 - 29 \times 10^4}$$

$$= \frac{-60 \times 10^4}{27 \times 10^4} = -2.222$$

As 2θ is negative, hence it lies in 2nd and 4th quadrant.

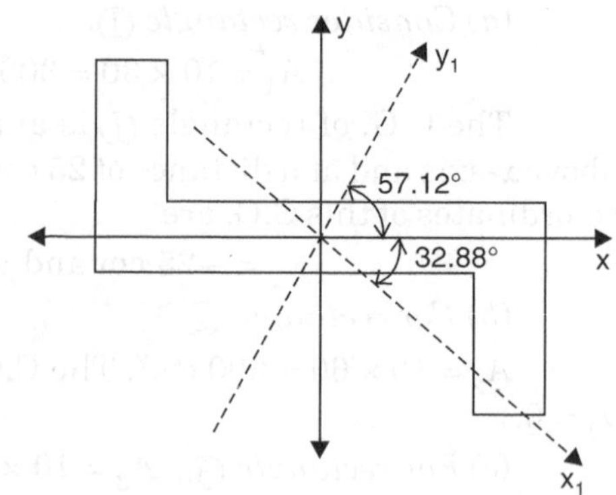

Fig. 3.41(e)

$$\therefore \quad 2\theta = \tan^{-1}(-2.222)$$
$$= -65.77° \text{ and } 114.23°$$

or
$$\theta = -32.88° \text{ and } 57.12°$$

The +ve angle is taken anti-clockwise and –ve angle is taken clockwise to the existing axes x and y. The principal axes are shown as x_1 and y_1 in Fig. 3.41 (*e*). The moment of inertia along these axes is the principal moment of inertia. Hence by substituting $\theta = -32.88°$ and $57.12°$, in equations (3.24) and (3.25), we get principal moment of inertia.

$$\therefore \quad (I_{x_1x_1})_{min.}^{max.} = \frac{I_{xx} + I_{yy}}{2} + \frac{I_{xx} - I_{yy}}{2} \cos 2\theta - I_{xy} \sin 2\theta$$

$$= \frac{29 \times 10^4 + 56 \times 10^4}{2} + \frac{29 \times 10^4 - 56 \times 10^4}{2} \times \cos(-2 \times 32.88) - (-30 \times 10^4) \sin(-2 \times 32.88)$$

$$[\because \ \theta = -32.88°]$$

$$= 42.5 \times 10^4 - 13.5 \times 10^4 \times 0.41 + 30 \times 10^4 \times (-0.912)$$

$$= 42.5 \times 10^4 - 5.535 \times 10^4 - 27.36 \times 10^4$$

$$= 9.605 \times 10^4 \text{ cm}^4$$

and

$$(I_{y_1y_1})_{min.}^{max.} = \frac{I_{xx} + I_{yy}}{2} - \frac{I_{xx} - I_{yy}}{2} \cos 2\theta + I_{yy} \sin 2\theta$$

$$= 42.5 \times 10^4 + 5.535 \times 10^4 + 27.36 \times 10^4 = 75.395 \times 10^4 \text{ cm}^4$$

Hence principal moment of inertia are

$I_{max.}$ = **75.395 × 10⁴ cm⁴. Ans.**

$I_{min.}$ = **9.605 × 10⁴ cm⁴. Ans.**

Alternate Method

The principal moments of inertia can also be obtained by

$$I_{\substack{max.\\min.}} = \frac{I_{xx} + I_{yy}}{2} \pm \sqrt{\frac{(I_{xx} - I_{yy})^2}{2} + I_{xy}^{\ 2}}$$

$$= \frac{29 \times 10^4 + 56 \times 10^4}{2} \pm \sqrt{\frac{(29 \times 10^4 - 56 \times 10^4)^2}{2} + (-30 \times 10^4)^2}$$

$$= 42.5 \times 10^4 \pm \sqrt{(-13.5 \times 10^4)^2 + (-30 \times 10^4)^2}$$

$$= 42.5 \times 10^4 \pm 10^4 \times 32.89$$

$$= (42.5 + 32.89) \times 10^4 \text{ and } (42.5 - 32.89) \times 10^4$$

$$= 75.39 \times 10^4 \text{ and } 9.61 \times 10^4 \text{ cm}^4$$

3.14. MASS MOMENT OF INERTIA

Consider a body of mass M as shown in Fig. 3.42.

Let x = Distance of the centre of gravity of mass M from axis OY

y = Distance of the C.G. of mass M from axis OX

Then moment of the mass about the axis $OY = M \, . \, x$

The above equation is known as *first moment* of mass about the axis OY.

If the moment of mass given by the above equation is again multiplied by the perpendicular distance between the C.G. of the mass and axis OY, then the quantity $(M \, . \, x) \, . \, x$

$= M \cdot x^2$ is known as *second* moment of mass about the axis OY. This second moment of the mass (*i.e.* quantity $M \cdot x^2$) is known as mass moment of inertia about the axis OY.

Similarly, the second moment of mass or mass moment of inertia about the axis OX

$$= (M \cdot y) \cdot y = M \cdot y^2$$

Hence *the product of the mass and the square of the distance of the centre of gravity of the mass from an axis is known as the mass moment of inertia about that axis*. Mass moment of inertia is represented by I_m. Hence mass moment of inertia about the axis OX is represented by $(I_m)_{xx}$ whereas about the axis OY by $(I_m)_{yy}$.

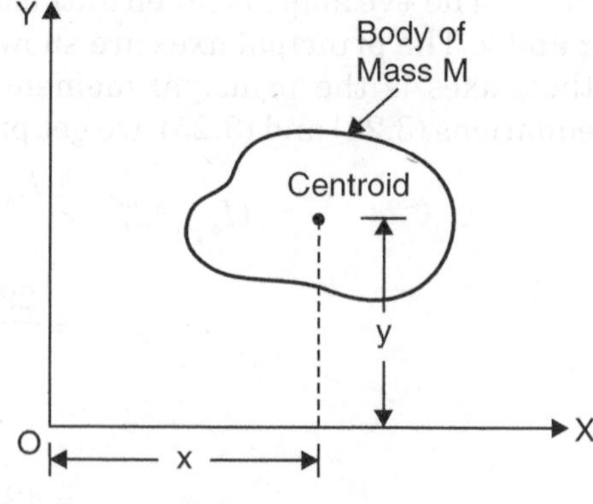

Fig. 3.42

Consider a body which is split up into small masses m_1, m_2, m_3 ...... etc. Let the C.G. of the small areas from a given axis be at a distance of r_1, r_2, r_3 ...... etc. as shown in Fig. 3.43. Then mass moment of inertia of the body about the given axis is given by

$$I_m = m_1 r_1^2 + m_2 r_2^2 + m_3 r_3^2 + \ldots\ldots$$
$$= \Sigma m r^2$$

If small masses are large in number then the summation in the above equation can be replaced by integration. Let the small masses are replaced by dm instead of 'm', then the above equation can be written as

$$I_m = \int r^2 \, dm \qquad \ldots(3.28)$$

Fig. 3.43

3.15. MASS MOMENT OF INERTIA FOR RECTANGLE SECTION, PRISM, SPHERE ETC. FROM FIRST PRINCIPAL

The mass moment of inertia of the following bodies will be determined by the method of integration :

1. Mass moment of inertia of a rectangular plate,
2. Mass moment of inertia of a circular plate.
3. Mass moment of inertia of a hollow circular cylinder.
4. Mass moment of inertia of a right circular cone of base radius R, height H and mass M about its axis.

3.15.1. Mass Moment of Inertia of a Rectangular Plate

(*a*) **Mass moment of inertia of a rectangular plate about *X-X* axis passing through the C.G. of the plate.** Fig. 3.44 shows a rectangular plate of width b, depth 'd' and uniform thickness 't'. Consider a small element of width 'b' at a distance 'y' from X-X axis as shown in Fig. 3.45.

Here X-X axis is the horizontal line passing through the C.G. of the plate.

Area of the element $= b \times dy$

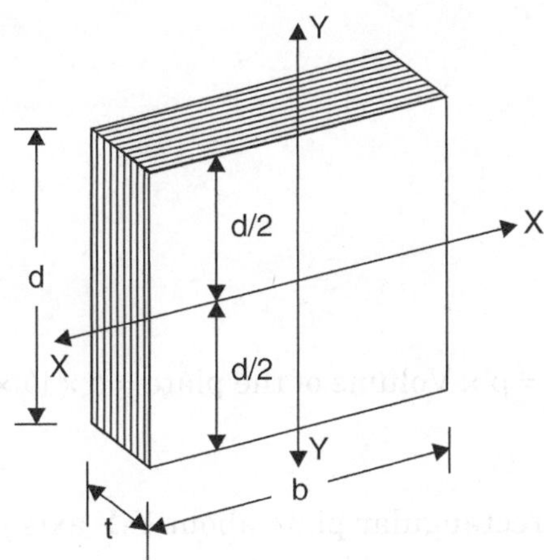

Fig. 3.44

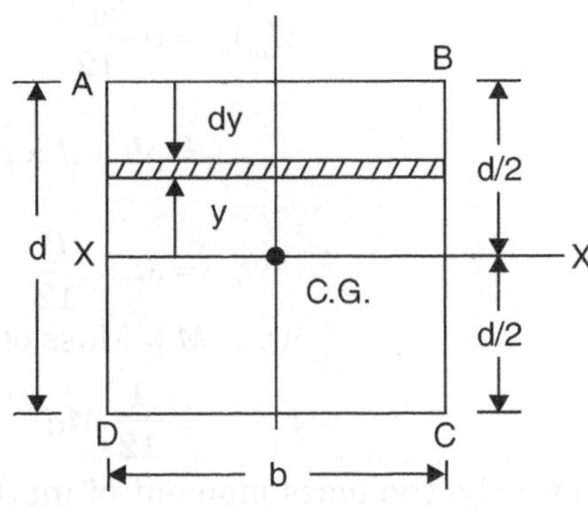

Fig. 3.45

$\therefore$ Mass of the element = Density × Volume of element

$= \rho \times$ [Area × thickness of element]

$= \rho \times [b \times dy \times t]$ [$\because$ ρ = Density and t = thickness]

$= \rho bt\, dy$

Mass moment of inertia of the element about X-X axis

= Mass of element × y^2

$= (\rho bt\, dy) \times y^2 = \rho bt\, y^2\, dy$

Mass moment of inertia of the plate will be obtained by integrating the above equation between the limits $-\frac{d}{2}$ to $\frac{d}{2}$.

$$\therefore \quad (I_m)_{xx} = \int_{-d/2}^{d/2} \rho bt\, y^2\, dy = \rho\, bt \int_{-d/2}^{d/2} y^2\, dy$$

[$\because$ ρ, b, t are constant and can be taken outside the integral sign]

$$= \rho bt \left[\frac{y^3}{3}\right]_{-d/3}^{d/2} = \frac{\rho bt}{3}\left[\left(\frac{d}{2}\right)^3 - \left(-\frac{d}{2}\right)^3\right]$$

$$= \frac{\rho bt}{3}\left[\frac{d^3}{8} - \left(-\frac{d^3}{8}\right)\right] = \frac{\rho bt}{3}\left[\frac{d^3}{8} + \frac{d^3}{8}\right] = \frac{\rho bt}{3} \times \frac{2d^3}{8}$$

$$= \frac{\rho bt}{3}\, d^3 = \rho \times t\, \frac{bd^3}{12} \qquad \text{...(3.29)}$$

But $\frac{bd^3}{12}$ is the moment of inertia of the area of the rectangular section about X-X axis. This moment of inertia of the area is represented by I_{xx}.

$$\therefore \quad (I_m)_{xx} = \rho \times t \times I_{xx} \qquad \text{...(3.30)}$$

where $(I_m)_{xx}$ = Mass moment of inertia of the plate about X-X axis passing through C.G. of the plate.

I_{xx} = Moment of inertia of the area of the plate about X-X axis.

Again from equation (3.5), we have

$$(I_m)_{xx} = \rho \frac{bd^3}{12}$$

$$= \rho b \times d \times t \times \frac{d^2}{12}$$

$$= M \times \frac{d^2}{12}$$

($\because$ M = Mass of the plate = ρ × Volume of the plate = $\rho \times [b \times d \times t]$)

$$= \frac{1}{12} Md^2 \qquad ...(3.31)$$

Similarly, the mass moment of inertia of the rectangular plate about *Y-Y* axis passing through the C.G. of the plate is given by

$$(I_m)_{yy} = \frac{1}{12} Mb^2. \qquad ...(3.32)$$

(*b*) **Mass moment of inertia of the rectangular plate about a line passing through the base.** Fig. 3.46 shows a rectangular plate *ABCD*, having width = *b*, depth = *d* and uniform thickness = *t*. We want to find the mass moment of inertia of the rectangular plate about the line *CD*, which is the base of the plate. Consider a rectangular elementary strip of width *b*, thickness *t* and depth '*dy*' at a distance *y* from the line *CD* as shown in Fig. 3.46.

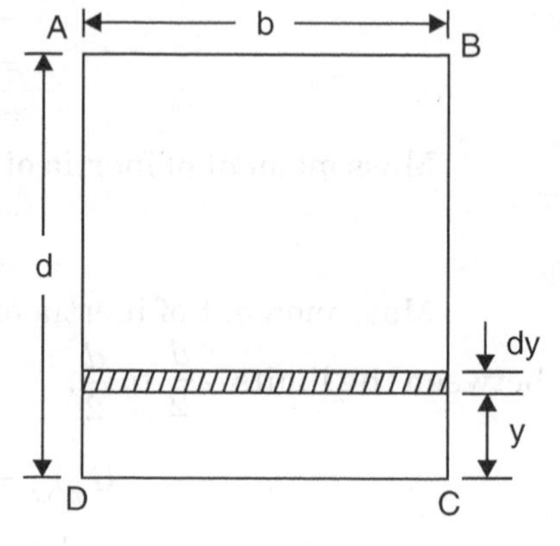

Fig. 3.46

Area of strip, $dA = b \,.\, dy$

Volume of strip $= dA \times t = b \,.\, dy \,.\, t = b \,.\, t \,.\, dy$

Mass of the strip, dm = Density × Volume of strip

$= \rho(b \,.\, t \,.\, dy) = \rho \,.\, b \,.\, t \,.\, dy$

Mass moment of inertia of the strip about the line *CD*

= Mass of strip . y^2

$= dm \,.\, y^2 = y^2 \,.\, dm$

Mass moment of inertia of the whole rectangular plate about the line *CD* is obtained by integrating the above equation between the limits 0 to *d*.

$\therefore$ Mass moment of inertia of the rectangular plate about the line *CD*

$$= \int_0^d y^2 \,.\, dm = \int_0^d y^2 \,(\rho \,.\, b \,.\, t \,.\, dy) \qquad [\because\ dm = \rho \,.\, b \,.\, t \,.\, dy]$$

$$= \rho \,.\, b \,.\, t \int_0^d y^2\, dy \qquad [\because\ \rho, b \text{ and } t \text{ are constant}]$$

$$= \rho \,.\, b \,.\, t \,.\left[\frac{y^3}{3}\right]_0^d = \rho \,.\, b \,.\, t \,.\, \frac{d^3}{3} = \rho \,.\, b \,.\, t \,.\, d \,.\, \frac{d^2}{3}$$

$$= \frac{M \,.\, d^2}{3} \quad [\because\ \rho \,.\, b \,.\, t \,.\, d = \text{Mass of rectangular plate} = M] \qquad ...(3.33)$$

(*c*) **Mass moment of inertia of a hollow rectangular plate.** Fig. 3.47 shows a hollow rectangular plate in which *ABCD* is the main plate and *EFGH* is the cut-out section.

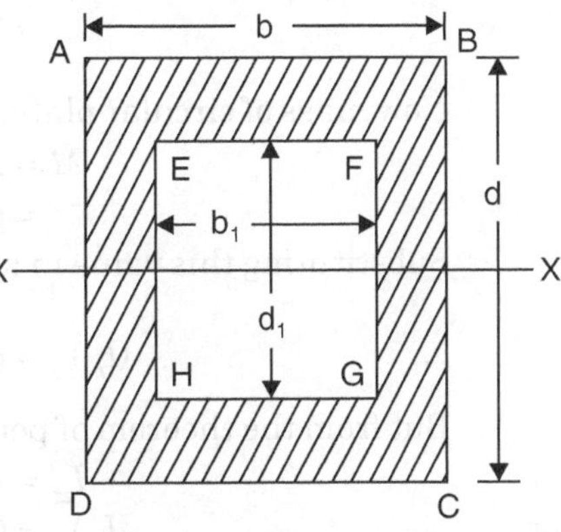

Fig. 3.47

The mass moment of inertia of the main plate *ABCD* about *X-X* is given by equation

$$= \frac{1}{12} Md^2$$

The mass moment of inertia of the cut-out section *EFGH* about *X-X* axis

$$= \frac{1}{12} md_1^2$$

where M = Mass of main plate *ABCD*

$= \rho . b . d . t$

m = Mass of the cut-out section *EFGH*

$= \rho . b_1 . d_1 . t$

Then mass moment of inertia of hollow rectangular plate about *X-X* axis is given by

$$(I_m)_{xx} = \frac{1}{12} Md^2 - \frac{1}{12} md_1^2 \quad ...(3.34)$$

3.15.2. Mass Moment of Inertia of a Circular Plate. Fig. 3.48 shows a circular plate of radius *R* and thickness *t* with *O* as centre. Consider an elementary circular ring of radius '*r*' and width *dr* as shown in Fig. 3.48 (*a*).

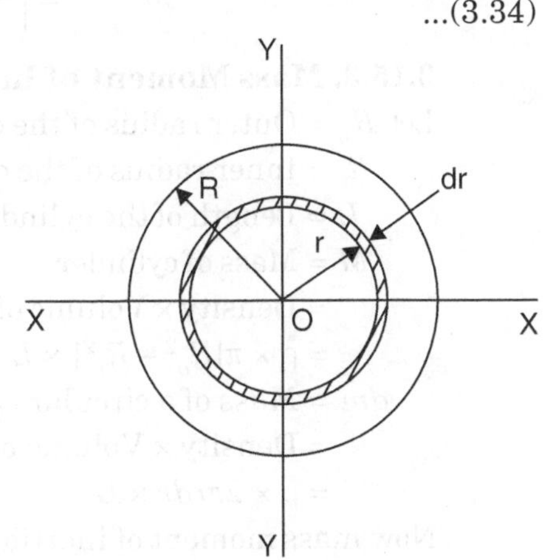

Fig. 3.48

Area of ring, $dA = 2\pi r . dr$

Volume of ring $= \text{Area of ring} \times t = dA . t$

$= 2\pi r . dr . t$

Mass of ring,

$$dm = \text{Density} \times \text{Volume of ring}$$

$$= \rho(2\pi r . dr . t)$$

In this case first find the mass moment of inertia about an axis passing through *O* and perpendicular to the plane of the paper *i.e.* about axis *Z-Z*.

∴ Mass moment of inertia of the circular ring about axis *Z-Z*

$= (\text{Mass of ring}) \times (\text{radius of ring})^2$

$= dm \times r^2 = (\rho . 2\pi r\, dr . t) \times r^2 = \rho . t . 2\pi r^3\, dr$

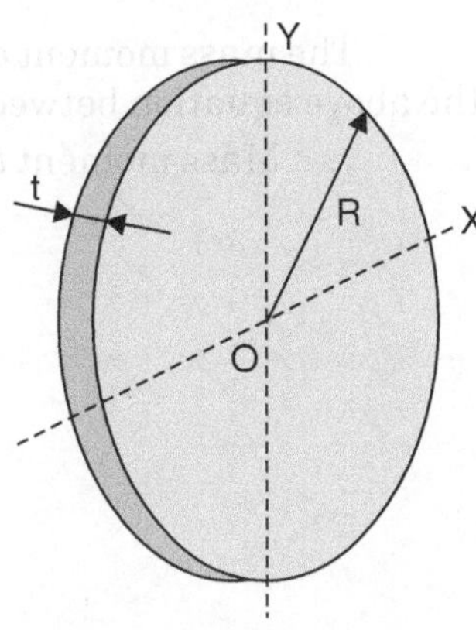

Fig. 3.48(a)

The mass moment of inertia of the whole circular plate will be obtained by integrating the above equation between the limits *O* to *R*.

∴ Mass moment of inertia of circular plate about *Z-Z* axis is given by

$$(I_m)_{zz} = \int_O^R \rho . t . 2\pi r^3\, dr = 2\pi . \rho . t \int_O^R r^3\, dr$$

$$= 2\pi\rho \,.\, t \left[\frac{r^4}{4}\right]_O^R = 2\pi \,.\, \rho \,.\, t \,.\, \frac{R^4}{4} = \pi \,.\, \rho \,.\, t \,.\, \frac{R^4}{2}$$

Now mass of circular plate,

$$M = \rho \times \text{Volume of plate}$$

$$= \rho \times \pi R^2 \times t \qquad [\text{Volume of plate} = \text{Area} \times t = \pi R^2 \times t]$$

Substituting this value in above equation, we get

$$\therefore \qquad (I_m)_{zz} = \rho \times \pi R^2 \times t \times \frac{R^2}{2} = \frac{MR^2}{2} \qquad \text{...(3.35)}$$

But from the theorem of perpendicular axis given by equation (3.7), we have

$$I_{zz} = I_{xx} + I_{yy}$$

or

$$(I_m)_{zz} = (I_m)_{xx} + (I_m)_{yy}$$

And due to symmetry, we have $(I_m)_{xx} = (I_m)_{yy}$

$$\therefore \qquad (I_m)_{xx} = (I_m)_{yy} = (I_m)_{zz}/2$$

$$= \left(\frac{MR^2}{2}\right)/2 = \frac{MR^2}{4} \qquad \text{...(3.36)}$$

3.15.3. Mass Moment of Inertia of a Hollow Circular Cylinder

Let R_o = Outer radius of the cylinder

R_i = Inner radius of the cylinder

L = Length of the cylinder

M = Mass of cylinder

= Density × Volume of cylinder

$= \rho \times \pi[R_o^2 - R_i^2] \times L$ ...(*i*)

dm = Mass of a circular ring of radius 'r', width 'dr' and length L [Refer to Fig. 3.48]

= Density × Volume of ring = ρ × Area of ring × L

$= \rho \times 2\pi r dr \times L$

Now mass moment of inertia of the circular ring about Z-Z axis

= Mass of ring × (radius)²

$= (\rho \times 2\pi r dr \times L) \times r^2$

The mass moment of inertia of the hollow circular cylinder will be obtained by integrating the above equation between the limits R_i to R_O.

∴ Mass moment of inertia of the hollow circular cylinder about Z-Z axis is given by,

$$(I_m)_{zz} = \int_{R_i}^{R_o} (\rho \times 2\pi r\, dr \,.\, L)\, r^2$$

$$= \rho \times 2\pi \times L \int_{R_i}^{R_o} r^3\, dr = \rho \times 2\pi \times L \left[\frac{r^4}{4}\right]_{R_i}^{R_o}$$

$$= \rho \times 2\pi \times L \times \left[\frac{R_o^4 - R_i^4}{4}\right]$$

$$= \rho \times 2\pi \times L \times \left[\frac{R_o^2 - R_i^2}{4}\right] [R_o^2 + R_i^2]$$

$$[\because \quad R_o^4 - R_i^4 = (R_o^2 - R_i^2)(R_o^2 + R_i^2)]$$

$$= \rho \times \pi[R_o^2 - R_i^2] \times L \times \frac{(R_o^2 + R_i^2)}{2}$$

$$= \frac{M(R_o^2 + R_i^2)}{2} \qquad [\because \quad \rho \times \pi \times (R_o^2 - R_i^2) = M]$$

Now $\qquad (I_m)_{xx} = (I_m)_{yy} = \frac{(I_m)_{zz}}{2} = \frac{M(R_o^2 + R_i^2)}{4}$

3.15.4. Mass Moment of Inertia of a Right Circular Cone of Base Radius R, Height H and Mass M about its Axis

Let R = Radius of the base of the cone,

H = Height of the cone,

M = Mass of the cone

$$= \text{Density} \times \text{Volume of cone} = \rho \times \frac{1}{3}\pi R^2 \times H$$

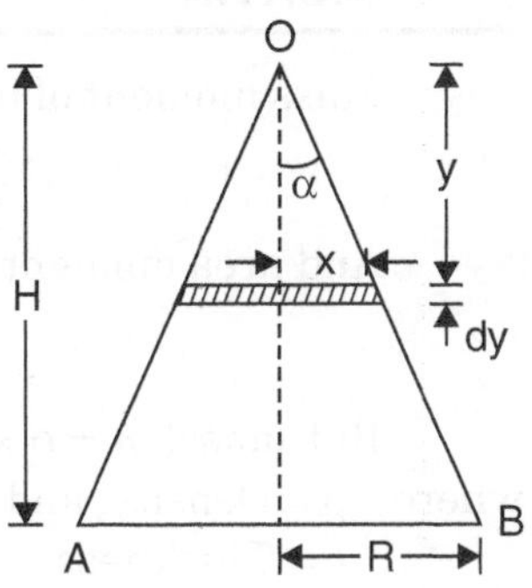

Fig. 3.49

Consider an elemental plate of thickness dy and of radius x at a distance y from the vertex (as shown in Fig. 3.49).

We have, $\quad \tan\alpha = \frac{x}{y} = \frac{R}{H} \quad \therefore \quad x = \frac{R}{H} \times y$

Mass of the elemental plate,

$$dm = \rho \times \text{Volume}$$

$$= \rho \times (\pi x^2 \times dy)$$

$$= \rho \times \left[\pi \frac{R^2 \times y^2}{H^2} \times dy\right] \qquad \left[\because \quad x = \frac{R \times y}{H}\right]$$

The mass moment of inertia of the circular elemental plate about the axis of the cone (here axis of the cone is Z-Z axis of the circular elemental plate) is given by equation (3.35) as

$$(I_m)_{zz} = \frac{\text{Mass of plate} \times \text{radius}^2}{2}$$

$$= \frac{(dm) \times r^2}{2} = \frac{dm \times x^2}{2} \qquad (\because \quad r = x)$$

$$= \left[\rho \times \frac{\pi R^2 y^2}{H^2} \times dy\right] \times \frac{x^2}{2} \qquad \left[\because \quad dm = \rho \times \frac{\pi R y}{H^2} dy\right]$$

$$= \frac{\rho \times \pi R^2 y^2}{H^2} \times dy \times \left[\frac{R^2 y^2}{H^2}\right] \times \frac{1}{2} \qquad \left[\because \quad x = \frac{Ry}{H}\right]$$

$$= \frac{\rho \times \pi r^4 \times y^4}{2H^4} dy$$

Now the total mass moment of inertia of the circular cone will be obtained by integrating the above equation between the limits O to H.

$$\therefore \quad (I_m)_{zz} = \int_O^H \frac{\rho\pi R^4 \times y^4}{2H^4} dy = \frac{\rho\pi R^4}{2H^4} \times \left[\frac{y^5}{5}\right]_O^H$$

$$= \frac{\rho\pi R^4}{2H^4} \times \frac{H^5}{5} = \frac{\rho\pi R^4 \times H}{2 \times 5}$$

But mass of cone, $M = \dfrac{\rho\pi R^4 \times H}{3}$

$$\therefore \quad (I_m)_{zz} = \frac{\rho\pi R^2 \times H}{3} \times \frac{R^2 \times 3}{10} = M \times \frac{3}{10} R^2 = \frac{3}{10} MR^2 \qquad ...(3.37)$$

3.16. RELATION BETWEEN MASS MOMENT OF INERTIA AND AREA MOMENT OF INERTIA

Mass moment of inertia is given by

$$I_m = \int r^2 dm$$

And area moment of inertia is given by

$$I = \int r^2 dA$$

But mass, $m = \rho \times A \times t$

where ρ = Density, and

t = Thickness

Hence

$$I_m = \int r^2 dm$$

$$= \int r^2 d\ (\rho A \times t)$$

$$= \rho \times t \int r^2 dA.$$

HIGHLIGHTS

1. The point, through which the whole weight of the body acts, is known as centre of gravity.
2. The point, at which the total area of a plane figure is assumed to be concentrated, is known as centroid of that area. The centroid and centre of gravity are at the same point.
3. The centre of gravity of a uniform rod lies at its middle point.
4. The C.G. of a triangle lies at a point where the three medians of a triangle meet.
5. The C.G. of a parallelogram or a rectangle is at a point where its diagonals meet each other.
6. The C.G. of a circle lies at its centre.
7. The C.G. of a body consisting of different areas is given by

$$\bar{x} = \frac{a_1x_1 + a_2x_2 + a_3x_3 + \ldots\ldots}{a_1 + a_2 + a_3 + \ldots\ldots} \quad \text{and} \quad \bar{y} = \frac{a_1y_1 + a_2y_2 + a_3y_3 + \ldots\ldots}{a_1 + a_2 + a_3 + \ldots\ldots}$$

where $\bar{x}$ and $\bar{y}$ = Co-ordinates of the C.G. of the body from axis of reference

$a_1, a_2, a_3, \ldots\ldots$ = Different areas of the sections of the body

$x_1, x_2, x_3, \ldots\ldots$ = Distances of the C.G. of the areas $a_1, a_2, a_3, \ldots\ldots$ from Y-axis

$y_1, y_2, y_3, \ldots\ldots$ = Distances of the C.G. of the areas $a_1, a_2, a_3, \ldots\ldots$ from X-axis.

8. If a given section is symmetrical about X-X axis or Y-Y axis, the C.G. of the section will lie on the axis symmetry.

9. The moment of inertia of an area (or mass) about an axis is the product of area (or mass) and square of the distance of the C.G. of the area (or mass) from that axis. It is represented by I.

10. Radius of gyration of a body (or a given lamina) is the distance from an axis of reference where the whole mass (or area) of the given body is assumed to be concentrated so as not to alter the moment of inertia about the given axis. It is represented by k. Mathematically, $k = \sqrt{\frac{I}{A}}$.

11. According to theorem of perpendicular axis $I_{ZZ} = I_{XX} + I_{YY}$ where I_{XX} and I_{YY} = Moment of inertia of a plane section about two mutually perpendicular axes X-X and Y-Y in the plane of the section, I_{ZZ} = Moment of inertia of the section perpendicular to the plane and passing through the intersection of X-X and Y-Y axes.

12. According to the theorem of parallel axis $I_{AB} = I_G + Ah^2$ where

I_G = Moment of inertia of a given area about an axis passing through C.G. of the area

I_{AB} = Moment of inertia of the given area about an axis AB, which is parallel to the axis passing through G

h = Distance between the axis passing through G and axis AB

A = Area of the section.

13. Moment of inertia of a rectangular section :

(*i*) about an horizontal axis passing through C.G. $= \dfrac{bd^3}{12}$

(*ii*) about an horizontal axis passing through base $= \dfrac{bd^3}{12}$.

14. Moment of inertia of a circular section $= \dfrac{\pi D^4}{64}$.

15. Moment of inertia of a triangular section :

(*i*) about the base $= \dfrac{bh^3}{12}$

(*ii*) about an axis passing through C.G. and parallel to the base $= \dfrac{bh^3}{36}$.

where b = Base width, and h = Height of the triangle.

16. The C.G. of an area by integration method is given by

$$\bar{x} = \frac{\int x^* \, dA}{\int dA} \quad \text{and} \quad \bar{y} = \frac{\int x^* \, dA}{\int dA}$$

where x^* = Distance of C.G. of area dA from y-axis

y^* = Distance of C.G. of area dA from x-axis.

17. The C.G. of a straight or curved line is given by

$$\bar{x} = \frac{\int x^* \, dL}{\int dL} \quad \text{and} \quad \bar{y} = \frac{\int y^* \, dL}{\int dL}.$$

EXERCISE 3

(A) Theoretical Questions

1. Define centre of gravity and centroid.
2. Derive an expression for the centre of gravity of a plane area using method of moments.
3. What do you understand by axes of reference ?
4. Define the terms : moment of inertia and radius of gyration.
5. State the theorem of perpendicular axis. How will you prove this theorem ?
6. State and prove the theorem of parallel axis.
7. Find an expression for the moment of inertia of a rectangular section :
   (*i*) about an horizontal axis passing through the C.G. of the rectangular section, and
   (*ii*) about an horizontal axis passing through the base of the rectangular section.
   (*AMIE Summer, 1985*)
8. Prove that the moment of inertia of a circular section about an horizontal axis (in the plane of the circular section) and passing through the C.G. of the section is given by $\frac{\pi D^4}{64}$.
9. Prove that moment of inertia of a triangular section about the base of the section

$$= \frac{bh^3}{12}$$

where b = Base of triangular section, and
h = Height of triangular section.

10. Derive an expression for the moment of inertia of a triangular section about an axis passing through the C.G. of the section and parallel to the base.
11. Show that $I_O = I_G + Ah^2$, where I_G is the moment of inertia of a lamina about an axis through its centroid and lying in its plane and h is the distance from the centroid to a parallel axis in the same plane about which its moment of inertia is I_O, A being the area of the lamina.
12. State and prove the parallel axes theorem on moment of inertia for a plane area.
13. Prove that the moment of area of any plane figure about a line passing through its centroid is zero.

(B) Numerical Problems

1. Find the centre of gravity of the T-section shown in Fig. 3.50. **[Ans.** 8.272 cm]

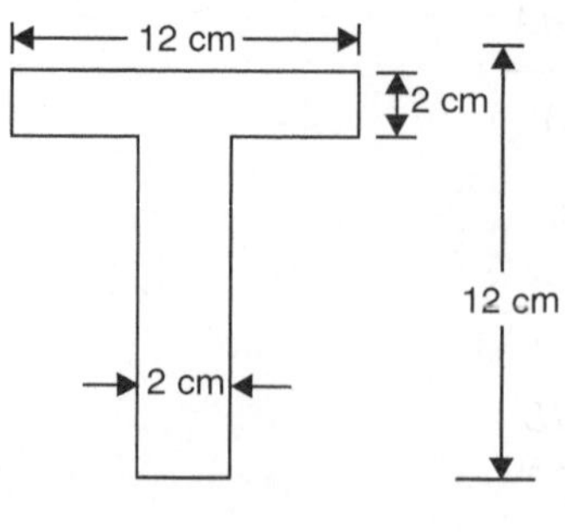

Fig. 3.50

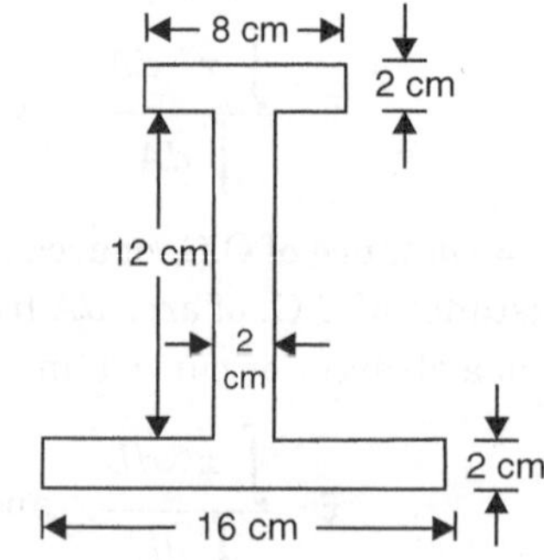

Fig. 3.51

2. Find the centre of gravity of the I-section shown in Fig. 3.51. **[Ans.** 6.44 cm]

3. Find the centre of gravity of the *L*-section shown in Fig. 3.52. [**Ans.** $\bar{x}$ = 1.857, $\bar{y}$ = 3.857]

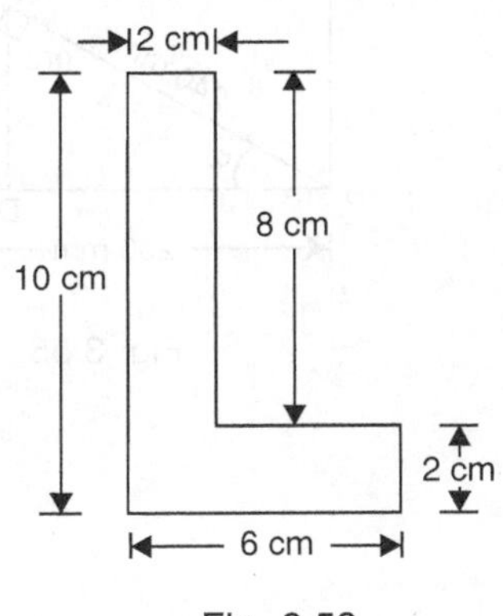

Fig. 3.52

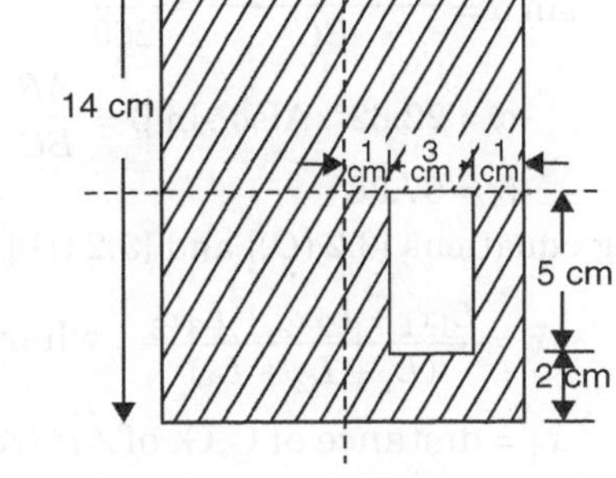

Fig. 3.53

4. From a rectangular lamina *ABCD* 10 cm × 14 cm a rectangular hole of 3 cm × 5 cm is cut as shown in Fig. 3.53. Find the centre of gravity of the remainder lamina. [**Ans.** $\bar{x}$ = 4.7 cm, $\bar{y}$ = 6.444 cm]

5. For the *T*-section shown in Fig. 3.54, determine the moment of inertia of the section about the horizontal and vertical axes, passing through the centre of gravity of the section. [**Ans.** 567.38 cm^4, 294.67 cm^4]

6. For the *I*-section shown in Fig. 3.51, find the moment of inertia about the centroidal axis *X-X* perpendicular to the web. [**Ans.** 2481.76 cm^4]

7. Locate the C.G. of the area shown in Fig. 3.54 with respect to co-ordinate axes. All dimensions are in mm.

[**Hint.** $a_1 = 10 \times 30 = 300$ mm^2, $x_1 = 5$ mm, $y_1 = 15$.

$a_2 = 40 \times 10 = 400$ mm^2, $x_2 = 10 + 20$
$= 30$ mm, $y_2 = 5$ mm

$a_3 = 10 \times 20 = 200$ mm^2, $x_3 = 5$ mm, $y_3 = -10$ mm

$a_4 = 10 \times 10 = 100$ mm^2, $x_4 = 45$ mm,

$y_4 = 10 + 5 = 15$ mm

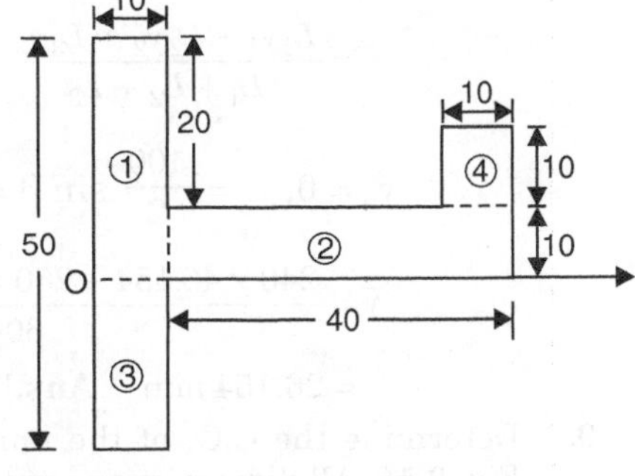

Fig. 3.54

$$\bar{x} = \frac{a_1x_1 + a_2x_2 + a_3x_3 + a_4x_4}{(a_1 + a_2 + a_3 + a_4)}$$

$$= \frac{1500 + 12000 + 1000 + 4500}{1000}$$

$= 1.5 + 12 + 1 + 4.5 = 19$ mm. **Ans.**

$$\bar{y} = \frac{a_1y_1 + a_2y_2 + a_3y_3 + a_4y_4}{(a_1 + a_2 + a_3 + a_4)} = \frac{4500 + 2000 - 2000 + 1500}{1000}$$

$= 4.5 + 2 - 2 + 1.5 = 6$ mm. **Ans.**]

8. A thin homogeneous wire is bent into a triangular shape *ABC* such that *AB* = 240 mm, *BC* = 260 mm and *AC* = 100 mm. Locate the C.G. of the wire with respect to co-ordinate axes. Angle at *A* is right angle.

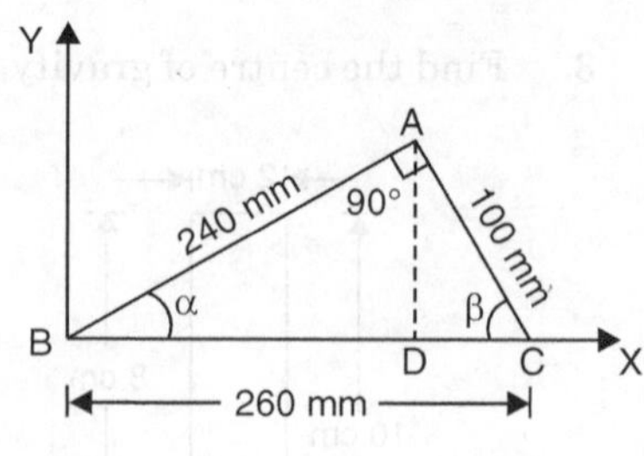

Fig. 3.55

[**Hint.** First determine angles α and β. Use sine rule

$$\frac{BC}{\sin 90^\circ} = \frac{AC}{\sin \alpha} = \frac{AB}{\sin \beta}.$$

$$\therefore \quad \sin \alpha = \frac{AC \times \sin 90^\circ}{BC} = \frac{100}{260}$$

$$\therefore \quad \alpha = 22.62^\circ. \text{ Also } \sin \beta = \frac{AB}{BC} \times \sin 90^\circ = \frac{240}{260}$$

$$\therefore \quad \beta = 67.38^\circ$$

Using equations [3.2 (*C*)] and [3.2 (*D*)]

$$\bar{x} = \frac{L_1x_1 + L_2x_2 + L_3x_3}{(L_1 + L_2 + L_3)}, \text{ where } L_1 = AB = 240,$$

x_1 = distance of C.G. of AB from y-axis

$$= \frac{240}{2} \times \cos \alpha = 120 \times \cos 22.62^\circ = 110.77 \text{ mm}$$

$L_2 = BC = 260$ mm, x_2 = Distance of C.G. of BC from y-axis = 130 mm

$L_3 = AC = 100$ mm, x_3 = Distance of C.G. of AC from y-axis

$$= BD + \frac{100}{2} \cos \beta = 240 \cos \alpha + 50 \cos \beta$$

$$= 240 \times \cos 22.62^\circ + 50 \cos 67.38^\circ = 240.77 \text{ mm}$$

$$\therefore \quad \bar{x} = \frac{240 \times 110.77 + 260 \times 130 + 100 \times 240.77}{240 + 260 + 100} = 140.77 \text{ mm. } \textbf{Ans.}$$

$$\bar{y} = \frac{L_1y_1 + L_2y_2 + L_3y_3}{L_1 + L_2 + L_3}, \quad \text{where } y_1 = \frac{240}{2} \sin \alpha = 120 \times \sin 22.62^\circ = 46.154$$

$$y_2 = 0, y_3 = \frac{100}{2} \sin \beta = 50 \sin 67.38^\circ = 48.154$$

$$\therefore \quad \bar{y} = \frac{240 \times 46.154 + 260 \times 0 + 100 \times 46.154}{600}$$

$= 26.154$ mm. **Ans.**]

9. Determine the C.G. of the uniform plane lamina shown in Fig. 3.56. All dimensions are in cm.

[**Hint.** The figure is symmetrical about *Y-Y* axis.

$$\therefore \quad \bar{y} = \frac{a_1y_1 + a_2y_2 + a_3y_3 + a_4y_4}{a_1 + a_2 + a_3 + a_4}$$

where $a_1 = 40 \times 30 = 1200 \text{ cm}^2, y_1 = \frac{30}{2} = 15 \text{ cm},$

$$a_2 = 30 \times 20 = 600 \text{ cm}^2, y_2 = 30 + \frac{30}{2} = 45 \text{ cm}$$

$$a_3 = -\frac{\pi \times 10^2}{2} = -50\pi, y_3 = \frac{4r}{3\pi} = \frac{4 \times 10}{3\pi} = \frac{40}{3\pi}$$

$$a_4 = -\frac{20 \times 10}{2} = -100, y_4 = 60 - \frac{10}{3} = \frac{170}{3}$$

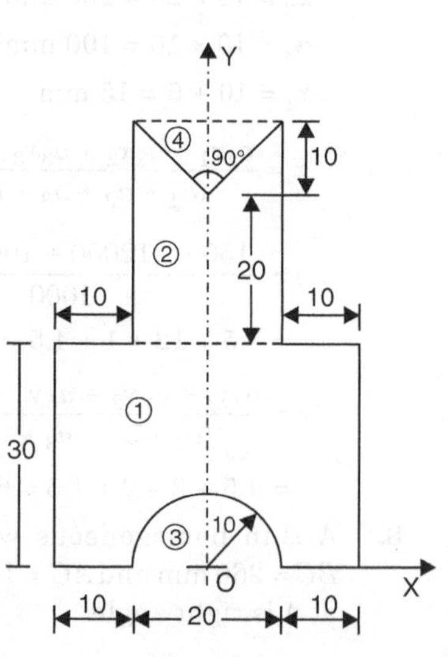

Fig. 3.56

$$\therefore \quad \bar{y} = \frac{1200 \times 15 + 600 \times 45 - 50\pi \times \dfrac{40}{3\pi} - 100 \times \dfrac{170}{3}}{1200 + 600 - 50\pi - 100}$$

$$= \frac{18000 + 27000 - 666.7 - 5666.7}{1700 - 50\pi} = \frac{38666.6}{1542.92}$$

= 25.06 cm from origin 0. **Ans.**]

10. From a circular plate of diameter 100 mm a circular part of diameter 50 mm is cut as shown in Fig. 3.57. Find the centroid of the remainder. (*U.P. Tech. University, 2002–2003*)

[**Hint.** Fig. 3.57 is symmetrical about x-axis. Hence centroid lies on x-axis.

$\therefore \quad \bar{y} = 0.6$. The value of $\bar{x}$ is given by $\bar{x} = \dfrac{a_1 x_1 + a_2 x_2}{a_1 - a_2}$

But $\quad a_1 = \dfrac{\pi}{4} \times 100^2 = 7853.98 \text{ mm}^2, x_1 = \dfrac{100}{2} = 50 \text{ mm}$

$$a_2 = -\left(\frac{\pi}{4} \times 50^2\right) = -1963.5 \text{ mm}^2, x_2 = 100 - 25 = 75 \text{ mm}$$

$$\therefore \quad \bar{x} = \frac{7853.98 \times 50 - 1963.5 \times 75}{7853.98 - 1963.5} = 41.67 \text{ mm}$$

Hence centroid is at (41.67 mm, 0). **Ans.**]

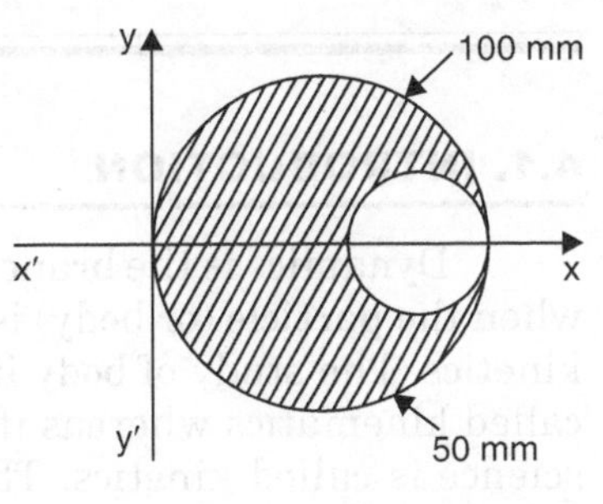

Fig. 3.57

4

Dynamics of Particles

4.1. INTRODUCTION

Dynamics is the branch of science which deals with the study of a particle (or of a body), when the particle (or body) is in motion. The dynamics is further divided into kinematics and kinetics. The study of body in motion, when forces which cause motion are not considered, is called kinematics whereas if the forces are considered for the body in motion, that branch of science is called kinetics. This chapter deals with velocity, acceleration, their relationship, relative motion, curvilinear motion, Newton's laws, work energy equation of particles, impulse and momentum and impact of elastic bodies.

4.2. VELOCITY

Velocity is defined as the rate of change of displacement of a body moving in a straight line. It is measured in metre per second. Velocity is a vector* quantity. It is denoted by v.

Let s = Distance travelled by a body in a straight line

t = Time taken to travel the distance.

Then velocity of the body $= \dfrac{s}{t}$.

Mathematically, it can also be written as

Velocity, $$v = \frac{ds}{dt} \qquad ...(4.1)$$

4.3. ACCELERATION

Acceleration is defined as the rate of change of velocity of a body. It is measured in metre per second square and is written as m/sec^2 in M.K.S. system and m/s^2 in S.I. system.

$$\therefore \quad \text{Acceleration} = \frac{\text{Change of velocity}}{\text{Time}}$$

Acceleration is denoted by 'a'. Acceleration is also a vector* quantity. Mathematically, it can also be written as

Acceleration, $$a = \frac{dv}{dt} \qquad ...(4.2)$$

$$= \frac{d}{dt}\left(\frac{ds}{dt}\right) \qquad \left(\because \; v = \frac{ds}{dt} \text{ from equation (4.1)}\right)$$

*A vector quantity is one which is represented in magnitude and direction.

$$= \frac{d^2s}{dt^2} \quad ...(4.2A)$$

But $\frac{dv}{dt}$ can also be written as

$$\frac{dv}{dt} = \frac{dv}{ds} \cdot \frac{ds}{dt} = \frac{dv}{ds} \cdot v \qquad \left(\because \frac{ds}{dt} = v\right)$$

But $\frac{dv}{dt}$ = Acceleration = a

$$\therefore \quad a = \frac{dv}{ds} \cdot v \quad ...(4.3)$$

4.4. RELATIONSHIP OF VELOCITY, ACCELERATION AND DISPLACEMENT

The relationship of velocity, acceleration and displacement is obtained from equations of motion in a straight line.

Let us consider a body, which is moving in a straight line.

Let u = Initial velocity of the body in m/s
v = Final velocity of the body in m/s
t = Time in seconds, during which velocity changes from u to v
s = Distance travelled by the body in time t
a = Acceleration of the body.

4.4.1. Equation for Final Velocity

Change of velocity = Final velocity – Initial velocity
$= (v - u)$

Rate of change of velocity $= \frac{\text{Change of velocity}}{\text{Time}} = \frac{(v-u)}{t}$.

But rate of change of velocity = Acceleration = a

$$\therefore \quad a = \frac{(v-u)}{t} \quad ...(4.4)$$

or $at = v - u$

$$\therefore \quad v = u + at \quad ...(4.5)$$

4.4.2. Equation of Motion for Distance Covered (s)

We know average velocity $= \frac{\text{Initial velocity + Final velocity}}{2}$

$= \frac{u+v}{2}$.

Distance covered, s = Average velocity × Time

$= \frac{(u+v)}{2} \times t$.

Substituting the value of v from equation (4.5),

$$s = \frac{(u+u+at)}{2} \times t = \frac{(2u+at)}{2} \times t = \left(u + \frac{1}{2}at\right) \times t$$

$$\therefore \quad s = ut + \tfrac{1}{2}at^2 \quad ...(4.6)$$

4.4.3. Derivation of $v^2 - u^2 = 2as$. The value of 't' from equation (4.5) is given as

$$t = \frac{(v-u)}{a}.$$

Substituting the value of 't' in equation (4.6), we get

$$s = u\left(\frac{v-u}{a}\right) + \frac{1}{2}a\left(\frac{v-u}{a}\right)^2$$

$$= (v-u)\left[\frac{u}{a} + \frac{1}{2a}(v-u)\right] = (v-u)\left[\frac{(2u+v-u)}{2a}\right]$$

$$= \frac{(v-u)(v+u)}{2a} = \frac{v^2-u^2}{2a}$$

or $$v^2 - u^2 = 2as \qquad ...(4.7)$$

In equations (4.5), (4.6) and (4.7), a is the linear acceleration. If instead of acceleration the retardation is given, then the value of 'a' in these equations should be taken negative. Retardation is –ve of acceleration.

2nd Method for deriving equations of motion for a body moving in a straight line

The equations of motion of a body moving in a straight line may also be derived by integration as given below :

1. Derivation of $s = ut + \frac{1}{2}at^2$

Let a body is moving with a uniform acceleration 'a'.

Then from equation (4.2), we have

$$\frac{d^2s}{dt^2} = a \qquad \text{or} \qquad \frac{d}{dt}\left(\frac{ds}{dt}\right) = a$$

or $$d\left(\frac{ds}{dt}\right) = a\,dt.$$

Integrating the above equation,

$$\int d\left(\frac{ds}{dt}\right) = \int a\,dt \qquad \text{or} \qquad \frac{ds}{dt} = at + C_1 \qquad ...(i)$$

where C_1 is the constant of integration.

But $$\frac{ds}{dt} = \text{Velocity at any instant}$$

when $t = 0$, the velocity is known as initial velocity which is represented by u.

$\therefore$ at $t = 0$, $\frac{ds}{dt}$ = Initial velocity = u

Substituting these values in equation (i).

$$u = a \times 0 + C_1$$

$\therefore$ $$C_1 = u.$$

Substituting the value of C_1 in equation (i),

$$\frac{ds}{dt} = at + u \qquad ...(ii)$$

Integrating the above equation, we get

$$s = \frac{at^2}{2} + ut + C_2 \qquad \text{...}(iii)$$

when C_2 is another constant of integration.

When $t = 0$, $s = 0$.

Substituting these values in equation (*iii*), we get

$$0 = \frac{a}{2} \times 0 + u \times 0 + C_2 = 0 + 0 + C_2$$

$$\therefore \quad C_2 = 0.$$

Substituting this value of C_2 in equation (*iii*), we get

$$s = \frac{1}{2} at^2 + ut$$

the above equation is the same as equation (4.6).

2. Derivation of v = u + at.

From equation (*ii*), we have $\frac{ds}{dt} = at + u$.

But $\frac{ds}{dt}$ represents the velocity at any time. After the time '*t*' the velocity is known as final velocity, which is represented by v.

$$\therefore \quad \frac{ds}{dt} \text{ after time '}t\text{'} = \text{Final velocity} = v.$$

Substituting the value of $\frac{ds}{dt} = v$, we get

$$v = at + u.$$

3. Derivation of $v^2 = u^2 + 2as$.

From equation (4.3), acceleration '*a*' is given by

$$a = \frac{v\,dv}{ds}$$

or

$$\frac{v\,dv}{ds} = a \quad \text{or} \quad v\,dv = a\,ds.$$

Integrating, we get

$$\frac{v^2}{2} = as + C_3 \qquad \text{...}(iv)$$

where C_3 is constant of integration.

When $s = 0$, the velocity is known as initial velocity.

$\therefore$ At $s = 0$, $\quad v = u$.

Substituting these values in equation (*iv*), we get

$$\frac{u^2}{2} = a \times 0 + C_3$$

$$\therefore \quad C_3 = \frac{u^2}{2}.$$

Substituting this value in equation (*iv*), we get

$$\frac{v^2}{2} = as + \frac{u^2}{2} \qquad \text{or} \quad v^2 = 2as + u^2.$$

Problem 4.1. *A body is moving with a velocity of 2 m/s. After 4 seconds the velocity of the body becomes 5 m/s. Find the acceleration of the body.*

Sol. Given :

Initial velocity, $u = 2$ m/s

Final velocity, $v = 5$ m/s

Time, $t = 4$ sec

Let a = Acceleration of the body.

Using equation (4.5), $v = u + at$

or $5.0 = 2.0 + a \times 4.0$

$\therefore$ $a = \frac{(5-2)}{4} = \frac{3}{4} = \mathbf{0.75\ m/s^2}$. **Ans.**

Problem 4.2. *A car is moving with a velocity of 15 m/s. The car is brought to rest by applying brakes in 5 seconds. Determine :*

(i) The retardation, and

(ii) Distance travelled by the car after applying brakes.

Sol. Given :

Initial velocity, $u = 15$ m/s

Final velocity, $v = 0$ ($\because$ Car has been brought to rest)

Time, $t = 5$ sec

(*i*) Let a = Acceleration of the car

Using equation (4.5), $v = u + at$

or $0 = 15 + a \times 5$

$\therefore$ $a = \frac{-15}{5} = -3$ m/s^2

$\therefore$ Retardation $= -a = -(-3) = \mathbf{3\ m/s^2}$. **Ans.**

(*ii*) Let s = Distance travelled by the car after applying brakes.

Using equation (4.6), we have

$$s = ut + \tfrac{1}{2}at^2$$

$$= 15 \times 5 + \tfrac{1}{2}(-3) \times 5^2 \qquad (\because a = -3 \text{ m/s}^2)$$

$$= 75 - \frac{3}{2} \times 25 = 75 - \frac{75}{2} = \frac{75}{2} = \mathbf{37.5\ m}. \quad \textbf{Ans.}$$

Problem 4.3. *A bullet, moving at the rate of 250 m/s, is fired into a log of wood. The bullet penetrates to a depth of 40 cm. If the bullet moving with the same velocity is fired into a similar piece of wood 20 cm thick, with what velocity would it emerge. Take the resistance to be uniform in both the cases.*

Sol. First Case. Given :

Initial velocity, $u = 250$ m/s

Final velocity, $v = 0$ (as bullet is not coming out of the wood)

Distance moved by bullet before coming to rest,

$$s = 40 = 0.40 \text{ m}.$$

First calculate the retardation of the bullet.

Let a = Acceleration of the bullet.

Using the equation (4.7)$v^2 - u^2 = 2as$

or $$0 - 250^2 = 2a \times 0.40$$

$$\therefore \quad a = \frac{-250 \times 250}{2 \times 0.40} = -\mathbf{78125\ m/s^2} \quad \textbf{Ans.}$$

Second Case

Initial velocity, $u = 250$ m/s (same as before)

Distance moved, $s = 20$ cm $= 0.20$ m.

As the wood is similar and also the resistance is uniform, the acceleration will be same as in the first case.

$\therefore$ Acceleration, $a = -78125$ m/s^2.

Let v = Final velocity with which bullet would come out from the wood.

Using the equation (4.7),

$$v^2 - u^2 = 2as$$

or $$v^2 - 250^2 = 2 \times (-78125) \times 0.20 \qquad (\because \quad a = -78125 \text{ m/s}^2)$$

or $$v^2 = 250^2 - 2 \times 78125 \times 0.20 = 62500 - 31250 = 31250$$

$$\therefore \quad v = \sqrt{31250} = \mathbf{176.77\ m/s.} \quad \textbf{Ans.}$$

Problem 4.4. *Three marks A, B and C at a distance of 100 m each are made along a straight road. A car starting from rest and with uniform acceleration passes the mark A and takes 10 seconds to reach B and further 8 seconds to reach the mark C. Calculate (i) the magnitude of the acceleration of the car, (ii) the velocity of the car at A, (iii) the velocity of car at B, and (iv) the distance of the mark A from the starting point.* (AMIE Winter, 1977)

Sol. Given :

Three marks A, B and C along a straight road are shown in Fig. 4.1, such that

Distance AB = 100 m

Distance BC = 100 m

A car starts from rest and with uniform acceleration.

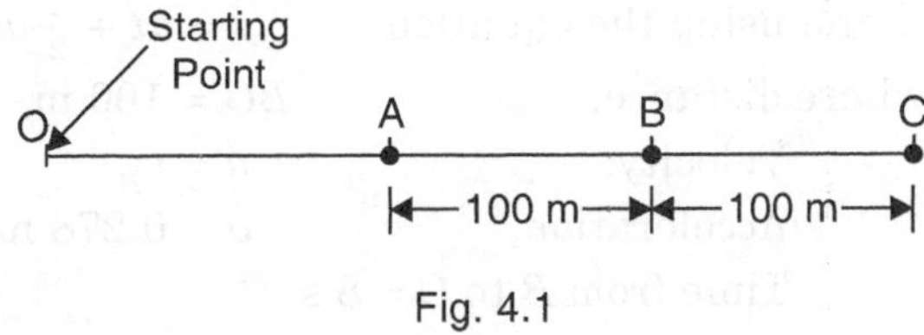

Fig. 4.1

Let the car starts from the point O

Time taken by car from A to B,

$$t_1 = 10 \text{ s}$$

Time taken by car from B to C,

$$t_2 = 8 \text{ s}$$

Let a = Acceleration of the car.

v_A = Velocity of the car at A, and

v_B = Velocity of the car at B.

For the motion of the car from A to B

Using the equation, $s = ut + \frac{1}{2}at^2$

where distance $s = 100$ m

Time $t = 10$ s

Velocity $u = v_A$

$\therefore \quad 100 = v_A \times 10 + \frac{1}{2} a \times 10^2$

$100 = 10v_A + 50a. \quad \text{...}(i)$

For the motion of the car from A to C

Using the equation, $s = ut + \frac{1}{2} at^2$

where distance AC = Distance AB + Distance BC = 100 + 100 = 200 m.

Time t = Time from A to B + Time from B to C

= 10 + 8 = 18 s

Velocity = v_A

$\therefore \quad 200 = v_A \times 18 + \frac{1}{2} a \times 18^2$

or $\quad 200 = 18v_A + 162a \quad \text{...}(ii)$

In equations (*i*) and (*ii*), there are two unknowns v_A and a. These equations are solved for v_A and a as given below :

Multiplying equation (*i*) by 9 and equation (*ii*) by 5, we get

$900 = 90v_A + 450a \quad \text{...}(iii)$

$1000 = 90v_A + 810a \quad \text{...}(iv)$

Subtracting equation (*iii*) from equation (*iv*), we have

$100 = (810 - 450)\, a = 360a$

(*i*) $\therefore \quad a = \frac{100}{360} = \frac{5}{18} = \mathbf{0.278\ m/s^2}$. **Ans.**

Substituting the value of a in equation (*i*), we get

$100 = 10v_A + 50 \times 0.278 = 10v_A + 13.9$

$\therefore \quad v_A = \frac{(100 - 13.9)}{10} = \frac{86.1}{10} = \mathbf{8.61\ m/s}$. **Ans.**

(*ii*) Velocity of car at A, v_A = **8.61 m/s. Ans.**

(*iii*) Velocity (v_B) of car at B is obtained from considering the motion of the car from B to C and using the equation $s = ut + \frac{1}{2} at^2$

where distance, $BC = 100$ m

Velocity, $u = v_B$

Acceleration, $a = 0.278$ m/s^2

Time from B to C = 8 s

$\therefore \quad 100 = v_B \times 8 + \frac{1}{2} \times .278 \times 8^2 = 8v_B + 8.896$

$\therefore \quad v_B = \frac{(100 - 8.896)}{8} = \mathbf{11.388\ m/s}$. **Ans.**

(*iv*) Distance (s) of the mark A from the starting point O is obtained by considering the motion of the car from O to A and using the equation $v^2 - u^2 = 2as$

where u = Velocity at 0 = 0

v = Velocity at $A = v_A = 8.61$ m/s

a = Acceleration of the car = 0.278 m/s^2

s = Distance OA

$\therefore \quad 8.61^2 - 0 = 2 \times .278 \times s$

$$\therefore \quad s = \frac{8.61^2}{2 \times .278} = \textbf{133.33 m.} \quad \textbf{Ans.}$$

Problem 4.5. *A burglar's car had a start with an acceleration of 2 m/s^2. A police vigilant party came after 10 seconds and continued to chase the burglar's car with a uniform velocity of 40 m/s. Find the time taken, in which the police van will overtake the car.*

Sol. Given :

Initial velocity of burglar's car, $u = 0$

Acceleration of burglar's car, $a = 2$ m/s^2.

Let t = Time taken by the police van in overtaking the burglar's car.

As the police party came after 10 seconds, hence burglar's car will be in motion for $(t + 10)$ seconds.

The uniform velocity of police van = 40 m/s.

When the police party will overtake the burglar's car, then the distance travelled by the police party and by the burglar's car will be same.

The distance travelled by police party = Uniform velocity × Time = $40 \times t$ ...(*i*)

The distance travelled by burglar's car in $(t + 10)$ second is obtained by using $s = ut + \frac{1}{2} at^2$ in which t is replaced by $(t + 10)$.

$\therefore \quad$ Distance travelled $= u\,(t + 10) + \frac{1}{2}\, a\,(t + 10)^2$

$= 0 \times (t + 10) + \frac{1}{2} \times 2\,(t + 10)^2 = (t + 10)^2$

$= [t^2 + 100 + 20t]$ ...(*ii*)

Equating the two distances given by equations (*i*) and (*ii*), we get

$$40t = [t^2 + 100 + 20t]$$

or $\quad t^2 + 100 + 20t - 40t = 0 \quad$ or $\quad t^2 + 100 - 20t = 0$

or $\quad (t - 10)^2 = 0 \quad$ or $\quad t - 10 = 0$

$\therefore \quad t = \textbf{10 s.} \quad$ **Ans.**

Problem 4.6. *On a straight road, a smuggler's car passes a police station with a uniform velocity of 10 m/s. After 10 seconds, a police party follows in pursuit in a jeep with a uniform acceleration of 1 m/s^2. Find the time necessary for the jeep to catch up with the smuggler's car.*

(AMIE Summer, 1988)

Sol. Given :

Uniform velocity of smuggler's car = 10 m/s

Uniform acceleration of police jeep = 1 m/s^2

Let S = Distance from police station to the point where police jeep catch up the smuggler's car.

t = Time taken by smuggler's car to cover distance S.

t' = Time taken by police party to catch up with the smuggler's car *or* to cover up the distance S metre.

$= (t - 10)$ seconds.

(*i*) *First consider the motion of smuggler's car*

Uniform velocity, $u = 10$ m/s

Distance S covered in time 't' is given by, $S = ut = 10 \times t$ ...(*i*)

(*ii*) *Now consider the motion of police jeep*

Acceleration, $a = 1 \text{ m/s}^2$

Initial velocity, $u = 0$ (jeep starts from rest)

Distance $= S$

Time, $t' = (t - 10)$

Using the relation, $S = ut' + \frac{1}{2} at'^2$ ($\because$ Here time $= t'$)

$$= 0 \times (t - 10) + \frac{1}{2} \times 1 \times (t - 10)^2 = 0 + \frac{1}{2}(t - 10)^2$$

$$= \frac{1}{2}(t^2 + 100 - 20t) \qquad ...(ii)$$

Equating the two values of S given by equations (*i*) and (*ii*), we get $\frac{1}{2}(t^2 + 100 - 20t) = 10t$

or $t^2 + 100 - 20t = 20t$

or $t^2 - 40t + 100 = 0$

The solution of the above quadratic equation is

$$t = \frac{40 \pm \sqrt{40^2 - 4 \times 100}}{2} = \frac{40 \pm \sqrt{1600 - 400}}{2} = \frac{40 \pm 34.64}{2}$$

$$= \frac{74.64}{2} \text{ and } \frac{5.36}{2} = 37.32 \text{ seconds and } 2.68 \text{ seconds}$$

The time 2.68 seconds is impossible, as jeep starts after $(t - 10)$ seconds *i.e.*, $2.68 - 10 = -7.32$ seconds. This gives absurd result.

$\therefore$ The time necessary for the jeep to catch up with smuggler's car is given by

$t' = (t - 10)$ seconds $= (37.32 - 10) =$ **27.32 seconds. Ans.**

4.4.4. Distance Travelled in the nth second

Let u = Initial velocity of a body

a = Acceleration of the body

S_n = Distance covered in n seconds

S_{n-1} = Distance covered in $(n - 1)$ seconds

Then distance travelled in the nth seconds

= Distance travelled in n seconds – Distance travelled in $(n - 1)$ seconds

$= S_n - S_{n-1}$.

Distance travelled in n seconds is obtained by substituting $t = n$ in equation

$$s = ut + \tfrac{1}{2} at^2.$$

$$\therefore \quad S_n = u \times n + \tfrac{1}{2} an^2.$$

Similarly $S_{n-1} = u(n - 1) + \frac{1}{2} a(n - 1)^2$ (Put $t = n - 1$ in $s = ut + \frac{1}{2} at^2$)

∴ Distance travelled in the nth seconds

$$= S_n - S_{n-1}$$

$$= (un + \frac{1}{2}an^2) - [u(n-1) + \frac{1}{2}a(n-1)^2]$$

$$= un + \frac{1}{2}an^2 - [un - u + \frac{1}{2}a(n^2 + 1 - 2n)]$$

$$= un + \frac{1}{2}an^2 - un + u - \frac{1}{2}an^2 - \frac{1}{2}a + \frac{1}{2}a \times 2n$$

$$= an + u - \frac{1}{2}a = u + \frac{a}{2}(2n - 1) \quad ...(4.8)$$

Problem 4.7. *A body is moving with uniform acceleration and covers 15 m in fifth second and 25 m in 10th second. Determine :*

(i) the initial velocity of the body, and

(ii) acceleration of the body.

Sol. Given :

Distance covered in 5th second = 15 m.

Distance covered in 10th second = 25 m.

Let u = Initial velocity, and

a = Acceleration of the body.

Using equation (4.8) for the distance covered in nth second $= u + \frac{a}{2}(2n - 1)$.

∴ Distance covered in 5th second

$$= u + \frac{a}{2}[2 \times 5 - 1] \quad (\because n = 5)$$

or
$$15 = u + \frac{9a}{2} \quad ...(i)$$

Distance covered in 10th second

$$= u + \frac{a}{2}[2 \times 10 - 1] \quad (\because n = 10)$$

or
$$25 = u + \frac{19a}{2} \quad ...(ii)$$

Subtracting equation (*i*) from (*ii*), we get

$$25 - 15 = \left(u + \frac{19a}{2}\right) - \left(u + \frac{9a}{2}\right)$$

or
$$10 = \frac{19a}{2} - \frac{9a}{2} = \frac{10a}{2}$$

∴ a = **2 m/s²**. **Ans.**

Substituting the value of 'a' in equation (*i*),

$$15 = u + \frac{9}{2} \times 2 = u + 9$$

∴ $u = 15 - 9 =$ **6 m/s. Ans.**

4.4.5. Equation of Motions due to Gravity. The acceleration due to gravity is 'g'. Hence when a body falls, the equation of motions given by equations (4.5), (4.6) and (4.7) are modified by substituting 'g' in place of 'a'. But when the body is moving vertically up the acceleration due to gravity is acting in the opposite direction. In that case the equations are modified by substituting $(-g)$ in place of a. The value of g is taken as 981 cm/s^2 or 9.81 m/s^2. The distance 'S' is replaced by 'h'. Hence the equations of motions due to gravity in the downward directions and upward directions becomes as :

1. For downward motion	2. For upward motion
$a = +g$	$a = -g$
$v = u + gt$	$v = u - gt$
$S = h = ut + \frac{1}{2}gt^2$	$h = ut - \frac{1}{2}gt^2$
$v^2 - u^2 = 2gh.$	$v^2 - u^2 = -2gh.$

4.4.6. Points to be Remembered

(*i*) If a body starts from rest, its initial velocity is zero, *i.e.*, $u = 0$.

(*ii*) If a body comes to rest, its final velocity is zero, *i.e.*, $v = 0$.

(*iii*) If a body is projected vertically upwards, the final velocity of the body at the highest point is zero, *i.e.*, $v = 0$.

(*iv*) If a body starts moving vertically downwards, its initial velocity is zero, *i.e.*, $u = 0$.

(*v*) Acceleration due to gravity is taken positive when a body is moving vertically downwards. But if the body is moving vertically upwards, the acceleration due to gravity is taken negative.

Problem 4.8. *A tower is 90 m in height. A particle is dropped from the top of the tower and at the same time another particle is projected upward from the foot of the tower. Both the particle meet at a height of 30 m. Find the velocity, with which the second particle is projected upward.*

Sol. Given :

Height of tower, $h = 90$ m.

Height at which both particle meet,

$$h_1 = 30 \text{ m}.$$

Distance travelled by first particle,

$$S_1 = h - h_1 = 90 - 30 = 60 \text{ m}.$$

Distance travelled by the second particle,

$$S_2 = h_1 = 30 \text{ m}.$$

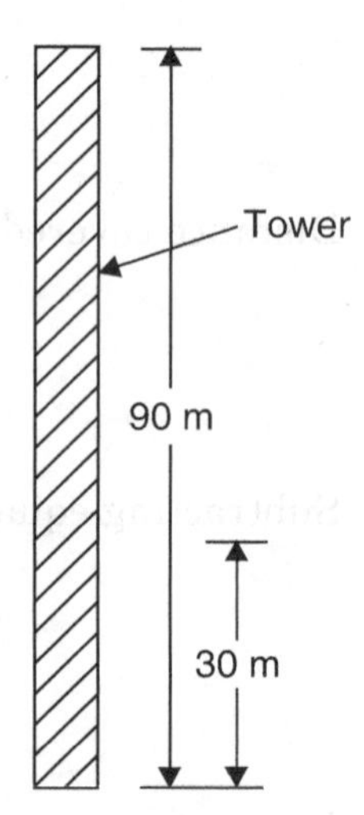

Fig. 4.2

For the first particle, we have (particle is moving down).

Initial velocity, $u = 0$

Distance, $S_1 = 60$ m

Acceleration due to gravity, $g = 9.81$ m/s^2.

Let the two particles meet after an interval of 't' seconds.

Using the equation,

$$S_1 = ut + \frac{1}{2}gt^2 \quad \text{or} \quad 60 = 0 \times t + \frac{1}{2} \times 9.81 \times t^2$$

$$\therefore \quad t = \sqrt{\frac{60 \times 2}{9.81}} = 3.497 \text{ s}.$$

For the second particle, we have (particle is moving up)

Time, $t = 3.497$ s (the time is same)

Distance, $S_2 = 30$ m

Acceleration, $g = -9.81\ m/s^2$

Let u = Initial velocity with which the second particle should be projected up.

Using the equation,

$$S_2 = ut + \frac{1}{2}gt^2 \quad \text{or} \quad 30 = u \times 3.497 + \frac{1}{2}(-9.81) \times 3.497^2$$

$$= 3.497\,u - 59.98$$

$$\therefore \quad u = \frac{30 + 59.98}{3.497} = \mathbf{25.73\ m/s.} \quad \textbf{Ans.}$$

Problem 4.9. *An aeroplane is flying horizontally at a height of 800 m. A bomb is released from the aeroplane when the speed of aeroplane is 600 kmph (kilometre per hour). Determine the time required for the bomb to reach the ground and the horizontal distance travelled by the bomb during flight.*

Sol. Given :

Height of aeroplane, $h = 800$ m

Speed of aeroplane, $U = 600$ kmph $= 600 \times 1000$ m per hour

$$= \frac{600 \times 1000}{60 \times 60} = 166.67\ m/s.$$

The aeroplane is moving in the horizontal direction, when the bomb is released. This means the velocity of bomb, initially will be in the horizontal direction only. This velocity is equal to the velocity of aeroplane, *i.e.*, 166.67 m/s. But at the same time, the bomb is attracted downwards due to gravitational effect. The initial velocity of the bomb in the downward direction is zero.

1st Case

Consider the motion of bomb in the downward direction.

Initial velocity in the downward direction, $u = 0$

Height of bomb, when it is released, $h = 800$ m

Acceleration due to gravity, $g = 9.81\ m/s^2$.

Let t = Time taken by bomb to reach the ground.

Using the equation, $h = ut + \frac{1}{2}gt^2$

$$\therefore \quad 800 = 0 \times t + \frac{1}{2} \times 9.81 \times t^2 = 4.905t^2$$

$$\therefore \quad t = \sqrt{\frac{800}{4.905}} = 12.77\ s.$$

$\therefore$ Time required to reach the ground = **12.77 s. Ans.**

2nd Case

Horizontal distance travelled = Horizontal velocity of bomb × Time

$= 166.67 \times t = 166.67 \times 12.77 =$ **2128.37 m. Ans.**

Problem 4.10. *A stone dropped into a well is heard to strike the water after 4 seconds. Find the depth of the well, if the velocity of sound is 350 m/sec.* (AMIE Summer, 1976)

Sol. Given :

Initial velocity of stone, $u = 0$.

Time taken by the stone to reach the bottom of well and time taken by the sound to reach from the bottom of the well to the top of the well = 4 s.

Velocity of sound = 350 m/s.

Let h = Depth of well

t_1 = Time taken by the stone to reach the bottom of well

t_2 = Time taken by the sound to reach from bottom of the well to the top of the well.

Then $$t_1 + t_2 = 4 \text{ s.} \qquad ...(i)$$

Consider the motion of the stone

Using the equation, $h = ut + \frac{1}{2} gt^2$

where $u = 0$, $t = t_1$, $g = 9.81$ m/s².

$$\therefore \quad h = 0 \times t_1 + \frac{1}{2} \times 9.81 \times t_1^2 = 0 + 4.905t_1^2 = 4.905t_1^2. \qquad ...(ii)$$

Consider the motion of sound

When the stone strikes the water in the well the sound, produced by the stone, travels with uniform velocity of 350 m/s in all directions. Hence the time taken by the sound to reach the top of the well

$$= \frac{\text{Depth of well}}{\text{Velocity of sound}}$$

or $$t_2 = \frac{h}{350} = \frac{4.905t_1^2}{350} \qquad (\because \text{ From equation } (ii), h = 4.905t_1^2)$$

Substituting the value of t_2 in equation (i), we get

$$t_1 + \frac{4.905t_1^2}{350} = 4 \quad \text{or} \quad 350t_1 + 4.905t_1^2 = 350 \times 4 = 1400$$

or $$4.905t_1^2 + 350t_1 - 1400 = 0 \qquad ...(iii)$$

The above equation is a quadratic equation, whose roots are given as,

$$t_1 = \frac{-350 \pm \sqrt{350^2 - 4(4.905)(-1400)}}{2(4.905)}$$

$$= \frac{-350 \pm \sqrt{122500 + 27468}}{9.81}$$

$$= \frac{-350 \pm 387.26}{9.81} = \frac{-350 + 387.26}{9.81} \text{ and}$$

$$= \frac{-350 - 387.26}{9.81} = 3.798 \text{ and } -75.154$$

$$\therefore \quad t_1 = 3.798 \text{ s.} \qquad \text{(–ve root is not possible)}$$

Substituting this value in equation (ii), we get

$$h = 4.905 \times 3.798^2 = 70.75 \text{ m.}$$

∴ Depth of well = **70.75 m. Ans.**

Problem 4.11. *A stone is thrown vertically upwards with a velocity of 19.6 m/s from the top of a tower 24.5 m high. Calculate :*

(i) Time required for the stone to reach the ground.

(ii) Velocity of the stone in its downward travel at the point in the same level as the point of projection.

(iii) The maximum height to which the stone will rise in its flight. (AMIE Summer, 1977)

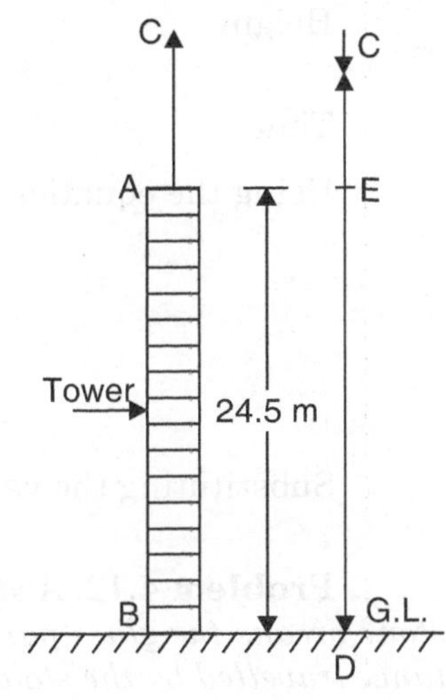

Fig. 4.3

Sol. (*i*) Given :

Initial velocity of stone, $u = 19.6$ m/s

Height of tower, $h = 24.5$ m

Let AB is a tower and stone is thrown upwards from A with a velocity 19.6 m/s. The stone will reach the maximum height and after that will start moving downwards. Let the highest point to which the stone reaches is C as shown in Fig. 4.3. At the point C the velocity of the stone will be zero.

The total time taken by the stone to reach the ground is equal to the time taken by the stone to reach from A to C and then from C to D.

Let us first calculate the maximum height (*i.e.*, AC) to which the stone will rise.

Let $AC = h_1$ = Maximum height to which stone will rise.

For the motion of stone in the upward direction

$u = 19.6$ m/s, $V = 0$, Distance $AC = h_1$

$g = -9.80$ m/s^2

Now using the equation

$$V^2 - u^2 = 2gh$$

$$0 - (19.6)^2 = -2 \times 9.80 \times h_1$$

$$\therefore \quad h_1 = \frac{(19.6)^2}{2 \times 9.80} = 19.6 \text{ m.}$$

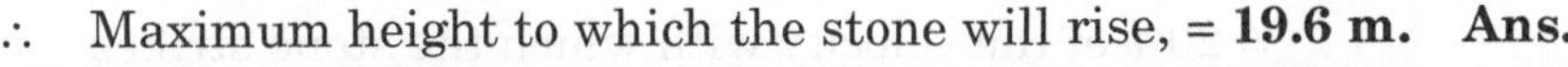

∴ Maximum height to which the stone will rise, = **19.6 m. Ans.**

(*ii*) The velocity of the stone when it is moving downward from C, at a point (E) which is in the same level as A is equal to the velocity with which the stone is thrown upward from A, *i.e.*,

= **19.6 m/s. Ans.**

(*iii*) **Time required for the stone to reach the ground**

Let t_1 = Time for the stone to move from A to C

t_2 = Time for the stone to move from C to D

t = Total time for the stone to reach the ground

Then $t = t_1 + t_2$ ...(*i*)

When the stone moves from A to C, we have

$u = 19.6$ m/s, $v = 0$

$AC = h_1 = 19.6$ m, $g = -9.8$ m/s^2.

Time $t = t_1$.

Using the equation, $v = u + gt$, we have

$$0 = 19.6 + (-9.8) \times t_1 = 19.6 - 9.8t_1$$

$$\therefore \quad t_1 = \frac{19.6}{9.8} = 2 \text{ s.}$$

When the stone moves from C to D, we have

$u = 0$, $g = 9.8$ m/s^2

Height, CD = Height BA + Height AC

$= 24.5 + 19.6 = 44.1$ m

Time, $t = t_2$.

Using the equation, $h = ut + \frac{1}{2} gt^2$, we have

$$44.1 = 0 \times t_2 + \frac{1}{2} \times 9.8 \times t_2^2 \qquad (\because \; h = 44.1 \text{ m and } t = t_2)$$

$$= 0 + 4.9t_2^2$$

$$\therefore \qquad t_2 = \sqrt{\frac{44.1}{4.9}} = \sqrt{9} = 3 \text{ s.}$$

Substituting the values of t_1 and t_2 in equation (*i*), we get

$$t = t_1 + t_2 = 2 + 3 = \mathbf{5\ s.} \quad \textbf{Ans.}$$

Problem 4.12. *A stone is dropped from a height. After falling 5 seconds from rest, the stone breaks the glass pane and in breaking, the stone loses 20% of its velocity. Find the distance travelled by the stone in the next second. Take g = 9.81 m/s².*

Sol. Given :

Initial velocity, $u = 0$

Time taken by stone in striking the glass pane = 5 s.

Acceleration due to gravity, $g = 9.81$ m/s².

Let v = the velocity with which stone strikes the pane.

Using $v = u + gt$, we get

$$v = 0 + 9.81 \times 5 = 49.05 \text{ m/s.}$$

Velocity lost in breaking the stone

= 20% of the velocity with which the stone strikes

$$= \frac{20}{100} \times 49.05 = 9.81 \text{ m/s}$$

$\therefore$ The velocity of the stone after breaking the glass pane

$$= 49.05 - 9.81 = 39.24 \text{ m/s.}$$

Now the distance travelled by the stone, after breaking the glass pane, is one second is given by

$$s = ut + \frac{1}{2} gt^2 = 39.24 \times 1 + \frac{1}{2} \times 9.81 \times 1^2$$

($\because$ $t = 1$ sec and initial velocity of stone after breaking = 39.24 m/s)

$$= 29.24 + 4.905 = \mathbf{44.145\ m.} \quad \textbf{Ans.}$$

Problem 4.13. *An object falls from rest from an unknown height. In the last second of its motion, the object travels a distance of 53.90 m. If g = 9.80 m/s², determine :*

(i) the height from which the object falls and

(ii) total time taken by the object in falling.

Sol. Given :

Initial velocity of the object, $u = 0$

Distance travelled in the last second = 53.90

Acceleration due to gravity, $g = 9.80$ m/s².

Let h = Height from which the object falls and

t = Total time of fall.

Then h is also equal to the distance travelled by the object time 't'

$\therefore$ $h = ut + \frac{1}{2}gt^2$

$= 0 \times t + \frac{1}{2} \times 9.80 \times t^2$ $[\because \ u = 0]$

$= 4.9t^2$ ...(i)

If the distance travelled in $(t - 1)$ second is h', then we have

$h' = u(t-1) + \frac{1}{2}g(t-1)^2$ {In equation $s = ut + \frac{1}{2}gt^2$, substitute $t = (t - 1)$}

$= 0 \times (t-1) + \frac{1}{2} \times 9.80\,(t-1)^2$ $(\because \ u = 0)$

$= 4.9\,(t-1)^2$ ...(ii)

$\therefore$ Distance travelled by the object in the last second of its motion

$= h - h' = 4.9t^2 - 4.9(t-1)^2$

$= 4.9[t^2 - (t-1)^2] = 4.9[t^2 - (t^2 + 1 - 2t)]$

$= 4.9[t^2 - t^2 - 1 + 2t] = 4.9[2t - 1]$

But according to the given condition, the distance travelled by the object in the last second of its motion = 53.90.

$\therefore$ $4.9[2t - 1] = 53.90$

$\therefore$ $2t - 1 = \frac{53.90}{4.90} = 11.0$

$\therefore$ $t = \frac{11+1}{2} = \frac{12}{2} = \mathbf{6\ s.}$ **Ans.**

Substituting the value of t in equation (i), we get

$h = 4.9 \times t^2 = 4.9 \times 6^2 = \mathbf{176.4\ m.}$ **Ans.**

Problem 4.14. *Find the height of a tower from the top of which an object falls freely and during the last second of its motion, the object travels a distance equal to 2/3 of the height of the tower. Taking g = 9.80 m/s².*

Sol. Given :

Distance travelled in the last one second = (2/3) of the height of tower

Acceleration due to gravity, $g = 9.80$ m/s^2

Initial velocity of the object, $u = 0$

Let h = Height of the tower

t = Total time of fall

Then distance travelled during last second $= \frac{2}{3} \times h$.

The distance travelled in time 't' is equal to the height of the tower h.

$\therefore$ $h = ut + \frac{1}{2}gt^2$

$= 0 \times t + \frac{1}{2} \times 9.80 \times t^2$ $(\because \ u = 0)$

$= 4.9t^2$ ...(i)

If the distance travelled in $(t - 1)$ second is h', then we have

$h' = u(t-1) + \frac{1}{2}g(t-1)^2$ [Replace t by $t - 1$ in $S = ut + \frac{1}{2}gt^2$]

$= 0 \times (t-1) + \frac{1}{2} \times 9.80\,(t-1)^2 = 4.9(t-1)^2$ ...(ii)

But $h - h' = \frac{2}{3} \times h$ (given)

Substituting the values of h and h' from equations (*i*) and (*ii*), we get

$$4.9t^2 - 4.9\,(t-1)^2 = \frac{2}{3} \times 4.9t^2$$

Cancelling 4.9 to both sides,

$$t^2 - (t-1)^2 = \frac{2}{3}\,t^2 \quad \text{or} \quad t^2 - (t^2 + 1 - 2t) = \frac{2}{3}\,t^2$$

or $$2t - 1 = \frac{2}{3}\,t^2 \quad \text{or} \quad 6t - 3 = 2t^2$$

or $$2t^2 - 6t + 3 = 0$$

The above equation is a quadratic equation. The solution of the equation is

$$t = \frac{6 \pm \sqrt{6^2 - 4 \times 2 \times 3}}{4} = \frac{6 \pm \sqrt{36-24}}{4} = \frac{6 \pm \sqrt{12}}{4}$$

$$= \frac{6 \pm 3.464}{4} = \frac{9.464}{4} \text{ and } \frac{6 - 3.464}{4} = 2.366 \text{ s and } 0.634 \text{ s.}$$

But t is greater than unity, according to the data given in the problem.

$\therefore$ $t = 2.366$ s.

Substituting the value of t in equation (*i*), we get

$$h = 4.9 \times t^2 = 4.9 \times 2.366^2 = 27.429 \text{ m}$$

$\therefore$ Height of the tower = h = **27.429 m. Ans.**

Problem 4.15. *An object is thrown vertically upward with a velocity of 30 m/s. Four second later, a second object is project vertically upward with a velocity of 40 m/s. Determine :*

(i) the time (after the first object is thrown) when the two objects will meet each other in air.

(ii) the height from the earth at which the two objects will meet.

Sol. Given :

Initial velocity of 1st object, $u_1 = 30$ m/s.

Initial velocity of 2nd object, $u_2 = 40$ m/s.

Let t = Time (after the first object is thrown) when the two objects will meet

h = Height at which the two objects will meet.

The height 'h' is covered by the first object in 't' seconds whereas the second object covers the same height in $(t - 4)$ seconds.

For the first object

$$h = ut - \tfrac{1}{2}\,gt^2 \qquad \text{(–ve sign as object is moving upward)}$$

$$= 30 \times t - \tfrac{1}{2} \times 9.81 \times t^2$$

$$= 30t - 4.905t^2 \qquad ...(i)$$

For the second object

$$h = u_2(t-4) - \tfrac{1}{2}\,g\,(t-4)^2 \qquad \text{(replace } t \text{ by } (t-4) \text{ in } S = ut - \tfrac{1}{2}\,gt^2)$$

$$= 40(t-4) - \tfrac{1}{2} \times 9.81(t-4)^2$$

$$= 40(t-4) - 4.905(t-4)^2 \qquad ...(ii)$$

Equating equations (*i*) and (*ii*), we get

$$30t - 4.905t^2 = 40(t-4) - 4.905(t-4)^2$$

or $$30t - 4.905t^2 - 40(t-4) + 4.905(t-4)^2 = 0$$

or $$30t - 40(t-4) + 4.905[(t-4)^2 - t^2)] = 0$$

or $$30t - 40t + 160 + 4.905[(t^2 + 16 - 8t - t^2) = 0$$

or $$-10t + 160 + 4.905[16 - 8t] = 0$$

or $$-10t + 160 + 78.48 - 39.24t = 0$$

or $$49.24t = 238.48$$

$\therefore$ $$t = \frac{238.48}{49.24} = \mathbf{4.843\ s.}\quad \textbf{Ans.}$$

Substituting the value of 't' in equation (*i*), we get

$$h = 30 \times 4.843 - 4.905 \times 4.843^2 = 145.29 - 115.045$$

$$= \mathbf{30.245\ m.}\quad \textbf{Ans.}$$

Problem 4.16. *A particle is dropped from the top of a tower 100 m high. Another particle is projected upwards at the same time from the foot of the tower, and meets the first particle at a height of 30 m. Find the velocity with which the second particle is projected upwards. Take $g = 9.8\ m/s^2$.*

Sol. Given :

Height of tower = 100 m

Initial velocity of 1st particle, $u_1 = 0$

Height from the foot of the tower at which the two particles meet = 30 m.

$\therefore$ Distance travelled vertically up by the second particle, $S_2 = 30$ m.

Distance travelled vertically down by the 1st particle,

$$S_1 = 100 - 30 = 70 \text{ m.}$$

Let t = Time, when the two particles meet

u_2 = Initial velocity of the second particle, projected vertically up.

Case of the 1st particle. The particle is dropped from the top of the tower.

Using $S = ut + \frac{1}{2}gt^2$

$\therefore$ $S_1 = 0 \times t + \frac{1}{2} \times 9.80t^2$ or $70 = 4.90t^2$

$\therefore$ $$t = \sqrt{\frac{70}{4.9}} = 3.78 \text{ s.}$$

Case of 2nd particle. The particle is projected vertically up.

Using $S = ut - \frac{1}{2}gt^2$ (g is –ve in this case)

or $S_2 = ut - \frac{1}{2} \times 9.80t^2$

or $30 = u \times 3.98 - 4.90 \times (3.78)^2$ $(\because\ t = 3.78 \text{ s})$

$= 3.78u - 70$

or $30 + 70 = 3.78u$ or $u = \frac{100}{3.78} = \mathbf{26.45\ m/s.}$ **Ans.**

Problem 4.17. *A body is dropped from an unknown height. The ratio of the distances covered by the falling body in the last and last but one second of its fall is $\frac{4}{3}$. If $g = 9.8\ m/s^2$, determine :*

(i) the height from which the body is dropped and

(ii) the velocity with which the body will reach the ground.

Sol. (*i*) Given :

Initial velocity of the body, $u = 0$

Ratio of distances covered in last and last but one second of the fall of the body = $\frac{4}{3}$.

Acceleration due to gravity, $g = 9.80 \text{ m/s}^2$.

Let h = Height from which the body is dropped

t = Time of the fall of the body

h = Distance covered in $(t - 1)$ second

h_2 = Distance covered in $(t - 2)$ second.

Then distance covered in last second $= h - h_1$ and distance covered in the last but one second $= h_1 - h_2$.

Now the distance covered in 't' second is given by

$$h = ut + \frac{1}{2}gt^2$$

$$= 0 \times t + \frac{1}{2}gt^2 \qquad (\because \ u = 0)$$

$$= \frac{1}{2}gt^2 \qquad ...(i)$$

Similarly, the distance covered in $(t - 1)$ second is given by

$$h_1 = \frac{1}{2}g(t-1)^2 \qquad [\because \ u = 0 \text{ and replace } t \text{ by } (t-1)]$$

and the distance covered in $(t - 2)$ second is given by

$$h_2 = \frac{1}{2}g(t-2)^2$$

But the distance covered in last second

$$= h - h_1 = \frac{1}{2}gt^2 - \frac{1}{2}g(t-1)^2$$

$$= \frac{1}{2}g[t^2 - (t-1)^2] = \frac{1}{2}g[t^2 - (t^2 + 1 - 2t)]$$

$$= \frac{1}{2}g[2t - 1]$$

and distance covered in the last but one second

$$= h_1 - h_2 = \frac{1}{2}g(t-1)^2 - \frac{1}{2}g(t-2)^2$$

$$= \frac{1}{2}g[(t-1)^2 - (t-2)^2] = \frac{1}{2}g[t^2 + 1 - 2t - (t^2 + 4 - 4t)]$$

$$= \frac{1}{2}g[2t - 3]$$

But $$\frac{\text{Distance covered in last second}}{\text{Distance covered in last but one second}} = \frac{4}{3} \qquad \text{(Given)}$$

$$\therefore \quad \frac{\frac{1}{2}g[2t-1]}{\frac{1}{2}g[2t-3]} = \frac{4}{3}$$

or $$\frac{(2t-1)}{(2t-3)} = \frac{4}{3} \qquad \text{or} \qquad 6t - 3 = 8t - 12$$

or $$2t = 9, \ t = \frac{9}{2} = 4.5 \text{ seconds.}$$

Substituting the value of t in equation (i), we get

$$h = \frac{1}{2}gt^2 = \frac{1}{2} \times 9.80 \times 4.5^2 \text{ m}$$

$$= \mathbf{99.225 \text{ m}.} \quad \textbf{Ans.}$$

(ii) Let v = Velocity with which the body will reach the ground.

Then using $v = u + gt$, we get

$$v = 0 + 9.80 \times 4.5 \qquad (\because \quad u = 0)$$

$$= \mathbf{44.1 \text{ m/s}.} \quad \textbf{Ans.}$$

Problem 4.18. *A body falling freely under the action of gravity passes two points 20 m apart vertically in 0.4 seconds. From what height, above the higher point, did the body start to fall ? Take g = 9.8 m/s².*

Sol. Given :

Initial velocity, $u = 0$

Let the body starts from A and passes the two points B and C, 20 m apart in 0.4 seconds as shown in Fig. 4.4.

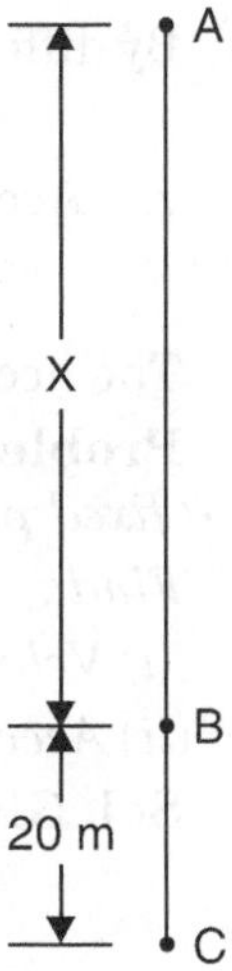

Let distance $AB = x$ m

Then distance $AC = (x + 20)$ m

If t = Time taken by the body to travel from A to B.

Then $(t + 0.4)$ = Time taken by the body to travel from A to C.

The distance 'x' travelled by the body in time 't' is given by

$$x = ut + \frac{1}{2}gt^2$$

$$= 0 \times t + \frac{1}{2}gt^2 \qquad (\because \quad u = 0)$$

$$= \frac{1}{2}gt^2 \qquad \ldots(i)$$

A

X

B

20 m

C

Fig. 4.4

The distance $(x + 20)$ travelled by the body in time $(t + 0.4)$ second is given by

$$(x + 20) = u \times (t + 0.4) + \frac{1}{2}g(t + 0.4)^2$$

$$= 0 \times (t + 0.4) + \frac{1}{2}g(t + 0.4)^2$$

$$= \frac{1}{2}g(t^2 + 0.4^2 + 0.8t) \qquad \ldots(ii)$$

Subtracting equation (i) from equation (ii),

$$(x + 20) - x = \frac{1}{2}g(t^2 + 0.4^2 + 0.8t) - \frac{1}{2}gt^2$$

or

$$20 = \frac{1}{2}g[t^2 + 0.16 + 0.8t - t^2]$$

$$= \frac{9.8}{2}[0.8t + 0.16] \quad \text{or} \quad \frac{20 \times 2}{9.8} = 0.8t + 0.16$$

or

$$0.8t = \frac{20 \times 2}{9.8} - 0.16 = 4.08 - 0.16 = 3.92$$

$$\therefore \quad t = \frac{3.92}{0.80} = 4.9 \text{ s.}$$

Substituting the value of t in equation (i), we get

$$x = \frac{1}{2}g \times (4.9)^2 = \frac{1}{2} \times 9.8 \times (4.9)^2 = \mathbf{117.649 \text{ m}.} \quad \textbf{Ans.}$$

4.4.7. Velocity and Acceleration of a Body Moving in a Straight Line by Differentiation. Let the equation of motion of a body moving in a straight line is given in terms of displacement (s) and time (t) as $s = t^3 + 2t^2 + 5t + 6$ ...(i)

By differentiating the above equation with respect to time, we get

Velocity, *i.e.*, $\frac{ds}{dt} = v$

$$\therefore \quad v = \frac{ds}{dt} = \frac{d}{dt}(t^3 + 2t^2 + 5t + 6)$$

$$= 3t^2 + 4t + 5 \quad \text{...}(ii)$$

If the velocity at start is to be calculated, the value of $t = 0$ is to be substituted in equation (ii).

By differentiating the equation (ii) with respect to time, we get acceleration, *i.e.*, $\frac{dv}{dt} = a$.

$$\therefore \quad \text{Acceleration,} \quad a = \frac{dv}{dt} = \frac{d}{dt}(3t^2 + 4t + 5)$$

$$= 6t + 4 \quad \text{...}(iii)$$

The acceleration at start will be obtained by substituting $t = 0$, in equation (iii).

Problem 4.19. *A particle moves along a straight line so that its displacement in metre from a fixed point is given by, $s = t^3 + 3t^2 + 4t + 5$.*

Find :

(i) *Velocity at start and after 4 seconds*

(ii) *Acceleration at start and after 4 seconds.*

Sol. Given :

$$s = t^3 + 3t^2 + 4t + 5$$

$$\text{Velocity,} \quad v = \frac{ds}{dt} = \frac{ds}{dt}(t^3 + 3t^2 + 4t + 5)$$

$$= 3t^2 + 6t + 4 \quad \text{...}(i)$$

(i) Velocity at start will be obtained, if $t = 0$ is substituted in equation (i),

$$\therefore \quad v(\text{at } t = 0) = 3 \times 0^2 + 6 \times 0 + 4 = 0 + 0 + 4$$

$$= \textbf{4 m/s.} \quad \textbf{Ans.}$$

Velocity after four seconds. Substitute $t = 4$ in equation (i)

$$v(\text{at } t = 4) = 3 \times 4^2 + 6 \times 4 + 4$$

$$= 48 + 24 + 4 = \textbf{76 m/s.} \quad \textbf{Ans.}$$

(ii) Differentiating equation (i) with respect to time 't', we get acceleration.

$$\therefore \quad \text{Acceleration,} \quad a = \frac{dv}{dt} = \frac{d}{dt}(3t^2 + 6t + 4)$$

$$= 6t + 6. \quad \text{...}(ii)$$

Acceleration at start will be obtained, if $t = 0$ is substituted in equation (ii).

$$\therefore \quad a(\text{at } t = 0) = 6 \times 0 + 6 = 0 + 6$$

$$= \textbf{6 m/s}^2\textbf{.} \quad \textbf{Ans.}$$

Acceleration after four seconds. Substitute $t = 4$ in equation (ii),

$$\therefore \quad a(\text{at } t = 4) = 6 \times 4 + 6 = 24 + 6$$

$$= \textbf{30 m/s}^2\textbf{.} \quad \textbf{Ans.}$$

Problem 4.20. *The equation of motion of a particle moving in a straight line is given by*

$$s = 18t + 3t^2 - 2t^3$$

where s is the total distance covered from the starting point is metres at the end of t seconds. Find :

(*i*) *the velocity and acceleration at start,*

(*ii*) *the time, when the particle reaches its maximum velocity and*

(*iii*) *the maximum velocity of the particle.* (AMIE Winter, 1982)

Sol. Given :

$$s = 18t + 3t^2 - 2t^3.$$

Differentiating both sides with respect to t,

$$\frac{ds}{dt} = 18 + 3 \times 2t - 2 \times 3t^2 = 18 + 6t - 6t^2.$$

But from equation (4.1), we know $\frac{ds}{dt} = v$

$\therefore$ Velocity, $v = 18 + 6t - 6t^2.$ ...(*i*)

(*i*) At start, $t = 0$. Hence velocity at start will be obtained by substituting $t = 0$ in equation (*i*),

$\therefore$ Velocity at start,

$$v = 18 + 6 \times 0 - 6 \times 0^2 = 18 + 0 - 0$$

$$= \mathbf{18\ m/s.}\quad \textbf{Ans.}$$

Differentiating both sides of equation (*i*) with respect to t,

$$\frac{dv}{dt} = 6 - 6 \times 2t = 6 - 12t \quad ...(ii)$$

But from equation (4.2), we know that $\frac{dv}{dt}$ = Acceleration = a

$\therefore$ $a = 6 - 12t$ ...(*iii*)

At start, $t = 0$. Hence acceleration at start will be obtained by substituting $t = 0$ in equation (*iii*).

$\therefore$ Acceleration at start = $6 - 12 \times 0 =$ **6 m/s²**. **Ans.**

(*ii*) The velocity will be maximum, when the differentiation of the velocity with respect to time is equal to zero or $\frac{dv}{dt} = 0$. But $\frac{dv}{dt}$ is given by equation (*ii*), hence equating the equation (*ii*) to zero for maximum velocity, we get

$$6 - 12t = 0 \quad \text{or} \quad 12t = 6$$

or
$$t = \frac{6}{12} = \frac{\mathbf{1}}{\mathbf{2}}\ \textbf{s.}\quad \textbf{Ans.}$$

(*iii*) From equation (*i*), velocity is given as

$$v = 18 + 6t - 6t^2$$

For maximum velocity, $t = \frac{1}{2}$.

Hence substituting $t = \frac{1}{2}$ in the above equation, we get maximum velocity as

$$v_{max} = 18 + 6 \times \tfrac{1}{2} - 6 \times \left(\tfrac{1}{2}\right)^2 = 18 + 3 - \tfrac{6}{4}$$

$$= 21 - 1.5 = \mathbf{19.5\ m/s.} \quad \textbf{Ans.}$$

4.4.8. Velocity and Displacement of a Body Moving in a Straight Line by Integration. Let the equation of motion of a body moving in a straight line is given in terms of acceleration (a) and time (t) as

$$a = t^3 + 2t^2 + 4t + 5 \qquad \text{...(i)}$$

But $$a = \frac{dv}{dt}$$

$\therefore$ $$\frac{dv}{dt} = t^3 + 2t^2 + 4t + 5$$

Integrating the above equation, we get

$$\int dv = \int (t^3 + 2t^2 + 4t + 5)dt$$

or $$v = \left(\frac{t^4}{4} + \frac{2t^3}{3} + \frac{4t^2}{2} + 5t\right) + C_1 \qquad \text{...(ii)}$$

where C_1 is the constant of integration and its value is obtained from given condition. Let the condition is given that at $t = 1$, the velocity is 10 m/s.

Substituting this condition in equation (*ii*), we get

$$10 = \frac{1^4}{4} + \frac{2 \times 1^3}{3} + \frac{4 \times 1^2}{2} + 5 \times 1 + C_1$$

$$= \frac{1}{4} + \frac{2}{3} + 2 + 5 + C_1$$

$$= 0.25 + 0.67 + 7 + C_1 = 7.92 + C_1$$

$\therefore$ $$C_1 = 10 - 7.92 = 2.08.$$

Substituting the value of C_1 in equation (*ii*), we get the velocity at any time t,

$$v = \frac{t^4}{4} + \frac{2t^3}{3} + \frac{4t^2}{2} + 5t + 2.08 \qquad \text{...(iii)}$$

But $$v = \frac{ds}{dt}$$

$\therefore$ $$\frac{ds}{dt} = \frac{t^4}{4} + \frac{2t^3}{3} + 2t^2 + 5t + 2.08$$

Integrating the above equation, we get

$$\int ds = \int \left(\frac{t^4}{4} + \frac{2t^3}{3} + 2t^2 + 5t + 2.08\right) dt$$

or $$s = \frac{t^5}{4 \times 5} + \frac{2t^4}{3 \times 4} + \frac{2t^3}{3} + \frac{5t^2}{2} + 2.08t + C_2 \qquad \text{...(iv)}$$

where C_2 is another constant of integration. The value of C_2 is also obtained from given condition. If at $t = 1$, the displacement (s) is given as equal to 10 m. Substituting this condition in equation (*iv*), we get

$$10 = \frac{1^5}{20} + \frac{2 \times 1^4}{12} + \frac{2 \times 1^3}{3} + \frac{5 \times 1^2}{2} + 2.08 \times 1 + C_2$$

or
$$10 = \frac{1}{20} + \frac{1}{6} + \frac{2}{3} + 2.5 + 2.08 + C_2$$

$$= 0.05 + 0.1667 + 0.667 + 2.5 + 2.08 + C_2 = 5.4637 + C_2$$

$\therefore$ $C_2 = 10 - 5.4637 = 4.5363.$

Substituting the value of C_2 in equation (*iv*), we get the displacement at any time t as

$$s = \frac{t^5}{20} + \frac{2t^4}{12} + \frac{2t^3}{3} + \frac{5t^2}{2} + 2.08t + 4.5363 \quad ...(v)$$

If it is required to find the velocity and displacement at any time 't', then equation (*iii*) and (*v*) are used.

Problem 4.21. *A particle moves along a straight line with an acceleration described by the equation $a = -8s^{-2}$ where a is in m/s² and s in m. When t = 1 s, s = 4 m and v = 2 m/s. Determine acceleration when t = 2 s.* (Osmania, 1989)

Sol. Given :

Acceleration, $a = -8s^{-2}$

When $t = 1$ s, $s = 4$ m and $v = 2$ m/s

Find acceleration when $t = 2$ s.

The acceleration in terms of velocity and displacement is given by equation (4.3),

$$a = \frac{v\,dv}{ds}$$

But $a = -8s^2$ (given)

$\therefore$ $$-8s^{-2} = \frac{v\,dv}{ds} \quad \text{or} \quad vdv = -8s^{-2}\,ds$$

Integrating the above equation, we get

$$\int v\,dv = \int -8s^{-2}\,ds \quad \frac{v^2}{2} = -8\frac{s^{-2+1}}{-2+1} + C_1$$

or
$$\frac{v^2}{2} = \frac{8}{s} + C_1 \quad ...(i)$$

where C_1 is the constant of integration and its value is obtained from given condition *i.e.*, $s = 4$ m and $v = 2$ m/s when $t = 1$ s.

Substituting $v = 2$ m/s and $s = 4$ m in equation (*i*),

$$\frac{2^2}{2} = \frac{8}{4} + C_1 \quad \text{or} \quad 2 = 2 + C_1$$

or $C_1 = 0$

Substituting $C_1 = 0$ in equation (*i*), we get

$$\frac{v^2}{2} = \frac{8}{s} \quad \text{or} \quad v^2 = \frac{4}{s} \quad \text{or} \quad v = \frac{4}{\sqrt{s}} \quad ...(ii)$$

But $$v = \frac{ds}{dt}$$

Substituting the value of v in equation (*ii*), we get

$$\frac{ds}{dt} = \frac{4}{\sqrt{s}} \quad \text{or} \quad \sqrt{s}\, ds = 4dt$$

Integrating, we get

$$\int \sqrt{s}\, ds = \int 4\, dt \quad \text{or} \quad \frac{s^{\frac{1}{2}+1}}{\frac{1}{2}+1} = 4t + C_2$$

or $$\frac{2}{3} s^{3/2} = 4t + C_2 \qquad \text{...(iii)}$$

when $t = 1$ s, $s = 4$ m.

Substituting these values in equation (*iii*), we get

$$\frac{2}{3} \times 4^{3/2} = 4 \times 1 + C_2 \quad \text{or} \quad \frac{16}{3} = 4 + C_2$$

or $$C_2 = \frac{16}{3} - 4 = \frac{16 - 12}{3} = \frac{4}{3}$$

Substituting the value of C_2 in equation (*iii*), we get

$$\frac{2}{3} s^{3/2} = 4t\ \frac{4}{3}.$$

When $t = 2$ s, the above equation becomes as

$$\frac{4}{3} s^{3/2} = 4 \times 2 + \frac{4}{3} = 8 + \frac{4}{3} = \frac{28}{3}$$

or $$s^{3/2} = \frac{28}{3} \times \frac{3}{2} = 14 \qquad \text{or} \quad s = 14^{2/3} = 5.81$$

Now acceleration, $a = -8s^{-2} = -8 \times 5.81^{-2} =$ **0.2373 m/s²**. **Ans.**

Problem 4.22. *A particle moves along a straight line with an acceleration described by the equation a = (4t² –2), where a is in m/s² and t is in seconds. What t = 0, the particle is located 2 m to the left of origin and when t = 2 seconds, the particle is 20 m to left of origin. Determine the position of the particle, when t = 4 seconds.* (AMIE Summer, 1988)

Sol. Given :

Acceleration, $a = 4t^2 - 2$

(If left of origin is considered –ve)

When $t = 0$, distance $s = -2$ m

When $t = 2$ seconds, distance $s = -20$ m

Find distance s when $t = 4$ seconds

Acceleration in terms of distance is given by,

$$a = \frac{d^2s}{dt^2}$$

But $a = 4t^2 - 2$

Equating the two values of acceleration, we get

$$\frac{d^2s}{dt^2} = 4t^2 - 2$$

Integrating the above equation, we get

$$\frac{ds}{dt} = 4 \times \frac{t^3}{3} - 2t + C_1 \qquad ...(i)$$

Integrating the above equation again,

$$s = \frac{4}{3} \times \frac{t^4}{4} - \frac{2t^2}{2} + C_1 t + C_2$$

or

$$s = \frac{t^4}{3} - t^2 + C_1 t + C_2 \qquad ...(ii)$$

where C_1 and C_2 are constant of integrations and their values are obtained from given conditions

(*i*) When $t = 0$, $s = -2$ m.

Substituting these values in equation (*ii*), we get

$$-2 = \frac{1}{3} \times 0 - 0 + C_1 \times 0 + C_2 = 0 - 0 + 0 + C_2$$

$$\therefore \quad C_2 = -2$$

Substituting the value of C_2 in equation (*ii*), we get

$$s = \frac{t^4}{3} - t^2 + C_1 t - 2 \qquad ...(iii)$$

(*ii*) When $t = 2$ seconds, $s = -20$ m.

Substituting these values in equation (*iii*), we get

$$-20 = \frac{1}{3} \times 2^4 - 2^2 + C_1 \times 2 - 2 = \frac{16}{3} - 4 + 2C_1 - 2$$

or

$$2C_1 = -20 - \frac{16}{3} + 4 + 2 = \frac{-60 - 16 + 12 + 6}{3} = \frac{-58}{3} = -19.33$$

$$\therefore \quad C_1 = -\frac{19.33}{2} = -9.665$$

Substituting this value in equation (*iii*),

$$s = \frac{t^4}{3} - t^2 - 9.665t - 2$$

When $t = 4$ seconds,

$$s = \frac{1}{3} \times 4^4 - 4^2 - 9.665 \times 4 - 2 = \frac{256}{3} - 16 - 38.66 - 2$$

$$= 85.33 - 16 - 38.66 - 2 = 85.33 - 56.66$$

$= \mathbf{28.67}$ (Right of origin). **Ans.**

(+ve sign shows that particle will be right of origin.)

Problem 4.23. *A particle moves along a straight line with a velocity given by the equation*

$$v = 2t^3 - t^2 - 2t + 4$$

where v is the velocity in m/s and t is time is seconds. When t = 2 seconds, the particle is found to be at a distance of 10 m from a station A. Determine :

(i) the acceleration and (ii) displacement of the particle after 6 seconds.

(Osmania, 1998)

Sol. Given :

Velocity, $v = 2t^3 - t^2 - 2t + 4$

When $t = 2$ seconds, distance $s = 10$ m

Find : (*i*) acceleration and (*ii*) displacement when $t = 6$ seconds

(*i*) *Acceleration when $t = 6$ seconds*

Acceleration, $$a = \frac{dv}{dt} = \frac{d}{dt}(2t^3 - t^2 - 2t + 4)$$

$$= 6t^2 - 2t - 2 = 6 \times 6^2 - 2 \times 6 - 2 \qquad (\because \ t = 6 \text{ seconds})$$

$$= 216 - 12 - 2 = \mathbf{202\ m/s^2}. \ \textbf{Ans.}$$

(*ii*) *Displacement when $t = 6$ seconds*

The velocity in terms of displacement is given by

$$v = \frac{ds}{dt} \quad \therefore \quad 2t^3 - t^2 - 2t + 4 = \frac{ds}{dt}$$

or $$ds = (2t^3 - t^2 - 2t + 4)\, dt$$

Integrating, $$\int ds = \int (2t^3 - t^2 - 2t + 4)\, dt$$

or $$s = \frac{2t^4}{4} - \frac{t^3}{3} - \frac{2t^2}{2} + 4t + C \qquad \ldots(i)$$

where C is a constant of integration and its value is obtained from the given condition *i.e.*, $s = 10$ m when $t = 2$ seconds.

Substituting these values in equation (*i*), we get

$$10 = \frac{2}{4} \times 2^4 - \frac{2^3}{3} - \frac{2}{2} \times 2^2 + 4 \times 2 + C = 8 - \frac{8}{3} - 4 + 8 + C$$

or $$C = 10 - 8 + \frac{8}{3} + 4 - 8 \quad \text{or} \quad = -2 + \frac{8}{3} = \frac{-6+8}{3} = \frac{2}{3}$$

Substituting the value of C in equation (*i*), we get

$$s = \frac{2t^4}{4} - \frac{t^3}{3} - \frac{2t^2}{2} + 4t + \frac{2}{3}$$

To find the displacement, when $t = 6$ s, substitute $t = 6$ s in the above equation.

$\therefore$ $$s = \frac{2}{4} \times 6^4 - \frac{6^3}{3} - \frac{2}{2} \times 6^2 + 4 \times 6 + \frac{2}{3}$$

$$= 648 - 72 - 36 + 24 + \frac{2}{3} = \mathbf{564.67\ m.} \ \textbf{Ans.}$$

Problem 4.24. *A body moves along a straight line and its acceleration a which varies with t is given by a = 2 – 3t. After 5 seconds from start of observations its velocity is observed to be 20 m/s. After 10 seconds from start of observations the body was 85 metres from the origin.*

(a) Determine its acceleration, velocity and distance from the origin at the start of observations.

(b) Determine the time after start of observation in which the velocity becomes zero and its distance from the origin.

(c) Describe the motion diagrammatically. (AMIE Summer, 1975)

Sol. Given :

Acceleration, $a = 2 - 3t$ ...(*i*)

At $t = 5$ seconds, $v = 20$ m/s

At $t = 10$ seconds, $s = 85$ m.

(*a*) (*i*) Acceleration at start will be obtained by substituting $t = 0$, in equation (*i*) as,

$a = 2 - 3 \times 0 = 2 - 0 =$ **2 m/s²**. **Ans.**

(*ii*) The equation (*i*) can be written as

$$\frac{dv}{dt} = 2 - 3t. \qquad \left(\because \; a = \frac{dv}{dt}\right)$$

Integrating the above equation,

$$\int dv = \int (2 - 3t)\, dt \quad \text{or} \quad v = 2t - \frac{3t^2}{2} + C_1 \quad \text{...(ii)}$$

where C_1 is the constant of integration.

At $t = 5$ seconds, velocity $v = 20$ m/s. Substituting these values in equation (*ii*), we get

$$20 = 2 \times 5 - \frac{3 \times 5^2}{2} + C_1$$

or $$20 = 10 - 37.5 + C_1 = -27.5 + C_1$$

or $$C_1 = 20 + 27.5 = 47.5$$

Substituting the value of C_1 in equation (*ii*), we get

$$v = 2t - \frac{3t^2}{2} + 47.5 \quad \text{...(iii)}$$

The velocity at start will be obtained by substituting $t = 0$, in equation (*iii*) as

$$v = 2 \times 0 - \frac{3}{2} \times 0^2 + 47.5 = 0 - 0 + 47.5 = \mathbf{47.5\ m/s.} \quad \textbf{Ans.}$$

(*iii*) As $v = \frac{ds}{dt}$, the equation (*iii*) can be written as

$$\frac{ds}{dt} = 2t - \frac{3t^2}{2} + 47.5$$

Integrating the above equation, we get

$$\int ds = \int \left(2t - \frac{3t^2}{2} + 47.5\right) dt$$

or $$s = \frac{2t^2}{2} - \frac{3t^3}{3 \times 2} + 47.5t + C_2 \quad \text{...(iv)}$$

where C_2 is another constant of integration.

At $t = 10$ seconds, distance $s = 85$ m is given. Substituting these values in equation (*iv*), we get

$$85 = \frac{2 \times 10^2}{2} - \frac{3 \times 10^3}{6} + 47.5 \times 10 + C_2$$

$$= 100 - 500 + 47.5 + C_2 = 75 + C_2$$

$$\therefore \quad C_2 = 85 - 75 = 10.$$

Substituting the value of C_2 in equation (*iv*), we get

$$s = t^2 - \frac{t^3}{2} + 47.5t + 10 \qquad \text{...}(v)$$

The distance at start will be obtained by substituting $t = 0$, in equation (*v*) as

$$s = 0^2 - \frac{0^3}{2} + 47.5 \times 0 + 10$$

$$= 0 - 0 + 0 + 10 = \mathbf{10\ m.} \quad \textbf{Ans.}$$

(*b*) The equation (*iii*) gives the velocity at any time t. To find the time after start of observation in which velocity becomes zero, substitute $v = 0$ in equation (*iii*).

$$\therefore \quad 0 = 2t - \frac{3t^2}{2} + 47.5 \qquad \text{or} \qquad 0 = 4t - 3t^2 + 95$$

or $\quad 3t^2 - 4t - 95 = 0$

The above equation is a quadratic equation. Its solution is given by

$$t = \frac{4 \pm \sqrt{4^2 + 4 \times 3 \times 95}}{2 \times 3} = \frac{4 \pm \sqrt{16 + 1140}}{6} = \frac{4 \pm \sqrt{1156}}{6}$$

$$= \frac{4 \pm 34}{6} = \frac{38}{6} \text{ and } \frac{-30}{6} = \mathbf{6.33.} \text{ (Neglecting –ve root).} \quad \textbf{Ans.}$$

The distance from origin after $t = 6.33$ second is obtained by substituting this value of t in equation (*v*).

$$\therefore \quad s = 6.33^2 - \frac{6.33^3}{2} + 47.5 \times 6.33 + 10$$

$$= 40.069 - 126.818 + 300.675 + 10 = \mathbf{223.926\ m.} \quad \textbf{Ans.}$$

(*c*) *Description of motion diagrammatically*

The velocities for different times in sec is obtained from equation (*iii*). They are given as

Time (t)	0	1	2	3	4	5	6	7	8	9
Velocity (v)	47.5	47	45.5	40	31.5	20	3.5	– 12	– 32.5	– 56
Acceleration (a)	2	– 1	– 4	– 7	– 10	– 13	– 16	– 19	– 22	– 25
Distance (s)	10	58	105	184						

Fig. 4.5 shows the graph of velocity and time.

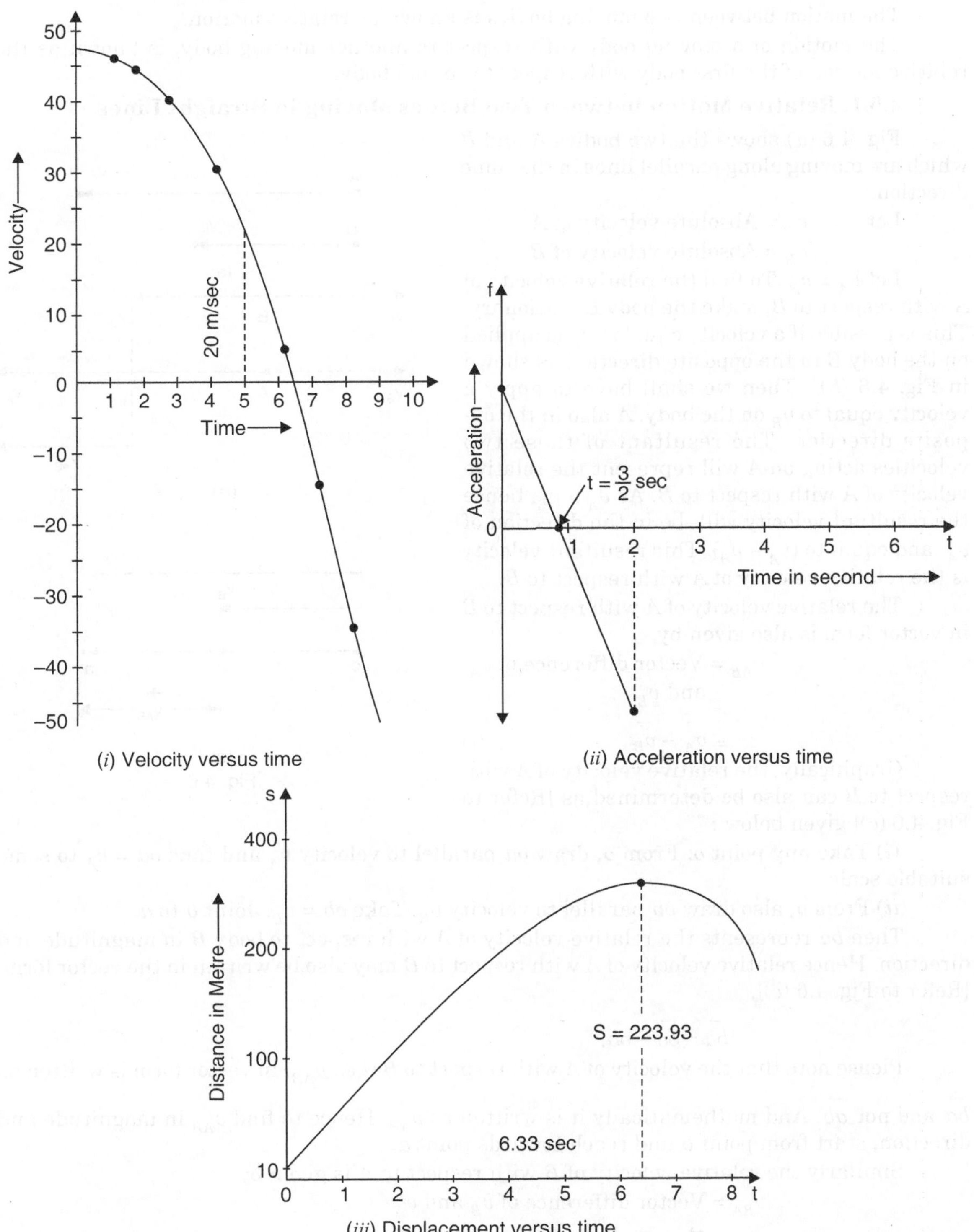

(*i*) Velocity versus time

(*ii*) Acceleration versus time

(*iii*) Displacement versus time

Fig. 4.5

4.5. RELATIVE MOTION

The motion between two moving bodies is known as relative motion.

The motion of a moving body with respect to another moving body, is known as the relative motion of the first body with respect to second body.

4.5.1. Relative Motion between Two Bodies Moving in Straight Lines

Fig. 4.6 (*a*) shows the two bodies *A* and *B* which are moving along parallel lines in the same direction.

Let v_A = Absolute velocity of *A*

v_B = Absolute velocity of *B*

Let $v_A > v_B$. To find the relative velocity of *A* with respect to *B*, make the body *B* stationary. This is possible if a velocity equal to v_B is applied on the body *B* in the opposite direction as shown in Fig. 4.6 (*b*). Then we shall have to apply a velocity equal to v_B on the body. *A* also in the opposite direction. The resultant of these two velocities acting on *A* will represent the relative velocity of *A* with respect to *B*. As $v_A > v_B$, hence the resultant velocity will be in the direction of v_A and equal to $(v_A - v_B)$. This resultant velocity is the relative velocity of *A* with respect to *B*.

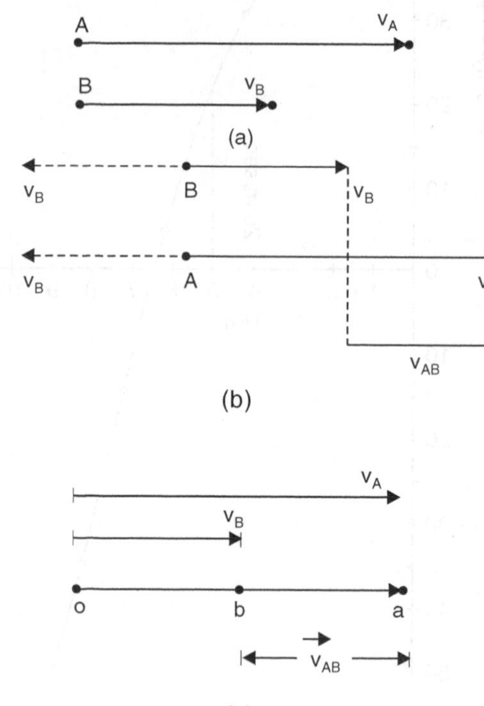

Fig. 4.6

The relative velocity of *A* with respect to *B* in vector form is also given by,

$$v_{AB} = \text{Vector difference of } v_A \text{ and } v_B$$

$$= \vec{v}_A - \vec{v}_B \qquad \text{...}(i)$$

Graphically, the relative velocity of *A* with respect to *B* can also be determined as [Refer to Fig. 4.6 (*c*)] given below :

(*i*) Take any point *o*. From *o*, draw *oa* parallel to velocity v_A and take $oa = v_A$ to some suitable scale.

(*ii*) From *o*, also draw *ob* parallel to velocity v_B. Take $ob = v_B$. Joint *b* to *a*.

Then *ba* represents the relative velocity of *A* with respect to body *B* in magnitude and direction. Hence relative velocity of *A* with respect to *B* may also be written in the vector form. [Refer to Fig. 4.6 (*c*)].

$$\vec{ba} = \vec{oa} - \vec{ob}.$$

Please note that the velocity of *A* with respect to *B* (*i.e.*, v_{AB}) in vector form is written as $\vec{ba}$ and not $\vec{ab}$. And mathematically it is written as v_{AB}. Hence to find v_{AB} in magnitude and direction, start from point *b* and reach towards point *a*.

Similarly the relative velocity of *B* with respect to *A* is given by

$$v_{BA} = \text{Vector difference of } v_B \text{ and } v_A$$

$$= \vec{v}_B - \vec{v}_A . \qquad \text{...}(ii)$$

The relative velocity of B with respect to A in vector form is written as

$$\vec{ab} = \vec{ob} - \vec{oa}.$$

From equations (*i*) and (*ii*), it is clear that the relative velocity of point A with respect to $B(v_{AB})$ and the relative velocity of point B with respect to $A(v_{BA})$ are equal in magnitude but opposite in direction *i.e.*,

$$v_{AB} = -\,v_{BA} \quad \text{or} \quad \vec{ba} = -\,\vec{ab}.$$

4.5.2. Relative Motion between Two Bodies Moving Along Inclined Lines

Fig. 4.7(*a*) shows the two bodies A and B moving along inclined lines with absolute velocities v_A and v_B respectively. To find the relative velocity of A with respect to B, make the body B stationary. This is possible if a velocity equal to v_B is applied on the body B in the opposite direction as shown in Fig. 4.7 (*b*). Then we shall have to apply a velocity equal to v_B on the body A also in opposite direction.

The resultant gives the relative velocity of A with respect to B as shown in Fig. 4.7 (*b*).

The vector oa in Fig. 4.7 (*c*) represents the velocity v_A in magnitude and direction, whereas the vector ob represents the velocity v_B in magnitude and direction. Join points b and a. Then relative velocity of A with respect to B is written as v_{AB} and is given by

$$v_{AB} = \text{Vector } ba = \vec{v_A} - \vec{v_B}$$

and relative velocity of B with respect to A is given by,

$$v_{BA} = \text{Vector } ab.$$

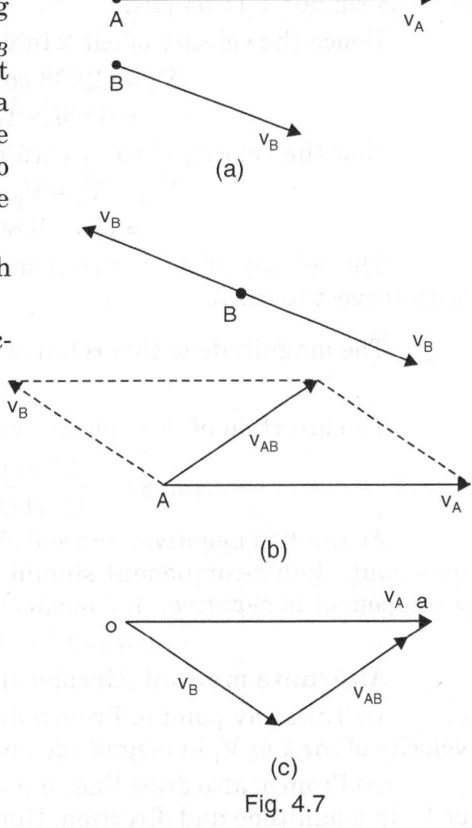

Fig. 4.7

Problem 4.25. *Two roads which are at an angle of 60°, intersect at a point O. A car 1 is moving with a velocity of 72 km/hr away from point O along x-axis at an instant of time. At the same time another car 2 is moving away from point O with a velocity of 48 km/hr along the road 2.*

Determine the velocity of car 1 with respect to car 2.

Fig. 4.8

Sol. Given :

Velocity of car 1 along x-axis = 72 km/hr

$$= \frac{72 \times 1000}{60 \times 60} = 20 \text{ m/s}$$

The car 1 is moving along x-axis, there is no component of velocity along y-axis. Hence the velocity of car 1 in vector form is written as

$$\mathbf{V}_1 = 20i$$

(As velocity is along x-axis only, hence only i component exists)

Velocity of car 2 along road 2, which is at an angle of 60° with x-axis = 48 km/hr.

$$= \frac{48 \times 1000}{60 \times 60} = 13.33 \text{ m/s}$$

The components of this velocity along x-axis = 13.33 cos 60° = 6.66 m/s and along y-axis = 13.33 sin 60° = 11.54 m/s.

Hence the velocity of car 2 in vector form is written as

$$\mathbf{V}_2 = (13.33 \cos 60°)i + (13.33 \sin 60°)j$$
$$= 6.66i + 11.54j$$

Now the velocity of car 1 with respect to car 2 is given by

$$\mathbf{V}_{12} = \mathbf{V}_1 - \mathbf{V}_2 = (20i) - (6.66i + 11.54j)$$
$$= (20 - 6.66)\, i - 11.54j = \mathbf{13.34i - 11.54j}. \quad \textbf{Ans.}$$

The velocity of car 1 with respect to car 2 is also known as the relative velocity of car 1 with respect to car 2.

The magnitude of this relative velocity $= \sqrt{13.34^2 + 11.54^2} = \sqrt{177.95 + 133.17}$

$$= \mathbf{17.64 \text{ m/s}}. \quad \textbf{Ans.}$$

The direction of this relative velocity is given by,

$$\tan \theta = \frac{-11.54\,(y\text{-axis component})}{13.34\,(x\text{-axis component})} = -0.865$$

As tan θ is negative, hence θ should be in second or fourth quadrant. If θ is in second quadrant, then x-component should be – ve and y-component should be positive. But here y-component is negative, this means θ lies in fourth quadrant.

$$\therefore \quad \theta = \tan^{-1}(-0.865) = 40.86° \quad \text{or} \quad (360 - 40.86) = \mathbf{319.14°}. \quad \textbf{Ans.}$$

Alternate method (Graphical Method)

(*i*) Take any point *o*. From *o* draw a line *oa* parallel to the velocity of car 1 or V_1 in magnitude and direction. Cut *oa* = 20 m/s.

(*ii*) From *o*, also draw line *ob* parallel to the velocity of car 2 or V_2 in magnitude and direction. Cut *ob* = 13.33 m/s. Join *b* to *a*. Then vector *ba* is the relative velocity of car 1 with respect to car 2.

By measurement vector *ba* = 17.6 m/s and angle α = 40.86°.

$$\therefore \quad \theta = 360 - 40.86 = \mathbf{319.14°}. \quad \textbf{Ans.}$$

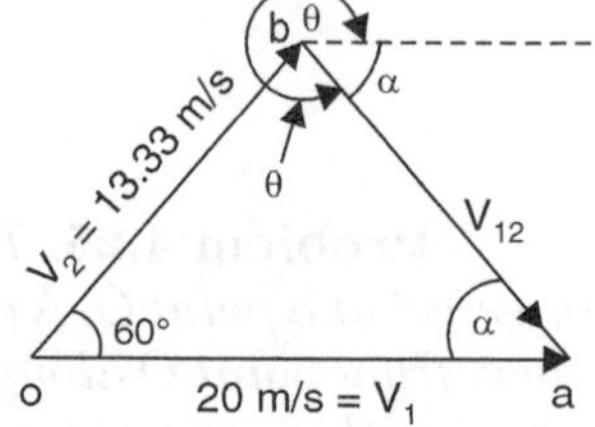

Fig. 4.8 (a)

Note. For graphical method both the velocities should be directed either away from the origin (point *o*) or towards the origin.

Problem 4.26. *If in the previous problem, the car 1 is at a distance of 100 m from O along x-axis and is moving with a speed of 72 km/hr towards O at the time of observation and car 2 is moving with a speed of 48 km/hr. along road 2 and away from the origin as shown in Fig. 4.9, then find the velocity of car 1 with respect to car 2.*

Sol. Given :

Distance of car 1 from O along x-axis = 100 m

Speed of car 1 = 72 km/hr = $\frac{72 \times 1000}{60 \times 60}$ = 20 m/s

$\therefore$ Velocity of car 1 = 20 m/s

The car 1 is moving along x-axis and towards origin.

Hence the velocity of car 1 in vector form is written as

$\mathbf{V}_1 = -20i$ [–ve sign is due to opposite direction of car 1]

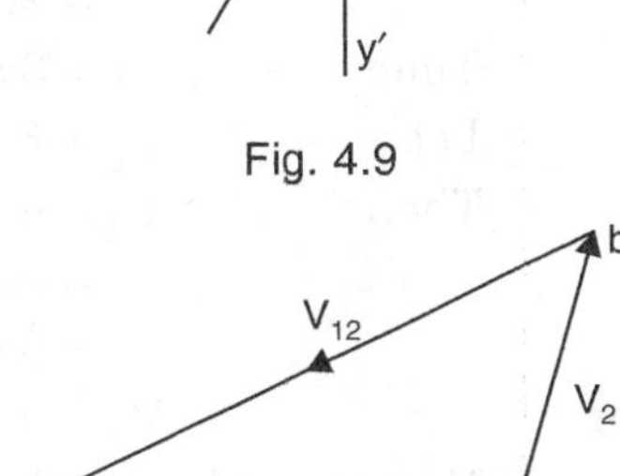

Fig. 4.9

Velocity of car 2 = 48 km/hr = $\frac{48 \times 1000}{60 \times 60}$ = 13.33 m/s

The velocity of car 2 in vector form is written as

$\mathbf{V}_2 = 13.33\,[\cos 60i + \sin 60j] = 6.66i + 11.54j$

The velocity of car 1 with respect to car 2 is given by,

$\mathbf{V}_{12} = \mathbf{V}_1 - \mathbf{V}_2 = -20i - [6.66i + 11.54j]$

$= \mathbf{-26.66i - 11.54j}$. **Ans.**

Fig. 4.9 (a)

Graphical Method

(*i*) Take any point o. From o draw a line oa parallel to velocity V_1 in magnitude and direction. Take oa = 20 m/s.

(*ii*) From o, also draw a line ob parallel to velocity V_2 in magnitude and direction. Take ob = 13.33 m/s. Now vector ba represents the velocity of 1 with respect to 2 in magnitude and direction. Measure vector ba.

Problem 4.27. *At an instant of time, a car 1 at a distance of 120 m from point o along x-axis moving at a velocity of 72 km / hr decelerates at 1.5 m / s² as it approaches the point o. At the same instant, a car 2, is moving along a road which is at an angle of 60° with the road on which car 1 is moving. This car is moving away from o with a velocity of 48 km / hr and accelerates at 2 m / s² as shown in Fig. 4.10. Find the velocity of car 1 with respect to car 2 after three seconds.*

Sol. Given :

Car 1. Initial velocity, u_1 = 72 km/hr = $\frac{72 \times 1000}{60 \times 60}$ = 20 m/s

$= -20i$ m/s (in vector notation)

(–ve sign is due to car 1 is moving towards o along x-axis)

Deceleration = 1.5 m/s² (towards o)

or Acceleration, a = 1.5 m/s² (away from o along x-axis)

$= 1.5i$ m/s² (in vector notation)

(+ve sign is as the acceleration is acting along x-axis away from o)

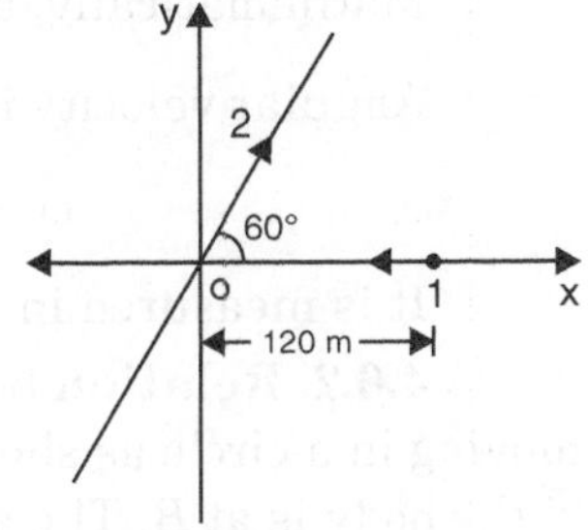

Fig. 4.10

Time, t = 3 seconds

Let v_1 = Final velocity of car 1 after 3 seconds

Then $v_1 = u_1 + at = (-20i) + (1.5i) \times 3$

($\because$ $u_1 = -20i$; $a_1 = 1.5i$)

$= -20i + 4.5i = -15.5i$

$\therefore$ $\mathbf{V}_1 = -15.5i$

[Final velocity of car 1, in vector notation]

Car 2. Initial velocity, $u_2 = 48$ km/hr $= \frac{48 \times 1000}{60 \times 60} = 13.33$ m/s

This velocity is at an angle of 60° with x-axis. Hence in vector notation, it is given by

$$u_2 = 13.33 (\cos 60°)i + 13.33 (\sin 60°)j$$
$$= 6.67i + 11.54j$$

Acceleration, $a = 2$ m/s^2. This acceleration is at an angle of 60° with x-axis. Hence this acceleration in vector form is given by,

$$\mathbf{a} = 2(\cos 60°)i + 2 \sin (60°)j = 1.0i + 1.732j$$

Time $t = 3s$

Let v_2 = Final velocity of car 2 after 3 seconds.

Then $v_2 = u_2 + at$

$$= (6.67i = 11.54j) + (1.0i + 1.732j) \times 3$$
$$= 6.67i + 11.54j + 3i + 5.196j = 9.67i + 16.736j$$

or $\mathbf{V_2} = 9.67i + 16.736j$

Now the velocity of car 1 with respect to the velocity of car 2 after 3 seconds

$$V_{12} = \mathbf{V_1} - \mathbf{V_2} = -15.5i - [9.67i + 16.736j]$$
$$= \mathbf{-25.17i - 16.736j \text{ m/s.}} \quad \textbf{Ans.}$$

4.6. CURVILINEAR MOTION

The motion of a body along a circular path is known as circular motion. In circular motion, the centre of rotation remains fixed. The examples of the bodies moving in a circular path are : shafts, flywheels, pulleys etc., rotating about their geometric axis.

4.6.1. Angular Velocity. It is defined as the rate of change of angular displacement of a body. Angular displacement is always measured in terms of angle covered by the body from the initial position.

Let a body is moving along a circular path as shown in Fig. 4.11. Let initially the body is at A and after time 't', the body is at B. Let $\angle AOB = \theta$.

Then angular displacement = $\angle AOB = \theta$.

Time taken = t.

$\therefore$ Angular velocity

$$= \frac{\text{Angular displacement}}{\text{Time}} = \frac{\theta}{t}$$

Mathematically, angular velocity $= \frac{d\theta}{dt}$.

Angular velocity is denoted by the symbol, ω.

$$\therefore \quad \omega = \frac{d\theta}{dt} \quad ...(4.9)$$

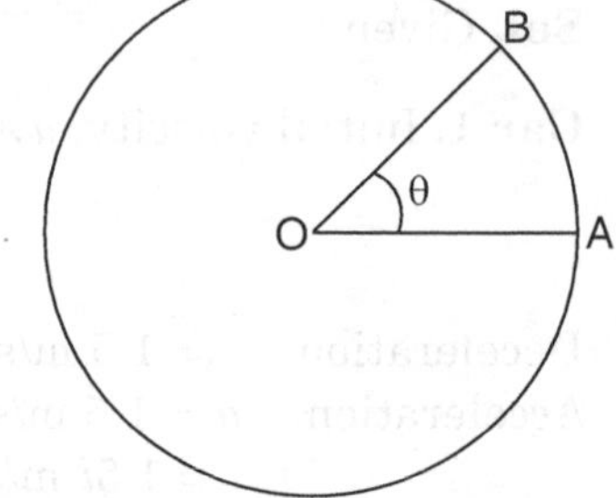

Fig. 4.11. Body moving in a circle.

It is measured in radians per second and is written as rad/s.

4.6.2. Relation between Linear Velocity and Angular Velocity. Consider the body moving in a circle as shown in Fig. 4.11. The initial position of the body is at A and after time 't' the body is at B. The angle AOB is equal to θ.

Angular velocity $= \frac{\theta}{t}$.

Let V = linear velocity = $\frac{\text{Linear displacement}}{\text{Time}}$

But linear displacement = Arc $AB = OA \times \theta = r \times \theta$ ($\because$ OA = radius of circle = r)

$$\therefore \quad V = \frac{r \times \theta}{t} = r \times \text{Angular velocity} \qquad \left(\because \frac{\theta}{t} = \text{Angular velocity}\right)$$

$$= r \times \omega \qquad \text{...(4.10)}$$

where ω = Angular velocity.

4.6.3. Angular Acceleration. It is defined as the rate of change of angular velocity. It is measured in radians per second per second and written as rad/s^2. Mathematically, angular acceleration (α) is given as

$$\alpha = \text{Rate of change of angular velocity}$$

$$= \frac{d\omega}{dt} = \frac{d}{dt}\left(\frac{d\theta}{dt}\right) \qquad \left(\because \omega = \frac{d\theta}{dt}\right)$$

$$= \frac{d^2\theta}{dt^2} \qquad \text{...(4.11)}$$

Also $$\frac{d\omega}{dt} = \frac{d\omega}{d\theta} \times \frac{d\theta}{dt} = \frac{d\omega}{d\theta} \times \omega \qquad \left(\because \frac{d\theta}{dt} = \omega\right)$$

But $$\frac{d\omega}{dt} = \text{angular acceleration} = \alpha$$

$$\therefore \quad \alpha = \frac{d\omega}{dt} = \omega \frac{d\omega}{d\theta} \qquad \text{...(4.12)}$$

4.6.4. Relation between Linear Acceleration and Angular Acceleration. From equation (4.10), we have $V = \omega \times r$.

Differentiating the above equation w.r.t. 't'

$$\frac{dV}{dt} = \frac{d}{dt}(\omega r)$$

or $$= r\frac{d\omega}{dt} \qquad (\because \quad r = \text{constant})$$

But $$\frac{dV}{dt} = \text{linear acceleration} = a$$

$$\frac{d\omega}{dt} = \text{angular acceleration} = \alpha$$

Substituting these values in the above equation, we get $a = r\alpha$ ...(4.13)

$\therefore$ Linear acceleration is equal to 'r' times the angular acceleration.

4.7. EQUATIONS OF MOTIONS ALONG A CIRCULAR PATH

Consider a body, which is moving along a circular path.

Let ω_0 = Final angular velocity of the body in radians per second

ω = Final angular velocity in radians per second

t = Time in second during which angular velocity changes from ω_0 to ω

θ = Angular displacement or angle traversed in radians

α = Angular acceleration in radians/s^2.

From Art. 4.6.3, we know that angular acceleration (α) is the rate of change of angular velocity. Hence

$$\alpha = \text{Rate of change of angular velocity} = \frac{\text{Change of velocity}}{\text{Time}}$$

$$= \frac{(\text{Final angular velocity} - \text{Initial angular velocity})}{\text{Time}}$$

$$= \frac{(\omega - \omega_0)}{t} \quad \text{...(4.14)}$$

or $\alpha t = \omega - \omega_0$

$\therefore$ $$\omega = \omega_0 + \alpha t \quad \text{...(4.15)}$$

4.7.1. Equation for Angular Displacement (θ). We know average angular velocity

$$= \frac{(\text{Initial} + \text{Final angular velocity})}{2} = \frac{\omega_0 + \omega}{\omega}.$$

$\therefore$ Angular displacement, θ = (Average angular velocity) × Time

$$= \frac{(\omega_0 + \omega)}{2} \times t.$$

Substituting the value of ω from equation (4.15) in the above equation, we get

$$\theta = \frac{(\omega_0 + \omega_0 + \alpha t)}{2} \times t = \frac{(2\omega_0 + \alpha t)}{2} \times t$$

$$= (\omega_0 + \tfrac{1}{2}\alpha t)t = \omega_0 t + \tfrac{1}{2}\alpha t^2$$

$$\theta = \omega_0 t + \tfrac{1}{2}\alpha t^2 \quad \text{...(4.16)}$$

4.7.2. Derivation of the Angular Displacement in terms of Initial and Final Angular Velocities. The final angular velocity is given by the equation (4.15)

$$\omega = \omega_0 + \alpha t \quad \text{or} \quad \frac{\omega - \omega_0}{\alpha} = t$$

Substituting the above value of t in the equation (4.16), we get

$$\theta = \omega_0 \times \left(\frac{\omega - \omega_0}{\alpha}\right) + \frac{1}{2}\alpha\left(\frac{\omega - \omega_0}{\alpha}\right)^2 = (\omega - \omega_0)\left[\frac{\omega_0}{\alpha} + \frac{1}{2\alpha}(\omega - \omega_0)\right]$$

$$= (\omega - \omega_0)\left[\frac{2\omega_0 + \omega - \omega_0}{2\alpha}\right] = (\omega - \omega_0)\left(\frac{\omega + \omega_0}{2\alpha}\right) = \frac{\omega^2 - {\omega_0}^2}{2\alpha}$$

or $$\omega^2 - {\omega_0}^2 = 2\alpha\theta \quad \text{...(4.17)}$$

Angular displacement in the nth second $= \omega_0 + \left(\dfrac{2n - 1}{2}\right)\alpha.$

4.7.3. Relationship between r.p.m. (N) and Angular Velocity (ω). Let N = r.p.m. of a body ω = angular velocity of the body.

Now the number of revolution in one minute or 60 seconds = N.

$\therefore$ Number of revolution in one second $= \dfrac{N}{60}$.

But in one revolution, the body covers an angle equal to 360° or 2π radians.

$\therefore$ Angle covered by the body in one second

= Angle covered in one revolution × Number of revolution per second

$$= 2\pi \times \frac{N}{60} = \frac{2\pi N}{60}.$$

But angle covered per second = Angular velocity = ω.

$$\therefore \quad \omega = \frac{2\pi N}{60} \qquad ...(4.18)$$

Problem 4.28. *A body is rotating with an angular velocity of 5 radians / s. After 4 seconds, the angular velocity of the body becomes 13 radians / s. Determine the angular acceleration of the body.*

Sol. Given :

Initial angular velocity, $\omega_0 = 5$ rad/s

Final angular velocity, $\omega = 13$ rad/s

Time, $t = 4$ s

Let α = Angular acceleration of the body.

Using equation (4.15),

$$\omega = \omega_0 + \alpha t \quad \text{or} \quad 13 = 5 + \alpha \times 4$$

$$\therefore \quad \alpha = \frac{(13-5)}{4} = \frac{8}{4} = \mathbf{2.0\ rad/s^2}. \quad \textbf{Ans.}$$

Problem 4.29. *A wheel, rotating about a fixed axis at 20 r.p.m., is uniformly accelerated for 70 seconds, during which time it makes 50 revolution. Find : (i) Angular velocity at the end of this interval, and (ii) Time required for the speed to reach 100 revolution per minute.* (AMIE Summer, 1977)

Sol. Given :

Initial r.p.m. of wheel, $N_0 = 20$

∴ Initial angular velocity,

$$\omega_0 = \frac{2\pi N_0}{60} = \frac{2\pi \times 20}{60} = \frac{2\pi}{3} \text{ rad/s.}$$

Time, $t = 70$ s

Number of revolution in 70 s = 50

∴ Angular displacement, θ = Angular displacement in one revolution × Number of revolution

$$= 2\pi \times 50 = 100\pi \text{ radians.}$$

(*i*) **Angular velocity at the end of 70 s.**

Let ω = Angular velocity at the end of 70 s

α = Angular acceleration.

Using equation (4.16),

$$\theta = \omega_0 t + \tfrac{1}{2}\alpha t^2$$

$$100\pi = \frac{2\pi}{3} \times 70 + \frac{1}{2}\alpha \times (70)^2 = 46.67\pi + 2450\alpha$$

$$\therefore \quad \alpha = \frac{100\pi - 46.67\pi}{2450} = 0.0684 \text{ rad/s}^2.$$

From equation (4.15), we have

$$\omega = \omega_0 + \alpha t = \frac{2\pi}{3} + 0.0684 \times 70 = 2.0944 + 4.788$$

$$= \mathbf{6.8824\ rad/s.} \quad \textbf{Ans.}$$

(*ii*) **Time required for the speed to reach 100 r.p.m.**

Final speed, $N = 100$ r.p.m.

$\therefore$ Final angular velocity, $\omega = \dfrac{2\pi N}{60} = \dfrac{2\pi \times 100}{60} = 13.472$ rad/s.

Using equation (4.15), we have $\omega = \omega_0 + \alpha \times t$

$$\therefore \qquad 10.472 = \frac{2\pi}{3} + 0.0684 \times t = 2.0944 + .0684t$$

$$\therefore \qquad t = \frac{(10.472 - 2.0944)}{0.0684} = 122.48 \text{ s} = \mathbf{2 \text{ min } 2.48 \text{ s.}} \quad \textbf{Ans.}$$

Problem 4.30. *A wheel is rotating about its axis with a constant angular acceleration of 1 rad/s². If the initial and final angular velocities are 5.25 rad/s and 13.5 rad/s, determine the total angle turned through during the time interval this change of angular velocity took place.* (AMIE Summer, 1975)

Sol. Given :

Constant angular acceleration, $\alpha = 1$ rad/s^2

Initial angular velocity, $\omega_0 = 5.25$ rad/s

Final angular velocity, $\omega = 10.50$ rad/s

Let the total angle turned $= \theta$

Using equation (4.17),

$$\omega^2 - \omega_0^2 = 2\alpha\theta \qquad \text{or} \qquad 13.50^2 - 5.25^2 = 2 \times 1 \times \theta$$

or $$110.25 - 27.56 = 2\theta$$

$$\therefore \qquad \theta = \frac{(110.25 - 27.56)}{2} = \frac{82.69}{2} = \mathbf{41.35 \text{ rad.}} \quad \textbf{Ans.}$$

Problem 4.31. (*a*) *A flywheel starts rotating from rest and is given an acceleration of 1 rad/s². Find the angular velocity and speed in r.p.m. after 1.5 minutes.*

(*b*) *If the flywheel is brought to rest with a uniform angular retardation of 0.5 rad/s², determine the time taken by the flywheel in seconds to come to rest.*

Sol. Given :

(*a*) Initial angular velocity, $\omega_0 = 0$

Angular acceleration, $\alpha = 1$ rad/s^2

Time, $t = 1.5$ min $= 1.5 \times 60 = 90$ s

Let final angular velocity, $= \omega$.

Using equation (4.15), $\omega = \omega_0 + \alpha t$

$= 0 + 1 \times 90 =$ **90 rad/s. Ans.**

Speed in r.p.m. is given by equation (4.18) as, $\omega = \dfrac{2\pi N}{60}$

$$\therefore \qquad N = \frac{60\omega}{2\pi} = \frac{60 \times 90}{2\pi} = \mathbf{859.44 \text{ r.p.m.}} \quad \textbf{Ans.}$$

(*b*) In this case, we have

Initial angular velocity, $\omega_0 = 90$ rad/s

Final angular velocity, $\omega = 0$

Angular retardation, $\alpha = 0.5$ rad/s^2

$\alpha = -0.5$ rad/s^2

Let the time taken by the flywheel in coming to rest $= t_1$.

Using equation (4.15), $\omega = \omega_0 + \alpha t$

or $0 = 90 + (-0.5) \times t_1$ $\quad (\because \alpha = -0.5 \text{ rad/s}^2)$

$= 90 - 0.5 \times t_1$

$\therefore \quad t_1 = \dfrac{90}{0.5} = 180 \text{ s} = \mathbf{3 \text{ min.}}$ **Ans.**

Problem 4.32. *A flywheel is rotating at 200 r.p.m. and after 10 seconds it is rotating at 160 r.p.m. If the retardation is uniform, determine member of revolution made by the flywheel and the time taken by the flywheel before it comes to rest from the speed of 200 r.p.m.*

Sol. Given :

Initial speed, $N_0 = 200$ r.p.m.

$\therefore$ Initial angular velocity, $\omega_0 = \dfrac{2\pi N_0}{60} = \dfrac{2\pi \times 200}{60} = 20.94$ rad/s.

Final speed, $N = 160$ r.p.m.

$\therefore$ Final angular velocity, $\omega = \dfrac{2\pi N}{60} = \dfrac{2\pi \times 160}{60} = 16.75$ rad/s.

Time, $t = 10$ s.

(*i*) **Number of revolutions made by flywheel before it stops.**

First determine the uniform retardation. Using equation (4.15), we have

$\omega = \omega_0 + \alpha t \quad$ or $\quad 16.75 = 20.94 + \alpha \times 10$

$\therefore \quad \alpha = \dfrac{(16.75 - 20.94)}{10} = -0.419 \text{ rad/s}^2.$

To find the number of revolutions made by the flywheel before coming to rest from 200 r.p.m. (or 20.94 rad/s), we have

Initial angular velocity, $\omega_0 = 20.94$ rad/s

Final angular velocity, $\omega = 0$

Angular retardation, $\alpha = -0.419$ rad/s^2

Now using the equation (4.17),

$\omega^2 - \omega_0^2 = 2\alpha\theta$

or $0 - 20.94^2 = 2 \times (-0.419) \times \theta$

$\therefore \quad \theta = \dfrac{-20.94^2}{-2 \times 0.419} = 523.25$ rad.

Number of revolution $= \dfrac{\text{Total angular displacement in radians}}{\text{Angle covered in one revolution in radian}}$

$= \dfrac{523.25}{2\pi} = \mathbf{83.27}$ **revolution. Ans.**

(*ii*) **Time taken by wheel before it comes to rest from 200 r.p.m.**

Initial angular velocity, $\omega_0 = 20.94$ rad/s

Final angular velocity, $\omega = 0$

Angular retardation, $\alpha = -0.419$ rad/s^2

Using the equation (4.15), $\omega = \omega_0 + \alpha t$

or $0 = 20.94 + (-0.419) \times t$

$$\therefore \qquad t = \frac{-20.94}{-0.419} = \mathbf{49.97\ s.} \quad \textbf{Ans.}$$

Problem 4.33. *The angle of rotation of a body is given by the equation*

$$\theta = 2t^3 - 5t^2 + 8t + 6$$

where θ is expressed in radians and t in seconds. Determine :

(i) angular velocity and

(ii) angular acceleration of the body

when t = 0 and t = 4 seconds.

Sol. Given :

Angular displacement $\theta = 2t^3 - 5t^2 + 8t + 6$.

Angular velocity (ω) is obtained differentiating the above equation with respect to time (t).

$$\therefore \qquad \omega = \frac{d\theta}{dt} = \frac{d}{dt}(2t^3 - 5t^2 + 8t + 6) = 6t^2 - 10t + 8 \qquad ...(i)$$

Angular acceleration (α) is obtained by differentiating the equation (i) with respect to time (t).

$$\therefore \qquad \alpha = \frac{d\omega}{dt} = \frac{d}{dt}(6t^2 - 10t + 8) = 12t - 10 \qquad ...(ii)$$

When t = 0

Angular velocity (ω) and angular acceleration (α), when $t = 0$, are obtained by substituting $t = 0$ in equations (i) and (ii) respectively.

$$\therefore \qquad \omega = 6 \times 0 - 10 \times 0 + 8 = 0 - 0 + 8 = \mathbf{8\ radian/s.} \quad \textbf{Ans.}$$

and $$\alpha = 12 \times 0 - 10 = \mathbf{-10\ radian/s^2.} \quad \textbf{Ans.}$$

When t = 4 seconds

Substituting $t = 4$ seconds in equations (i) and (ii), we get

$$\omega = 6 \times 4^2 - 10 \times 4 + 8 = 96 - 40 + 8 = \mathbf{64\ radian/s.} \quad \textbf{Ans.}$$

and $$\alpha = 12 \times 4 - 10 = 48 - 10 = \mathbf{38\ radian/s^2.} \quad \textbf{Ans.}$$

Problem 4.34. *The initial angular velocity of a rotating body is 2 rad/s and initial angular acceleration is zero. The rotation of the body is according to the relation α = 3t² – 3. Find : (i) angular velocity, and (ii) angular displacement when t = 5 seconds. Consider the angular displacement in radians and time in seconds.*

Sol. Given :

Initial angular velocity, $\omega_0 = 2$ rad/s

This means that when $t = 0$, the angular velocity is 2 rad/s

Initial angular acceleration, $\alpha = 0$

The law of rotation, $\alpha = 3t^2 - 3$.

Find : (i) angular velocity (ω), and (ii) angular displacement θ when $t = 5$ s.

(i) Angular velocity (ω)

We know, $$\alpha = \frac{d\omega}{dt} \quad \text{or} \quad d\omega = \alpha dt$$

$$= (3t^2 - 3)\,dt \qquad (\because \alpha = 3t^2 - 3)$$

Integrating, we get

$$\int d\omega = \int (3t^2 - 3)dt \qquad \text{or} \qquad \omega = \frac{3t^3}{3} - 3t + C \qquad ...(i)$$

where C is a constant of integration. Its value is obtained from the given condition *i.e.*, $\omega = 2$ rad/s when $t = 0$.

Substituting these values in equation (*i*), we get

$$2 = 0 - 0 + C \qquad \text{or} \qquad C = 2$$

Substituting the value of C in equation (*i*),

$$\omega = t^3 - 3t + 2$$

When $t = 5$ s, angular velocity becomes

$$\omega = 5^3 - 3 \times 5 + 2 = 125 - 15 + 2 = \mathbf{112\ rad/s.} \quad \textbf{Ans.}$$

(*ii*) *Angular displacement* (θ)

We know that $\qquad \omega = \dfrac{d\theta}{dt} \qquad \text{or} \qquad d\theta = \omega.dt$

$$= (t^3 - 3t + 2)dt \qquad (\because \quad \omega = t^3 - 3t + 2)$$

Integrating, we get

$$\int d\theta = \int (t^2 - 3t + 2)dt$$

or

$$\theta = \frac{t^4}{4} - \frac{3t^2}{4} + 2t + C_1 \qquad ...(ii)$$

where C_1 is a constant of integration. Its value is obtained from the condition, $\theta = 0$ when $t = 0$.

Substituting these values in equation (*ii*), we get

$$0 = 0 - 0 + 0 + C_1 \qquad \text{or} \qquad C_1 = 0$$

Now equation (*ii*) becomes as

$$\theta = \frac{t^4}{4} - \frac{3t^2}{2} + 2t$$

When $t = 5$ seconds, the angular displacement becomes as

$$\theta = \frac{5^4}{4} - \frac{3 \times 5^2}{2} + 2 \times 5$$

$$= 156.25 - 37.5 + 10 = \mathbf{128.75\ radians.} \quad \textbf{Ans.}$$

Problem 4.35. *The angle of rotation of a body is given as a function of time by the equation,*

$$\theta = \theta_0 + at + bt^2$$

where θ_0 initial angular displacement, a and b are constants.

Obtain general expressions for : (a) the angular velocity and (b) the angular acceleration of the body.

If the initial angular velocity be 3π radian per second and after two seconds the angular velocity is 8π radian per second, determine the constants a and b.

Sol. Given :

$$\theta = \theta_0 + at + bt^2.$$

Angular velocity (ω) is obtained by differentiating the above equation with respect to time (t).

$$\therefore \qquad \omega = \frac{d\theta}{dt} = \frac{d}{dt}(\theta_0 + at + bt^2) = a + 2bt \qquad ...(i)$$

$$\left(\because \quad \theta_0 \text{ is constant and hence } \frac{d\theta_0}{dt} = 0\right)$$

(*i*) When $t = 0$, $\omega = 3\pi$

Substituting these values in equation (*i*), we get

$$3\pi = a + 2b \times 0 = a$$

$\therefore$ $a = 3\pi$. **Ans.**

(*ii*) When $t = 2$ seconds, $\omega = 8\pi$ radians.

Substituting these values in $\omega = a + 2bt$, we get

$$8\pi = a + 2b \times 2 = a + 4b = 3\pi + 4b \qquad (\because \; a = 3\pi)$$

$\therefore$ $4b = 8\pi - 3\pi = 5\pi$

or $b = \dfrac{5\pi}{4} = \mathbf{1.25\pi}$. **Ans.**

(*a*) *General expression for angular velocity*

General expression for angular velocity is obtained by substituting the values of a and b in equation (*i*).

$$\therefore \quad \omega = a + 2bt$$

$$= 3\pi + 2 \times 1.25\pi \times t \qquad (\because \; a = 3\pi, b = 1.25\pi)$$

$$= \mathbf{3\pi + 2.5\pi t}. \quad \textbf{Ans.}$$

(*b*) *General expression for angular acceleration*

Angular acceleration (α) is obtained by differentiating the equation (*i*) with respect to time (*t*).

$$\therefore \quad \alpha = \frac{d\omega}{dt} = \frac{d}{dt}(a + 2bt) = 2b \qquad \left(\because \; a \text{ is constant, hence } \frac{da}{dt} = 0\right)$$

$$= 2 \times 1.25\pi \qquad (\because \; b = 1.25\pi)$$

$$= \mathbf{2.50\pi}. \quad \textbf{Ans.}$$

4.8. NEWTON'S LAWS

When a body is at rest or moving in a straight line or rotating about an axis, the body obeys certain laws of motion. These laws are called Newton's laws of motion. There are three laws of motion. These laws for linear motions and angular motions are given in the next articles.

4.8.1. Newton's Laws for Linear Motion

First law. It states that a body continues in its state of rest or of uniform motion in a straight line unless it is compelled by an external force to change that state.

Second law. It states that the rate of change of momentum of a body is proportional to the external force applied on the body and takes place in the direction of the force.

Third law. It states that to every action, there is always an equal and opposite reaction.

Before discussing Newton's laws of motion, let us define certain terms like mass, weight and momentum.

1. Mass. The quantity of matter contained in a body is known as mass of the body. Mass is a scalar quantity. In C.G.S. units, the mass is expressed in gramme (gm) whereas in S.I. units the mass is expressed in kilogramme (kg).

2. Weight. Weight of a body is defined as the force, by which the body is attracted towards the centre of the earth. Mathematically weight of a body is given by

$$\text{Weight} = \text{Mass} \times \text{Acceleration due to gravity} = \text{Mass} \times g \qquad ...(4.19)$$

If mass is taken in kilogram (kg) and acceleration due to gravity in metre per second square (m/s^2), then weight is expressed in newton (N). But if mass is taken in gramme (gm) and acceleration due to gravity in centimetre per second square (cm/s^2), then weight is expressed in dyne. The relation between newton (N) and dyne is given as

One Newton = 10^5 dyne.

3. Momentum. The product of the mass of a body and its velocity is known as momentum of the body. Momentum is a vector quantity. Mathematically, momentum is given by

Momentum = Mass × Velocity.

4.8.2. Newton's First Law of Motion. It consists of two parts. First part states that a body continues in its state of rest unless it is compelled by an external force to change that state. A book lying on a table remains at rest, unless it is lifted by some external force.

Second part states that a body continues in its state of uniform motion in a straight line unless it is compelled by an external force to change that state. In actual practice, we see that when a body is moving with a uniform velocity in a straight line, the body does not continue in its state of uniform motion but comes to rest after some time. This is due to frictional force acting on the body. For an ideal case (*i.e.*, when there is no frictional force acting on the body), the body will continue to move with uniform velocity in a straight line, unless compelled by an external force to change that state.

4.8.3. Newton's Second Law of Motion. This law enables us to measure a force. Let a body of mass 'm' is moving with a velocity 'u' along a straight line. It is acted upon by a force F and the velocity of the body becomes v in time t. Then we have

u = Initial velocity of the body,

v = Final velocity of the body,

m = Mass of the body,

a = Uniform linear acceleration,

F = Force acting on the body, which changes the velocity u to v in time t,

t = Time in second to change the velocity from u to v.

Initial momentum of the body

= Mass × Initial velocity = $m \times u$

Final momentum of the body = $m \times v$.

∴ Change of momentum

= Final momentum – Initial momentum = $mv - mu = m(v - u)$

$$\text{Rate of change of momentum} = \frac{\text{Change of momentum}}{\text{Time}} = \frac{m(v-u)}{t} \quad \ldots(i)$$

But from equation (4.4),

$$\frac{v-u}{t} = a \quad (\textit{i.e., } \text{linear acceleration})$$

Substituting the value of $\left(\frac{v-u}{t}\right)$ in equation (*i*), we get

Rate of change of momentum = $m \times a$.

But according to Newton's second law of motion, the rate of change of momentum is directly proportional to the external force acting on the body.

$$\therefore \quad F \propto m \times a \quad \text{or} \quad F = k \times m \times a \quad \ldots(ii)$$

where k is a constant of proportionality.

In equation (*ii*), k and m (mass of a body) are constants for a given body and hence force acting on a body is proportional to the acceleration produced by the force. This means that for a given body, greater force products greater acceleration while a smaller force produces smaller acceleration. The acceleration produced will be zero if no force is applied on the body.

Two important conclusions are drawn from the first two Newton's laws of motion :

(*i*) There will be no acceleration, if no external force is applied on the body. This means the body will continue in its state of existing uniform motion in a straight line.

(*ii*) Force applied on the body is proportional to the product of mass of the body and the acceleration produced by the force.

4.8.4. Unit of Force. Let us first define a 'unit force'. A unit force can be suitably defined so as to make the value of k in equation (*ii*) equal to one. A unit force (*i.e.*, Force = 1.0) is that which produces unit acceleration on an unit mass. Then by substituting $F = 1.0$, $m = 1.0$ and $a = 1.0$ in equation (*ii*) (*i.e.*, $F = k \times m \times a$), we get

$$1 = k \times 1 \times 1 \quad \text{or} \quad k = 1.$$

Substituting the value of $k = 1$, in equation (*ii*), we get

$$F = m \times a \qquad \text{...(4.20)}$$

(*i*) If mass (m) = 1 kg and acceleration produced (a) = 1 m/s^2 the unit of force is known as newton (which is written as N). Thus newton is defined as that force which acts on a body of mass one kg and produces an acceleration of 1 m/s^2 in the direction of force. Newton is the unit of force in S.I. system.

$$\therefore \quad 1 \text{ N} = 1 \text{ kg} \times 1 \text{ m/s}^2 = 1 \text{ kg-m/s}^2.$$

(*ii*) If mass (m) = 1 gm and acceleration (a) = 1 cm/s^2, then unit of force is known as 'dyne'. Thus a dyne may be defined as that force which acts on a body of mass one gm and produces an acceleration of 1 cm/s^2. Dyne is the unit of force in C.G.S. system.

$$\therefore \quad 1 \text{ dyne} = 1 \text{ gm} \times 1 \text{ cm/s}^2 = \text{gm cm/s}^2$$

By definition,

$$1 \text{ N} = 1 \text{ (kg)} \times 1 \text{ (m/s}^2) = 1 \times 1000 \text{ (gm)} \times 1 \times 100 \text{ (cm/s}^2) = 10^5 \text{ (gm cm/s}^2)$$

$$= 10^5 \text{ dyne} \qquad (\because \ 1 \text{ dyne} = 1 \text{ gm} \times 1 \text{ cm/s}^2 = \text{gm cm/s}^2)$$

Note. (*i*) The body will have acceleration if the external force is acting on the body in the direction of motion of the body.

(*ii*) The body will have retardation if the external force is acting opposite to the direction of motion of the body.

Problem 4.36. *A force of unknown magnitude acts on a body of mass 150 kg and produces an acceleration of 3 m/s^2 in the direction of force. Find the force.*

Sol. Given :

Mass of the body, $m = 150$ kg

Acceleration, $f = 3$ m/s^2

The force is given by equation (4.20),

Hence $F = m \times a = 150 \text{ (kg)} \times 3 \text{ (m/s}^2)$

$= 450 \text{ (kg-m/s}^2) =$ **450 N. Ans.** $(\because \ \text{kg-m/s}^2 = \text{N})$

Problem 4.37. *A force of 100 N acts on a body having a mass of 4 kg for 10 seconds. If the initial velocity of the body is 5 m/s, determine :*

(*i*) *acceleration produced in the direction of force, and*

(*ii*) *distance moved by the body in 10 seconds.*

Sol. Given :

Force, $F = 100$ N ; Mass, $m = 4$ kg

Time, $t = 10$ second ; Initial velocity, $u = 5$ m/s

Let a = Acceleration produced in the direction of force

s = Distance travelled by body in 10 seconds.

(*i*) Using equation (4.20), we get

$$F = m \times a \quad \text{or} \quad 100 = 4 \times a$$

$$\therefore \quad a = \frac{100}{4} = \mathbf{25\ m/s^2}. \quad \textbf{Ans.}$$

(*ii*) The distance moved is given by

$$s = ut + \tfrac{1}{2}at^2 = 5 \times 10 + \tfrac{1}{2} \times 25 \times 10^2$$

$$= 50 + 1250 = \mathbf{1300\ m}. \quad \textbf{Ans.}$$

Problem 4.38. *The weight of a body on earth is 980 N. If the acceleration due to gravity on earth = 9.80 m/s², what will be weight of the body on :*

(*i*) *the moon, where gravitational acceleration is 1.6 m/s², and*

(*ii*) *the sun, where gravitational acceleration is 270 m/s².*

Sol. Given :

Weight of body on earth, $W = 980$ N

Acceleration due to gravity on earth, $g = 9.80$ m/s².

First calculate the mass of the body on earth. Using equation (4.19), we have

$$\text{Weight} = \text{Mass} \times g \quad \text{or} \quad 980 = \text{Mass} \times 9.80$$

$$\therefore \quad \text{Mass} = \frac{980}{9.80} = 100 \text{ kg.}$$

The mass of the body will remain the same on moon as well as on sun.

(*i*) *Weight of the body on moon,* where $g = 1.6$ m/s²

Using equation (4.19),

$$\text{Weight} = \text{Mass} \times g = 100 \times 1.6 = \mathbf{160\ N}. \quad \textbf{Ans.}$$

(*ii*) *Weight of the body on sun,* where $g = 270$ m/s²

$$\text{Weight} = \text{Mass} \times g = 100 \times 270 = \mathbf{27000\ N}. \quad \textbf{Ans.}$$

Problem 4.39. *A force of 200 N acts on a body having mass of 300 kg for 90 seconds. If the initial velocity of the body is 20 m/s, determine the final velocity of the body :*

(*i*) *when the force acts in the direction of motion, and*

(*ii*) *when the force acts in the opposite direction of the motion.*

Sol. Given :

Force, $F = 200$ N

Mass, $m = 300$ kg

Time, $t = 90$ seconds

Initial velocity, $u = 20$ m/s.

Using equation (4.20), we have

$$F = m \times a \quad \text{or} \quad 200 = 300 \times a$$

$$\therefore \quad a = \frac{200}{300} = \frac{2}{3} \text{ m/s}^2.$$

(*i*) *Final velocity when the force acts in the direction of motion.*

When the force acts in the direction of motion, the body will have acceleration. The final velocity (v) is given by

$$v = u + at = 20 + \frac{2}{3} \times 90 = 20 + 60 = \textbf{80 m/s.} \quad \textbf{Ans.}$$

(*ii*) *Final velocity when the force acts in the opposite direction of motion.*

When the force acts in the opposite direction of motion, the body will have retardation. The final velocity is given by

$$v = u - at \qquad \text{(–ve sign shows retardation)}$$

$$= 20 - \frac{2}{3} \times 90 = 20 - 60 = \textbf{– 40 m/s.} \quad \textbf{Ans.}$$

(–ve sign shows that the body will be moving in the opposite direction).

Problem 4.40. *A body of mass 15 kg falls on the ground from a height of 19.6 m. The body penetrates into the ground. Find the distance through which the body will penetrate into the ground, if the resistance by the ground to penetration is constant and equal to 4900 N. Take* $g = 9.8\ m/s^2$.

Sol. Given :

Mass, $m = 15$ kg

Height of body from ground, $h = 19.6$ m

Resistance to penetration, $F^* = 4900$ N

Acceleration due to gravity, $g = 9.8$ m/s^2.

Let us first consider the motion of the body from a height of 19.6 m to the ground surface.

Initial velocity of the body, $u = 0$

Final velocity of the body, when it reaches the ground = v.

Using the equation,

$$v^2 - u^2 = 2gh \quad \text{or} \quad v^2 - 0 = 2 \times 9.8 \times 19.6$$

or

$$v = \sqrt{2 \times 9.8 \times 19.6} = 19.6 \text{ m/s.}$$

When the body is penetrating into the ground, the resistance to penetration is acting in the upward direction ($\because$ Resistance always acts in the opposite direction of motion of the body). But the weight of the body is acting in the downward direction.

Weight of the body is given by equation (4.19),

Weight of the body = Mass × g = 15 × 9.80 N = 147 N

Upward resistance to penetration = 4900 N

$\therefore$ Net force acting in the upward direction,

$$F = 4900 - 147 = 4753 \text{ N.}$$

As the net force on the body is acting in the opposite direction to the motion of the body, this force will produce retardation.

Using equation (4.20), we have

$$F = m \times a \quad \text{or} \quad 4753 = 15 \times a$$

$$\therefore \quad a = \frac{4753}{15} = 316.866 \text{ m/s}^2.$$

Distance through which body will penetrate into the ground.

Consider the motion of the body from the ground to the point of penetration into the ground.

Let the distance of penetration = s

Final velocity, $v = 0$

Initial velocity, u = Velocity of the body on the ground = 19.6 m/s

Retardation, $a = 316.866 \text{ m/s}^2$.

Using the relation,

$$v^2 - u^2 = -2a \times s \quad \text{(–ve sign is taken due to retardation)}$$

$$\therefore \quad 0 - 19.6^2 = -2 \times 316.866 \times s$$

$$\therefore \quad s = \frac{19.6 \times 19.6}{2 \times 316.866} = 0.606 \text{ m} = \mathbf{60.6 \text{ cm.}} \quad \textbf{Ans.}$$

Problem 4.41. *A man weighing 637 N dives into a swimming pool from a tower of height 19.6 m. He was found to go down in water by 2 m and then started rising. Find the average resistance of water. Neglect the resistance of air.* (AMIE Summer, 1982)

Sol. Given :

Weight of the man = 637 N

Height of tower, $h = 19.6$ m

Distance travelled by man from the water surface into the water = 2 m.

First consider the motion of the man from the top of the tower to the water surface of the swimming pool.

Initial velocity of man, $u = 0$

Acceleration due to gravity, $g = 9.8 \text{ m/s}^2$.

Let the final velocity of the man, when he reaches the water surface = v.

Now using the relation,

$$v^2 - u^2 = 2gh \quad \text{or} \quad v^2 - 0^2 = 2 \times 9.80 \times 19.60$$

$$\therefore \quad v = \sqrt{2 \times 9.80 \times 19.60} = 19.6 \text{ m/s}.$$

Now consider the motion of the man from the water surface of the swimming pool up to the point, from where the man started rising.

Distance traversed, $s = 2$ m

Initial velocity of the man on the water surface

$u = 19.6$ m/s

Final velocity, $v = 0$.

As the velocity of the man becomes zero after travelling a distance 2 m inside the water, hence a force due to water resistance is acting in the opposite direction to the motion of the man. This force will produce retardation.

Let a = Retardation due to water resistance

$\therefore$ Using the relation, $v^2 - u^2 = -2as$ (–ve sign is taken due to retardation)

or $$0^2 - 19.6^2 = -2 \times a \times 2$$

$$\therefore \quad a = \frac{19.6 \times 19.6}{2 \times 2} = 96.04 \text{ m/s}^2.$$

Average resistance of water

Let F^* = Average resistance of water acting on the man in the upward direction.

Weight of man = 637 N acting in the downward direction.

$\therefore$ Net force acting on the man in the upward direction

$$= F^* - \text{Weight of man} = (F^* - 637) \text{ N}.$$

But the net force acting on the man must be equal to the product of mass of the man and retardation.

$$\therefore \quad (F^* - 637) = m \times a \qquad (\because \; F = m \times a) \quad ...(i)$$

But mass of man,

$$m = \frac{\text{Weight of man}}{g} \qquad \text{[using equation (4.19)]}$$

$$= \frac{637}{9.8} = 65 \text{ kg} \quad \text{and} \quad a = 96.04 \text{ m/s}^2.$$

Substituting the values of m and a in equation (i), we get

$$(F^* - 637) = 65 \times 96.04$$

$$\therefore \quad F^* = 65 \times 96.04 + 637 = \mathbf{6879.6 \text{ N.}} \quad \textbf{Ans.}$$

Problem 4.42. *A bullet of mass 81 gm and moving with a velocity of 300 m/s is fired into a log of wood and it penetrates to a depth of 10 cm. If the bullet moving with the same velocity, were fired into a similar piece of wood 5 cm thick, with what velocity would it emerge ? Find also the force of resistance, assuming it to be uniform.* (AMIE Winter, 1981)

Sol. Given :

Mass of bullet, $m = 81 \text{ gm} = \frac{81}{1000} \text{ kg} = 0.081 \text{ kg}$

Initial velocity of bullet, $u = 300$ m/s

Distance travelled, $s = 10 \text{ cm} = 0.10$ m

Final velocity, $v = 0$.

As the force of resistance is acting in the opposite direction of motion of bullet, hence force of resistance will produce retardation on the bullet.

Let a = Retardation

Now using the relation,

$$v^2 - u^2 = -2as \qquad \text{(–ve sign is taken due to retardation)}$$

or

$$0^2 - 300^2 = -2 \times a \times 0.1$$

$$\therefore \quad a = \frac{300 \times 300}{2 \times 0.1} = 450000 \text{ m/s}^2.$$

Let F = Force of resistance offered by wood to the bullet.

Using equation (4.20), we get

$$F^* = m \times a = 0.081 \times 450000 = \mathbf{36450 \text{ N.}} \quad \textbf{Ans.}$$

Velocity of the bullet with which the bullet will come out from a piece of wood of 5 cm thick.

Let v = velocity with which the bullet emerges from the piece of wood of 5 cm thick.

Initial velocity, $u = 300$ m/s.

As the resistance offered by wood is uniform, hence retardation will be same as before.

$$\therefore \quad a = 450000 \text{ m/s}^2.$$

Distance travelled, $r = 5 \text{ cm} = 0.05$ m

Using the relation,

$$v^2 - u^2 = -2as \quad \text{(–ve sign is taken due to retardation)}$$

$\therefore$ $v^2 - 300^2 = -2 \times 450000 \times 0.05$ or $v^2 = 300 \times 300 - 45000$

$= 90000 - 45000 = 45000$

$\therefore$ $v = \sqrt{45000} =$ **212.132 m/s. Ans.**

Problem 4.43. *A car, moving on a straight level road, skidded for a total distance of 60 metres after the brakes were applied. Determine the speed of the car, just before the brakes were applied, if the co-efficient of friction between the car tyres and the road is 0.4. Take $g = 9.80\ m/s^2$.*

Sol. Given :

Let u = Velocity of car just before applying the brakes.

Final velocity of car, $v = 0$

Distance travelled, $s = 60$ m

Co-efficient of friction between car tyres and road,

$\mu = 0.4$

Let W = Weight of car in Newton

R = Normal reaction

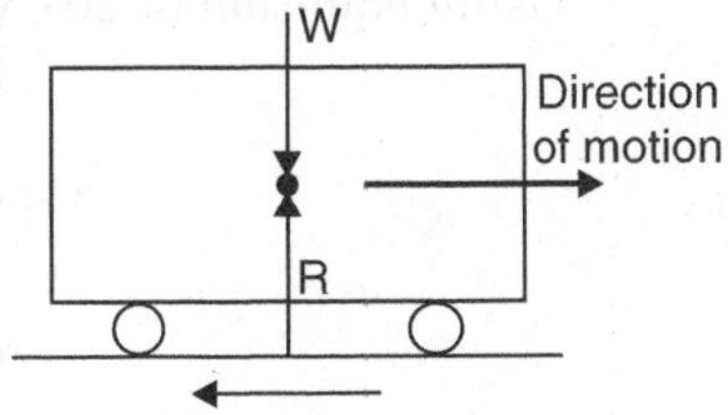

Fig. 4.12

$\therefore$ Mass of car $m = \dfrac{W}{g}$

Now frictional resistance

$= \mu R = \mu W$ $(\because R = W)$

$= 0.4\ W$ Newton

$\therefore$ $F = 0.4\ W.$

As frictional force is acting in the opposite direction of motion, hence the frictional force will produce retardation.

Using equation (4.20),

Force = Mass × a ...(*i*)

$\therefore$ $0.4\ W = \dfrac{W}{g} \times a$ or $a = \dfrac{0.4\ W \times g}{W} = 0.4 \times 9.80 = 3.92\ \text{m/s}^2$

$\therefore$ Retardation $= 3.92\ \text{m/s}^2$.

Now using the relation,

$$v^2 - u^2 = -2as \quad \text{(–ve sign is due to retardation)}$$

$\therefore$ $0^2 - u^2 = -2 \times 3.92 \times 60$

$\therefore$ $u = \sqrt{2 \times 3.92 \times 60} = \sqrt{470.4} = 21.688$ m/s

$= \dfrac{21.688}{1000} \times 60 \times 60$ km/hour = **78.077 km/hour. Ans.**

Problem 4.44. *The tractive force, exerted by a railway car weighing 50 kN, is 2000 N. If the frictional resistance is 5 N per kN of the railway car's weight, determine the acceleration when the railway car is moving on a level track.*

Sol. Given :

Tractive force exerted by railway car,

$F_1 = 2000$ N

Weight of car, $W = 50 \text{ kN} = 50 \times 1000 \text{ N}$

$\therefore$ Mass of car, $m = \frac{W}{g} = \frac{50 \times 1000}{9.81}$ kg

Frictional resistance, $F_2 = 5$ N per kN of car's weight

$= 5 \text{ N} \times \text{Weight of car in kN} = 5 \times 50 \text{ N} = 250 \text{ N}.$

The tractive force is acting in the direction of motion, while frictional resistance is acting in opposite direction of motion.

$\therefore$ Net force in the direction of motion,

$$F = F_1 - F_2 = 2000 - 250 = 1750 \text{ N}.$$

As the net force is acting in the direction of motion, it will produce acceleration.

Let a = Acceleration produced

Using equation (4.20), we have

$$F = m \times a$$

or $$1750 = \frac{50 \times 1000}{9.81} \times a \qquad \left(\because \; m = \frac{50 \times 1000}{9.81} \right)$$

$$\therefore \quad a = \frac{1750 \times 9.81}{50 \times 1000} = \mathbf{0.343 \; m/s^2}. \quad \textbf{Ans.}$$

Problem 4.45. *A train of weight 1960 kN starts from rest and attains a speed of 120 km/hr in 5 minute. If the frictional resistance of the track is 10 N per kN of the train's weight, find the average pull required. Take g = 9.8 m/s².*

Sol. Given :

Weight of train, $W = 1960 \text{ kN} = 1960 \times 1000 \text{ N}$

$\therefore$ Mass of train, $m = \frac{W}{g} = \frac{1960 \times 1000}{9.80}$ kg

Initial velocity of train, $u = 0$

Final velocity, $v = 120 \text{ km/hr} = \frac{120 \times 1000}{60 \times 60} \text{ m/s} = \frac{100}{3} \text{ m/s}$

Time in second, $t = 5 \times 60 = 300$ seconds

Let a = Acceleration of train

Using the relation,

$$v = u + a \times t \quad \text{or} \quad \frac{100}{3} = 0 + a \times 300 = 300\,a$$

$$\therefore \quad a = \frac{100}{3} \times \frac{1}{300} = \frac{1}{9} \text{ m/s}^2$$

Let F^* = Average pull required in newton

F_1 = Frictional resistance in newton

F = Net force acting on the engine in the direction of motion.

Then $F = F^* - F_1$ ...(*i*)

But F_1 = 10 N per kN of train's weight

$= 10 \text{ N} \times \text{Weight of train in kN} = 10 \times 1960 = 19600 \text{ N}.$

Substituting this value in equation (*i*), we get

$$F = (F^* - 19600)$$

Now using equation (4.20), we get

$$F = m \times a$$

or $$(F^* - 19600) = \frac{1960 \times 1000}{9.80} \times \frac{1}{9} \qquad \left(\because\ m = \frac{1960 \times 1000}{9.81} \text{ and } a = \frac{1}{9}\right)$$

$$= 22222.22 \text{ N}$$

$$\therefore \quad F^* = 22222.22 + 19600 = 41822.22 \text{ N} = \mathbf{41.822\ kN.} \quad \textbf{Ans.}$$

4.8.5. Motion on an Inclined Smooth Surface. Fig. 4.13 shows a body of weight W, sliding down on a smooth inclined plane.

Let θ = Angle made by inclined plane with horizontal

W = Weight of the body

a = Acceleration of the body

m = Mass of the body

$= \dfrac{W}{g}$.

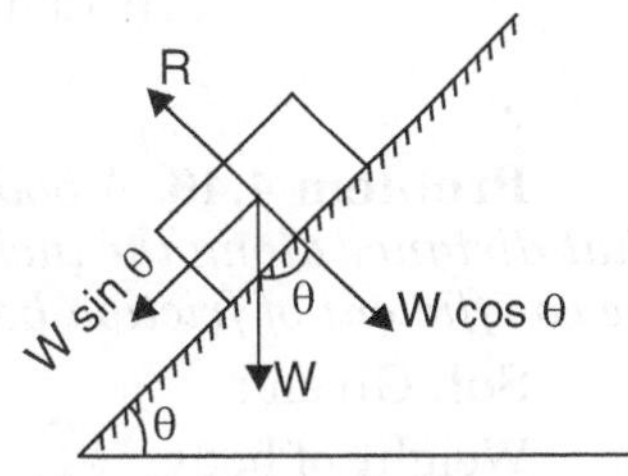

Fig. 4.13

As the surface of the plane is smooth, hence the frictional force will be zero. Hence the forces acting on the body are its own weight W and reaction R of the plane. The resolved part of W perpendicular to the plane is $W \cos \theta$, which is balanced by R, while the resolved part parallel to the plane is $W \sin \theta$, which produces acceleration down the plane. This force is responsible for the movement of the body down the plane.

$\therefore$ Net force acting on the body down the pane is

$$F = W \sin \theta$$

Now using the equation (4.20), we have

$$F = m \times a.$$

Substituting the values of F and m in the above equation, we get

$$W \sin \theta = \frac{W}{g} \times a$$

$$\therefore \quad a = g \sin \theta \qquad \text{...(4.21)}$$

If the body is moving up the plane, the corresponding acceleration will be $- g \sin \theta$.

4.8.6. Motion on an Inclined Rough Surface. Fig. 4.14 shows a body of weight W, sliding down the rough inclined surface.

Let a = Acceleration of the body

m = Mass of the body $= \dfrac{W}{g}$

θ = Inclination of the plane with horizontal

μ = Co-efficient of friction

F_1 = Force of friction.

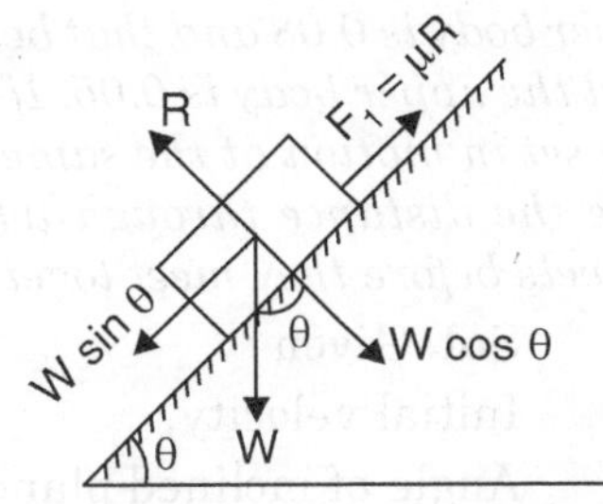

Fig. 4.14

As the body is moving down the plane, the force of friction will be acting up the plane as shown in Fig. 4.14.

Now force of friction, $F_1 = \mu R$

$= \mu \times W \cos\theta$ $\qquad (\because \ R = W\cos\theta)$

Force acting down the plane, $F_2 = W \sin\theta$

$\therefore$ Net force acting on the body down the plane,

$$F = F_2 - F_1 = W\sin\theta - \mu W \cos\theta.$$

Now using the equation (4.20), $F = m \times a$

or $$(W\sin\theta - \mu W\cos\theta) = \frac{W}{g} \times a \qquad \left(\because \ m = \frac{W}{g}\right)$$

or $$W(\sin\theta - \mu\cos\theta) = \frac{W}{g} \times a \quad \text{or} \quad a = W(\sin\theta - \mu\cos\theta) \times \frac{g}{W}$$

$\therefore$ $$a = g(\sin\theta - \mu\cos\theta). \qquad \text{...(4.22)}$$

Problem 4.46. *A body of weight 200 N is initially stationary on a 45° inclined plane. What distance along the inclined plane must the body slide, before it reaches a speed of 2 m/s. The co-efficient of friction between the body and the plane = 0.1.*

Sol. Given :

Weight of body, $W = 200$ N

$\therefore$ Mass of body, $m = \dfrac{W}{g} = \dfrac{200}{9.81}$ kg

Angle of plane, $\theta = 45°$

Initial velocity, $u = 0$

Final velocity, $v = 2$ m/s

Co-efficient of friction, $\mu = 0.1$.

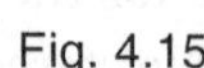
Fig. 4.15

The acceleration of the body is given by equation (4.22) as

$$a = g[\sin\theta - \mu\cos\theta] = 9.81[\sin 45° - 0.1 \cos 45°]$$
$$= 9.81[0.707 - 0.1 \times .707] = 6.242 \text{ m/s}^2.$$

Now using the relation,

$$v^2 - u^2 = 2as \quad \text{or} \quad 2^2 - 0^2 = 2 \times 6.242 \times s$$

$\therefore$ $$s = \frac{2 \times 2}{2 \times 6.242} = 0.32 \text{ m} = \mathbf{32 \text{ cm}}. \quad \textbf{Ans.}$$

Problem 4.47. *Two bodies directly in line and 10 m apart are held stationary on an inclined plane having inclination of 20°. The co-efficient of friction between the plane and lower body is 0.08 and that between the plane and the upper body is 0.05. If both the bodies are set in motion at the same instant, calculate the distance through which each body travels before they meet together.*

Sol. Given :

Initial velocity, $u = 0$

Angle of inclined plane, $\theta = 20°$

Co-efficient of friction between the plane and lower body, $\mu_1 = 0.08$.

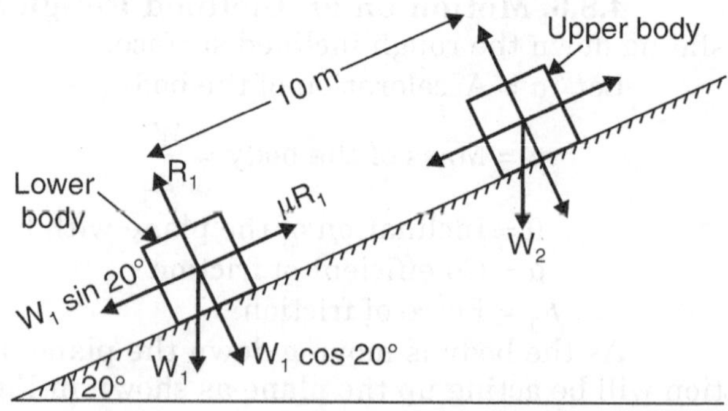

Fig. 4.16

Co-efficient of friction between the plane and upper body,

$\mu_2 = 0.05.$

Distance between two bodies,

$d = 10$ m.

Let the two bodies meet after 't' seconds.

If s_1 = Distance travelled by lower body in time 't', and

s_2 = Distance travelled by upper body in time 't'.

Then $s_2 = s_1 + 10$ ...(*i*)

Let a_1 = Acceleration of the lower body down the plane, and

a_2 = Acceleration of the upper body down the plane.

But the acceleration of a body moving down a rough plane is given by equation (4.22).

$$a = g(\sin \theta - \mu \cos \theta).$$

∴ Acceleration of lower body,

$$\begin{aligned} a_1 &= g(\sin \theta - \mu_1 \cos \theta) \\ &= 9.81(\sin 20° - 0.08 \times \cos 20°) \quad (\because \mu_1 = 0.08, \theta = 20°) \\ &= 9.81(0.342 - 0.08 \times 0.9397) \\ &= 9.81(0.342 - 0.0752) = 2.617 \text{ m/s}^2. \end{aligned}$$

Acceleration of upper body,

$$\begin{aligned} a_2 &= g(\sin \theta - \mu_2 \cos \theta) \\ &= 9.81(\sin 20° - 0.05 \cos 20°) \quad (\because \mu_2 = .05) \\ &= 9.81 (0.342 - 0.05 \times 0.9397) \\ &= 9.81(0.342 - 0.04698) = 2.894 \text{ m/s}^2. \end{aligned}$$

The distance travelled by the body is given by

$$s = ut + \frac{1}{2} at^2$$

∴ Distance travelled by the lower body is given by

$$\begin{aligned} s_1 &= ut + \frac{1}{2} a_1 t^2 \\ &= 0 \times t + \frac{1}{2} \times 2.617 \times t^2 \quad (\because u = 0 \text{ and } a_1 = 2.617) \\ &= \frac{1}{2} \times 2.617 \times t^2 = 1.3085\, t^2 \quad \text{...(ii)} \end{aligned}$$

Distance travelled by the upper body is given by

$$\begin{aligned} s_2 &= ut + \frac{1}{2} a_2 t^2 = 0 \times t + \frac{1}{2} \times 2.894 \times t^2 \\ &= \frac{1}{2} \times 2.894 \times t^2 = 1.447\, t^2 \quad \text{...(iii)} \end{aligned}$$

Substituting the values of s_1 and s_2 in equation (*i*), we get

$$1.447\, t^2 = 1.3085\, t^2 + 10$$

or $1.447\, t^2 - 1.3085\, t^2 = 10$ or $0.1385\, t^2 = 10$

$$\therefore \quad t = \sqrt{\frac{10}{.1385}} = \sqrt{72.202} = 8.497 \text{ s.}$$

Substituting the value of t in equations (*ii*) and (*iii*), we get

$s_1 = 1.3085 \times 8.497^2 =$ **94.476 m. Ans.**

and $s_2 = 1.447 \times 8.497^2 =$ **104.476 m. Ans.**

Problem 4.48. *A truck weighing 6 kN just moves freely (engine is not running) at 36 kilometre per hour down a slope of 1 in 40, the road resistance at this speed just being sufficient to prevent any acceleration. Find the road resistance per kN weight of truck.*

What power will the engine have to exert to run up the same slope at double the speed when the road resistance remains the same ?

Sol. Given :

Weight of truck, $W = 6 \text{ kN} = 6 \times 1000 \text{ N} = 6000 \text{ N}$

Speed of truck, $u = 36 \text{ km/hr} = \dfrac{36 \times 1000}{60 \times 60} = 10 \text{ m/s}$

Slope of the road $= 1 \text{ in } 40$

$$\therefore \quad \sin\theta = \frac{1}{40}$$

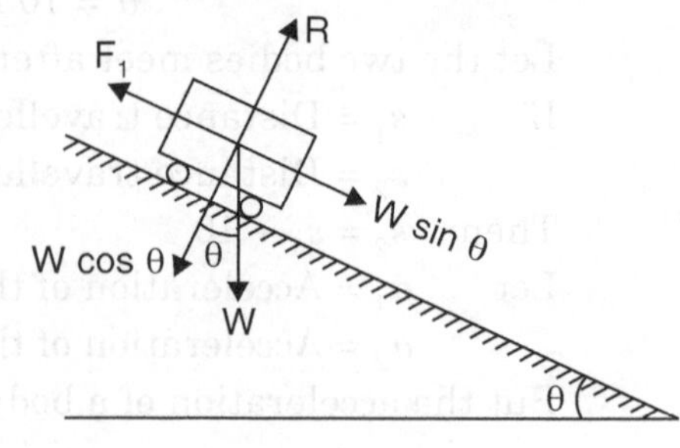

Fig. 4.17

1st Case. Fig. 4.17 shows the position of the truck, when it is moving down the plane. The road resistance (F_1) is acting in the opposite direction of the motion of the truck. The truck is not having any acceleration and hence it is moving with a constant velocity of 10 m/s. Hence the net force on the truck in the direction of motion should be zero. But net force on the truck in the direction of motion (see Fig. 4.17).

$$= W \sin\theta - F_1$$

$$= 6000 \times \frac{1}{40} - F_1 \qquad \left\{\because W = 6000 \text{ N}, \sin\theta = \frac{1}{40}\right\}$$

$\therefore$ According to the given condition,

$$6000 \times \frac{1}{40} - F_1 = 0$$

$$\therefore \quad F_1 = 6000 \times \frac{1}{40} = \frac{6000}{40} = 150 \text{ N}$$

$\therefore$ Frictional force per tonne weight of truck

$$= \frac{F_1}{\text{Weight of truck in kN}} = \frac{150}{6} = \mathbf{25\ N.} \quad \textbf{Ans.}$$

2nd Case. Truck is moving up an inclined plane of slope 1 in 40 with double speed. Road resistance is same.

$\therefore$ Speed of truck $= 2u = 2 \times 10 = 20$ m/s

$$\sin\theta = \frac{1}{40}.$$

Road resistance $= F_1 = 150$ N.

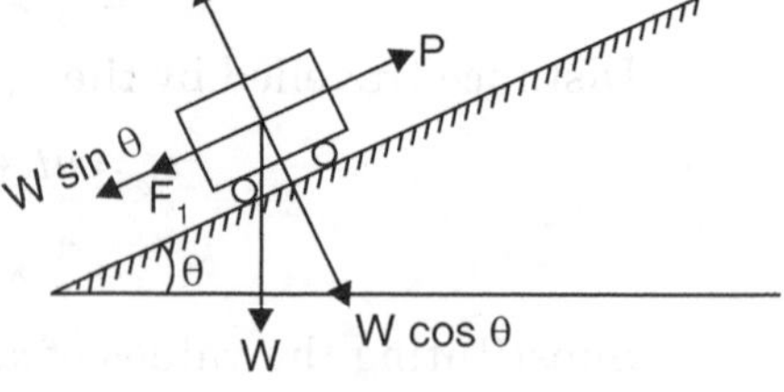

Fig. 4.18

The truck is moving at constant speed of 20 m/s up the plane.

Let P = Force exerted by engine up the plane.

As the truck is moving with uniform speed, the net force on the truck along the plane should be zero (see Fig. 4.18).

or $\quad P - W \sin \theta - F_1 = 0$

$$\therefore \quad P = W \sin \theta + F_1 = 6000 \times \frac{1}{40} + 150 = 150 + 150 = 300 \text{ N}$$

*Power exerted by engine

$$= \frac{(\text{Force exerted by engine in N}) \times \text{Speed of engine in m/s}}{1000}$$

$$= \frac{(300 \text{ N}) \times 20 \text{ m/s}}{1000} = \frac{300 \times 20}{1000} = \mathbf{6 \text{ kW}}. \quad \textbf{Ans.}$$

Problem 4.49. *A train weighing 200 kN moves down a slope of 1 in 150 at 18 km / hour and engine develops 3.5 kW. What power will be required to pull the train up the slope at the same speed ?*

Sol. Given :

Weight of train, $\quad W = 200 \text{ kN} = 200 \times 1000 \text{ N}$

Slope of the track $\quad = 1 \text{ in } 150$

$$\therefore \quad \sin \theta = \frac{1}{150}$$

Speed of train, $\quad u = 18 \text{ km/hr} = \frac{18 \times 1000}{60 \times 60} \text{ m/s} = 5 \text{ m/s}$

Power developed by engine = 3.5 kW.

1st Case. Train is moving down the plane as shown in Fig. 4.19 with a constant velocity and hence it will not have any acceleration. So the net force acting on the train in the direction of motion should be zero. But the forces acting on the train in the direction of motion are :

(*i*) $W \sin \theta$ in the direction of motion

(*ii*) Force of friction F in the opposite direction of motion

(*iii*) Force (P) exerted by engine* in the direction of motion.

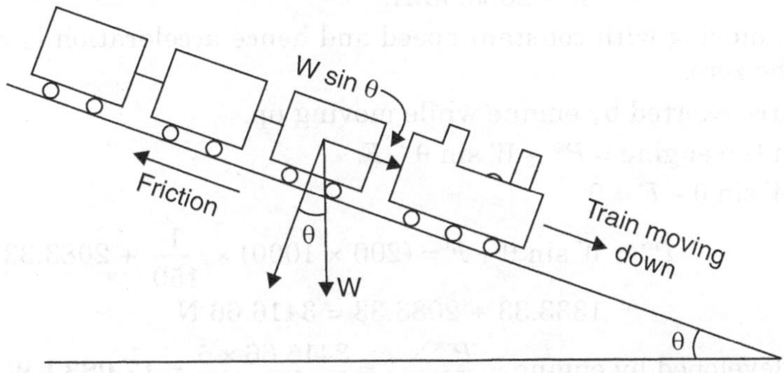

Fig. 4.19

Now the net force acting on engine in the direction of motion

$$= W \sin \theta - F + P$$

$$\therefore \quad W \sin \theta - F + P = 0$$

$$\text{*Power in kW} = \frac{\text{Force in N} \times \text{Speed in m/s}}{1000}.$$

or $$(200 \times 1000) \times \frac{1}{150} - F + P = 0 \quad \text{or} \quad 1333.33 - F + P = 0 \quad ...(i)$$

But power developed by engine

$$= \frac{(\text{Force exerted by engine in N}) \times u}{1000} \quad \text{or} \quad 3.5 = \frac{P \times 5}{1000}$$

$$\therefore \quad P = \frac{3.5 \times 1000}{5} = 750 \text{ N.}$$

Substituting the value of P in equation (i), we get

$$1333.33 - F + 750 = 0$$

or $$F = 1333.33 + 750 = 2083.33 \text{ N.}$$

2nd Case. Train is moving up the same plane at the same speed as shown in Fig. 4.20. As the plane is same and hence frictional force will remain same.

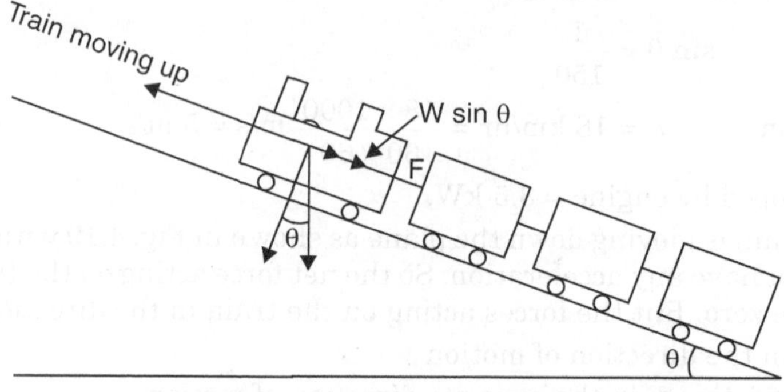

Fig. 4.20

$$\therefore \quad F = 2083.33 \text{ N.}$$

The train is moving with constant speed and hence acceleration is zero. And net force on engine should be zero.

Let P^* = Force exerted by engine while moving up.

Net force on the engine = $P^* - W \sin \theta - F$.

$$\therefore \quad P^* - W \sin \theta - F = 0$$

or $$P^* = W \sin \theta + F = (200 \times 1000) \times \frac{1}{150} + 2083.33$$

$$= 1333.33 + 2083.33 = 3416.66 \text{ N.}$$

$\therefore$ Power developed by engine $= \dfrac{P^* \times u}{1000} = \dfrac{3416.66 \times 5}{1000}$ = **17.083 kW. Ans.**

Problem 4.50. *At a mine, a train of small dump cars is pulled along a level track from A to B as shown in Fig. 4.21 and then up a 5% grade. At C, 100 m up the grade the last car*

---

*In problem 4.48, when the truck was moving down the plane engine was not running and hence force exerted by engine was zero. In this case engine is working and hence force will be exerted by engine.

becomes detached while its velocity is 24 km/hour. If the car weighs 20 kN and the track resistance is 8 N per kN, determine :

(i) How far up the grade will the car continue before stopping at D ?

(ii) Will the car stop at D or move again down the grade ?

(iii) If the car moves again down the grade, what will be the distance travelled by the car beyond B on the level track before stopping at E ? Take g = 9.8 m/s².

Sol. Given :

Inclination of BD = 5% grade

This means if θ is the inclination of the track BD to the horizontal, then

$$\sin\theta = \frac{5}{100} = 0.05$$

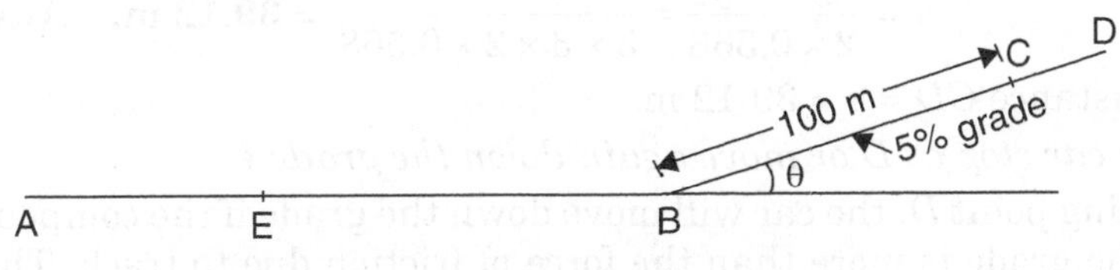

Fig. 4.21

Distance, $BC = 100$ m

Velocity at $C = 24 \text{ km/hr} = \frac{24 \times 1000}{60 \times 60} \text{ m/s} = \frac{20}{3} \text{ m/s}.$

Weight of car, $W = 20 \text{ kN} = 20 \times 1000 \text{ N}$

Mass of car, $m = \frac{W}{g} = \frac{20 \times 1000}{9.81}$ kg

Frictional resistance due to track,

$= 8$ N per kN weight of car $= 8 \text{ N} \times 20 = 160$ N.

(i) How far up the grade from C will the car continue before stopping at D ?

Consider the motion from C to D

Initial velocity of car at C, $u = \frac{20}{3}$ m/s

Final velocity of car at D, $v = 0$

Let s = Distance from C to D

a = Retardation of the car from C to D.

Component of the weight of the train along the grade

$$= W \times \sin\theta = 20 \times 1000 \times \frac{5}{100} = 1000 \text{ N}.$$

When the car is moving up the grade, the frictional resistance due to track and the component of the weight of the train along the grade are acting in the direction opposite to the direction of motion of the car.

∴ Total retarding force against motion

= Frictional resistance due to track + Component of the weight of train along the grade

= 160 + 1000 = 1160 N

But total retarding force = Mass × Retardation $(\because \quad F = m \times a)$

or $$1160 = \left(\frac{20 \times 1000}{9.8}\right) \times a \qquad \left(\because \text{ Mass} = \frac{W}{g} = \frac{20 \times 1000}{9.80}\right)$$

$$\therefore \quad a = \frac{1160 \times 9.8}{20 \times 1000} = 0.568 \text{ m/s}^2.$$

Now using the relation,

$$v^2 = u^2 - 2as \qquad \text{(–ve sign due to retardation)}$$

or $$0 = \left(\frac{20}{3}\right)^2 - 2 \times 0.568 \times s$$

or $$s = \frac{\frac{20}{3} \times \frac{20}{3}}{2 \times 0.568} = \frac{20 \times 20}{3 \times 3 \times 2 \times 0.568} = \textbf{39.12 m.} \quad \textbf{Ans.}$$

or Distance $CD = s = 39.12$ m.

(ii) Will the car stop at D or move again down the grade ?

After reaching point D, the car will move down the grade if the component of the weight of the car along the grade is more than the force of friction due to track. The force of friction along the grade is acting up the plane whereas the component of the weight of the car along the grade is acting down the plane.

But force of friction = 160 N

and component of the weight of the car along the grade = 1000 N.

As the component of the weight of the car along the grade is more than the force of friction, the car will move down from D. **Ans.**

(iii) Distance travelled by car from B to E on level track while moving down the grade.

First consider the motion from point D to point B.

Initial velocity at D, $u = 0$

Final velocity at $B = v$ (assumed)

Distance BD, s = Distance CD + Distance BC

= 39.12 + 100 = 139.12 m.

Let a' = Acceleration when car is moving from D to B.

When the car is moving down from D to B, the frictional resistance of 160 N is acting upward while the component of the weight of the car along the grade is acting downward. Hence the net downward force along the grade which produces acceleration a' is (1000 N – 160 N) or 840 N.

$\therefore$ Net force down the grade = 840 N.

Now using the relation, Force = Mass × Acceleration

or $$840 = \left(\frac{20 \times 1000}{9.80}\right) \times a' \qquad \left(\because \text{ Mass} = \frac{W}{g} = \frac{20 \times 1000}{9.80}\right)$$

$$\therefore \quad a' = \frac{840 \times 9.80}{20 \times 1000} = 0.4116 \text{ m/s}^2$$

Now using the relation,

$$v^2 = u^2 + 2a's \qquad \text{(where } a' = \text{acceleration)}$$

or
$$v^2 = 0 + 2 \times 0.4116 \times 139.12$$

$\therefore$
$$v = \sqrt{2 \times 0.4116 \times 139.12} = 10.70 \text{ m/s.}$$

$\therefore$ Velocity at point $B = 10.70$ m/s.

Consider the motion from B to E

Now, from B to E the track is level. The weight of the car is acting vertically downward. The component of the weight of the car along the horizontal plane is zero. The only force acting on the car, when it is moving from B to E, is frictional force. This frictional force of 160 N will oppose the motion of the car and finally the car will stop at E. The frictional force will produce retardation.

Let a^* = Retardation from B to E

Using, Force = Mass × Retardation

$\therefore$
$$160 = \left(\frac{20 \times 1000}{9.80}\right) \times a^*$$

$\therefore$
$$a^* = \frac{160 \times 9.80}{20 \times 1000} = 0.0784 \text{ m/s}^2.$$

For motion from B to E

Initial velocity at $B = 10.70$ m/s

Final velocity at $E = 0$

Retardation, $a^* = 0.0784$.

Let s^* = Distance from B to E.

Now using
$$v^2 = u^2 - 2a^*s^* \qquad \text{(–ve sign is due to retardation)}$$

$$0 = 10.70^2 - 2 \times 0.78 \times s^*$$

$\therefore$
$$s^* = \frac{10.70 \times 10.70}{2 \times 0.078} = 733.91 \text{ m}$$

$\therefore$ Distance BE = **733.91 m. Ans.**

4.8.7. Analysis of Lift Motion. Fig. 4.22 shows a lift (elevator or cage) carrying some weight and moving with a uniform acceleration.

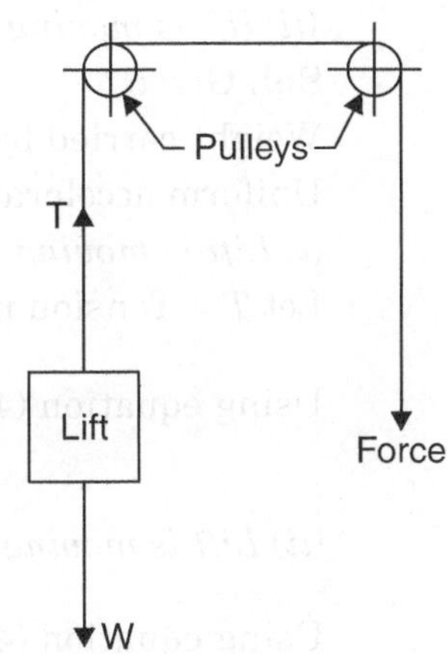

Fig. 4.22

Let W = Weight carried by the lift

m = Mass carried by lift = $\dfrac{W}{g}$

a = Uniform acceleration of the lift

T = Tension in cable supporting the lift. This is also called the *reaction of the lift.

The lift may be moving upwards or moving downwards.

*Reaction of lift is equal to the tension (T) in cable supporting the lift.

1st Case. Let the lift is moving upwards as shown in Fig. 4.22 (*a*). The weight carried by lift is acting downwards while the tension in the cable is acting upwards. As the lift is moving up, the net force which is equal to $(T - W)$ is acting upwards.

$\therefore$ Net force in upward direction = $T - W$.

This net force produces an acceleration 'a'

Hence using,

Net force = Mass × Acceleration

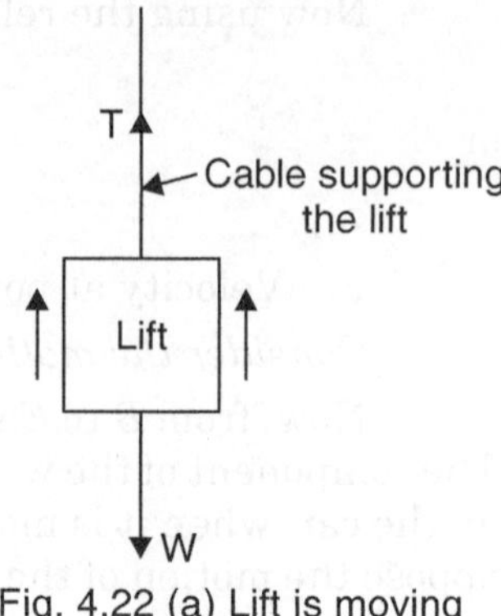

Fig. 4.22 (a) Lift is moving upwards

or $$(T - W) = \left(\frac{W}{g}\right) \times a \qquad \left(\because \text{ Mass} = \frac{W}{g}\right)$$

or $$T = W + \frac{W}{g} \times a = W\left(1 + \frac{a}{g}\right) \qquad ...(4.23)$$

2nd Case. As the lift is moving downwards as shown in Fig. 4.22 (*b*), the net force is acting downwards. Hence in this case W is more than T (tension in string).

$\therefore$ Net force in downward direction = $(W - T)$.

This net force produces an acceleration 'a'.

Hence using, Net force = Mass × Acceleration

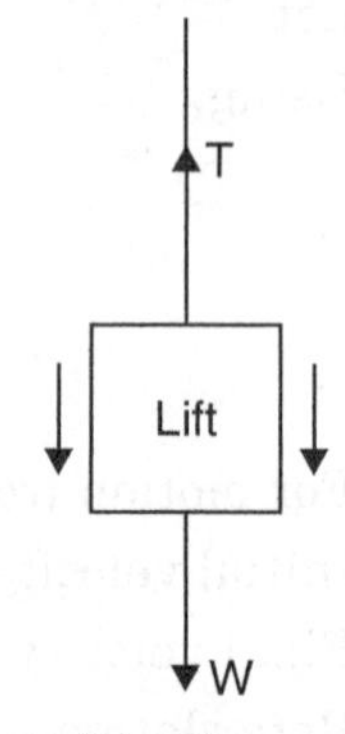

Fig. 4.22 (b) Lift is moving downwards

or $$(W - T) = \frac{W}{g} \times a$$

or $$T = W - \frac{W}{g} \times a$$

$$= W\left(1 - \frac{a}{g}\right) \qquad ...(4.24)$$

Problem 4.51. *A lift carries a weight of 100 N and is moving with a uniform acceleration of 2.45 m/s². Determine the tension in the cables supporting the lift, when :*

(*i*) *lift is moving upwards, and*

(*ii*) *lift is moving downwards. Take g = 9.80 m/s².*

Sol. Given :

Weight carried by lift, $W = 100$ N.

Uniform acceleration, $a = 2.45$ m/s².

(*i*) *Lift is moving upwards*

Let T = Tension in the cables supporting the lift.

Using equation (4.23), $T = W\left(1 + \frac{a}{g}\right) = 100\left(1 + \frac{2.45}{9.80}\right)$

$= 100\ (1.25) =$ **125 N. Ans.**

(*ii*) *Lift is moving downwards*

Using equation (4.24), $T = W\left(1 - \frac{a}{g}\right) = 100\left(1 - \frac{2.45}{9.80}\right) = 100(1 - 0.25)$

$= 100 \times 0.75 =$ **75 N. Ans.**

Problem 4.52. *A lift has an upward acceleration of 1.225 m / s². What pressure will a man weighing 500 N exert on the floor of the lift ? What pressure would he exert if the lift had an acceleration of 1.225 m / s² downwards ? What upward acceleration would cause his weight to exert a pressure of 600 N on the floor ? Take g = 9.8 m / s².*

Sol. Given :

Upward acceleration, $a = 1.225 \text{ m/s}^2$

Weight of man, $W = 500$ N.

1st Case. The lift is moving up with an acceleration of 1.225 m/s². The pressure exerted by a man on the floor of the lift is equal to the reaction of the lift and it is the same as the tension in the cables supporting the lift.

Let T = Tension in the cables supporting the lift or the reaction of the lift or the pressure exerted by the man on the floor of the lift.

Using equation (4.23) for the lift moving upwards, we have

$$T = W\left(1+\frac{a}{g}\right) = 500\left(1+\frac{1.225}{9.80}\right) = 500(1 + 0.125) = \mathbf{562.5\ N.}\quad \textbf{Ans.}$$

2nd Case. The lift is moving downwards with an acceleration of 1.225 m/s². The pressure exerted by the man on the floor of the lift is equal to the reaction of the lift and it is the same as the tension in the cables supporting the lift.

Using equation (4.24),

$$T = W\left(1-\frac{a}{g}\right) = 500\left(1-\frac{1.225}{9.80}\right) = 500(1- 0.125) = \mathbf{437.5\ N.}\quad \textbf{Ans.}$$

3rd Case. The lift is moving upwards with an unknown acceleration.

Let a = Acceleration upwards

T = Pressure exerted by man on floor of lift = 600 N

W = Weight of man = 500 N.

Using equation (4.23), we get

$$T = W\left(1+\frac{a}{g}\right) \quad \text{or} \quad 600 = 500\left(1+\frac{a}{9.80}\right)$$

or $$\frac{600}{500} = 1+\frac{a}{9.80} \quad \text{or} \quad 1.2 = 1 + \frac{a}{9.80}$$

or $$a = (1.2 - 1.0) \times 9.80 = 0.2 \times 9.80 = \mathbf{1.96\ m/s^2.}\quad \textbf{Ans.}$$

Problem 4.53. *An elevator weighs 2500 N and is moving vertically downwards with a constant acceleration. Write the equation for the elevator cable tension. Starting from rest it travels a distance of 35 metres during an interval of 10 seconds. Find the cable tension during this time. Neglect all other resistances to motion. What are the limits of cable tension ?*

(AMIE Summer, 1979)

Sol. Given :

Weight of elevator, $W = 2500$ N

Initial velocity, $u = 0$

Distance travelled, $s = 35$ m

Time, $t = 10$ sec.

Let T = Tension in the cable supporting the elevator in N.

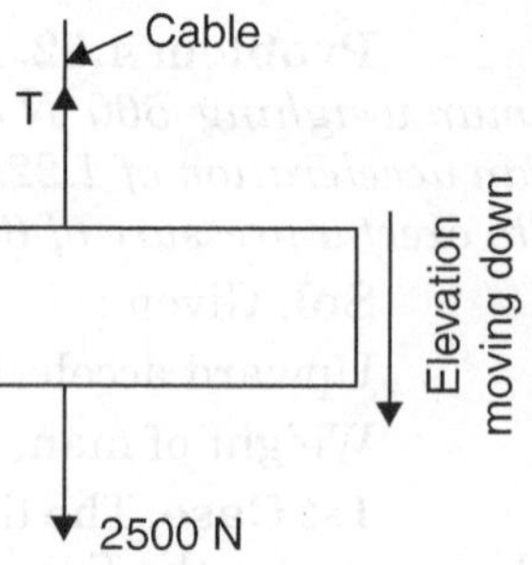

1st Part. The equation for the elevator cable tension is obtained as given below : (see Fig. 4.23). The elevator is moving down.

$\therefore$ Net acceleration force in the downward direction

$$= (W - T) = (2500 - T) \text{ N}.$$

The net accelerating force produces an acceleration 'a' in the downward direction.

Hence using the relation,

Net force = Mass × Acceleration.

Cable

T

Elevation moving down

2500 N

Fig. 4.23

$$(2500 - T) = \frac{2500}{9.81} \times a \qquad \left(\because \text{ Mass} = \frac{W}{g} = \frac{2500}{9.81}\right)$$

or
$$T = 2500 - \frac{2500}{9.81} \times a$$

$$\therefore \quad T = 2500\left(1 - \frac{a}{9.81}\right)\text{N} \quad \text{...}(i)$$

Hence the above equation (*i*) represents the equation for the elevator cable tension when the elevator is moving downwards.

2nd Part. Limits of cable tension are obtained from equation (*i*) as given below :

(*i*) When $a = 0$, and this value is substituted in equation (*i*), the value of T is obtained as

$$T = 2500\left(1 - \frac{0}{9.81}\right) = 2500 \text{ N}.$$

(*ii*) When $a = 9.81 \text{ m/s}^2$, and this value is substituted in equation (*i*), the value of T is obtained as

$$T = 2500\left(1 - \frac{9.81}{9.81}\right) = 2500\,(1 - 1) = 0$$

$\therefore$ Limits of cable tension (T) are :

At $a = 0,$ $T = 2500$ N

At $a = 9.81,$ $T = 0.$ **Ans.**

3rd Part. $s = 35$ m, $u = 0$ and $t = 10$ seconds.

Using the relation, $s = ut + \frac{1}{2}at^2$

or
$$35 = 0 \times 10 + \tfrac{1}{2} \times a \times 10^2 = 0 + 50\,a$$

$$\therefore \quad a = \frac{35}{50} = 0.7 \text{ m/s}^2.$$

Substituting this value of f in equation (*i*), we get

$$T = 2500\left(1 - \frac{0.7}{9.81}\right) = \mathbf{2321.61\ N.} \quad \textbf{Ans.}$$

Problem 4.54. *A cage, carrying 10 men each weighing 500 N, starts moving downwards from rest in a mine vertical shaft. The cage attains a speed of 12 metres/s in 20 metres. Find the pressure exerted by each man on the floor of the cage. Take g = 9.80 m/s².*

Sol. Given :

Weight of one man, $= 500$ N

Total weight of 10 men on the cage, $W = 500 \times 10 = 5000$ N

Initial velocity of cage, $u = 0$

Final velocity of cage, $v = 12$ m/s

Distance travelled, $s = 20$ m

Acceleration due to gravity, $g = 9.80$ m/s^2.

The pressure exerted by the man on the floor of the cage will be same as the tension produced in the cables supporting the cage.

Let T = Total tension produced by 10 men in the cables.

a = Acceleration of the cage.

The acceleration will be obtained by using the relation,

$$v^2 - u^2 = 2as \quad \text{or} \quad 12^2 - 0 = 2a \times 20$$

$$\therefore \quad a = \frac{12 \times 12}{2 \times 20} = 3.6 \text{ m/s}^2.$$

The cage is moving downwards, hence using equation (4.24),

$$T = W\left(1 - \frac{a}{g}\right) \quad \text{or} \quad T = 5000\left(1 - \frac{3.6}{9.8}\right)$$

$$= 5000\,(1 - 0.3673) = 3163.26 \text{ N}.$$

$\therefore$ Tension produced by one man in the cable,

$$= \frac{T}{10} = \frac{3163.26}{10} = 316.326 \text{ N}.$$

But tension produced by each man in the cable is the same as the pressure exerted by each man on the floor of the cage.

$\therefore$ Pressure exerted by each man on the floor = **316.326 N. Ans.**

Problem 4.55. *An elevator weighing 5000 N is ascending with an acceleration of 3 m/s^2. During this ascent its operator whose weight is 700 N is standing on the scales placed on the floor. What is the scale reading ? What will be the total tension in the cables of the elevator during this motion ?* (AMIE Summer, 1982)

Sol. Given :

Weight of the elevator, $W_1 = 5000$ N

Acceleration of elevator, $a = 3$ m/s^2

Weight of the operator, $W_2 = 700$ N

When the operator is standing on the scale, placed on the floor of the elevator, the reading of the scale will be equal to the reaction (R) offered by the floor on the operator.

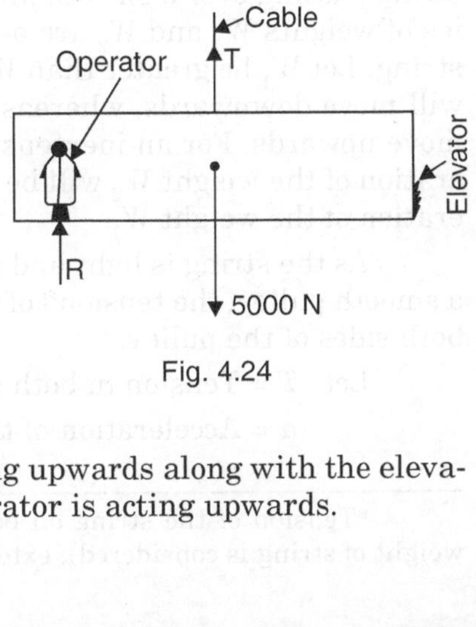

Hence let R = Reaction offered by floor on operator. This is also equal to the reading of scale.

T = Total tension in the cables of elevator.

Fig. 4.24

Consider the motion of operator. The operator is moving upwards along with the elevator with an acceleration $a = 3$ m/s^2. The net force on the operator is acting upwards.

∴ Net upward force on operator

= Reaction offered by floor on operator – Weight of operator

= $(R - 700)$

Mass of operator $= \dfrac{\text{Weight of operator}}{g} = \dfrac{700}{g}$

But, Net force = Mass × Acceleration

$$\therefore \quad (R - 700) = \frac{700}{9.8} \times 3 \qquad (\because \text{ Acceleration} = 3 \text{ m/s}^2)$$

$$\therefore \quad R = 700 + \frac{700}{9.8} \times 3 = 700 + 214.28 = \mathbf{914.28\ N.}\quad \textbf{Ans.}$$

Scale reading is equal to the reaction, R

∴ Scale reading = **914.28 N. Ans.**

Total tension in the cables of elevator.

Let T = Total tension in the cables of elevator

W = Total weight (*i.e.*, weight of elevator + weight of operator)

= 5000 + 700 = 5700 N.

As the elevator with the operator is moving upwards with an acceleration $f = 3$ m/s², the net force will be acting on the elevator and operator in the upward direction.

∴ Net upward force on elevator and operator

= Total tension in the cables – Total weight of elevator and operator

= $(T - 5700)$

Mass of elevator and operator $= \dfrac{\text{Total weight}}{g} = \dfrac{5700}{9.80}$

But, Net force = Mass × Acceleration

$$\therefore \quad (T - 5700) = \frac{5700}{9.80} \times 3$$

$$\therefore \quad T = 5700 + \frac{5700}{9.80} \times 3 = 5700 + 1745 = \mathbf{7445\ N.}\quad \textbf{Ans.}$$

4.8.8. Analysis of the Motion of Two Bodies Connected by a String. Fig. 4.25 shows a light and inextensible string passing over a smooth and weightless pulley. Two bodies of weights W_1 and W_2 are attached to the two ends of the string. Let W_1 be greater than W_2. As $W_1 > W_2$, the weight W_1 will move downwards, whereas the smaller weight (W_2) will move upwards. For an inextensible string, the upward acceleration of the weight W_2 will be equal to the downward acceleration of the weight W_1.

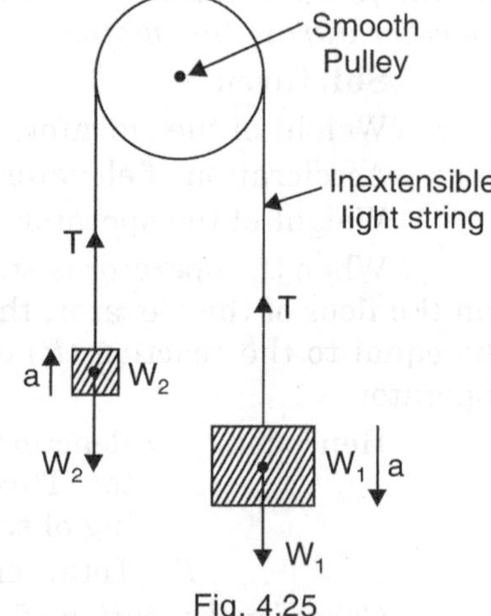

Fig. 4.25

As the string is light and inextensible and passing over a smooth pulley, the tension* of the string will be the same on both sides of the pulley.

Let T = Tension in both strings,

a = Acceleration of the bodies.

*Tension of the string on both sides of the pulley will be different if the string is heavy (*i.e.*, weight of string is considered), extensible and does not pass over a smooth pulley.

Consider the motion of weight, W_1. The weight W_1 is moving downwards with an acceleration a. The forces acting on W_1 are (*i*) its weight W_1 acting downwards and (*ii*) tension T acting upwards. As the weight W_1 is moving downwards, hence net force on the weight W_1 is acting downwards.

$\therefore$ Net downward force = $(W_1 - T)$

But net force = Mass × Acceleration

$$\therefore \quad (W_1 - T) = \frac{W_1}{g} \times a \qquad \left(\because \text{ Mass} = \frac{\text{Weight}}{g} = \frac{W_1}{g}\right) \qquad ...(i)$$

Now consider the motion of weight, W_2. The forces acting on W_2 are : (*i*) its weight W_2 acting downwards and (*ii*) tension T acting upwards. But the weight W_2 is moving upwards, hence net force on weight W_2, is acting upwards.

$\therefore$ Net upward force = $(T - W_2)$

But net upward force = Mass × Acceleration

$$\text{or} \quad (T - W_2) = \frac{W_2}{g} \times a \qquad \left(\because \text{ Mass} = \frac{W_2}{g}\right) \qquad ...(ii)$$

Adding equations (*i*) and (*ii*), we get

$$(W_1 - W_2) = \frac{a}{g}(W_1 + W_2)$$

$$\text{or} \quad a = \frac{g(W_1 - W_2)}{(W_1 + W_2)} \qquad ...(4.25)$$

Equation (4.25) is used for finding the acceleration. If the value of this acceleration is substituted either in equation (*i*) or in equation (*ii*) the value of tension (T) is obtained.

Hence substituting the value of 'a' in equation (*ii*), we get

$$(W_1 - T) = \frac{W_1}{g} \times g \frac{(W_1 - W_2)}{(W_1 + W_2)} = \frac{W_1(W_1 - W_2)}{(W_1 + W_2)} \qquad \text{(Cancelling } g)$$

$$\text{or} \quad T = W_1 - \frac{W_1(W_1 - W_2)}{(W_1 + W_2)} = W_1\left[1 - \frac{(W_1 - W_2)}{(W_1 + W_2)}\right]$$

$$= W_1\left[\frac{W_1 + W_2 - W_1 + W_2}{(W_1 + W_2)}\right] = \frac{2W_1W_2}{(W_1 + W_2)} \qquad ...(4.26)$$

Problem 4.56. *Two bodies of weight 50 N and 30 N are connected to the two ends of a light inextensible string. The string is passing over a smooth pulley. Determine :*

(i) The acceleration of the system, and

(ii) Tension in the string. Take $g = 9.80\ m/s^2$.

Sol. Given :

Heavier weight, $W_1 = 50$ N

Lighter weight, $W_2 = 30$ N

Let a = Acceleration of the system, and

T = Tension in the string.

Refer to Fig. 4.26 (*a*) *i.e.*, *FBD* of 50 N.

Net force = Mass × Acceleration

$$(50 - T) = \left(\frac{50}{g}\right) \times a \qquad ...(i)$$

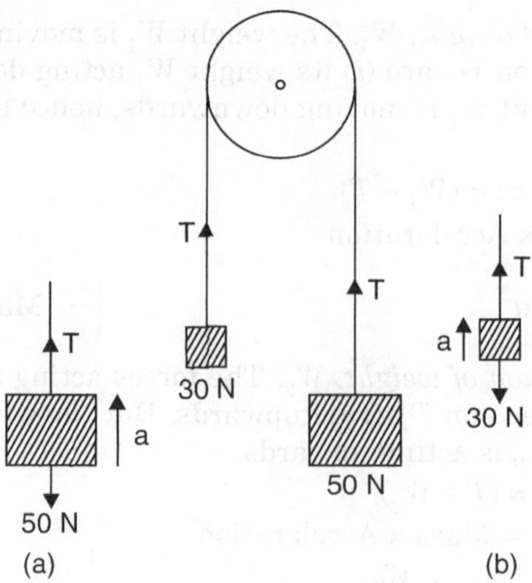

Fig. 4.26

Refer to Fig. 4.26 (*b*).

$$\text{Net force} = \text{Mass} \times \text{Acceleration}$$

$$(T - 30) = \left(\frac{30}{g}\right) \times a \qquad ...(ii)$$

Adding equations (*i*) and (*ii*), we get the value of a

$$20 = \frac{80}{g} \times a \quad \therefore \quad a = \frac{20 \times g}{80} = \frac{g}{4} = \frac{9.8}{4} = \mathbf{2.45\ m/s^2.}\ \ \textbf{Ans.}$$

Substitute this value in equation (*ii*), we get tension, T.

$$\therefore \qquad T - 30 = \frac{30}{g} \times 2.45$$

$$\therefore \qquad T = 30 + \frac{30}{9.8} \times 2.45$$

$$= 30 + 7.5 = \mathbf{37.5\ N.}\ \ \textbf{Ans.}$$

Problem 4.57. *Two bodies of different weights are connected to the two ends of a light inextensible string, which passes over a smooth pulley. If the acceleration of the system is 3 m/s² and bigger weight is 60 N, determine :*

(i) The smaller weight, and

(ii) Tension in the string. Take g = 9.80 m/s².

Sol. Given :

Acceleration, $a = 3$ m/s²

Bigger weight, $W_1 = 60$ N

Let W_2 = Smaller weight, and

T = Tension in the string.

(*i*) Using equation (4.25), we get

$$a = \frac{g(W_1 - W_2)}{(W_1 + W_2)} \quad \text{or} \quad 3 = \frac{9.80(60 - W_2)}{(60 + W_2)}$$

or

$$3(60 + W_2) = 9.80(60 - W_2)$$

or $$180 + 3W_2 = 9.80 \times 60 - 9.80\ W_2$$

or $$3W_2 + 9.80\ W_2 = 9.80 \times 60 - 180 \quad \text{or} \quad 12.80\ W_2 = 408$$

$$\therefore \quad W_2 = \frac{408}{12.80} = \mathbf{31.875\ N.} \quad \textbf{Ans.}$$

(*ii*) Tension in the string is obtained from equation (4.26),

$$T = \frac{2W_1W_2}{(W_1 + W_2)}$$

$$= \frac{2 \times 60 \times 31.875}{(60 + 31.875)} = \frac{2 \times 60 \times 31.875}{91.875} = \mathbf{41.632\ N.} \quad \textbf{Ans.}$$

Problem 4.58. *A pulley whose axis passes through the centre O, carries load as shown in Fig. 4.27. Neglecting the inertia of the pulley and assuming that the cord is inextensible, determine the acceleration of the block A. How much weight should be added to or taken away from the block A, if the acceleration of the block A is required to be g/3.0 downwards ?*

(AMIE Summer, 1981)

Sol. Given :

Bigger load, $W_1 = 700$ N

Smaller load, $W_2 = 500$ N

Let a = Acceleration of block A or the acceleration of the system.

Using equation (4.25),

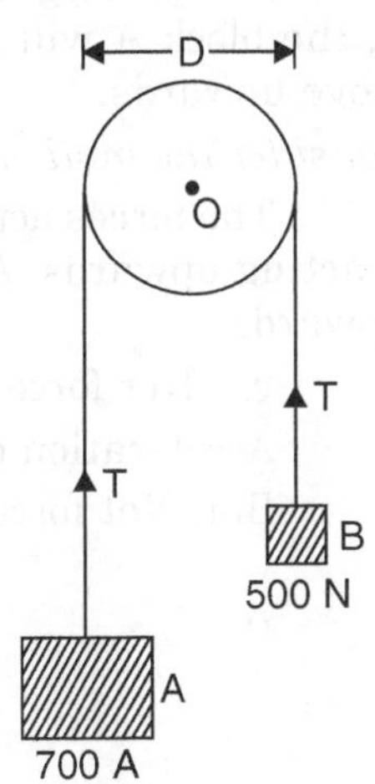

Fig. 4.27

$$a = \frac{g(W_1 - W_2)}{(W_1 + W_2)}$$

$$= \frac{g(700 - 500)}{(700 + 500)} = g\,\frac{200}{1200} = \frac{\mathbf{g}}{\mathbf{6}}. \quad \textbf{Ans.}$$

How much weight should be added to or taken away from the block A (i.e., from bigger load 700 N) when acceleration of bigger load is g / 3.0 downwards.

Let W_1^* = Total weight of block A when acceleration is $g/3.0$

$$a = \frac{g}{3} \quad W_2 = 500 \text{ N}$$

Using equation (4.25), $$a = \frac{g(W_1^* - W_2)}{(W_1^* + W_2)} \qquad (\because \text{ Here } W_1 = W_1^*)$$

or $$\frac{g}{3} = \frac{g(W_1^* - 500)}{(W_1^* + 500)}$$

or $$\frac{1}{3} = \frac{(W_1^* - 500)}{(W_1^* + 500)} \qquad \text{(Cancelling } g \text{ on both sides)}$$

or $$3(W_1^* - 500) = W_1^* + 500 \quad \text{or} \quad 3W_1^* - 1500 = W_1^* + 500$$

or $$2W_1^* = 2000$$

$$\therefore \quad W_1^* = \frac{2000}{2} = 1000 \text{ N.}$$

As W_1^* is more than W_1. Hence the weight must be added to the block A.

$\therefore$ Weight added $= W_1^* - W_1 = 1000 - 700 =$ **300 N. Ans.**

Problem 4.59. *Determine the tensions in the strings and accelerations of blocks A and B weighing 150 N and 50 N connected by a string and a frictionless and weightless pulley as shown in Fig. 4.28.* (AMIE Nov., 1970)

Sol. Given :

Weight of block A = 150 N

Weight of block B = 50 N.

As the pulley is smooth, the tension in the string will be same throughout.

Let T = Tension in the string

a = Acceleration of block B.

Then acceleration of block A will be equal to half the acceleration of block B.

$\therefore$ Acceleration of block $A = \dfrac{a}{2}$.

As the weight of block A is more than the weight of block B, the block A will move downwards whereas the block B will move upwards.

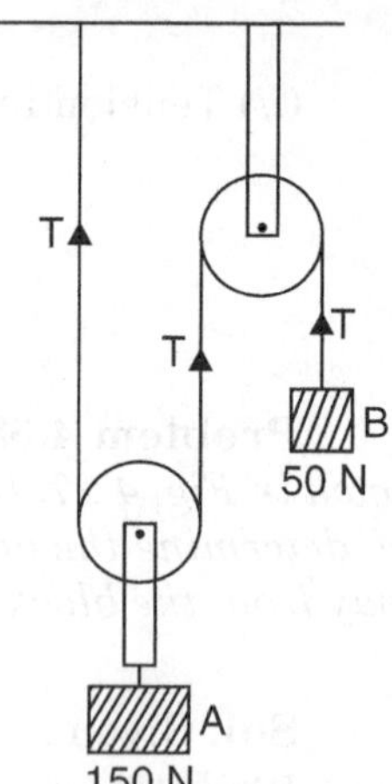

Fig. 4.28

Consider the motion of block B

The forces acting on block B are : (*i*) its weight 50 N acting downwards, and (*ii*) tension T acting upwards. As the block B is moving upwards, the net force on the block B is acting upwards.

$\therefore$ Net force on block $B = T - 50$

Acceleration of block $B = a$.

But, Net force = Mass × Acceleration

$$\therefore \quad T - 50 = \frac{50}{9.81} \times a \qquad \left(\because \text{ Mass} = \frac{\text{Weight}}{g} = \frac{50}{9.81}\right)$$

or
$$T = 50 + \frac{50}{9.81} \times a \qquad ...(i)$$

Consider the motion of block A

Two portions of the string are supporting A, so that the upward force on A is $2T$ and the downward force is 150 N. As the block A is moving downwards, the net force on the block A is acting downwards.

$\therefore$ Net force on block $A = 150 - 2\,T$.

Acceleration of block A = Half the acceleration of block $B = \dfrac{a}{2}$.

Using net force = Mass × Acceleration

$$150 - 2T = \frac{150}{9.81} \times \frac{a}{2} \qquad \left(\because \text{ Mass} = \frac{50}{9.81}\right)$$

$$2T = 150 - \frac{150}{9.81} \times \frac{a}{2} \qquad ...(ii)$$

Substituting the value of T from equation (*i*) into equation (*ii*), we get

$$2\left(50 + \frac{50}{9.81} \times a\right) = 150 - \frac{150}{9.81} \times \frac{a}{2}$$

or
$$100 + \frac{100}{9.81}\, a = 150 - \frac{75}{9.81} \times a$$

or
$$\left(\frac{100}{9.81} + \frac{75}{9.81}\right) \times a = 150 - 100 = 50 \quad \text{or} \quad \frac{175}{9.81} \times a = 50$$

$\therefore$
$$a = \frac{50 \times 9.81}{175} = 2.803 \text{ m/s}^2 \quad \text{...(iii)}$$

$\therefore$ Acceleration of block B = **2.803 m/s²**. **Ans.**

And acceleration of block $A = \frac{2.803}{2}$ = **1.4015 m/s²**. **Ans.**

Substituting the value of 'a' from equation (*iii*) into equation (*i*), we get

$$T = 50 + \frac{50}{9.81} \times 2.803 = 50 + 14.286 = \mathbf{64.286\ N.} \quad \textbf{Ans.}$$

Problem 4.60. *A system of weights connected by strings passing over pulleys A and B is shown in Fig. 4.29. Find the acceleration of the three weights, assuming weightless strings and ideal conditions for pulleys.* (AMIE Summer, 1977)

Sol. Given :

As the strings are weightless and ideal conditions prevail, hence the tensions in the string passing over pulley A will be same. The tensions in the string passing over pulley B will also be same. But the tensions in the strings passing over pulley A and over pulley B will be different as shown in Fig. 4.29.

Let T_1 = Tension in the string passing over pulley A, and

T_2 = Tension in the string passing over pulley B.

One end of the string passing over pulley A is connected to a weight of 15 N, and the other end is connected to pulley B. As the weight 15 N is more than the weights (6 + 4 = 10 N), hence weight 15 N will move downwards, whereas pulley B will move upwards. The acceleration of the weight 15 N and of the pulley B will be same.

Let a = Acceleration of 15 N weight in downward direction.

Then acceleration of pulley B in upward direction is also equal to 'a'.

The string passing over pulley B, has weights of 6 N and 4 N. Hence the weight 6 N will move downwards, whereas the weight 4 N will move upwards.

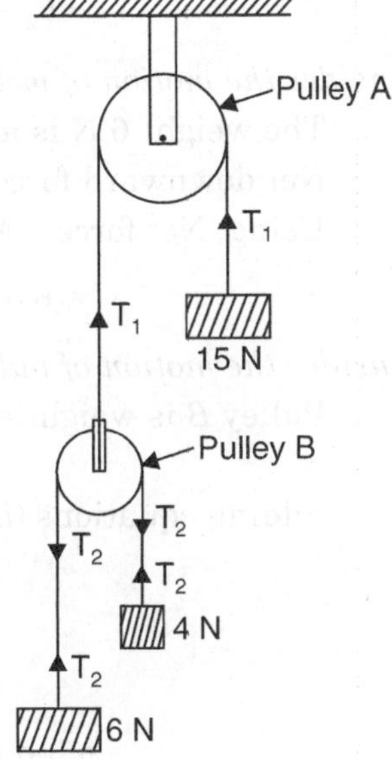

Fig. 4.29

Let 'a_1' = Acceleration of 6 N downwards with respect to pulley B.

Then acceleration of weight of 4 N with respect to pulley B

= a_1 in the upward direction.

$\therefore$ Absolute acceleration of weight 4 N.

$= $ Acceleration of 4 N w.r.t. pulley B + Acceleration of pulley B.

$= a_1 + a$ (upwards)

(As both the accelerations are in upward directions, total acceleration will be sum of the two accelerations).

Absolute acceleration of weight 6 N

$= $ Acceleration of 6 N w.r.t. pulley B + Acceleration of pulley B

$= (a_1 - a)$ downwards.

(As a_1 is acting downward whereas a is acting upward, hence total acceleration in the downward direction $= a_1 - a$).

Consider the motion of weight 15 N

The forces acting are : (*i*) 15 N downwards, and (*ii*) tension T_1 upwards. As the weight 15 N is moving downwards, the net force is acting downwards.

$\therefore$ Net downward force $= (15 - T_1)$.

Using, Net force = Mass × Acceleration

$$\therefore \quad (15 - T_1) = \frac{15}{9.80} \times a \qquad \left(\because \text{ Mass} = \frac{15}{g}\right) \qquad ...(i)$$

Consider the motion of weight 4 N

The weight 4 N is moving upwards with an acceleration of $(a_1 + a)$.

Net force on weight 4 N in upward direction $= (T_2 - 4)$.

Using, Net force = Mass × Acceleration

$$\therefore \quad (T_2 - 4) = \frac{4}{9.8} \times (a_1 + a) \qquad ...(ii)$$

Consider the motion of weight 6 N

The weight 6 N is moving downwards with an acceleration of $(a_1 - a)$.

Net downward force on weight 6 N $= 6 - T_2$.

Using, Net force = Mass × Acceleration

$$\therefore \quad 6 - T_2 = \frac{6}{9.8} \times (a_1 - a) \qquad ...(iii)$$

Consider the motion of pulley B

Pulley B is weightless,

$$\therefore \quad T_1 = 2T_2 \qquad ...(iv)$$

Adding equations (*ii*) and (*iii*), we get

$$2 = \frac{4}{9.8}(a_1 + a) + \frac{6}{9.8}(a_1 - a)$$

$$= \frac{4a_1 + 4a + 6a_1 - 6a}{9.8} = \frac{10a_1 - 2a}{9.8}$$

or $\quad 2 \times 9.80 = 2(5a_1 - a) \quad$ or $\quad 9.80 = 5a_1 - a \qquad ...(v)$

Multiplying equation (*ii*) by 2, we get

$$2T_2 - 8 = \frac{8}{9.80}(a_1 + a)$$

But from equation (*iv*), $2T_2 = T_1$. Substituting the value of $2T_2 = T_1$ in the above equation, we get

$$T_1 - 8 = \frac{8}{9.80}(a_1 + a) \quad ...(vi)$$

Adding equations (*i*) and (*vi*), we get

$$15 - 8 = \frac{15}{9.8} \times a + \frac{8}{9.80}(a_1 + a)$$

or
$$7 = \frac{15a + 8a_1 + 8a}{9.80} = \frac{23a + 8a_1}{9.80}$$

or
$$7 \times 9.80 = 23a + 8a_1 \quad ...(vii)$$

Multiplying equation (*v*) by 23, we get

$$23 \times 9.80 = -23a + 5 \times 23a_1 \quad ...(viii)$$

Adding equations (*vii*) and (*viii*), we get

$$7 \times 9.80 + 23 \times 9.80 = 8a_1 + 115\,a_1 \quad \text{or} \quad 30 \times 9.80 = 123a_1$$

$$\therefore \quad a_1 = \frac{30 \times 9.80}{123} = 2.39 \text{ m/s}^2.$$

Substituting the value of a_1 in equation (*v*), we get

$$9.80 = 5 \times 2.39 - a_1 \quad \text{or} \quad a = 5 \times 2.39 - 9.80$$

$$= 11.95 - 9.80 = 2.15 \text{ m/s}^2$$

$\therefore$ Acceleration of weight 15 N = a = **2.15 m/s²**. **Ans.**

Acceleration of weight 6 N = $a_1 - a = 2.39 - 2.15$ = **0.24 m/s²**. **Ans.**

Acceleration of weight 4 N = $a_1 + a = 2.39 + 2.15$ = **4.54 m/s²**. **Ans.**

4.8.9. Analysis of the Motion of two Bodies Connected by a String when one Body is Lying on a Horizontal Surface and other is Hanging Free

1. First case, when the horizontal surface is smooth and the string is passing over a smooth pulley. Fig. 4.30 shows the two weights W_1 and W_2 connected by a light inextensible string, passing over a smooth pulley. The weight W_2 is placed on a smooth horizontal surface, whereas the weight W_1 is hanging free.

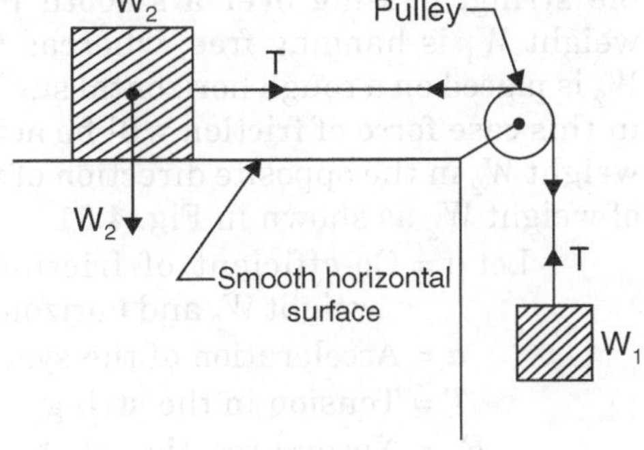

Fig. 4.30

The weight W_1 is moving downwards, whereas the weight W_2 is moving on smooth horizontal surface. The velocity and acceleration of W_1 will be same as that of W_2.

As the string is light and inextensible and passing over a smooth pulley, the tensions of the string will be same on both sides of the pulley.

Let T = Tension in the string

a = Acceleration of the weight W_1 and also of W_2

(*i*) *Consider the motion of the hanging weight* W_1

The weight W_1 is moving downwards with an acceleration a. The forces acting on W_1 are : (*i*) its weight W_1 acting downwards, and (*ii*) tension T acting upwards.

∴ Net downward force = $(W_1 - T)$.

Using, Net force = Mass × Acceleration

$$\therefore \quad (W_1 - T) = \frac{W_1}{g} \times a \quad \left(\because \text{ Mass} = \frac{\text{Weight}}{g}\right) \quad \ldots(i)$$

(ii) Consider the motion of weight W_2

The weight W_2 is moving on the horizontal surface with an acceleration of 'a'. As the weight W_2 is moving in the horizontal direction, the only force causing the motion is T. The weight W_2 is acting downwards and hence the component of this weight in horizontal direction is $W_2 \cos 90°$, which is zero.

Using, Force = Mass × Acceleration

or
$$T = \frac{W_2}{g} \times a \quad \left(\because \text{ Mass} = \frac{\text{Weight}}{g}\right) \quad \ldots(ii)$$

Adding equations (i) and (ii),

$$W_1 = \frac{W_1}{g} \times a + \frac{W_2}{g} \times a = \frac{a}{g}\,[W_1 + W_2]$$

$$\therefore \quad a = \frac{g \times W_1}{(W_1 + W_2)} \quad \ldots(4.27)$$

Equation (4.27) gives the acceleration of the system.

To find the tension (T) in the string, substitute the value of 'a' in equation (ii).

$$\therefore \quad T = \frac{W_2}{g} \times \frac{g \times W_1}{(W_1 + W_2)} = \frac{W_1 W_2}{(W_1 + W_2)}. \quad \ldots(4.28)$$

Equation (4.28) gives the tension in the string.

2. Second case when the horizontal surface is rough and string is passing over a smooth pulley. Fig. 4.31 shows the two weights W_1 and W_2 connected by a light inextensible string, passing over a smooth pulley. The weight W_1 is hanging free, whereas the weight W_2 is placed on a rough horizontal surface. Hence in this case force of friction will be acting on the weight W_2 in the opposite direction of the motion of weight W_2 as shown in Fig. 4.31.

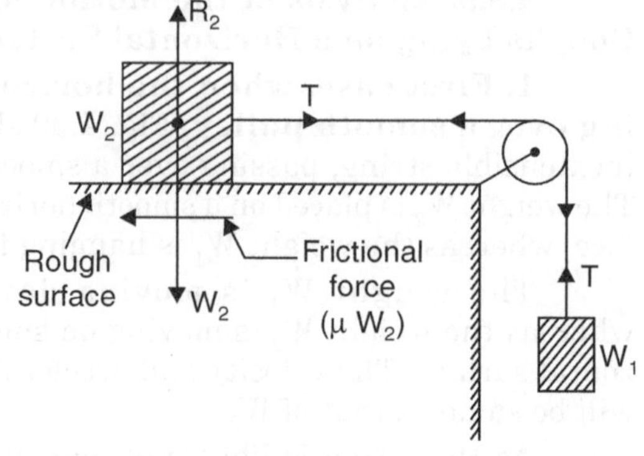

Fig. 4.31

Let μ = Co-efficient of friction between weight W_2 and horizontal surface

a = Acceleration of the system

T = Tension in the string

R_2 = Normal reaction at the horizontal rough surface = W_2

∴ Force of friction = μR_2

$= \mu W_2 \quad (\because R_2 = W_2) \quad \ldots(i)$

(i) Consider the motion of the hanging weight, W_1

The weight W_1 is moving downwards with an acceleration 'a'. The net downward force acting on weight $W_1 = (W_1 - T)$.

Using, Net force = Mass × Acceleration

$$W_1 - T = \frac{W_1}{g} \times a \qquad \left(\because \quad \text{Mass} = \frac{\text{Weight}}{g}\right) \qquad ...(ii)$$

(ii) Consider the motion of the weight, W_2.

The weight W_2 is moving on the rough horizontal surface towards right with an acceleration 'a'. The forces acting in the horizontal directions are : (*i*) tension (T) towards right, and (*ii*) force of friction = $\mu R_2 = \mu W_2$ towards left.

∴ Net horizontal force towards right = $T - \mu W_2$.

Using, Net force = Mass × Acceleration

or
$$T - \mu W_2 = \frac{W_2}{g} \times a \qquad ...(iii)$$

Adding equations (*ii*) and (*iii*),

$$W_1 - \mu W_2 = \frac{a}{g}(W_1 + W_2)$$

$$\therefore \quad a = \frac{g(W_1 - \mu W_2)}{(W_1 + W_2)} \text{ m/s}^2 \qquad ...(4.29)$$

Equation (4.29) is used to find the acceleration.

To find the tension T, substitute the value of 'a' in equation (*ii*).

$$\therefore \quad W_1 - T = \frac{W_1}{g} \times \frac{g(W_1 - \mu W_2)}{(W_1 + W_2)} = \frac{W_1(W_1 - \mu W_2)}{(W_1 + W_2)}$$

$$\therefore \quad T = W_1 - \frac{W_1(W_1 - \mu W_2)}{(W_1 + W_2)} = W_1\left[1 - \frac{(W_2 - \mu W_2)}{(W_1 + W_2)}\right]$$

$$= W_1\left[\frac{W_1 + W_2 - (W_1 - \mu W_2)}{(W_1 + W_2)}\right] = \frac{W_1}{(W_1 + W_2)}(W_1 + W_2 - W_1 + \mu W_2)$$

$$= \frac{W_1(W_2 + \mu W_2)}{(W_1 + W_2)} = \frac{W_1 W_2(1 + \mu)}{(W_1 + W_2)}. \qquad ...(4.30)$$

Problem 4.61. *Two bodies of weight 20 N and 10 N are connected to the two ends of a light inextensible string, passing over a smooth pulley. The weight of 20 N is placed on a smooth horizontal surface while the weight of 10 N is hanging free in air. Find :*

(i) the acceleration of the system, and

(ii) the tension in the string. Take $g = 9.81\ m/s^2$.

Sol. Given :

Weight placed on horizontal surface, $W_2 = 20$ N

Weight hanging free in air, $W_1 = 10$ N

Acceleration due to gravity, $g = 9.81$ m/s²

Let a = Acceleration of the system, and

T = Tension in the string.

The horizontal surface is smooth. Hence the acceleration and tension are obtained by using equations (4.27) and (4.28).

Using equation (4.27) for acceleration, we have

$$a = \frac{gW_1}{(W_1 + W_2)} = \frac{9.81 \times 10}{(10 + 20)} = \mathbf{3.27\ m/s^2}. \quad \textbf{Ans.}$$

Using equation (4.28) for tension, we have

$$T = \frac{W_1 W_2}{(W_1 + W_2)} = \frac{10 \times 20}{(10 + 20)} = \frac{200}{30} = \mathbf{6.67\ N.}\quad \textbf{Ans.}$$

Problem 4.62. *If in the problem 4.61, the horizontal surface is a rough one, having co-efficient of friction between the weight 20 N and the plane surface equal to 0.3, determine :*

(i) the acceleration of the system, and

(ii) the tension in the string.

Sol. Given : (From Problem 4.61)

$$W_1 = 10\text{ N},\ W_2 = 20\text{ N},\ g = 9.81\text{ m/s}^2.$$

Co-efficient of friction between weight 20 N and horizontal surface,

$\mu = 0.3$

Let a = Acceleration of the system, and

T = Tension in the string.

For a rough horizontal surface, the acceleration and tension are obtained by using equations (4.29) and (4.30).

Using equation (4.29) for acceleration, we have

$$a = \frac{g(W_1 - \mu W_2)}{(W_1 + W_2)} = \frac{9.81(10 - 0.3 \times 20)}{(10 + 20)} = \frac{9.81 \times 4}{30} = \mathbf{1.308\ m/s^2.}\quad \textbf{Ans.}$$

Using equation (4.30) for tension, we have

$$T = \frac{W_1 W_2 (1 + \mu)}{(W_1 + W_2)} = \frac{10 \times 20(1 + 0.3)}{(10 + 20)} = \frac{200 \times 1.3}{30} = \mathbf{8.66\ N.}\quad \textbf{Ans.}$$

Problem 4.63. *Two blocks shown in Fig. 4.32 (a) have weights A = 20 N and B = 10 N and coefficient of friction between the block A and the horizontal plane is μ = 0.25. If the system is released from the rest and the block B falls through a vertical distance of 2 m, what is the velocity attained by block B. Neglect the friction in the pulley and the extension of the string.*

Sol. Given :

Weight of block A, $W_2 = 20$ N

Weight of block B, $W_1 = 10$ N

Coefficient of friction, $\mu = 0.25$

Distance moved by block B, $s = 2$ m

Initial velocity of block B, $u = 0$.

Let v = Final velocity of block B

a = Acceleration of the system or acceleration of block B.

Fig. 4.32 (a)

Using equation (4.29) for acceleration, we have

$$a = \frac{g(W_1 - \mu W_2)}{(W_1 + W_2)}$$

$$= \frac{9.81(10 - 0.25 \times 20)}{(10 + 20)} = \frac{9.81 \times 5}{30} = 1.635\text{ m/s}^2.$$

Now using the relation,

$$v^2 = u^2 + 2as = 0 + 2 \times 1.635 \times 2 = 6.54$$

$$\therefore \quad v = \sqrt{6.54} = \mathbf{2.557\ m/s.}\quad \textbf{Ans.}$$

Problem 4.64. *Two bodies of weights 10 N and 1.5 N are connected to the two ends of a light inextensible string, passing over a smooth pulley. The weight 10 N is placed on a rough horizontal surface while the weight of 1.5 N is hanging vertically in air. Initially the friction between the weight 10 N and the table is just sufficient to prevent motion. If an additional weight of 0.5 N is added to the weight 1.5 N, determine :*

(i) the acceleration of the two weights, and

(ii) tension in the string after adding additional weight of 0.5 N to the weight 1.5 N. Take $g = 9.80\ m/s^2$.

Sol. Given :

Weight placed on rough horizontal surface,

$$W_2 = 10\text{ N}$$

Weight hanging free in air, $W_1{}^* = 1.5$ N

Additional weight added $= 0.5$ N

∴ Total weight hanging in air in second case

$W_1 = 1.5 + 0.5 = 2.0$ N.

Let T = Tension in string when hanging weight is 1.5 N

R = Normal reaction between the weight 10 N and the table surface

T_1 = Tension in string when hanging weight is 2.0 N

a = Acceleration of the system when hanging weight is 2.0 N

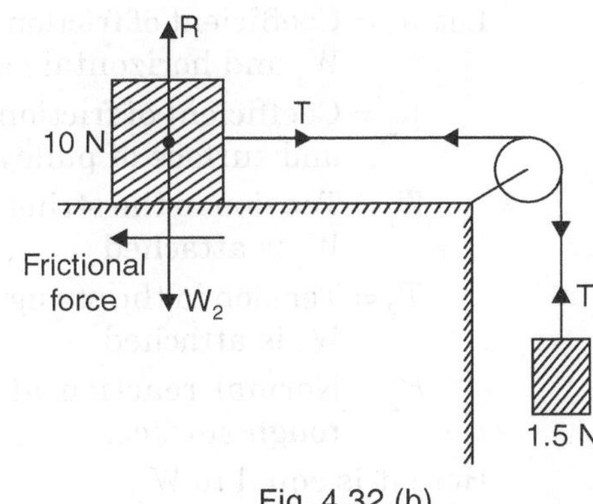

Fig. 4.32 (b)

Initially, the friction between the weight 10 N and the table is just sufficient to prevent motion.

∴ Max. Frictional Force, $F = T = 1.5$ N.

But frictional force, $F = \mu R$

or $1.5 = \mu \times 10$ $\quad (\because \ R = W_2 = 10\text{ N})$

$$\therefore \quad \mu = \frac{1.5}{10} = 0.15$$

where μ is the coefficient of friction.

When an additional weight 0.5 N is added to the hanging weight 1.5 N, the system starts moving with an acceleration '*a*'.

(i) Acceleration of the two weights

Using equation (4.29), we have

$$a = \frac{g(W_1 - \mu W_2)}{(W_1 + W_2)} = \frac{9.80(2.0 - 0.15 \times 10)}{(2.0 + 10)}$$

$$= \frac{9.8 \times 0.5}{12} = \mathbf{0.408\ m/s^2.} \quad \textbf{Ans.}$$

(ii) Tension in the string

Using equation (4.30), we have

$$T_1 = \frac{W_1 W_2 (1 + \mu)}{(W_1 + W_2)} = \frac{2 \times 10(1 + 0.15)}{(2 + 10)} \quad (\because \ \mu = 0.15)$$

$$= \frac{20 \times 1.15}{12} = \mathbf{1.916\ N.} \quad \textbf{Ans.}$$

3. Third case when the horizontal surface is rough and the string passes over a rough pulley. Fig. 4.33 shows the two weights W_1 and W_2 connected by a string passing over a *rough* pulley. The weight W_1 is hanging free, whereas the weight W_2 is placed on a rough horizontal surface. Hence in this case force of friction will be acting on the weight W_2 in the opposite direction of the motion of weight W_2 as shown in Fig. 4.33. As the string is passing over a rough pulley, the tensions on both sides of the string will not be same.

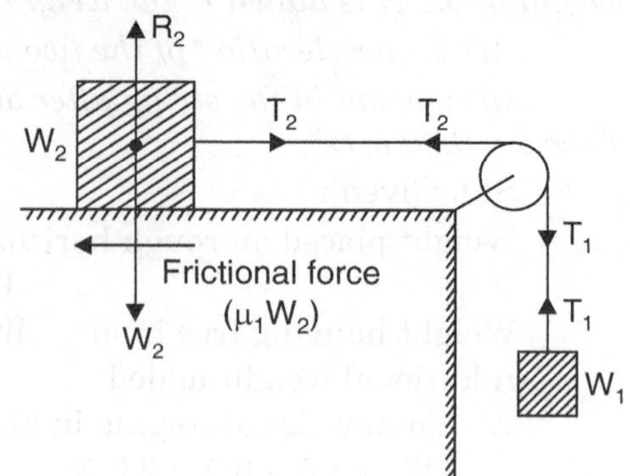

Fig. 4.33

Let μ_1 = Coefficient of friction between weight W_2 and horizontal surface

μ_2 = Coefficient of friction between string and surface of pulley

T_1 = Tension in the string to which weight W_1 is attached

T_2 = Tension in the string to which weight W_2 is attached.

R_2 = Normal reaction at the horizontal rough surface.

Here it is equal to W_2.

$\therefore$ Force of friction $= \mu_1 R_2 = \mu_1 W_2$ $\quad (\because R_2 = W_2)$ ...(*i*)

(*i*) *Consider the motion of the hanging weight,* W_1

The weight W_1 is moving downward with an acceleration 'a'. The net downward force acting on weight W_1 is equal to $(W_1 - T_1)$.

But, Net force = Mass × Acceleration

$$\therefore \quad (W_1 - T_1) = \left(\frac{W_1}{g}\right) \times a \qquad \left(\because \text{ Mass} = \frac{\text{Weight}}{g}\right) \qquad ...(ii)$$

(*ii*) *Consider the motion of the weight,* W_2

The weight W_2 is moving on the rough horizontal surface towards right with an acceleration 'a'. The forces acting in the horizontal direction are : (*i*) Tension T_2 towards right and (*ii*) force of friction $= \mu_1 R_2 = \mu_1 W_2$ towards left.

$\therefore$ Net force towards right $= (T_2 - \mu_1 W_2)$

Now, Net force = Mass × Acceleration

$$(T_2 - \mu_1 W_2) = \left(\frac{W_2}{g}\right) \times a \qquad ...(iii)$$

In equations (*ii*) and (*iii*), there are three unknowns *i.e.*, T_1, T_2 and 'a'. We require one more equation. This equation is given by,

$$\frac{T_1}{T_2} = e^{\mu_1 \times \theta} = e^{\mu_2 \times \theta} \qquad ...(iv)$$

In equation (*iv*), the angle θ* is angle of lap and is used in radians. In this particular case, from the geometry of Fig. 4.33, the value of $\theta = 90° = \frac{\pi}{2}$ radians.

From the three equations [*i.e.*, equations (*ii*), (*iii*) and (*iv*)], three unknown (*i.e.*, T_1, T_2 and acceleration 'a') can be obtained.

Problem 4.65. *Two bodies of weight 1500 N and 1000 N are connected to the two ends of a string, passing over a rough pulley. The weight of 1500 N is placed on a rough horizontal surface while the weight 1000 N is hanging free in air. If the coefficient of kinetic friction is 0.20 at all contact surfaces, find :*

(*i*) *the acceleration of the system, and*

(*ii*) *tensions in the string.*

Sol. Given :

Weight of body A, $W_2 = 1500$ N

Weight of body B, $W_1 = 1000$ N

Coefficient of kinetic friction for all contact surfaces, $\mu = \mu_1 = \mu_2 = 0.2$

Let a = Acceleration of the system

T_1 = Tension in the string to which weight 1000 N is attached

T_2 = Tension in the string to which weight 1500 N is attached

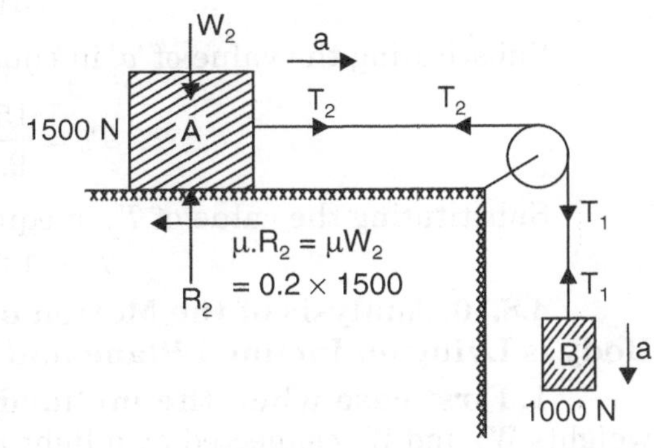

Fig. 4.33 (a)

We know that

$$\frac{T_1}{T_2} = e^{\mu\theta} = e^{0.2 \times \frac{\pi}{2}} = 1.3691 \qquad \text{where } \mu = \mu_1 = \mu_2 = 0.2,\ \theta = 90° = \frac{\pi}{2} \text{ radians}$$

$$\therefore \quad T_1 = 1.3691 T_2 \qquad ...(i)$$

Now consider the motion of body B. The body B is moving downwards with an acceleration 'a' Net force = Mass × Acceleration

$$(1000 - T_1) = \frac{1000}{9.81} \times a$$

or

$$1000 - 1.3691 T_2 = \frac{1000}{9.81} \times a \qquad (\because\ T_1 = 1.3691 T_2) \qquad ...(ii)$$

Now consider the motion of body A.

The body A is moving towards right with an acceleration 'a'. Net force = Mass × Acceleration

$$(T_2 - \mu W_2) = \left(\frac{W_2}{g}\right) \times a \qquad \text{or} \qquad T_2 - 0.2 \times 1500 = \frac{1500}{9.81} \times a$$

or

$$T_2 - 300 = \frac{1500}{9.81} \times a \qquad ...(iii)$$

or

$$T_2 = \frac{1500}{9.81} \times a + 300 \qquad ...(iv)$$

Substituting the value of T_2 in equation (*ii*), we get

$$1000 - 1.3691 \left(\frac{1500}{9.81} \times a + 300\right) = \frac{1000}{9.81} \times a$$

or

$$1000 - 1.3691 \times \frac{1500}{9.81} \times a - 1.3691 \times 300 = \frac{1000}{9.81} \times a$$

or

$$1000 - 1.3691 \times 300 = \frac{1000}{9.81} \times a + \frac{1.3691 \times 1500}{9.81} \times a$$

or $$1000 - 410.73 = \left(\frac{1000 + 1.3691 \times 1500}{9.81}\right) \times a$$

or $$589.27 = 315.287 \times a$$

$$\therefore \quad a = \frac{589.27}{311.287} = \mathbf{1.89\ m/s^2.} \quad \textbf{Ans.}$$

Substituting the value of '*a*' in equation (*iv*), we get

$$T_2 = \frac{1500}{9.81} \times 1.89 + 300 = \mathbf{594.6\ N.} \quad \textbf{Ans.}$$

Substituting the value of T_2 in equation (*i*), we get

$$T_1 = 1.3691 \times 594.6 = \mathbf{N.} \quad \textbf{Ans.}$$

4.8.10. Analysis of the Motion of Two Bodies Connected by a String when one Body is Lying on Inclined Plane and the other is Hanging Free in Air

1. First case when the inclined surface is smooth. Fig. 4.34 shows two bodies of weights W_1 and W_2 connected by a light inextensible string, which passes over a smooth and weightless pulley. The weight W_2 is placed on an inclined smooth plane of inclination θ with the horizontal, whereas the weight W_1 is hanging free in air.

As the inclined plane is smooth, hence the friction between the weight W_2 and the inclined plane will be neglected. When the weight W_1 is moving downwards, the weight W_2 will be moving upwards along the inclined plane. The velocity and acceleration of the weight W_1 will be same as that of weight W_2. Since the pulley is smooth and string is light and inextensible, the tension* in the string on both sides of the pulley will be same.

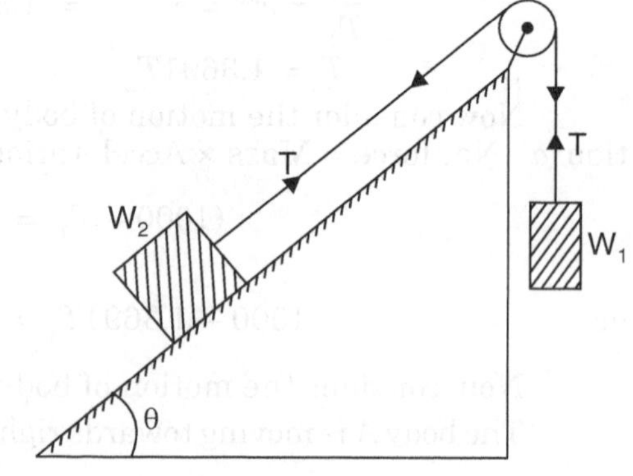

Fig. 4.34

Let a = Acceleration of the system *i.e.*, acceleration of weight W_1 as well as acceleration of weight W_2,

T = Tension in the string,

θ = Inclination of the inclined plane.

Consider the motion of weight W_1. The weight W_1 is moving downwards with an acceleration '*a*'. The forces acting on W_1 are : (*i*) its weight W_1 acting downwards, and (*ii*) tension T acting upwards.

∴ Net downwards force = $W_1 - T$

But net downwards force = Mass × Acceleration.

$$\therefore \quad (W_1 - T) = \frac{W_1}{g} \times a \qquad \left(\because \text{mass} = \frac{W_1}{g}\right) \qquad \ldots(i)$$

*If the string is inextensible and passes over a smooth pulley, then tension on both sides of the pulley will be same. The tension is independent of the inclination.

Now consider the motion of Weight W_2. The weight W_2 is moving upwards along the inclined plane with an acceleration a. The forces acting on W_2 along the plane are shown in Fig. 4.35. They are :

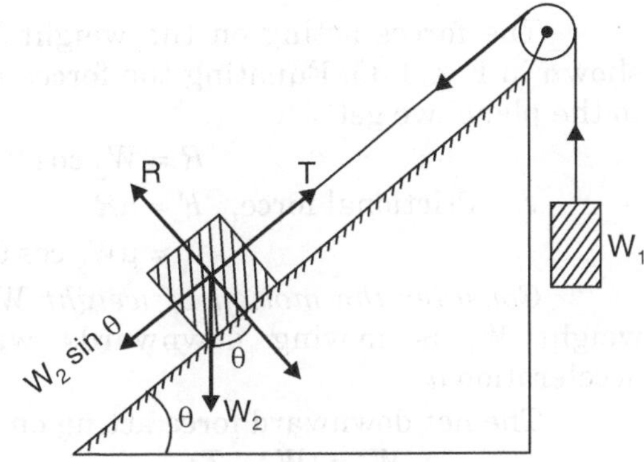

Fig. 4.35

(*i*) $W_2 \sin\theta$ downwards, and

(*ii*) Tension T upwards.

∴ Net force acting on W_2 along the plane in the upward direction $= T - W_2 \sin\theta$.

But, Net force = Mass × Acceleration

$$\therefore \quad T - W_2 \sin\theta = \frac{W_2}{g} \times a$$

$$\left(\because \text{ Mass} = \frac{\text{Weight}}{g}\right) \quad \ldots(ii)$$

Adding equations (*i*) and (*ii*)

$$W_1 - W_2 \sin\theta = \frac{W_1}{g} \times a + \frac{W_2}{g} \times a = \frac{a}{g}(W_1 + W_2)$$

$$\therefore \quad a = \frac{g(W_1 - W_2 \sin\theta)}{(W_1 + W_2)} \text{ m/s}^2 \quad \ldots(4.31)$$

Hence equation (4.31) is used for finding the acceleration of the system.

To find the value of tension T, in the string, the value of a from equation (4.31) is substituted in equation (*i*).

$$\therefore \quad W_1 - T = \frac{W_1}{g} \times g \, \frac{(W_1 - W_2 \sin\theta)}{(W_1 + W_2)}$$

$$= \frac{W_1(W_1 - W_2 \sin\theta)}{(W_1 + W_2)} \quad \text{(cancelling `}g\text{')}$$

or

$$T = W_1 - \frac{W_1(W_1 - W_2 \sin\theta)}{(W_1 + W_2)} = W_1\left[1 - \frac{(W_1 - W_2 \sin\theta)}{W_1 + W_2}\right]$$

$$= W_1\left[\frac{W_1 + W_2 - W_1 + W_2 \sin\theta}{W_1 + W_2}\right] = \frac{W_1[W_2 + W_2 \sin\theta]}{(W_1 + W_2)}$$

$$= \frac{W_1 W_2 (1 + \sin\theta)}{W_1 + W_2} \quad \ldots(4.32)$$

Equation (4.32) is used for finding tension T in the string.

2. Second case when the inclined surface is rough. As the surface of the inclined plane is not smooth, hence a force of friction equal to μR will be acting on the weight W_2 in the opposite direction of motion of weight W_2 as shown in Fig. 4.36.

Let μ = Coefficient of friction between the weight W_2 and inclined surface,

θ = Angle of the inclination of the plane,

a = Acceleration of the system,

T = Tension in the string,

R = Normal reaction acting on W_2.

The forces acting on the weight W_2 are shown in Fig. 4.36. Equating the forces normal to the plane, we get

$$R = W_2 \cos\theta.$$

$\therefore$ Frictional force, $F = \mu R$

$$= \mu W_2 \cos\theta \quad ...(i)$$

Consider the motion of weight W_1. The weight W_1 is moving downwards with an acceleration a.

The net downward force acting on $W_1 = (W_1 - T)$.

Mass of weight $W_1 = \dfrac{W_1}{g}$.

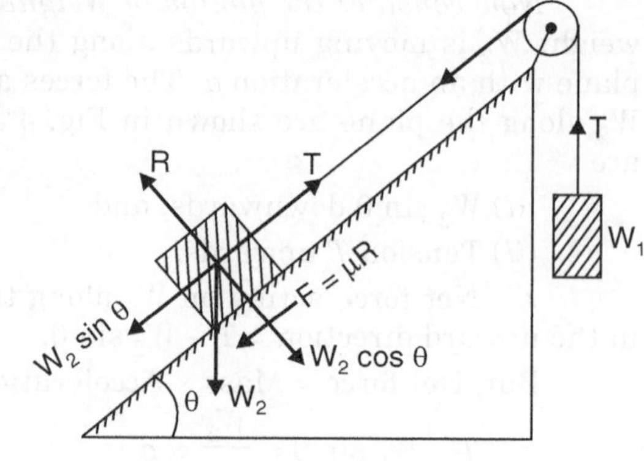

Fig. 4.36

But, Net force = Mass × Acceleration

$$\therefore \quad (W_1 - T) = \frac{W_1}{g} \times a \quad ...(ii)$$

Now consider the motion of weight W_2. The weight W_2 is moving upwards along the inclined plane with an acceleration a. The net upward force along the inclined plane acting on weight W_2

$$= T - W_2 \sin\theta - \mu R$$

$$= T - W_2 \sin\theta - \mu W_2 \cos\theta \qquad (\because \text{ From } (i), \mu R = \mu W_2 \cos\theta)$$

Mass of weight, $W_2 = \dfrac{W_2}{g}$

Using, Net force = Mass × Acceleration

$$\therefore \quad T - W_2 \sin\theta - \mu W_2 \cos\theta = \frac{W_2}{g} \times a \quad ...(iii)$$

Adding equations (*ii*) and (*iii*), we get

$$W_1 - W_2 \sin\theta - \mu W_2 \cos\theta = \frac{W_1}{g} \times a + \frac{W_2}{g} \times a = \frac{a}{g}[W_1 + W_2]$$

$$\therefore \quad a = \frac{g(W_1 - W_2 \sin\theta - \mu W_2 \cos\theta)}{(W_1 + W_2)} \quad ...(4.33)$$

Equation (4.33) is used for finding acceleration of the system. To find tension T in the string, the value of 'a' from equation (4.33) is substituted in equation (*ii*).

$$\therefore \quad W_1 - T = \frac{W_1}{g} \times \frac{g(W_1 - W_2 \sin\theta - \mu W_2 \cos\theta)}{(W_1 + W_2)}$$

$$= \frac{W_1(W_1 - W_2 \sin\theta - \mu W_2 \cos\theta)}{(W_1 + W_2)}$$

$$\therefore \quad T = W_1 - \frac{W_1(W_1 - W_2 \sin\theta - \mu W_2 \cos\theta)}{(W_1 + W_2)}$$

$$= W_1\left[1 - \frac{(W_1 - W_2 \sin\theta - \mu W_2 \cos\theta)}{(W_1 + W_2)}\right]$$

$$= W_1 \left[\frac{W_1 + W_2 - W_1 + W_2 \sin \theta + \mu W_2 \cos \theta}{(W_1 + W_2)} \right]$$

$$= \frac{W_1}{(W_1 + W_2)} (W_2 + W_2 \sin \theta + \mu W_2 \cos \theta)$$

$$= \frac{W_1 W_2}{(W_1 + W_2)} (1 + \sin \theta + \mu \cos \theta) \qquad ...(4.34)$$

Equation (4.34) is used for finding tension in the string.

Problem 4.66. *Two bodies of weights 40 N and 15 N are connected to the two ends of a light inextensible string, which passes over a smooth pulley. The weight 40 N is placed on a smooth inclined plane, while the weight 15 N is hanging free in air. If the angle of the plane is 15°, determine :*

(i) acceleration of the system, and

(ii) tension in the string. Take $g = 9.80\ m/s^2$.

Sol. Given :

Weight placed on inclined plane, $W_2 = 40$ N

Weight hanging free in air, $W_1 = 15$ N

Angle of inclination, $\theta = 15°$

Acceleration due to gravity, $g = 9.80$ m/s^2

Let a = Acceleration of the system

T = Tension in the string.

The inclined surface is smooth. Hence the acceleration and tension are obtained by using equations (4.31) and (4.32).

(*i*) Using equation (4.31) for acceleration,

$$a = \frac{g(W_1 - W_2 \sin \theta)}{(W_1 + W_2)}$$

$$= \frac{9.80(15 - 40 \sin 15°)}{(15 + 40)} = \frac{9.80(15 - 40 \times 0.2588)}{55}$$

$$= \frac{9.80 \times 4.684}{55} = \mathbf{0.828\ m/s^2.} \quad \textbf{Ans.}$$

(*ii*) Using equation (4.32) for tension,

$$T = \frac{W_1 W_2 (1 + \sin \theta)}{(W_1 + W_2)}$$

$$= \frac{15 \times 40(1 + .2588)}{(15 + 40)} = \frac{15 \times 40 \times 1.2588}{55} = \mathbf{13.732\ N.} \quad \textbf{Ans.}$$

Problem 4.67. *If in the problem 4.65, the inclined surface is a rough one, having a co-efficient of friction between the weight 40 N and the inclined surface equal to 0.2 and the weight hanging free is 25 N, determine :*

(i) the acceleration of the system,

(ii) the tension in the string, and

(iii) the distance moved by the weight 25 N in 3 seconds starting from rest. Take $g = 9.80\ m/s^2$.

Sol. Given : (From problem 4.65)

$W_1 = 25$ N, $W_2 = 40$ N, $g = 9.80$ m/s^2

$\theta = 15°$.

Coefficient of friction, $\mu = 0.2$.

Let a = Acceleration of the system,

T = Tension in the string, and

s = Distance travelled by weight 15 N in 3 seconds starting from rest.

As the inclined surface is a rough one, hence acceleration and tension are obtained by using equations (4.33) and (4.34).

(*i*) Using equation (4.33) for acceleration,

$$a = \frac{g(W_1 - W_2 \sin\theta - \mu W_2 \cos\theta)}{(W_1 + W_2)}$$

$$= \frac{9.80(25 - 40 \times \sin 15° - 0.2 \times 40 \times \cos 15°)}{(25 + 40)} \qquad (\because \; W_1 = 25 \text{ N})$$

$$= \frac{9.80(25 - 40 \times .2588 - 0.2 \times 40 \times 0.9659)}{65}$$

$$= \frac{9.80(25 - 10.352 - 7.727)}{65} = \frac{9.80 \times 6.921}{65} = \mathbf{1.043 \text{ m/s}^2}. \text{ **Ans.**}$$

(*ii*) Using equation (4.34) for tension,

$$T = \frac{W_1 W_2}{(W_1 + W_2)}(1 + \sin\theta + \mu\cos\theta)$$

$$= \frac{25 \times 40}{(25 + 40)}(1 + \sin 15° + 0.2 \cos 15°)$$

$$= \frac{1000}{65}(1 + 0.2588 + 0.2 \times 0.9659)$$

$$= \frac{1000}{65} \times 1.45198 = \mathbf{22.338 \text{ N}}. \text{ **Ans.**}$$

(*iii*) *Distance moved by the weight 25 N in 3 seconds, starting from rest.*

Initial velocity, $u = 0$

Time, $t = 3$ seconds

Acceleration, $a = 1.043$ m/s^2

Then distance s is given by

$$s = ut + \frac{1}{2}at^2 = 0 \times 3 + \frac{1}{2} \times 1.043 \times 3^2$$

$$= 0 + 4.6935 = \mathbf{4.6935 \text{ m}}. \text{ **Ans.**}$$

4.9. MOMENTUM AND ANGULAR MOMENTUM (OR MOMENT OF MOMENTUM)

4.9.1. Momentum. The product of mass and velocity of a body is known as momentum of the body. If '*m*' is the mass of a body and *v* is the velocity of the body, the momentum of the body is equal to $m \times v$.

4.9.2. Moment of Momentum or Angular Momentum. The product of mass moment of inertia and angular velocity of a rotating body is known as moment of momentum or angular momentum.

If ω = Angular velocity of a body rotating about an axis, and
I = Moment of inertia of the body about the axis.

Then angular momentum = $\omega \times I$. ...(4.35)

The equation (4.35) is derived as given below :

Consider a body of mass 'm' rotating in a circle about its centre O.

Let dm = Mass of the elementary strip

r = Radius of the mass 'dm'

ω = Angular velocity of the body or angular velocity of the mass 'dm'. (As angular velocity is constant, hence the angular velocity of the body will be same as angular velocity of mass 'dm')

v = Linear velocity of mass 'dm' = $\omega \times r$.

Now momentum of elementary mass

= Elementary mass × Velocity

$= dm \times v$

$= dm \times \omega r$ $(\because \quad v = \omega r)$

Moment of momentum of elementary mass 'dm' about O

= Momentum × Radius

$= (dm \times \omega r) \times r$

$= dm \times \omega r^2$ ...(*i*)

The moment of momentum of the entire mass about O is obtained by integrating the above equation.

$\therefore$ Moment of momentum of the entire mass

$= \int dm \times \omega r^2$

$= \omega \int r^2\, dm$ (ω is constant and can be taken outside the integral sign)

But $\int r^2\, dm$ = moment of inertia of the whole body about $O = I$.

Substituting this value, we get

$\therefore$ Moment of momentum of the entire mass

$= \omega \times I$ or $I \times \omega$ ...(4.36)

But moment of momentum is also known as **angular momentum.**

4.9.3. Moment of Inertia of a Circular Section. Consider a circular section as shown in Fig. 4.37.

Let R = Radius of the section

m = Mass per unit area of the section

Then total mass,

M = Area of the section × $m = \pi R^2 \times m$.

Consider an elementary circular ring of thickness 'dr' at a radius r.

Mass of the circular ring

= m × Area of circular ring

$= m \times 2\pi r \times dr$.

Moment of inertia of the circular ring about an axis passing through O and perpendicular to the plane of paper

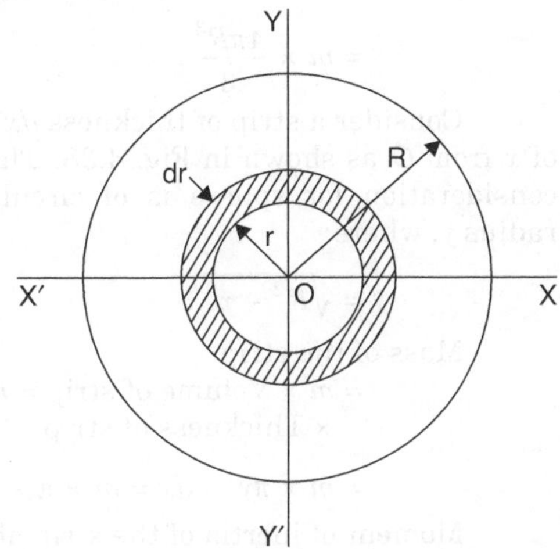

Fig. 4.37

$= \text{(Mass of circular ring)} \times \text{Radius}^2$

$= m \times 2\pi r\, dr \times r^2 = m \times 2\pi r^3 \times dr$ ...(*i*)

Moment of inertia of the whole circular section is obtained by integrating equation (*i*) between the limits 0 to R.

Hence moment of inertia of the whole section,

$$I_{ZZ} = \int_0^R m \times 2\pi r^3 \times dr$$

$$= m.2\pi \int_0^R r^3\, dr = m \times 2\pi \times \left[\frac{r^4}{4}\right]_0^R$$

$$= 2m \times \frac{\pi R^4}{4} = m \times \frac{\pi R^4}{2} = m \times \pi R^2 \times \frac{R^2}{2}$$

$$= M\frac{R^2}{2} \qquad (\because \; m \times \pi R^2 = M) \quad ...(4.37)$$

But from the theorem of perpendicular axis, the moment of inertia about X-X or Y-Y axis,

$$I_{XX} = I_{YY} = \frac{I_{ZZ}}{2} = \frac{MR^2}{2 \times 2} = \frac{MR^2}{4} \quad ...(4.38)$$

4.9.4. Moment of Inertia of Solid Cylinder. The moment of inertia of a solid cylinder is also given by equation (4.37). However, in case of a solid cylinder, M is equal to the mass of the solid cylinder.

4.9.5. Moment of Inertia of a Solid Sphere. Consider a solid sphere of radius R with O as centre, as shown in Fig. 4.38.

Let m = Mass of sphere per unit volume

M = Total mass of the sphere

$= m \times$ Volume of sphere

$= m \times \dfrac{4\pi R^3}{3}$.

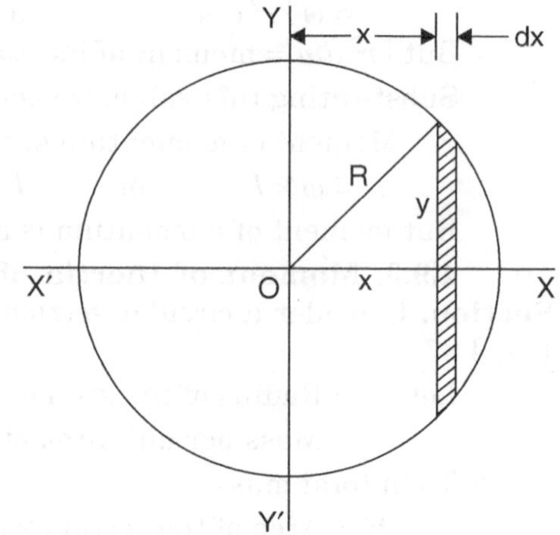

Fig. 4.38

Consider a strip of thickness dx at a distance of x from O as shown in Fig. 4.38. The strip under consideration for sphere is of circular section of radius y, where

$$y = \sqrt{R^2 - x^2}.$$

Mass of the strip

$= m \times$ Volume of strip $= m \times$ Area of strip $\times$ Thickness of strip

$= m \times \pi y^2 \times dx = m \times \pi \times (R^2 - x^2) \times dx \qquad (\because \; y = \sqrt{R^2 - x^2})$

Moment of inertia of the strip about X-X axis

$= \text{Mass of the strip} \times \dfrac{\text{Radius}^2}{2}$

$$= m\pi(R^2 - x^2)\,dx \times \frac{y^2}{2} = m\pi(R^2 - x^2) \times dx \times \frac{(R^2 - x^2)}{2}$$

$$= \frac{m\pi}{2} \times (R^2 - x^2)^2\,dx = \frac{m\pi}{2} \times (R^4 + x^4 - 2R^2x^2)dx$$

The mass moment of inertia of the whole sphere is obtained by integrating the above equation between the limits $-R$ to $+R$. Therefore,

$$I = \int_{-R}^{+R} \frac{m\pi}{2}(R^4 + x^4 - 2R^2x^2)\,dx = \frac{m\pi}{2}\int_{-R}^{+R}(R^4 + x^4 - 2R^2x^2)\,dx$$

$$= \frac{m\pi}{2}\left[R^4x + \frac{x^5}{5} - \frac{2R^2x^3}{3}\right]_{-R}^{+R}$$

$$= \frac{m\pi}{2}\left[R^4 \times R + \frac{R^5}{5} - \frac{2R^2 \times R^3}{3} - \left\{R^4(-R) + \frac{(-R)^5}{5} - \frac{2R^2(-R)^3}{3}\right\}\right]$$

$$= \frac{m\pi}{2}\left[R^5 + \frac{R^5}{5} - \frac{2R^5}{3} - \left\{-R^5 - \frac{R^5}{5} + \frac{2R^5}{3}\right\}\right]^3$$

$$= \frac{m\pi}{2}\left[2R^5 + \frac{2R^5}{5} - \frac{4R^5}{3}\right] = \frac{m\pi}{2}\left[\frac{30 + 6 - 20}{15}\right]R$$

$$= \frac{m\pi}{2} \times \frac{16}{15}R^5 = \frac{8}{15}m\pi R^5 = \frac{2}{5} \times \frac{4}{3}m\pi R^3 \times R^2$$

$$= \frac{2}{5} \times M \times R^2 \qquad \left(\because \frac{4}{3}m\pi R^3 = M\right)$$

$$= \frac{2MR^2}{5} \qquad \text{...(4.39)}$$

The moment of inertia of the sphere about every axis is same as sphere is symmetrical.

4.9.6. Radius of Gyration (k). It is defined as the distance of a body from the axis of reference where whole mass of the body is assumed to be concentrated so as not to alter the moment of inertia about the given axis.

The moment of inertia of a body in terms of radius of gyration is given by

$$I = Mk^2$$

$\therefore$ Radius of gyration, $$k = \sqrt{\frac{I}{M}}. \qquad \text{...(4.40)}$$

(A) Radius of gyration for circular lamina

From equation (4.37), moment of inertia of a circular section about the central axis is given as,

$$I_{ZZ} = \frac{MR^2}{2} \qquad \text{where } R = \text{Radius and } M = \text{Mass}$$

But moment of inertia in terms of radius of gyration is given by

$$I_{ZZ} = Mk^2.$$

Equating the two values of I_{ZZ},

$$\therefore \qquad \frac{MR^2}{2} = Mk^2$$

$$\therefore \qquad k^2 = \frac{R^2}{2} \quad \text{or} \quad k = \frac{R}{\sqrt{2}}. \qquad ...(4.41)$$

(B) Radius of gyration of a solid sphere

From equation (4.39), the moment of inertia of a solid sphere about any axis is given as,

$$I = \frac{2MR^2}{5}$$

But moment of inertia in terms of radius of gyration is given by

$$I = Mk^2$$

$\therefore$ Equating the two values of I.

$$\therefore \qquad Mk^2 = \frac{2M}{5} R^2$$

$$\therefore \qquad k = \sqrt{\frac{2R^2}{5}} = R \times \sqrt{\frac{2}{5}} = 0.6324\,R. \qquad ...(4.42)$$

Problem 4.68. *Calculate the moment of inertia and radius of gyration with respect to the axis of rotation of : (i) a circular lamina of diameter 60 cm and of mass 0.001 kg/cm², (ii) a circular cylinder of diameter 80 cm of height 15 cm and of mass 0.002 kg/cm³, and (iii) a solid sphere of diameter 40 cm and of mass 0.0015 kg/cm³.*

Sol. Given :

(*i*) **For a circular lamina :**

Diameter, $D = 60$ cm

$\therefore$ Radius, $R = \frac{D}{2} = \frac{60}{2} = 30$ cm

Mass per unit area, $m = 0.001$ kg/cm^2

$\therefore$ Total mass, $M = m \times$ Area of circular lamina

$= 0.001 \times \pi R^2 = 0.001 \times \pi \times 30^2 = 2.827$ kg.

Moment of inertia of the circular section about the axis of rotation is given by equation (4.37) as,

$$I_{ZZ} = \frac{MR^2}{2} = \frac{2.827}{2} \times (30)^2 = \mathbf{1272.15\ kg/cm^2}. \quad \textbf{Ans.}$$

Radius of gyration for circular section is given by equation (4.41),

$$k = \frac{R}{\sqrt{2}} = \frac{30}{\sqrt{2}} = \mathbf{21.21\ cm.} \quad \textbf{Ans.}$$

(*ii*) **For a circular cylinder**

Diameter, $D = 80$ cm

$\therefore$ Radius, $R = \frac{D}{2} = \frac{80}{2} = 40$ cm

Height, $h = 15$ cm.

Mass per unit volume, $m = 0.002$ kg/cm^3

$\therefore$ Total mass, $M = m \times$ Volume of the cylinder

$= m \times \pi R^2 \times h$

$= 0.002 \times \pi \times 40^2 \times 15 = 150.796$ kg.

Moment of inertia of a circular cylinder about the axis of rotation is given by equation (4.37) as

$$I_{ZZ} = \frac{MR^2}{2} = \frac{150.796 \times 40^2}{2} = \mathbf{120636.8\ kg/cm^2}. \quad \textbf{Ans.}$$

Radius of gyration for the circular cylinder is given by equation (4.41) as

$$k = \frac{R}{\sqrt{2}} = \frac{40}{\sqrt{2}} = \mathbf{28.28\ cm.} \quad \textbf{Ans.}$$

(*iii*) **For a solid sphere**

Diameter, $D = 40$ cm

$\therefore$ Radius, $R = \frac{D}{2} = \frac{40}{2} = 20$ cm

Mass per unit volume, $m = 0.0015$ kg/cm^3

$\therefore$ Total mass, $M = m \times$ Volume of the sphere

$$= m \times \frac{4\pi R^3}{3} = \frac{0.0015 \times 4 \times \pi \times 20^3}{3} \text{ kg} = 50.265 \text{ kg.}$$

Moment of inertia of the solid sphere is given by equation (4.39) as

$$I = \frac{2}{5} MR^2 = \frac{2}{5} \times 50.265 \times 20^2 \text{ kg/cm}^2 = \mathbf{8042.4\ kg/cm^2}. \quad \textbf{Ans.}$$

Radius of gyration of a solid sphere is given by equation (4.42) as,

$$k = 0.6324\, R = 0.6324 \times 20 = \mathbf{12.648\ cm.} \quad \textbf{Ans.}$$

Problem 4.69. *Calculate the moment of inertia and radius of gyration of a grindstone 90 cm in diameter and 10 cm thick with respect to its axis of rotation. Stone mass is 0.0026 kg/cm^3.*

(AMIE Summer, 1978)

Sol. Given :

Diameter of grindstone, $D = 90$ cm

$\therefore$ Radius, $R = \frac{90}{2} = 45$ cm

Thickness of grindstone, $t = 10$ cm

Mass per unit volume, $m = 0.0026$ kg/cm^3

$\therefore$ Total mass, $M = m \times$ Volume of grindstone

$= m \times \pi R^2 \times t$

$= 0.0026 \times \pi \times 45^2 \times 10 = 165.4$ kg.

Moment of inertia (I_{ZZ}) of grindstone (which is similar to cylinder) about the axis of rotation is given by equation (4.37) as

$$I_{ZZ} = \frac{MR^2}{2} = \frac{165.4 \times 45^2}{2} = \mathbf{167467.5\ kg/cm^2}. \quad \textbf{Ans.}$$

Radius of gyration (k) is given by equation (4.41) as

$$k = \frac{R}{\sqrt{2}} = \frac{45}{\sqrt{2}} = \mathbf{31.819\ cm.} \quad \textbf{Ans.}$$

4.10. LAWS FOR ROTARY MOTION

First law. It states that a body continues in its state of rest or of rotation about an axis with constant angular velocity unless it is compelled by an external torque to change the state.

In actual practice, we see that when a body is rotating about an axis with constant angular velocity, the body does not continue in its state of rotation, but comes to rest after some time. This is due to air resistance and friction between the body and its bearing or axis. If these forces had been absent, the body would have gone on rotating indefinitely.

Second law. It states that the rate of change of angular momentum of a rotating body is proportional to the external torque applied on the body and takes place in the direction of the torque.

Consider a body of moment of inertia I rotating with an angular velocity ω_0. It is acted upon a torque T and the angular velocity of the body becomes ω. Then we have

ω_0 = Initial angular velocity,

ω = Final angular velocity,

I = Moment of inertia of the body = Mk^2,

M = Mass of the body and k = radius of gyration. At this radius the whole mass of the body is assumed to concentrate in rotation.

T = Torque acting on the body,

t = Time in seconds.

Initial angular momentum of the body

= Moment of Inertia × Initial angular velocity = $I \times \omega_0 = I\omega_0$

Final angular momentum

$= I \times \omega = I\omega$

∴ Change of angular momentum

= Final angular momentum – Initial angular momentum

$= I\omega - I\omega_0 = I(\omega - \omega_0)$

∴ Rate of change of angular momentum

$$= \frac{\text{Change of momentum}}{\text{Time}} = \frac{I(\omega - \omega_0)}{t}.$$

But from equation (4.14), we have

$$\frac{(\omega - \omega_0)}{t} = \alpha = \text{Angular acceleration.}$$

∴ Rate of change of angular momentum $= I \times \alpha = I\alpha$.

But according to second law of motion for rotation, the rate of change of angular momentum is directly proportional to the external torque applied on the body.

∴ Torque ∝ Rate of change of angular momentum

or $\quad T \propto I\alpha \quad \therefore \quad T = kI\alpha \quad$ ...(*i*)

where k is a constant of proportionality.

If we define unit torque as that which produces unit angular acceleration on unit moment of inertia, then we have $T = 1$, $I = 1$ and $\alpha = 1$. Substituting these values in equation (*i*), we get

$$1 = k \times 1 \times 1 = k.$$

Substituting this value of $k = 1$ in equation (*i*), we get

$$T = I\alpha \quad \text{...(4.43)}$$

Units of Torque

The units of torque depend upon the units of I, where I = moment of inertia = Mk^2 and k is the radius of gyration from the axis of the rotation. It is at this radius that the whole mass of the rotating body is assumed to be concentrated. If I is in kg m^2, then T is in Newton-metre written as Nm.

Note. If the value of M is taken in kg, then the unit of torque is in N m. However if the value of M is taken as $\frac{W}{g}$, where W is in kgf, then unit of torque will be in kgf-m.

Third law. It states that to every torque there is always an equal and opposite torque.

4.10.1. Kinetic Energy due to Rotation. Consider a rigid body rotating about O as shown in Fig. 4.39.

Let ω = Angular velocity of the body

dm = Elementary mass of the body

r = Radius of elementary mass from O

V = Tangential velocity of elementary mass

$= \omega \times r.$

K.E. of the elementary mass

$= \frac{1}{2} \times \text{Mass} \times \text{Velocity}^2 = \frac{1}{2} \times dm \times V^2.$

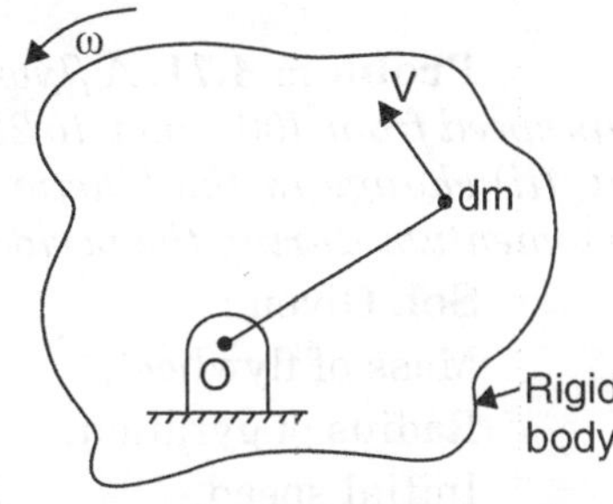

Fig. 4.39. Rigid body rotating about O

K.E. of the whole body is obtained by integrating the above equation. Hence K.E. of the body

$$= \int \frac{1}{2}\, dm \times V^2 = \frac{1}{2} \int dm \times (\omega r)^2 \qquad (\because \; V = \omega r)$$

$$= \frac{1}{2} \int \omega^2 r^2 dm = \frac{1}{2} \omega^2 \int r^2\, dm. \qquad (\because \; \omega \text{ is a constant})$$

But $\int r^2\, dm = I$ = Moment of Inertia of the body about O.

$$\therefore \quad \text{K.E. of the body} = \frac{1}{2} \omega^2 I. \qquad \text{...(4.44)}$$

If mass is taken in kg, the unit of K.E. will be Nm.

4.10.2. Total Kinetic Energy of a Body. Equation (4.44) gives only the K.E. of the body due to rotation. But if the body possesses both the motion of translation as well as the motion of rotation, then the total kinetic energy of body is equal to the K.E. due to rotation plus K.E. due to translation. If the linear velocity of the body is V, then K.E. due to linear velocity = $\frac{1}{2} mV^2$.

$$\therefore \quad \text{Total Kinetic Energy} = \frac{1}{2} I\omega^2 + \frac{1}{2} mv^2. \qquad \text{...(4.45)}$$

Problem 4.70. *Find the angular acceleration of a flywheel of an engine, which weighs 1000 N and has a radius of gyration 0.5 m if the wheel is subjected to a torque of 1200 Nm. Take g = 9.80 m/s^2.*

Sol. Given :

Weight of flywheel, $W = 1000$ N

$\therefore$ Mass of the flywheel, $M = \frac{W}{g} = \frac{1000}{9.80}$ kg $\qquad (\because \; g = 9.80 \text{ m/s}^2)$

Radius of gyration, $k = 0.5$ m

Torque, $T = 1200$ Nm.

Let α = Angular acceleration

Using equation (4.43), $T = I \times \alpha$ ...(*i*)

where I = Moment of inertia

$$= Mk^2 = \frac{1000}{9.80} \times (0.5)^2 = \mathbf{25.51\ kg/m^2}.$$

Substituting the value of I, and T in equation (*i*), we get

$$1200 = 25.51 \times \alpha$$

$$\therefore \quad \alpha = \frac{120}{25.51} = \mathbf{47.04\ radians/s^2.\ Ans.}$$

Problem 4.71. *A flywheel, of mass 5000 kg and having a radius of gyration 1 m, loses its speed from 400 r.p.m. to 280 r.p.m. in 2 minutes. Calculate (i) the retarding torque acting on it, (ii) change in the kinetic energy during the above period, and (iii) change in its angular momentum during the period.* (AMIE Winter, 1977)

Sol. Given :

Mass of flywheel, $M = 5000$ kg

Radius of gyration, $k = 1$ m

Initial speed, $N_0 = 400$ r.p.m.

$\therefore$ Initial angular velocity, $\omega_0 = \dfrac{2\pi N_0}{60} = \dfrac{2\pi \times 400}{60} = 41.88$ rad/s.

Final speed, $N = 280$ r.p.m.

$\therefore$ Final angular velocity, $\omega = \dfrac{2\pi N}{60} = \dfrac{2\pi \times 280}{60} = 29.32$ rad/s.

Time, $t = 2$ min $= 2 \times 60 = 120$ seconds.

(*i*) *Retarding torque acting on the flywheel*

Using equation (4.43), torque (T) is given by

$$T = I\alpha \quad \text{...(i)}$$

where I = Moment of inertia $= Mk^2 = 5000 \times 1^2 = 5000$ kg/m^2

α = Angular acceleration and from equation (4.14)

$$= \frac{(\omega - \omega_0)}{t} = \frac{(29.32 - 41.88)}{120} = 0.1046 \text{ rad/s}^2$$

T = Torque in Nm.

Substituting the values of I and α in equation (*i*), we get

$$T = 5000 \times (-0.1046) = -523.0 \text{ Nm.}$$

(–ve sign means the torque is retarding torque)

$\therefore$ Retarding torque = **523 Nm. Ans.**

(*ii*) *Change in kinetic energy*

Change in Kinetic Energy = Final K.E. – Initial K.E. ...(*ii*)

But the kinetic energy of a rotating body is given by equation (4.44)

$$\text{K.E.} = \frac{1}{2}\omega^2 I.$$

∴ Final kinetic energy,

$$E = \frac{1}{2}\omega^2 I = \frac{1}{2} \times 29.32^2 \times 5000 \text{ Nm} \qquad (\because \omega = 29.32)$$

$$= 2149156 \text{ Nm}$$

Initial K.E., $E_0 = \frac{1}{2}\omega_0^2 I = \frac{1}{2} \times 41.88^2 \times 5000 \qquad (\because \omega_0 = 41.88)$

$$= 4384836 \text{ Nm.}$$

∴ Change in Kinetic Energy

$= 2149156 - 4384836 = -$ **2235680 Nm. Ans.**

The –ve sign shows that K.E. is decreasing.

(iii) Change in angular momentum

Using equation (4.36),

Angular momentum $= \omega I$ or $I\omega$

∴ Initial angular momentum $= I\omega_0 = 5000 \times 41.88 = 209400$ Nm/s.

Final angular momentum $= I \times \omega = 5000 \times 29.32 = 146600$ N m/s.

∴ Change in angular momentum = Final – Initial angular momentum

$= 146600 - 209400 = -62800$ N m/s.

= **62800 N m/s Numerically. Ans.**

Problem 4.72. *A uniform homogeneous cylinder rolls without slip along a horizontal level surface with a translational velocity of 20 cm/s. If its weight is 0.1 N and its radius is 10 cm, what is its total kinetic energy ?* (AMIE Summer, 1976)

Sol. Given :

Linear velocity, $V = 20$ cm/s $= 0.20$ m/s

Weight of cylinder, $W = 0.1$ N

∴ Mass, $M = \frac{W}{g} = \frac{0.1}{9.81}$ kg

Radius, $R = 10$ cm $= 0.1$ m

Total K.E. is given by equation (4.45)

$$= \frac{1}{2}I\omega^2 + \frac{1}{2}mV^2 \qquad \ldots(i)$$

where I = Moment of inertia of solid cylinder

$$= \frac{MR^2}{2} \qquad \text{[see equation (4.37)]}$$

$$= \frac{0.1}{9.80} \times \frac{0.1^2}{2} = 0.000051$$

ω = Angular velocity

$$= \frac{V}{R} = \frac{0.20}{0.10} = 2 \text{ rad/s} \qquad (\because V = \omega R)$$

Substituting these values in equation (*i*), we get

Total K.E. $= \frac{1}{2} \times 0.000051 \times 2^2 + \frac{1}{2} \times \frac{0.1}{9.81} \times 0.2^2$

$= 0.000102 + 0.000204 =$ **0.000306 Nm. Ans.**

Problem 4.73. *The radius of gyration of a flywheel, which weighs 6 kN, is 50 cm. If the wheel starts from rest and attains a speed of 200 r.p.m. in 2 minutes, determine the average torque exerted on the flywheel.*

Sol. Given :

Radius of gyration, $k = 50 \text{ cm} = 0.50 \text{ m}$

Weight of flywheel, $W = 6 \text{ kN} = 6 \times 1000 = 6000 \text{ N}$

$\therefore$ Mass, $M = \dfrac{W}{g} = \dfrac{6000}{9.80} \text{ kg}$

Initial r.p.m., $N_0 = 0$

$\therefore$ Initial angular velocity, $\omega_0 = \dfrac{2\pi N_0}{60} = 0$

Final r.p.m. $N = 200$ r.p.m.

$\therefore$ Final angular velocity, $\omega = \dfrac{2\pi N}{60} = \dfrac{2\pi \times 200}{60} = 20.94 \text{ rad/s.}$

$\therefore$ Time, $t = 2 \text{ min} = 2 \times 60 = 120 \text{ seconds.}$

The torque exerted is given by equation (4.43) as

$$T = I\alpha \qquad \text{...}(i)$$

where I = Moment of inertia $= Mk^2 = \dfrac{6000}{9.80} = 0.5^2 = 306.12$

α = Constant angular acceleration and from equation (4.14)

$$= \frac{(\omega - \omega_0)}{t} = \frac{20.94 - 0}{120} = 0.1745 \text{ rad/s}^2$$

Substituting these values in equation (*i*), we get average torque,

$$T = 306.12 \times 0.1745 = \mathbf{53.42 \text{ Nm.}} \quad \textbf{Ans.}$$

4.10.3. Rotation Due to a Weight W Attached to one End of a String Passing Over a Pulley of Weight W_0. Fig. 4.40 shows a weight W, attached to one end of an inextensible string passing over a pulley. The other end of the string is attached to the periphery of the pulley. Let the weight W is moving downwards. It is required to find the acceleration with which the weight W will move down and the tension P in the string.

Let W = Weight attached to the string,

W_0 = Weight of pulley,

R = Radius of the pulley,

P = Tension in string,

a = Acceleration with which the weight W moves down,

I = Moment of Inertia of the pulley about the axis of rotation, and

α = Angular acceleration of the pulley.

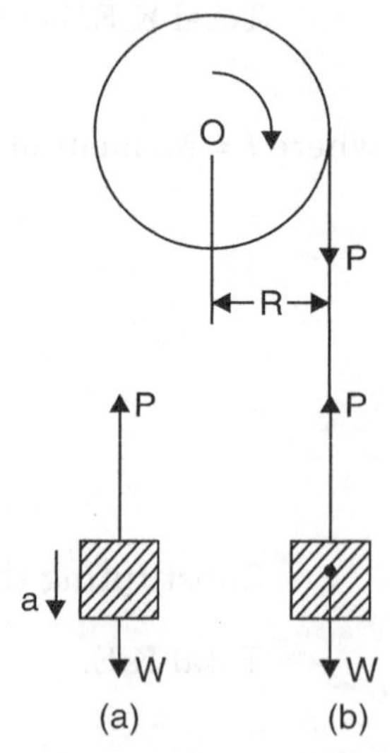

Fig. 4.40

The weight W is moving downwards with an acceleration a, hence net force on this weight is acting downwards. [Refer to Fig. 4.40 (*a*)].

$\therefore$ Net downward force on the weight $W = (W - P)$.

Using, Net force = Mass × Acceleration

$$\therefore \quad W - P = \frac{W}{g} \times a \qquad \left(\because \text{ Mass} = \frac{\text{Weight}}{g}\right) \quad ...(i)$$

Now consider the rotation of pulley

The torque on the pulley is given by equation (4.43), as

$$T = I \times \alpha \quad ...(ii)$$

where I = Moment of inertia of the pulley about the axis of rotation.

If pulley is considered as a solid disc, then moment of inertia of pulley is given by equation (4.37).

$$\therefore \quad I = \frac{MR^2}{2} \qquad (\because \text{ Solid disc is like a cylinder})$$

In the above equation M = Mass of the pulley

$$= \frac{\text{Weight of the pulley}}{g} = \frac{W_0}{g}$$

$$\therefore \quad I = \frac{W_0}{g} \times \frac{R^2}{2}$$

Substituting this value of I in equation (*ii*), we get

$$T = \left(\frac{W_0}{g} \times \frac{R^2}{2}\right) \times \alpha. \quad ...(iii)$$

But the relation between angular acceleration (α) and linear acceleration is given by (eqn. 4.13)* as

$$a = \alpha \times R$$

$$\therefore \quad \alpha = \frac{a}{R}.$$

Substituting the value of α in equation (*iii*), we get

$$T = \left(\frac{W_0}{g} \times \frac{R^2}{2}\right) \times \frac{a}{R} \quad ...(iv)$$

But the torque at the centre of the pulley is due to the tension P** and its magnitude is given by

$$T = P \times R.$$

Substituting this value of T in equation (*iii*), we get

$$P \times R = \left(\frac{W_0}{g} \times \frac{R^2}{2}\right) \times \frac{a}{R} \quad \text{or} \quad P = \frac{W_0}{2g} \times a. \quad ...(v)$$

Adding equations (*i*) and (*v*), we get

$$W = \frac{W}{g} \times a + \frac{W_0}{2g} \times a = \frac{a}{g}\left[W + \frac{W_0}{2}\right]$$

---

*See on page 245.

**If the weight W is not moving, even then the torque at the centre = $P \times R$. But in that case $W = P$ and hence $T = W \times R$. When weight W is moving, $W \neq P$.

$$\therefore \qquad a = \frac{gW}{\left(W + \frac{W_0}{2}\right)} \text{ m/s}^2 \qquad \text{...(4.46)}$$

To find the tension (P) in the string, the value of a is substituted in equation (i).

$$\therefore \qquad W - P = \frac{W}{g} \times \frac{gW}{\left(W + \frac{W_0}{2}\right)} = \frac{W^2}{\left(W + \frac{W_0}{2}\right)}$$

$$\therefore \qquad P = W - \frac{W^2}{W + \frac{W_0}{2}} = W - \frac{2W^2}{(2W + W_0)}$$

$$= \frac{2W^2 + WW_0 - 2W^2}{(2W + W_0)} = \frac{WW_0}{(2W + W_0)} \qquad \text{...(4.47)}$$

Problem 4.74. *A weight of 5 N is suspended by a light rope wound round a pulley of weight 50 N and radius 30 cm, the other end of the rope being fixed to the periphery of the pulley. If the weight is moving downwards, determine :*

(*i*) *Acceleration of the weight 5 N, and*

(*ii*) *Tension in the string. Take g = 9.80 m / s²*.

Sol. Given :

Weight suspended by the rope, $W = 5$ N

Weight of pulley, $W_0 = 50$ N

Radius of pulley, $R = 30$ cm $= 0.30^*$ m

Acceleration due to gravity, $g = 9.80$ m/s²

Let a = Acceleration of weight 5 N, and

P = Tension in the string.

(*i*) *Consider the motion of weight 5 N.* [Refer to Fig. 4.40 (*c*)]. Weight 5 N is moving downwards,

$\therefore$ Net force = Mass × acceleration

$$\text{or} \qquad (5 - P) = \left(\frac{5}{9.8}\right) \times a \qquad \text{...}(i)$$

(*ii*) *Consider the rotation of the pulley,*

Net torque $= I \times \alpha$

$$\text{or} \qquad P \times R = I \times \alpha$$

where $I = \dfrac{M_0 R^2}{2}$ and $\alpha = \dfrac{a}{R}$

$$\therefore \qquad P \times R = \frac{M_0 \times R^2}{2} \times \frac{a}{R}$$

$$\text{or} \qquad P \times R = \frac{M_0}{2} \times R \times a$$

$$\text{or} \qquad P = \frac{M_0}{2} \times a \qquad \text{But } M_0 = \frac{W_0}{g} = \frac{50}{9.8}$$

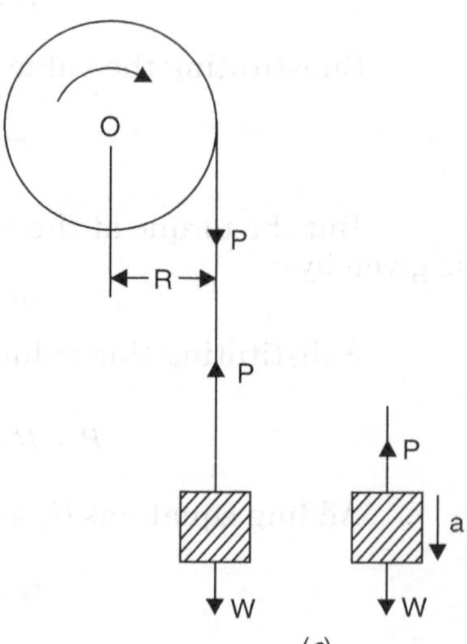

Fig. 4.40

*Please, note that the value of R is extra data. If has not been used.

$\therefore \qquad P = \frac{50}{9.8} \times \frac{1}{2} \times a$

or $\qquad P = \frac{25}{9.8} \times a$ ...(*ii*)

Adding equations (*i*) and (*ii*), we get the value of acceleration a.

$\therefore \qquad 5 = \frac{(5+25)}{9.8} \times a$

$\therefore \qquad a = \frac{5 \times 9.8}{30} = \frac{9.8}{6} = \mathbf{1.633\ m/s^2}$. **Ans.**

To get the value of tension P, substitute this value of a in equation (*ii*).

$\therefore \qquad P = \frac{25}{9.8} \times 1.633 = \mathbf{4.166\ N}$. **Ans.**

4.10.4. Rotation Due to Weights Attached to the Two Ends of a String, which Passes Over a Rough Pulley of Weight W_0. Fig. 4.41 shows the two weights W_1 and W_2 attached to the two ends of a string, which passes over a rough pulley of radius R. As pulley is rough and having certain weight, the tensions on both sides of the string will not be same. If $W_1 > W_2$, the weight W_1 will move downwards whereas the weight W_2 will move upwards with the same acceleration.

Let a = Acceleration of the system ;

T_1 = Tension in the string to which weight W_1 is attached ;

T_2 = Tension in the string to which weight W_2 is attached ;

R = Radius of the pulley ;

I = Moment of inertia of the pulley about the axis of rotation ;

α = Angular acceleration ; and

W_0 = Weight of the pulley.

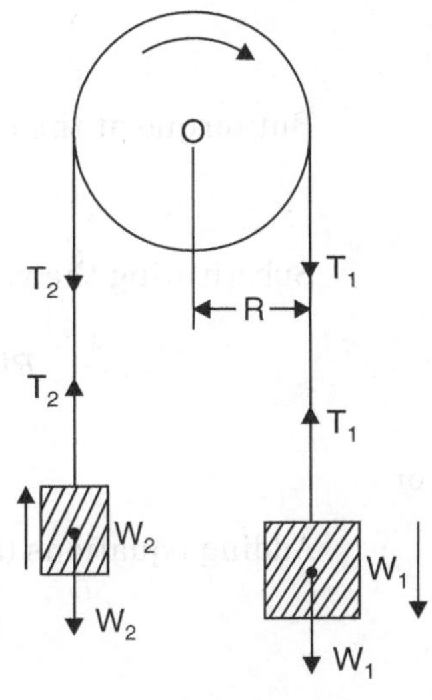

Fig. 4.41

Consider the motion of weight W_1. The weight W_1 is moving downwards with an acceleration a.

$\therefore$ The net downward force on weight $W_1 = (W_1 - T_1)$.

Mass of weight, $\qquad W_1 = \frac{W_1}{g}$.

Now using, Net force = Mass × Acceleration

or $\qquad (W_1 - T_1) = \frac{W_1}{g} \times a.$ ...(*i*)

Now consider the motion of weight W_2. The weight W_2 is moving upwards with an acceleration a.

$\therefore$ Net upward force on weight $W_2 = (T_2 - W_2)$.

Mass of weight, $\qquad W_2 = \frac{W_2}{g}$

Using, Net force = Mass × Acceleration

or $$(T_2 - W_2) = \frac{W_2}{g} \times a. \quad ...(ii)$$

Now consider the rotation of the pulley. The pulley is rotating with an angular acceleration α.

But angular acceleration, $\alpha = \frac{a}{R}$. [Using equation (4.13)]

If the pulley is considered as a solid disc, then moment of inertia of the pulley is given by equation (4.37) as

$$I = \frac{MR^2}{2} \qquad (\because \text{ Solid disc is like a cylinder})$$

where M = Mass of pulley $= \frac{\text{Weight of pulley}}{g} = \frac{W_0}{g}$.

$$\therefore \quad I = \frac{W_0}{g} \frac{R^2}{2}.$$

The torque at the centre of the pulley is given by equation (4.43) as

$$T = I \times \alpha = \frac{W_0}{g} \frac{R^2}{2} \times \frac{a}{R} \qquad \left(\because \alpha = \frac{a}{R}\right) \quad ...(iii)$$

But torque at the centre of the pulley

= Torque due to T_1 – Torque due to T_2

$= T_1 \times R - T_2 \times R = R[T_1 - T_2]$.

Substituting the value of torque in equation (*iii*), we get

$$R[T_1 - T_2] = \frac{W_0}{g} \frac{R^2}{2} \times \frac{a}{R}$$

or $$T_1 - T_2 = \frac{W_0}{2g} \times a. \quad ...(iv)$$

Adding equations (*i*), (*ii*) and (*iv*), we get

$$W_1 - W_2 = \frac{W_1}{g} \times a + \frac{W_2}{g} \times a + \frac{W_0}{2g} \times a = \frac{a}{g}\left[W_1 + W_2 + \frac{W_0}{2}\right]$$

$$\therefore \quad a = \frac{g[W_1 - W_2]}{\left(W_1 + W_2 + \frac{W_0}{2}\right)} \quad ...(4.48)$$

To find the tensions T_1 and T_2, substitute the value of a in equations (*i*) and (*ii*).

Substituting the value of a in equation (*i*), we get

$$W_1 - T_1 = \frac{W_1}{g} \times \frac{g(W_1 - W_2)}{\left(W_1 + W_2 + \frac{W_0}{2}\right)} = \frac{W_1(W_1 - W_2)}{\left(W_1 + W_2 + \frac{W_0}{2}\right)}$$

$$\therefore \quad T_1 = W - \frac{W_1(W_1 - W_2)}{\left(W_1 + W_2 + \frac{W_0}{2}\right)} = W_1\left[1 - \frac{(W_1 - W_2)}{\left(W_1 + W_2 + \frac{W_0}{2}\right)}\right]$$

$$= W_1 \left[\frac{W_1 + W_2 + \frac{W_0}{2} - W_1 + W_2}{W_1 + W_2 + \frac{W_0}{2}} \right] = W_1 \left[\frac{2W_2 + \frac{W_0}{2}}{W_1 + W_2 + \frac{W_0}{2}} \right] \quad \text{...(4.49)}$$

Substituting the value of a in equation (*ii*), we get

$$T_2 - W_2 = \frac{W_2}{g} \times \frac{g(W_1 - W_2)}{\left(W_1 + W_2 + \frac{W_0}{2}\right)} = \frac{W_2(W_1 - W_2)}{\left(W_1 + W_2 + \frac{W_0}{2}\right)}$$

$$\therefore \quad T_2 = W_2 + \frac{W_2(W_1 - W_2)}{\left(W_1 + W_2 + \frac{W_0}{2}\right)} = W_2 \left[1 + \frac{W_1 - W_2}{\left(W_1 + W_2 + \frac{W_0}{2}\right)} \right]$$

$$= W_2 \left[\frac{W_1 + W_2 + \frac{W_0}{2} + W_1 - W_2}{\left(W_1 + W_2 + \frac{W_0}{2}\right)} \right] = \frac{W_2 \left[2W_1 + \frac{W_0}{2} \right]}{\left(W_1 + W_2 + \frac{W_0}{2}\right)} \quad \text{...(4.50)}$$

Problem 4.75. *Two blocks weighing 100 N and 40 N are supported at the ends of a rope of negligible weight which is passing over the rough surface of a pulley mounted on a horizontal axle. The pulley may be assumed as a solid disc with a weight of 50 N. Friction in the bearings of the pulley may be neglected. Find the tension on the two parts of the two ropes and the linear acceleration of the blocks.* (AMIE Summer, 1984)

Sol. Given :

Bigger weight, $W_1 = 100$ N

Smaller weight, $W_2 = 40$ N

Weight of pulley, $W_0 = 50$ N

Pulley is given as a solid disc.

Let a = Linear acceleration of the system

T_1 = Tension in the rope to which bigger weight is attached

T_2 = Tension in the rope to which smaller weight is attached.

(*i*) *Consider the motion of weight 100 N* [Refer to Fig. 4.42 (*a*)]

$$(100 - T_1) = \frac{100}{9.81} \times a \quad \text{...}(i)$$

(*ii*) *Consider the motion of weight 40 N* [Refer to Fig. 4.42 (*b*)]

$$(T_2 - 40) = \frac{40}{9.81} \times a \quad \text{...}(ii)$$

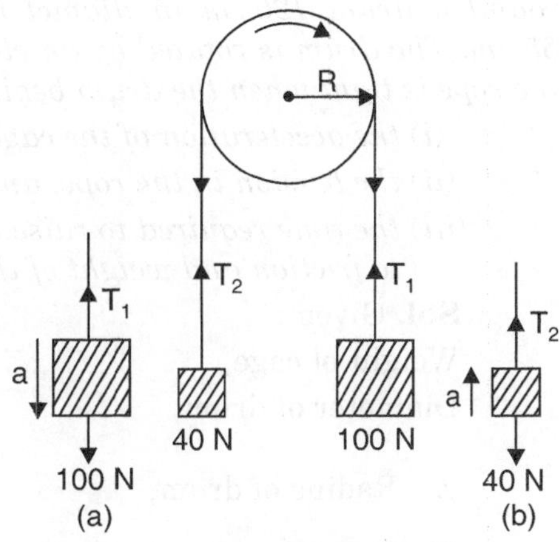

Fig. 4.42

(*iii*) *Consider the rotation of the pulley*

Net torque = $I \times \alpha$

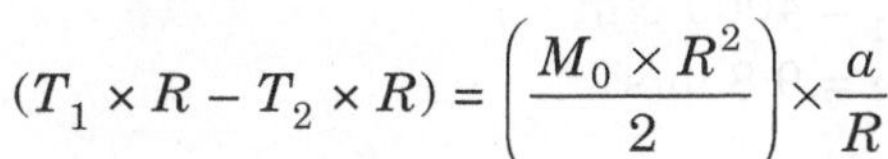

or $$(T_1 \times R - T_2 \times R) = \left(\frac{M_0 \times R^2}{2}\right) \times \frac{a}{R} \qquad \left[\because \; I = \frac{M_0 \times R^2}{2} \text{ and } \alpha = \frac{a}{R} \right]$$

or $$(T_1 - T_2) \times R = \frac{M_0}{2} \times R \times a$$

or $$(T_1 - T_2) = \frac{M_0}{2} \times a \quad \text{But } M_0 = \frac{W_0}{g} = \frac{50}{9.81}$$

$$\therefore \quad (T_1 - T_2) = \frac{50}{2 \times 9.81} \times a$$

or $$(T_1 - T_2) = \frac{25}{9.81} \times a \qquad ...(iii)$$

Add equations (*i*), (*ii*) and (*iii*) to get the value of a.

$$\therefore \quad 100 - 40 = \left(\frac{100 + 40 + 25}{9.81}\right) \times a$$

or $$60 = \frac{165}{9.81} \times a$$

$$\therefore \quad a = \frac{60 \times 9.81}{165} = \mathbf{3.567\ m/s^2}.\quad \textbf{Ans.}$$

Substituting the value of a in equations (*i*) and (*ii*), we get the values of T_1 and T_2.

From equation (*i*), we get

$$(100 - T_1) = \frac{100}{9.81} \times 3.567$$

or $$100 - \frac{100}{9.81} \times 3.567 = T_1$$

$$\therefore \quad T_1 = 100 - 36.36 = \mathbf{33.64\ N}.\quad \textbf{Ans.}$$

From equation (*ii*), we get

$$T_2 - 40 = \frac{40}{9.81} \times 3.567 = 14.54$$

$$\therefore \quad T_2 = 40 + 14.54 = \mathbf{54.54\ N}.\quad \textbf{Ans.}$$

Problem 4.76. *A cage weighing 2940 N is raised by a rope, one end of which is wrapped round a drum 120 cm in diameter and weighing 735 N and having a radius of gyration of 55 cm. The drum is rotated by an electric motor, which exerts constant torque of 4000 Nm. If the rope is tight when the drum begins to rotate, determine :*

(*i*) *the acceleration of the cage,*

(*ii*) *the tension in the rope, and*

(*iii*) *the time required to raise the cage 20 m from ground. Neglect the loss of energy due to friction and weight of the rope. Take g = 9.8 m/s².* (AMIE Winter, 1982)

Sol. Given :

Weight of cage, $W = 2940$ N

Diameter of drum, $D = 120$ cm $= 1.20$ m

$\therefore$ Radius of drum, $R = \frac{1.20}{2} = 0.6$ m

Weight of drum, $W_0 = 735$ N

Radius of gyration of drum, $k = 55$ cm $= 0.55$ m

Constant torque exerted by motor, $T_1 = 4000$ Nm

Acceleration due to gravity, $g = 9.8$ m/s^2

Let a = Acceleration of the cage, and

P = Tension in the rope.

(*i*) *Consider the motion of the cage.* The cage is moving upward with an acceleration a. The net force acting on the cage in the upward direction is equal to tension in the rope minus weight of the cage.

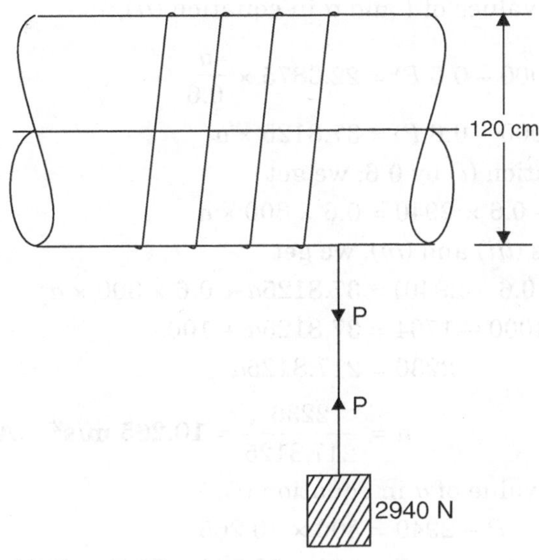

Fig. 4.43

$\therefore$ Net upward force $= (P - 2940)$

Mass of the cage $= \dfrac{\text{Weight}}{g} = \dfrac{2940}{9.8} = 300$ kg

Using, Net force = Mass × Acceleration

or $(P - 2940) = 300 \times a.$ ...(*i*)

(*ii*) *Now consider the rotation of the drum.* The drum is rotated by an electric motor, which exerts constant torque of 4000 N m in the upward direction so that cage may be lifted. But the torque on the drum due to tension P (which is equal to $P \times R$ or $P \times 0.6$ N m) is acting downwards.

$\therefore$ Net torque at the centre of the drum in the upward direction

= Torque exerted by motor – Torque due to P

= $(4000 - 0.6\,P)$ Nm.

But the net torque on drum is also given by equation (4.43).

Net torque $= I \times \alpha$

or $(4000 - 0.6\,P) = I \times \alpha$ ...(*ii*)

where I = Moment of inertia of the drum

= Mass of drum × k^2

$= \dfrac{735}{9.80} \times 0.55^2 \quad \left(\because \text{ Mass of drum} = \dfrac{\text{Weight of drum}}{g} = \dfrac{735}{9.80}\right)$

= 22.6875.

and α = Angular acceleration

$$= \frac{a}{R} \quad \text{[see equation (4.13)]}$$

$$= \frac{a}{0.6} \quad (\because R = 0.6 \text{ m})$$

Substituting the values of I and α in equation (*ii*),

$$(4000 - 0.6\,P) = 22.6875 \times \frac{a}{0.6}$$

or
$$(4000 - 0.6\,P) = 37.8125 \times a \quad ...(iii)$$

Multiplying equation (*i*) by 0.6, we get

$$0.6\,P - 0.6 \times 2940 = 0.6 \times 300 \times a \quad ...(iv)$$

Adding equations (*iii*) and (*iv*), we get

$$(4000 - 0.6 \times 2940) = 37.8125a + 0.6 \times 300 \times a$$

or
$$4000 - 1764 = 37.8125a + 100a$$

or
$$2236 = 217.8125a$$

$$\therefore \quad a = \frac{2236}{217.8125} = \mathbf{10.265\ m/s^2}. \quad \textbf{Ans.}$$

Substituting the value of a in equation (*i*),

$$P - 2940 = 300 \times 10.265$$

or
$$P = 300 \times 10.265 + 2940 = \mathbf{6019.5\ N}. \quad \textbf{Ans.}$$

(*iii*) *Time required to raise the cage 20 m from ground.*

Let t = Time required

Now, initial velocity of cage, $u = 0$

Acceleration, $a = 10.265\ \text{m/s}^2$

Distance, $s = 20$ m

Using equation, $s = ut + \frac{1}{2}at^2$

or
$$20 = 0 \times t + \frac{1}{2} \times 10.265 \times t^2 = \frac{1}{2} \times 10.265t^2$$

$$\therefore \quad t = \sqrt{\frac{2 \times 20}{10.265}} = \mathbf{1.974\ s}. \quad \textbf{Ans.}$$

Problem 4.77. *A homogeneous solid cylinder of weight 100 N whose axis is horizontal rotates about its axis, in frictionless bearings under the action of the weight of a 10 N block which is carried by a rope wrapped round the cylinder. What will be the angular velocity of the cylinder two seconds after the motion starts ? Assume diameter of the cylinder as 100 cm.*

(AMIE Summer, 1983)

Sol. Given :

Weight of cylinder, $W_0 = 100$ N

Weight of block, $W = 10$ N

Diameter of cylinder, $D = 100$ cm = 1 m

$\therefore$ Radius of cylinder, $R = \frac{1}{2} = 0.5$ m

The cylinder is rotating due to the action of weight of 10 N attached to the rope. This means that when weight 10 N is moving downwards, the cylinder is rotating.

Let a = Linear acceleration of the system

P = Tension in the rope

α = Angular acceleration

$= \frac{a}{R} = \frac{a}{0.5}$

$\therefore \quad a = 0.5 \times \alpha.$

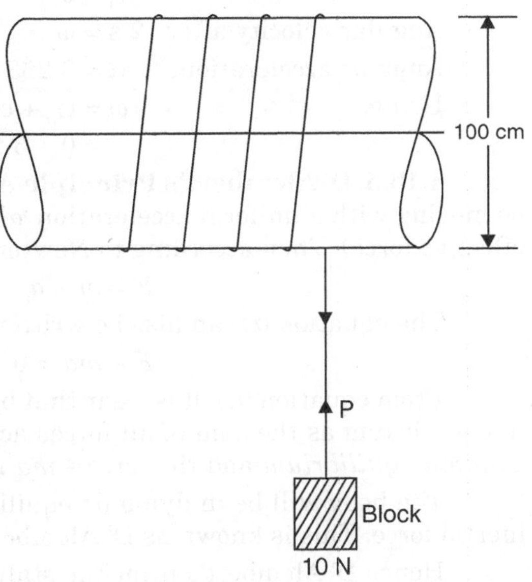

Fig. 4.44

Consider the motion of block of weight 10 N. The block is moving downwards with an acceleration a.

$\therefore$ Net downwards force on block

$= (10 - P)$

Mass of block

$$= \frac{\text{Weight of block}}{g} = \frac{10}{9.80}$$

But, Net force = Mass × Acceleration

$$\therefore \quad (10 - P) = \frac{10}{9.80} \times a$$

$$= \frac{10}{9.80} \times 0.5 \times \alpha$$

$$(\because \quad a = 0.5 \times \alpha) \qquad ...(i)$$

Now consider the rotation of cylinder

Moment of inertia of cylinder about the axis of rotation,

$$I = \frac{MR^2}{2}, \text{ where } M = \text{Mass of cylinder} = \frac{100}{g}$$

$$= \frac{100}{g} \times \frac{(0.5)^2}{2} = \frac{100 \times 0.25}{9.8 \times 2}.$$

Net torque at the centre of cylinder,

$$T = P \times R = P \times 0.5 = 0.5 P.$$

But torque is given by (4.43),

$$T = I \times \alpha$$

$$\therefore \quad 0.5 P = I \times \alpha \qquad (\because \quad T = 0.5 P)$$

$$= \frac{100 \times 0.25}{9.8 \times 2} \times \alpha \qquad \left(\because \quad I = \frac{100 \times 0.25}{9.8 \times 2}\right)$$

Multiplying the above equation by 2, we get

$$0.5 \times 2 \times P = \frac{100 \times .25}{9.8 \times 2} \times 2 \times \alpha$$

or

$$P = \frac{25}{9.8} \alpha \qquad ...(ii)$$

Adding equations (*i*) and (*ii*), we get

$$10 = \frac{10}{9.8} \times 0.5 \times \alpha + \frac{25}{9.8} \alpha = 0.51 \alpha + 2.551 \alpha = 3.06 \alpha$$

$\therefore \qquad \alpha = \frac{10}{3.06} = 3.268 \text{ rad/s}^2.$

Now, initial angular velocity of cylinder,

$\omega_0 = 0$ (cylinder starts from rest)

Angular velocity after 2 s = ω

Angular acceleration, $\alpha = 3.268$

Using, $\omega = \omega_0 + \alpha t$

$= 0 + 3.268 \times 2 =$ **6.536 rad/s. Ans.**

4.10.5. D' Alembert's Principle Applicable to Plane Motion. Let a body of mass 'm' be moving with a uniform acceleration 'a' under the action of external force F (also known as effective force). Then according to Newton's seconds law of motion, we have

$$F = m \times a \qquad ...(i)$$

The equation (i) can also be written as

$$F - ma = 0 \qquad ...(ii)$$

From equation (ii), it is clear that by applying a force $-ma$ on the body, the body will be in equilibrium as the sum of all forces acting on the body is zero. Such equilibrium is called *dynamic equilibrium* and the force $-ma$ is called **D' Alembert force** or *inertia force.*

The body will be in dynamic equilibrium under the action of external force F and the inertia force. This is known as D' Alembert's principle.

Hence D' Alembert's principle states that the net external forces acting on the system and the resultant inertia force are in equilibrium.

The inertia force on a body = Mass of the body × Acceleration.

Inertia force acts in the opposite direction of motion of the body and passes through the C.G. of the body. The above method is considerably quicker for problems where a number of bodies are connected as this method does not require the consideration of tension in the string.

To illustrate this principle, consider the following two cases :

1. *Analysis of the motion of two bodies connected by a string.* (Refer to Art. 4.8.8 on page 274 where this problem is done by another method). Two bodies of weights W_1 and W_2 are connected to the two ends of a light inextensible string passing over a smooth pulley as shown in Fig. 4.45. The weight W_1 is more than weight W_2.

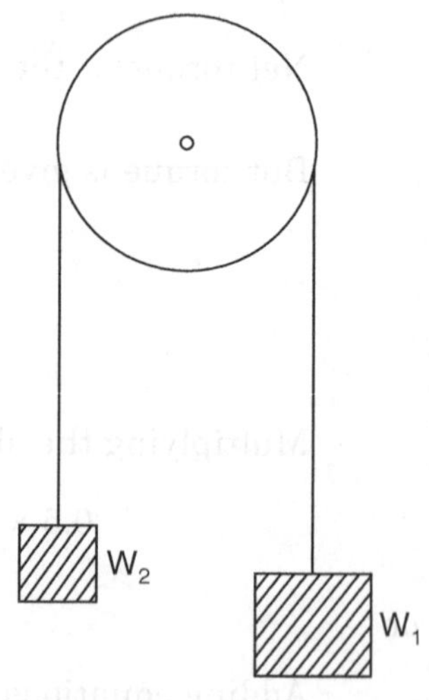

Fig. 4.45

Let a = Acceleration of the system.

Now the net external force acting on the system in the direction of motion

$$= (W_1 - W_2). \qquad ...(i)$$

Inertia force on weight W_1

$= -$ Mass of W_1 × Acceleration

$$= -\frac{W_1}{g} \times a \qquad \left(\because \text{ Mass of } W_1 = \frac{W_1}{g}\right)$$

Inertia force on weight W_2

$= -$ Mass of W_2 × Acceleration

$$= -\frac{W_2}{g} \times a$$

∴ Resultant inertia force

$$= -\frac{W_1}{g} \times a + \left(-\frac{W_2}{g} \times a\right) = -\frac{a}{g}(W_1 + W_2).$$

But according to D' Alembert's Principle the net external force acting on the system and the resultant inertia force should be in equilibrium.

∴ Net external force + Resultant inertia force = 0

or $$(W_1 - W_2) - \frac{a}{g}(W_1 + W_2) = 0$$

or $$(W_1 - W_2) = \frac{a}{g}(W_1 + W_2)$$

∴ $$a = \frac{g(W_1 - W_2)}{(W_1 + W_2)}.$$

The acceleration given by the above equation is the same as given by equation (4.25). But the above method is more simple.

2. *Analysis of motion of two bodies connected by a string when one body is lying on a smooth horizontal surface and other is hanging free.* (Refer to Art. 4.8.9 on page 281.)

Let a = Acceleration of the system.

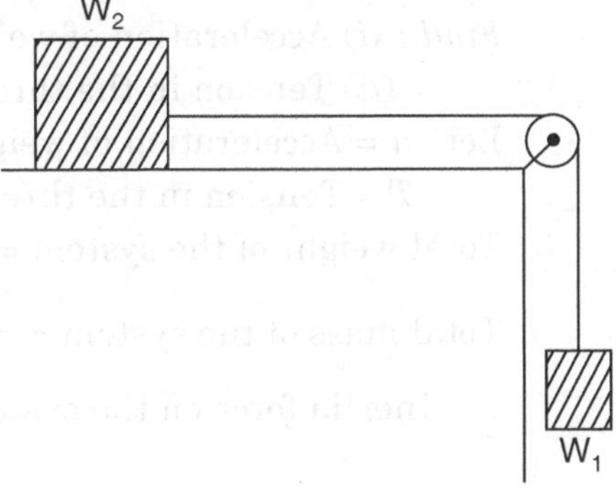

Fig. 4.45 (a)

Net external force acting on the system shown in Fig. 4.45 (a) in the direction of motion = W_1.

(Kindly note here, the weight W_2 is not acting in the direction of motion).

Motion of W_2 is horizontal whereas W_2 is acting downwards. Motion of W_1 is vertically downwards and also W_1 is acting downwards.

Inertia force on weight W_1

= – (Mass of W_1) × (Acceleration)

$$= -\frac{W_1}{g} \times a \qquad \left(\because \text{ Mass} = \frac{W_1}{g}\right)$$

Inertia force on weight W_2

= – (Mass of W_2) × Acceleration

$$= -\frac{W_2}{g} \times a.$$

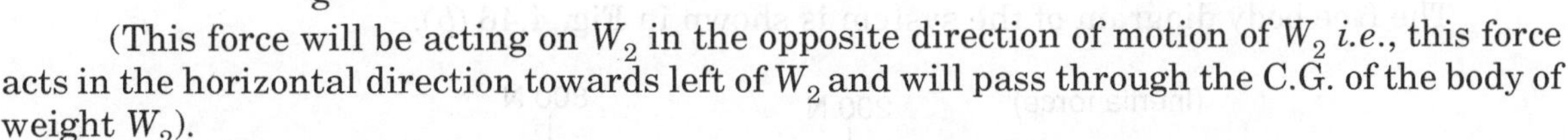

(This force will be acting on W_2 in the opposite direction of motion of W_2 *i.e.*, this force acts in the horizontal direction towards left of W_2 and will pass through the C.G. of the body of weight W_2).

∴ Resultant inertia force

$$= -\frac{W_1}{g} a + \left(-\frac{W_2}{g} a\right) = -\frac{a}{g}[W_1 + W_2].$$

But according to D' Alembert's principle, the net external force acting on the system in the direction of motion and the resultant inertia forces should be in equilibrium.

$\therefore$ Net external force + Resultant inertia force = 0.

or $$W_1 + \left[-\frac{a}{g}(W_1 + W_2) \right] = 0$$

or $$W_1 = \frac{a}{g}[W_1 + W_2] \qquad \therefore \quad a = \frac{gW_1}{(W_1 + W_2)}.$$

The above equation is the same as equation (4.27).

Problem 4.78. *Two weights 800 N and 200 N are connected by a thread and move along a rough horizontal plane under the action of a force 400 N applied to the first weight of 800 N as shown in Fig. 4.46 (a). The coefficient of friction between the sliding surfaces of the weights and the plane is 0.3. Determine the acceleration of the weights and the tension in the thread using D' Alembert's principle.*

Sol. Given :

Weights, $W_1 = 200$ N

$W_2 = 800$ N

Force applied, $F = 400$ N

Coefficient of friction, $\mu = 0.3$

Find : (*i*) Acceleration of weights

(*ii*) Tension in the threads

800 N
200 N
Thread
400 N
Rough Horizontal Plane

Fig. 4.46 (a)

Let a = Acceleration of weights and

T = Tension in the threads.

Total weight of the system = $W_1 + W_2 = 200 + 800 = 1000$ N

$$\text{Total mass of the system} = \frac{\text{Total weight}}{g} = \frac{1000}{g} \text{ kg}$$

$\therefore$ Inertia force on the system

$$= -\text{(Mass of system)} \times \text{Acceleration of the system}$$

$$= -\left(\frac{1000}{g}\right) \times a \text{ Newton.}$$

Force of friction on the system = μ × Total weight

$= 0.3 \times (1000) = 300$ N ($\leftarrow$)

This is also acting in the backward direction.

The free body diagram of the system is shown in Fig. 4.46 (*b*).

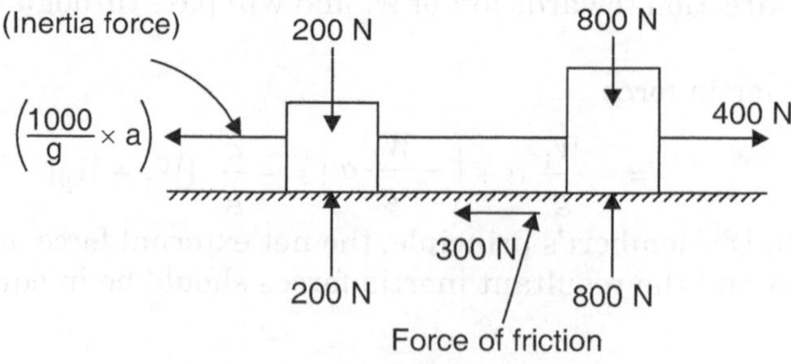

Fig. 4.46 (b)

By D' Alembert principle, the sum of the inertia force and the forces on the system is zero.

Force applied + Frictional force + Inertia force = 0

$$\therefore \quad 400 - 300 - \frac{1000}{g} \times a = 0 \quad \text{or} \quad 100 = \frac{1000 \times a}{g}$$

or $$a = \frac{100 \times g}{1000} = \frac{g}{10} = \frac{9.80}{10} = \mathbf{0.98\ m/s^2}. \quad \textbf{Ans.}$$

Tension in the thread

Let T = Tension in the thread

Consider the motion of 200 N weight.

The free body diagram of 200 N weight is shown in Fig. 4.46 (*c*).

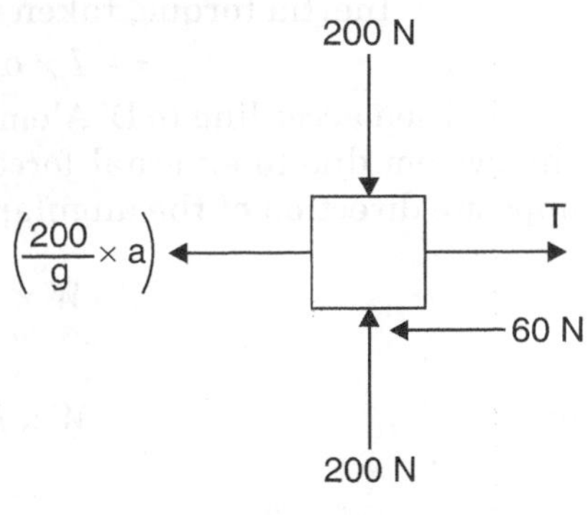

Fig. 4.46 (c)

Force of friction = μ × 200 = 0.3 × 200 = 60 N

Inertia force on it = Mass × Acceleration

$$= -\left(\frac{200}{g}\right) \times a = \frac{200}{g} \times \frac{g}{10} = -\ 20\ \text{N}$$

By D' Alembert's principle, $T - 60 - 20 = 0$

$$\therefore \quad T = 60 + 20 = \mathbf{80\ N}. \quad \textbf{Ans.}$$

4.10.6. D' Alembert's Principle Applicable to Rotary Motion. It states that when external torques (also called active torques) act on a system having rotating motion, then the algebraic sum of all the torques acting on the system due to external forces and inertia forces including the inertia torques (taken in the opposite direction of the angular acceleration) is zero.

To illustrate the above principle, consider the following two cases :

1. *Rotation due to a weight W attached to one end of a string, passing over a pulley of weight W_0.* (Refer to Art. 4.10.3 on page 302 where this problem is done by another method).

A weight W is attached to one end of an inextensible string, which passes over a pulley of weight W_0, the other end of the string is attached to the periphery of the pulley as shown in Fig. 4.47. When the weight W moves downwards, the rotation to the pulley is caused in clockwise direction.

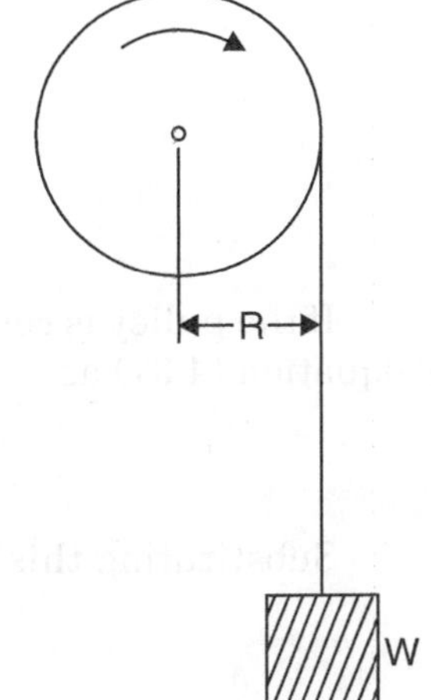

Fig. 4.47

Let a = Linear acceleration of the weight W

α = Angular acceleration of the pulley

R = Radius of the pulley

I = Moment of inertia of the pulley about the axis of rotation.

From equation (4.13), we know

Linear acceleration = Angular acceleration × Radius.

$$\therefore \quad a = \alpha \times R$$

Inertia force on weight W

= – (Mass of W) × Acceleration

$$= -\left(\frac{W}{g}\right) \times a = -\left(\frac{W}{g}\right) \times \alpha \times R \qquad (\because \ a = \alpha \times R)$$

Now torque at the centre of the pulley due to external force = $W \times R$.

Torque due to inertia force = (Inertia force) × R

$$= \left(-\frac{W}{g} \times \alpha \times R\right) \times R = -\frac{W}{g} \times \alpha \times R^2.$$

Torque at the centre of the pulley due to angular acceleration is given by equation (4.43)

$$= I \times \alpha.$$

∴ Inertia torque, taken in opposite direction of the angular acceleration

$$= -I \times \alpha.$$

Then according to D' Alembert's principle, the algebraic sum of all the torques acting on the system due to external forces and inertia forces including inertia torques (taken in the opposite direction of the angular acceleration) should be zero.

$$\therefore \quad W \times R - \frac{W}{g} \times \alpha \times R^2 - I \times \alpha = 0$$

or

$$W \times R = \frac{W}{g} \times \alpha \times R^2 + I \times \alpha = \alpha \left[\frac{W}{g} R^2 + I\right]$$

$$\therefore \quad \alpha = \left(\frac{W \times R}{\frac{W}{g} R^2 + I}\right) \quad \ldots(i)$$

If linear acceleration 'a' is to be determined, then

$$\alpha = \frac{a}{R}.$$

∴ Equation (i) becomes as

$$\frac{a}{R} = \frac{W \times R}{\frac{W}{g} R^2 + I}$$

$$\therefore \quad a = \frac{WR^2}{\frac{W}{g} R^2 + I} \quad \ldots(ii)$$

If the pulley is considered to be a solid disc, then moment of inertia of the pulley is given by equation (4.37) as

$$I = \frac{MR^2}{2}, \text{ where } M = \text{Mass of pulley} = \left(\frac{W_0}{g}\right) = \left(\frac{W_0}{g}\right)\frac{R^2}{2}.$$

Substituting this value in equation (ii), we get

$$a = \frac{WR^2}{\frac{W}{g} R^2 + \frac{W_0}{g} \frac{R^2}{2}} = \frac{gW}{\left(W + \frac{W_0}{2}\right)}$$

The above result is same as given by equation (4.46).

2. *Rotation due to weights attached to the two ends of a string, which passes over a rough pulley of weight* W_0 (Refer to Art. 4.10.4 on page 305).

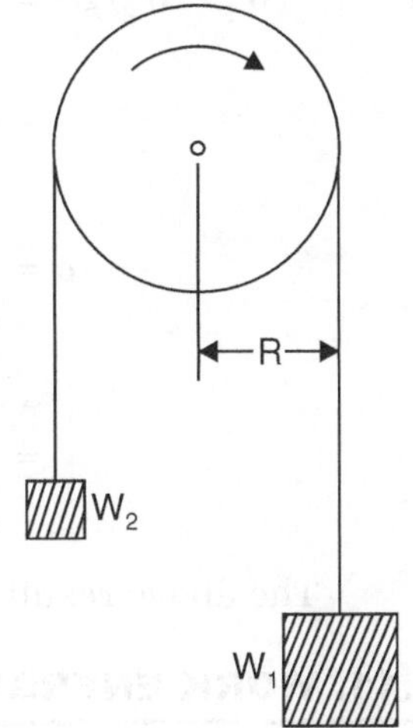

Fig. 4.48

Let R = Radius of pulley

a = Linear acceleration of the system

α = Angular acceleration of the pulley

W_1, W_2 = Two weights attached

W_0 = Weight of pulley

I = Moment of inertia of the pulley

$$= \frac{MR^2}{2} \text{ if pulley is considered a solid disc}$$

$$= \frac{W_0}{g}\frac{R^2}{2} \qquad \left(\because \text{ Mass of pulley} = \frac{\text{Weight of pulley}}{g}\right)$$

Consider $W_1 > W_2$, so that the rotation of the pulley is caused in clockwise direction.

Resultant torque at the centre of the pulley caused by the weights

$$= W_1 \times R - W_2 R = (W_1 - W_2) \times R.$$

Inertia force on weight W_1

$$= -(\text{Mass of } W_1) \times \text{Acceleration} = -\frac{W_1}{g} \times a.$$

$\therefore$ Torque due to reversed force on weight W_1

$$= \left(-\frac{W_1}{g} \times a\right) \times R$$

Inertia force on weight W_2

$$= -\frac{W_2}{g} \times a$$

$\therefore$ Torque due to reversed effective force on weight W_2

$$= -\frac{W_2}{g} \times a \times R.$$

Torque due to angular acceleration on pulley = $I \times \alpha$ [See equation (4.43)].

$\therefore$ Inertia torque taken in opposite direction = $-I \times \alpha$.

Now according to D' Alembert's principle, the algebraic sum of all the torques acting on the system due to external forces and inertia forces including inertia torques should be zero.

$$\therefore \quad (W_1 - W_2) \times R - \frac{W_1}{g} \times a \times R - \frac{W_2}{g} \times a \times R - I \times \alpha = 0$$

or
$$(W_1 - W_2) \times R - \frac{W_1}{g} \times a \times R - \frac{W_2}{g} \times a \times R - I \times \frac{a}{R} = 0 \qquad \left(\because \alpha = \frac{a}{R}\right)$$

or $$(W_1 - W_2) R^2 = \frac{W_1}{g} \times a \times R^2 + \frac{W_2}{g} a \times R^2 + I \times a$$

$$= a \left(\frac{W_1 R^2}{g} + \frac{W_2 R^2}{g} + I \right)$$

$$a = \frac{(W_1 - W_2) \times R^2}{\left(\frac{W_1 R^2}{g} + \frac{W_2 R^2}{g} + I \right)} = \frac{(W_1 - W_2) \times R^2}{\left(\frac{W_1 R^2}{g} + \frac{W_2 R^2}{g} + \frac{W_0}{g} \frac{R^2}{2} \right)} \qquad \left(\because \; I = \frac{W_0}{g} \frac{R^2}{2} \right)$$

$$= \frac{g(W_1 - W_2)}{\left(W_1 + W_2 + \frac{W_0}{2} \right)}$$

The above result is same as given by equation (4.48).

4.11. WORK ENERGY EQUATION

The various important methods of solving kinetic problems of rigid bodies are :

1. Newton's laws of motion
2. Work and energy method
3. Impulse-momentum method.

Solving kinetic problems by using Newton's laws of motion, have been described in previous articles. From second law of motion ($F = m \times a$), the acceleration of the rigid body can be obtained. After knowing acceleration, the velocity and displacement of the body as a function of time can be determined. Hence Newton's second law relates the force, mass and acceleration.

The kinetic problems can also be solved by work-energy method. The work-energy approach relates force, velocity and displacements. Work and energy are scalar quantities and hence can be added algebraically. Secondly in work-energy method the velocity and displacement of the rigid body can be determined directly without knowing acceleration of the body. Thirdly for the connected bodies, the motion can be obtained without drawing free-body diagram of each body.

The work energy equation states that the total work done by all the forces acting on a particle moving from one position to other, is equal to change of the kinetic energy of the particle. Mathematically it is written as :

$$(\Sigma F_x) \,.\, S = \frac{1}{2} m \,(v^2 - u^2)$$

where ΣF_x = Sum of all the forces acting on the particle

v = Final velocity of particle

u = Initial velocity of particle.

To understand fully the work energy equation, let us first define work and energy.

4.11.1. Work and Energy. Work is defined as the product of force and displacement of the body on which the force is acting. The force and displacement should be in the same direction. *Energy* is defined as the capacity to do work. The energy exists in many forms like, mechanical, electrical, heat, chemical and light etc. But in engineering mechanics, we only consider mechanical energy. This article deals with work and energy.

Work. As defined above, work is the product of force and distance. The distance should be in the direction of the force. If a force P is acting on a body and the body moves a distance of S in the direction of the force, as shown in Fig. 4.49 (*a*), then the work done on the body is given by :

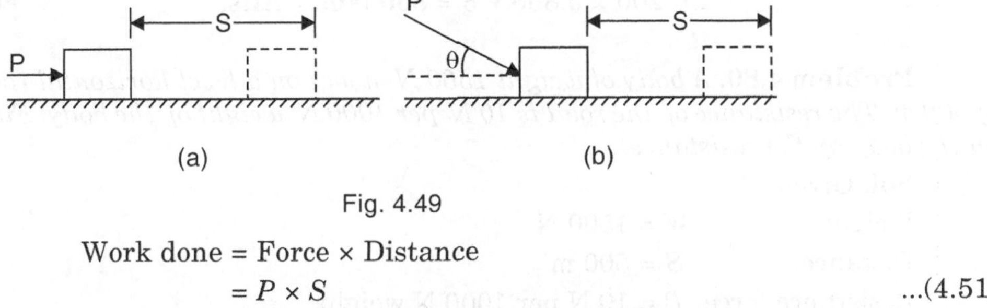

Fig. 4.49

$$\text{Work done} = \text{Force} \times \text{Distance}$$
$$= P \times S \qquad ...(4.51)$$

But if the force acting on the body and the distance moved are not in the same direction as shown in Fig. 4.49 (*b*), then the work done on the body is given by :

$$\text{Work done} = \text{Component of force in the direction of motion} \times \text{Distance}$$
$$= P \cos \theta \times S \qquad ...(4.52)$$

where θ is the angle made by the force with the direction of motion.

If $\theta = 90°$, $\cos \theta = 0$ and hence work done will be zero. Hence if the force and displacement are at right angles, then work done will be zero. If a man of weight W is moving on a horizontal road and the distance moved by man is S, then the work done by the weight of man will be zero. This is due to the fact that W and S are at right angle.

Units of Work. Work is the product of force and distance. In S.I. system, force is expressed in Newton (N) whereas the distance is expressed in metre (m). Hence the unit of work is Nm.

The work done by a force of 1 N on a body which moves a distance of 1 m, is called 1 Nm. This is also known as one Joule (J). Hence one Joule is the work done by a force of 1 N on a body whose displacement is 1 m.

Problem 4.79. *Find the work done in drawing a body :*

(*i*) *Weighing 500 N through a distance of 5 m along a horizontal surface by a horizontal force of 250 N.*

(*ii*) *Weighing 500 N through a distance of 5 m along a horizontal surface by a force of 200 N whose line of action makes an angle of 30° with the horizontal.*

Sol. Given :

Weight, $W = 500$ N

Distance, $S = 5$ m

Force, $P = 250$ N

(*i*) The forces acting on the body are shown in Fig. 4.50 (*a*).

Work done = Force × Distance

= 250 × 5 = **1250 Nm. Ans.**

(Note that here the force applied and not the weight of the body is taken into account).

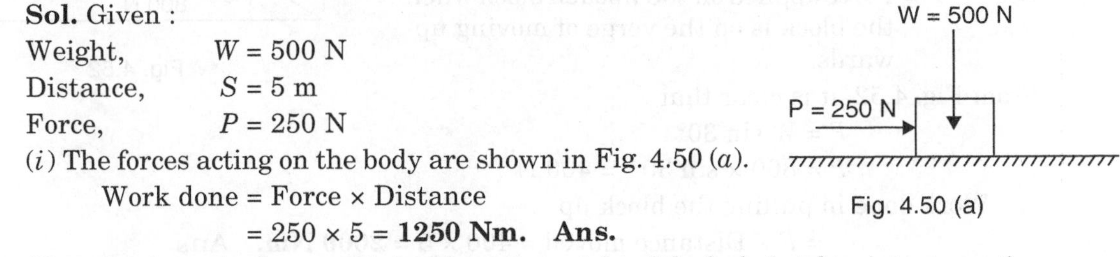

Fig. 4.50 (a)

(*ii*) $W = 500$ N, $P = 200$ N, $\theta = 30°$

[Refer to Fig. 4.50 (*b*)]

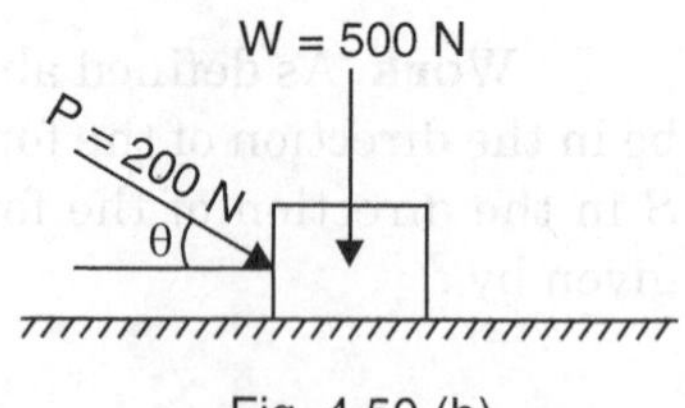

Fig. 4.50 (b)

Here the work done is given by equation (4.52).

$\therefore$ Work done $= P \cos \theta \times$ Distance

$= (200 \times \cos 30°) \times 5$

$= 200 \times 0.866 \times 5 =$ **866 Nm. Ans.**

Problem 4.80. *A body of weight 1500 N moves on a level horizontal road for a distance of 500 m. The resistance of the road is 10 N per 1000 N weight of the body. Find the work done on the body by the resistance.*

Sol. Given :

Weight, $W = 1500$ N

Distance, $S = 500$ m

Resistance force, $P = 10$ N per 1000 N weight

$$= \frac{10}{1000} \times W = \frac{10}{1000} \times 1500 = 15 \text{ N}$$

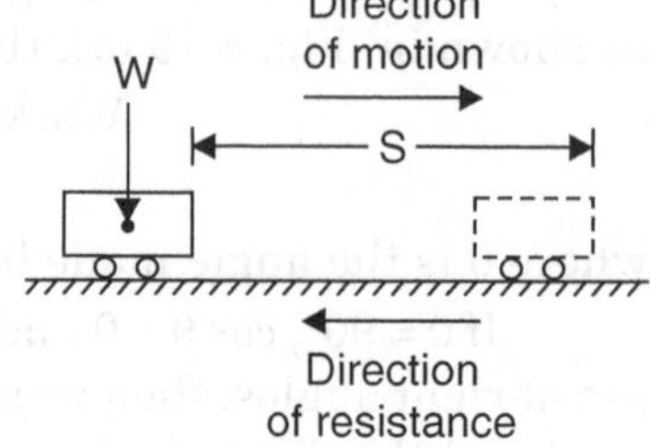

Fig. 4.51

The work done by weight of the body will be zero as W and S are at right angles. The work will be done only by resistance force as this force and displacement are in the same line. (Though here resistance force and displacement are in opposite direction and hence work done will be –ve).

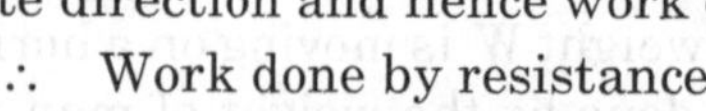

$\therefore$ Work done by resistance

= Resistance force × Distance

$= P \times S = 15 \times 500$ Nm = **7500 Nm. Ans.**

Problem 4.81. *A block of wood of weight 800 N is placed on a smooth inclined plane which makes an angle of 30° with the horizontal. Find the work done in pulling the block up for a length of 5 m.*

Sol. Given :

Weight, $W = 800$ N

Angle, $\theta = 30°$

Distance along the plane moved by block = 5 m.

The inclined plane is given smooth. Hence coefficient of friction will be zero. This means the force of friction will also be zero.

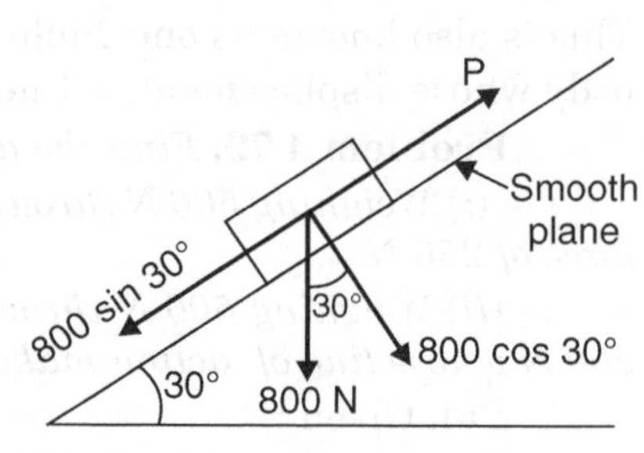

Fig. 4.52

Let P = Force applied on the wooden block when the block is on the verge of moving upwards.

From Fig. 4.52, it is clear that

$P = W \sin 30°$

$\therefore$ $P = 800 \times \sin 30° = 400$ N

$\therefore$ Work done in pulling the block up

$= P \times$ Distance moved $= 400 \times 5 =$ **2000 Nm. Ans.**

Problem 4.82. *If in Problem 4.81, the inclined plane is rough and the coefficient of friction between the block and plane is 0.3, then find the work done in pulling the block up for the length of 5 m.*

Sol. Given :

From Problem 4.81, $W = 800$ N, $\theta = 30°$, $S = 5$ m

Coefficient of friction, $\mu = 0.3$

Let P = Force applied so that the block starts moving up the plane.

The forces acting on the block are shown in Fig. 4.53.

For equilibrium of the block

$$R = 800 \cos 30°$$

(Force normal to inclined plane)

$\therefore$ Force of friction,

$$F = \mu R = 0.3 \times 800 \cos 30° = 207.846 \text{ N}$$

Consider forces along the plane,

$$P = 800 \sin 30° + F$$
$$= 400 + 207.846 = 607.846 \text{ N}$$

$\therefore$ Work done $= P \times S$

$$= 607.846 \times 5 = \mathbf{3039.23 \text{ Nm.}} \quad \textbf{Ans.}$$

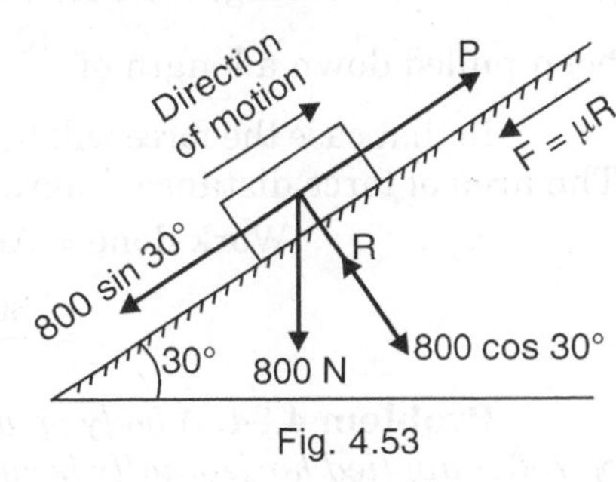

Fig. 4.53

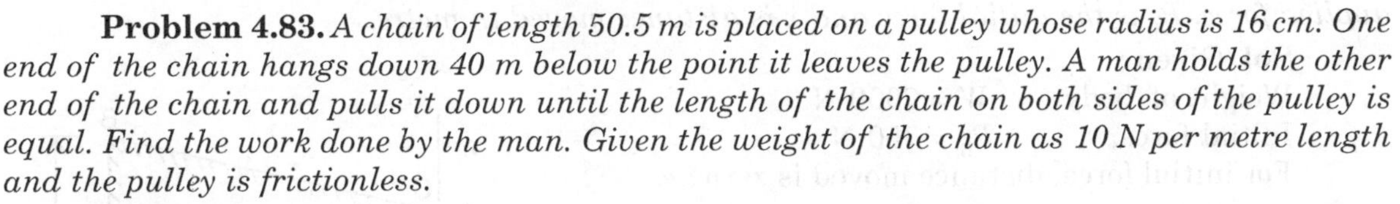

Problem 4.83. *A chain of length 50.5 m is placed on a pulley whose radius is 16 cm. One end of the chain hangs down 40 m below the point it leaves the pulley. A man holds the other end of the chain and pulls it down until the length of the chain on both sides of the pulley is equal. Find the work done by the man. Given the weight of the chain as 10 N per metre length and the pulley is frictionless.*

Sol. Given :

Total length of chain, $L = 50.5$ m

Radius of pulley, $R = 16$ cm $= 0.16$ m

Length of chain between A and $C = 40$ m

Weight of chain $= 10$ N per m length

$= 10 \times$ Total length of chain

$= 10 \times 50.5 = 505$ N

Length BD = Total length of chain − Length AB − Length AC

$= 50.5 - \frac{1}{2}$ of circumference of pulley $- 40$

$= 50.5 - \frac{1}{2}(2\pi R) - 40 = 50.5 - \frac{1}{2}(2\pi \times 0.16) - 40$

$= 50.5 - 0.5 - 40 = 10$ m

Now weight of chain

$AC = 10 \times$ length $AC = 10 \times 40 = 400$ N

Weight of chain

$BD = 10 \times$ length $BD = 10 \times 10 = 100$ N

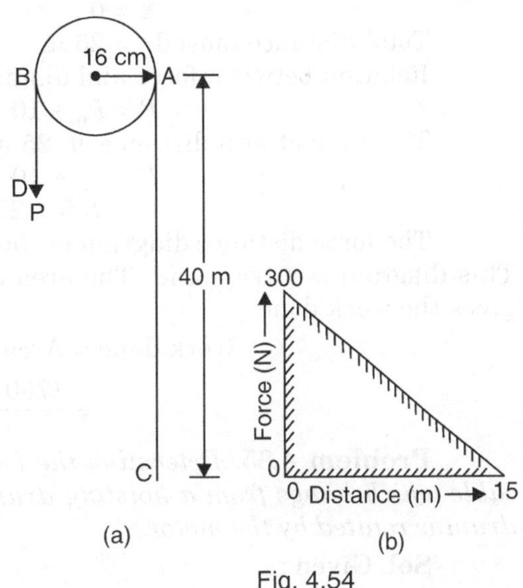

Fig. 4.54

Let P = Force applied at point D by the man

The force P applied at D in the beginning is given by the relation :

P + Weight of chain BD = Weight of chain AC

or $P + 100 = 400$

$\therefore$ $P = 400 - 100$

$= 300$ N

When the chain is pulled, the length *BD* of the chain goes on gradually increasing and length *AC* of the chain goes on gradually decreasing. The force required to pull the chain, which is equal to the difference of the weights of the chain on either side of the pulley, goes on gradually decreasing, till the length of the chain on both the sides is equal (after chain has been pulled down a length of $\frac{40 - 10}{2} = \frac{30}{2} = 15$ m).

In that case the force will become zero. The force distance diagram is shown in Fig. 4.54 (*b*). The area of force distance diagram represents the work done.

$\therefore$ Work done = Area of the curve

$$= \frac{300 \times 15}{2} = \textbf{2250 Nm.} \quad \textbf{Ans.}$$

Problem 4.84. *A body of weight 2000 N is placed on a horizontal surface. An initial force of 750 N applied horizontally is sufficient just to move the body. The applied force is not constant but varies uniformly with the distance moved by the body. If the body moves a distance of 25 m and force applied at any distance is given by relation $P = P_0 + 10x$, find the work done by the applied force. P_0 is the initial force and x is distance moved in metre.*

Sol. Given :

Weight of body, $W = 2000$ N

Initial force, $P_0 = 750$ N

For initial force, distance moved is zero *i.e.*,

$x = 0$

Total distance moved $= 25$ m

Relation between force and distance,

$P = P_0 + 10\,x$

The force after a distance of 25 m,

$P = P_0 + 10 \times 25$

$= 750 + 250 = 1000$ N

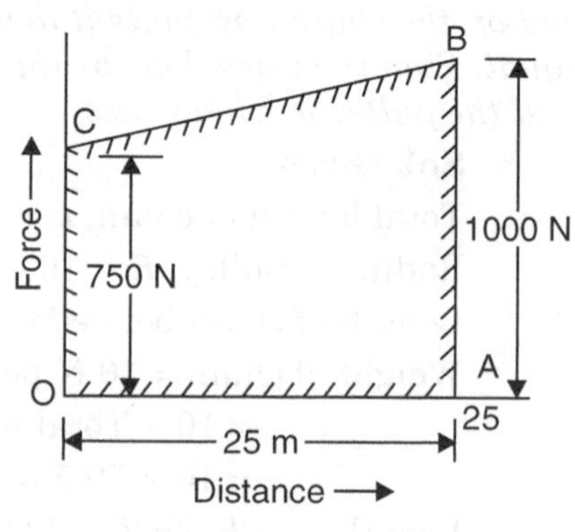

Fig. 4.55

The force distance diagram is shown in Fig. 4.55. This diagram is a trapezoid. The area of this diagram, gives the work done.

$\therefore$ Work done = Area of trapezoidal

$$= \frac{(750 + 1000)}{2} \times 25 = \textbf{21875 Nm.} \quad \textbf{Ans.}$$

Problem 4.85. *Determine the work done by electric motor in winding up a uniform cable which hangs from a hoisting drum if its free length is 10 m and it weighs 500 N. The drum is rotated by the motor.* (AMIE Summer, 1984)

Sol. Given :

Free length of cable, $L = 10$ m

Weight of 10 m length cable $= 500$ N

Weight of cable per m length $= \frac{500}{10} = 50$ N

Refer to Fig. 4.56.

Consider an element of length dx at a distance of x m from free end. This element is to be raised to a height of $(10 - x)$ m. The work done on this element is equal to the product of its weight and the height through which it is raised.

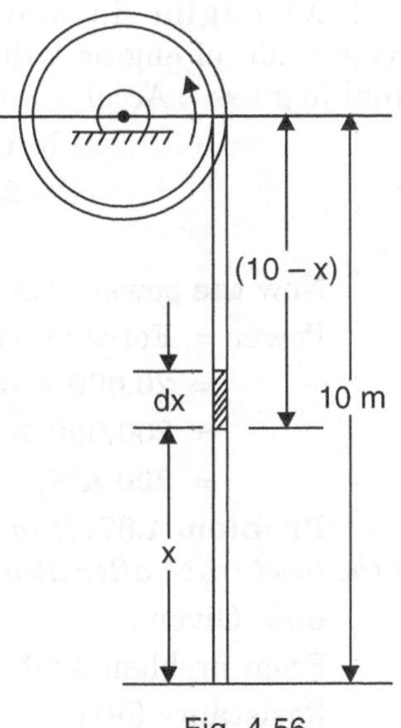

Fig. 4.56

Weight of element = Weight of cable per m length × length of element

$$= 50 \times dx = 50dx$$

Work done on the element,

$$dW = \text{Weight of element} \times \text{Height through which it is raised}$$

$$= (50dx) \times (10 - x) = 50 \times (10 - x)dx$$

Total work done will be obtained by integrating the above equation from 0 to 10 m.

$$\therefore \quad W = \int_0^{10} 50\,(10 - x)\,dx$$

$$= 50\left[10x - \frac{x^2}{2}\right]_0^{10}$$

$$= 50\left[10 \times 10 - \frac{10^2}{2}\right] = 50\,[100 - 50] = 50 \times 50 = \mathbf{2500\ Nm.} \quad \textbf{Ans.}$$

4.11.2. Power. The rate of doing work is known as power. Hence, power can be obtained by dividing the total work done by time. Or in other words, the power is the work done per second. The unit of power in S.I. system is Nm/s or Watt. It is denoted by W. The power in kilowatts is obtained by dividing watt by 1000.

$$\therefore \quad \text{Power} = \text{Work done per second}$$

$$= \frac{\text{Force} \times \text{Distance}}{\text{Time}} = \text{Force} \times \frac{\text{Distance}}{\text{Time}} = \text{Force} \times \text{Velocity} \qquad ...(4.53)$$

The force and velocity should be in the same direction.

Problem 4.86. *A train of weight 2000 kN is pulled by an engine on a level track at a constant speed of 36 kilometre per hour. The resistance due to friction is 10 N per kN of the trains weight. Find the power of the engine.*

Sol. Given :

Weight of train, $W = 2000$ kN

Speed of train, $v = 36$ km/hr

$$= \frac{36 \times 1000}{60 \times 60} = 10 \text{ m/s}$$

Resistance due to friction,

$$F = 10 \text{ N per kN weight of train}$$

$$= 10 \times \text{weight of train in kN} = 10 \times 2000 = 20{,}000 \text{ N}$$

Let P = Force exerted by the engine to pull the train at constant speed.

$\therefore$ Net force in the direction of motion $= P - F = (P - 20{,}000)$ N

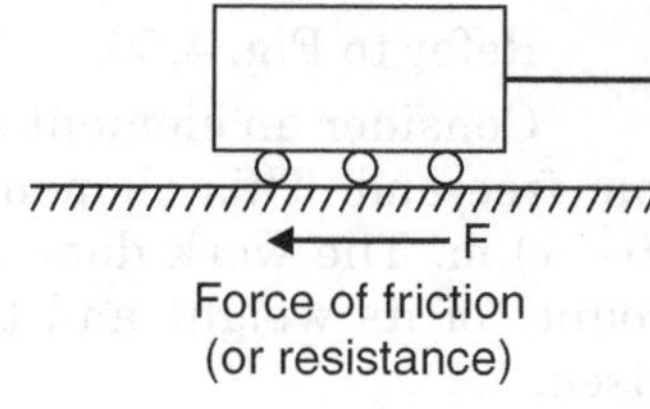

Fig. 4.57

As engine is moving with uniform speed, the acceleration of engine will be zero. The net force, which is equal to mass × Acceleration, will also be zero.

$\therefore$ Net force = 0

or $P - 20{,}000 = 0$

or $P = 20{,}000$ N

Now the power is given by equation (4.53), as

Power = Force exerted by engine × Velocity

= 20,000 × 10 Nm/s

= 200,000 W (∵ Watt, W = Nm/s)

= **200 kW. Ans.**

Problem 4.87. *If in problem 4.86, the train is to move with an acceleration of 0.5 m/s² on the level track after attaining a speed of 36 km/hr, then find the power of the engine.*

Sol. Given :

From problem 4.86, $W = 2000$ kN

Resistance force, $F = 20{,}000$ N

Velocity, $v = 10$ m/s

Acceleration, $a = 0.5 \text{ m/s}^2$

Now, mass of the train, $m = \dfrac{W}{g} = \dfrac{2000 \text{ kN}}{9.81 \text{ m/s}^2}$

$$= \frac{2000 \times 1000 \text{ N}}{9.81 \text{ m/s}^2} = \frac{2000 \times 1000}{9.81} \text{ kg}$$

Let P = Force exerted by engine, when train is moving with an acceleration of 0.5 m/s²

Now using the relation,

Net force = Mass × Acceleration

or $(P - F) = m \times a$ or $F - 20{,}000 = \left(\dfrac{2000 \times 1000}{9.81}\right) \times 0.5$

$$\therefore \quad P = 20{,}000 + \frac{2000 \times 1000}{9.81} \times 0.5 = 20{,}000 + 101936.1 = 121936.8 \text{ N}$$

$\therefore$ Power = Force exerted by engine × Velocity = 121936.8 × 10 Nm/s

= 1219368 W = **1219.368 kW. Ans.**

Problem 4.88. *A train of weight 1500 kN, is ascending (i.e., going upwards) a slope of 1 in 100 with a uniform speed of 36 km/hr. Find the power exerted by the engine, if the road resistance is 5 N per kN weight of the train.*

Sol. Given :

Weight of train, $W = 1500$ kN

= 1500 × 1000 N

Slope* = 1 in 100

$\therefore \quad \sin\theta = \dfrac{1}{100}$

* When slope is very small, then $\theta = \tan\theta = \sin\theta$.

Speed, $v = 36$ km/hr

$$= \frac{36 \times 1000}{60 \times 60} \text{ m/s} = 10 \text{ m/s}$$

Resistance force (or Force of friction)

$F = 5$ N per kN weight of train

$= 5 \text{ N} \times$ Weight of train in kN

$= 5 \times 1500 = 7500$ N

Let F = Force exerted by the engine.

Fig. 4.58

Fig. 4.58 shows the various forces acting on the train, when it is moving up. The net force acting along the plane in the upward direction

$$= P - W \sin \theta - F$$

$$\therefore \quad \text{Net force} = P - W \sin \theta - F$$

$$= P - (1500 \times 1000) \times \frac{1}{100} - 7500 = P - 15000 - 7500$$

But the train is moving with uniform speed, and hence acceleration of the train is zero. If acceleration is zero, then net force acting on the engine should also be zero.

$$\therefore \quad P - 15000 - 7500 = 0$$

or $$P = 15000 + 7500 = 22500 \text{ N}$$

$\therefore$ Power exerted by engine = Force exerted by engine × Velocity

$= 22500 \times 10$ Nm/s $= 225000$ W $=$ **225 kW. Ans.**

Problem 4.89. *A train of weight 2000 kN moves down a slope of 1 in 150 at 18 km/hr and engine develops a power of 35 kW. If the train is pulled up at the same speed, what power will be required to pull the train.*

Sol. Given :

Weight, $W = 2000 \text{ kN} = 2000 \times 1000$ N

Slope $= 1$ in 150

$$\therefore \quad \sin \theta = \frac{1}{150}$$

Speed of train, $$v = 18 \text{ km/hr} = \frac{18 \times 1000}{60 \times 60} = 5 \text{ m/s}$$

Power developed by engine $= 35 \text{ kW} = 35 \times 1000 \text{ W} = 35000$ W.

1st Case. Train is moving down the plane as shown in Fig. 4.59 with a constant velocity and hence it will not have any acceleration. So the net force acting on the train in the direction of motion should be zero. But the forces acting on the train are :

(*i*) $W \sin \theta$ in the direction of motion ;

(*ii*) Force of friction (F) in the opposite direction of motion ;

(*iii*) Force (P) exerted by engine in the direction of motion.

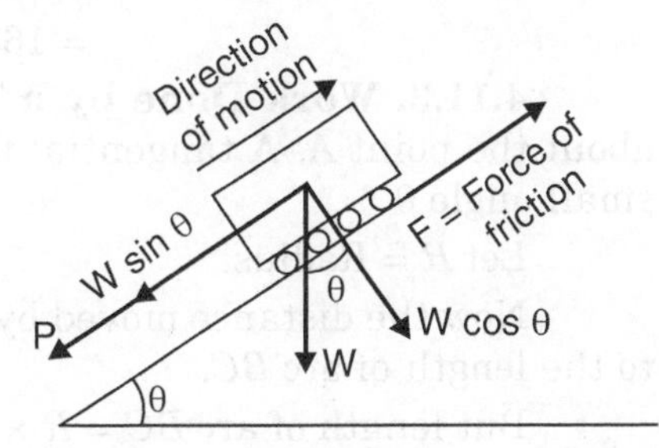

Fig. 4.59

Now net force acting on the engine in the direction of motion

$$= W \sin\theta - F + P = (2000 \times 1000) \times \frac{1}{150} - F + P$$

$$= 13333.3 - F + P \qquad ...(i)$$

The power developed by engine

$$= (\text{Force exerted by engine}) \times \text{Velocity}$$

$$= P \times 5 \quad \text{or} \quad 35{,}000 = 5 \times P$$

$$\therefore \quad P = \frac{35{,}000}{5} = 7000 \text{ N}$$

Substituting the value of P in equation (i), we get

Net force $= 13333.3 - F + 7000$

But net force in the direction of motion is zero.

$\therefore$ $13333.3 - F + 7000 = 0$ or $13333.3 + 7000 = F$ or $20333.3 = F$

$$\therefore \quad F = 20333.3 \text{ N} \qquad ...(ii)$$

2nd Case. The train is moving up the same plane with the same speed as shown in Fig. 4.60.

As the plane is same and hence frictional force will be same.

Let P^* = Force exerted by engine while moving up

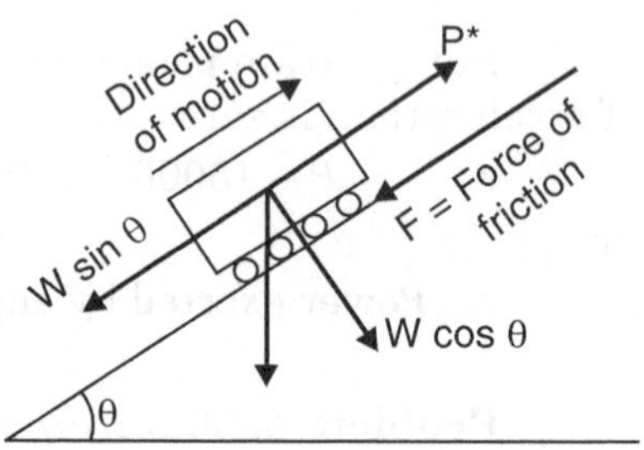

Net force in the direction of motion

$$= P^* - W \sin\theta - F$$

$$= P^* - (2000 \times 1000) \times \frac{1}{150} - 20333.3$$

$$= P^* - 13333.3 - 20333.3$$

$$= P^* - 33666.6.$$

Direction of motion
P^*
F = Force of friction
$W \sin\theta$
$W \cos\theta$
θ

Fig. 4.60

The train is moving with constant speed and hence acceleration is zero. This means the net force will also be zero.

$$\therefore \quad P^* - 33666.6 = 0$$

or

$$P^* = 33666.6 \text{ N}$$

$\therefore$ Power developed by engine

$$= \text{Force exerted by engine} \times \text{Speed}$$

$$= P^* \times v = 33666.6 \times 5 \text{ Nm/s}$$

$= 168333$ W $=$ **168.333 kW. Ans.**

4.11.3. Work Done by a Torque. Fig. 4.61 shows a circular body, which can rotate about the point A. A tangential force P is applied at point B and the body moves through a small angle θ.

Let R = Radius.

Now the distance moved by the force P in rotating the body through an angle θ is equal to the length of arc BC.

But length of arc $BC = R \times \theta$.

As the body has moved a distance $R \times \theta$, the work will be done by the force on the body.

$\therefore$ Work done

$= \text{Force} \times \text{Distance moved}$

$= P \times R \times \theta = (P \times R) \times \theta$

But $P \times R$ is equal to torque *(T)*.

$\therefore$ Work done $= T \times \theta$ ...(4.54)

In the above equation, θ should be taken in radians.

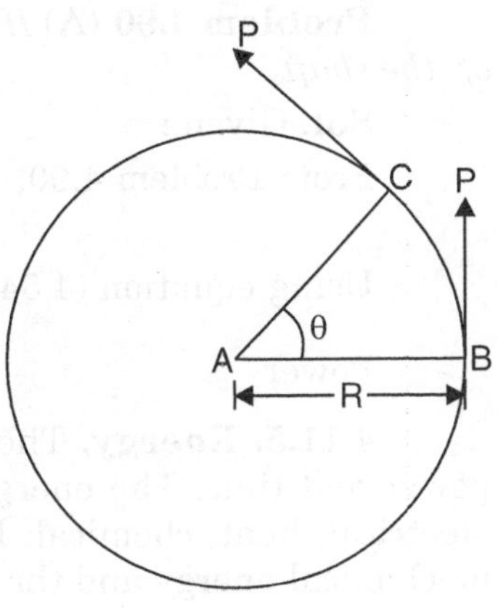

Fig. 4.61

4.11.4. Power Developed by a Torque

Let T = torque

N = r.p.m. (or revolution per minute)

Consider a body acted upon by a torque (T) and rotating at N r.p.m.

Now angle turned in one revolution $= 2\pi$ or 360°.

$\therefore$ Angle turned in N revolution $= 2\pi \times N$.

But N revolutions are performed in one minute.

$\therefore$ Angle turned in one minute $= 2\pi N$

or Angle turned in one second $= \dfrac{2\pi N}{60}$

Now using equation (4.54), we get

Work done $= T \times \theta$

or Work done per second $= T \times$ Angle turned in one second

$$= T \times \frac{2\pi N}{60} \text{ Nm/s}$$

But, power = Work done per second

$$= T \times \frac{2\pi N}{60} = \frac{2\pi NT}{60} \text{ W.} \qquad \text{...(4.54A)}$$

Problem 4.90. *A tangential force of 1800 N is acting on a shaft of diameter 10 mm. Find the work done by the force for one revolution of the shaft.*

Sol. Given :

Force, $P = 1800$ N

Dia. of shaft, $D = 10 \text{ mm} = \dfrac{10}{1000} = 0.01$ m

$\therefore$ Radius of shaft, $R = \dfrac{0.01}{2} = 0.005$ m

Angle turned for one revolution $= 2\pi$ radians

$\therefore$ $\theta = 2\pi$ radians

Torque, $T = P \times R = 1800 \times 0.005 = 9.0$ Nm

Work done is given by equation (4.54A).

$\therefore$ Work done $= T \times \theta$

$= 9 \times 2\pi$ Nm $=$ **56.54 Nm. Ans.**

Problem 4.90 (A) *If in problem 4.90, the shaft is rotating at 200 r.p.m., find the power of the shaft.*

Sol. Given :

From Problem 4.90, $P = 1800$ N, $R = 0.005$ m, $T = 9$ Nm

$N = 200$ r.p.m.

Using equation (4.54A), we get

$$\text{Power} = \frac{2\pi NT}{60}\ W = \frac{2\pi \times 200 \times 9}{60} = \mathbf{188.49\ W.}\quad \textbf{Ans.}$$

4.11.5. Energy. The capacity of doing work is known as energy. It is the product of power and time. The energy is expressed in Nm. It exists in many forms *i.e.,* mechanical, electrical, heat, chemical, light etc. In engineering mechanics, we are only concerned with mechanical energy and the same will be dealt with.

Mechanical energy consists of the following two types :

1. Potential energy (or position energy or datum energy)
2. Kinetic energy.

Depending upon the state of rest or that of motion, a body may possess anyone or both of the above energies.

1. Potential energy. Potential energy is also known as position energy or datum energy. It is the energy by *virtue of position* of a body with respect to any given reference or datum. A weight W lying on the top of a tower of height h possesses a potential energy of $W \times h$ with respect to the ground, as the weight W is capable of doing $W \times h$ work if it falls on the ground.

A compressed spring has potential energy, because it can do work in recovering its original shape. Similarly, compressed air also possesses potential energy because it is capable of doing work when allowed to expand.

2. Kinetic energy. The energy possessed by a body by virtue of its velocity (or its motion) is known as kinetic energy. It is represented by K.E.

The expression for the kinetic energy is derived as follows : Consider a body of mass m starting from rest. Let it be subjected to an accelerating force F and after covering a distance S, its velocity becomes v.

∴ Initial velocity, $u = 0$

Now work done on the body = Force × Distance

$$= F \times S \qquad \ldots(i)$$

But Force = Mass × Acceleration

$$\therefore \quad F = m \times a$$

Substituting the value of F in equation (*i*),

$$\text{Work done} = m \times a \times S$$

$$= m \times (a \times S) \qquad \ldots(ii)$$

But from equation of motion, we have

$$v^2 - u^2 = 2a \times S \quad \text{or} \quad v^2 - 0^2 = 2a \times S \qquad (\because u = 0)$$

or
$$a \times S = \frac{v^2}{2}$$

Substituting the value of $a \times S$ in equation (*ii*),

$$\text{Work done} = m \times \frac{v^2}{2}$$

But work done on the body is equal to K.E. possessed by the body.

$\therefore$ K.E. $= \frac{1}{2} mv^2$.

4.11.6. Derivation of Work-Energy Equation. The work energy principle states that the total work done by all forces is equal to change of kinetic energy. This is proved as below :

We know that, $F = m \times a$. ...(*i*)

where F = Resultant of all forces acting on a body,

m = Mass of the body, and

a = Acceleration* in the direction of resultant force

$= v \dfrac{dv}{ds}$

Substituting the value of a in equation (*i*), we get

$$F = m \times \left(v \frac{dv}{ds}\right) \quad \text{or} \quad F \cdot ds = m \times v\, dv \qquad ...(ii)$$

But $F \cdot ds$ is the work done by the resultant force F in displacing the body by a small distance ds. The total work done by the resultant force F in displacing the body by a distance S is obtained by integrating the above equation (*ii*).

Hence integrating equation (*ii*) on both sides, we get

$$\int_0^S F \cdot ds = \int_u^v m \times v \times dv$$

where u is the initial velocity when distance is zero and v is the final velocity when distance is S.

$$\therefore \quad F \cdot S = m \cdot \left[\frac{v^2}{2}\right]_u^v = \frac{m}{2}[v^2 - u^2] = \frac{mv^2}{2} - \frac{mu^2}{2}$$

or Work done by resultant force = change in kinetic energy

The above equation is known as work-energy equation.

Note. Some of the forces acting on a body donot do work when the body is displaced. These forces are : (*i*) the forces acting on a fixed point such as support reactions of pin ball and socket connections about which the body rotates, (*ii*) the forces acting perpendicular to direction of motion such as normal reaction of the surface over which the body is moving and weight of body if body moves in a horizontal direction.

Problem 4.91. *A body of mass 2 kg is moving with a velocity of 50 m/s. What will be the kinetic energy of the body ?*

Sol. Given :

Mass, m = 2 kg

Velocity, v = 50 m/s

K.E. is given by $\frac{1}{2} mv^2$.

$$\therefore \quad \text{K.E.} = \frac{1}{2} mv^2 = \frac{1}{2} \times 2 \times 50^2 \text{ kg} \left(\frac{\text{m}}{\text{s}}\right)^2 = 2500 \text{ kg} \cdot \frac{\text{m}}{\text{s}^2} \cdot \text{m}$$

$$\left(\because \ \text{kg}\frac{\text{m}}{\text{s}^2} = \text{N}\right)$$

= **2500 Nm. Ans.**

---

*Acceleration, $a = \dfrac{dv}{dt}$. But $v = \dfrac{ds}{dt}$. Hence $a = \dfrac{dv}{dt} = \dfrac{dv}{ds} \cdot \dfrac{ds}{dt} = \dfrac{dv}{ds} \cdot v$ or $a = v \dfrac{dv}{ds}$.

Problem 4.92. *A bullet of mass 81 gm and moving with a velocity of 300 m/s is fired into a block of wood and it penetrates to a depth of 10 cm. If the bullet moving with the same velocity, were fired into a similar piece of wood 5 cm thick, with what velocity would it emerge ? Find also the force of resistance, assuming it to be uniform.* (AMIE Winter, 1981)

Sol. Given :

Mass of bullet, $m = 81 \text{ gm} = 0.081 \text{ kg}$

Initial velocity of bullet, $u = 300$ m/s

First Case

Penetration of bullet, $S = 10 \text{ cm} = 0.1 \text{ m}$

Final velocity of bullet, $v = 0$

Let P = Force of resistance in Newton (N)

Fig. 4.62

The force of resistance will be acting in the opposite direction of motion of bullet. The work done by the force of resistance

= Force of resistance × Depth of penetration

$= (-P) \times S = (-P) \times 0.1$ Nm

$= -0.1P$ (–ve sign is due to opposite direction of force)

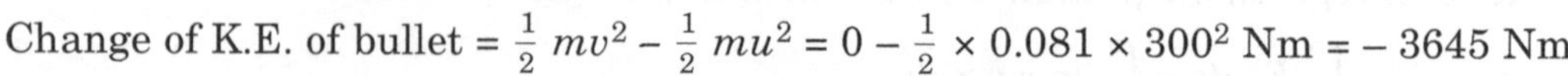

Change of K.E. of bullet $= \frac{1}{2} mv^2 - \frac{1}{2} mu^2 = 0 - \frac{1}{2} \times 0.081 \times 300^2 \text{ Nm} = -3645 \text{ Nm}$

But the change of K.E. of the bullet is equal to work done

Change of K.E of bullet = Work done by force of resistance

or $$-3645 = -0.1P$$

$\therefore$ $$P = \frac{3645}{0.1} = 36450 \text{ N}$$

Second Case. *Velocity of bullet after 5 cm penetration*

The force of resistance will be same *i.e.*, $P = 36450$ N.

Now depth of penetration,

$$S = 5 \text{ cm} = 0.05 \text{ m}$$

$\therefore$ Work done by force of resistance

$$= -P \times S = -36450 \times 0.05 \text{ Nm}$$

(–ve sign is due to opposite direction of force)

$$= -1822.5 \text{ Nm}$$

Let v_1 = Velocity of bullet after 5 cm penetration

Change of K.E. of bullet after penetration

$$= \frac{1}{2} mv_1^2 - \frac{1}{2} mu^2$$

$$= \frac{1}{2} \times 0.081 \times v_1^2 - \frac{1}{2} \times 0.081 \times 300^2 \quad (\because u = 300 \text{ m/s})$$

But the change of K.E. is equal to work done

$$\therefore \frac{1}{2} \times 0.081 \times v_1^2 - \frac{1}{2} \times 0.081 \times 300^2 = -1822.5$$

or $$\frac{1}{2} \times 0.081 \times v_1^2 - 3645 = -1822.5$$

or $$\frac{1}{2} \times 0.081\, v_1^2 = 3645 - 1822.5 \quad \text{or} \quad \frac{1}{2} \times 0.081\, v_1^2 = 1822.5$$

or $$v_1 = \left(\frac{1822.5 \times 2}{0.081}\right)^{1/2} = 212.132 \text{ m/s.}$$

4.11.7. Law of Conservation of Energy. It states that the energy can neither be created nor destroyed though it can be transformed from one form to another form. The second statement of this law is :

"The total energy possessed by a body remains constant provided no energy is added to or taken from it."

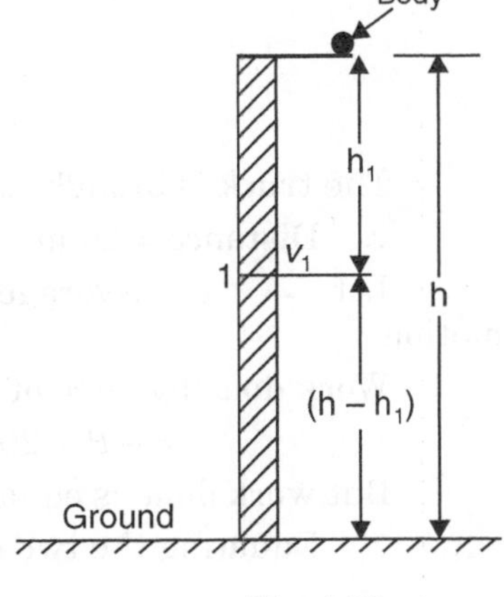

Fig. 4.63

Fig. 4.63 shows a body resting on the top of a tower.

Let W = Weight of body

m = Mass of body = $\frac{W}{g}$

h = Height of tower

Let us find the potential energy and K.E. of the body at the top of the tower.

Potential energy of the body with respect to ground

$= W \times h$ Nm ...(i)

K.E. of the body

$= \frac{1}{2} mv^2 = 0$ $(\because \; v = 0)$

$\therefore$ Total energy at the top of tower

= Potential energy + Kinetic energy

$= Wh + 0 = Wh$

Suppose the body falls down by a height h_1 to the position 1 as shown in Fig. 4.63. Let v_1 is the velocity at position 1.

Using equation $v^2 - u^2 = 2gh$, we get

$$v_1^2 - 0^2 = 2gh_1 \qquad (\because \; u = 0, h = h_1)$$

$$v_1^2 = 2gh_1$$

$\therefore$ K.E. of the body at position 1

$$= \frac{1}{2} mv_1^2 = \frac{1}{2} m \times 2gh_1 \qquad (\because \; v_1^2 = 2gh_1)$$

$$= mgh_1 = Wh_1 \qquad (\because \; W = mg)$$

Potential energy at position 1

$= W \times$ Height of body at 1 w.r.t. ground $= W \times (h - h_1)$

$\therefore$ Total energy at position 1

= Potential energy + Kinetic energy

$= W(h - h_1) + Wh_1 = Wh - Wh_1 + Wh_1 = Wh$

And it is the same energy, the body was possessing at the top of the tower.

Similarly, it can be proved that total energy of the body on the ground will be equal to 'Wh'.

Problem 4.93. *A truck of weight 20 kN is travelling at 45 km/hr on a level road. It is brought to rest in 20 metre, what is the average force of resistance acting on the truck.*

Sol. Given :

Weight of truck, $W = 20 \text{ kN} = 20 \times 1000 = 20{,}000$ N

Speed of truck, $u = 45$ km/hr $= \frac{45 \times 1000}{60 \times 60}$ m/s = 12.5 m/s

∴ Initial velocity of truck, $u = 12.5$ m/s

Final velocity of truck, $v = 0$

Change of K.E. of the truck, $= \frac{1}{2} mv^2 - \frac{1}{2} mu^2 = 0 - \frac{1}{2} mu^2$

$= -\frac{1}{2} mu^2$, where $m = \frac{W}{g} = \frac{20,000}{9.81}$ kg

$$= -\frac{1}{2} \times \left(\frac{20,000}{9.81}\right) \times 12.5^2 \text{ Nm} \quad ...(i)$$

The truck is brought to rest, in a distance of 20 m.

∴ Distance = 20 m

Let P = Average force of resistance which is acting in the opposite direction of motion.

Work done by force of resistance = $(-P) \times$ Distance.

$= -P \times 20 = -20P$ ...(*ii*)

But work done is equal to change of K.E.

∴ Equating the two equations (*i*) and (*ii*),

$$-20\,P = -\frac{1}{2} \times \left(\frac{20000}{9.81}\right) \times 12.5^2 \quad \text{or} \quad P = \frac{\frac{1}{2} \times \left(\frac{20000}{9.81}\right) \times 12.5^2}{20} = \mathbf{7963.8\ N.} \quad \textbf{Ans.}$$

Problem 4.94. *A car of weight 9810 N accelerates from rest to a speed of 45 km/hr in a distance of 50 m against a resistance of 100 N. Find the average driving force acting on the car. Using the average force, find the greatest power developed by the engine.*

Sol. Given :

Weight of car, $W = 9810$ N

∴ Mass of car, $m = \frac{W}{g} = \frac{9810}{9.81} = 1000$ kg

Initial velocity of car, $u = 0$

Velocity attained by car, $v = 45 \text{ km/hr} = \frac{45 \times 1000}{60 \times 60} = 12.5$ m/s

∴ Final velocity of car, $v = 12.5$ m/s

Distance = 50 m

Resistance, $P = 100$ N

Change in K.E. of car $= \frac{1}{2} mv^2 - \frac{1}{2} mu^2 = \frac{1}{2} mv^2$ $(\because \; u = 0)$

$= \frac{1}{2} \times 1000 \times 12.5^2$ Nm = 78125 Nm ...(*i*)

Let P^* = Average driving force exerted by engine.

The average driving force is acting in the direction of motion, whereas the resistance is acting opposite to the direction of motion.

*The work done by resultant force on the body must be equal to change of K.E. But change of K.E. $= \frac{1}{2}mv^2 - \frac{1}{2}mu^2$. Here $u = 0$ and hence change in K.E. $= \frac{1}{2}mv^2$.

∴ Resultant force on the car

$= P^* - P$

$= (P^* - 100)$ N

Work done by the resultant force

= (Resultant force) × Distance

$= (P^* - 100) \times 50$ Nm ...(*ii*)

**The change in K.E. of car must be equal to the work done by the resultant force.*

∴ Equating equations (*i*) and (*ii*),

$$78125 = (P^* - 100) \times 50$$

or $$P^* = \frac{78125}{50} + 100 = 1562.5 + 100 = \mathbf{1662.5\ N.}\quad \textbf{Ans.}$$

Power developed by engine = Driving force × speed of car

$= P^* \times v = 1662.5 \times 12.5$ Nm/s

= 20781.25 W = **20.781 kW. Ans.**

Problem 4.95. *A body weighing 196.2 N slides up a 30° inclined plane under the action of an applied force 300 N acting parallel to the inclined plane. The coefficient of friction, μ is equal to 0.2. The body moves from rest. Determine :*

(*i*) *acceleration of the body,*
(*ii*) *distance travelled by body in four seconds,*
(*iii*) *velocity of body after four seconds,*
(*iv*) *kinetic energy of the body after four seconds,*
(*v*) *work done on the body in four seconds,*
(*vi*) *momentum of the body after four seconds,*
(*vii*) *impulse applied in four seconds.*

Sol. Given :

Weight, $W = 196.2$ N

∴ Mass, $m = \frac{W}{g} = \frac{196.2}{9.81} = 20$ kg

Applied force = 300 N

Angle of inclination, $\theta = 30$

Co-efficient of friction, $\mu = 0.2$

As body moves from rest, hence initial velocity (*u*) will be zero.

∴ $u = 0$

Fig. 4.64 (*a*) shows the free-body diagram. The net force in the direction of motion is given by

F = Applied force $- W \sin\theta - \mu R$

$= 300 - 196.2 \times \sin 30° - 0.2 \times W \cos\theta$

$(\because\ R = W\cos\theta)$

$= 300 - 98.1 - 0.2 \times 196.2 \times \cos 30°$

$= 300 - 98.1 - 33.98 = 167.92$ N

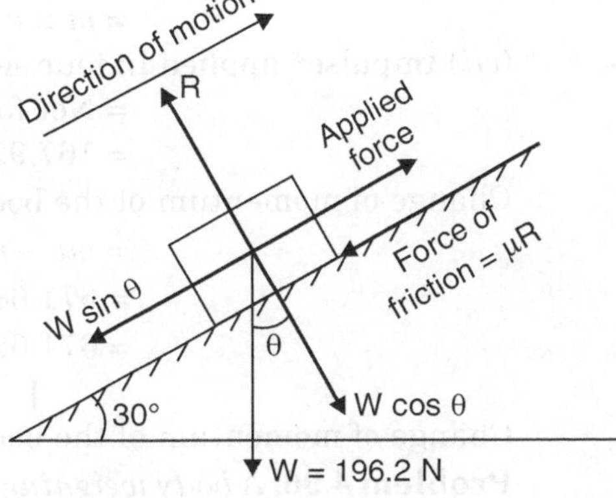

Fig. 4.64 (a)

(*i*) Let a = Acceleration of the body

But, Net force = Mass × Acceleration

or $$F = m \times a$$

or $$167.92 = 20 \times a$$

$\therefore$ $$a = \frac{167.92}{20} = \mathbf{8.396\ m/s^2.\ \ Ans.}$$

(*ii*) Let s = distance travelled in four seconds

Using $$s = ut + \frac{1}{2}at^2$$

$$= 0 \times 4 + \frac{1}{2} \times 8.396 \times 4^2 = \mathbf{67.168\ m.\ \ Ans.} \qquad (\because\ u = 0)$$

(*iii*) Let v = velocity after 4 seconds

$\therefore$ $$v = u + at$$

$\therefore$ $$= 0 + 8.396 \times 4 = \mathbf{33.584\ m/s.\ \ Ans.}$$

(*iv*) The kinetic energy after 4 seconds is given by

$$\text{K.E.} = \frac{1}{2}mv^2 \qquad \text{where } v = \text{velocity after 4 seconds}$$

$$= \frac{1}{2} \times 20 \times 33.584^2 = \mathbf{11278.8\ Nm.\ \ Ans.}$$

(*v*) Work done on the body in four seconds

= Net force × distance moved in four seconds

= 167.92 × 67.168 Nm = **11278.8 Nm. Ans.**

The work done on the body is equal to the change of kinetic energy of the body as proved below :

$$\text{Change of K.E.} = \frac{1}{2}mv^2 - \frac{1}{2}mu^2$$

$$= \frac{1}{2} \times 20 \times 33.584^2 - \frac{1}{2} \times 20 \times 0^2 \qquad (\because\ u = 0)$$

$$= 11278.8 - 0 = 11278.8 \text{ Nm.}$$

(*vi*) Momentum of the body after four seconds

$= m \times v = 20 \times 33.584 =$ **671.68 kg m/s. Ans.**

(*vii*) Impulse* applied in four seconds

$= \text{Net force} \times \text{time} = F \times 4$ $\qquad (\because\ \text{time} = 4 \text{ s})$

= 167.92 × 4 = **671.68 N s. Ans.**

Change of momentum of the body

$= mv - mu = 20 \times 33.584 - 20 \times 0$

= 671.68 kg m/s

= 671.68 Ns

$[\because\ \text{kg m/s} = \text{kg m/s} \times \text{s/s} = (\text{kg} \times \text{m/s}^2) \times \text{s} = \text{N} \times \text{s} = \text{N s}]$

Change of momentum of the body is equal to the impulse of the body.

Problem 4.96. *A body weighing 20 N is projected up a 20° inclined plane with a velocity of 12 m/s, coefficient of friction is μ = 0.15. Find : (i) The maximum distance S, that the body will move up the inclined plane, (ii) Velocity of the body when it returns to its original position.*

---

*Impulse = $F \times dt$

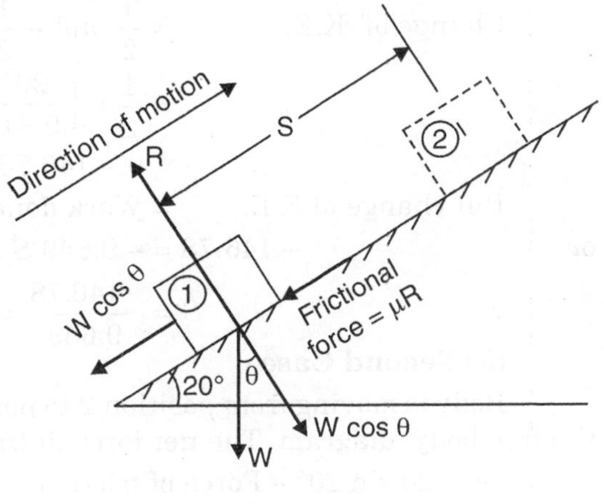

Sol. Given :

Weight, $W = 20$ N

Angle, $\theta = 20°$

Initial velocity, $u = 12$ m/s

Coefficient of friction,

$\mu = 0.15$

Mass, $m = \frac{W}{g} = \frac{20}{9.81}$ kg.

Let S = Maximum distance moved by the body up the inclined plane.

(*i*) **First Case**

Body is moving up the inclined plane from position 1 to position 2 where velocity is zero.

At the maximum distance, the velocity of the body (*i.e.*, final velocity) will become zero.

$\therefore$ Initial velocity, $u = 12$ m/s

Final velocity, $v = 0$

Force of friction, $F = \mu R = \mu \times W \cos 20°$ $\quad (\because\ R = W \cos 20°)$

$= 0.15 \times 20 \times \cos 20°$

Fig. 4.64 (b)

Fig. 4.64 (*b*) shows the free-body diagram. The net force in the direction of motion is given by,

$$F = -W \sin \theta - \mu R$$

$$= -20 \times \sin 20° - 0.15 \times W \cos \theta \quad (\because\ R = W \cos \theta)$$

$$= -20 \times 0.342 - 0.15 \times 20 \times \cos 20°$$

$$= -6.84 - 2.819 = -9.659 \text{ N}$$

(Negative sign means the net force is acting opposite to the direction of motion.)*

Using, Net force = Mass × Acceleration

or $$-9.659 = \left(\frac{20}{9.81}\right) \times a$$

$\therefore$ $$a = \frac{-9.659 \times 9.81}{20} = -4.737 \text{ m/s}^2$$

Now using $v^2 - u^2 = 2aS$

or $$0^2 - 12^2 = 2 \times (-4.737) \times S \quad \text{or} \quad \frac{-144}{-2 \times 4.737} = S$$

or $$S = \frac{144}{2 \times 4.737} = \mathbf{15.19 \text{ m.}} \quad \textbf{Ans.}$$

Alternate Method

Work done on the body = Net force × Distance moved

$= -9.659 \times S = -9.659\, S$ Nm

*The net force is acting downwards, but the body is moving upwards. This is due to the fact that initially some external force was acting on the body upwards. This external force projected the body with a velocity of 12 m/s upwards. After that the external force was removed.

Change of K.E. $= \frac{1}{2} mv^2 - \frac{1}{2} mu^2$

$= \frac{1}{2} \times \left(\frac{20}{9.81}\right) \times 0^2 - \frac{1}{2} \times \left(\frac{20}{9.81}\right) \times 12^2$ $(\because v = 0)$

$= 0 - 146.78 = -146.78$ Nm

But change of K.E. = Work done on the body

or $-146.78 = -9.659\, S$

$\therefore$ $S = \frac{-146.78}{9.659} =$ **15.196 m. Ans.**

(*ii*) **Second Case**

Body is moving from position 2 to position 1 down the inclined plane. Fig. 4.64 (*c*) shows the free-body diagram. The net force in the direction of motion is given by

$F = 20 \sin 20° -$ Force of friction

$= 20 \sin 20° - \mu R$

$= 20 \sin 20° - 0.15 \times 20 \cos 20°$

$(\because R = 20 \cos 20°)$

$= 6.84 - 2.819 = 4.021$ N

Distance between position 2 and position 1 is 15.196 m

$\therefore$ $S = 15.196$ m

$\therefore$ Work done on the body $= F \times S$

$= 4.021 \times 15.196$ Nm

$= 61.103$ Nm

Fig. 4.64 (c)

Change in K.E. $= \frac{1}{2} mV_1^2 - \frac{1}{2} mV_2^2$

where V_2 = Velocity at position 2 and equal to zero

V_1 = Velocity at position 1

$\therefore$ Change in K.E. $= \frac{1}{2} mV_1^2 - 0 = \frac{1}{2} mV_1^2$

$= \frac{1}{2}\left(\frac{20}{9.81}\right) \times V_1^2$ $\left(\because m = \frac{W}{g} = \frac{20}{9.81}\right)$

But change in K.E. = Work done on the body

$\frac{1}{2} \times \frac{20}{9.81} \times V_1^2 = 61.103$

$\therefore$ $V_1^2 = \frac{61.103 \times 2 \times 9.81}{20} = 59.94$

$\therefore$ $V_1 = \sqrt{59.94} =$ **7.742 m/s. Ans.**

Problem 4.97. *A bullet of mass 25 gm, moving horizontal with a velocity of 600 m/s strikes a wooden block of mass 5 kg resting on a rough horizontal surface. The bullet after striking the block remains burried in the block and both travel a distance of 90 cm, before coming to rest. Determine (i) average resistance between block and horizontal surface, (ii) coefficient of friction between block and horizontal surface.*

Sol. Given :

Mass of bullet, $m_1 = 25$ gm $= 0.025$ kg

Initial velocity of bullet, $u_1 = 600$ m/s

Mass of wooden block, $m_2 = 5$ kg

Initial velocity of block, $u_2 = 0$

Total mass of bullet and block, $M = m_1 + m_2$

$= (0.025 + 5) = 5.025$ kg

Let V = common velocity of bullet and block after impact

The law of conservation of momentum gives :

Total momentum before impact = Total momentum after impact

or $(m_1u_1 + m_2u_2) = M \times V$ or $0.025 \times 600 + 5 \times 0 = 5.025 \times V$

$$\therefore \quad V = \frac{0.025 \times 600}{5.025} = 2.985 \text{ m/s}$$

(i) Average resistance between block and horizontal surface

If we apply the principle of work and energy to the system of block and bullet, then the initial velocity of block and bullet will be their common velocity after impact and final velocity of the system will be zero.

Let P = Average frictional resistance

S = Distance travelled by block and bullet

= 90 cm = 0.9 m (given)

Initial velocity of block and bullet, $V_i = V = 2.985$ m/s

After travelling a distance of 90 cm, the block and bullet come to rest.

Hence final velocity of block and bullet, $V_f = 0$

Change of K.E. of bullet and block

$$= \frac{1}{2}M(V_f^2 - V_i^2) = \frac{1}{2} \times 5.025\,(0 - 2.985^2)$$

$$= -\frac{1}{2}(5.025) \times 2.985^2 \text{ Nm} \qquad ...(i)$$

Work done by frictional resistance

= Frictional resistance × Distance

$= (-P) \times 0.9$ Nm. ...(*ii*)

(–ve sign is due to opposite direction of frictional resistance)

The work done by the frictional resistance is equal to the change of K.E.

Equating the two equations (*i*) and (*ii*), we get

$$-P \times 0.9 = -\frac{1}{2} \times 5.025 \times 2.985^2$$

or

$$P = \frac{1}{2} \times \frac{5.025 \times 2.985^2}{0.9}$$

= **24.874 N. Ans.**

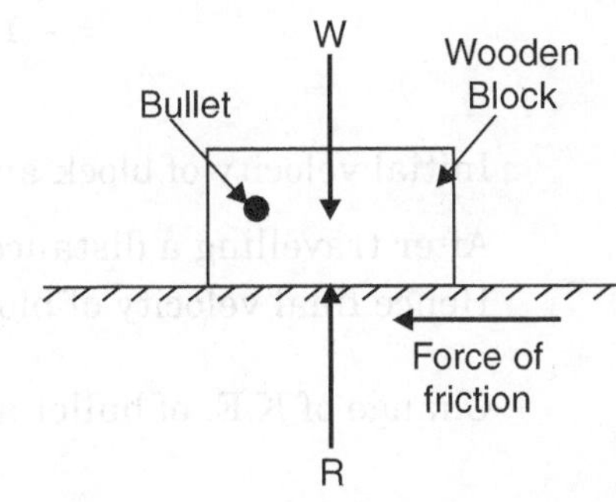

Fig. 4.65

(ii) To find coefficient of friction

Let μ = Coefficient of friction

R = Normal reaction

W = Total weight of bullet and block.

Refer to Fig. 4.65

The frictional force calculated above

$= P = 24.874$ N

W = Total weight = (Total mass) $\times g = M \times g = 5.025 \times 9.81$ N

$\therefore \quad R = W = 5.025 \times 9.81$ N

Now using the equation,

Force of friction = $\mu \times R$

$$\therefore \quad \mu = \frac{\text{Force of friction}}{\text{R}} = \frac{24.874}{5.025 \times 9.81} = \mathbf{0.5046.} \quad \textbf{Ans.}$$

Problem 4.98. *A 10 gm bullet is shot horizontally in a wood block of mass 1 kg. The bullet gets embedded in the block and the block is displaced on a rough horizontal table ($\mu = 0.2$) through 1 m. What was the velocity of bullet ?*

Sol. Given :

Mass of bullet, $m_1 = 10$ gm $= 0.01$ kg

Mass of block, $m_2 = 1$ kg

Distance travelled by block and bullet, $S = 1$ m

Value of $\mu = 0.2$

Let u_1 = Initial velocity of bullet

u_2 = Initial velocity of block which is equal to zero

V = Common velocity of bullet and block after impact

M = Total mass of bullet and block = $0.01 + 1 = 1.01$ kg

From the law of conservation of momentum, we get

Total momentum before impact = Total momentum after impact

or $(m_1u_1 + m_2u_2) = M \times V$

or $(0.01 \times u_1 + 1 \times 0) = 1.01 \times V$ or $0.01\, u_1 = 1.01\, V$

$$\therefore \quad V = \frac{0.01u_1}{1.01} \qquad ...(i)$$

Now force of friction = μR (R is normal reaction)

$= \mu W$ (Here $R = W$)

$= \mu \times M \times g$ ($\because \; W = M \times g$)

$= 0.2 \times 1.01 \times 9.81$ N

Work done by force of friction = Force of friction $\times$ Distance moved

$= -\,0.2 \times 1.01 \times 9.81 \times 1$ Nm

$= -\,1.9816$ Nm ...(ii)

(–ve sign is due to opposite direction of force of friction)

Initial velocity of block and bullet, $V_i = V = \dfrac{0.01u_1}{1.01}$

After travelling a distance of 1 m, the block and bullet come to rest.

Hence final velocity of block and bullet, $V_f = 0$

$$\text{Change of K.E. of bullet and block} = \frac{1}{2}M[v_f^2 - v_i^2] = \frac{1}{2} \times 1.01 \left[0 - \left(\frac{0.01u_1}{1.01}\right)^2\right]$$

$$= -\,\frac{1}{2}\,(1.01) \times \left[\left(\frac{0.01u_1}{1.01}\right)^2\right] \qquad ...(iii)$$

But the work done by force of friction is equal to change of K.E. of block and bullet.

Equating the work done by force of friction to the change of K.E. of bullet and block *i.e.*, equating equations (*ii*) and (*iii*), we get

$$-1.9816 = -\frac{1}{2} \times 1.01 \times \left(\frac{0.01u_1}{1.01}\right)^2$$

$$\left(\frac{0.01u_1}{1.01}\right)^2 = \frac{1.9816 \times 2}{1.01} \left(\frac{0.01u_1}{1.01}\right) = \sqrt{\frac{1.9816 \times 2}{1.01}} = 1.98 \text{ m/s}$$

or $$0.01\, u_1 = 1.01 \times 1.98$$

or $$u_1 = \frac{1.01 \times 1.98}{0.01} = \mathbf{199.98 \simeq 200 \text{ m/s.}} \quad \textbf{Ans.}$$

4.11.8. Energy Lost by a Body Falling on Another Body and to Calculate the Resistance Offered by the Ground. Fig. 4.66 shows a body of mass M falling on another body of mass m, from a height h. The mass m penetrates into the ground through a distance S before coming to rest.

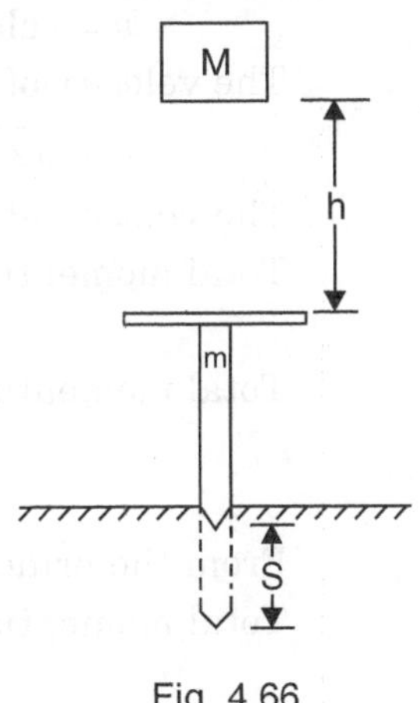

Fig. 4.66

Let v = velocity of mass M before striking the mass m

$= \sqrt{2gh}$

V = Velocity of both masses M and m after impact (*i.e.*, common velocity)

The velocity of mass m before impact is zero. Applying law of conservation of momentum :

Total momentum of M and m before impact = Total momentum of both masses after impact

or $$M \times v + m \times 0 = (M + m) \times V$$

or $$M \times \sqrt{2gh} = (M + m) \times V$$

$\therefore$ $$V = \frac{M \times \sqrt{2gh}}{(M + m)} \quad \ldots(i)$$

Resistance offered by the ground

The K.E. of the system (*i.e.*, K.E. of both masses) just after the impact

$$= \frac{1}{2}(\text{Total mass of system}) \times V^2 = \frac{1}{2}\,[M + m] \times V^2 \text{ Nm}$$

and potential energy of the system just after the impact

$$= (\text{Total weight of system}) \times \text{Vertical depth of penetration}$$

$$= [(M \times m) \times g] \times S \text{ Nm}$$

Total energy of the system

$$= \text{K.E. of system} + \text{P.E. of system}$$

$$= \frac{1}{2}\,[M + m] \times V^2 + (M + m) \times g \times S$$

This total energy is lost in penetrating the mass m into the ground.

$\therefore$ Work done during penetration of mass m into ground

$$= \text{Resistance of ground} \times \text{Depth of penetration} = R \times S$$

Now total energy lost is equal to work done during penetration.

$\therefore$ $$\frac{1}{2}[M + m] \times V^2 + (M + m) \times g \times S = R \times S$$

From the above equation, the resistance R can be calculated.

Problem 4.99. *A hammer of mass 1500 kg drops from a height of 60 cm on a pile of mass 750 kg. Find :*

(*i*) *the common velocity after impact assuming plastic impact.*

(*ii*) *the average resistance of the ground if the pile comes to rest after penetrating 5 cm into the ground.* (AMIE Winter, 1984)

Sol. Given :

Mass of hammer, $M = 1500$ kg

Height from which the hammer drops,

$h = 60$ cm $= 0.6$ m

Mass of pile, $m = 750$ kg

Depth of penetration, $S = 5$ cm $= 0.05$ m.

(*i*) Let V = common velocity after impact

Plastic impact means that after impact, both (hammer and pile) masses will move together with a common velocity.

v = velocity of the hammer before impact

The velocity of hammer after falling through a height of 0.6 m from rest, is given by

$$v = \sqrt{2gh} = \sqrt{2 \times 9.81 \times 0.6} = 3.43 \text{ m/s}$$

The velocity of pile before impact is zero.

Total momentum of hammer and pile just before impact

$= M \times v + m \times 0 = M \times v = 1500 \times 3.43 = 5145$ kg-m/s

Total momentum of hammer and pile just after impact

= (Total mass of hammer and pile) × Common velocity

$= (M + m) \times V = (1500 + 750) \times V = 2250\, V$ kg-m/s

From the principle of conservation of momentum, we know

Total momentum before impact

= Total momentum after impact

or $5145 = 2250 \times V$

$\therefore$ $V = \dfrac{5145}{2250} =$ **2.29 m/s. Ans.**

(*ii*) Let R = Average resistance of ground to the penetration.

Now first of all calculate the total energy of the system (*i.e.*, hammer and pile) just after the impact. This total energy is the sum of potential energy and kinetic energy.

K.E. of the system (hammer and pile) just after the impact

$$= \frac{1}{2}(M + m)\, V^2 = \frac{1}{2}(1500 + 750) \times 2.29^2 = 5899.6 \text{ Nm}$$

Loss of potential energy of system in moving a distance S

= Total weight of system × S = $[(M + m) \times g] \times S$

$= (1500 + 750) \times 9.81 \times 0.05$ $(\because\ S = 0.05 \text{ m})$

$= 1103.625$ Nm

$\therefore$ Total energy loss of the system

= Loss of K.E. of system + Loss of P.E.

$= 5899.6 + 1103.625 = 7003.225$ Nm

This total energy lost is used in penetrating the pile into the ground.

The work done during penetration of the file

$= \text{Resistance of ground} \times \text{Depth of penetration} = R \times 0.05$

Equating the work done during penetration to the total energy lost, we get

Work done = Total energy lost

$R \times 0.05 = 7003.225$

$\therefore \quad R = \dfrac{7003.225}{0.05} = \mathbf{140064.5\ N.}$ **Ans.**

Problem 4.100. *A hammer of mass 750 kg drops from a height of 1.2 m on a pile of mass 200 kg. Find the depth of penetration of the pile into the ground, if the average resistance of the ground is 79 kN. Assume the impact between the hammer and pile to be plastic.*

Sol. Given :

Mass of hammer, $M = 750$ kg

Height through which the hammer drops, $h = 1.2$ m

Mass of pile, $m = 200$ kg

Average resistance of ground, $R = 79\ \text{kN} = 79 \times 1000\ \text{N} = 79000\ \text{N}$

Impact between hammer and pile is plastic.

Let S = Depth of penetration into the ground

V = Common velocity after impact

v = Velocity of hammer before impact.

Let us first find the common velocity.

The velocity of hammer after falling through a height of 1.2 m from rest is given by,

$$v = \sqrt{2gh} = \sqrt{2 \times 9.81 \times 1.2} = 4.85 \text{ m/s}$$

Total momentum of hammer and pile just before impact

$= M \times v + m \times 0$ ($\because$ Pile is at rest before impact)

$= 750 \times 4.85 = 3637.5$ kg m/s

Total momentum of hammer and pile just after impact

$= (M + m) \times \text{Common velocity} = (750 + 200) \times V = 950\ V$ kg m/s

Equating the total momentum after and before impact,

$950\ V = 3637.5$

$\therefore \quad V = \dfrac{3637.5}{950} = 3.829$ m/s

Now let us find the total energy of the system just after impact.

K.E. of the hammer and pile just after impact

$= \frac{1}{2}[M + m] \times V^2 = \frac{1}{2}[750 + 200] \times 3.829^2 = 6964.09$ Nm

P.E. of system lost in moving a depth of S

$= \text{Total weight} \times \text{Vertical depth } S$

$= (M + m) \times g \times S = (750 + 200) \times 9.81 \times S = 9319.5\ S$ Nm

$\therefore$ Total energy of the system = (K.E. + P.E.) of the system

$= 6964.09 + 9319.5\ S$

Work done during penetration of pile

$= \text{Resistance of ground} \times \text{Depth of penetration}$

$= R \times S = 79000\ S$ Nm

Work done during penetration of pile is equal to the total energy lost of the system.

$\therefore$ $79000S = 6964.09 + 9319.5\,S$ or $(79000 - 9319.5)S = 6964.09$

$\therefore$ $S = \dfrac{6964.09}{(79000 - 9319.5)} = \mathbf{0.0999 \simeq 0.1\ m.}$ **Ans.**

Problem 4.101. *A hammer of mass 400 kg falls through a height of 3 m on a pile of negligible mass. If it drives the pile 1 m into the ground, find the average resistance of the ground for penetration.*

Sol. Given :

Mass of hammer, $M = 400$ kg

Height through which hammer falls, $h = 3$ m

Depth of penetration, $S = 1$ m

Mass of pile is negligible *i.e.*, $m = 0$

Let R = Average resistance of the ground

The velocity of hammer before striking the pile is given by,

$$v = \sqrt{2gh} = \sqrt{2 \times 9.81 \times 3} = 7.672 \text{ m/s}$$

The velocity* of hammer just after striking the pile will also be equal to 7.672 m/s. Hence $V = v = 7.672$ m/s.

The K.E. of the system $= \frac{1}{2}(M + m)V^2 = \frac{1}{2}MV^2$

$= \frac{1}{2} \times 400 \times 7.672^2 = 11771.9$ Nm

The P.E. of the system lost in moving a depth of 1 m

$= (M + m)g \times S = 400 \times 9.81 \times 1$ $(\because \ m = 0)$

$= 3924$ Nm

Total energy of the system $= 11771.9 + 3924 = 15695.9$ Nm

Work done during penetration of pile

= Resistance of the ground × Depth of penetration

$= R \times S = R \times 1$ Nm

But from principle of conservation of energy

Work done = Total energy lost

$R \times 1 = 15695.9$

$\therefore$ $\mathbf{R = 15695.9\ N.}$ **Ans.**

Problem 4.102. *A mass of 10 kg is dropped through 2 cm on a spring of stiffness $k = 100$ N/cm. Working from the first principles, find the maximum displacement of the spring.*

Sol. Given :

Mass = 10 kg

Stiffness, $k = 100$ N/cm

Height through which mass 10 kg is dropped

= 2 cm

Let x = Maximum displacement of the spring when mass 10 kg is dropped as shown in position (1) of Fig. 4.67.

*Let V = Common velocity. Here $m = 0$. Now $m \times v + M \times 0 = (M + m) \times V$.

or $M \times v = (M + 0) \times V$ or $M \times v = M \times V$ or $v = V$.

Considering the reference line *A-A*, let us find the total mechanical energy of the system for position (1) and position (2).

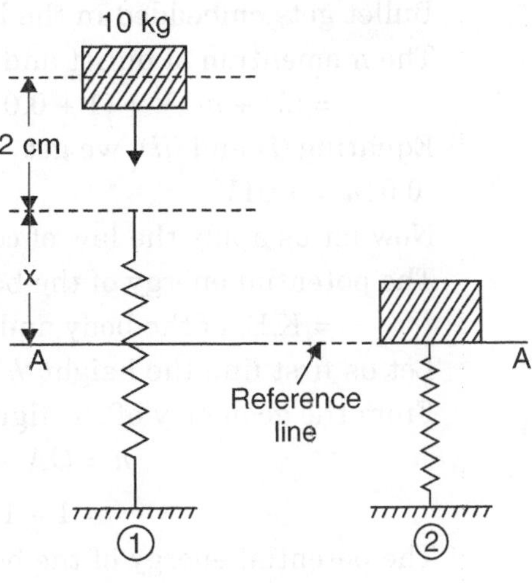

Fig. 4.67

For position (1) (with reference to line *A-A*)

$$\text{P.E.} = m \times g \times h$$
$$= 10 \times 9.81 \times (x + 2) \text{ N cm}$$
$$\text{K.E.} = 0$$
$$\therefore \quad \text{P.E.} + \text{K.E.} = 10 \times 9.81 \times (x + 2) + 0$$
$$= 10 \times 9.81\,(x + 2) \text{ N cm} \quad ...(i)$$

For position (2)

$$\text{P.E. of mass} = 0$$
$$\text{P.E. of spring} = \frac{1}{2}\,kx^2$$
$$= \frac{1}{2} \times 100 \times x^2 = 50\,x^2 \text{ N cm}$$
$$\text{K.E.} = 0$$

∴ Total mechanical energy of the system

$$= 0 + 50x^2 + 0 = 50x^2 \text{ N cm} \quad ...(ii)$$

But according to principle of total mechanical energy, the total energy of a system remains constant.

Equating equations (*i*) and (*ii*), we get

$$10 \times 9.81 \times (x + 2) = 50x^2 \quad \text{or} \quad 9.81\,(x + 2) = 5x^2$$

or $$9.81x + 19.62 = 5x^2 \quad \text{or} \quad 5x^2 - 9.81x - 19.62 = 0$$

$$\therefore \quad x = \frac{9.81 \pm \sqrt{9.81^2 + 4 \times 5 \times 19.62}}{2 \times 5} = \frac{9.81 \pm \sqrt{96.236 + 392.4}}{10}$$
$$= \frac{9.81 + 22.1}{10} \text{ (Neglecting negative value which is not possible)}$$
$$= \mathbf{3.19 \text{ cm.}} \quad \textbf{Ans.}$$

Problem 4.103. *A bullet of mass 10 gm is fired into a body of mass 1 kg, which is suspended by a string 1 m long. The bullet gets embedded in the body and due to the impact, the body swings through an angle of 18.2°. Find the velocity of the bullet.*

Sol. Given :

Mass of bullet, $m = 10$ gm $= 0.01$ kg

Mass of body, $M = 1$ kg

Length of string, $L = 1$ m

Angle of swing, $\theta = 18.2°$

Let, u = Velocity of bullet

V = Common velocity of bullet and body

Initial velocity of the body is zero.

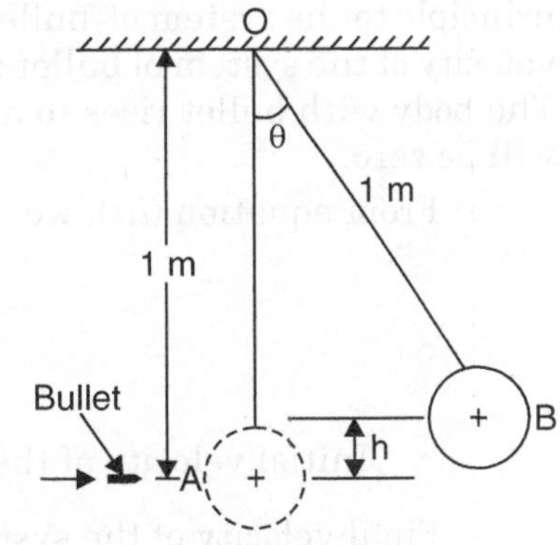

Fig. 4.68

The momentum of the bullet and body before impact

$$= m \times u + M \times \text{Initial velocity of body}$$
$$= 0.01 \times u + 1 \times 0 = 0.01u \text{ kg m/s} \quad ...(i)$$

Bullet gets embedded in the body. Both of them move together with common velocity.

The momentum of bullet and body after impact

$= (M + m)\,V = (1 + 0.01)\,V = 1.01\,V$ kg m/s ...(*ii*)

Equating (*i*) and (*ii*), we get

$0.01u = 1.01V$ ...(*iii*)

Now let us apply the law of conservation of energy.

The potential energy of the body and bullet at B

= K.E. of the body and bullet after impact.

Let us first find the height 'h' to which the body with bullet rises after impact.

From the geometry of the figure,

$h = OA - OB \cos\theta$ (But $OA = OB = 1$ m)

$= 1 - 1 \times \cos 18.2° = 1 - 0.9499 \simeq 0.05$ m

The potential energy of the body and bullet at B

= Total weight of bullet and body × h = $[(M + m) \times g\,] \times h$

$= (1 + .01) \times 9.81 \times 0.05 = 0.4954$ Nm ...(*iv*)

K.E. of the body and bullet after impact

$= \frac{1}{2}(M + m) \times V^2 = \frac{1}{2}(1 + .01) \times V^2$

$= 0.505\,V^2$ Nm ...(*v*)

But P.E. of body and bullet at B

= K.E. of body and bullet after impact

∴ $0.4954 = 0.505\,V^2$

or $V = \sqrt{\dfrac{0.4954}{0.505}} = 0.99$ m/s

Substituting this value in equation (*iii*), we get

$0.01\,u = 1.01 \times 0.99$

∴ $u = \dfrac{1.01 \times 0.99}{0.01} =$ **99.99 m/s. Ans.**

∴ Velocity of bullet = 99.99 m/s.

Alternate Method

This question can also be solved by alternate method. Let us apply the work and energy principle to the system of bullet and a body of given mass suspended by the spring. The initial velocity of the system of bullet and body is the common velocity (V) of the system after impact. The body with bullet rises to a height 'h' after impact. Hence the final velocity of the system will be zero.

From equation (*iii*), we have,

$$0.01\,u = 1.01\,V$$

∴ $$V = \frac{0.01u}{1.01}$$

∴ Initial velocity of the system, $V_i = V = \dfrac{0.01u}{1.01}$ m/s

Final velocity of the system, $V_f = 0$

Total mass of the system $= (M + m) = 1 + 0.01 = 1.01$ kg

Change of K.E. of the system $= \frac{1}{2}(M + m)[V_f^2 - V_i^2]$

$$= \frac{1}{2}(1 + 0.01)\left[0 - \left(\frac{0.01u}{1.01}\right)^2\right]$$

$$= -\frac{1}{2} \times 1.01\left(\frac{0.01}{1.01}u\right)^2 = -\frac{0.0001}{2 \times 1.01}u^2 \quad ...(A)$$

The work done will be due to the rise of the weight of the system. As the weight of the system is acting downwards, but the displacement of the weight of the system is upwards, hence work done will be –ve.

∴ Work done = – (Weight of system) × rise in height

$= -(M + m) \times g \times h$ [∵ Weight = Mass × g]

But the value of 'h' = 0.05 m (already calculated)

∴ Work done = – (1 + 0.01) × 9.81 × 0.05 ...(B)

But according to the principal of work and energy, the total work done on the system must be equal to the change of K.E. of the system.

Hence equating the two values given by equations (A) and (B), we get

$$-\frac{0.0001}{2 \times 1.01}u^2 = -1.01 \times 9.81 \times 0.05$$

$$\therefore \quad u^2 = \frac{2 \times 1.01 \times 1.01 \times 9.81 \times 0.05}{0.0001}$$

or $$u = \sqrt{\frac{2 \times 1.01 \times 1.01 \times 9.81 \times 0.05}{0.0001}} \simeq \textbf{100 m/s.} \quad \textbf{Ans.}$$

Problem 4.104. *Bullet A of mass 30 gm moving with a velocity of 483 m/s is fired into a body of mass 10 kg, which is suspended by a string 0.8 m long. Due to this impact, the body swings through some angle. Find the angle through which the body swings when the bullet gets embedded in the body.*

Sol. Given :

Mass of bullet, m = 30 gm = 0.03 kg

Velocity of bullet, u = 483 m/s

Mass of body, M = 10 kg

Initial velocity of body is zero.

Length of string, L = 0.8 m

Let θ = Angle of swing.

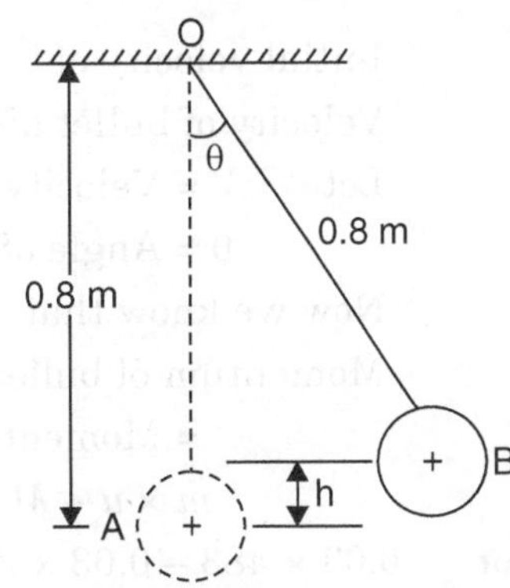

Fig. 4.68 (a)

From the geometry of Fig. 4.68 (a), we have

$h = 0.8 - 0.8 \cos \theta$ ...(i)

Momentum of bullet and body before impact

$= m \times u + M \times 0$

$= m \times u = 0.03 \times 483$

$= 14.49$ kg m/s ...(ii)

The bullet (after impact), gets embedded into the body and so both of them move with a common velocity (V).

∴ Momentum of bullet and body after impact

$$= (M + m)V = (10 + 0.03)\ V = 10.03\ V \text{ kg-m/s} \quad ...(iii)$$

Equating the two momentum given by (*ii*) and (*iii*),

$$14.49 = 10.03\ V$$

$$\therefore \quad V = \frac{14.49}{10.03} = 1.4446 \text{ m/s}$$

The K.E. of the bullet and body after impact

$$= \frac{1}{2}(M + m)\ V^2 = \frac{1}{2}(10 + 0.03) \times 1.4446^2$$

$$= \frac{1}{2} \times 10.03 \times 1.4446^2 = 10.4656 \text{ Nm} \quad ...(iv)$$

P.E. of the body and bullet at B

$$= (\text{Total weight of bullet and body}) \times h = (M + m) \times g \times h$$

$$= (10.03) \times 9.81 \times [0.8 - 0.8 \cos \theta] \quad ...(v)$$

But P.E. of the body and bullet at B

= K.E. of body and bullet after impact

or $$10.03 \times 9.81 \times [0.8 - 0.8 \cos \theta] = 10.4656$$

or $$(0.8 - 0.8 \cos \theta) = \frac{10.4656}{10.03 \times 9.81} = 0.1063 \quad \text{or} \quad 0.8 - 0.1063 = 0.8 \cos \theta$$

or $$0.6937 = 0.8 \cos \theta \quad \text{or} \quad \cos \theta = \frac{0.6937}{0.8} = 0.867$$

$$\therefore \quad \theta = \cos^{-1} 0.867 = \mathbf{29.87°}. \quad \textbf{Ans.}$$

Problem 4.105. *If in Problem 4.104 the bullet, instead of getting embedded into the body, escapes from the other end of the body with a velocity of 96.5 m/s, find the angle through which the body will swing.*

Sol. Given :

From Problem 4.104, we have

$$m = 0.03 \text{ kg},\ u = 483 \text{ m/s},\ M = 10 \text{ kg},\ L = 0.8 \text{ m}$$

Initial velocity of body is zero.

Velocity of bullet after impact, $v = 96.5$ m/s

Let V = Velocity of body after impact and

θ = Angle of swing.

Now we know that

Momentum of bullet and body before impact

= Momentum of bullet and body after impact

$$m \times u + M \times 0 = m \times v + M \times V \quad \text{or} \quad 0.03 \times 483 = 0.03 \times 96.5 + 10 \times V$$

or $$0.03 \times 483 - 0.03 \times 96.5 = 10V \quad \text{or} \quad 14.49 - 2.895 = 10\ V$$

or $$V = \frac{(14.49 - 2.895)}{10} = 1.1595 \text{ m/s}$$

Due to this velocity the body will swing. The *total K.E. of the body* (after impact) will be used for lifting the body *only* for a height 'h'.

$\therefore$ K.E. of the body after impact

$= \text{P.E. of the body at } B$

or $\frac{1}{2} MV^2 = (M \times g) \times h$

or $h = \frac{V^2}{2g} = \frac{1.1595^2}{2 \times 9.81} = 0.0685$ m

But from the geometry of Fig. 4.68 (*b*), we have

$h = L - L \cos \theta$

$= 0.8 - 0.8 \cos \theta \quad (\because L = 0.8 \text{ m})$

or $0.0685 = 0.8 - 0.8 \cos \theta \quad (\because h = 0.0685 \text{ m})$

or $0.8 \cos \theta = 0.8 - 0.0685 = 0.7315$

$\therefore \cos \theta = \frac{0.7315}{0.8} = 0.9143$

$\therefore \theta = \cos^{-1} 0.9143 =$ **23.89°. Ans.**

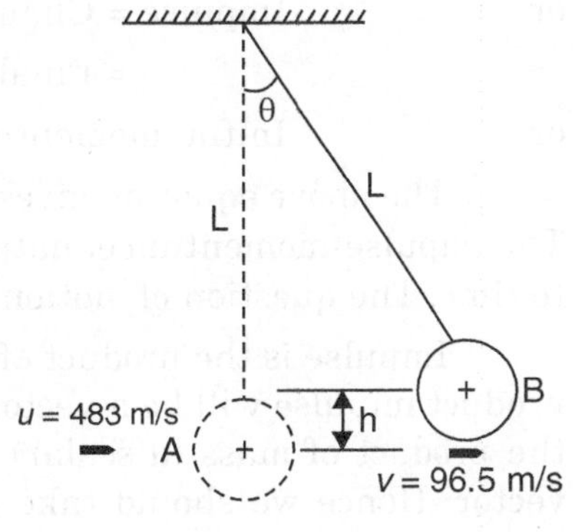

Fig. 4.68 (b)

4.12. IMPULSE AND MOMENTUM

To solve kinetic problems of rigid body. Impulse-momentum method is also used. This method which relates force, velocity and time, is based on the integration of equation of motion with respect to time.

In this method also there is no need of determing the acceleration of the body.

Let us first define the impulse and momentum.

Let F = Net force acting on a rigid body in the direction of motion, through the C.G. of the body

m = Mass of the rigid body, and

a = Acceleration of the body.

Then we have

$$F = m \times a = m \times \frac{dv}{dt} \quad \left[\because \text{ Acceleration is rate of change of velocity } i.e., a = \frac{dv}{dt}\right]$$

or $F \times dt = m \times dv$

Integrating the above equation, we get

$$\int_{t_1}^{t_2} F \times dt = \int_{v_1}^{v_2} m \times dv = m(v_2 - v_1) \qquad ...(4.55)$$

If time-interval is very small, then $\int_{t_1}^{t_2} F \times dt$ is known as impulse. *Hence impulse is the product of force and time when time is very small.* Momentum is the product of mass and velocity.

$$\therefore \quad \text{Impulse} = \int_{t_1}^{t_2} F \times dt = m(v_2 - v_1) = mv_2 - mv_1$$

or Impulse = Change of momentum [∵ Momentum = Mass × Velocity]

= Final momentum – Initial momentum

or Initial momentum + Impulse = Final momentum

The above equation gives the relation between impulse and momentum of a rigid body. The impulse-momentum equation is based on integrating the equation of motion with respect to time. The question of motion relates force, velocity and time.

Impulse is the product of force and time. As force is a vector and time is a scalar, their product impulse will be a vector. Similarly, the momentum will be a vector as momentum is the product of mass (a scalar) and velocity (a vector). Also change of momentum will be a vector. Hence we should take care of the direction in solving the kinetic equation. Hence equation (4.55) for rectangular co-ordinates will be written as

$$\int_{t_1}^{t_2} F_x \times dt = m\,(V_{2x} - V_{1x})$$

and

$$\int_{t_1}^{t_2} F_y \times dt = m\,(V_{2y} - V_{1y})$$

If $t_1 = 0$ and $t_2 = t$, then the above equations are

$$F_x \times t = m\,(V_{2x} - V_{1x}) \qquad ...(4.55A)$$

$$F_y \times t = m\,(V_{2y} - V_{1y}) \qquad ...(4.55B)$$

where F_x = Impulsive force in x-direction

F_y = Impulsive force in y-direction

V_{1x} and V_{2x} = Initial and final velocities in x-direction and

V_{1y} and V_{2y} = Initial and final velocities in y-direciton

F = Resultant impulsive force

$$= \sqrt{F_x^2 + F_y^2} \qquad ...(4.55C)$$

Note. (*i*) The impulse-momentum approach is particularly convenient in situations when forces act for very small interval of time as in an impact or sudden below.

(*ii*) In satellite motion, a combination of impulse-momentum method and work-energy method is used.

(*iii*) The unit of impulse is Ns (*i.e.*, Newton second) in S.I. system.

Problem 4.105A. *A football of mass 200 gm is at rest. A player kicks the ball which moves with a velocity of 20 m/s at an angle of 30° with respect to ground level. Find the force exerted by the player on the ball if duration of strike is 0.02 seconds.*

Sol. Given :

Mass, m = 200 gm = 0.2 kg

Duration of strike, t = 0.02 s

Initially the ball is at rest. Hence $V_{1x} = 0$ and $V_{1y} = 0$

The ball leaves with a velocity of 20 m/s at an angle of 30°. Refer to Fig. 4.69

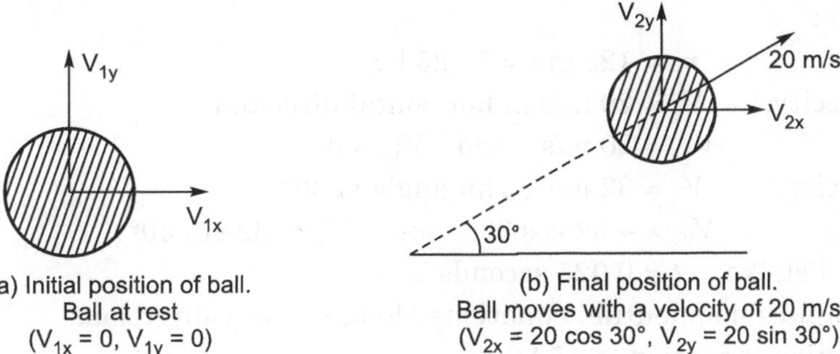

Fig. 4.69

Writing impulse-momentum equation along x- and y-directions, we get [Refer to equations (4.55A) and (4.55B)].

For x-direction

$$F_x \times t = m\,(V_{2x} - V_{1x}), \quad \text{where} \quad V_{2x} = 20 \cos 30° \text{ and } V_{1x} = 0$$

or $$F_x \times 0.02 = 0.2\,(20 \cos 30° - 0)$$

or $$F_x = \frac{0.2(20 \cos 30°)}{0.02} = 173.2 \text{ N}$$

For y-direction

$$F_y \times t = m\,(V_{2y} - V_{1y}), \quad \text{where} \quad V_{2y} = 20 \sin 30° \text{ and } V_{1y} = 0$$

or $$F_y \times 0.02 = 0.2\,(20 \sin 30° - 0)$$

or $$F_y = \frac{0.2 \times 20 \sin 30°}{0.02} = 100 \text{ N}$$

Hence the resultant impulse force exerted by the player on the ball,

$$F = \sqrt{F_x^2 + F_y^2} = \sqrt{173.2^2 + 100^2}$$

$$= 199.99 \simeq \mathbf{200\ N.} \quad \textbf{Ans.}$$

Problem 4.105B. *Fig. 4.70 shows a ball of mass 125 gm, moving in a horizontal direction towards a bat with a velocity of 20 m/s. The ball attains a velocity of 32 m/s at an angle of 30° with respect to x-axis, when hit by a bat. Find the impulsive force exerted by the bat on the ball if duration of strike is 0.025 second.*

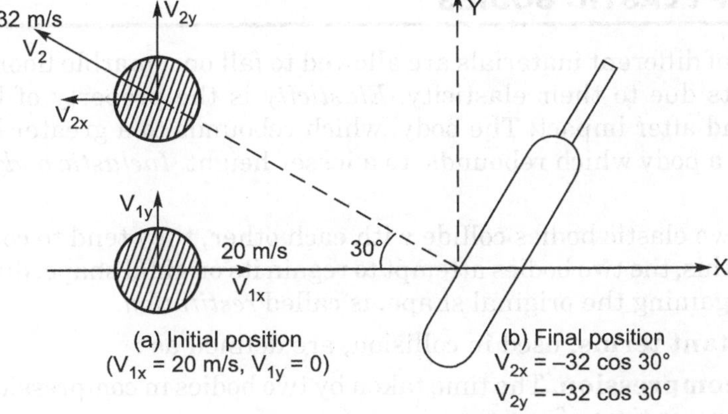

Fig. 4.70

Sol. Given :

Mass, $m = 125 \text{ gm} = 0.125 \text{ kg}$

Initial velocity, $V_1 = 20$ m/s in horizontal direciton

$\therefore$ $V_{1x} = 20$ m/s and $V_{1y} = 0$

Final velocity, $V_2 = 32$ m/s at an angle of 30°

$\therefore$ $V_{2x} = -32 \cos 30°$ and $V_{2y} = 32 \sin 30°$

Duration of strike, $t = 0.025$ seconds

Writing impulse-momentum equations along x- and y-directions.

For x-direction [equation (4.55A)]

$$F_x \times t = m(V_{2x} - V_{1x})$$

or $F_x \times 0.025 = 0.125\,(-32 \cos 30° - 20)$ ($\because$ V_{2x} is in opposite direction hence –ve)

or $$F_x = \frac{+0.125\,(-32 \cos 30° - 20)}{0.025} = -238.56 \text{ N}$$

For y-direction

$$F_y \times t = m(V_{2y} - V_{1y})$$ [equation (4.55B)]

or $F_y \times 0.025 = 0.125\,(32 \sin 30° - 0)$

or $$F_y = \frac{0.125 \times 32 \sin 30°}{0.025} = 80 \text{ N}$$

$\therefore$ Resultant average impulsive force exerted by bat on the ball [Refer to Fig. 4.70 (c)]

$$F = \sqrt{F_x^2 + F_y^2}$$

$$= \sqrt{(-238.56)^2 + 80^2} = \mathbf{251.6\ N.}\quad \textbf{Ans.}$$

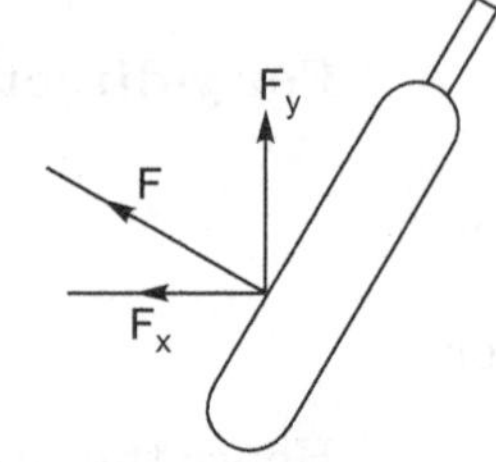

Fig. 4.70 (c)

Note. (i) The impulse-momentum approach is particularly convenient in situations when forces act for very small interval of time as in an impact or a sudden blow.

(ii) In satellite motion, a combination of impulse-momentum method and work-energy method is used.

(iii) The unit of impulse is Ns ($i.e.$, Newton second) in S.I. system.

4.13. IMPACT OF ELASTIC BODIES

If the balls of different materials are allowed to fall on a marble floor, they will rebound to different heights due to their elasticity. *Elasticity* is the property of bodies by virtue of which they rebound after impact. The body, which rebounds to a greater height is known as more elastic, than a body which rebounds to a lesser height. *Inelastic body* is one which does not rebound at all.

Whenever two elastic bodies collide with each other, they tend to compress each other. Immediately after this, the two bodies attempt to regain its original shape, due to their elasticity. This process, of regaining the original shape, is called *restitution.*

The important terms, used in collision, are defined as :

1. Time of compression. The time taken by two bodies in compression, after the instant of collision, is known as time of compression.

2. Time of restitution. The time taken by two bodies to regain the original shape, after compression, is known as time of restitution.

3. Time of collision. The sum of time of compression and time of restitution is known as time of collision or period of collision or period of impact.

4. Law of conservation of momentum. It states that if the resultant of the external forces acting on a system is zero, the momentum of the system remains constant. This means that the total momentum of the system before collision is equal to the total momentum of the system after collision. The system may consist of one body or two bodies or more.

4.13.1. Types of Impacts. Impact means the collision of two bodies which occurs in a *very small interval of time* and during which the two bodies exert a *very large force* on each other. The important types of impacts are :

(*i*) Direct impact

(*ii*) Indirect (oblique) impact.

4.13.2. Direct Impact of Two Bodies. The two bodies A and B are moving in a horizontal line before collision with velocities u_1 and u_2 in the same direction *i.e.*, along x-axis as shown in Fig. 4.71 (*a*) . If $u_1 > u_2$, the body A will strike the body B and collision will take place. Let C is the point of collision of the two bodies as shown in Fig. 4.71 (*b*). The point C is also known as the point of contact. The line joining the centres of these two bodies and passing through the point of contact is known as *line of impact.* Hence here the line O_1-C-O_2 is called line of impact.

The collision between two bodies is known as *direct impact* if the two bodies before impact, are moving along the line of impact.

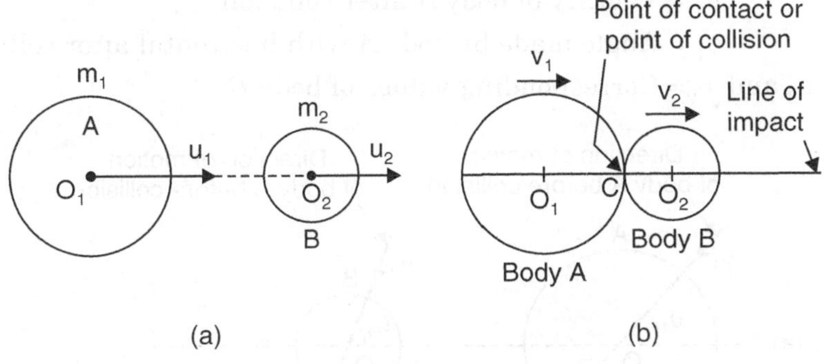

Fig. 4.71

The two bodies shown in Fig. 4.71 are having a direct impact.

Let m_1 = Mass of the body A

u_1 = Initial velocity of body A, *i.e.*, the velocity of body A before collision along x-axis.

v_1 = Final velocity of body A (after collision) along x-axis.

m_2, u_2 and v_2 are the mass of body B, velocity of body B before collision and velocity of the body B after collision along x-axis respectively.

The momentum of the body A before collision

$$= \text{Mass} \times \text{Velocity} = m_1u_1 \text{ kg m/s.}$$

The momentum of the body B before collision

$$= m_2 \times u_2 = m_2u_2 \text{ kg m/s.}$$

∴ Total initial momentum (*i.e.*, momentum before collision)

$= m_1u_1 + m_2u_2$ kg m/s ...(*i*)

Similarly, total final momentum (*i.e.*, momentum after collision)

= Mass of body *A* × Final velocity of *A* + Mass of body *B* × Final velocity of body *B*

$= m_1v_1 + m_2v_2$

But according to the law of conservation of momentum,

Total initial momentum = Total final momentum

or $m_1u_1 + m_2u_2 = m_1v_1 + m_2v_2.$

4.13.3. Indirect Impact of Two Bodies. Fig. 4.72 (*a*) shows the two bodies *A* and *B*, moving with a velocity of u_1 and u_2 in different direction. The velocity of body *A* makes an angle θ_1 with horizontal direction whereas the velocity of body *B* makes an angle θ_2 with horizontal direction. As the two bodies are moving in different direction, the collision between two bodies takes place as shown in Fig. 4.72 (*b*). The point *C* is the point of contact. And line joining the centres O_1, O_2 with point *C* is known as *line of impact.*

If the two bodies, before impact are not moving along the line of impact, the collision is known as indirect impact.

Here the two bodies *A* and *B* are not moving along the line of impact before collision, hence the collision is known as indirect impact (or oblique impact).

Let m_1 = Mass of body *A*

u_1 = Initial velocity of *A*

θ_1 = Angle made by body *A* with horizontal before collision

v_1 = Velocity of body *A* after collision

ϕ_1 = Angle made by body *A* with horizontal after collision.

m_2, u_2, θ_2, v_2 and ϕ_2= Corresponding values of body *B*.

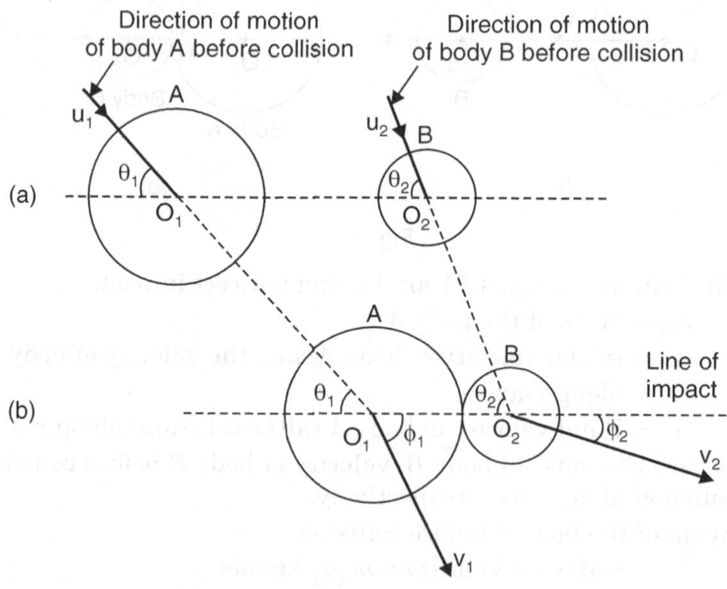

Fig. 4.72

Now according to the law of conservation of momentum the total initial momentum in any direction must be equal to final momentum in that direction.

Now initial momentum (*i.e.*, momentum before collision) of body A in x-direction

$$= \text{Mass of } A \times \text{Initial velocity of } A \text{ in } x\text{-direction} = m_1 \times u_1 \cos\theta_1$$

Initial momentum of body B in x-direction

$$= \text{Mass of } B \times \text{Initial velocity of } B \text{ in } x\text{-direction} = m_2 \times u_2 \cos\theta_2$$

$\therefore$ Total initial momentum in x-direction

$$= m_1 \times u_1 \cos\theta_1 + m_2 \times u_2 \cos\theta_2 \quad ...(i)$$

Final momentum of body A (*i.e.*, momentum after collision) in x-direction

$$= m_1 \times \text{Final velocity of } A \text{ in } x\text{-direction} = m_1 \times v_1 \cos\phi_1$$

Similarly, final momentum of B in x-direction

$$= m_2 \times \text{Final velocity of } B \text{ in } x\text{-direction} = m_2 \times v_2 \cos\phi_2$$

$\therefore$ Total final momentum in x-direction

$$= m_1 \times v_1 \cos\phi_1 + m_2 \times v_2 \cos\phi_2 \quad ...(ii)$$

Equating the initial momentum to final momentum given by equations (*i*) and (*ii*), we get

$$m_1 \times u_1 \cos\theta_1 + m_2 \times u_2 \cos\theta_2 = m_1 \times v_1 \cos\phi_1 + m_2 \times v_2 \cos\phi_2.$$

4.13.4. Coefficient of Restitution. It is defined as the ratio of velocity of separation (of the two moving bodies which collide with each other) to their velocity of approach. It is also defined as the ratio of the relative velocities of colliding bodies after impact to their relative velocity before impact. It is denoted by symbol 'e'. The relative velocities are measured along the line of impact, which is the common normal to the colliding surfaces.

According to Newton's Law of collision of elastic bodies, "*the velocity of separation, of the two moving bodies which collide with each other, bears a constant ratio to their velocity of approach*". And the constant of proportionality is known as coefficient of restitution.

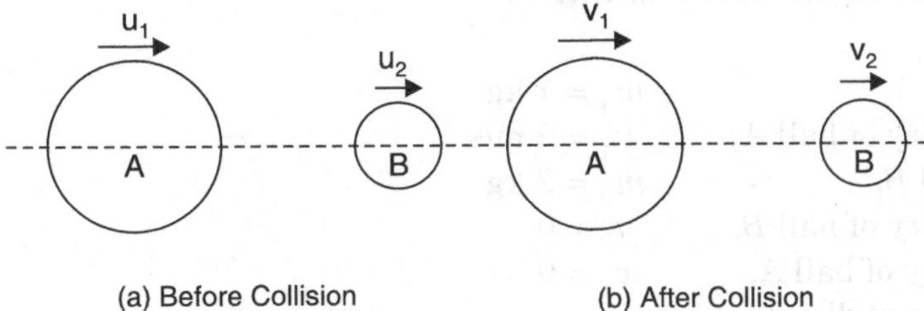

Fig. 4.73

Fig. 4.73 shows two bodies A and B.

Let u_1 = Velocity of A before collision along x-axis

v_1 = Velocity of A after collision along x-axis

u_2 = Velocity of B before collision along x-axis

v_2 = Velocity of B after collision along x-axis.

The body A will collide with body B if velocity of A is more than that of B. Hence velocity of approach (or relative velocity of colliding bodies before impact)

$$= \text{Initial velocity of } A - \text{Initial velocity of } B = (u_1 - u_2)$$

After collision, the separation of the two bodies will take place if final velocity of B is more than that of A.

Hence velocity of separation (or relative velocity of colliding bodies after impact)

$$= \text{Final velocity of } B - \text{Final velocity of } A = (v_2 - v_1)$$

Now according to Newton's Law of collision of elastic bodies,

Velocity of separation $\propto$ Velocity of approach

or $$(v_2 - v_1) \propto (u_1 - u_2)$$

or $$(v_2 - v_1) = e \times (u_1 - u_2)$$

where e is the constant of proportionality and is known as coefficient of restitution.

$$\therefore \quad e = \frac{(v_2 - v_1)}{(u_1 - u_2)} \quad \text{...(4.56)}$$

For most of bodies, the value of e lies between 0 and 1. For perfectly elastic bodies $e = 1$ and for perfectly plastic bodies $e = 0$.

To determine the velocities after the impact (*i.e.*, velocity v_1 and v_2) the equation (4.56) is not sufficient to determine the two unknowns. One more equation is needed. This equation is law of conservation of momentum *i.e.*, total initial momentum is equal to total final momentum.

Note 1. The velocity of approach or separation of the two bodies, which are moving in the same direction before or after the impact, is the difference of their velocities.

Note 2. For the two bodies moving in the opposite direction, the velocity of approach or separation is the sum of their velocities.

Note 3. Newton's law of collision of elastic bodies, also holds good for indirect impact, *i.e.*,

$$(v_2 \cos \phi_2 - v_1 \cos \phi_1) = e\,(u_1 \cos \theta_1 - u_2 \cos \theta_2) \quad \text{...(4.57)}$$

Note 4. If the velocity in a direction is +ve then the velocity in the opposite direction will be –ve.

Problem 4.106. *Ball A of mass 1 kg moving with a velocity of 2 m/s, strikes directly on a ball B of mass 2 kg at rest. The ball A, after striking, comes to rest. Find the velocity of ball B after striking and coefficient of restitution.*

Sol. Given:

Mass of ball A, $m_1 = 1$ kg

Initial velocity of ball A, $u_1 = 2$ m/s

Mass of ball B, $m_2 = 2$ kg

Initial velocity of ball B, $u_2 = 0$

Final velocity of ball A, $v_1 = 0$

This is a case of direct impact.

Let v_2 = Velocity of ball B after impact, and

e = Coefficient of restitution.

Total initial momentum $= m_1u_1 + m_2u_2 = 1 \times 2 + 2 \times 0 = 2$ kg m/s.

Total final momentum $= m_1v_1 + m_2v_2 = 1 \times 0 + 2 \times v_2 = 2v_2$ kg m/s.

According to the law of conservation of momentum,

Total initial momentum = Total final momentum

$$\therefore \quad 2 = 2 \times v_2$$

$$\therefore \quad v_2 = \frac{2}{2} = \mathbf{1\ m/s.}\quad \textbf{Ans.}$$

To find the coefficient of restitution, the equation (4.56) is used

$$\therefore \quad e = \frac{(v_2 - v_1)}{(u_1 - u_2)} = \frac{(1-0)}{(2-0)} = \frac{\mathbf{1}}{\mathbf{2}}. \quad \textbf{Ans.}$$

Problem 4.107. *A body of mass 50 kg, moving with a velocity of 6 m/s, collides directly with a stationary body of mass 30 kg. If the two bodies become coupled so that they move on together after the impact, what is their common velocity ?*

Sol. Given :

Mass of first body, $m_1 = 50$ kg

Initial velocity of first body, $u_1 = 6$ m/s

Mass of second body, $m_2 = 30$ kg

Initial velocity of second body, $u_2 = 0$.

Total mass of two bodies $= (m_1 + m_2) = (50 + 30) = 80$ kg.

Let V = Common velocity of the two bodies after impact.

Total momentum before impact $= m_1u_1 + m_2u_2 = 50 \times 6 + 30 \times 0 = 300$ kg m/s

Total momentum after impact $= (m_1 + m_2) \times V = (50 + 30) \times V = 80\ V$ kg m/s

But total momentum before impact = Total momentum after impact

or $300 = 80 \times V$

$$\therefore \quad V = \frac{300}{80} = \mathbf{3.75\ m/s.} \quad \textbf{Ans.}$$

Problem 4.108. *A bullet of mass 50 gm is fired into a freely suspended target to mass 5 kg. On impact, the target moves with a velocity of 7 m/s along with the bullet in the direction of firing. Find the velocity of bullet.*

Sol. Given :

Mass of bullet, $m_1 = 50 \text{ gm} = \frac{50}{1000} = 0.05$ kg

Mass of target, $m_2 = 5$ kg

Initial velocity of bullet $= u_1$

Initial velocity of target, $u_2 = 0$

Total mass of bullet and target $= 5 + 0.05 = 5.05$ kg

Final velocity of bullet and target $= 7$ m/s

Total initial momentum (*i.e.*, momentum before impact)

$$= m_1 \times u_1 + m_2 \times u_2 = 0.05 \times u_1 + 5 \times 0$$
$$= 0.05\ u_1 \text{ kg m/s}$$

Total final momentum (*i.e.*, momentum after impact)

= Total mass × Common velocity = (5.05) × 7 kg m/s

Equating the initial momentum to final momentum, we get

$$0.05\ u_1 = 5.05 \times 7 \quad \therefore \quad u_1 = \frac{5.05 \times 7}{0.05} = \mathbf{707\ m/s.} \quad \textbf{Ans.}$$

Problem 4.109. *A ball of mass 20 kg moving with a velocity of 5 m/s strikes directly another ball of mass 10 kg moving in the opposite direction with a velocity of 10 m/s. If the coefficient of restitution is equal to 5/6, then determine the velocity of each ball after impact.*

Sol. Given :

Mass of first ball, $m_1 = 20$ kg

Initial velocity of first ball, $u_2 = 5$ m/s

Mass of second ball, $m_2 = 10$ kg
Initial velocity of second ball, $u_2 = -10$ m/s

(Negative sign is due to opposite direction)

Co-efficient of restitution, $e = \frac{5}{6}$

Let v_1 = Velocity of first ball after impact
v_2 = Velocity of second ball after impact

Total momentum before impact $= m_1u_1 + m_2u_2 = 20 \times 5 + 10 \times (-10) = 100 - 100 = 0$
Total momentum after impact $= m_1v_1 + m_2v_2 = 20\, v_1 + 10\, v_2$
Equating the total momentum after impact and before impact, we get

$$20v_1 + 10v_2 = 0 \quad \text{or} \quad v_1 = \frac{-10v_2}{10} = -\frac{1}{2}v_2 \qquad ...(i)$$

Now using equation (4.56), we get

$$e = \frac{(v_2 - v_1)}{(u_1 - u_2)} \quad \text{or} \quad v_2 - v_1 = e\,(u_1 - u_2)$$

$$= \frac{5}{6}[5 - (-10)] \qquad (\because \; u_2 = -10)$$

$$= \frac{5}{6} \times 15 = \frac{25}{2} \qquad ...(ii)$$

Substituting the value of v_1 from equation (*i*) into equation (*ii*), we get

$$v_2 - \left(\frac{-v_2}{2}\right) = \frac{25}{2} \quad \text{or} \quad \frac{3v_2}{2} = \frac{25}{2}$$

or
$$v_2 = \frac{25}{3} = \textbf{8.33 m/s.} \quad \textbf{Ans.}$$

Substituting the value of v_2 in equation (*i*), we get

$$v_1 = -\frac{1}{2} \times 8.33 = \mathbf{-4.165\ m/s.} \quad \textbf{Ans.}$$

(Minus sign indicates that the velocity of first ball after impact will be in opposite direction).

Problem 4.110. *Prove that the two perfect elastic bodies of equal masses exchange velocities in the case of direct central impact.*

Sol. Given :

Two perfect elastic bodies of equal mass

Let m_1 = Mass of first body
m_2 = Mass of second body
u_1 = Velocity of first body before impact
u_2 = Velocity of second body before impact
v_1 = Velocity of first body after impact
v_2 = Velocity of second body after impact

Given : $m_1 = m_2$
and $e = 1$ (for perfect elastic body)

Using equation (4.56), $e = \frac{(v_2 - v_1)}{(u_1 - u_2)}$ or $1 = \frac{(v_2 - v_1)}{(u_1 - u_2)}$

$$\therefore \quad u_1 - u_2 = v_2 - v_1 \qquad ...(i)$$

According to the law of conservation of momentum,

Total initial momentum = Total final momentum

or $$m_1u_1 + m_2u_2 = m_1v_1 + m_2v_2$$

But $$m_1 = m_2 \qquad \text{(Given)}$$

$$\therefore \qquad u_1 + u_2 = v_1 + v_2 \qquad \text{...}(ii)$$

Adding equations (*i*) and (*ii*), we get

$$2u_1 = 2v_2 \quad \therefore \quad u_1 = v_2 \qquad \text{...}(iii)$$

Substituting the value of u_1 in equation (*ii*), we get

$$v_2 + u_2 = v_1 + v_2 \quad \text{or} \quad u_2 = v_1 \qquad \text{...}(iv)$$

From equation (*iii*), it is clear that the velocity of first body before impact is equal to the velocity of the second body after impact.

Equation (*iv*) shows that the velocity of second body before impact is equal to the velocity of first body after impact.

Hence the two perfect elastic bodies of equal masses exchange velocities after impact.

Problem 4.111. *Three perfectly elastic balls A, B and C of masses 2 kg, 6 kg and 12 kg are moving in the same direction with velocities 12 m/s, 4 m/s and 2 m/s respectively. If the ball A strikes with the ball B, which in turn, strikes with the ball C, prove that the balls A and B will be brought to rest by the impact.*

Sol. Given:

For perfectly elastic balls, $e = 1$

Mass of first ball, $m_1 = 2$ kg

Mass of second ball, $m_2 = 6$ kg

Mass of third ball, $m_3 = 12$ kg

Initial velocity of first ball, $u_1 = 12$ m/s

Initial velocity of second ball, $u_2 = 4$ m/s

Initial velocity of third ball, $u_3 = 2$ m/s

(*a*) *First of all consider the impact of first and second ball*

Let v_1 = Final velocity of first ball after impact

v_2 = Final velocity of second ball after impact

Total momentum of 1st and 2nd ball before impact

= Total momentum of 1st and 2nd ball after impact

$$\therefore \qquad m_1u_1 + m_2u_2 = m_1v_1 + m_2v_2$$

or $$2 \times 12 + 6 \times 4 = 2 \times v_1 + 6 \times v_2$$

or $$48 = 2v_1 + 6v_2 \quad \text{or} \quad v_1 + 3v_2 = 24 \qquad \text{...}(i)$$

Now using equation (4.56), $e = \dfrac{(v_2 - v_1)}{(u_1 - u_2)}$

or $$(v_2 - v_1) = e\,(u_1 - u_2) = 1(\,12 - 4\,) \qquad (\because\ e = 1)$$

or $$(v_2 - v_1) = 8 \qquad \text{...}(ii)$$

Adding equations (*i*) and (*ii*), we get

$$4v_2 = 24 + 8 = 32$$

$$\therefore \qquad v_2 = \frac{32}{4} = 8 \text{ m/s.}$$

Substituting this value of v_2 in equation (*ii*),

$$8 - v_1 = 8 \quad \text{or} \quad v_1 = 0.$$

The velocity of ball A after impact is zero. Hence the ball A will be brought to rest by the impact of A and B. **Ans.**

The velocity of ball B after impact with A will be ($v_2 = 8$ m/s).

(*b*) *Now consider the impact of second and third ball.* The second ball is now moving with a velocity of 8 m/s and strikes the third ball which is moving with a velocity of 2 m/s.

Now Initial velocity of second ball, $u_2^* = 8$ m/s

Initial velocity of third ball, $u_3 = 2$ m/s

Let v_2^* = New final velocity of second ball after striking the third ball.

v_3 = Final velocity of third ball

Law of conservation of momentum gives :

Total momentum of 2nd and 3rd ball before impact

= Total momentum of 2nd and 3rd ball after impact

or $$m_2 \times u_2^* + m_3 u_3 = m_2 \times v_2^* + m_3 \times v_3$$

or $$6 \times 8 + 12 \times 2 = 6 \times v_2^* + 12 \times v_3$$

or $$48 + 24 = 6v_2^* + 12v_3 \quad \text{or} \quad 72 = 6v_2^* + 12v_3$$

or $$6v_2^* + 12v_3 = 72 \quad \text{or} \quad v_2^* + 2v_3 = 12 \qquad \ldots(iii)$$

Using equation (4.56), we get

$$e = \frac{\text{Velocity of separation}}{\text{Velocity of approach}} = \frac{(v_3 - v_2^*)}{(u_2^* - u_3)}$$

or $$(v_3 - v_2^*) = e\,(u_2^* - u_3) = 1(8 - 2) = 6$$

$\therefore$ $$v_3 - v_2^* = 6 \qquad \ldots(iv)$$

Adding equations (*iii*) and (*iv*), we get

$$3v_3 = 12 + 6 = 18$$

$\therefore$ $$v_3 = \frac{18}{3} = 6 \text{ m/s.}$$

Substituting the value of v_3 in equation (*iv*),

$$6 - v_2^* = 6 \quad \text{or} \quad v_2^* = 6 - 6 = 0.$$

Hence the velocity of second ball, after impact with third ball, is zero. Hence the second ball will also be brought to rest after impact with third ball C. **Ans.**

Problems on Indirect Impact

Problem 4.112. *A ball of mass 1 kg, moving with a velocity of 6 m/s, strikes on a ball of mass 2 kg moving with a velocity of 2 m/s. At the instant of impact, the velocities of the two balls are parallel and inclined at 30° to the line joining their centres. If coefficient of restitution is* $\frac{1}{2}$, *find:*

(*i*) *the velocity and direction in which the 1 kg ball will move after impact.*

(*ii*) *the velocity and direction in which the 2 kg ball will move after impact.*

Sol. Given :

Mass of first ball,

$$m_1 = 1 \text{ kg}$$

Mass of second ball,

$$m_2 = 2 \text{ kg}$$

Initial velocity of first ball,

$$u_1 = 6 \text{ m/s}$$

Initial velocity of second ball,

$$u_2 = 2 \text{ m/s}$$

Angle made by first ball with line of impact, $\theta_1 = 30°$

Angle made by second ball with line of impact, $\theta_2 = 30°$

Coefficient of restitution,

$$e = \frac{1}{2}$$

Fig. 4.74

Let v_1 = Velocity of first ball after impact

v_2 = Velocity of second ball after impact

ϕ_1 = Angle made by first ball after impact with line of impact

ϕ_2 = Angle made by second ball after impact with line of impact.

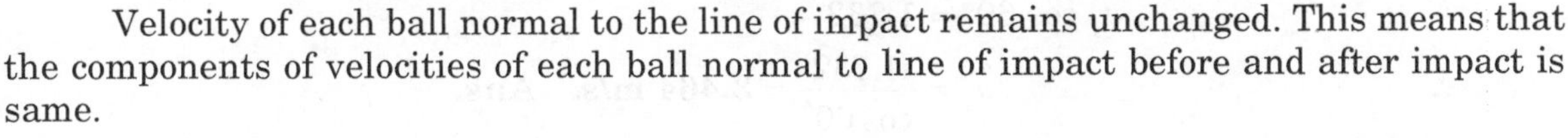

Velocity of each ball normal to the line of impact remains unchanged. This means that the components of velocities of each ball normal to line of impact before and after impact is same.

$\therefore$ *For ball A,*

Initial normal component = Final normal component

or $$6 \sin 30° = v_1 \sin \phi_1 \quad \text{or} \quad v_1 \sin \phi_1 = 6 \sin 30° = 3 \qquad ...(i)$$

For ball B,

Final normal component = Initial normal component

$$v_2 \sin \phi_2 = 2 \sin 30° = 1 \qquad ...(ii)$$

According to law of conservation of momentum,

Total initial momentum along the line of impact

= Total final momentum along the line of impact.

or $$m_1 \times u_1 \cos \theta_1 + m_2 \times u_2 \cos \theta_2 = m_1 \times v_1 \cos \phi_1 + m_2 \times v_2 \cos \phi_2$$

or $$1 \times 6 \cos 30° + 2 \times 2 \cos 30° = 1 \times v_1 \cos \phi_1 + 2 \times v_2 \cos \phi_1$$

or $$6 \times 0.866 + 4 \times 0.866 = v_1 \cos \phi_2 + 2v_2 \cos \phi_2$$

or $$8.66 = v_1 \cos \phi_1 + 2v_2 \cos \phi_2$$

or $$v_1 \cos \phi_1 + 2v_2 \cos \phi_2 = 8.66 \qquad ...(iii)$$

The coefficient of restitution for this case is given by the equation

$$(v_2 \cos \phi_2 - v_1 \cos \phi_1) = e\,(u_1 \cos \theta_1 - u_2 \cos \theta_2)$$

$$= \frac{1}{2}\,[6 \times \cos 30° - 2 \times \cos 30°] = 2 \times \cos 30° = 1.732$$

$\therefore$ $$v_2 \cos \phi_2 - v_1 \cos \phi_1 = 1.732 \qquad ...(iv)$$

Adding equations (*iii*) and (*iv*),

$$3v_2 \cos \phi_2 = 8.66 + 1.732 = 10.392$$

or $$v_2 \cos \phi_2 = \frac{10.392}{3} = 3.464 \qquad ...(v)$$

Substituting the value of $v_2 \cos \phi_2$ in equation (*iii*),

$$v_1 \cos \phi_1 \times 2 \times 3.464 = 8.66$$

or $$v_1 \cos \phi_1 = 8.66 - 2 \times 3.464 = 1.732 \quad ...(vi)$$

Dividing equation (*i*) by equation (*vi*), we get

$$\tan \phi_1 = \frac{3}{1.732} = 1.732$$

$$\therefore \quad \phi_1 = \tan^{-1} 1.732 = \mathbf{60^\circ.} \quad \textbf{Ans.}$$

Dividing equation (*ii*) by equation (*v*), we get

$$\tan \phi_2 = \frac{1}{3.464} = 0.2886$$

$$\therefore \quad \phi_2 = \tan^{-1} 0.2886 = \mathbf{16.1^\circ.} \quad \textbf{Ans.}$$

Substituting the value of ϕ_2 in equation (*v*),

$$v_2 \cos 16.1^\circ = 3.464$$

$$\therefore \quad v_2 = \frac{3.464}{\cos 16.1^\circ} = \mathbf{3.6 \text{ m/s.}} \quad \textbf{Ans.}$$

Substituting the value of ϕ_1 in equation (*vi*),

$$v_1 \cos 60^\circ = 1.732$$

$$\therefore \quad v_1 = \frac{1.732}{\cos 60^\circ} = \mathbf{3.464 \text{ m/s.}} \quad \textbf{Ans.}$$

Problem 4.113. *A ball of mass 2 kg, moving with a velocity of 20 m/s, strikes on a ball of mass 2 kg moving with a velocity of 30 m/s. At the instant of impact, the velocities of the balls are inclined at an angle of 30° and 60° to the line joining their centres, as shown in Fig. 4.75. If coefficient of restitution is 0.9, then find :*

(*i*) *the magnitude and direction of first ball after impact*

(*ii*) *the magnitude and direction of second ball after impact.*

Sol. Given :

Mass of first ball,

$$m_1 = 2 \text{ kg}$$

Mass of second ball,

$$m_2 = 2 \text{ kg}$$

Initial velocity of first ball,

$$u_1 = 20 \text{ m/s}$$

Initial velocity of second ball,

$$u_2 = 30 \text{ m/s}$$

Angle made by first ball with line of impact,

$$\theta_1 = 30^\circ$$

Angle made by second ball with line of impact,

$$\theta_2 = 60^\circ$$

Coefficient of restitution,

$$e = 0.9$$

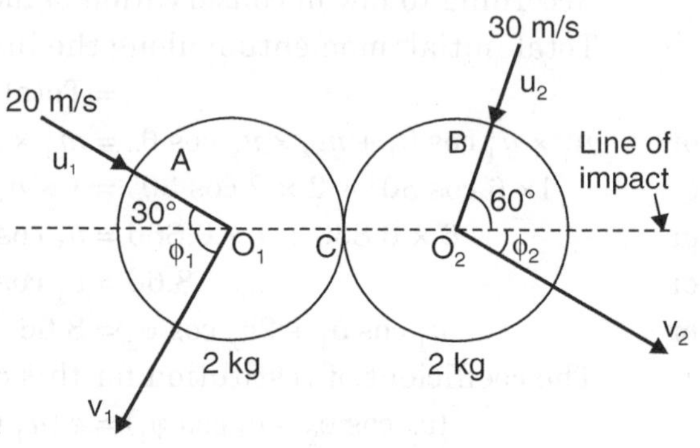

Fig. 4.75

Let v_1 = Velocity of first ball after impact

v_2 = Velocity of second ball after impact

ϕ_1 = Angle made by first ball with line of impact after impact

ϕ_2 = Angle made by second ball after impact.

The components of velocity of each ball perpendicular to the line of impact before and after impact is same.

$\therefore$ *For ball A,*

Final normal component = Initial normal component

$$v_1 \sin \phi_1 = 20 \sin 30° = 10 \quad ...(i)$$

For ball B,

Final normal component = Initial normal component

$$v_2 \sin \phi_2 = 30 \sin 60° = 25.98 \quad ...(ii)$$

Now according to the law of conservation of momentum

Total momentum along the line of impact before collision

= Total momentum along the line of impact after collision

$$m_1 \times u_1 \cos \theta_1 + m_2 \times u_2 \cos \theta_2 = m_1 \times v_1 \cos \phi_1 + m_2 \times v_2 \cos \phi_2$$

Here $u_1 \cos \theta_1 = 20 \times \cos 30° = 17.32$

$u_2 \cos \theta_2 = -30 \times \cos 60° = -15$

(Minus sign is due to opposite direction)

Substituting these values in the above equation

$$2 \times 17.32 + 2 \times (-15) = 2 \times v_1 \cos \phi_1 + 2 \times v_2 \cos \phi_2$$

or $$17.32 - 15 = v_1 \cos \phi_1 + v_2 \cos \phi_2$$

or $$2.32 = v_1 \cos \phi_1 + v_2 \cos \phi_2$$

or $$v_1 \cos \phi_1 + v_2 \cos \phi_2 = 2.32 \quad ...(iii)$$

The coefficient of restitution for the indirect impact between two bodies is given by

$$v_2 \cos \phi_2 - v_1 \cos \phi_1 = e\,[u_1 \cos \theta_1 - u_2 \cos \theta_2] = 0.9\,[17.32 - (-15)]$$

$(\because \; u_1 \cos \theta_1 = 17.32 \text{ and } u_2 \cos \theta_2 = -15)$

$$= 0.9 \times 32.32$$

or $$v_2 \cos \phi_2 - v_1 \cos \phi_1 = 29.08 \quad ...(iv)$$

Adding equations (*iii*) and (*iv*), we get

$$2v_2 \cos \phi_2 = 2.32 + 29.03 = 31.40$$

or $$v_2 \cos \phi_2 = \frac{31.40}{2} = 15.7 \quad ...(v)$$

Substituting the above value in equation (*iii*), we get

$$v_1 \cos \phi_1 + 15.7 = 2.32$$

or $$v_1 \cos \phi_1 = 2.32 - 15.7 = -13.38$$

Minus sign shows that final velocity of first ball will be in the opposite direction. Numerically,

$$v_1 \cos \phi_1 = 13.38$$

$\therefore$ $$v_1 \cos \phi_1 = 13.38 \quad ...(vi)$$

Dividing equation (*i*) by equation (*vi*), we get

$$\frac{v_1 \sin \phi_1}{v_1 \cos \phi_1} = \frac{10}{13.38} \quad \text{or} \quad \tan \phi_1 = 0.7474$$

$$\therefore \quad \phi_1 = \tan^{-1} 0.7474 = \mathbf{36.77°.} \quad \textbf{Ans.}$$

Substituting this value of ϕ_1 in equation (*vi*),

$$v_1 \cos 36.77° = 13.38$$

$$\therefore \quad v_1 = \frac{13.38}{\cos 36.77} = \mathbf{16.7\ m/s.} \quad \textbf{Ans.}$$

Dividing equation (*ii*) by equation (*v*),

$$\frac{v_2 \sin \phi_2}{v_2 \cos \phi_2} = \frac{25.98}{15.7} = 1.6547 \quad \text{or} \quad \tan \phi_2 = 1.6547$$

$$\therefore \quad \phi_2 = \tan^{-1} 1.6547 = \mathbf{58.85°.} \quad \textbf{Ans.}$$

Substituting this value of ϕ_2 in equation (*ii*),

$$v_2 \sin 58.85° = 25.98$$

$$\therefore \quad v_2 = \frac{25.98}{\sin 58.85} = \mathbf{30.35\ m/s.} \quad \textbf{Ans.}$$

4.13.5. Loss of Kinetic Energy During Impact. When the two bodies collide with each other, the loss of kinetic energy takes place due to impact. This loss of energy may be obtained by finding out the kinetic energy of the two bodies before and after the impact. The loss of kinetic energy during impact is equal to the difference of the two kinetic energies.

Consider two bodies *A* and *B* having a direct impact.

Let m_1 = Mass of the first body,

u_1 = Velocity of first body before impact

v_1 = Velocity of first body after impact

m_2, u_2 and v_2 = Corresponding values of mass, initial velocity and final velocity of second body

E_L = Loss of energy

e = Coefficient of restitution.

Kinetic energy of first body before impact

$$= \frac{1}{2} \times m_1 \times u_1^2$$

Kinetic energy of second body before impact

$$= \frac{1}{2} m_2 u_2^2$$

$\therefore$ Total K.E. of the two bodies before impact

$$= \frac{1}{2} m_1 u_1^2 + \frac{1}{2} m_2 u_2^2 \qquad ...(i)$$

Similarly, total K.E. of the two bodies after impact

$$= \frac{1}{2} m_1 v_1^2 + \frac{1}{2} m_2 v_2^2 \qquad ...(ii)$$

$\therefore$ Loss of K.E., during impact

= Total K.E. before impact – Total K.E. after impact

$$= \left(\frac{1}{2} m_1 u_1^2 + \frac{1}{2} m_2 u_2^2\right) - \left(\frac{1}{2} m_1 v_1^2 + \frac{1}{2} m_2 v_2^2\right).$$

Problem 4.114. *A vehicle of mass 600 kg and moving with a velocity of 12 m/s strikes another vehicle of mass 400 kg, moving at 9 m/s in the same direction. Both the vehicles get coupled together due to impact. Find the common velocity with which the two vehicles will move. Also find the loss of kinetic energy due to impact.*

Sol. Given :

Mass of first vehicle, $m_1 = 600$ kg

Initial velocity of first vehicle, $u_1 = 12$ m/s

Mass of second vehicle, $m_2 = 400$ kg

Initial velocity of second vehicle, $u_2 = 9$ m/s

When the two vehicles get coupled, then total mass of the two vehicles,

$$M = m_1 + m_2 = 600 + 400 = 1000 \text{ kg}$$

(*i*) Let V = Common velocity of the two vehicles after impact

Total momentum before impact

$$= m_1u_1 + m_2u_2 = 600 \times 12 + 400 \times 9$$

$$= 7200 + 3600 = 10800 \text{ kg m/s} \qquad ...(i)$$

Total momentum after impact

$$= (m_1 + m_2) \times V = (600 + 400) \times V = 1000\, V \text{ kg m/s} \qquad ...(ii)$$

According to law of conservation of momentum,

Total momentum before impact = Total momentum after impact

or $$10800 = 1000 \times V$$

$\therefore$ $$V = \frac{10800}{1000} = \mathbf{10.8 \text{ m/s.}} \quad \textbf{Ans.}$$

(*ii*) Let E_L = Loss of kinetic energy due to impact

Total K.E. before impact

$$= \frac{1}{2}\, m_1u_1^2 + \frac{1}{2}\, m_2u_2^2 = \frac{1}{2} \times 600 \times 12^2 + \frac{1}{2} \times 400 \times 9^2$$

$$= \frac{118800}{2} \text{ Nm}$$

Total K.E. after impact $= \frac{1}{2} MV^2 = \frac{1}{2} \times 1000 \times 10.8^2 = \frac{116640}{2}$ Nm

$\therefore$ Loss of K.E. = Total K.E. before impact – Total K.E. after impact

$$= \frac{118800}{2} - \frac{116640}{2} = \frac{2160}{2} = \mathbf{1080 \text{ Nm.}} \quad \textbf{Ans.}$$

Problem 4.115. *A bullet of mass 100 gm is fixed into a freely suspended target of mass 10 kg. Due to impact, the bullet gets embedded in the target and the target with bullet moves with a velocity of 7 m/s. Find the velocity of the bullet and the loss of kinetic energy.*

Sol. Given :

Mass of bullet, $m_1 = 100 \text{ gm} = \frac{100}{1000} = 0.1$ kg

Mass of target, $m_2 = 10$ kg

Common velocity of bullet and target = 7 m/s

Initial velocity of target, $u_2 = 0$

Let u_1 = Initial velocity of bullet

E_L = Loss of kinetic energy due to impact

Total momentum before impact

$= m_1u_1 + m_2u_2 = 0.1 \times u_1 + 10 \times 0 = 0.1\ u_1$ ...(*i*)

Total momentum after impact

$= (m_1 + m_2) \times$ Common velocity

$= (0.1 + 10) \times 7 = 70.7$ kg m/s ...(*ii*)

Equating the total momentum before impact to the total momentum after impact, we get

$$0.1\ u_1 = 70.7$$

$$\therefore \quad u_1 = \frac{7.07}{0.1} = \mathbf{707\ m/s.} \quad \textbf{Ans.}$$

To find the loss of kinetic energy due to impact, find the total K.E. before impact and after impact.

Total K.E. before impact $= \frac{1}{2}\ m_1u_1^2 + \frac{1}{2}\ m_2u_2^2$

$= \frac{1}{2}\ (0.1) \times 707^2 + \frac{1}{2} \times 10 \times 0^2 = 24992.45$ Nm

Total K.E. after impact $= \frac{1}{2}\ MV^2$

$= \frac{1}{2}\ (m_1 + m_2)V^2$ $(V = 7$ m/s)

$= \frac{1}{2}\ (0.1 + 10) \times 7^2 = 247.45$ Nm

$\therefore$ Loss of K.E. = K.E. before impact – K.E. after impact

$= 24992.45 - 247.45 =$ **24745 Nm. Ans.**

Problem 4.116. *A ball of mass 100 gm strikes directly another ball of same mass, which is at rest. The first ball is reduced to rest by the impact. Find the loss of kinetic energy if the coefficient of restitution is* $1/\sqrt{2}$.

Sol. Given :

Mass of first ball, $m_1 = 100$ gm $= 0.1$ kg

Mass of second ball, $m_2 = 100$ gm $= 0.1$ kg

Initial velocity of second ball, $u_2 = 0$

Final velocity of first ball, $v_1 = 0$

Coefficient of restitution, $e = \frac{1}{\sqrt{2}}$

Let u_1 = Initial velocity of first ball

v_1 = Final velocity of second ball

The coefficient of restitution for direct impact is given by equation (4.56), as

$$e = \frac{(v_2 - v_1)}{(u_1 - u_2)} \quad \text{or} \quad \frac{1}{\sqrt{2}} = \frac{(v_2 - 0)}{(u_1 - 0)} \qquad (\because\ v_1 = 0 \text{ and } u_2 = 0)$$

$$= \frac{v_2}{u_1}$$

$$\therefore \quad u_1 = \sqrt{2}\ v_2 \qquad ...(i)$$

Now K.E. before impact $= \frac{1}{2} m_1u_1^2 + \frac{1}{2} m_2u_2^2$

$= \frac{1}{2}(0.1)\ u_1^2 + \frac{1}{2}\ (0.1) \times 0^2 = 0.05u_1^2$

$= 0.05\,(\sqrt{2}\,v_2)^2$ $\quad (\because \; u_1 = \sqrt{2}\,v_2)$

$= 0.1\,v_2^2$

K.E. after impact $= \frac{1}{2}m_1v_1^2 + \frac{1}{2}m_2v_2^2$

$= \frac{1}{2}(0.1) \times 0^2 + \frac{1}{2}(0.1)v_2^2 = 0.05\,v_2^2$

$\therefore$ Loss of K.E. = K.E. before impact – K.E. after impact

$= 0.1\,v_2^2 - 0.05\,v_2^2 = \mathbf{0.05\,v_2^2}$. **Ans.**

Problem 4.117. *Two bodies, of masses m_1 and m_2 and of initial velocities u_1 and u_2, are having a direct impact. Find an expression for the loss of kinetic energy due to impact in terms of masses of two bodies and their velocities before impact.*

Sol. Given :

First Body

Mass $= m_1$

Initial velocity $= u_1$

Final velocity $= v_1$

Second Body

Mass $= m_2$

Initial velocity $= u_2$

Final velocity $= v_2$

Total K.E. of two bodies before impact

$$= \frac{1}{2}m_1u_1^2 + \frac{1}{2}m_2u_2^2$$

Total K.E. of two bodies after impact

$$= \frac{1}{2}m_1v_1^2 + \frac{1}{2}m_2v_2^2$$

Loss of K.E. during impact

$$= \left(\frac{1}{2}m_1u_1^2 + \frac{1}{2}m_2u_2^2\right) - \left(\frac{1}{2}m_1v_1^2 + \frac{1}{2}m_2v_2^2\right)$$

$$= \frac{1}{2}[(m_1u_1^2 + m_2u_2^2) - (m_1v_1^2 + m_2v_2^2)]$$

Multiplying the numerator and denominator by $(m_1 + m_2)$

$\therefore$ Loss of K.E.

$$= \frac{1}{2(m_1+m_2)}\,[(m_1 + m_2)\,(m_1u_1^2 + m_2u_2^2) - (m_1v_1^2 + m_2v_2^2\,)\,(m_1 + m_2\,)]$$

$$= \frac{1}{2(m_1+m_2)}\,[(m_1^2u_1^2 + m_1m_2u_2^2 + m_1m_2u_1^2 + m_2^2u_2^2) - (m_1^2v_1^2 + m_1m_2v_2^2 + m_1m_2v_1^2 + m_2^2v_2^2)$$

$$= \frac{1}{2(m_1+m_2)}\,[\{m_1^2u_1^2 + m_2^2u_2^2 + m_1m_2\,(u_1^2 + u_2^2)\} - \{m_1^2v_1^2 + m_2^2v_2^2 + m_1m_2(v_1^2 + v_2^2)\}]$$

$$= \frac{1}{2(m_1+m_2)}\,[\{(m_1u_1 + m_2u_2)^2 - 2m_1m_2u_1u_2 + m_1m_2(u_1^2 + u_2^2)\} - \{(m_1v_1 + m_2v_2)^2 - 2m_1m_2v_1v_2 + m_1m_2\,(v_1^2 + v_2^2)\}]$$

Now $m_1m_2\,(u_1^2 + u_2^2) = m_1m_2(u_1 - u_2)^2 + 2m_1m_2u_1u_2$

Also $m_1m_2(v_1^2 + v_2^2) = m_1m_2(v_1 - v_2)^2 + 2m_1m_2v_1v_2$

$$\therefore \quad \text{Loss of K.E.} = \frac{1}{2(m_1 + m_2)} [\{(m_1u_1 + m_2u_2)^2 - 2m_1m_2u_1u_2 + m_1m_2(u_1 - u_2)^2 + 2m_1m_2u_1u_2\} - \{(m_1v_1 + m_2v_2)^2 - 2m_1m_2v_1v_2 + m_1m_2(v_1 - v_2)^2 + 2m_1m_2v_1v_2\}]$$

$$= \frac{1}{2(m_1 + m_2)} [\{(m_1u_1 + m_2u_2)^2 + m_1m_2(u_1 - u_2)^2\} - \{(m_1v_1 + m_2v_2)^2 + m_1m_2(v_1 - v_2)^2\}]$$

Now according to law of conservation of momentum,

Total momentum before impact

= Total momentum after impact

$$(m_1u_1 + m_2u_2) = (m_1v_1 + m_2v_2)$$

Squaring both sides, we get

$$(m_1u_1 + m_2u_2)^2 = (m_1v_1 + m_2v_2)^2$$

Substituting the value of $(m_1v_1 + m_2v_2)^2$ in the loss of energy equation, we get

$$\text{Loss of K.E.} = \frac{1}{2(m_1 + m_2)} [\{(m_1u_1 + m_2u_2)^2 + m_1m_2(u_1 - u_2)^2\} - \{(m_1u_1 + m_2u_2)^2 + m_1m_2(v_1 - v_2)^2\}]$$

$$= \frac{1}{2(m_1 + m_2)} [m_1m_2(u_1 - u_2)^2 - m_1m_2\,(v_1 - v_2)^2]$$

The coefficient of restitution is given by equation (4.56), as

$$e = \frac{v_2 - v_1}{u_1 - u_2} \quad \text{or} \quad v_2 - v_1 = e\,(u_1 - u_2)$$

Squaring both sides, we get

$$(v_2 - v_1)^2 = e^2\,(u_1 - u_2)^2 \quad \text{or} \quad (v_1 - v_2)^2 = e^2\,(u_1 - u_2)^2$$

Substituting the value of $(v_1 - v_2)^2$ in the loss of energy equation, we get

$$\text{Loss of K.E.} = \frac{1}{2(m_1 + m_2)} [m_1m_2(u_1 - u_2)^2 - m_1m_2 \times e^2(u_1 - u_2)^2]$$

$$= \frac{m_1\,m_2(u_1 - u_2)^2}{2(m_1 + m_2)} [1 - e^2]$$

The above equation gives the loss of K.E. due to direct impact in terms of masses, initial velocity and coefficient of restitution.

Problem 4.118. *The coefficient of restitution between two spheres of masses 1 kg and 5 kg is 0.75. The sphere of mass 1 kg, moving with a velocity of 3 m/s, strikes the sphere of mass 5 kg moving in the same direction with a velocity of 60 cm/s. Find the velocities of the two spheres after impact and also loss of kinetic energy during impact.*

Sol. Given :

Coefficient of restitution = 0.75

Mass of 1st sphere, $m_1 = 1$ kg

Mass of 2nd sphere, $m_2 = 5$ kg

Initial velocity of 1st sphere, $u_1 = 3$ m/s

Initial velocity of 2nd sphere, $u_2 = 60$ cm/s = 0.6 m/s

Let v_1 = Final velocity of 1st sphere

v_2 = Final velocity of 2nd sphere

Law of conservation of momentum gives :

Total momentum before impact

= Total momentum after impact

$\therefore \quad m_1 u_1 + m_2 u_2 = m_1 v_1 + m_2 v_2$

or $\quad 1 \times 3 + 5 \times 0.6 = 1 \times v_1 + 5v_2$

or $\quad 3 + 3 = v_1 + 5v_2 \quad$ or $\quad v_1 + 5v_2 = 6 \quad$ ...(i)

Now using equation (4.56), we get

$$e = \frac{(v_2 - v_1)}{(u_1 - u_2)} \quad \text{or} \quad v_2 - v_1 = e\,(u_1 - u_2)$$

$$= 0.75\,(3 - 0.6) = 0.75 \times 2.4 = 1.8$$

$\therefore \quad v_2 - v_1 = 1.8 \quad$ ...(ii)

Adding equations (i) and (ii), we get

$$6v_2 = 6 + 1.8 = 7.8$$

$\therefore \quad v_2 = \dfrac{7.8}{6} = \mathbf{1.3\ m/s.}$ **Ans.**

Substituting this value in equation (ii), we get

$$1.3 - v_1 = 1.8 \quad \text{or} \quad -v_1 = 1.8 - 1.3 = 0.5$$

$\therefore \quad v_1 = \mathbf{-\,0.5\ m/s.}$ **Ans.**

Minus sign indicates that the direction of the first sphere after impact is reversed.

Loss of K.E. during impact

= Total K.E. before impact – Total K.E. after impact

$$= (\tfrac{1}{2} m_1 u_1^2 + \tfrac{1}{2} m_2 u_2^2) - (\tfrac{1}{2} m_1 v_1^2 + \tfrac{1}{2} m_2 v_2^2)$$

$$= [\tfrac{1}{2} \times 1 \times 3^2 + \tfrac{1}{2} \times 5 \times (0.6^2)] - [\tfrac{1}{2} \times 1 \times (-\,0.5^2) + \tfrac{1}{2} \times 5 \times 1.3^2]$$

$(\because \ v_1 = -\,0.5)$

$$= [4.5 + 0.9] - [0.125 + 4.225] = 5.4 - 4.35 = \mathbf{1.05\ Nm.} \quad \textbf{Ans.}$$

Second Method

The loss of K.E. during direct impact can also be obtained by using the expression derived in problem 4.117.

$$\therefore \quad \text{Loss of K.E.} = \frac{m_1 m_2 (u_1 - u_2)^2}{2(m_1 + m_2)} (1 - e^2) = \frac{1 \times 5 (3 - 0.6)^2}{2(1 + 5)} (1 - 0.75^2)$$

$$= \frac{5 \times 2.4^2}{12} (1 - 0.5625) = \mathbf{1.05\ Nm.} \quad \textbf{Ans.}$$

4.13.6. Impact of a Body on a Fixed Plane. There are two types of impact of a body on a fixed plane. These are :

(i) Direct impact of a body on a fixed plane, and

(ii) Indirect impact of a body on a fixed plane.

4.13.7. Direct Impact of a Body on a Fixed Plane. The fixed plane is at rest before impact and after impact. The mass of the fixed plane is very large. Consider a body having a direct impact on a fixed plane.

Let u = Initial velocity of the body,

v = Final velocity of the body

e = Coefficient of restitution.

The velocity of approach between body and fixed plane

= Initial velocity of body – velocity of fixed plane = $u - 0 = u$.

The velocity of separation

= Final velocity of body – Velocity of fixed plane = $v - 0 = v$.

∴ Coefficient of restitution is given by,

$$e = \frac{\text{Velocity of separation}}{\text{Velocity of approach}} = \frac{v}{u} \quad \text{or} \quad v = e \times u \qquad ...(4.58)$$

Important Cases :

(a) *A body is falling on a floor from some height*

Let H = Height from which the body is falling

u = Velocity with which the body strikes the floor

This velocity will be obtained by using the equation

(Final vel.)^2 – (Initial vel.)^2 = $2g \times H$

Initial velocity of the body = 0

Final velocity of the body = u

$$\therefore \quad u^2 - 0 = 2gH$$

or

$$u = \sqrt{2gH}. \qquad ...(4.59)$$

(b) *A body is first projected upwards and then strikes the floor.* When the body is projected upwards with some initial velocity, the body will reach the greatest height and then will return to the ground with the same velocity with which it was projected.

Problem 4.119. *A ball of mass 500 gm is dropped on a horizontal floor from a height of 18 m. The ball rebounds due to impact with horizontal floor to a height of 8 m. Find the coefficient of restitution between the floor and the ball.*

Sol. Given :

Mass of the ball, m = 500 gm = 0.5 kg

Height from which the ball is dropped,

$$H = 18 \text{ m}$$

The velocity with which the ball strikes the floor is given by equation (4.59),

$$\therefore \quad u = \sqrt{2gH} = \sqrt{2 \times 9.81 \times 18} = \sqrt{36 \times 9.81} = 6 \times \sqrt{9.81} \text{ m/s.}$$

Height to which the ball rebounds = 8 m

or

$$h = 8 \text{ m}$$

This means that after the impact, the ball rises to a height of 8 m. At the highest point, the velocity will be zero. The velocity of the ball after impact, will be obtained by using the relation

(Final vel.)^2 – (Initial vel.)^2= $2gh$

Let v = velocity of the ball after impact

Then $v = \sqrt{2gh} = \sqrt{2 \times 9.81 \times 8} = 4 \times \sqrt{9.81}$

The coefficient of restitution is given by equation (4.58), as

$$v = e \times u \quad \text{or} \quad 4 \times \sqrt{9.81} = e \times 6\sqrt{9.81}$$

$$\therefore \qquad e = \frac{4 \times \sqrt{9.81}}{6 \times \sqrt{9.81}} = \frac{\mathbf{2}}{\mathbf{3}}. \quad \textbf{Ans.}$$

Problem 4.120. *A ball is dropped on a horizontal floor from which it rebounds to a height of 16 m. If the coefficient of restitution between the floor and the ball is 0.8, find the height from which the ball was dropped.*

Sol. Given :

Height after impact, $h = 16$ m

Coefficient of restitution, $e = 0.8$

Let H = Height from which ball was dropped.

The velocity with which the ball strikes the floor,

$$u = \sqrt{2gH}$$

The velocity with which the ball rebounds,

$$v = \sqrt{2gh} = \sqrt{2g \times 16}$$

The coefficient of restitution is given by equation (4.58) as

$$v = e \times u \quad \text{or} \quad \sqrt{2g \times 16} = 0.8 \times \sqrt{2gH}$$

Squaring both sides, we get

$$2g \times 16 = 0.8^2 \times 2gH \quad \text{or} \quad 2g \times 16 = 0.64 \times 2g \times H$$

or
$$H = \frac{2g \times 16}{0.64 \times 2g} = \frac{16}{0.64} = \mathbf{25\ m.} \quad \textbf{Ans.}$$

Problem 4.121. *A ball is dropped from a height of 13.5 m on a smooth horizontal floor, from which it rebounds to a height of 9 m. The ball again strikes the floor and rises to an unknown height. Find :*

(a) Unknown height of the second rebound, and

(b) The coefficient of restitution.

Sol. Given :

Height from which ball is dropped,

$$H = 13.5 \text{ m}$$

Height to which the ball rebounds in the first case,

$$h = 9 \text{ m}$$

$\therefore$ The velocity with which the ball strikes the floor for the first time is given by

$$u = \sqrt{2gH} = \sqrt{2 \times 9.81 \times 13.5}$$

The velocity of the ball after first impact is given by,

$$v = \sqrt{2gh} = \sqrt{2 \times 9.81 \times 9}$$

The coefficient of restitution is given by equation (4.58) as

$$v = e \times u \quad \text{or} \quad \sqrt{2 \times 9.81 \times 9} = e \times \sqrt{2 \times 9.81 \times 13.5}$$

or
$$e = \frac{\sqrt{2 \times 9.81 \times 9}}{\sqrt{2 \times 9.81 \times 13.5}} = \sqrt{\frac{9}{13.5}} = \mathbf{0.816.} \quad \textbf{Ans.}$$

To find the height of the second rebound

Let h_2 = Height of second rebound

Height from which ball strikes the floor for the second time,

$h = 9$ m

$\therefore$ The velocity, with which the ball strikes the floor second time, is given by

$$u_1 = \sqrt{2gh} = \sqrt{2 \times 9.81 \times 9}$$

The velocity with which the ball rebounds second time is given by,

$$v_1 = \sqrt{2gh_2} = \sqrt{2 \times 9.81 \times h_2}$$

Again using equation (4.58), as

$$v_1 = e \times u_1 \quad \text{or} \quad \sqrt{2 \times 9.81 \times h_2} = 0.816 \times \sqrt{2 \times 9.81 \times 9}$$

Squaring, we get

$$2 \times 9.81 \times h_2 = 0.816^2 \times 2 \times 9.81 \times 9 \quad \text{or} \quad h_2 = 0.816^2 \times 9 = \mathbf{6\ m.} \quad \textbf{Ans.}$$

4.13.8. Indirect Impact of a body on a fixed plane. Fig. 4.76 shows a body striking a fixed plane at certain angle. This type of impact is known as indirect impact. The point C is known as point of contact. The line joining the centre of the body with the point of contact is known as *line of impact.*

Let u = Velocity of body before impact

v = Velocity of body after impact

θ = Angle made by the initial velocity with line of impact

ϕ = Angle made by the final velocity with line of impact

e = Coefficient of restitution.

The component of initial velocity along the line of impact will cause the direct impact with the fixed plane. The components of the initial velocity and final velocity normal to the line of impact will remain unchanged.

Here velocity of approach $= u \cos \theta$

and velocity of separation $= v \cos \phi$

But, velocity of separation

$= e \times$ velocity of approach

or $v \cos \phi = e \times u \cos \theta$. ...(4.60)

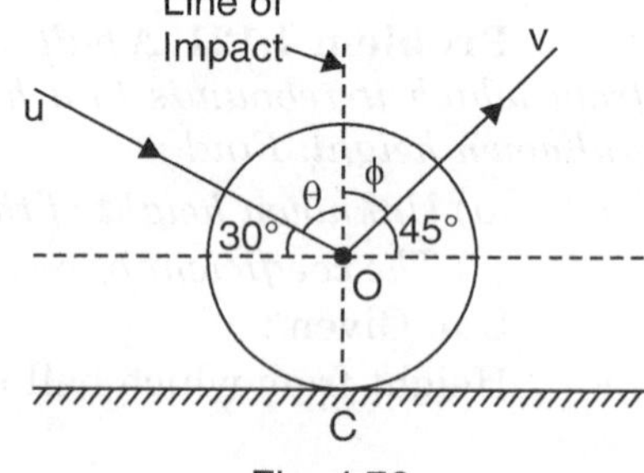

Fig. 4.76

Problem 4.122. *A ball is thrown against a wall with a velocity 10 m/s making an angle of 30° with the wall as shown in Fig. 4.77. If the coefficient of restitution = 0.5, find :*

(i) Direction of the ball after impact, and

(ii) Velocity of the ball after impact.

Sol. Given :

Initial velocity of ball, $u = 10$ m/s.

Angle made by initial velocity with wall = 30°.

$\therefore$ Angle made by initial velocity with line of impact,

$\theta = 90 - 30 = 60°$

Co-efficient of restitution, $e = 0.5$

Let v = Final velocity of the ball,

and θ = Angle made by final velocity with line of impact.

The components of the initial velocity and final velocity at right angles to the line of impact is same.

$\therefore \quad u \sin \theta = v \sin \phi \quad$ or $\quad 10 \sin 60° = v \sin \phi$

or $\quad 10 \times 0.866 = v \sin \phi \quad$ or $\quad v \sin \phi = 8.66 \quad ...(i)$

Now using equation (4.60),

$$v \cos \phi = e \times u \cos \theta = 0.5 \times 10 \times \cos 60° = 2.5 \quad ...(ii)$$

Divide equation (*i*) by equation (*ii*),

$$\frac{v \sin \phi}{v \cos \phi} = \frac{8.66}{2.5}$$

$$\therefore \quad \tan \phi = \frac{8.66}{2.5} = 3.464$$

$$\therefore \quad \phi = \tan^{-1} 3.464 = \mathbf{73.9°}. \quad \textbf{Ans.}$$

Substituting the value of ϕ in equation (*i*), we get

$$v \sin 73.9° = 8.66$$

$$\therefore \quad v = \frac{8.66}{\sin (73.9)} = \mathbf{9.013 \text{ m/s}}. \quad \textbf{Ans.}$$

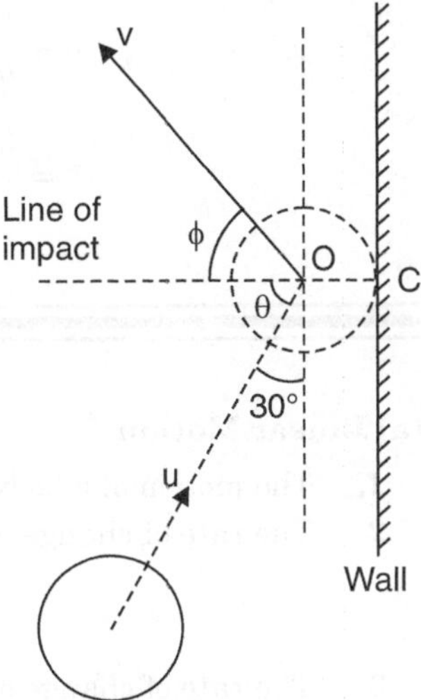

Fig. 4.77

Problem 4.123. *A ball moving with a velocity of 5 m/s strikes on a fixed plane, making an angle of 30° with the fixed plane. If the angle made by the velocity of the ball after impact is 45° with the fixed plane, then find :*

(*i*) *Velocity of the body after impact, and*

(*ii*) *Coefficient of restitution between the ball and fixed plane.*

Sol. Given :

Initial velocity of the body,

$$u = 5 \text{ m/s}$$

Angle made by initial velocity with the fixed plane

$$= 30°.$$

$\therefore$ Angle made by initial velocity with line of impact, $\quad \theta = 90° - 30° = 60°$

Angle made by final velocity with the fixed plane

$$= 45°.$$

$\therefore$ Angle made by final velocity with the line of impact, $\quad \phi = 90 - 45 = 45°$

Let $\quad v$ = Final velocity, and

$\quad e$ = Coefficient of restitution.

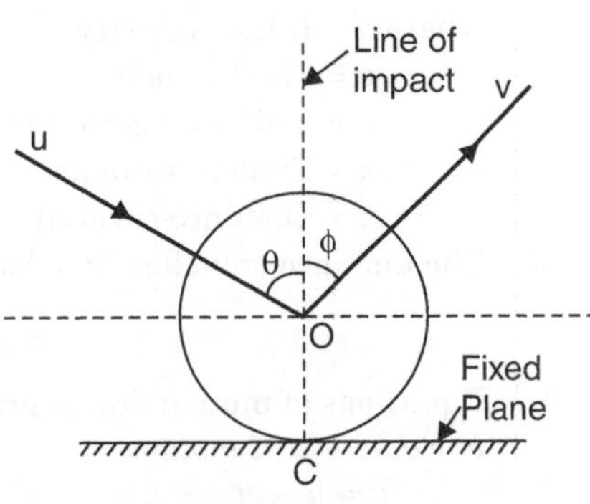

Fig. 4.78

The components of the initial velocity and final velocity at right angles to the line of impact is same.

$\therefore \quad u \sin \theta = v \sin \phi$

or $$v = \frac{u \sin \theta}{\sin \phi} = \frac{5 \sin 30°}{\sin 45°} = \frac{5 \times 0.5}{0.707} = \mathbf{3.536 \text{ m/s}}. \quad \textbf{Ans.}$$

For the indirect impact on a fixed plane, the coefficient of restitution is given by equation (4.60).

$\therefore \quad v \cos \phi = e \times u \cos \theta$

$$\therefore \qquad e = \frac{v \cos \phi}{u \cos \theta} = \frac{3.536 \times \cos 45°}{5 \times \cos 60°} \qquad [\because \phi = 45°, \theta = 60°]$$

$$= \frac{3.536 \times 0.707}{5 \times 0.05} = \mathbf{0.9999 \approx 1.0.} \quad \textbf{Ans.}$$

HIGHLIGHTS

(a) Linear Motion

1. The motion of a body in a straight line is called linear motion.
2. The rate of change of displacement of a body is known as velocity. It is given by

$$v = \frac{s}{t} \quad \text{or} \quad \frac{ds}{dt}.$$

3. The rate of change of velocity is known as acceleration. It is represented by 'a'. The acceleration is given by

$$a = \frac{dv}{dt} \quad \text{..... in terms of velocity}$$

$$= \frac{d^2s}{dt^2} \quad \text{..... in terms of distance}$$

$$= \frac{v\,dv}{ds} \quad \text{..... in terms of velocity and distance.}$$

4. When a body is moving with a uniform acceleration, then equations of motions are :

(i) $v = u + at$ (ii) $s = ut + \frac{1}{2}at^2$

(iii) $v^2 - u^2 = 2as$

where u = Initial velocity
v = Final velocity
a = Uniform acceleration
t = Time in seconds
s = Distance covered.

5. The distance travelled by a body moving with uniform acceleration in nth second

$$= u + \frac{a}{2}(2n - 1).$$

6. Equations of motion due to gravity in the downward and upward directions are given as :

For downward motion
$v = u + gt$
$h = s = ut + \frac{1}{2}gt^2$
$v^2 - u^2 = 2gh$

For upward motion
$v = u - gt$
$h = ut - \frac{1}{2}gt^2$
$v^2 - u^2 = -2gh.$

7. If a body starts from rest, its initial velocity is zero, *i.e.*, $u = 0$ whereas if a body comes to rest, the final velocity of the body will be zero *i.e.*, $v = 0$.
8. If a body is projected vertically upwards, the final velocity of the body at the highest point is zero.
9. If the equation of motion of a body moving in a straight line is given in terms of displacement (s) and time (t), the velocity of the body is obtained by differentiating the displacement with respect to time. And acceleration is obtained by differentiating the velocity with respect to time.

10. If the equation of motion of a body moving in a straight line is given in terms of acceleration (a) and time (t), then velocity is obtained by integrating the acceleration. The constant of integration for velocity is obtained from the given condition in terms of velocity and time.
The displacement is obtained by integrating the velocity. There will be another constant of integration. This constant of integration is also obtained from given condition in terms of displacement and time.

(b) Laws of Motion

11. The quantity of matter contained in a body is known as mass of the body.

12. The weight of a body is defined as the force by which the body is attracted towards the centre of earth. Weight of the body is given by

$$\text{Weight} = \text{Mass} \times g.$$

13. Momentum of a body is the product of the mass and its velocity.

14. The external force acting on a body is directly proportional to the rate of change of momentum in the same direction.

$$\therefore \qquad F = m \times a$$

where m = Mass of the body, and a = Acceleration of the body.

15. A unit force is one, which produces unit acceleration on unit mass.

16. Newton is that force which acts on a body of mass one kilogram and produces an acceleration of 1 m/s^2 in the direction of force.

17. Dyne is that force which acts on a body of mass one gram and produces an acceleration of 1 cm/s^2 in the direction of force.

18. The relation between newton and dyne is given by

$$1 \text{ N} = 10^5 \text{ dyne}.$$

19. If the external force acts on the body in the direction of motion of the body, the body will have acceleration. But if the external force acts opposite to the direction of motion of the body, the body will have retardation.

20. When a body moves down the inclined smooth plane, the acceleration of the body is given by

$$a = g \sin \theta$$

where θ = Inclination of the plane with horizontal.

21. If the body moves up the inclined smooth plane, then the acceleration of the body is given by

$$a = -g \sin \theta.$$

22. The acceleration of a body moving down a rough inclined plane is given by

$$a = g[\sin \theta - \mu \cos \theta]$$

where μ = Coefficient of friction between the body and inclined plane.

23. When a man is standing on a scale, which is placed on the floor of a moving elevator (or lift), the reading of the scale will be equal to the reaction (R) offered by the floor on the man.

24. The product of moment of inertia (I) and angular velocity (ω) of a rotating body is known as angular momentum.
Mathematically, angular momentum = $I \times \omega$.

25. Moment of inertia about an axis of rotation is given by

$$I = \frac{MR^2}{2} \qquad \text{......For a circular section and solid cylinder}$$

$$= \frac{2}{5} MR^2 \qquad \text{......For a solid sphere}$$

where M = Mass of a circular section, or a solid cylinder or a solid sphere.

26. Volume of a solid sphere is given by,

$$\text{Volume} = \frac{4}{3} \pi R^3.$$

27. The radius of gyration (k) is given by

$$k = \frac{R}{\sqrt{2}} \qquad \text{......For a circular lamina}$$

$$= \frac{R}{\sqrt{2}} \qquad \text{......For a solid cylinder}$$

$$= 0.6324\ R \qquad \text{......For a solid sphere.}$$

28. The rate of change of Angular momentum of a rotating body is proportional to the external torque applied on the body and takes place in the direction of the torque.

$\therefore \qquad T = I \times \alpha$

where I = Moment of inertia = Mk^2

α = Angular acceleration.

If in I, the value of M is taken in kg, the unit of torque will be in N m. But if the value of M is taken as $\frac{W}{g}$, where W is in N, then unit of torque will also be in N m.

29. Kinetic energy due to rotation of a body is given by

$$\text{K.E.} = \frac{1}{2} I \omega^2.$$

30. If a body is having motion of translation as well as motion of rotation, then total kinetic energy is given by,

$$\text{Total K.E.} = \frac{1}{2} mv^2 + \frac{1}{2} I\omega^2.$$

(c) Work and Energy

31. Work is the product of force and distance. Energy is the capacity to do work. The work done is given by

Work done = $P \times S$ ... When force and distance are in the same direction

= $P \cos \theta \times S$ When force acts at an angle θ with the direction of displacement.

32. The units of work is N m (or Joule). Hence one Joule is the work done by a force of 1 N when displacement is 1 m.

33. The area of force-distance curve represents the work done.

34. Power is the rate of doing work. Or power is the work done per second. The unit of power is N m/s or Watt (W)

Power = Force × Velocity

The force and velocity should be in the same direction.

35. Work done on a rotating body, by a torque (T) is given by

Work done = $T \times \theta$

where θ = angular displacement in radians.

And the power is given by

$$\text{Power} = \frac{2\pi NT}{60} \text{ watts.}$$

36. The product of power and time is known as Energy. The unit of energy is N m. Energy is also defined as capacity to do work. Though there are many forms of energies, yet potential energy and kinetic energy are important from the subject point of view.

37. The energy, by virtue of position of a body with respect to any given reference, is known as potential energy and it is equal to $W \times h$, where W = weight of body and h is height of body from the reference line.

38. The energy possessed by a body by virtue of its velocity (or motion) is known as kinetic energy. K.E. is given by

$$\text{K.E.} = \frac{1}{2} mv^2.$$

39. Law of conservation of energy states, "Energy can neither be created nor destroyed, though it can be transformed from one form into another form".

40. The resistance offered by the ground against penetration of a body into ground is obtained by equating the work done against penetration to the total energy lost by the system. Total energy means the loss of K.E. + loss of P.E. of the system.

(d) Impact of Elastic Body

41. According to the law of conservation of momentum, the total momentum of a system before collision is equal to the total momentum of the system after collision.

42. Line of impact is obtained by joining the centres of the two bodies with the point of contact.

43. The collision between two bodies, which are moving along the line of impact before collision, is known as direct impact.

44. The collision between two bodies, which are moving at some angle to the line of impact before collision, is known as indirect impact.

45. For a direct impact between two bodies, we have

$$m_1 u_1 + m_2 u_2 = m_1 v_1 + m_2 v_2$$

where m_1 = Mass of the first body
u_1 = Velocity of the first body before impact
v_1 = Velocity of the first body after impact
m_2 = Mass of the second body
u_2 = Velocity of the second body before impact
v_2 = Velocity of the second body after impact

46. For an indirect impact between two bodies, we have

$$m_1 \times u_1 \cos \theta_1 + m_2 \times u_2 \cos \theta_2 = m_1 \times v_1 \cos \phi_1 + m_2 \times v_2 \cos \phi_2$$

where θ_1 = Angle made by initial velocity of the first body with line of impact.
θ_2 = Angle made by the initial velocity of the second body with line of impact.
ϕ_1 = Angle made by final velocity of the first body with line of impact.
ϕ_2 = Angle made by final velocity of the second body with line of impact.

47. When the two moving bodies collide with each other, the ratio of their velocity of separation to their velocity of approach bears a constant ratio. This ratio is known as coefficient of restitution.

$\therefore$ Coefficient of restitution (e)

$$= \frac{\text{Velocity of separation}}{\text{Velocity of approach}} = \frac{\text{Relative velocity after impact}}{\text{Relative velocity before impact}}$$

48. The coefficient of restitution is given by

$$e = \frac{(v_2 - v_1)}{(u_1 - u_2)} \quad \text{... For direct impact}$$

$$= \frac{(v_2 \cos \phi_2 - v_1 \cos \phi_1)}{(u_1 \cos \theta_1 - u_2 \cos \theta_2)} \quad \text{...For indirect impact}$$

where u_1, u_2, v_1, v_2, ϕ and θ are having their usual values.

49. The value of e lies between 0 and 1.0.

If $e = 0$ ...body is perfectly plastic.
$= 1$ ...body is perfectly elastic.

50. The loss of kinetic energy during impact

= Total K.E. before impact – Total K.E. after impact.

51. The loss of K.E. during impact in terms of masses, initial velocities of two bodies and coefficient of restitution is given by

$$E_1 = \frac{m_1 m_2 (u_1 - u_2)^2}{2(m_1 + m_2)} [1 - e^2].$$

EXERCISE 4

(A) Theoretical Questions

(a) Linear Motion

1. Define the terms : velocity and acceleration.
2. (*a*) What do you mean by linear motion ? Give some examples of linear motion.
   (*b*) Distinguish between uniform motion and uniformly accelerated motion.
   (*AMIE Summer, 1988*)
3. Acceleration can be expressed as

$$a = \frac{d^2 s}{dt^2} \quad \text{and} \quad a = \frac{u\,dv}{ds}$$

   Derive the above two expressions.
4. Derive the following equations of motion of a body moving in a straight line with uniform acceleration :

   (*i*) $v = u + at$ (*ii*) $s = ut + \frac{1}{2}at^2$ (*iii*) $v^2 - u^2 = 2as.$
5. Derive a relation for the distance travelled, by a body in the *n*th second.
6. The distance travelled by a body moving with uniform acceleration in the *n*th second is equal to

   (*a*) $un + \frac{1}{2}an^2$ (*b*) $2u + a(2n - 1)$

   (*c*) $u + \frac{a}{2}(2n - 1)$ (*d*) $u + 2a(n - 1).$ **[Ans.** (*c*)**]**
7. If a body is projected vertically upwards, then at the highest point the final velocity is

   (*a*) maximum (*b*) zero

   (*c*) equal to '*g*' (*d*) none of the above. **[Ans.** (*b*)**]**
8. The acceleration of a body is given by

   (*a*) $\frac{dv}{ds}$ (*b*) $\frac{ds}{dt}$

   (*c*) $\frac{s\,dv}{ds}$ (*d*) $\frac{v\,dv}{ds}$. **[Ans.** (*d*)**]**
9. The equation of motion of a body moving in a straight line is given in terms of displacement (*s*) and time (*t*). How will you determine :

   (*i*) the velocity at any time *t*, (*ii*) acceleration at any time *t*,

   (*iii*) velocity at start, (*iv*) acceleration at start,

   (*v*) the time when the body is having maximum velocity and

   (*vi*) maximum velocity.
10. The equation of motion of a body moving in a straight line is given in terms of acceleration (*f*) and time (*t*). How will you determine :

    (*i*) velocity at any time, and (*ii*) displacement at any time.

(b) Laws of Motion

11. Define the terms : Mass of a body and weight of a body. What is the relationship between the two ?
12. Explain the terms : Momentum of a body and angular momentum of a body.
13. Define and explain the Newton's Laws of motion for linear motion and rotational motion.

14. Derive the relation, $F = ma$
where m = Mass, a = Acceleration, and F = Force acting on a body.
15. Explain the laws of motion for a body in rotational motion. (*AMIE Winter, 1981*)
16. State Newton's laws of motion and explain them by giving an example in each case. (*AMIE Winter, 1982*)
17. Define a unit force, Newton and dyne. What is the relationship between a newton and a dyne ?
18. Prove that the acceleration of a body, which moves down a smooth inclined plane is given by

$$a = g \sin \theta$$

where θ = Inclination of the plane with the horizontal.
19. The acceleration of a body moving down a rough inclined plane is given by

$$a = g (\sin \theta - \mu \cos \theta).$$

Prove the above relation.
20. State D' Alembert's Principle giving equations expressing the above principle for a rigid body in plane motion.
21. State the equations expressing the dynamics of a rigid body in rotation about a fixed axis.
22. Deduce the relation,

Kinetic energy $= \frac{1}{2} I\omega^2$

where I = Moment of inertia about the axis of rotation, and
ω = Angular velocity.
23. Prove that the angular momentum of a rotating body is equal to the product of moment of inertia and angular velocity.
24. Determine the moment of inertia of the following bodies :
(*i*) A circular section, (*ii*) A solid cylinder, and (*iii*) A solid sphere.
25. Find an expression for the radius of gyration for the following bodies :
(*i*) a circular lamina, (*ii*) a solid cylinder, and (*iii*) a solid sphere.
26. Deduce the relation,
Torque, $T = I \times \alpha$
where I = Moment of inertia about the axis of rotation, and
α = Angular acceleration.
27. A body is having motion of translation as well as motion of rotation. How will you determine the total kinetic energy of the body ?

(c) Work and Energy

28. Define the following terms :
Work, energy, momentum, impulse and moment of momentum.
29. What are the units of work ? What is the relation between work done and power ? Also mention the unit of power.
30. Find an expression for the work done and power of rotating body when a torque T is acting on the body.
31. Explain the principle of conservation of energy.
32. Mention the various forms of mechanical energy known to you and define them.
33. State the law of conservation of momentum.
34. What do you understand by 'Potential energy' and 'Kinetic energy' of a body ?
35. Prove that the K.E. of a moving body is expressed as :

K.E. $= \frac{1}{2} mv^2$ where m = mass of body and v = velocity of body.

36. State the law of conservation of energy. Give proof of this law taking mechanical energy only into account.

(d) Impact of Elastic Body

37. Define and explain the following terms :
Collision of elastic bodies, law of conservation of momentum and law of conservation of energy.

38. (*a*) What do you mean by direct impact and indirect impact ?
(*b*) Define the coefficient of restitution.

39. Define the coefficient of restitution. Two bodies are having direct impact. Find an expression for the coefficient of restitution in terms of initial and final velocities of the two bodies.

40. How will you find the loss of energy during impact between two moving bodies, which are having direct impact ?

41. Prove that the loss of energy due to direct impact between two bodies of masses, m_1, m_2 of initial velocities u_1, u_2 of final velocities v_1, v_2 and of coefficient of restitution e is given by

$$E_L = \frac{m_1 m_2}{2(m_1 + m_2)} (u_1 - u_2)^2 (1 - e^2).$$

42. A ball strikes directly on a similar ball which is at rest. The first ball comes to rest by the impact. If half of the initial kinetic energy is lost by impact, then prove that the coefficient of restitution will be equal to 0.707.

43. A body of mass m is allowed to fall from a height 'H' on a floor. After the impact, the body rises to a height of 'h' (or body rebounds to a height of h), then prove that the coefficient of restitution will be equal to $\sqrt{h/H}$.

44. What is the difference between the impact of the two bodies and the impact of a body on a fixed plane ?

45. A heavy elastic ball drops from the ceiling of a room, and after rebounding twice from the floor reaches a height equal to one-half that of ceiling. Prove that the coefficient of restitution is $(1/2)^{1/4}$.

46. Find an expression for coefficient of restitution in case of a body which is having indirect impact on a fixed plane.

47. Prove that the two elastic bodies of equal masses exchange velocities in the case of direct central impact.

(B) Numerical Problems

(a) Linear Motion

1. A body is moving with a velocity of 3 m/s. After five seconds the velocity of the body becomes 13 m/s. Find the acceleration of the body. **[Ans.** 2 m/s^2]

2. A car is moving with a velocity of 20 m/s. The car is brought to rest by applying brakes in 4 seconds. Determine : (*i*) the retardation and (*ii*) distance travelled by the car after applying brakes. **[Ans.** – 5 m/s^2, 40 m/s]

3. A bullet, moving at the rate of 200 m/s, is fired into a log of wood. The bullet penetrates to a depth of 50 cm. If the bullet moving with the same velocity is fired into a similar piece of wood 25 cm thick, with what velocity would it emerge. Take the resistance to be uniform in both the cases. **[Ans.** 141.4 m/s]

4. Three marks A, B and C at distance of 84 m each are made along a straight road. A car starting from rest and with uniform acceleration passes, the mark A and takes 8 seconds to reach B and further 6 seconds to reach the mark C. Calculate : (*i*) acceleration of the car, (*ii*) the velocity of car at A, (*iii*) the velocity of car at B, and (*iv*) the distance of the mark A from the starting point. **[Ans.** (*i*) 0.5 m/s^2, (*ii*) 8.5 m/s, (*iii*) 12.5 m/s, (*iv*) 72.25 m]

5. A burglar's car had a start with an acceleration of 2 m/s^2. A police vigilant party came after 5 seconds and continued to chase the burglar's car with a uniform velocity of 27 m/s. Find the time taken, in which the police van will overtake the car. **[Ans.** 5 s]

6. A body is moving with uniform acceleration and covers 20 m in 4th second and 30 m in 8th second. Determine : (*i*) the initial velocity of the body and (*ii*) acceleration of the body. **[Ans.** (*i*) 11.25 m/s, (*ii*) 2.5 m/s^2]

7. A tower is 100 m in height. A particle is dropped from the top of the tower and at the same time another particle is projected upward from the foot of the tower. Both the particles meet at a height of 40 m. Find the velocity with which the second particle is projected upward. **[Ans.** 28.59 m/s]

8. An aeroplane is flying horizontally at a height of 490 m. A bomb is released from the aeroplane when the speed of the aeroplane is 360 kilometre per hour (kmph). Determine the time required for the bomb to reach the ground and the horizontal distance travelled by the bomb during flight. **[Ans.** 10 s, 1000 m]

9. A stone dropped into a well is heard to strike the water after 6 seconds. Find the depth of the well, if the velocity of sound is 350 m/s. **[Ans.** 151.63 m]

10. A stone is thrown vertically upwards with a velocity of 29.40 m/s from the top of a tower 49 m high. Calculate : (*i*) time required for the stone to reach the ground, (*ii*) velocity of the stone in its downward travel at the point in the same level as the point of projection, and (*iii*) the maximum height to which the stone will rise in its flight. **[Ans.** (*i*) 7.358 s, (*ii*) 29.4 m/s, (*iii*) 44.1 m]

11. A stone is dropped from a height. After falling four seconds from rest, the stone breaks the glass pane and in breaking, the stone loses 25% of its velocity. Find the distance travelled by the stone in the next second. Take g = 9.80 m/s^2. **[Ans.** 34.3 m/s]

12. An object falls from rest from an unknown height. In the last second of its motion, the object travels a distance of 50 m. If g = 9.80 m/s^2, determine : (*i*) the height from which the object falls, and (*ii*) total time taken by the object in falling. **[Ans.** (*i*) 153.664 m, (*ii*) 5.6 s]

13. A body is falling freely under the action of gravity passes two points 10 m apart vertically in 0.2 seconds. From what height, above the higher point, did the body start to fall ? Take g = 9.8 m/s^2. **[Ans.** 122.5 m]

14. A particle moves along a straight line so that its displacement in metre from a fixed point is given by, $s = 2t^3 + 4t^2 - 6t + 8$. Find :
(*i*) velocity at start, (*ii*) velocity after 5 seconds, (*iii*) acceleration at start and (*iv*) acceleration after 5 seconds] **[Ans.** (*i*) 6 m/s, (*ii*) 184 m/s, (*iii*) 8 m/s^2, (*iv*) 68 m/s^2]

15. The equation of motion of a particle moving in a straight line is given by

$$s = 15t + 3t^2 - t^3$$

where s is the distance covered from the starting point in metre at the end of t seconds. Find :
(*i*) the velocity and acceleration at start,
(*ii*) the time, when the particle reaches its maximum velocity, and
(*iii*) the maximum velocity of the particle. **[Ans.** (*i*) 15 m/s, 6 m/s^2, (*ii*) t = 1 s, (*iii*) 18 m/s]

16. A body moves along a straight line and its acceleration (a) which varies with time (t) is given by : $a = 4 - 5t$. After 4 seconds from start of observations its velocity is observed to be 16 m/s. After 8 seconds from start of observations, the body was 70 m from the origin. (*i*) Determine its acceleration, velocity and distance from the origin at the start of observations, (*ii*) Determine the time after start of observation in which the velocity becomes zero and its distance from the origin. **[Ans.** (*i*) 4 m/s^2, 40 m/s, 48.67 m, (*ii*) 8.4 s, 31.87 m]

17. The velocity of a particle moving along x-axis is defined by $v = kx^3 - 4x^2 + 6x$ where v is in m/s and x in m and k is a constant. If k = 2, compute the acceleration when x = 2 m. Also find the smallest value of k that will make acceleration = 16 m/s^2 at x = 3 m. *(Osmania, 1993)*

[**Hint.** $a = \dfrac{v\,dv}{ds}$ or $= \dfrac{v\,dv}{dx}$. But $v = kx^3 - 4x^2 + 6x$ and $\dfrac{dv}{dx} = 3kx^2 - 8x + 6$ $\therefore$ $a = (kx^3 - 4x^2 + 6x)(3kx^2 - 8x + 6)$

When $k = 2$ and $x = 2$ then $a = (2.2^3 - 4.2^2 + 6.2)(3.2.2^2 - 8.2 + 6)$ or $a = (16 - 16 + 12)(24 - 16 + 6)$ $= 12 \times 14 = 168$ m/s^2. **Ans.**]

(b) Laws of Motion

18. Find the force acting on a body of mass 100 kg and producing an acceleration of 2 m/s^2 in its direction. [**Ans.** 200 N]

19. A force of 450 N acts on a body having a mass of 150 kg for 5 seconds. If the initial velocity of the body is 10 m/s, determine :
(*i*) acceleration produced in the direction of force, and
(*ii*) distance moved by the body in 4 seconds. [**Ans.** (*i*) 3 m/s^2, (*ii*) 64 m]

20. The weight of a body on earth is 490 N. If the acceleration due to gravity on earth = 9.8 m/s^2, what will be the weight of the body on :
(*i*) the moon where gravitational acceleration is 1.5 m/s^2, and
(*ii*) the sun, where gravitational acceleration is 300 m/s^2. [**Ans.** (*i*) 75 N, (*ii*) 15000 N]

21. A force of 300 N acts on a body of mass 150 kg for 30 seconds. If the initial velocity of the body is 25 m/s, determine the final velocity of the body, when the force :
(*i*) acts in the direction of motion, and
(*ii*) acts in the opposite direction of motion. [**Ans.** (*i*) 85 m/s, (*ii*) – 35 m/s]

22. A body of mass 20 kg falls on a muddy ground from a height of 39.2 m. The body penetrates into the ground. Find the distance through which the body will penetrate into the ground, if the resistance by the ground to penetration is constant and equal to 980 N. Take g = 9.8 m/s^2. [**Ans.** 9.8 m]

23. A man weighing 588 N dives into a swimming pool from a tower of height 9.8 m. He was found to go down in water by 1.5 m and then started rising. Find the average resistance of water. [**Ans.** 452 kgf]

24. A bullet weighing 80 gmf and moving with a velocity of 350 m/sec is fired into a log of wood and it penetrates to a depth of 15 cm. If the bullet moving with the same velocity, were fired into a similar piece of wood 10 cm thick, with what velocity would it emerge ? Find also the force of resistance, assuming it to be uniform. [**Ans.** 202 m/s, 32666 N]

25. A car, moving on a straight level road, skidded for a total distance of 40 metres after the brakes were applied. Determine the speed of the car, just before the brakes were applied, if the co-efficient of friction between the car tyres and the road is 0.4. Take g = 9.80 m/s^2. [**Ans.** 63.75 km/hr]

26. The tractive force, exerted by a railway car weighing 40 kN, is 1600 N. If the frictional resistance is 4 N per kN of the railway car's weight, determine the acceleration when the railway car is moving on a level track. [**Ans.** 0.353 m/s^2]

27. A train of weight 2000 kN starts from rest and attains a speed of 100 km/hr in 4 minutes. If the frictional resistance of the track is 8 N per kN of the train's weight, find the average pull required. Take g = 9.8 m/sec^2. [**Ans.** 39.588 kN]

28. A body of weight 300 N is initially stationary on a 45° inclined plane. What distance along the inclined plane must the body slide, before it reaches a speed of 3 m/s. The coefficient of friction between the body and the plane is equal to 0.2. [**Ans.** 81.1 cm]

29. A truck weighing 5 tonnes just moves freely (engine is not running) at a speed of 40 kilometre per hour down a slope of 1 in 45, the road resistance at this speed just being sufficient to prevent any acceleration. Find the road resistance per tonne weight of truck.
What power will the engine have to exert to run up the same slope at double the speed when the road resistance remains the same. [**Ans.** 22.22 kgf, 65.83]

30. A train weighing 150 tonnes moves down a slope of 1 in 200 at 20 km/hr and engine develops 40 h.p. What horse power will be required to pull the train up the slope at the same speed ? **[Ans.** 150.99]

31. A lift carries a weight of 110 N and is moving with a uniform acceleration of 3 m/s^2. Determine the tension in the cables supporting the lift, when
(*i*) lift is moving upwards, and
(*ii*) lift is moving downwards. Take g = 9.80 m/s^2. [**Ans.** (*i*) 143.67 N, (*ii*) 76.34 N]

32. A lift has an upward acceleration of 1.5 m/s^2. What pressure will a man weighing 500 N exert on the floor of the lift ? What pressure would he exert if the lift had an acceleration of 1.5 m/s^2 downward ? Take g = 9.8 m/s^2. [**Ans.** 576.5 N, 423.5 N]

33. An elevator weighs 2000 N and is moving vertically downwards with a uniform acceleration. Write the equation for the elevator cable tension. Starting from rest it travels a distance of 30 m during an interval of 12 seconds. Find the cable tension during this time. Neglect all other resistances to motion. What are the limits of cable tension ?
[**Ans.** 1915.2 N, at f = 0, T = 2000 and at f = 9.81, T = 0]

34. An elevator weighing 6000 N is ascending with an acceleration of 2 m/s^2. During this ascent its operator whose weight is 600 N is standing on the scales placed on the floor. What is the scale reading ? What will be the total tension in the cables of the elevator during this motion ?
[**Ans.** 722.3 N, 7945.5 N]

35. Two bodies of weights 60 N and 40 N are connected to the two ends of a light inextensible string. The string is passing over a smooth pulley. Determine :
(*i*) the acceleration of the system, and
(*ii*) the tension in the string. Take g = 9.80 m/s^2. [**Ans.** (*i*) 1.96 m/s^2, (*ii*) 48 N]

36. Determine the tension in the strings and acceleration of blocks *A* and *B* weighing 200 N and 70 N connected by a string and a frictionless and weightless pulley as shown in Fig. 4.79.
[**Ans.** 87.49 N ; 2.452 and 1.226 m/s^2]

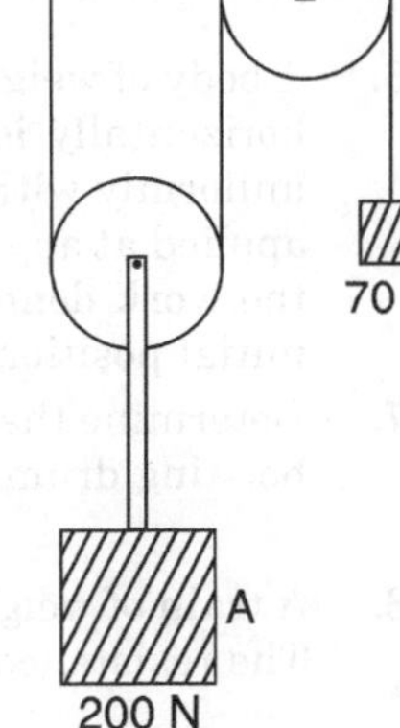

Fig. 4.79

37. Two bodies of weights 40 N and 20 N are connected to the two ends of a light inextensible string, passing over a smooth pulley. The weight of 40 N is placed on a smooth horizontal surface while the weight of 20 N is hanging free in air. Find :
(*i*) the acceleration of the system, and
(*ii*) the tension in the string. Take g = 9.81 m/sec^2.
[**Ans.** (*i*) 3.27 m/s^2, (*ii*) 13.33 N]

38. If in the problem 37, the horizontal surface is rough one, having a coefficient of friction between the weight 40 N and the horizontal surface equal to 0.3, determine :
(*i*) the acceleration of the system, and
(*ii*) the tension in the string. [**Ans.** (*i*) 1.308 m/s^2, (*ii*) 17.33 N]

39. Two bodies of weights 20 N and 4 N are connected to the two ends of a light inextensible string, passing over a smooth pulley. The weight 20 N is placed on a rough horizontal surface while the weight of 4 N is hanging vertically in air. Initially the friction between the weight 20 N and the horizontal surface is just sufficient to prevent motion. If an additional weight of 1 N is added to the weight, 4 N, determine :
(*i*) the acceleration of the two weights, and
(*ii*) tension in the string after adding additional weight of 1 N to the weight 4 N. Take g = 9.80 m/s^2.
[**Ans.** (*i*) 0.392 m/s^2, (*ii*) 4.8 N]

40. Two bodies of weights 60 N and 40 N are connected to the two ends of a light inextensible string, which passes over a smooth pulley. The weight 60 N is placed on a smooth inclined plane of angle of inclination 10°, while the weight 40 N is hanging free in air. Determine :

(*i*) acceleration of the system, and

(*ii*) tension in the string. Take $g = 9.80$ m/s^2. [**Ans.** (*i*) 2.89 m/s^2, (*ii*) 28.16 N]

(c) Work and Enegry

41. Find the work done in drawing a body :

(*i*) Weighing 1000 N through a distance 10 m along a horizontal surface by a horizontal force of 400 N.

(*ii*) Weighing 1000 N through a distance 10 m along a horizontal surface by a force of 400 N whose line of action makes an angle of 30° with horizontal.

[**Ans.** (*i*) 4000 N m, (*ii*) 3464 N m]

42. A body of weight 2000 N moves on a level horizontal rough road for a distance of 200 m. The resistance of the road is 10 N per 1000 N weight of the body. Find the work done by the resistance on the body. [**Ans.** 4000 N m]

43. A block of wood of weight 1200 N is placed on a smooth inclined plane which makes an angle of 30° with the horizontal. Find the work done in pulling the block up for a length of 8 m.

[**Ans.** 4800 N m]

44. If in question 43, the inclined plane is rough and coefficient of friction between the wooden block and plane is 0.3, then find the work done in pulling up for a length of 8 m. [**Ans.** 7293.92 N m]

45. A chain of length 40.5 m is placed on a pulley whose radius is 16 cm. One end of the chain hangs down 30 m below the point it leaves the pulley. A man holds the other end of the chain and pulls it down until the length of the chain on both sides of the pulley is equal. Find the work done by the man. Given the weight of the chain as 10 N per metre length and pulley is frictionless.

[**Ans.** 1000 N m]

46. A body of weight 1500 N is placed on a rough horizontal surface. An initial force of 600 N applied horizontally is sufficient just to move the body. The applied force is not constant but varies uniformly with the distance moved by the body. If the body moves a distance of 20 m and force applied at any distance x from the initial position is given by the relation $P = P_0 + 10x$, find the work done by the applied force, where P_0 = initial force and x is the distance moved from initial position. [**Ans.** 14000 N m]

47. Determine the work done by an electric motor in winding up a uniform cable which hangs from a hoisting drum if its free length is 20 m and weighs 800 N. The drum is rotated by the motor.

[**Ans.** 8000 N m]

48. A train of weight 4000 kN is pulled by an engine on a level track at a constant speed of 54 km/hr. The resistance due to friction is 5 N per kN weight of the train. Find the power of the engine.

[**Ans.** 300 kW]

49. If in the above question, the train is to move with an acceleration of 0.5 m/s^2 on the level track after attaining a speed of 54 km/hr, then find the power of the engine. [**Ans.** 3358.1 kW]

50. A train of weight 2000 kN, is ascending a slope of 1 in 100 with a uniform speed of 36 km/hr. Find the power exerted by the engine, if road resistance is 8 N per kN weight of the train.

[**Ans.** 288 kW]

51. A train of weight 2000 kN moves down a slope of 1 in 100 at 36 km/hr and engine develops a power of 70 kW. If the train is pulled up with the same speed, what power will be required to pull the train. [**Ans.** 470 kW]

52. A tangential force of 2700 N is acting on a shaft of diameter 20 mm. Find the work done by the force for one revolution of shaft. [**Ans** 169.62 N m]

53. If in the above question, the shaft is rotating at 400 r.p.m., find the power of the shaft. [**Ans.** 1130.94 W]

54. A body of mass 1 kg is moving with a velocity of 40 m/s. Find the K.E. of the body. [**Ans.** 800 N m]

55. A bullet of mass 25 gm is fired with a velocity of 400 m/s. What is the K.E. of the bullet ? If the bullet can penetrate 20 cm in a block of wood, what is the average resistance of wood ? What will be the exit velocity of the bullet, if fired into a similar block of wood 10 cm thick. [**Ans.** 2000 N m ; 1000 N ; 282.8 m/s]

56. A truck of mass 3000 kg is travelling at 36 km/hr on a level road. It is brought to rest in 15 m , what is the average force of resistance acting on the truck. [**Ans.** 10,000 N]

57. A car of weight 19620 N accelerates from rest to a speed of 36 km/hr in a distance of 40 m against a resistance of 500 N. Find the average driving force acting on the car. Using that average force, find the greatest power developed by the engine. [**Ans.** 3000 N, 30 kW]

58. A bullet travelling horizontally with a velocity of 600 m/s and weighing 0.25 N strikes a wooden block of weight 50 N resting on a rough horizontal floor. The coefficient of friction between the floor and the block $u = 0.5$. Find the distance through which the block is displaced from its initial position. [**Ans.** 0.90 m]

59. A hammer of mass 750 kg drops from a height of 1.2 m on a pile of mass 200 kg. Find : (*i*) the common velocity after impact assuming plastic impact and (*ii*) the average resistance of the ground if the pile comes to rest after penetrating 10 cm into the ground. [**Ans.** (*i*) 3.829 m/s ; (*ii*) 79 kN]

60. A hammer of mass 1500 kg drops from a height of 60 cm on a pile of mass 750 kg. Find the depth of penetration of the pile into the ground, if the average resistance of the ground is 140 kN. Assume the impact between the hammer and pile to be plastic. [**Ans.** 5.005 cm]

61. A hammer of mass 200 kg falls through a height of 4 m on a pile of negligible mass. If it drives the pile 50 cm into the ground, find the average resistance of the ground for penetration. [**Ans.** 17658 N]

62. A bullet of mass 30 gm is fired into a body of mass 10 kg , which is suspended by a string 0.8 m long. Due to this impact, the body swings through an angle of 30°. Find the velocity of bullet. [**Ans.** 483 m/s]

63. A bullet of mass 10 gm moving with a velocity of 100 m/s is fired into a body of mass 1 kg which is suspended by a string 1 m long. Due to the impact, the body swings through some angle. Find the angle of swing when the bullet gets embedded in the body. [**Ans.** 18.2°]

64. If in problem 63, the bullet, instead of getting embedded into the body, escapes from the other end of the body with a velocity of 20 m/s, find the angle through which the body will swing. [**Ans.** 14.7°]

65. If in problem 63, the bullet rebounds from the surface of the body with a velocity of 20 m/s, find the angle through which the body will swing. [**Ans.** 22.09°]

(d) Impact of Elastic Bodies

66. Ball *A* of mass 2 kg moving with a velocity of 4 m/s, strikes directly on a ball *B* of mass 4 kg at rest. The ball *A*, after striking, comes to rest. Find the velocity of ball *B* after striking and co-efficient of restitution. [**Ans.** 2 m/s, 0.5]

67. A ball of mass 1 kg moving with a velocity of 2 m/s, strikes directly on another ball of mass 2 kg at rest. If the coefficient of restitution between the two balls is 0.5, find the velocities of the two balls after impact. [**Ans.** 0, 1 m/s]

68. A vehicle of mass 600 kg and moving with a velocity of 12 m/s strikes another vehicle of mass 400 kg moving at 9 m/s in the same direction. Due to the impact both the vehicles get coupled and move together. Find the common velocity with which the two vehicles move after impact. [**Ans.** 10.8 m/s]

69. A body of mass 100 kg, moving with a velocity of 9 m/s, collides directly with a stationary body of mass 50 kg. If the two bodies become coupled so that they move on together after the impact, what is their common velocity. [**Ans.** 6 m/s]

70. A bullet of mass 100 gm is fired into a freely suspended target of mass 5 kg. On impact, the target along with the bullet moves with a velocity of 5 m/s in the direction of firing. Find the velocity of bullet. [**Ans.** 255 m/s]

71. A ball of mass 30 kg moving with a velocity of 4 m/s strikes directly another ball of mass 15 kg moving in the opposite direction with a velocity of 12 m/s. If the coefficient of restitution is equal to 5/6, then determine the velocity of each ball after impact. [**Ans.** $v_1 = -3.33$ m/s, $v_2 = 10$ m/s]

72. Three perfectly elastic balls *A*, *B* and *C* of masses 2 kg, 4 kg and 8 kg move in the same direction with velocities of 4 m/s, 1 m/s and 0.75 m/s respectively. If the ball *A* strikes with the ball *B*, which in turn strikes with the ball *C*, prove that the balls *A* and *B* will be brought to rest by the impacts.

73. A ball of mass 20 gm, moving with a velocity of 3 m/s strikes on a ball of mass 40 gm moving with a velocity of 1 m/s. The velocities of the two balls are parallel and inclined at 30° to the line joining their centres at the instant of impact. If the coefficient of restitution is 1/2, find :

(*i*) the direction and magnitude of velocity of ball of mass 20 gm after impact ; and

(*ii*) the direction and magnitude of velocity of ball of mass 40 gm.

[**Ans.** $\phi = 60°$, $v_1 = 1.732$ m/s, $\phi_2 = 16.1°$, $v_2 = 1.803$ m/s]

74. A vehicle of mass 600 kg and moving with a velocity of 10 m/s strikes another vehicle of mass 400 kg, moving at 5 m/s in the same direction. Both the vehicles get coupled together due to impact. Find the common velocity with which the two vehicles will move. Also find the loss of kinetic energy due to impact. [**Ans.** 8 m/s, 3000 N m]

75. A ball of mass 5 kg strikes directly on another ball of mass 5 kg which is at rest. If the coefficient of restitution between the two balls is equal to 0.707, then prove that the loss of kinetic energy due to impact is equal to half of the initial kinetic energy.

76. A sphere of mass 2 kg, moving at 4 m/s, strikes directly another sphere of mass 10 kg moving in the same line at 2 m/s. If the coefficient of restitution between the two spheres is 0.8, then find the loss of kinetic energy during impact. [**Ans.** 12 N m]

77. A ball of mass 1 kg is dropped on a horizontal floor from a height of 36 m. The ball rebounds due to impact to a height of 25 m. Find the value of coefficient of restitution between the floor and the ball. [**Ans.** 0.833]

78. A ball is dropped from an unknown height on a horizontal floor from which it rebounds to a height of 8 m. If the coefficient of restitution between the floor and the ball is 0.667, find the height from which the ball was dropped. [**Ans.** 18 m]

79. A ball is dropped from a height of 100 cm on a smooth floor, from which it rebounds to a height of 81 cm. Find :

(*i*) coefficient of restitution ; and

(*ii*) the expected height of the second rebound. [**Ans.** $e = 0.9$ and $h_2 = 65.61$ cm]

80. From what height, must a heavy elastic ball be dropped on a floor, so that after rebounding thrice it will reach a height of 16 metres. Take $e = \left(\frac{1}{2}\right)^{1/3}$. [**Ans.** 64 m]

81. A ball is dropped from a height of 16 m on a horizontal floor, from which it rebounds to a height of 9 m. Find the coefficient of restitution. [**Ans.** 0.75]

5

Friction and Elements of Rigid Body Dynamics

5.1. FRICTIONAL FORCE

When a solid body slides over a stationary solid body, a force is exerted at the surface of contact by the stationary body on the moving body. This force is called the *force of friction or frictional force* and is always acting in the direction opposite to the direction of motion. The property of the bodies by virtue of which a force is exerted by a stationary body on the moving body to resist the motion of the moving body is called *friction*. Friction acts parallel to the surface of contact and depends upon the nature of surface of contact.

5.2. LIMITING FORCE OF FRICTION AND DEFINITIONS OF CERTAIN TERMS

For defining the terms like coefficient of friction (μ) and angle of friction (ϕ), consider a solid body placed on a horizontal plane surface as shown in Fig. 5.1.

Let W = Weight of body acting through C.G. downward,

R = Normal reaction of body acting through C.G. upward,

P = Force acting on the body through C.G. and parallel to the horizontal surface.

If P is small, the body will not move as the force of friction acting on the body in the direction opposite to P will be more than P. But if the magnitude of P goes on increasing, a stage comes, when the solid body is on the point of motion. At this stage, the force of friction acting on the body is called *limiting force of friction*. The limiting force of friction is denoted by F.

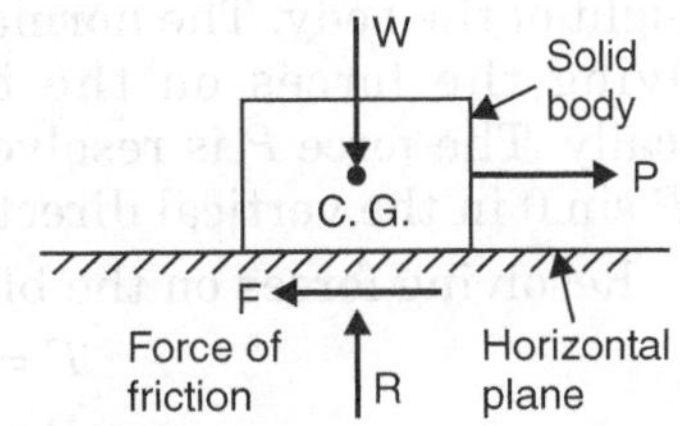

Fig. 5.1. Solid body on horizontal surface

Resolving the forces on the body vertically and horizontally, we get

$$R = W$$

$$F = P.$$

If the magnitude of P is further increased, the body will start moving. The force of friction, acting on the body when the body is moving, is called *kinetic friction*.

5.2.1. Coefficient of Friction (μ). It is defined as the ratio of the limiting force of friction (F) to the normal reaction (R) between two bodies. It is denoted by the symbol μ. Thus

$$\mu = \frac{\text{Limiting force of friction}}{\text{Normal reaction}} = \frac{F}{R}.$$

$$\therefore \quad F = \mu R \qquad ...(5.1)$$

5.2.2. Angle of Friction (ϕ). It is defined as the angle made by the resultant of the normal reaction (R) and the limiting force of friction (F) with the normal reaction (R). It is denoted by ϕ. Fig. 5.2 shows a solid body resting on a rough horizontal plane.

Let S = Resultant of the normal reaction (R) and limiting force of friction (F)

Then angle of friction = ϕ

= Angle between S and R

From Fig. 5.2, we have

$$\tan \phi = \frac{F}{R} = \frac{\mu R}{R} \quad [\because \ F = \mu R \text{ from } (5.1)]$$

$$= \mu = \text{Coefficient of friction} \qquad ...(5.2)$$

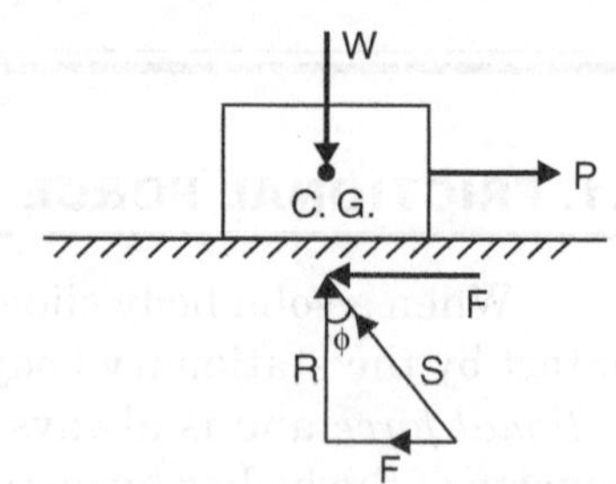

Fig. 5.2. Angle of friction

Thus the tangent of the angle of friction is equal to the coefficient of friction.

A block of weight W is placed on a rough horizontal plane surface as shown in Fig. 5.3 and a force P is applied at an angle θ with the horizontal such that the block just tends to move.

Let R = Normal reaction

μ = Coefficient of friction

F = Force of friction

$= \mu R$

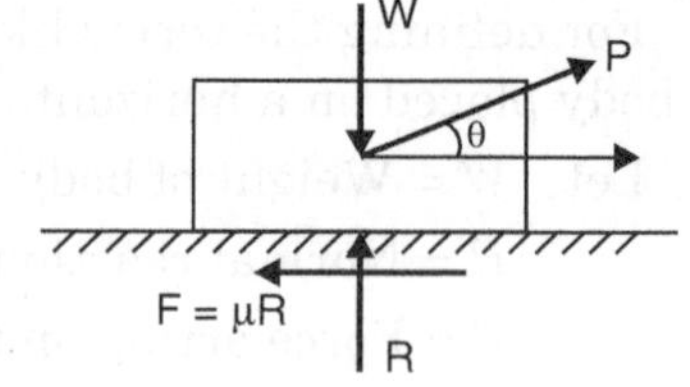

Fig. 5.3

In this case the normal reaction R will not be equal to weight of the body. The normal reaction is obtained by resolving the forces on the block horizontally and vertically. The force P is resolved in two components *i.e.*, $P \cos \theta$ in the horizontal direction and $P \sin \theta$ in the vertical direction.

Resolving forces on the block horizontally, we get

$$F = P \cos \theta$$

or

$$\mu R = P \cos \theta \qquad ...(i) \quad (\because \ F = \mu R)$$

Resolving forces on the block vertically, we get

$$R + P \sin \theta = W$$

$$\therefore \quad R = W - P \sin \theta \qquad ...(ii)$$

From equation (*ii*), it is clear that normal reaction is not equal to the weight of the block.

If in equation (*ii*), the values of W, P and θ are known, then value of normal reaction (R) can be obtained. This value of R can be substituted in equation (*i*) to determine the value of coefficient of friction μ.

Note. (*i*) The force of friction is always equal to μR (*i.e.*, $F = \mu R$).

(*ii*) The normal reaction (R) is not equal to the weight of the body always.

5.2.3. Cone of Friction. It is defined as the right circular cone with vertex at the point of contact of the two bodies (or surfaces), axis in the direction of normal reaction (R) and semi-vertical angle equal to angle of friction (ϕ). Fig. 5.4 shows the cone of friction in which,

O = Point of contact between two bodies

R = Normal reaction and also axis of the cone of friction

ϕ = Angle of friction.

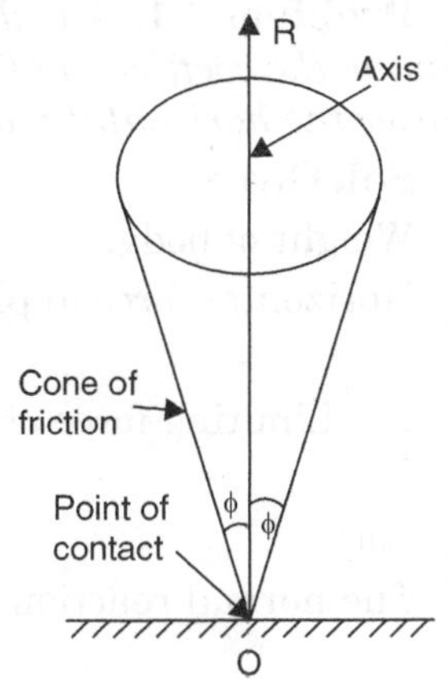

Fig. 5.4. Cone of friction

5.2.4. Types of Friction. The friction is divided into following two types depending upon the nature of the two surfaces in contact :

1. Static friction, and
2. Dynamic friction.

If the two surfaces, which are in contact, are at rest, the force experienced by one surface is called static friction. But if one surface starts moving and the other is at rest, the force experienced by the moving surface is called dynamic friction. If between the two surfaces, no lubrication (oil or grease) is used, the friction, that exists between two surfaces is called *'Solid Friction'* or *'Dry Friction'*. Solid friction or Dry friction is also known as *Coulomb friction.*

5.3. LAWS OF SLIDING OR COULOMB FRICTION

The friction, that exists between two surfaces which are not lubricated, is known as solid friction. The two surfaces may be at rest or one of the surface is moving and other surface is at rest. The following are the laws of solid friction :

1. The force of friction acts in the opposite direction in which the surface is having a tendency to move.
2. The force of friction is equal to the force applied to the surface, so long as the surface is at rest.
3. When the surface is on the point of motion, the force of friction is maximum and this maximum frictional force is called the limiting friction force.
4. The limiting frictional force bears a constant ratio to the normal reaction between two surfaces.
5. The limiting frictional force does not depend upon the shape and areas of the surfaces in contact.
6. The ratio between limiting friction and normal reaction is slightly less when the two surfaces are in motion.
7. The force of friction is independent of the velocity of sliding.

The above laws of solid friction are also called laws of static and dynamic friction.

5.4. SIMPLE CONTACT FRICTION EQUILIBRIUM ANALYSIS OF SIMPLE SYSTEMS WITH SLIDING FRICTION

The problems on simple contact friction consist of

(*i*) Simple contact friction on a horizontal surface,

(*ii*) Simple contact friction on an inclined plane, and

(*iii*) Simple contact friction on Ladder surface.

Problem 5.1. *A body of weight 100 Newtons is placed on a rough horizontal plane. Determine the coefficient of friction if a horizontal force of 60 Newtons just causes the body to slide over the horizontal plane.*

Sol. Given :

Weight of body, $W = 100$ N

Horizontal force applied,

$P = 60$ N

$\therefore$ Limiting force of friction,

$F = P = 60$ N

Let μ = Coefficient of friction.

The normal reaction of the body is given as

$R = W = 100$ N

Fig. 5.5

Using equation (5.1),

$$F = \mu R$$

or
$$\mu = \frac{F}{R} = \frac{60}{100} = \mathbf{0.6.} \quad \textbf{Ans.}$$

Problem 5.2. *A body of weight 200 N is placed on a rough horizontal plane. If the coefficient of friction between the body and the horizontal plane is 0.3, determine the horizontal force required to just slide the body on the plane.*

Sol. Given :

Weight of body, $W = 200$ N

Coefficient of friction, $\mu = 0.3$

Normal reaction, $R = W = 200$ N

Let F = horizontal force which causes the body to just slide over the plane.

Using equation (5.1),

$F = \mu R = 0.3 \times 200 = $ **60 N. Ans.**

Problem 5.3. *The force required to pull a body of weight 50 N on a rough horizontal plane is 15 N. Determine the coefficient of friction if the force is applied at an angle of 15° with the horizontal.*

Sol. Given :

Weight of the body, $W = 50$ N

Force applied, $p = 15$ N

Angle made by the force P, with horizontal, $\theta = 15°$

Let the coefficient of friction $= \mu$

Normal reaction $= R$

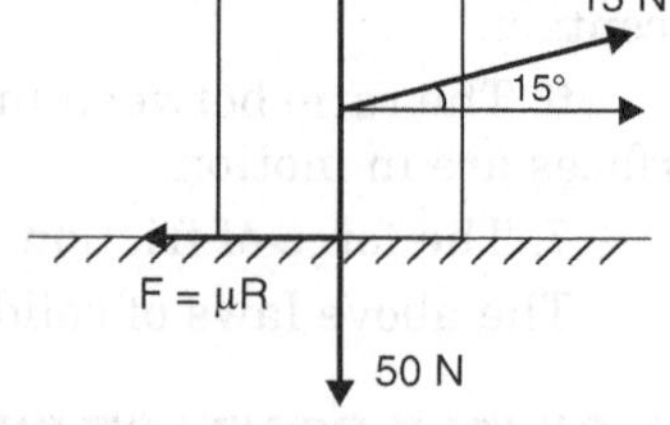

Fig. 5.6

When a force equal to 15 N is applied to the body at an angle 15° to the horizontal, the body is on the point of motion in the forward direction. Hence a force of friction equal to μR will be acting in the backward direction. The body is in equilibrium under the action of the forces shown in Fig. 5.6.

Resolving the forces along the plane, $\mu R = 15 \cos 15°$ ...(*i*)

Resolving the forces normal to the plane

$R + 15 \sin 15° = 50$

$\therefore \quad R = 50 - 15 \sin 15° = 50 - 15 \times 0.2588$

$= 46.12$ N

Substituting the value of R in equation (*i*), we get

$\mu \times 46.12 = 15 \cos 15°$

$$\therefore \quad \mu = \frac{15 \cos 15°}{46.12} = \frac{15 \times 0.9659}{46.12} = \mathbf{0.314.} \quad \textbf{Ans.}$$

Problem 5.4. *A body of weight 70 N is placed on a rough horizontal plane. To just move the body on the horizontal plane, a push of 20 N inclined at 20° to the horizontal plane is required. Find the coefficient of friction.*

Sol. Given :

Weight of body, $W = 70$ N

Force applied, $P = 20$ N

Inclination of P, $\theta = 20°$.

Let μ = Coefficient of friction

R = Normal reaction

F = Force of friction = μR.

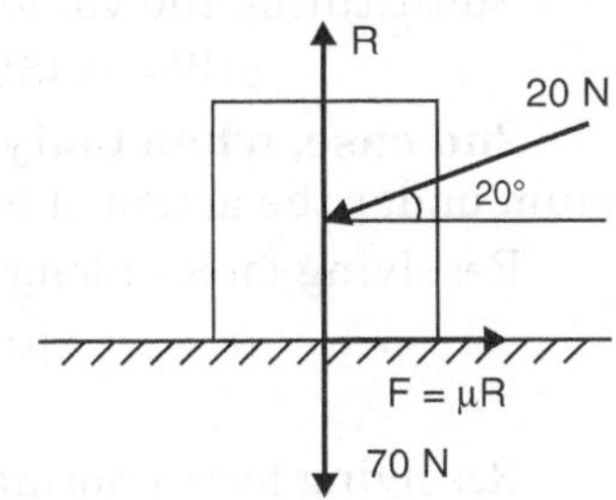

Fig. 5.7

When a push of 20 N at an angle 20° to the horizontal is applied to the body, the body is just to move towards left. Hence a force of friction $F = \mu R$, will be acting towards right as shown in Fig. 5.7.

Resolving forces along the plane, $\mu R = 20 \cos 20°$ ...(*i*)

Resolving forces normal to the plane, $R = 70 + 20 \sin 20°$

$= 70 + 20 \times 0.342 = 70 + 6.84 = 76.84$

Substituting the value of R in equation (*i*),

$\mu \times 76.84 = 20 \cos 20°$

$$\therefore \quad \mu = \frac{20 \cos 20°}{76.84} = \frac{20 \times 0.9397}{76.84} = \mathbf{0.244.} \quad \textbf{Ans.}$$

Problem 5.5. *A pull of 20 N, inclined at 25° to the horizontal plane, is required just to move a body placed on a rough horizontal plane. But the push required to move the body is 25 N. If the push is inclined at 25° to the horizontal, find the weight of the body and coefficient of friction.*

Sol. Given :

Pull required, $P = 20$ N

Inclination of pull, $\theta = 25°$

Push required, $P' = 25$ N

Inclination of push, $\theta' = 25°$

Let W = Weight of body

μ = Coefficient of friction

R = Normal reaction when body is pulled

R' = Normal reaction when body is pushed.

1st case, when body is pulled. The body is in equilibrium under the action of forces shown in Fig. 5.8.

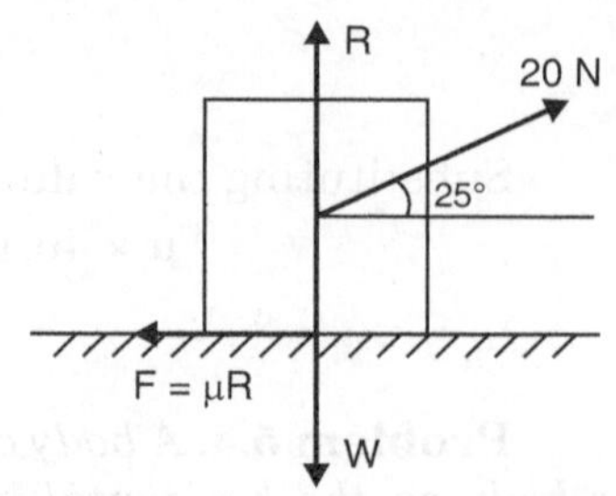

Fig. 5.8. Body is pulled

Resolving the forces along the plane,

$$\mu R = 20 \cos 25° = 20 \times 0.9063$$
$$= 18.126 \quad ...(i)$$

Resolving forces normal to the plane,

$$R + 20 \sin 25° = W$$
$$\therefore \quad R = W - 20 \sin 25°$$
$$= W - 20 \times 0.4226$$
$$= W - 8.452.$$

Substituting the value of R in equation (*i*),

$$\mu(W - 8.452) = 18.126 \quad ...(ii)$$

2nd case, when body is pushed. The body is in equilibrium under the action of forces shown in Fig. 5.9.

Fig. 5.9. Body is pushed

Resolving forces along the plane,

$$\mu R' = 25 \cos 25° = 25 \times 0.9063$$
$$= 22.657 \quad ...(iii)$$

Resolving forces normal to the plane,

$$R' = W + 25 \sin 25°$$
$$= W + 25 \times 0.4226$$
$$= W + 10.565$$

Substituting the value of R' in equation (*iii*),

$$\mu(W + 10.565) = 22.657 \quad ...(iv)$$

Dividing equation (*ii*) by equation (*iv*),

$$\frac{\mu(W - 8.452)}{\mu(W + 10.565)} = \frac{18.126}{22.656}$$

or $$\frac{(W - 8.452)}{(W + 10.565)} = \frac{18.126}{22.656}$$

or $$22.656\,(W - 8.452) = 18.126\,(W + 10.565)$$

or $$22.656\,W - 22.656 \times 8.452 = 18.126\,W + 18.126 \times 10.565$$

or $$22.656\,W - 18.126\,W = 18.126 \times 10.565 + 22.656 \times 8.452$$

or $$4.53\,W = 191.5 + 191.5 = 383$$

$$W = \frac{383}{4.53} = \mathbf{84.547.} \quad \textbf{Ans.}$$

Substituting the value of W in equation (*ii*),

$$\mu(84.547 - 8.452) = 18.126$$

$$\therefore \quad \mu = \frac{18.126}{(84.547 - 8.452)} = \frac{18.126}{76.095} = \mathbf{0.238.} \quad \textbf{Ans.}$$

Problem 5.6. *A block of weight W is placed on a rough horizontal plane surface as shown in Fig. 5.10 and a force P is applied at an angle θ with the horizontal such that the block just tends to move. Prove that the force P will be the least if the angle θ is equal to the angle of friction φ.*

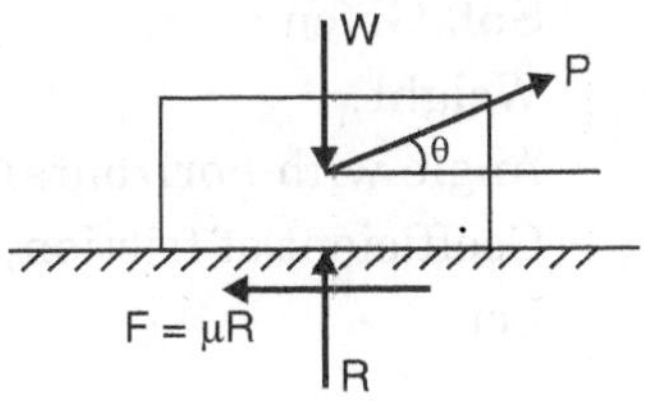

Sol. Given :

Weight of block $= W$

Force applied $= P$

Inclination of force $= \theta$

Let R = Normal reaction

μ = Coefficient of friction

F = Force of friction = μR.

Fig. 5.10

The forces acting on the block are shown in Fig. 5.10.

Resolving forces vertically, we get

$$R + P \sin \theta = W$$

or
$$R = W - P \sin \theta \quad \text{...}(i)$$

Resolving forces horizontally, we get

$$P \cos \theta = F = \mu R \qquad (\because \ F = \mu R) \quad \text{...}(ii)$$

Substituting the value of R from equation (i), the equation (ii) becomes as

$$P \cos \theta = \mu[W - P \sin \theta] \quad \text{...}(iii)$$

But from equation (5.2), we know

$$\mu = \tan \phi$$

where ϕ = angle of friction.

Substituting the value of μ in equation (iii), we get

$$P \cos \theta = \tan \phi \, (W - P \sin \theta)$$

$$= \frac{\sin \phi}{\cos \phi} (W - P \sin \theta)$$

or
$$P \cos \theta \cos \phi = W \sin \phi - P \sin \theta \sin \phi$$

or $P \cos \theta \cos \phi + P \sin \theta \sin \phi = W \sin \phi$

or $P(\cos \theta \cos \phi + \sin \theta \sin \phi) = W \sin \phi$

or
$$P \cos (\theta - \phi) = W \sin \phi \qquad [\because \ \cos \theta \cos \phi + \sin \theta \sin \phi = \cos (\theta - \phi)]$$

$$\therefore \quad P = \frac{W \sin \phi}{\cos (\theta - \phi)}$$

The force P will be least, if the denominator *i.e.*, $\cos (\theta - \phi)$ is maximum. But $\cos (\theta - \phi)$ will be maximum, if

$$\cos (\theta - \phi) = 1$$

or
$$\theta - \phi = 0$$

or
$$\theta = \phi$$

$$\therefore \quad P_{least} = W \sin \phi \text{ or } W \sin \theta.$$

Hence the force P will be least if the angle of inclination of P with the horizontal is equal to the angle of friction ϕ.

Problem 5.7. *A man wishing to slide a stone block of weight 1000 N over a horizontal concrete floor, ties a rope to the block and pulls it in a direction inclined upward at an angle of 20° to the horizontal. Calculate the minimum pull necessary to slide the block if the coefficient of friction $\mu = 0.6$. Calculate also the pull required if the inclination of the rope with the horizontal is equal to the angle of friction and prove that this is the least force required to slide the block.* (AMIE Winter, 1990)

Sol. Given :

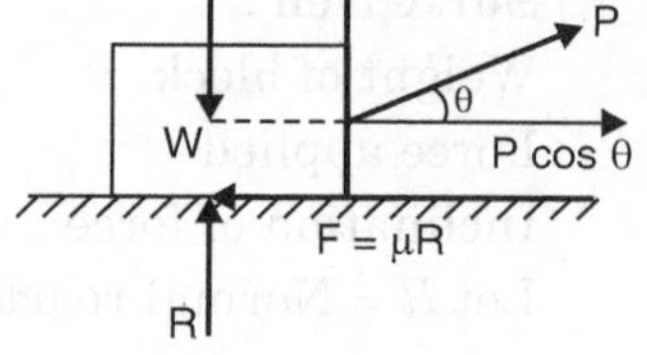

Fig. 5.10 (a)

Weight, $W = 1000$ N

Angle with horizontal, $\theta = 20°$

Coefficient of friction, $\mu = 0.6$

Let P = Force applied

R = Normal reaction

F = Force of friction = μR

The forces acting on the block are shown in Fig. 5.10 (*a*).

Resolving forces horizontally,

$$P \cos \theta = \mu R$$

or $$P \cos 20° = 0.6 \times R \quad ...(i)$$

Resolving forces vertically, $R + P \sin \theta = W$

or $$R + P \sin 20° = 1000$$

or $$R = 1000 - P \sin 20° \quad ...(ii)$$

Substituting the value of R in equation (*i*), we get

$$P \cos 20° = 0.6 (1000 - P \sin 20°)$$

$$= 600 - 0.6 P \sin 20°$$

or $$P \cos 20° + 0.6 P \sin 20° = 600$$

or $$P (\cos 20° + 0.6 \times \sin 20°) = 600 \quad ...(iii)$$

$$\therefore \quad P = \frac{600}{(\cos 20° + 0.6 \sin 20°)} = \frac{600}{(0.9397 + 0.6 \times 0.342)}$$

$$= \frac{600}{1.1449} = \textbf{524 N. Ans.}$$

Pull required if the inclination of the rope with the horizontal is equal to angle of friction.

Let ϕ = Angle of friction

= the angle made by rope with horizontal (given) = 20°

If in equation (*iii*), the angle 20° is replaced by angle ϕ, then we get the force required to pull the body as,

$$P(\cos \phi + 0.6 \sin \phi) = 600$$

$$\therefore \quad P = \frac{600}{(\cos \phi + 0.6 \sin \phi)} \quad ...(iv)$$

The force P will be minimum, if $(\cos \phi + 0.6 \sin \phi)$ is maximum. But $(\cos \phi + 0.6 \sin \phi)$ will be maximum if

$$\frac{d}{d\phi} (\cos \phi + 0.6 \sin \phi) = 0 \quad \text{or} \quad - \sin \phi + 0.6 \cos \phi = 0$$

or $$0.6 \cos \phi = \sin \phi \quad \text{or} \quad 0.6 = \frac{\sin \phi}{\cos \phi} = \tan \phi$$

But $0.6 = \mu$. Hence force P will be minimum if $\tan \phi = \mu = 0.6$ (Proved)

Now $$\tan \phi = 0.6$$

$$\therefore \quad \phi = \tan^{-1} 0.6 = 30.96°$$

Substituting this value of ϕ in equation (*iv*), we get

$$P = \frac{600}{(\cos 30.96° + 0.6 \sin 30.96°)} = \frac{600}{(0.8575 + 0.6 \times 0.5144)}$$

$$= \frac{600}{1.1661} = \textbf{514.5 N. Ans.}$$

Problem 5.8. *A uniform rod AB of length 50.8 cm weighing 100 N is resting on a rough horizontal surface whose coefficient of friction is 0.1. It is subjected to a force P applied horizontally at the end A in a direction perpendicular to its length. Determine the point O about which it would commence to turn.* [AMIE (S) 1982, (S) 1986]

Sol. Given :

Length of the rod, $AB = 50.8$ cm

Weight of the rod, $W = 100$ N

Coefficient of friction, $\mu = 0.1$

Force applied horizontally at $A = P$.

Let the rod commences to turn about point O, when a horizontal force P is applied at A as shown in Fig. 5.11. Let the distance of O from A is x cm.

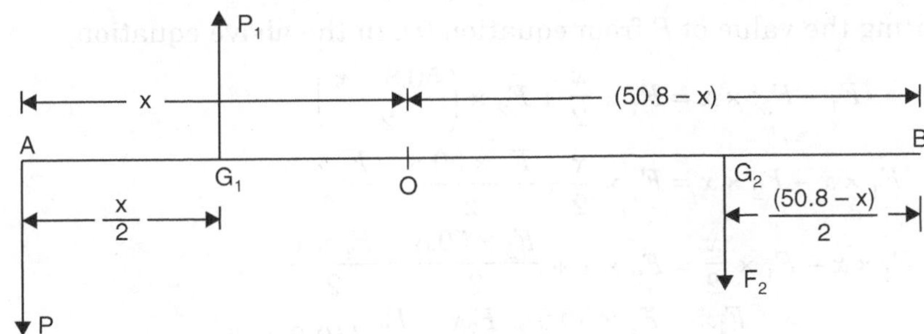

Fig. 5.11

$\therefore$ Distance $AO = x$ cm

Then $OB = AB - AO = (50.8 - x)$ cm

Let w = Weight per unit length of the rod

$\therefore$ Weight of length AO of the rod $= w \times$ Length AO

$= w \times x$ N.

This weight is acting at the middle point of AO. Let G_1 is the middle point of AO.

Weight of length BO of the rod $= w \times$ Length BO

$= w \times (50.8 - x)$ N.

This weight is acting at the middle point of BO. Let G_2 is the middle point of OB.

The distance $AG_1 = x/2$ and $BG_2 = \dfrac{(50.8 - x)}{2}$

When the force P is applied horizontally at A, the portion AO will commence to move in the same direction as that of P, while the portion OB will commence to move in the direction opposite to that of P. Hence force of friction, which acts in the opposite direction of the motion of the body, will act for the portion AO in the direction opposite to P whereas for the portion OB will act in the same direction as of P.

Let F_1 = Force of friction for portion AO

$= \mu \times$ weight of AO

$= \mu \times w \times x.$

This will act at point G_1

F_2 = Force of friction for portion BO

$= \mu \times$ weight of BO

$= \mu \times (50.8 - x) \times w.$

This will act at point G_2.

The directions of F_1 and F_2 are shown in Fig. 5.11. All the forces *i.e.*, P, F_1 and F_2 shown in Fig. 5.11, are acting in the horizontal plane (*i.e.*, in the plane of the paper).

Resolving forces in the direction perpendicular to AB, we get

$$P + F_2 = F_1$$

or $$P = F_1 - F_2 \quad \text{...(i)}$$

Taking moments of all forces about O, we get

$$P \times x = F_1 \times \frac{x}{2} + F_2 \times \left(\frac{50.8 - x}{2}\right)$$

Substituting the value of P from equation (*i*), in the above equation,

$$(F_1 - F_2) \times x = F_1 \times \frac{x}{2} + F_2 \times \left(\frac{50.8 - x}{2}\right)$$

or $$F_1 \times x - F_2 \times x = F_1 \times \frac{x}{2} + \frac{F_2 \times 50.8}{2} - \frac{F_2 \times x}{2}$$

or $$F_1 \times x - F_1 \times \frac{x}{2} = F_2 \times x + \frac{F_2 \times 50.8}{2} - \frac{F_2 \times x}{2}$$

or $$\frac{F_1 x}{2} = \frac{F_2 \times 50.8}{2} - \frac{F_2 x}{2} = \frac{F_2}{2}\,[50.8 + x].$$

Substituting the values of F_1 and F_2, we get

$$\frac{0.1wx \,.\, x}{2} = \frac{0.1(50.8 - x)\,w}{2}\,[50.8 + x] \quad (\because \ \mu = 0.1)$$

Cancelling $\frac{0.1w}{2}$ throughout, we have

$$x^2 = (50.8 - x)(50.8 + x) = (50.8)^2 - x^2$$

or $$x^2 + x^2 = (50.8)^2$$

or $$2x^2 = (50.8)^2 = 2580.64$$

$\therefore$ $$x^2 = \frac{2580.64}{2} = 1290.32$$

$\therefore$ $$x = \sqrt{1290.32} = \mathbf{35.92 \text{ cm.}} \quad \textbf{Ans.}$$

Hence the rod will commence to turn about the point O, which is at a distance of 35.92 cm from point A.

5.4.1. Angle of Repose. The angle of repose is defined as the maximum inclination of a plane at which a body remains in equilibrium over the inclined plane by the assistance of friction only.

Consider a body of weight W, resting on a rough inclined plane as shown in Fig. 5.12.

Let R = Normal reaction acting at right angle to the inclined plane.

α = Inclination of the plane with the horizontal

F = Frictional force acting upward along the plane.

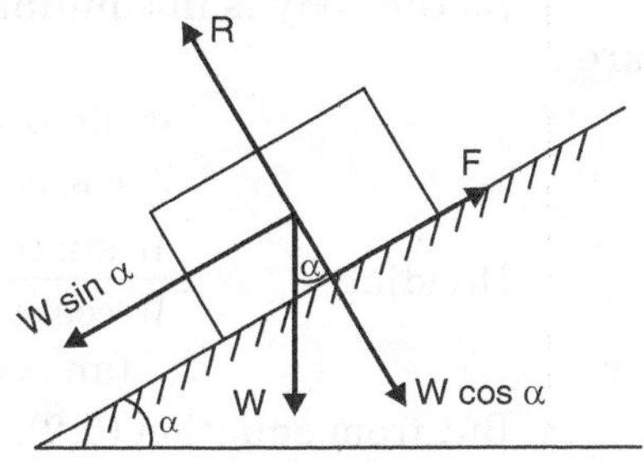

Fig. 5.12

Let the angle of inclination (α) be gradually increased, till the body just starts sliding down the plane. This angle of inclined plane, at which a body just begins to slide down the plane, is called angle of repose.

Resolving the forces along the plane, we get

$$W \sin \alpha = F \quad ...(i)$$

Resolving the forces normal to the plane, we get

$$W \cos \alpha = R \quad ...(ii)$$

Dividing equation (*i*) by equation (*ii*),

$$\frac{W \sin \alpha}{W \cos \alpha} = \frac{F}{R} \quad \text{or} \quad \tan \alpha = \frac{F}{R} \quad ...(iii)$$

But from equation (5.2), we know

$$\tan \phi = \frac{F}{R} \quad ...(iv)$$

where ϕ = angle of friction.

Hence from equations (*iii*) and (*iv*), we have

$$\tan \alpha = \tan \phi$$

or $$\alpha = \phi$$

or Angle of repose = Angle of friction.

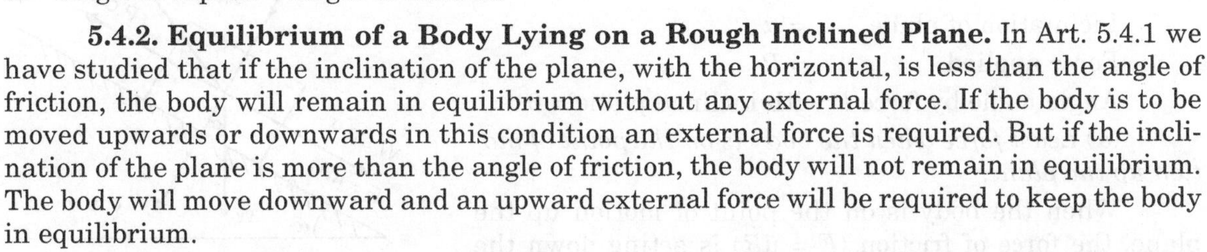

5.4.2. Equilibrium of a Body Lying on a Rough Inclined Plane. In Art. 5.4.1 we have studied that if the inclination of the plane, with the horizontal, is less than the angle of friction, the body will remain in equilibrium without any external force. If the body is to be moved upwards or downwards in this condition an external force is required. But if the inclination of the plane is more than the angle of friction, the body will not remain in equilibrium. The body will move downward and an upward external force will be required to keep the body in equilibrium.

Such problems are solved by resolving the forces along the plane and perpendicular to the planes. The force of friction (F), which is always equal to μR is acting opposite to the direction of motion of the body.

Problem 5.9. *Prove that the angle of friction (ϕ) is equal to the angle made by an inclined plane with the horizontal when a solid body, placed on the inclined plane, is about to slide down.*

Sol. A solid body of weight W is placed on an inclined plane AC as shown in Fig. 5.13.

Let α = Angle of the inclined plane AC with horizontal plane AB, such that the body just starts moving downward.

The body is in equilibrium under the action of following forces :

1. Weight of the body (W) acting vertically downwards.
2. Normal reaction (R), acting perpendicular to the inclined plane, AC.

3. The force of friction, $F = \mu R$, acting up the plane as the body is about to slide down the plane.

The weight, W can be resolved in two components, one along the plane and other perpendicular to the plane. The components are $W \sin \alpha$ and $W \cos \alpha$ respectively.

As the body is in equilibrium, the forces along and perpendicular to the inclined plane are :

$$W \sin \alpha = F = \mu R$$

$$W \cos \alpha = R$$

Dividing $$\frac{W \sin \alpha}{W \cos \alpha} = \frac{\mu R}{R} = \mu$$

or $$\tan \alpha = \mu$$

But from equation (5.2), we have

$$\tan \phi = \mu$$

where ϕ = angle of friction

$\therefore$ $$\tan \alpha = \tan \phi = \mu$$

or $$\alpha = \phi.$$

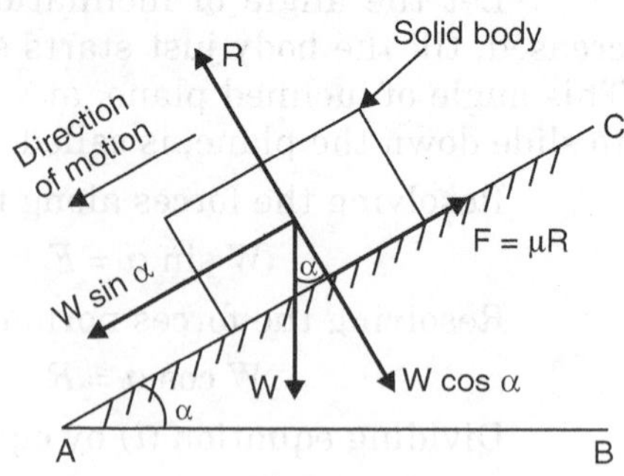

Fig. 5.13

The above relation shows that the angle of friction is equal to angle of the inclined plane when a solid body, placed on the inclined plane is about to slide down.

Problem 5.10. *Find the least force required to drag a body of weight W, placed on a rough inclined plane having inclination α to the horizontal. The force is applied to the body in such a way that it makes an angle θ to the inclined plane and the body is (a) on the point of motion up the plane and (b) on the point of motion down the plane.*

Sol. Given :

Weight of body $= W$

Inclination of plane $= \alpha$

Force applied $= P$

Angle made by force P with inclined plane = θ.

(*a*) *Least force when the body is on the point of motion up the plane.*

When the body is on the point of motion up the plane, the force of friction ($F = \mu R$) is acting down the plane. The body is in equilibrium under the action of following forces as shown in Fig. 5.14.

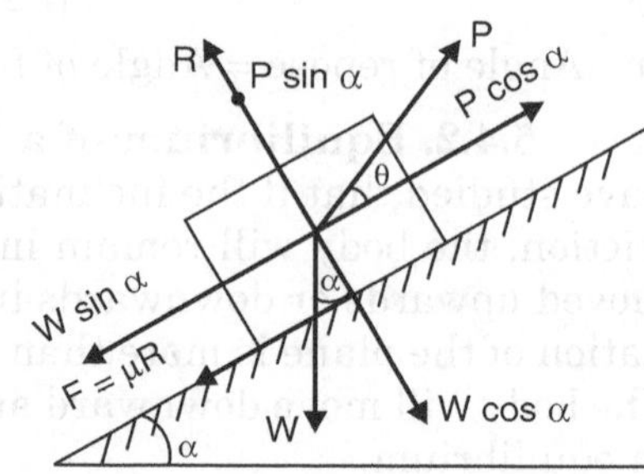

Fig. 5.14. Body moving up

1. Weight (W) of the body acting vertically downwards,
2. Normal reaction (R), perpendicular to the inclined plane,
3. The force of friction, $F = \mu R$ acting down the plane, and
4. Force P, inclined at an angle θ to the plane.

Resolving the forces along the plane, we get

$$W \sin \alpha + \mu R = P \cos \theta \quad \ldots(i)$$

Resolving force perpendicular to the inclined plane,

$$W \cos \alpha = R + P \sin \theta$$

or $$R = W \cos \alpha - P \sin \theta \quad \ldots(ii)$$

Substituting the value of R in equation (i), we get

$$W \sin \alpha + \mu (W \cos \alpha - P \sin \theta) = P \cos \theta$$

or $$W \sin \alpha + \mu W \cos \alpha - \mu P \sin \theta = P \cos \theta$$

or $$W \sin \alpha + \mu W \cos \alpha = P \cos \theta + \mu P \sin \theta$$

or $$W (\sin \alpha + \mu \cos \alpha) = P (\cos \theta + \mu \sin \theta)$$

$$\therefore \quad P = \frac{W(\sin \alpha + \mu \cos \alpha)}{(\cos \theta + \mu \sin \theta)}$$

$$= \frac{W(\sin \alpha + \tan \phi \cos \alpha)}{(\cos \theta + \tan \phi \sin \theta)} \quad (\because \ \mu = \tan \phi)$$

$$= \frac{W\left(\sin \alpha + \dfrac{\sin \phi}{\cos \phi} \cos \alpha\right)}{\left(\cos \theta + \dfrac{\sin \phi}{\cos \phi} \sin \theta\right)} = \frac{W(\sin \alpha \cos \phi + \sin \phi \cos \alpha)}{(\cos \theta \cos \phi + \sin \phi \sin \theta)}$$

$$= \frac{W(\sin \alpha + \phi)}{\cos (\theta - \phi)} \quad \text{...(5.3)}$$

The force P will be least if the denominator $\cos (\theta - \phi)$ is maximum. But the maximum value of $\cos (\theta - \phi)$ will be equal to one.

$$\therefore \quad \cos (\theta - \phi) = 1 = \cos 0°$$

$$\therefore \quad \theta - \phi = 0 \text{ or } \theta = \phi.$$

Substituting this value of $\theta = \phi$ in equation (5.3), we get

$$P_{min} = W \sin (\alpha + \theta) \quad \text{...(5.4)}$$

Thus the force P will be minimum if the angle of inclination of the force with the inclined plane is equal to the angle of friction.

(*b*) *Least force when the body is on the point of motion down the plane.*

When the body is on the point of motion down the plane, the force of friction $F = \mu R$ is acting up the plane as shown in Fig. 5.15.

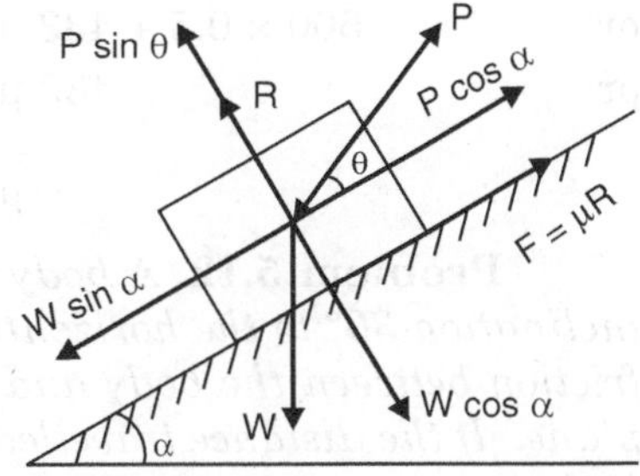

Fig. 5.15. Body moving down

Resolving the forces along the plane

$$W \sin \alpha = P \cos \theta + \mu R \quad \text{...}(iii)$$

Resolving forces perpendicular to the inclined plane

$$R + P \sin \theta = W \cos \alpha$$

or $$R = W \cos \alpha - P \sin \theta$$

Substituting the value of R in equation (iii), we get

$$W \sin \alpha = P \cos \theta + \mu (W \cos \alpha - P \sin \theta)$$

$$= P \cos \theta + \mu W \cos \alpha - \mu P \sin \theta$$

or $$W \sin \alpha - \mu W \cos \alpha = P \cos \theta - \mu P \sin \theta$$

or $$W [\sin \alpha - \mu \cos \alpha] = P [\cos \theta - \mu \sin \theta]$$

or $$P = \frac{W(\sin \alpha - \mu \cos \alpha)}{(\cos \theta + \mu \sin \theta)}$$

$$= \frac{W(\sin \alpha - \tan \phi \cos \alpha)}{(\cos \theta - \tan \phi \sin \theta)} \quad (\because \ \mu = \tan \phi)$$

$$= \frac{W\left(\sin\alpha - \dfrac{\sin\phi}{\cos\phi}\cos\alpha\right)}{\left(\cos\theta - \dfrac{\sin\phi}{\cos\phi}\sin\theta\right)} = \frac{W(\sin\alpha\cos\phi - \sin\phi\cos\alpha)}{(\cos\theta\cos\phi - \sin\phi\sin\theta)}$$

$$= \frac{W\sin(\alpha - \phi)}{\cos(\theta + \phi)} \quad \ldots(5.5)$$

Problem 5.11. *A body of weight 500 N is pulled up an inclined plane, by a force of 350 N. The inclination of the plane is 30° to the horizontal and the force is applied parallel to the plane. Determine the coefficient of friction.*

Sol. Given :

Weight of body, $W = 500$ N

Force applied, $P = 350$ N

Inclination, $\alpha = 30°$

Let μ = Coefficient of Q friction

R = Normal reaction

F = Force of friction = μR.

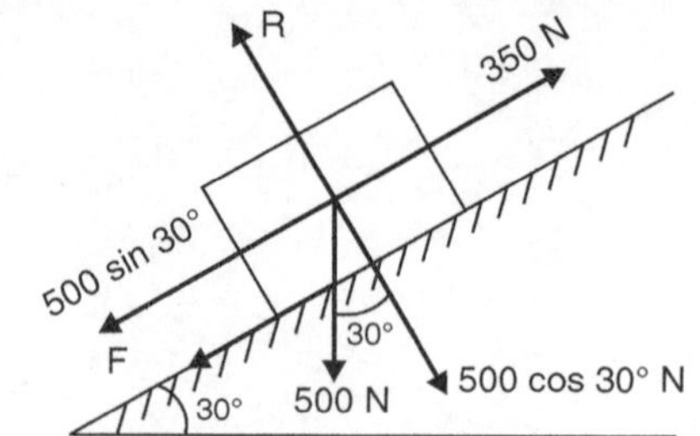

Fig. 5.16. Body moving up

The body is in equilibrium under the action of the forces shown in Fig. 5.16.

Resolving the forces along the plane,

$$500 \sin 30° + F = 350$$

or $$500 \sin 30° + \mu R = 350 \quad (\because \; F = \mu R) \quad \ldots(i)$$

Resolving forces normal to the plane,

$$R = 500 \cos 30° = 500 \times 0.866 = 433 \text{ N}$$

Substituting the value of R in equation (i), we get

$$500 \sin 30° + \mu \times 433 = 350$$

or $$500 \times 0.5 + 433\,\mu = 350$$

or $$433\,\mu = 350 - 500 \times 0.5 = 350 - 250 = 100$$

$\therefore$ $$\mu = \frac{100}{433} = \mathbf{0.23.} \quad \textbf{Ans.}$$

Problem 5.12. *A body of weight 450 N is pulled up along an inclined plane having inclination 30° to the horizontal at a steady speed. Find the force required if the coefficient of friction between the body and the plane is 0.25 and force is applied parallel to the inclined plane. If the distance travelled by the body is 10 m along the plane, find the work done on the body.*

Sol. Given :

Weight of body, $W = 450$ N

Inclination of plane, $\alpha = 30°$

Coefficient of friction, $\mu = 10.25$

Distance travelled by body $= 10$ m

Let the force required $= P$.

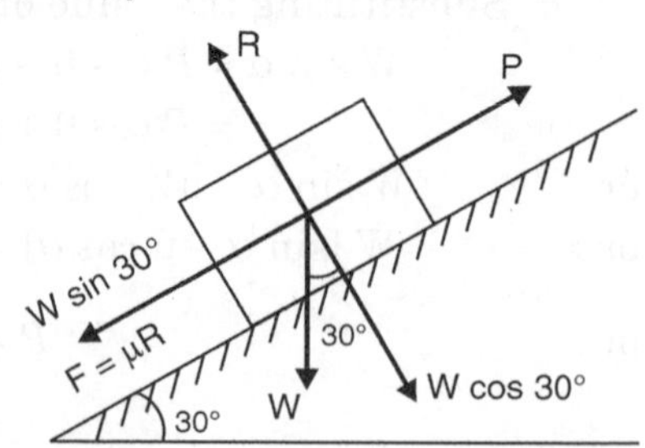

Fig. 5.17. Body moving up

The body is in equilibrium under the action of forces shown in Fig. 5.17.

Resolving forces along the plane,

$$P = W \sin 30° + \mu R = 450 \times 0.5 + 0.25 \times R$$

or $$P = 225 + 0.25\, R \quad ...(i)$$

Resolving forces normal to the plane,

$$R = W \cos 30° = 450 \times 0.866 = 389.7 \text{ N}$$

Substituting the value of R in equation (*i*),

$$P = 225 + 0.25 \times 389.7 = \mathbf{322.425\ N.}$$ **Ans.**

Work done on the body = Force × Distance travelled in the direction of force

= 322.525 × 10 Nm = 3224.25 Nm

= **3224.25 J** (where **J** = Joules = Nm). **Ans.**

Problem 5.13. *An effort of 200 N is required just to move a certain body up an inclined plane of angle 15°, the force acting parallel to the plane. If the angle of inclination of the plane is made 20°, the effort, required again, applied parallel to the plane, is found to be 230 N. Find the weight of the body and the coefficient of friction.* (AMIE Summer, 1980)

Sol. Given :

Effort required, $P_1 = 200$ N ; when inclination, $\theta_1 = 15°$

Effort required, $P_2 = 230$ N ; when inclination, $\theta_2 = 20°$.

In both the cases, the effort is applied parallel to the inclined plane and body is just to move up. Hence the force of friction ($F = \mu R$) will be acting downwards.

1st Case

$P_1 = 200$ N, $\theta_1 = 15°$

Let W = Weight of body,

μ = Coefficient of friction,

R_1 = Normal reaction, and

F_1 = Force of friction

$= \mu R_1$

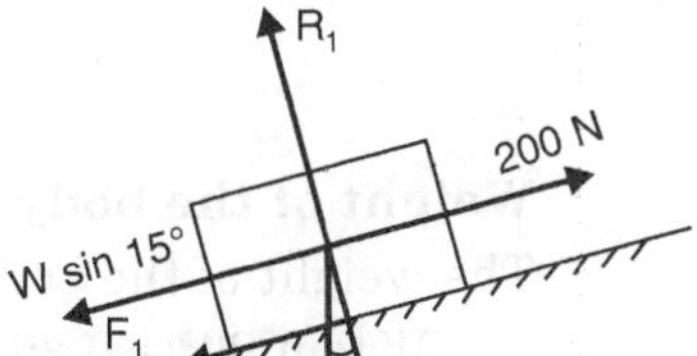

R₁
200 N
W sin 15°
F₁
15°
W
W cos 15°

Fig. 5.18

The body is in equilibrium under the action of the forces shown in Fig. 5.18.

Resolving the forces along the plane,

$$W \sin 15° + F_1 = 200$$

or $$W \sin 15° + \mu R_1 = 200 \qquad (\because F_1 = \mu R_1) \quad ...(i)$$

Resolving the forces normal to the plane,

$$R_1 = W \cos 15°$$

Substituting the value of R_1 in equation (*i*),

$$W \sin 15° + \mu \times W \cos 15° = 200$$

or $$W (\sin 15° + \mu \cos 15°) = 200 \quad ...(ii)$$

2nd Case

$P_2 = 230$ N and $\theta_2 = 20°$

Let R_2 = Normal reaction

F_2 = Force of friction

$= \mu R_2$

Resolving forces along the plane,

$$W \sin 20° + F_2 = 230$$

or $$W \sin 20° + \mu R_2 = 230 \quad ...(iii)$$

Resolving forces normal to the plane,

$$R_2 = W \cos 20°.$$

Substituting the value of R_2 in equation (*iii*), we get

$$W \sin 20° + \mu \times W \cos 20° = 230$$

or $$W [\sin 20° + \mu \cos 20°] = 230 \quad ...(iv)$$

Fig. 5.19

Dividing equation (*iv*) by equation (*ii*),

$$\frac{W(\sin 20° + \mu \cos 20°)}{W(\sin 15° + \mu \cos 15°)} = \frac{230}{200} = 1.15$$

or $$\frac{(\sin 20° + \mu \cos 20°)}{(\sin 15° + \mu \cos 15°)} = 1.15$$

or $$\sin 20° + \mu \cos 20° = 1.15 (\sin 15° + \mu \cos 15°)$$

$$= 1.15 \sin 15° + 1.15 \mu \cos 15°$$

or $$0.342 + \mu \times 0.9397 = 1.15 \times 0.2588 + 1.15 \times \mu \times 0.9659$$

$$= 0.2976 + 1.11 \mu$$

or $$0.342 - 0.2976 = 1.11 \mu - 0.9397 \mu$$

or $$0.0444 = 0.1703 \mu$$

$\therefore$ $$\mu = \frac{0.0444}{0.1703} = \mathbf{0.26.} \quad \textbf{Ans.}$$

Weight of the body (W)

The weight of the body is obtained by substituting the value of μ in equation (*iv*),

$$W (\sin 20° + 0.26 \times \cos 20°) = 230$$

or $$W (0.342 + 0.26 \times 0.9397) = 230$$

or $$W (0.342 + 0.2443) = 230$$

or $$0.5863 W = 230$$

$\therefore$ $$W = \frac{230}{0.5863} = \mathbf{392.3 \text{ N.}} \quad \textbf{Ans.}$$

Problem 5.14. *A rough inclined plane, coefficient of friction = μ, rises 1 cm for every 5 cm of its length. Calculate the effort require to drag a body weighing 100 N up the plane :*

(*i*) *when the effort is applied horizontally, and*

(*ii*) *when the effort is applied parallel to the plane.* (AMIE Winter, 1988)

Sol. Given :

Coefficient of friction = μ

Rise of plane is 1 cm for every 5 cm of its length. See Fig. 5.19 (*a*).

$\therefore$ $$\sin \theta = \frac{1}{5}$$

$$= 0.2$$

or $$\theta = \sin^{-1} 0.2$$

$$= 11.53°$$

Weight of body, $W = 100$ N.

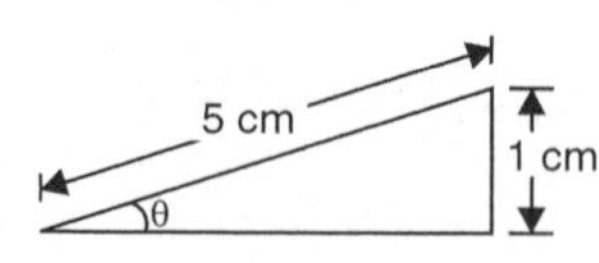

Fig. 5.19 (a)

(i) Find the effort when it is applied horizontally

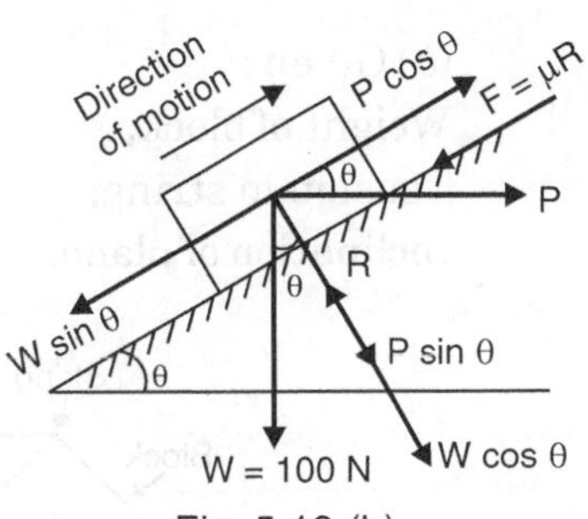

Fig. 5.19 (b)

The body is in equilibrium under the action of the forces shown in Fig. 5.19 (*b*).

Resolving the forces along the plane,

$$P \cos \theta = W \sin \theta + \mu R \quad \text{...(i)}$$

Resolving the forces normal to the plane,

$$R = P \sin \theta + W \cos \theta \quad \text{...(ii)}$$

Substituting the value of R in equation (*i*),

$$P \cos \theta = W \sin \theta + \mu \times (P \sin \theta + W \cos \theta)$$

$$= W \sin \theta + \mu P \sin \theta + \mu W \cos \theta$$

or $$P \cos \theta - \mu P \sin \theta = W \sin \theta + \mu W \cos \theta$$

or $$P (\cos \theta - \mu \sin \theta) = W (\sin \theta + \mu \cos \theta)$$

or $$P = \frac{W(\sin \theta + \mu \cos \theta)}{(\cos \theta - \mu \sin \theta)}$$

$$= \frac{100(\sin 11.53° + \mu \cos 11.53°)}{(\cos 11.53° - \mu \sin 11.53°)} = \frac{100(0.2 + \mu \times 0.9798)}{(0.9798 - \mu \times 0.2)}$$

$$= \mathbf{\frac{100(0.2 + 0.9798\mu)}{(0.9798 - 0.2\mu)}} \textbf{ N. Ans.}$$

(ii) Find the effort when it is applied parallel to the plane

Let P = Effort applied parallel to the plane. The body is in equilibrium under the action of forces shown in Fig. 5.19 (*c*).

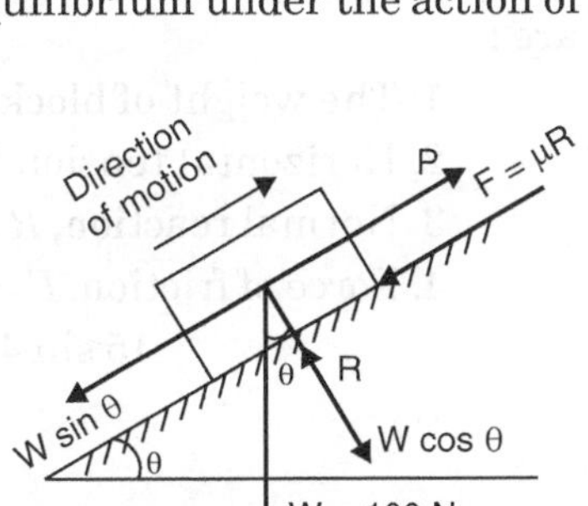

Fig. 5.19 (c)

Resolving the forces along the plane,

$$P = W \sin \theta + \mu \times R \quad \text{...(iii)}$$

Resolving the forces normal to the plane,

$$R = W \cos \theta \quad \text{...(iv)}$$

Substituting the value of R in equation (*iii*),

$$P = W \sin \theta + \mu \times W \cos \theta$$

$$= W (\sin \theta + \mu \cos \theta)$$

$$= 100 (\sin 11.53° + \mu \cos 11.53°)$$

$$= 100 (0.2 + \mu \times 0.9798)$$

$$= \mathbf{100 (0.2 + 0.9798 \times \mu)} \textbf{ N. Ans.}$$

Problem 5.15. *(a) Define coefficient of friction and limiting friction. (b) Block A weighing 15 N is a rectangular prism resting on a rough inclined plane as shown in Fig. 5.20. The block is tied up by a horizontal string which has a tension of 5 N. Find :*

(i) The frictional force on the block,

(ii) The normal reaction of the inclined plane, and

(iii) The coefficient of friction between the surfaces of contact. (AMIE Nov., 1967)

Sol. (*a*) For definition of coefficient of friction and limiting friction, please refer to Art. 5.2 and Art. 5.2.1.

(*b*) Given :

Weight of block, $W = 15$ N

Tension in string, $T = 5$ N

Inclination of plane, $\alpha = 45°$

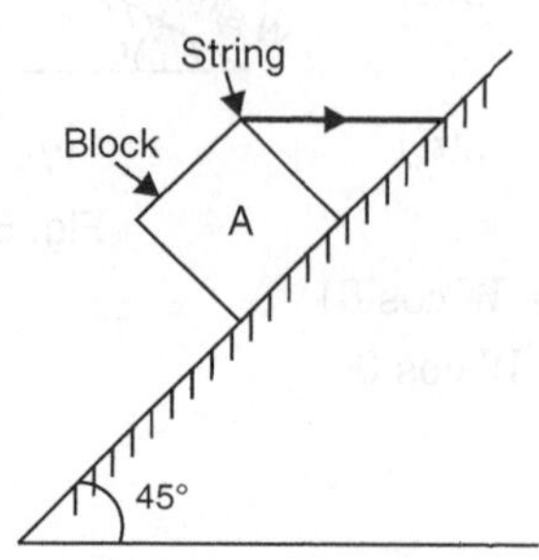

Fig. 5.20

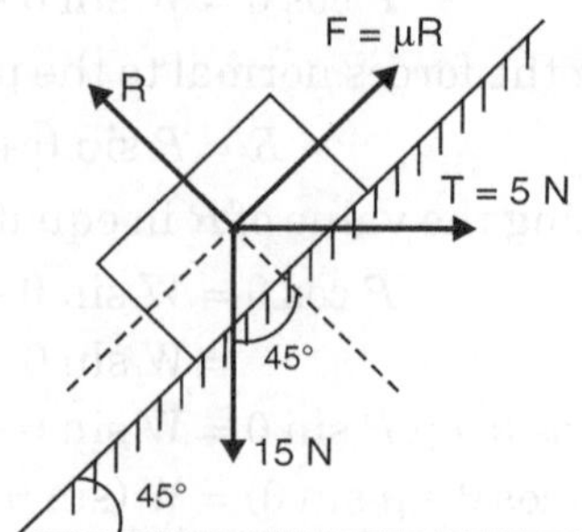

Fig. 5.21. Forces acting on the Block

Let F = Frictional force,

R = Normal reaction, and

μ = Coefficient of friction.

Since there is tension in string which means if string is removed, the block will slide down the plane. Hence force of friction will be acting in the upward direction.

The block A is in equilibrium under the action of the forces shown in Fig. 5.21. The forces are :

1. The weight of block, $W = 15$ N
2. Horizontal tension in the string, $T = 5$ N
3. Normal reaction, R
4. Force of friction, $F = \mu R$, acting upward. Resolving forces along the inclined plane,

$$15 \sin 45° = F + 5 \cos 45°$$

$$\therefore \quad F = 15 \sin 45° - 5 \cos 45° = 15 \times 0.707 - 5 \times 0.707$$

$$= 10 \times 0.707 = 7.07 \text{ N}$$

Resolving forces normal to inclined plane,

$$R = 15 \cos 45° + T \cos 45° = 15 \cos 45° + 5 \cos 45°$$

$$= 15 \times 0.707 + 5 \times 0.707 = 20 \times 0.707 = 14.14 \text{ N}$$

Using equation (5.1), we get

$$F = \mu R$$

$$\therefore \quad \mu = \frac{F}{R} = \frac{7.07}{14.14} = 0.5.$$

$\therefore$ (*i*) Frictional force on the block, F = **7.07 N. Ans.**

(*ii*) Normal reaction of the inclined plane, R = **14.14 N. Ans.**

(*iii*) Coefficient of friction, μ = **0.5. Ans.**

Problem 5.16. *Find the force required to drag a body of weight W, placed on a rough inclined plane having inclination α to the horizontal. The force P is applied to the body horizontally and the body is (a) on the point of motion up the plane, and (b) on the point of motion down the plane.*

Sol. Given :

Weight of the body $= W$

Inclination of the plane $= \alpha$

Force applied $= P$.

(a) Force required to drag the body when it is on the point of motion up the plane.

When the body is on the point of motion up the plane, the force of friction $F = \mu R$ is acting down the plane. The body is in equilibrium under the action of the forces shown in Fig. 5.22.

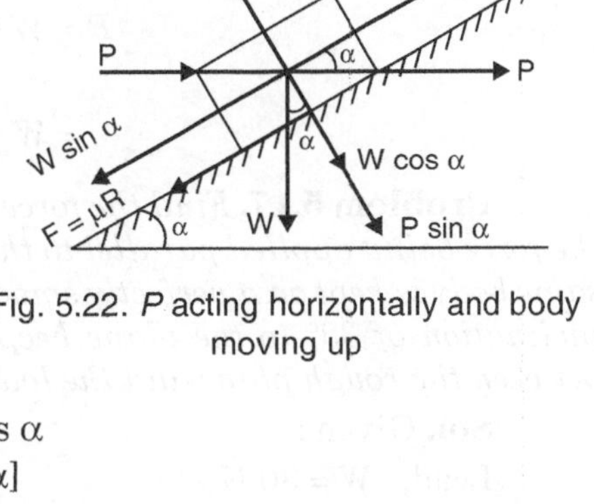

Fig. 5.22. *P* acting horizontally and body moving up

Resolving forces along the plane,

$$W \sin \alpha + \mu R = P \cos \alpha \quad ...(i)$$

Resolving forces normal to the inclined plane,

$$W \cos \alpha + P \sin \alpha = R \quad ...(ii)$$

Substituting the value of R in equation (*i*), we get

$$W \sin \alpha + \mu (W \cos \alpha + P \sin \alpha) = P \cos \alpha$$

or $$W \sin \alpha + \mu W \cos \alpha + \mu P \sin \alpha = P \cos \alpha$$

or $$P \cos \alpha - \mu P \sin \alpha = W \sin \alpha + \mu W \cos \alpha$$

or $$P [\cos \alpha - \mu \sin \alpha] = W [\sin \alpha + \mu \cos \alpha]$$

or $$P [\cos \alpha - \tan \phi \sin \alpha] = W [\sin \alpha + \tan \phi \cos \alpha] \quad (\because \quad \mu = \tan \phi)$$

or $$P \left[\cos \alpha - \frac{\sin \phi}{\cos \phi} \sin \alpha \right] = W \left[\sin \alpha + \frac{\sin \phi}{\cos \phi} \cos \alpha \right]$$

or $$P \left[\frac{\cos \alpha \cos \phi - \sin \phi \sin \alpha}{\cos \phi} \right] = W \left[\frac{\sin \alpha \cos \phi + \sin \phi \cos \alpha}{\cos \phi} \right]$$

Cancelling $\cos \phi$ on both sides,

$$P [\cos \alpha \cos \phi - \sin \phi \sin \alpha] = W [\sin \alpha \cos \phi + \sin \phi \cos \alpha]$$

$$\therefore \quad P = W \frac{(\sin \alpha \cos \phi + \sin \phi \cos \alpha)}{(\cos \alpha \cos \phi - \sin \phi \sin \alpha)}$$

$$= W \frac{\sin (\alpha + \phi)}{\cos (\alpha + \phi)}$$

or $$P = W \tan (\alpha + \phi) \quad ...(5.6)$$

(b) Force required to drag the body when it is on the point of motion down the plane.

In this case the force of friction $F = \mu R$ will be acting up the plane. The body is in equilibrium under the action of the forces shown in Fig. 5.23.

Resolving forces along the plane,

$$W \sin \alpha = P \cos \alpha + \mu R \quad ...(iii)$$

Resolving forces along the plane,

$$R = W \cos \alpha + P \sin \alpha \quad ...(iv)$$

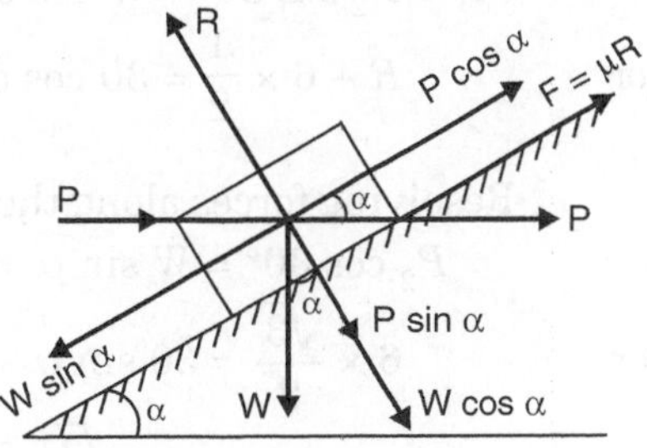

Fig. 5.23. Body moving down

Substituting the value of R in equation (*iii*), we get

$$W \sin \alpha = \text{P} \cos \alpha + \mu(W \cos \alpha + P \sin \alpha)$$

$$= P \cos \alpha + \mu W \cos \alpha + \mu P \sin \alpha$$

or $$P \cos \alpha + \mu P \sin \alpha = W \sin \alpha - \mu W \cos \alpha$$

or $$P(\cos \alpha + \alpha \sin \alpha) = W(\sin \alpha - \mu \cos \alpha)$$

or $$P(\cos \alpha + \tan \phi \sin \alpha) = W(\sin \alpha - \mu \cos \alpha) \qquad (\because \mu = \tan \phi)$$

or $$P\left(\cos \alpha + \frac{\sin \phi}{\cos \phi} \sin \alpha\right) = W\left(\sin \alpha - \frac{\sin \phi}{\cos \phi} \cos \alpha\right)$$

Multiplying by $\cos \phi$ to both sides, we get

$$P(\cos \phi \cos \alpha + \sin \phi \sin \alpha) = W(\sin \alpha \cos \phi - \sin \phi \cos \alpha)$$

$$P = W \frac{(\sin \alpha \cos \phi - \sin \phi \cos \alpha)}{(\cos \alpha \cos \phi + \sin \phi \sin \alpha)}$$

$$= W \frac{\sin (\alpha - \phi)}{\cos (\alpha - \phi)} = W \tan (\alpha - \phi) \qquad \text{...(5.7)}$$

Problem 5.17. *Find the force required to move a load of 30 N up a rough inclined plane, the force being applied parallel to the plane. The inclination of the plane is such that when the same body is kept on a perfectly smooth plane inclined at that angle, a force of 6 N applied at an inclination of 30° to the plane keeps the same in equilibrium. Assume coefficient of friction between the rough plane and the load is equal to 0.3.* (AMIE Winter, 1983)

Sol. Given :

Load, $W = 30$ N

Coefficient of friction between the rough plane and load,

$\mu = 0.3$

Let α = Inclination of the plane with horizontal

P_1 = Force required to move the load up a rough inclined plane, when the force is applied parallel to the plane.

The force applied when same body is kept on a smooth inclined plane, $P_2 = 6$ N.

Inclination of the force with the inclined plane, $\theta = 30°$.

1st Case. Consider the body of weight 30 N placed on a smooth inclined plane as shown in Fig. 5.24.

The forces acting on the body are :

(*i*) The weight (W = 30 N) vertically downward.

(*ii*) The force P_2 (= 6 N) at an angle of 30° with the inclined plane.

(*iii*) The normal reaction R.

Resolving forces normal to the inclined plane.

$$R + P_2 \sin 30° = W \cos \alpha$$

or $$R + 6 \times \frac{1}{2} = 30 \cos \alpha \qquad \text{...}(i)$$

$$(\because P_2 = 6 \text{ N}, W = 30 \text{ N})$$

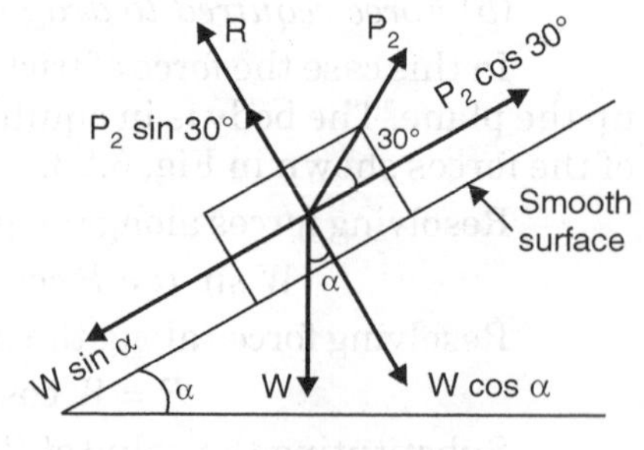

Fig. 5.24

Resolving forces along the inclined plane,

$$P_2 \cos 30° = W \sin \alpha$$

or $$6 \times \frac{\sqrt{3}}{2} = 30 \sin \alpha$$

$$\therefore \quad \sin \alpha = 6 \times \frac{\sqrt{3}}{2} \times \frac{1}{30} = \frac{\sqrt{3}}{10} = 0.1732$$

$$\therefore \quad \alpha = \mathbf{9.974°}.$$

2nd Case. The body of weight 30 N is placed on a rough inclined plane having inclination α (= 9.974°) with the horizontal as shown in Fig. 5.25.

The forces acting on the body are :

(*i*) The weight W (= 30 N) vertically downward.

(*ii*) The force P_1, parallel to the plane.

(*iii*) The normal reaction R^*.

(*iv*) Force of friction $F = \mu R^*$.

Fig. 5.25

Resolving the forces along the inclined plane,

$$W \sin \alpha + F = P_1$$

or $$30 \sin 9.974° + \mu R^* = P_1 \qquad (\because \; F = \mu R^* \text{ and } \alpha = 9.974°) \quad ...(ii)$$

Resolving the forces normal to the inclined plane

$$R^* = W \cos \alpha$$
$$= 30 \times \cos 9.974° \qquad (\because \; \alpha = 9.974°)$$
$$= 30 \times 9.9848 = 29.544.$$

Substituting the value of R^* in equation (*ii*), we get

$$30 \times \sin 9.974 + \mu \times 29.544 = P_1$$

or $$30 \times 0.1732 + 0.3 \times 29.544 = P_1$$

or $$5.196 + 8.8632 = P_1$$

or $$P_1 = \mathbf{14.059\ N.} \quad \textbf{Ans.}$$

Problem 5.18. *A cord connects two bodies of weights 300 N and 800 N. The two bodies are placed on an inclined plane and cord is parallel to inclined plane. The coefficients of friction for the weight of 400 N is 0.15 and that for 800 N is 0.4. Determine the inclination of the plane to the horizontal and the tension in the cord when the motion is about to take place, down the inclined plane. The body weighing 400 N is below the body weighing 800 N.*

Sol. Weight of first body, $W_1 = 400$ N

Weight of second body, $W_2 = 800$ N

Coefficient of friction for first body, $\mu_1 = 0.15$

Coefficient of friction for second body, $\mu_2 = 0.40$

Let α = Inclination of plane

T = Tension in cord

R_1 = Normal reaction for 1st body

R_2 = Normal reaction for 2nd body

F_1 = Force of friction between 1st body and plane = $\mu_1 R_1$

F_2 = Force of friction between 2nd body and plane = $\mu_2 R_2$.

As the motion of the two bodies is about to take place down the inclined plane, the force of frictions F_1 and F_2 will be acting upward. The two bodies are in equilibrium under the action of forces shown in Fig. 5.26.

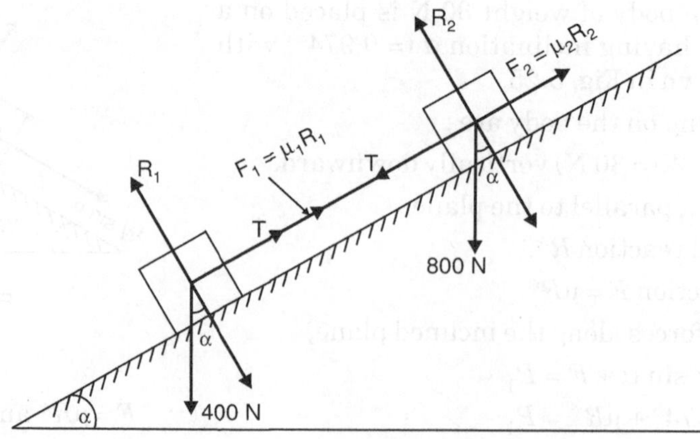

Fig. 5.26

Forces on the first body

Resolving forces along the plane,

$$400 \sin \alpha = T + F_1 = T + \mu_1 R_1 = T + 0.15\, R_1 \quad ...(i)$$

Resolving forces normal to the plane,

$$400 \cos \alpha = R_1$$

Substituting the value of R_1 in equation (*i*),

$$400 \sin \alpha = T + 0.15 \times 400 \cos \alpha$$
$$= T + 60 \cos \alpha$$

$$\therefore \quad T = 400 \sin \alpha - 60 \cos \alpha \quad ...(ii)$$

Forces on the second body

Resolving forces along the plane,

$$800 \sin \alpha + T = F_2 = \mu_2 R_2 = 0.40\, R_2 \quad ...(iii)$$

Resolving forces normal to the plane,

$$R_2 = 800 \cos \alpha$$

Substituting the values of R_2 in equation (*iii*),

$$800 \sin \alpha + T = 0.40 \times 800 \cos \alpha = 320 \cos \alpha$$

or $$T = 320 \cos \alpha - 800 \sin \alpha \quad ...(iv)$$

Equating the values of T, given by equations (*ii*) and (*iv*),

$$400 \sin \alpha - 60 \cos \alpha = 320 \cos \alpha - 800 \sin \alpha$$

or $$400 \sin \alpha + 800 \sin \alpha = 320 \cos \alpha + 60 \cos \alpha$$

or $$1200 \sin \alpha = 380 \cos \alpha$$

or $$\frac{\sin \alpha}{\cos \alpha} = \frac{380}{1200} = 0.3166 \quad \text{or} \quad \tan \alpha = 0.3166$$

$$\therefore \quad \alpha = \tan^{-1} 0.3166 = \mathbf{17.56°} \quad \text{or} \quad \mathbf{17° 34'}. \quad \textbf{Ans.}$$

Substituting the value of α in equation (*ii*),

$$T = 400 \sin 17.56° - 60 \cos 17.56°$$
$$= 400 \times 0.3017 - 60 \times 0.9534$$
$$= 120.68 - 57.20 = \mathbf{63.48\ N.} \quad \textbf{Ans.}$$

Problem 5.19. *Two blocks A and B are connected by a horizontal rod and are supported on two rough planes as shown in Fig. 5.27. If the weight of block B is 1500 N and coefficient of friction of block A and B are 0.25 and 0.35 respectively, find the smallest weight of block A for which the equilibrium can exist.*

Sol. Given : Weight of block B,

$$W_B = 1500 \text{ N}$$

Coefficient of friction for block A,

$$\mu_A = 0.25$$

Coefficient of friction for block B,

$$\mu_B = 0.35$$

Let the smallest weight of block A for equilibrium $= W_A$.

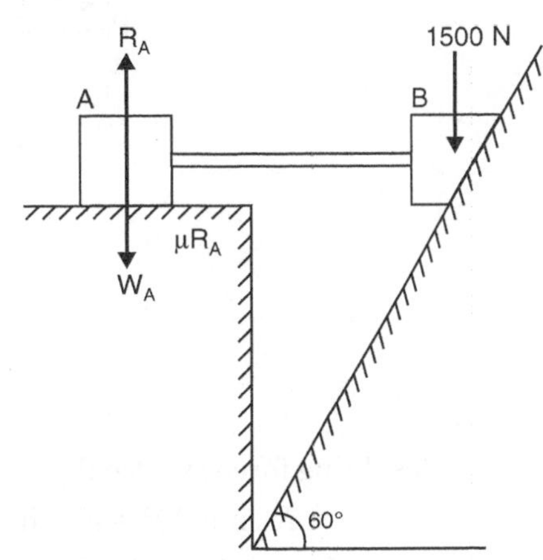

Fig. 5.27

If the weight of block A is less than the value required for equilibrium, the block B will slide downwards. But the block A and B are connected by a horizontal rod of fixed length. Now when block B starts moving in the downward direction, the block A starts moving towards left. Hence a force of friction F_A equal to $\mu_A R_A$ will be acting on block A towards right as shown in Fig. 5.27. On block B, the force of friction F_B equal to $\mu_B R_B$ will be acting in the upward direction.

For block A

Resolving force normal to plane, $R_A = W_A$.

Force of friction, $F_A = \mu_A R_A = \mu\, 0.25 \times W_A = 0.25\, W_A$.

This force will be transmitted to block B through rod AB.

For block B

The block B will be in equilibrium under the action of the forces, shown in Fig. 5.28. The forces are :

(*i*) The weight of block B = 1500 N acting vertically downwards.

(*ii*) The normal reaction R_B of the plane.

(*iii*) The horizontal force = $0.25\, W_A$, transmitted to block B through rod AB.

(*iv*) Force of friction $F_B = \mu_B R_B = 0.35\, R_B$ acting up the inclined plane.

In this case, the forces are resolved horizontally and vertically instead of along the inclined plane and normal to the plane. For an equilibrium state, the forces acting in any direction must be zero.

Resolving forces horizontally,

$$0.25\, W_A + F_B \cos 60° = R_B \cos 30°$$

$$0.25\, W_A + 0.35\, R_B \cos 60° = R_B \cos 30° \qquad (\because\ F_B = 0.35\, R_B)$$

$$0.25\, W_A + 0.35 \times 0.5\, R_B = R_B \times .866$$

$$0.25\, W_A + 0.175\, R_B = 0.866\, R_B$$

or

$$0.25\, W_A = 0.866\, R_B - 0.175\, R_B = 0.691\, R_B \qquad ...(i)$$

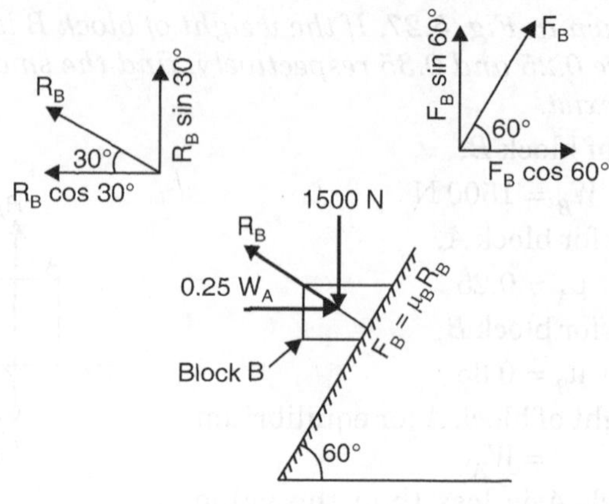

Fig. 5.28

Resolving forces vertically,

$$R_B \sin 30° + F_B \sin 60° = 1500$$

$$R_B \times 0.5 + 0.35\, R_B \times .866 = 1500 \qquad (\because \; F_B = 0.35\, R_B)$$

$$0.5\, R_B + 0.303\, R_B = 1500 = 0.803\, R_B = 1500$$

$$\therefore \qquad R_B = \frac{1500}{0.803} = 1868 \text{ N.}$$

Substituting this value of R_B in (*i*), we get

$$0.25\, W_A = 0.691 \times 1868$$

$$\therefore \qquad W_A = \frac{0.691 \times 1868}{0.25} = \mathbf{5163\ N.\quad Ans.}$$

Problem 5.20. *A block A weighing 100 N rests on a rough inclined plane whose inclination to the horizontal is 45°. This block is connected to another block B weighing 300 N resting on a rough horizontal plane, by a weightless rigid bar inclined at 30° to the horizontal. Find the horizontal force required to be applied to the block B to just move the block A in upward direction. Assume angle of limiting friction as 15° at all surfaces where there is a sliding.*

(AMIE Summer, 1984)

Sol. Given :

Weight of block *A*, $W_A = 100$ N

Inclination of plane, $\alpha = 45°$

Weight of block *B*, $W_B = 300$ N

Inclination of rigid bar with horizontal = 30°

Angle of limiting friction, $\phi = 15°$

$\therefore$ $\mu = \tan 15° = 0.2679.$

Let P = Horizontal force to be applied to block *B* to just move the block *A* in the upward direction

T = Thrust in the rigid bar.

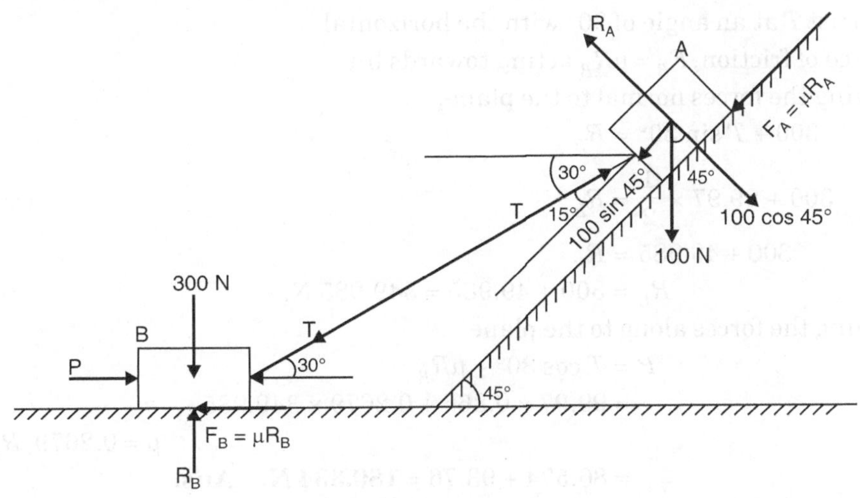

Fig. 5.29

Consider the equilibrium of block A

When the block A is at the point of moving up the plane, the forces acting on block A are :

(*i*) Weight (= 100 N) acting vertically downward

(*ii*) Normal reaction R_A

(*iii*) Force of friction ($F_A = \mu R_A$) acting down the plane

(*iv*) Thrust T in the rod at an angle of 30° to the horizontal and 15° with the inclined plane.

Resolving the forces along the plane,

$$100 \sin 45° + F_A = T \cos 15°$$

or $$100 \times 0.707 + \mu R_A = T \times 0.9659$$

or $$70.7 + 0.2679 R_A = 0.9659 T \qquad (\because \mu = 0.2679) \quad ...(i)$$

Resolving the forces normal to the plane

$$100 \cos 45° + T \sin 15° = R_A$$

or $$100 \times 0.707 + T \times 0.2588 = R_A$$

Substituting the value of R_A in equation (*i*), we get

$$70.7 + 0.2679\,(100 \times 0.707 + T \times 0.2588) = 0.9659\, T$$

or $$70.7 + 0.2679 \times 100 \times 0.707 + T \times 0.2588 \times 0.2679 = 0.9659\, T$$

or $$70.7 + 18.94 + 0.0693\, T = 0.9659\, T$$

or $$89.64 = 0.9659\, T - 0.0693\, T = 0.8966\, T$$

or $$T = \frac{89.64}{0.8966} = 99.97 \text{ N.}$$

Consider the equilibrium of block B

The forces acting on the block B are

(*i*) Weight equal to 300 N acting vertically downward

(*ii*) Normal reaction R_B

(*iii*) Horizontal force P

(*iv*) Thrust T at an angle of 30° with the horizontal

(*v*) Force of friction, $F_B = \mu R_B$ acting towards lift.

Resolving the forces normal to the plane,

$$300 + T \sin 30° = R_B$$

or $$300 + 99.97 \times \frac{1}{2} = R_B \qquad (\because \; T = 99.97)$$

or $$300 + 49.985 = R_B$$

$\therefore$ $$R_B = 300 + 49.985 = 349.985 \text{ N.}$$

Resolving the forces along to the plane,

$$P = T \cos 30° + \mu R_B$$
$$= 99.97 \times 0.866 + 0.2679 \times 349.985$$
$$(\because \; \mu = 0.2679, R_B = 349.985)$$
$$= 86.574 + 93.76 = \mathbf{180.334 \text{ N.}} \quad \textbf{Ans.}$$

Problem 5.21. *Referring to the Fig. 5.29 (a) given below, determine the least value of the force P to cause motion to impend rightwards. Assume the coefficient of friction under the blocks to be 0.2 and pulley to be frictionless.* (AMIE Winter, 1988)

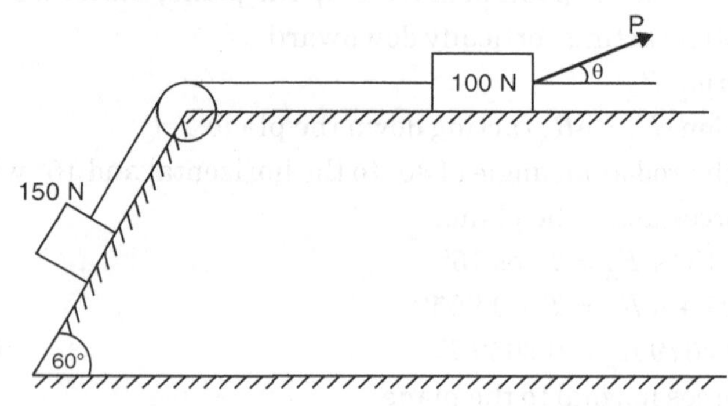

Fig. 5.29 (a)

Sol. Given :

Coefficient of friction under both blocks, $\mu = 0.2$

Pulley is frictionless. Motion of block of weight 100 N is towards right. Find least value of P.

1st Case

Consider the equilibrium of block of weight 150 N

As the block of weight 100 N tends to move rightwards, the block of weight 150 N will tend to move upwards. Hence force of friction will act downwards as shown in Fig. 5.29 (*b*).

Let T = Tension in the string

R = Normal reaction

F = Force of friction = μR

$= 0.2R$

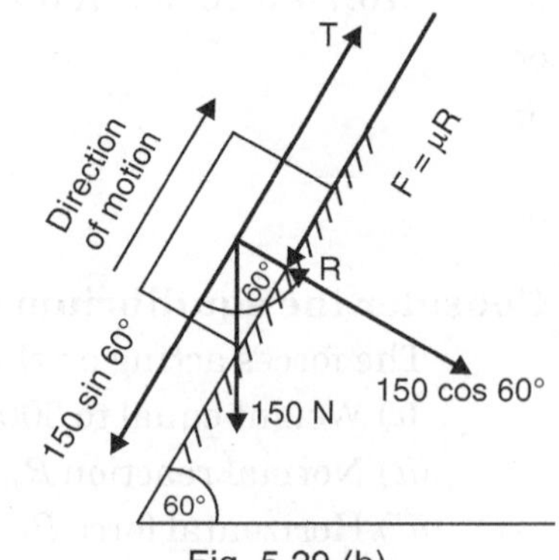

Fig. 5.29 (b)

The weight 150 N is acting vertically downwards. The body is in equilibrium under the action of forces shown in Fig. 5.29 (*b*). Resolving the forces along the plane,

$$T = 150 \sin 60° + \mu R \quad ...(i)$$

Resolving the forces normal to inclined plane,

$$R = 150 \cos 60° = 150 \times \frac{1}{2} = 75 \text{ N} \quad ...(ii)$$

Substituting the value of R in equation (*i*),

$$T = 150 \sin 60° + 0.2 \times 75 \quad (\because \mu = 0.2)$$

$$= 144.9 \text{ N}$$

2nd Case

Now consider the equilibrium of block of weight 100 N

The block of weight 100 N tends to move rightwards, hence force of friction will be acting towards left as shown in Fig. 5.29 (*c*). Also the pulley is frictionless, hence the tension in the string which is attached to the block of weight 100 N will be 144.9 N. The body will be in equilibrium under the action of forces shown in Fig. 5.29 (*c*).

Resolving forces along the plane, (*i.e.*, horizontally),

$$P \cos \theta = T + 0.2 R^*$$

$$= 144.9 + 0.2 R^* \quad ...(iii)$$

Fig. 5.29 (c)

Resolving the forces normal to the plane (*i.e.*, vertically)

$$R^* + P \sin \theta = 100$$

or $$R^* = 100 - P \sin \theta \quad ...(iv)$$

Substituting the value of R^* in equation (*iii*),

$$P \cos \theta = 144.9 + 0.2 \times (100 - P \sin \theta) = 144.9 + 20 - 0.2 P \sin \theta$$

or $$P \cos \theta + 0.2 P \sin \theta = 164.9$$

or $$P (\cos \theta + 0.2 \sin \theta) = 164.9$$

or $$P = \frac{164.9}{(\cos \theta + 0.2 \sin \theta)} \quad ...(v)$$

The force P will be minimum, if $(\cos \theta + 0.2 \sin \theta)$ is maximum. But $(\cos \theta + 0.2 \sin \theta)$ will be maximum if

$$\frac{d}{d\theta} (\cos \theta + 0.2 \sin \theta) = 0$$

or $$-\sin \theta + 0.2 \cos \theta = 0$$

or $$0.2 \cos \theta = \sin \theta$$

or $$0.2 = \frac{\sin \theta}{\cos \theta} = \tan \theta$$

or $$\theta = \tan^{-1} 0.2 = 11.309°$$

Substituting the value of θ in equation (*v*), the least value of P will be obtained.

$$\therefore \quad P_{(least)} = \frac{164.9}{(\cos 11.309° + 0.2 \times \sin 11.309°)} = \mathbf{161.88 \text{ N}}. \quad \textbf{Ans.}$$

Problem 5.22. *What should be the value of the angle θ in Fig. 5.29 (d) so that the motion of the 90 N block impends down the plane ? The coefficient of friction μ for all the surfaces is 1/3.*
(AMIE Summer, 1990)

Sol. Given :

Coefficient of friction for all surfaces, $\mu = 1/3$

Motion of weight 90 N impends down the plane. Find the value of θ.

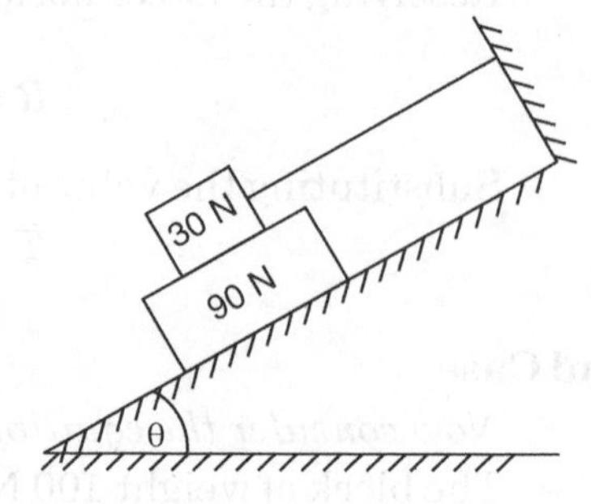

Fig. 5.29 (d)

First consider the equilibrium of weight 30 N.

As the weight 90 N tends to move downwards, there will be a rubbing action between the surfaces of weight 90 N and 30 N. Hence a force of friction will be acting between these two surfaces.

The weight 30 N is tied to a string, the other end of the string is fixed to the plane. When the weight 90 N tends to move downwards, the weight 30 N with respect to weight 90 N will move upwards. Hence the force of friction on the lower surface of the weight 30 N will act downward as shown in Fig. 5.29 (*e*). The weight 30 N will be in equilibrium under the action of the forces shown in Fig. 5.29 (*e*) in which

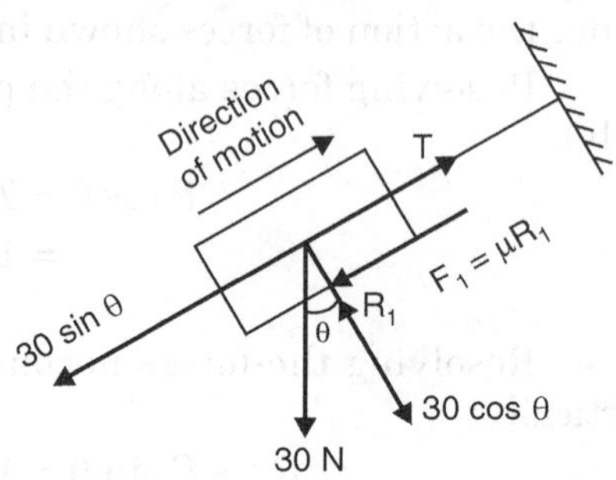

Fig. 5.29 (e)

T = Tension in the string

R_1 = Normal reaction on the lower surface of weight 30 N

F_1 = Force of friction = μR_1

Resolving the forces along the plane,

$$T = 30 \sin \theta + \mu R_1$$

$$= 30 \sin \theta + \frac{1}{3} R_1 \quad \text{...(i)}$$

Resolving the forces normal to the plane,

$$R_1 = 30 \cos \theta \quad \text{...(ii)}$$

Substituting the value of R_1 in equation (*i*),

$$T = 30 \sin \theta + \frac{1}{3} \times 30 \cos \theta$$

$$= 30 \sin \theta + 10 \cos \theta \quad \text{...(iii)}$$

Now consider the equilibrium of weight 90 N

The weight 90 N will be in equilibrium under the action of forces shown in Fig. 5.29 (*f*).

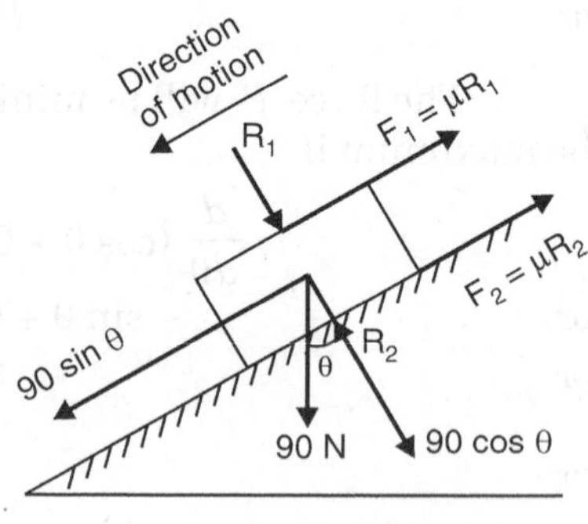

Fig. 5.29 (f)

Resolving the forces along the plane,

$$90 \sin \theta = \mu R_1 + \mu R_2$$

$$= \frac{1}{3} R_1 + \frac{1}{3} R_2$$

$$= \frac{1}{3} \times 30 \cos \theta + \frac{1}{3} R_2 \quad (\because \; R_1 = 30 \cos \theta)$$

$$= 10 \cos \theta + \frac{1}{3} R_2 \quad ...(iv)$$

Resolving the forces normal to the plane,

$$R_2 = R_1 + 90 \cos \theta$$

$$= 30 \cos \theta + 90 \cos \theta \quad (\because \ R_1 = 30 \cos \theta)$$

$$= 120 \cos \theta \quad ...(v)$$

Substituting the value of R_2 in equation (*iv*), we get

$$90 \sin \theta = 10 \cos \theta + \frac{1}{3} \times 120 \cos \theta$$

$$= 10 \cos \theta + 40 \cos \theta = 50 \cos \theta$$

$$\therefore \quad \frac{\sin \theta}{\cos \theta} = \frac{50}{90} = 0.5555$$

or $$\tan \theta = 0.5555$$

$$\therefore \quad \theta = \tan^{-1} 0.5555 = \mathbf{29.05°}. \quad \textbf{Ans.}$$

5.4.3. Analysis of Ladder Friction. Fig. 5.30 shows a ladder *AC* resting on the ground and leaning against a wall.

Let R_A = Reaction at *A*

R_C = Reaction at *C*

F_A = Force of friction at *A* $= \mu R_A$

F_C = Force of friction at *C* $= \mu R_C$.

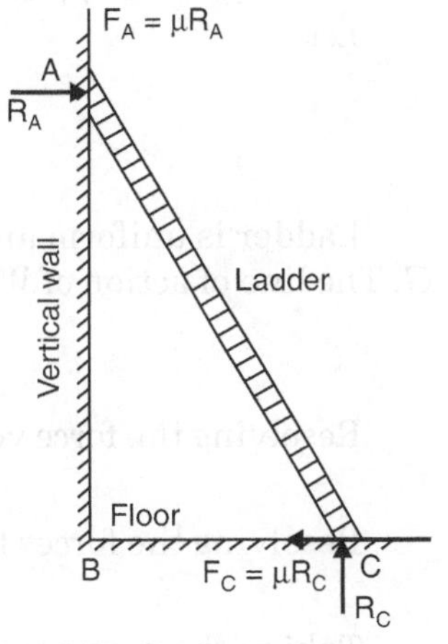

Fig. 5.30

Due to the self weight of the ladder or when some man stands on the ladder, the upper end *A* of the ladder tends to slip downwards, and hence the force of friction between the ladder and the vertical wall $F_A = \mu R_A$ will be acting upwards as shown in Fig. 5.30. Similarly, the lower end *C* of the ladder will tend to move towards right and hence a force of friction between ladder and floor $F_C = \mu R_C$ will be acting towards left.

For the equilibrium of the system, the algebraic sum of the horizontal and vertical components of the forces must be zero. Also the moments of all the forces about any point must be zero.

Note. If the vertical wall is smooth, there will be no force of friction between the ladder and the vertical wall.

Problem 5.23. *A uniform ladder of length 10 m and weighing 20 N is placed against a smooth vertical wall with its lower end 8 m from the wall. In this position the ladder is just to slip. Determine :*

(i) the coefficient of friction between the ladder and the floor, and

(ii) frictional force acting on the ladder at the point of contact between ladder and floor.

Sol. Given :

Weight of ladder, $W = 20$ N

Length of ladder, $AC = 10$ m

Distance of lower end of ladder from wall, *i.e.*,

$$BC = 8 \text{ m.}$$

In right-angled triangle ABC,

$$AB = \sqrt{AC^2 - BC^2}$$
$$= \sqrt{10^2 - 8^2} = \sqrt{100 - 64}$$
$$= \sqrt{36} = 6 \text{ m.}$$

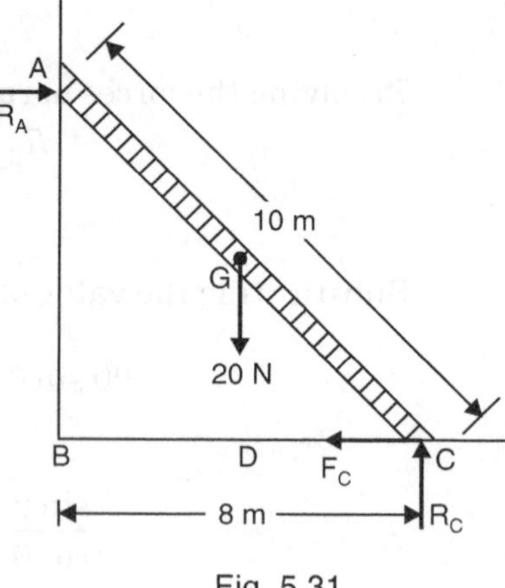

Fig. 5.31

As vertical wall is smooth, hence there will be no force of friction between ladder and wall.

In the position of the ladder shown in Fig. 5.31, the ladder is just to slip. Hence the lower end C of the ladder will tend to move towards right and hence a force of friction between ladder and floor (*i.e.*, $F_A = \mu R_C$) will be acting towards left.

(*i*) *Coefficient of friction between ladder and floor*

Let R_A = Reaction at A

R_C = Reaction at C

μ = Coefficient of friction between ladder and floor at C

F_C = Force of friction at $C = \mu R_C$

Ladder is uniform and the weight of ladder (W = 20 N) is acting at the middle point of AC at G. The line of action of W will pass through the middle point of BC. Hence distance

$$CD = \frac{1}{2} \times BC = \frac{8}{2} = 4.0 \text{ m.}$$

Resolving the force vertically, we get

$$R_C = 20 \text{ N}$$

Resolving the forces horizontally, we get

$$R_A = F_C = \mu \times R_C = \mu \times 20 = 20\,\mu \quad \text{...(i)}$$

Taking the moments of all forces about point C,

Clockwise moments = Anti-clockwise moments

$$R_A \times AB = 20 \times CD$$

(The moments of R_C and F_C about point C is zero)

or $$20\,\mu \times 6 = 20 \times 4 \qquad [\because R_A = 20\,\mu \text{ from equation } (i)]$$

$\therefore$ $$\mu = \frac{20 \times 4}{20 \times 6} = \mathbf{0.67.} \quad \textbf{Ans.}$$

(*ii*) *Frictional force acting at C is given as*

$$F_C = \mu \times R_C = 0.67 \times 20 = \mathbf{13.40\ N.} \quad \textbf{Ans.}$$

Problem 5.24. *A uniform ladder of length 13 m and weighing 25 N is placed against a smooth vertical wall with its lower end 5 m from the wall. The coefficient of friction between the ladder and the floor is 0.3. Show that the ladder will remain in equilibrium in this position. What is the frictional force acting on the ladder at the point of contact between the ladder and floor ?* (AMIE Winter, 1981)

Sol. Given :

Length of ladder, $L = AB = 13$ m

Weight of ladder, $W = 25$ N

Distance of lower end of ladder from wall,

$$AC = 5 \text{ m}$$

Coefficient of friction between ladder and floor,

$$\mu = 0.3$$

Vertical wall is smooth and hence there will be no force of friction between ladder and wall.

Let F_A = Limiting frictional force acting at $A = \mu R_A = 0.3\, R_A$

R_A = Normal reaction at A

R_B = Normal reaction at B

The ladder AB is placed as shown in Fig. 5.32. The weight of 25 N is acting at the middle point of AB (*i.e.*, at G) vertically downwards. If the ladder is not in equilibrium, it will start moving at A towards right and force of friction (F_A) will act towards left as shown in Fig. 5.32. The forces acting on the ladder are shown in Fig. 5.32.

Fig. 5.32

Equating the vertical forces,

$$R_A = 25 \text{ N}$$

Equating the horizontal forces, we have

$$R_B = F_A = \mu R_A = 0.3\, R_A = 0.3 \times 25 = 7.5 \text{ N} \qquad (\because\ R_A = 25 \text{ N})$$

$\therefore$ Maximum amount of force of friction available at A is

$$F_A = 7.5 \text{ N}.$$

To prove that ladder is in equilibrium for the given position

Let F_A' = Force of friction when ladder is in equilibrium

R_B' = Normal reaction at B when ladder is in equilibrium

R_A' = Normal reaction at A when ladder is in equilibrium

The ladder will be in equilibrium if F_A' is less than F_A.

From Fig. 5.32, ABC is a right-angled triangle. G is the middle point of AB and GD is normal to AC. Hence D is the middle point of AC.

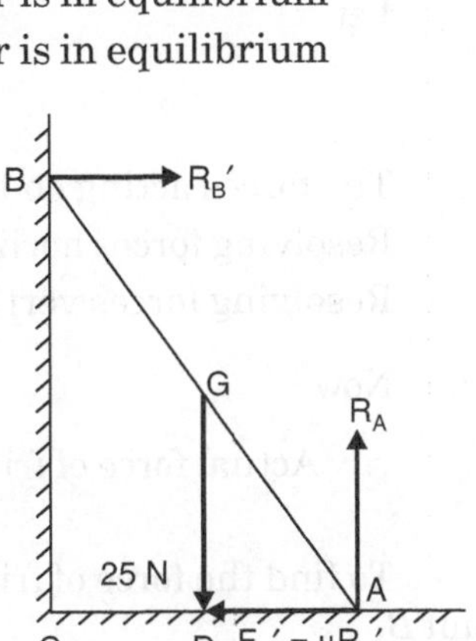

Fig. 5.33

$\therefore$ $AD = CD = 2.5$ m

Also $AB^2 = AC^2 + BC^2$

$$\therefore \quad BC = \sqrt{AB^2 - AC^2} = \sqrt{13^2 - 5^2}$$

$$= \sqrt{169 - 25}$$

$$= \sqrt{144} = 12 \text{ m}.$$

The ladder will be in equilibrium, if the moments of all forces (acting on the ladder) about any point is zero.

Taking moments of all forces shown in Fig. 5.33 about A,

$$25 \times AD = R_B' \times BC = R_B' \times 12$$

or

$$25 \times 2.5 = R_B' \times 12 \qquad (\because\ AD = 2.5)$$

$$\therefore \quad R_B' = \frac{2.5 \times 2.5}{12} = 5.21 \text{ N}$$

Equating horizontal forces,

$$F_A' = R_B' = 5.21 \text{ N}$$

Hence for equilibrium, the force of friction required is 5.21 N. But maximum amount of force of friction available is 7.5 N which is more than the required amount. Hence the ladder *will remain in equilibrium in the given position.*

Problem 5.25. *The weight of 14 m long bar as shown in Fig. 5.33 (a) is 600 N and it may be considered to be concentrated at a point 6 m from the bottom. It rests against a smooth vertical wall at A and on a rough horizontal floor at B. The coefficient of static friction between the bar and the floor is 1/3. Establish by calculations if the bar would stand in the 60° position as shown.* (AMIE Summer, 1989)

Sol. Given :

Length, $AB = 14$ m

Weight, $W = 600$ N

The weight is acting at D, which is 6 m from point B.

∴ Length $BD = 6$ m and length $AD = 8$ m

Vertical wall is smooth and hence $F_A = 0$

Coefficient of friction, $\mu = 1/3$

Angle $CBA = 60°$

Let us find :

(*i*) actual force of friction available at point B, and

(*ii*) force of friction required for equilibrium at point B.

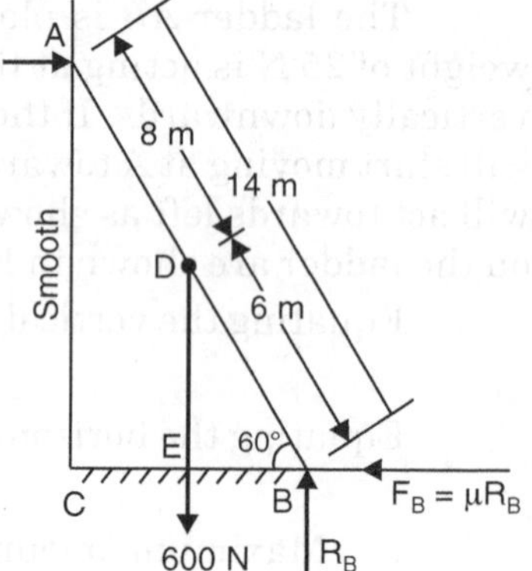

Fig. 5.33 (a)

The force of friction required for equilibrium will be obtained from moment equation *i.e.*, $\Sigma M = 0$. The actual force of friction available will be obtained from summation of forces in horizontal and vertical directions *i.e.*, $\Sigma F_x = 0$ and $\Sigma F_y = 0$.

Let R_A = Normal reaction at A

R_B = Normal reaction at B

F_B = Force of friction at $B = \mu R_B$

The forces acting on the ladder are shown in Fig. 5.33 (*a*).

Resolving forces horizontally, $R_A = F_B$ ...(*i*)

Resolving forces vertically, $R_B = 600$ N ...(*ii*)

Now $$F_B = \mu \times R_B = \frac{1}{3} \times 600 = 200 \text{ N}$$

∴ Actual force of friction available at point B

$$= F_B = 200 \text{ N}$$

To find the force of friction required for equilibrium, take the moments of all forces about point B.

∴ $$600 \times BE = R_A \times AC \quad ...(iii)$$

But $$BE = BD \cos 60° = 6 \times \frac{1}{2} = 3 \text{ m}$$

and $$AC = AB \sin 60° = 14 \times 0.866 = 12.124 \text{ m}$$

Substituting these values in equation (*iii*), we get

$$600 \times 3 = R_A \times 12.124$$

$$\therefore \qquad R_A = \frac{600 \times 3}{12.124} = 148.46 \text{ N}$$

But from equation (*i*),

$$F_B = R_A = 148.46 \text{ N}$$

∴ Force of friction required for equilibrium = 148.46 N.

As the actual force of friction available at B is 200 N, which is more than the force of friction required for equilibrium, the ladder will stand stable in the 60° position as shown in the given figure. **Ans.**

Problem 5.26. (S.I. Units). *A uniform ladder of weight 850 N and of length 6 m rests on a horizontal ground and leans against a smooth vertical wall. The angle made by the ladder with the horizontal is 65°. When a man of weight 750 N stands on the ladder at a distance 4 m from the top of the ladder, the ladder is at the point of sliding. Determine the coefficient of friction between the ladder and the floor.*

Sol. Given :

Weight of ladder, $W = 50$ N

Length of ladder, $L = AB = 6$ m

Angle made by ladder with horizontal,

$$\alpha = 65°$$

Weight of man, $W_1 = 750$ N

Distance of man from top of the ladder,

$$L_1 = 4 \text{ m}$$

∴ Distance of man from the foot of ladder,

$$L_2 = L - L_1 = 6 - 4 = 2 \text{ m}.$$

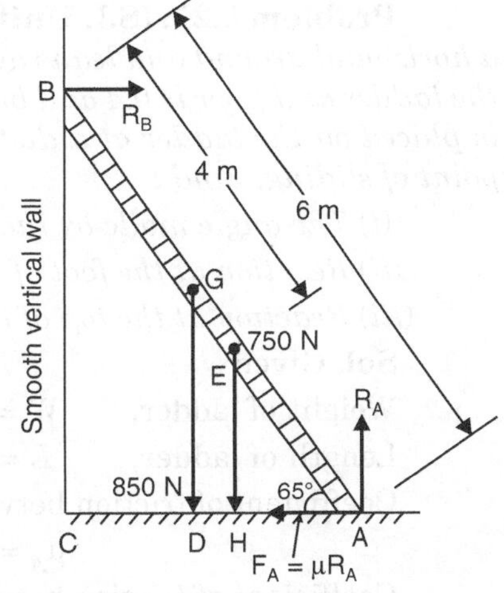

Fig. 5.34

Let μ = Coefficient of friction between the ladder and floor. Vertical wall is smooth and hence there will be no force of friction between the ladder and vertical wall.

Let AB is the ladder and G is the middle point of the ladder at which the weight 850 N is acting. The man of weight 750 N is standing at E. At this position, the ladder is at the point of sliding. This means that ladder at A will start moving towards right. Hence a force of friction $F_A = \mu R_A$ will be acting towards left as shown in Fig. 5.34.

Now R_A = Normal reaction at A

R_B = Normal reaction at B.

The forces acting on the ladder are shown in Fig. 5.34.

Resolving forces vertically,

$$R_A = 850 + 750 = 1600 \text{ N} \qquad \text{...(i)}$$

Resolving forces horizontally,

$$R_B = F_A = \mu R_A = \mu \times 1600 = 1600 \mu\text{N} \qquad \text{...(ii)}$$

In right-angled ΔABC, $BC = AB \sin 65° = 6 \sin 65°$

$$= 6 \times 0.9063 = 5.437 \text{ m}$$

$$AC = AB \cos 65° = 6 \cos 65° = 6 \times 0.4226 = 2.5357 \text{ m}$$

As G is the middle point of AB and GD is normal to AC,

$\therefore$ D is the middle point of AC

$$\therefore \quad AD = \frac{AC}{2} = \frac{2.5357}{2} = 1.267 \text{ m.}$$

Now $AH = AE \cos 65° = (AB - BE) \cos 65° = (6 - 4) \cos 65°$

$= 2 \times \cos 65° = 2 \times 0.4226 = 0.8452$ m.

Taking the moments of all forces about A, we get

$$R_B \times BC = 850 \times AD + 750 \times AH.$$

Substituting the values of BC, AD, AH and R_B from equation (*ii*),

$$1600\ \mu \times 5.437 = 850 \times 1.267 + 750 \times 0.8452$$

or $8699.2\ \mu = 1076.95 + 633.9 + 1710.85$

$$\therefore \quad \mu = \frac{1710.85}{8699.24} = \mathbf{0.199.} \quad \textbf{Ans.}$$

Problem 5.27. (S.I. Units). *A uniform ladder of weight 200 N of length 4.5 m rests on a horizontal ground and leans against a rough vertical wall. The coefficient of friction between the ladder and floor is 0.4 and between ladder and vertical wall is 0.2. When a weight of 900 N is placed on the ladder at a distance of 1.2 m from the top of the ladder, the ladder is at the point of sliding. Find :*

(*i*) *The angle made by the ladder with horizontal,*

(*ii*) *Reaction at the foot of the ladder, and*

(*iii*) *Reaction at the top of the ladder.*

Sol. Given :

Weight of ladder, $W = 200$ N

Length of ladder, $L = AB = 4.5$ m

Coefficient of friction between ladder and floor,

$\mu_A = 0.4.$

Coefficient of friction between ladder and wall,

$\mu_B = 0.2.$

Weight on ladder, $W_1 = 900$ N

Distance $BE = 1.2$ m.

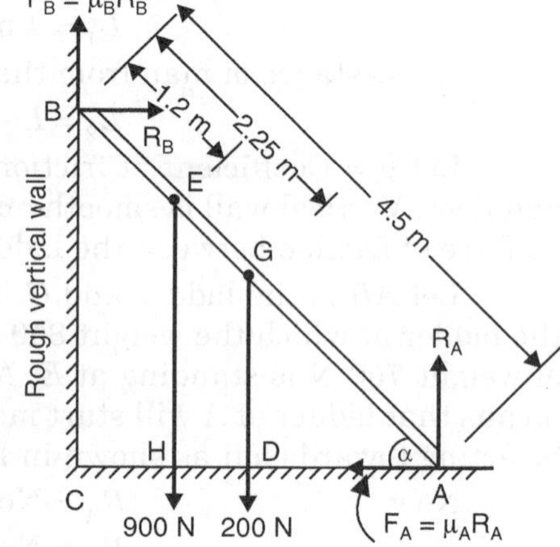

Fig. 5.35

Let α = Angle made by ladder with horizontal

R_A = Normal reaction at A

F_A = Force of friction at $A = \mu_A R_A$

R_B = Normal reaction at B

F_B = Force of friction at $B = \mu_B R_B$.

When the ladder AB is on the point of sliding, the foot of the ladder will move towards right and hence a force of friction $F_A = \mu_A R_A$ will be acting towards left. The top of the ladder will be moving downwards on the vertical wall and hence a force of friction $F_B = \mu_B R_B$ will be acting upwards as shown in Fig. 5.35.

From right-angled triangle ABC,

$$BC = AB \sin \alpha = 4.5 \sin \alpha \quad (\because \ AB = 4.5 \text{ m})$$

$$AC = AB \cos \alpha = 4.5 \cos \alpha$$

$$AD = AG \cos \alpha = \frac{AB}{2} \cos \alpha = \frac{4.5}{2} \cos \alpha$$

$$= 2.25 \cos \alpha \qquad \left(\because \; AG = \frac{AB}{2}\right)$$

$$AH = AE \cos \alpha = (AB - BE) \cos \alpha = (4.5 - 1.2) \cos \alpha$$

$$= 3.3 \cos \alpha.$$

Forces acting on the ladder are shown in Fig. 5.35.

Resolving forces vertically,

$$R_A + F_B = 900 + 200 = 1100 \text{ N}$$

or $$R_A + \mu_B R_B = 1100 \qquad (\because \; F_B = \mu_B R_B)$$

or $$R_A + 0.2 R_B = 1100 \qquad ...(i)$$

Resolving forces horizontally, $R_B = F_A = \mu_A R_A = 0.4 R_A$ ...(*ii*)

Substituting the value of R_B in equation (*i*), we get

$$R_A + 0.2 \times 0.4 R_A = 1100$$

or $$R_A + 0.08 R_A = 1100$$

or $$1.08 R_A = 1100$$

$$\therefore \quad R_A = \frac{1100}{1.08} = 1018.52 \text{ N} \qquad ...(iii)$$

Substituting this value of R_A in equation (*ii*), we get

$$R_B = 0.4 \times 1018.52 = 407.41 \text{ N} \qquad ...(iv)$$

Now taking moments of all forces about A, we get

$$200 \times AD + 900 \times AH = R_B \times BC + F_B \times AC$$

or $$200 \times 2.25 \cos \alpha + 900 \times 3.3 \cos \alpha = 407.41 \times 4.5 \sin \alpha + \mu_B R_B \times 4.5 \cos \alpha$$

or $$450 \cos \alpha + 2970 \cos \alpha = 1833.345 \sin \alpha + 0.2 \times 407.41 \times 4.5 \cos \alpha$$

$$(\because \; \mu_B = 0.2, R_B = 40.41)$$

or $$3420 \cos \alpha = 1833.345 \sin \alpha + 366.67 \cos \alpha$$

or $$3420 \cos \alpha - 366.67 \cos \alpha = 1833.345 \sin \alpha$$

or $$3053.33 \cos \alpha = 1833.345 \sin \alpha$$

or $$\frac{\sin \alpha}{\cos \alpha} = \frac{3053.33}{1833.345} = 1.665$$

or $$\tan \alpha = 1.665$$

$$\therefore \quad \alpha = \tan^{-1} 1.665 = \mathbf{59° 65'}. \quad \textbf{Ans.}$$

Problem 5.28. *A ladder 5 m long and 250 N weight is placed against a vertical wall in a position where its inclination to the vertical is 30°. A man weighing 800 N climbs the ladder. At what position will he induce slipping ? The coefficient of friction for both the contact surfaces of the ladder i.e., with the wall and the floor is 0.2.* (AMIE Summer, 1983)

Sol. Given :

Length of ladder, $AB = 5$ m

Angle made by ladder with vertical = 30°

$$\therefore \quad \angle ABC = 30°$$

and $$\angle BAC = 60°$$

Weight of ladder = 250 N

Weight of man = 800 N

Coefficient of friction between ladder and floor = 0.2.

Also coefficient of friction between ladder and wall = 0.2

Let x = The distance climbed by the man up the ladder from A when the ladder is on the point of slipping.

AB is the ladder and G is the middle point of the ladder at which the weight 250 N is acting. The man of weight 800 N is standing at E, at a distance x from A, when the ladder is on the point of slipping. This means the ladder at A will start moving towards right and a force of friction $F_A = \mu R_A$ will be acting towards left at A. The ladder at B will be moving downwards and hence a force of friction $F_B = \mu R_B$ will be acting upwards at B.

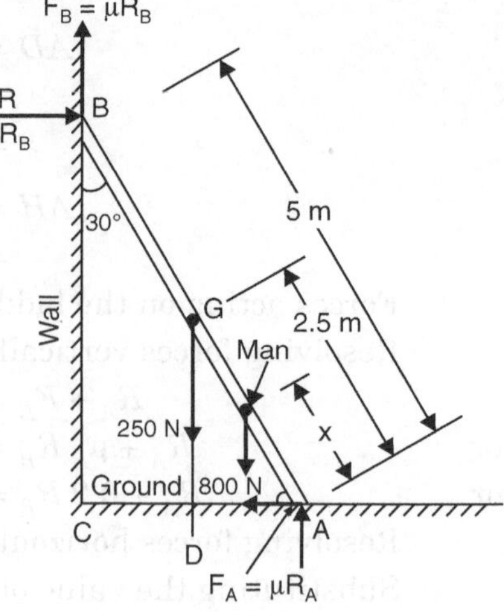

Fig. 5.36

Now R_A = Normal reaction at A

R_B = Normal reaction at B.

The forces acting on the ladder are shown in Fig. 5.36.

Resolving forces vertically, we get

$$R_A + \mu R_B = 250 + 800 = 1050$$

or $$R_A + 0.2\, R_B = 1050 \qquad (\because \ \mu = 0.2) \qquad ...(i)$$

Resolving forces horizontally, we get

$$R_B = \mu R_A = 0.2\, R_A \qquad ...(ii)$$

Substituting the value of R_B in equation (i), we get

$$R_A + 0.2\,(0.2\, R_A) = 1050$$

or $$R_A + 0.04\, R_A = 1050$$

or $$1.04\, R_A = 1050$$

$$\therefore \quad R_A = \frac{1050}{1.04} = 1009.6 \text{ N}$$

Substituting this value of R_A in equation (ii), we get

$$R_B = 0.2 \times 1009.6 = 201.92 \text{ N}$$

In right-angled triangle AGD,

$$AD = AG \cos 60° = 2.5 \times \frac{1}{2} = 1.25 \text{ m}$$

In right-angled triangle AEH,

$$AH = x \cos 60° = x \times \frac{1}{2} = \frac{x}{2}$$

Taking moments of all forces about A,

$$800 \times AH + 250 \times AD = R_B \times BC + F_B \times AC \qquad ...(iii)$$

where $$BC = AB \cos 30° = 5 \times 0.866 = 4.33 \text{ m}$$

$$AC = AB \cos 60° = 5 \times \frac{1}{2} = 2.5 \text{ m}$$

$$F_B = \mu R_B = 0.2 \times 201.92 = 40.384 \text{ N} \qquad (\because \ R_B = 20.192 \text{ N})$$

Substituting the above values in equation (*iii*), we get

$$800 \times \frac{x}{2} + 250 \times 1.25 = 201.92 \times 4.33 + 40.384 \times 2.5$$

or $$40x + 31.25 = 87.43 + 10.096$$

or $$40x = 87.43 + 10.096 - 31.25 = 66.276$$

$\therefore$ $$x = \frac{66.276}{40} = 1.6569 \simeq \mathbf{1.657\ m.}\quad \textbf{Ans.}$$

5.5. WEDGE FRICTION

A wedge is a piece of metal or wood which is usually of a triangular or trapezoidal in cross-section. It is used for either lifting loads or used for slight adjustments in the position of a body *i.e.*, for tightening fits or keys for shafts.

When lifting a heavy load the wedge is placed below the load and a horizontal force P is applied as shown in Fig. 5.37. If the force P is just sufficient to lift the load, the wedge will move towards left and load will move up. When the wedge moves towards left, the sliding of the surfaces AC and AB will take place. At the same time load moves up and sliding of the load takes place along GD. Thus for the wedge and load shown in Fig. 5.37 sliding takes place along surfaces AB, AC and DG. Hence there will be three normal reactions at AB, AC and DG.

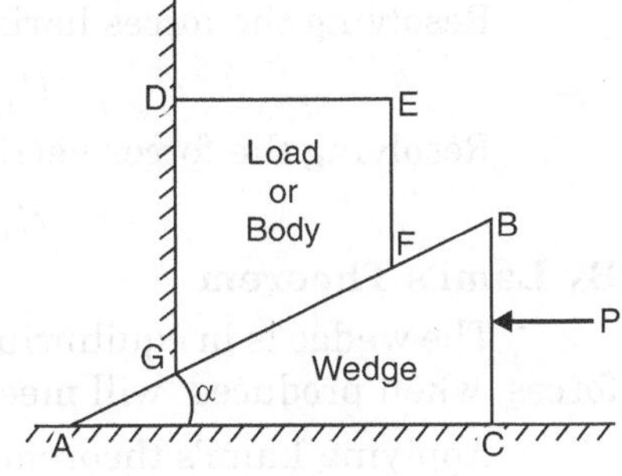

Fig. 5.37

The problems on wedges are generally the problems of equilibrium on inclined planes. Therefore, these problems are solved by equilibrium method or by applying Lami's Theorem.

Equilibrium Method. In this method, the equilibrium of the load (or the body placed on the wedge) and the equilibrium of the wedge are considered.

Equilibrium of Wedge

Consider the equilibrium of the wedge. The forces acting on the wedge are shown in Fig. 5.37(*a*). They are :

(*i*) The force P applied horizontally on face BC.

(*ii*) Reaction R_1 on the face AC (The reaction R_1 is the resultant of the normal reaction on the rubbing face AC and force of friction on surface AC). The reaction R_1 will be inclined at an angle ϕ_1 (when ϕ_1 is angle of friction) with the normal.

(*iii*) Reaction R_2 on the face AB (The reaction R_2 is the resultant of normal reaction on the rubbing face AB and force of friction on surface AB). The reaction R_2 will be inclined at an angle ϕ_2 with the normal.

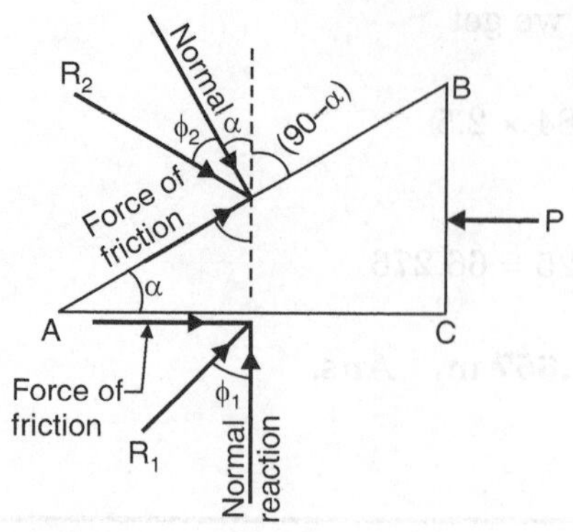

Fig. 5.37(*a*). Equilibrium of wedge Fig. 5.37(*b*)

When the force *P* is applied on the wedge, the surface *CA* will be moving towards left and hence force of friction on this surface will be acting towards right. Similarly, the force of friction on face *AB* will be acting from *A* to *B*. These forces are shown in Fig. 5.37(*a*).

Resolving the forces horizontally, we get

$$R_1 \sin \phi_1 + R_2 \sin (\phi_2 + \alpha) = P.$$

Resolving the forces vertically, we get

$$R_1 \cos \phi_1 = R_2 \cos (\phi_2 + \alpha).$$

By Lami's Theorem

The wedge is in equilibrium under the action of three forces namely R_1, R_2 and *P*. These forces, when produced, will meet at a point as shown in Fig. 5.37(*b*).

Applying Lami's theorem, we get

$$\frac{P}{\sin (180 - \phi_1 - \phi_2 - \alpha)} = \frac{R_1}{\sin (90 + \alpha + \phi_2)}$$

$$= \frac{R_2}{\sin (90 + \phi_1)}$$

Equilibrium of Body placed on the Wedge

The forces acting on the body are shown in Fig. 5.37(*c*). They are :

(*i*) The weight *W* of the body.

(*ii*) Reaction R_3 on the face *GD*. (The reaction R_3 is the resultant of the normal reaction N_3 on the rubbing face *GD* and force of friction on surface *GD*).

(*iii*) Reaction R_2 on the face *GF* (The reaction R_2 is the resultant of the normal reaction on the rubbing face *GF* and force of friction on surface *GF*).

These forces are shown in Fig. 5.37 (*c*).

Resolving the forces R_2, R_3 and *W* horizontally, we get

$$R_3 \cos \phi_3 = R_2 \sin (\alpha + \phi_2)$$

Resolving forces vertically,

$$W + R_3 \sin \phi_3 = R_2 \cos (\alpha + \phi_2)$$

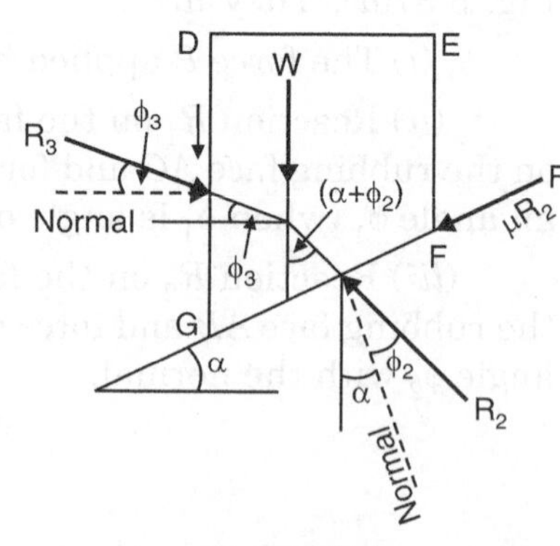

Fig. 5.37(*c*)

By Lami's Theorem

The forces R_3, R_2 and W are produced to meet at a point as shown in Fig. 5.37(*c*).

The body is in equilibrium under the action of these three forces.

Hence applying Lami's theorem, we get

$$\frac{W}{\sin(90+\phi_3+\alpha+\phi_2)} = \frac{R_2}{\sin(90-\phi_3)} = \frac{R_3}{\sin[180-(\alpha+\phi_1)]}.$$

Problem 5.29. *A block over lying a 10° wedge on a horizontal floor and leaning against a vertical wall and weighing 1500 N is to be raised by applying a horizontal force to the wedge. Assuming co-efficient of friction between all the surfaces in contact to be 0.3, determine the minimum horizontal force to be applied to raise the block.*

Sol. Given :

Angle of the wedge, $\alpha = 10°$

Weight of the block, $W = 1500$ N

Co-efficient of friction for all surfaces in contact,

$\mu = 0.3$

But $\mu = \tan\phi = 0.3$

$\therefore$ $\phi = \tan^{-1} 0.3 = 16° 42'$

Let P = Minimum horizontal force applied on wedge to raise the block.

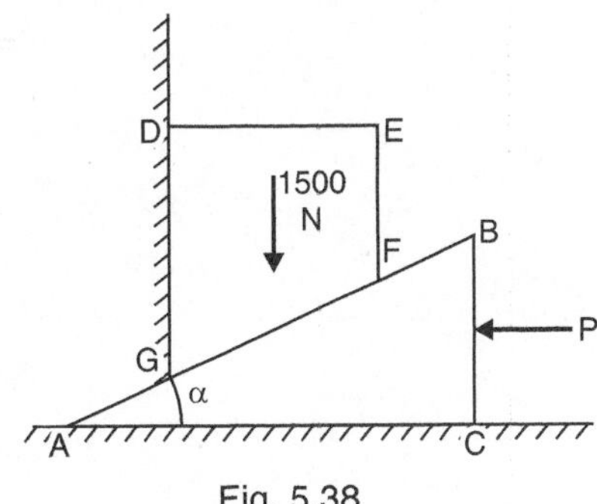

Fig. 5.38

Fig. 5.38 shows the block which is to be raised by the wedge.

1st Method

Considering the equilibrium of the block* as shown in Fig. 5.39(*a*).

Let R_2 = Reaction on face *GF*

R_3 = Reaction on face *GD*

Resolving the forces acting on the block horizontally, we get

$$R_3 \cos\phi = R_2 \sin(\alpha+\phi)$$

or $$R_3 \cos(16° 42') = R_2 \sin(10° + 16° 42')$$

or $$R_2 \times 0.9578 = R_3 \times 0.4493$$

$\therefore$ $$R_2 = \frac{0.9578}{0.4493} R_3 = 2.1317 R_3 \qquad ...(i)$$

Now resolving the forces vertically

$$R_3 \sin\phi + W = R_2 \cos(\alpha+\phi)$$

or $$R_3 \sin(16° 42') + 1500 = R_2 \cos(10 + 16° 42')$$

or $$R_3 \times 0.2874 + 1500 = R_2 \times 0.8934$$

$$= 2.1317 \times R_3 \times 0.8934 \qquad (\because \text{ From } (i), R_2 = 2.1317 R_3)$$

or $$0.2874 R_3 + 1500 = 1.9044 R_3$$

or $$1.9044 R_3 - 0.2874 R_3 = 1500$$

or $$(1.9044 - 0.2874) R_3 = 1500$$

or $$1.617 R_3 = 1500$$

---

*If the value of load is given, then consider the equilibrium of block first. But if the value of *P* is given, then consider the equilibrium of wedge first.

$$\therefore \qquad R_3 = \frac{1500}{1.617} = 927.54 \text{ N} \qquad \text{...(ii)}$$

From equation (*ii*), $R_2 = 2.1317\ R_3$

$$= 2.1317 \times 927.64 = 1977.45 \text{ N} \qquad \text{...(iii)}$$

Now considering the equilibrium of the wedge as shown in Fig. 5.39(*b*).

Resolving the forces vertically, we get

$$R_2 \cos(\alpha + \phi) = R_1 \cos \phi$$

or

$$R_1 = \frac{R_2 \cos(\alpha + \phi)}{\cos \phi}$$

$$= 1977.45 \times \frac{\cos(10^\circ + 16^\circ 42')}{\cos(16^\circ 42')}$$

$$= 1977.45 \times \frac{0.8943}{0.9578} = 1844.49 \text{ N}$$

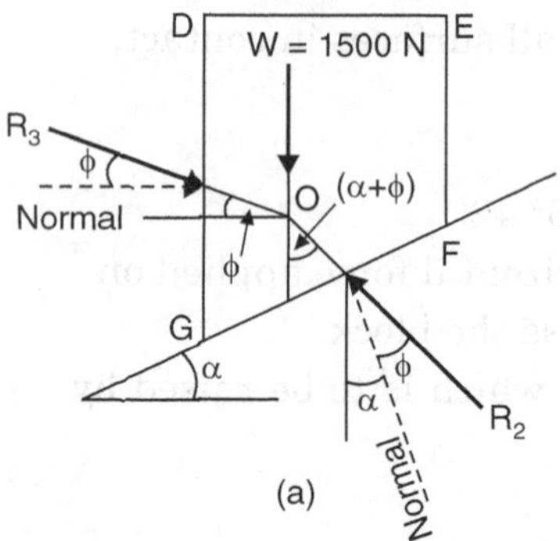

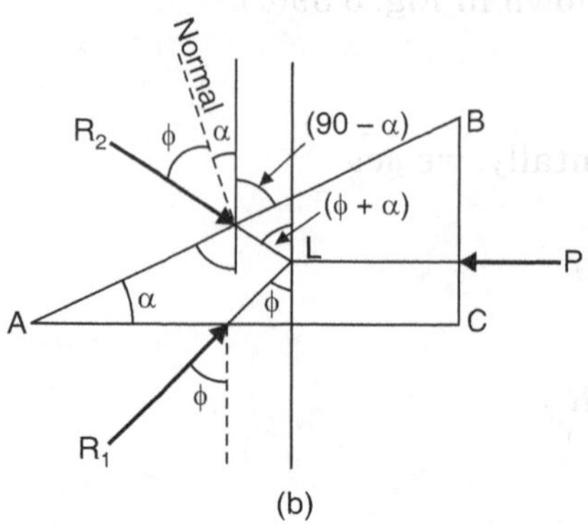

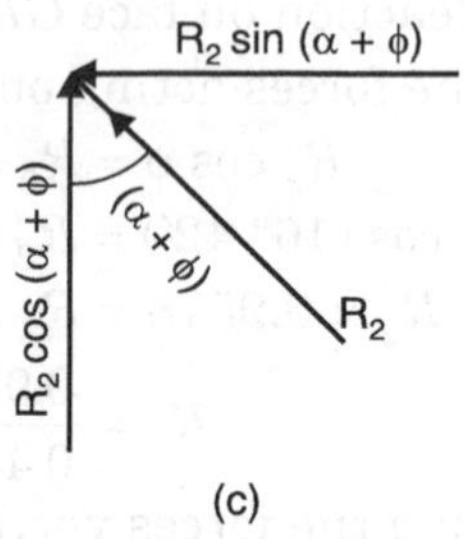

Fig. 5.39

Resolving the forces horizontally, we get

$$P = R_1 \sin \phi + R_2 \sin(\alpha + \phi)$$

$$= 1844.49 \times \sin(16^\circ 42') + 1977.45 \times \sin(10^\circ + 16^\circ 42')$$

$$= 1844.49 \times 0.2874 + 1977.45 \times 0.4493$$

$$= 530.106 + 888.468 = \mathbf{1418.57\ N.} \quad \textbf{Ans.}$$

2nd Method by Lami's Theorem

The block shown in Fig. 5.39 (*a*) is in equilibrium under the action of three forces W, R_3 and R_2. These forces when produced are meeting at the point O.

Applying Lami's theorem to the point O, we get

$$\frac{W}{\sin(\phi + 90 + \alpha + \phi)} = \frac{R_3}{\sin[180 - (\alpha + \phi)]} = \frac{R_2}{\sin(90 - \phi)}$$

or
$$\frac{1500}{\sin(16° 42' + 90 + 10° + 16° 42')} = \frac{R_2}{\sin[180 - (10° + 16° 42')]}$$

$$= \frac{R_2}{\sin(90 - 16° 42')}$$

or
$$\frac{1500}{\sin(133° 24')} = \frac{R_3}{\sin(153° 18')} = \frac{R_2}{\sin(72° 18')}$$

or
$$\frac{1500}{0.7265} = \frac{R_3}{0.4493} = \frac{R_2}{0.9578}$$

$\therefore$
$$R_3 = 1500 \times \frac{0.4493}{0.7265} = 927.66 \text{ N}$$

and
$$R_2 = 1500 \times \frac{0.9578}{0.7265} = 1977.56 \text{ N.}$$

Now consider the equilibrium of the wedge shown in Fig. 5.39(b). Three forces R_1, R_2 and P when produced are meeting at the point L. Applying Lami's theorem to the point L, we get

$$\frac{R_1}{\sin(90 + \alpha + \phi)} = \frac{R_2}{\sin(90 + \phi)} = \frac{P}{\sin[180 - \phi - (\phi + \alpha)]}$$

or
$$\frac{R_1}{\sin(90 + 10° + 16°42')} = \frac{R_2}{\sin(90 + 16°42')}$$

$$= \frac{P}{\sin[180 - 16° 42' - (16° 42' + 10°)]}$$

or From the last two parts, we get

$$\frac{R_2}{\sin(106°42')} = \frac{P}{\sin(136°36')}$$

$\therefore$
$$P = R_2 \times \frac{\sin(136° 36')}{\sin(106° 42')} = 1977.56 \times \frac{0.687}{0.9578}$$

$$= \mathbf{1418.44.} \quad \textbf{Ans.}$$

5.6. ROLLING RESISTANCE

Fig. 5.40 (a) shows a hard roller moving without slipping on a horizontal surface while supporting a load W at the centre. If the roller is moving with uniform velocity, due to horizontal force P, then some sort of resistance must be present so that the net force in the direction of uniform motion is zero [$\because$ Net force = Mass × Acc. As due to uniform motion of roller, acceleration is zero, hence net force is zero. But net force = P – resisting force]. The resistance to the motion of the roller in such cases is known as rolling resistance.

Hence rolling resistance is the resistance to the motion of the roller moving without slipping on a horizontal surface while supporting a load W at the centre.

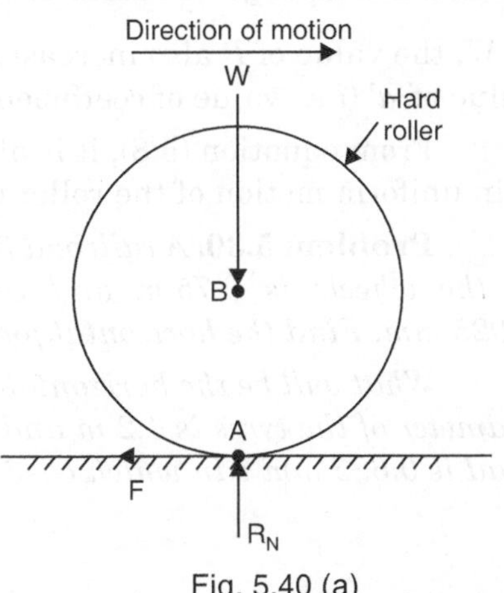

Fig. 5.40 (a)

The rolling resistance can be understood if we imagine the surface to be yielding as shown in Fig. 5.40 (*b*). The same result may be there if the roller is assumed to be yielding and moving over a rigid surface or both the roller and surface are yielding. The ground in front of the roller is depressed, causing the normal reaction (R_N) to act ahead of the line of action of the weight W as shown in Fig. 5.40 (*b*). Now for equilibrium of the roller, the three forces *i.e.*, W, P and R_N must be concurrent. As W and P intersect at B, hence R_N will also pass through B, making an angle ϕ with the vertical direction as shown in Fig. 5.40 (*b*).

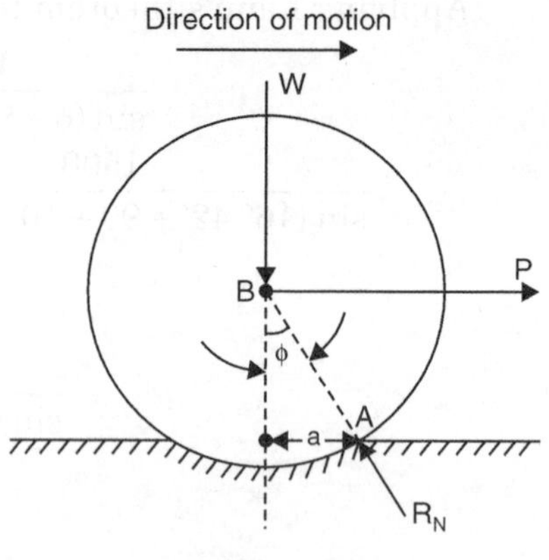

Direction of motion
W
P
B
φ
a
A
R_N

Fig. 5.40 (b)

Now for equilibrium of the roller,

$$\Sigma F_x = 0 \quad \text{or} \quad P = R_N \sin \phi \qquad \ldots(i)$$

and
$$\Sigma F_y = 0 \quad \text{or} \quad W = R_N \cos \phi \qquad \ldots(ii)$$

To eliminate R_N from equations (*i*) and (*ii*), divide (*ii*) by equation (*i*),

$$\therefore \qquad \frac{P}{W} = \frac{\sin \phi}{\cos \phi} = \tan \phi \qquad \ldots(iii)$$

Generally the angle ϕ is very small, hence $\tan \phi \simeq \sin \phi$. But $\sin \phi$ from Fig. 5.40 (*b*), is equal to $\frac{a}{R}$, where R = Radius of roller.

$$\therefore \qquad \tan \phi \simeq \sin \phi = \frac{a}{r}$$

Substituting the value of $\tan \phi$ in equation (*iii*), we get

$$\frac{P}{W} = \frac{a}{R}$$

or
$$P = \frac{W \times a}{R} \qquad \ldots(5.8)$$

The distance 'a' in the above equation is known as *coefficient of rolling resistance*. For a given material of given radius of roller, the ratio $\frac{P}{W}$ is constant. This means with the increase of W, the value of P also increases. Also for a given material of constant radius of roller, the value of 'a' (*i.e.*, value of coefficient of rolling resistance) is constant.

From equation (5.8), it is also clear that for a given load W, the force required to maintain uniform motion of the roller decreases as the radius of roller increases.

Problem 5.30. *A railroad freight car is having a weight of 80 metric tons. The diameter of the wheels is 0.75 m and coefficient of rolling resistance between wheel and track is 0.025 mm. Find the horizontal force required to maintain uniform speed.*

What will be the horizontal force for the truck and trailer of weight 80 metric tons if the diameter of the tyres is 1.2 m and coefficient of rolling resistance between the truck tyres and road is 0.625 mm ? In which case, the horizontal force is minimum ?

Sol. Given :

Railroad freight car

Weight, W_1 = 80 metric tons,
= 80 × 1000 kgf
= 80 × 1000 × 9.81 N

Dia., D_1 = 0.75 m = 0.75 × 10^3 mm

∴ Radius, $R_1 = \dfrac{0.75 \times 1000}{2}$ = 0.75 × 500 mm

Coefficient of rolling resistance,

a_1 = 0.025 mm.

Truck and trailer

W_2 = 80 metric tons
= 80 × 1000 × 9.81 N
(∵ 1 ton = 1000 kgf = 1000 × 9.81 N)

D_2 = 1.2 m = 1.2 × 10^3 mm

$R_2 = \dfrac{1.2 \times 1000}{2}$ = 1.2 × 500 mm

a_2 = 0.625 mm.

Using equation (5.8), we get

$$P_1 = \frac{W_1 \times a_1}{R_1} = \frac{(80 \times 1000 \times 9.81) \times 0.025}{0.75 \times 500}$$

= **52.32 N. Ans.**

$$P_2 = \frac{W_2 \times a_2}{R_2} = \frac{(80 \times 1000 \times 9.81) \times 0.625}{1.2 \times 500}$$

= **817.5 N. Ans.**

In case of railroad freight car, the horizontal force required to maintain uniform speed, is minimum. **Ans.**

5.7. TRANSLATION, ROTATION AND GENERAL PLANE MOTION

A body may have the following types of motion :

(*i*) Motion of translation,

(*ii*) Motion of rotation, and

(*iii*) Combination of motion of translation and motion of rotation.

5.7.1. Motion of Translation. If a body moves in such a way that all its particles move in parallel planes and travel the same distance, then the body is said to have a motion of translation. When a rigid body is in translation, all the points of the particles of the body have the same velocity and same acceleration at any particular instant. The motion of the rigid link *AB*, from its initial position *AB* to A_1B_1, shown in Fig. 5.41 is an example of motion of translation.

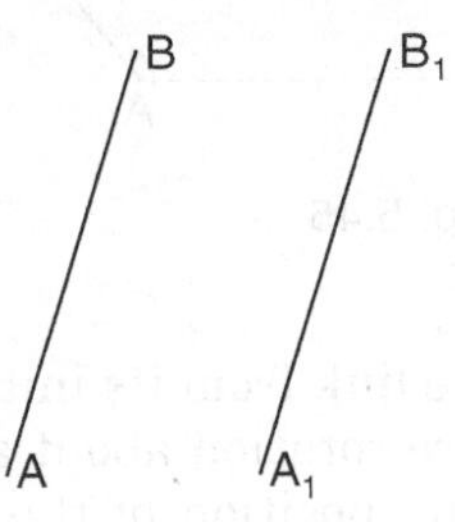

Fig. 5.41

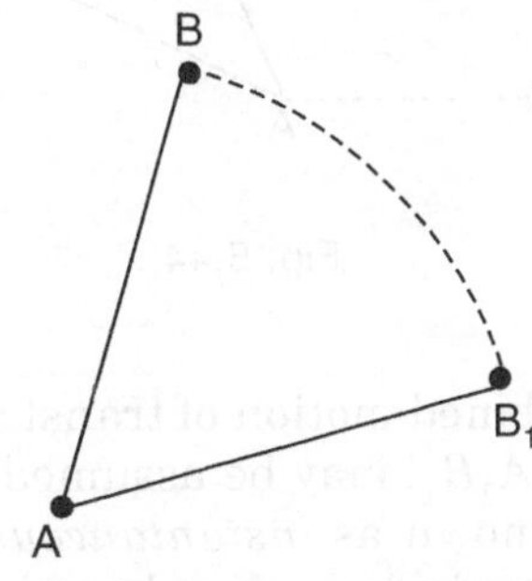

Fig. 5.42

5.7.2. Motion of Rotation. If a body rotates about a fixed point in such a way that all its particle move in a circular path, the body is said to have the motion of rotation. The fixed point about which the body rotates is called the point of rotation and the axis, passing through the fixed point, is known as axis of rotation. The particles, lying on the axis of rotation, have zero velocity and zero acceleration. The motion of the link AB, from its initial position AB to AB_1, shown in Fig. 5.42 is an example of motion of rotation.

5.7.3. General Plane Motion. General plane motion of a rigid body is the sum of the translation of the body and rotation of the body. The examples of general plane motion are rolling wheels connecting rod of a reciprocating engine, links joining rotating elements of a machine. Hence

General Plane motion = Motion of translation

+ Motion of rotation.

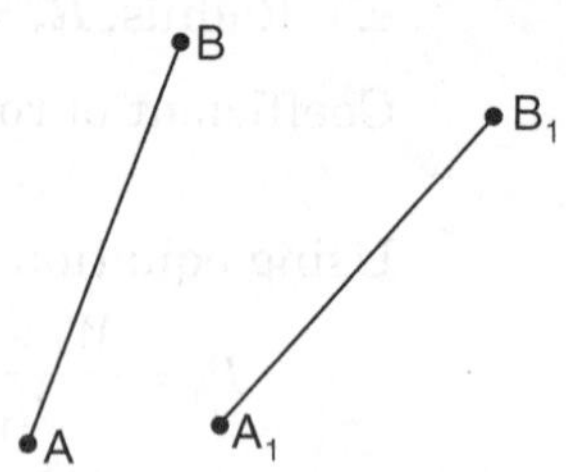

Fig. 5.43

Consider a link AB, which moves from its initial position AB to A_1B_1 in a short interval of time as shown in Fig. 5.43. The link has neither entire motion of translation nor entire rotation, but a combination of the two. The motion of the link from the position AB to the position A_1B_1 may be regarded as to consist of :

1st Case

(*i*) A motion of entire translation from the position AB to the position A_1B^* so that A_1B^* is parallel to AB as shown in Fig. 5.44, and

(*ii*) A motion of entire rotation about A_1 from the position A_1B^* to the position A_1B_1.

2nd Case

(*i*) A motion of entire rotation about A from the position AB to the position AB^* as shown in Fig. 5.45 and

(*ii*) A motion of entire translation from the position AB^* to the position A_1B_1.

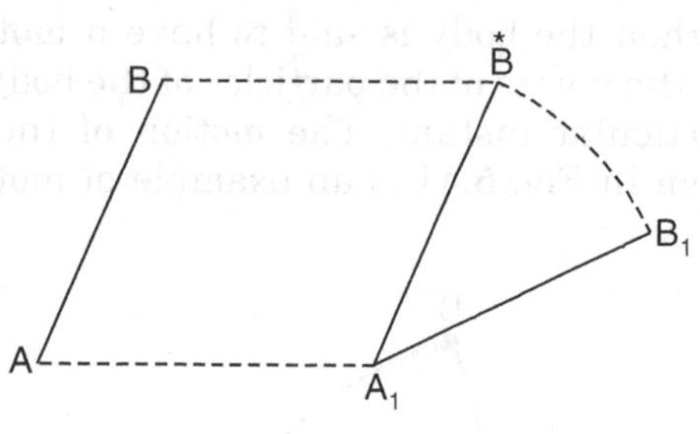

Fig. 5.44

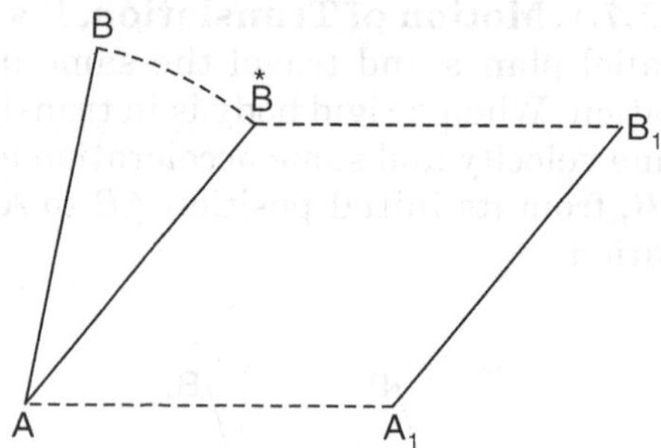

Fig. 5.45

3rd Case

The combined motion of translation and rotation of the link from its initial position AB to the position A_1B_1, may be assumed to be a motion of entire rotation about a certain point. This point is known as *instantaneous centre of rotation.* The position of the instantaneous centre of rotation is determined as given below :

Let the link AB in a short interval of time changes its position from AB to A_1B_1. The point A of the link has moved to point A_1 where the point B of the link has moved to the point B_1 as shown in Fig. 5.46. Draw the right bisectors of chord AA_1 and chord BB_1. Let CD is the right bisectors of AA_1 whereas EF is the right bisectors of BB_1. Let these two bisectors meet at the point O. Then the point O is the instantaneous centre of rotation of the link AB. This means the link AB as a whole has rotated about O.

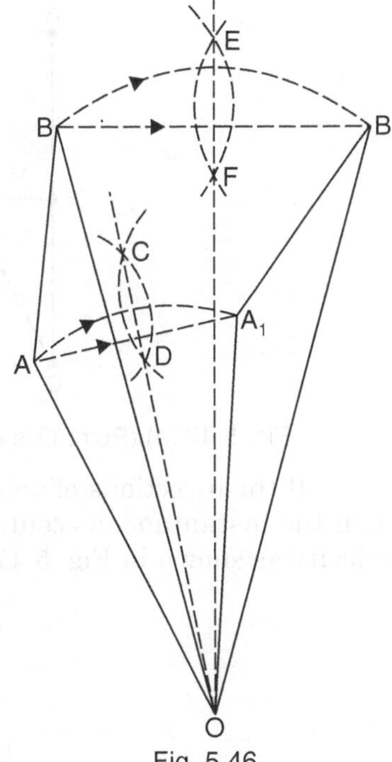

Fig. 5.46

Let V_A = Linear velocity of point A

V_B = Linear velocity of point B

ω = Angular velocity of link AB about O.

This means the angular velocity of point A and point B about O will be ω.

Now, we know

Linear velocity = Angular velocity × r

∴ Linear velocity of point A is given by

V_A = Angular velocity of point A about O × Distance of A from centre of rotation (*i.e.*, from O)

$= \omega \times OA$

$$\therefore \quad \omega = \frac{V_A}{OA} \qquad ...(i)$$

Similarly, the linear velocity of point B is given by

V_B = Angular velocity of point B about O × Distance of B from centre of rotation (*i.e.*, from O)

$= \omega \times BO$

$$\therefore \quad \omega = \frac{V_B}{BO} \qquad ...(ii)$$

Equating (*i*) and (*ii*), we get

$$\frac{V_A}{OA} = \frac{V_B}{BO} \quad \text{or} \quad \frac{V_A}{V_B} = \frac{OA}{BO} \qquad ...(5.9)$$

The direction of the velocity at A will be at right angle to OA whereas the velocity at B will be at right angle to BO.

Thus, if the directions of velocities at A and B are known, then the instantaneous centre of AB is obtained by drawing perpendiculars to the directions of the velocities at A and B. The point, where these two perpendiculars, meet is the instantaneous centre.

If the directions of velcoities at A and B are parallel but unequal in magnitude, then instantaneous centre of AB is obtained by determining the point of intersection of the line AB with the line joining the extremities of the vectors V_A and V_B as shown in Fig. 5.47 (*a*) and 5.47 (*b*). In Fig. 5.47 (*a*), the instantaneous centre O is outside the link AB whereas in Fig. 5.47 (*b*) the instantaneous centre O is on the link AB. Hence instantaneous centre may lie on the link or outside the link.

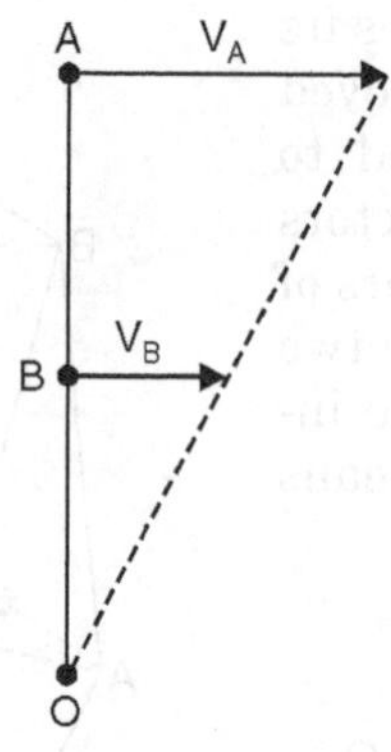

Fig. 5.47 (a) (Point *O* is outside the link *AB*)

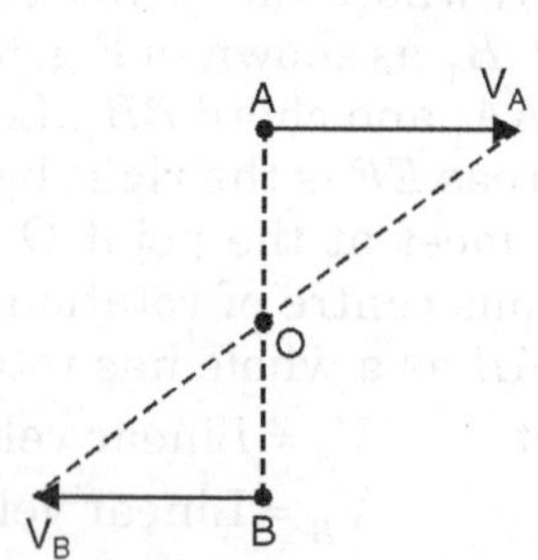

Fig. 5.47 (b) (Point *O* is on the link *AB*)

If the directions of velocities at *A* and *B* are parallel and also are equal in magnitude, then the instantaneous centre is at infinity and all the points of the link *AB* have the same velocity as shown in Fig. 5.47 (*c*). The two parallel lines will meet at infinity.

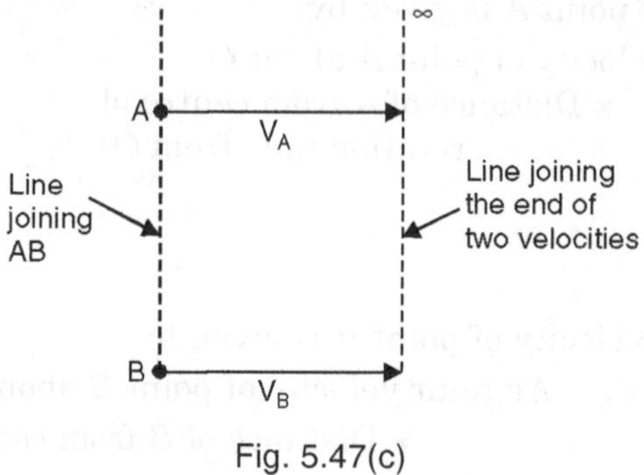

Fig. 5.47(c)

Note:

(*i*) The instantaneous centre is at rest and has zero velocity.

(*ii*) The instantaneous centre may be outside the body or within the body.

(*iii*) If any point on the body or outside the body has zero velocity, then that point will be the instantaneous centre for the body at that instant.

Problem 5.31. *A link AB is moving in a vertical plane. At a certain instant, when the link is inclined at 30° to the horizontal, the point A is moving horizontally at 4 m / s, while B is moving vertically upwards. Find the velocity of B.*

Sol. Given :

Inclination of link *AB* with horizontal = 30°

Horizontal velocity of *A*, V_A = 4 m/s

Let V_B = Velocity of *B* in the vertically upward direction.

Since the actual motions of *A* and *B* are in different directions and are known, the position of the instantaneous centre can be easily determined by drawing perpendiculars to the directions of motions at *A* and *B* as shown in Fig. 5.48.

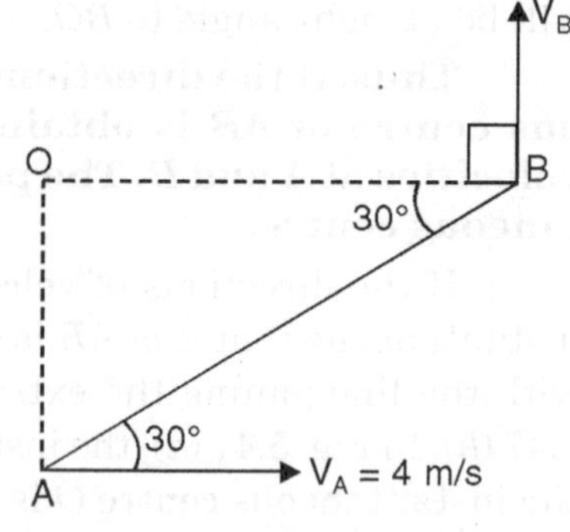

Fig. 5.48

AO is the perpendicular to the direction of motion of *A* whereas *BO* is the perpendicular to the direction of motion of *B*. The point *O*, at which these two perpendiculars meet, is the instantaneous centre of the link *AB*.

$$\left[\text{or} \quad V_A = \omega \times AO \text{ and } V_B = \omega \times BO \therefore \frac{V_A}{V_B} = \frac{AO}{BO}\right]$$

Now using equation (5.9), we get

$$\frac{V_A}{V_B} = \frac{OA}{BO} = \tan 30° \qquad \left(\because \text{ In } \Delta\, AOB, \tan 30° = \frac{AO}{BO}\right)$$

or

$$\frac{4}{V_B} = \frac{1}{\sqrt{3}}$$

$\therefore$ $V_B = 4 \times \sqrt{3} = 4 \times 1.732 =$ **6.928 m/s. Ans.**

Problem 5.32. *A cylinder of radius 1 m rolls without slipping along a horizontal plane AB. Its centre has uniform velocity of 20 m/s. Find the velocity of point E and F on the circumference of the cylinder shown in Fig. 5.49 (a).* (U.P. Tech. University, 2001–2002)

Sol. Given : $r = 1$ m, $V_C = 20$ m/s

As the cylinder rolls without slipping, hence the point of contact *D* with the horizontal plane *AB* at any instant will have zero velocity. The point on the cylinder, which has zero velocity is the instantaneous centre of rotation of the cylinder. Hence point *D* is the instantaneous centre at that instant. Hence the velocity of any other point on the cylinder will be equal to the angular velocity of cylinder multiplied by the distance of that point from instantaneous centre.

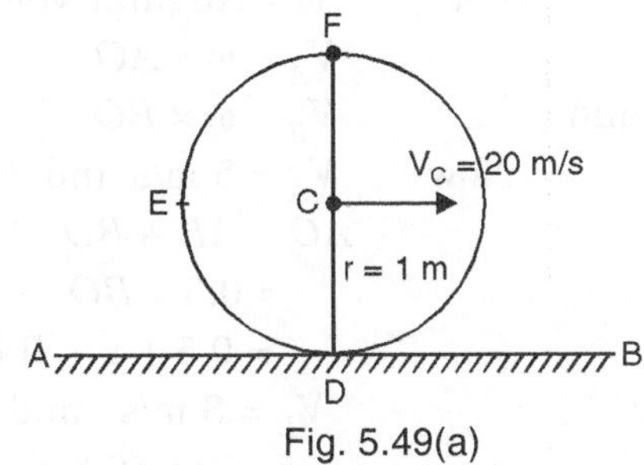

Fig. 5.49(a)

$\therefore$ Velocity of point *E*, $V_E = \omega \times DE$

where ω = Angular velocity of cylinder

and Velocity of point *F*, $V_F = \omega \times DF$

Also velocity of point *C*, $V_C = \omega \times DC$

But $V_C = 20$ m/s and $DC = r = 1$ m (given)

$\therefore$ $20 = \omega \times 1$

or $\omega = 20$ rad/s

$\therefore$ Velocity of *E*, $V_E = \omega \times DE = 20 \times DE$

where $DE = \sqrt{r^2 + r^2} = \sqrt{1^2 + 1^2} = \sqrt{2} = 1.414$ m

$\therefore$ $V_E = 20 \times 1.414 =$ **28.28 m/s**

Fig. 5.49(b)

The direction of velocity V_E is normal to the line *ED* as shown in Fig. 5.49 (*b*).

The velocity of point *F*,

$$V_F = \omega \times FD = 20 \times 2 \qquad (\because \ FD = 2r = 2)$$

= **40 m/s. Ans.**

The direction of velocity V_F will be normal to the line *FD* as shown in Fig. 5.49 (*b*).

Problem 5.33. *A cylindrical roller, 50 cm in diameter is in contact with two conveyor belts at its top and bottom as shown in Fig. 5.50 (a). If the belts run at the uniform speed of 5 m/s and 3 m/s, find linear velocity and angular velocity of the roller.*

(U.P. Tech. University, 2000–2001)

Sol. Given :

Dia. of roller $= 50$ cm

$\therefore$ Radius, $r = 25 \text{ cm} = 0.25 \text{ m}$

$V_A = 5 \text{ m/s}, \quad V_B = 3 \text{ m/s}$

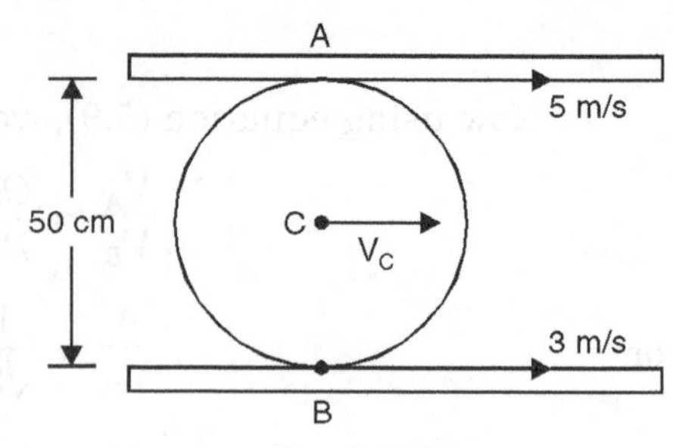

Fig. 5.50(a)

Let us first find the instantaneous centre (O). As the directions of velocities at A and B are parallel and unequal in magnitude, hence the instantaneous centre is obtained by determining the point of intersection of the line AB with the line joining the extremities of the vectors V_A and V_B as shown in Fig. 5.51.

Here $AL = V_A = 5 \text{ m/s}$

$BM = V_B = 3 \text{ m/s}$

Point O is the intersection of lines AB and LM.

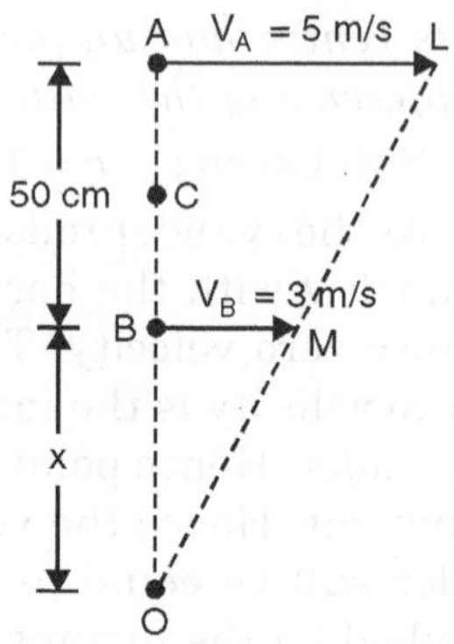

Fig. 5.51

Let ω = Angular velocity of AB or of roller. Then

$V_A = \omega \times AO$

and $V_B = \omega \times BO$

Now $V_A = 5$ m/s and

$AO = AB + BO$

$= 0.5 + BO$

$= 0.5 + x \quad$ [Let $BO = x$]

$V_B = 3$ m/s and $BO = x$

$$\therefore \quad 5 = \omega \times (0.5 + x) = 0.5 \times \omega + \omega \times x \quad \ldots(i)$$

$$3 = \omega \times x \quad \ldots(ii)$$

Subtracting equation (*ii*) from equation (*i*), we get

$$2 = 0.5 \times \omega \quad \text{or} \quad \omega = \frac{2}{0.5} = \mathbf{4 \text{ rad/s.}} \quad \textbf{Ans.}$$

$$\text{Now} \quad V_C = \omega \times OC = \omega(0.25 + x) \quad \ldots(iii) \left[\because \; OC = BC + OB = \frac{0.5}{2} + x = 0.25 + x\right]$$

But x from equation (*ii*),

$$x = \frac{3}{\omega} = \frac{3}{4} = 0.75 \quad (\because \; \omega = 4 \text{ rad/s})$$

Substituting the value of x in equation (*iii*), we get

$V_C = \omega(0.25 + 0.75)$. But $\omega = 4$ rad/s

$= 4(1) = \mathbf{4 \text{ m/s.}}$ **Ans.**

$\therefore$ Angular velocity of roller, $\omega = 4$ rad/s and linear velocity of roller, $V_C = 4$ m/s.

5.8. VELOCITY AND ACCELERATION FOR COMBINED MOTION OF TRANSLATION AND ROTATION OF RIGID BODIES

5.8.1. Analysis of Reciprocating Engine Mechanism. Fig. 5.52 shows a reciprocating engine mechanism in which AB is the connecting rod and BC is the crank. The crank BC is rotating at a uniform angular velocity in the clockwise direction about point C. The point A connected to the piston rod and connecting rod is having to and fro motion in the horizontal plane. The connecting rod is having a combined motion of translation and rotation.

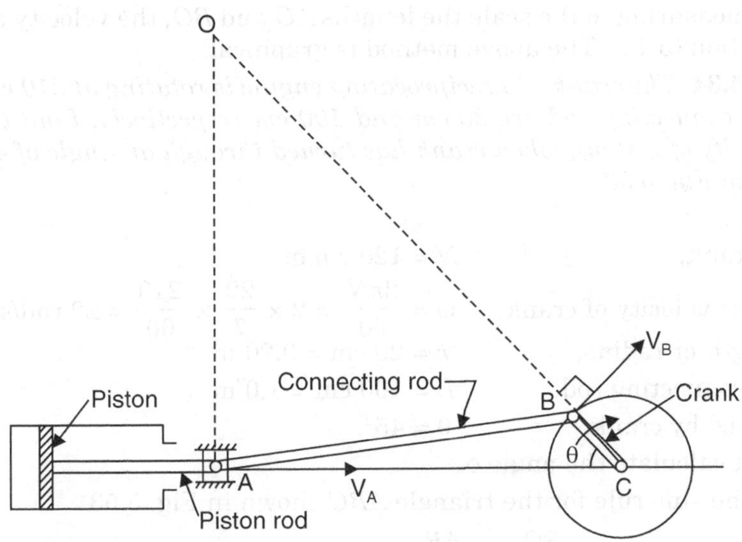

Fig. 5.52

Let N = Revolution of crank in r.p.m.

ω = Angular velocity of crank

r = Radius of crank, *i.e.*, length BC

L = Length of connecting rod AB.

Then $\omega = \dfrac{2\pi N}{60}$

Linear velocity of point $B = V_B = \omega \times BC = \omega \times r$ ...(*i*)

The direction of the linear velocity at B is along the tangent at B to the crank circle. Hence V_B will be acting at right angles to BC as shown in Fig. 5.52. The velocity at A (*i.e.*, V_A) is in the horizontal direction acting along AC for the given position of the crank. Since the directions of V_A and V_B are known, the position of the instantaneous centre can be determined by drawing perpendiculars to the directions of the velocities at A and B. The two perpendiculars meet at O, which is the instantaneous centre for the connecting rod AB. Thus for the given position, the connecting rod can be considered as having a motion of entire rotation about the instantaneous centre *i.e.*, about O.

Let ω_0 = Angular velocity of the connecting rod AB about O.

$\therefore \quad V_A = \omega_0 \times AO$ ...(*ii*)

and $\quad V_B = \omega_0 \times BO$ ...(*iii*)

Equating equations (*i*) and (*iii*), we get

$$\omega \times r = \omega_0 \times BO \quad \text{or} \quad \omega_0 = \frac{\omega \times r}{BO} \qquad ...(5.10)$$

As the values of ω and r are known, hence the value of ω_0 will be obtained if length BO is measured to the scale.

Substituting the value of ω_0 in equation (*ii*), we get the velocity of the point *A* as

$$V_A = \frac{\omega \times r}{BO} \times AO \quad \text{...(5.11)}$$

Hence by measuring to the scale the lengths *AO* and *BO*, the velocity at *A* can be determined from equation (5.11). The above method is graphical.

Problem 5.34. *The crank of a reciprocating engine is rotating at 210 r.p.m. The lengths of the crank and connecting rod are 20 cm and 100 cm respectively. Find the velocity of the point A (i.e., velocity of piston), when crank has turned through an angle of 45° with the horizontal as shown in Fig. 5.53.*

Sol. Given :

Speed of crank, $N = 120$ r.p.m.

$\therefore$ Angular velocity of crank, $\omega = \frac{2\pi N}{60} = 2 \times \frac{22}{7} \times \frac{210}{60} = 22$ rad/s.

Crank length or radius, $r = 20$ cm $= 0.20$ m

Length of connecting rod, $L = 100$ cm $= 1.0$ m

Angle turned by crank, $\theta = 45°$.

Let us first calculate the angle ϕ.

Applying the sine rule for the triangle *ABC* shown in Fig. 5.53

$$\frac{BC}{\sin \phi} = \frac{AB}{\sin \theta}$$

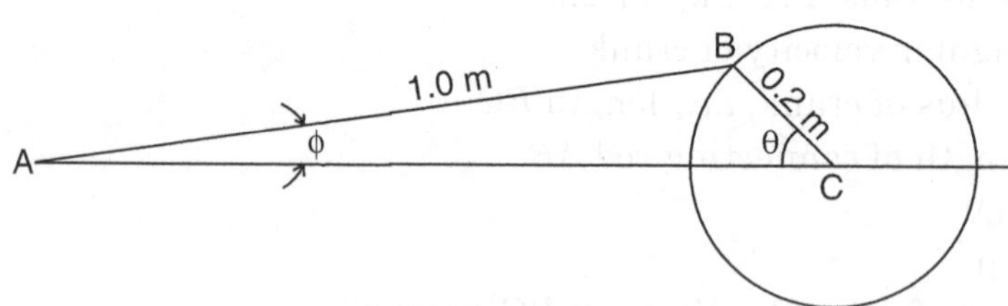

Fig. 5.53

$$\therefore \quad \sin \phi = \frac{BC \sin \theta}{AB} = \frac{0.2}{1.0} \times \sin 45° = \frac{0.2}{1.0} \times 0.7071 = 0.1414$$

$$\therefore \quad \phi = \sin^{-1} 0.1414 = 8.13°.$$

Graphical Method. The velocity of the point *A* is determined graphically as shown in Fig. 5.53 (*a*). The procedure of finding velocity graphically is given below :

1. With *C* as centre, draw a circle of radius 0.2 m.

2. Draw a line *BC* through *C*, making an angle of 45° with the horizontal and cutting the circle at *B*.

3. Draw a horizontal line *AC* through *C*.

4. With *B* as centre, cut-off *AB* equal to 1.0 m.

5. Produce the line *CB* and draw a normal to the line *AC* passing through *A*. The intersection of the normal through *A* and produced line *BC*, give the instantaneous centre *O* for the rod *AB*.

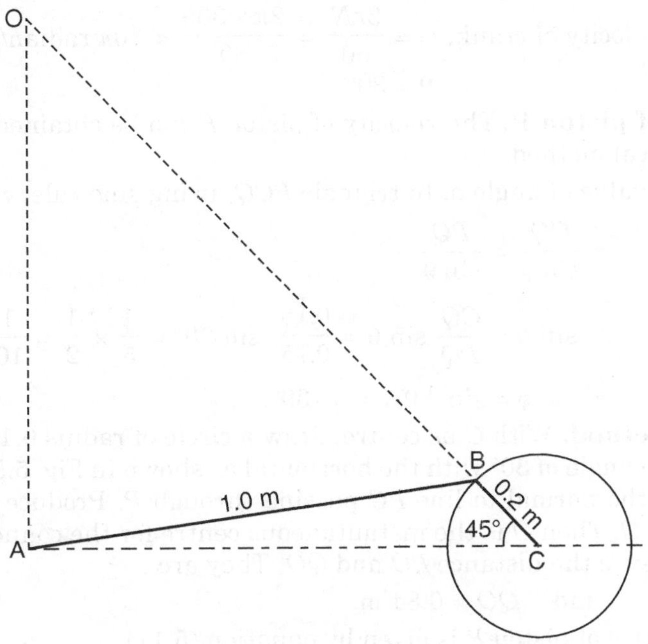

Fig. 5.53 (a)

Measuring lengths AO and BO from the diagram, we get

$$AO = 1.15 \text{ m} \quad \text{and} \quad BO = 1.41 \text{ m}.$$

Now using equation (5.28), we get

$$V_A = \frac{\omega \times r \times AO}{BO} = \frac{22 \times 0.2 \times 1.15}{1.41} = \mathbf{3.58 \text{ m/s.}} \quad \textbf{Ans.}$$

Problem 5.35. *A steam engine has a crank radius of length 15 cm and connecting rod of length 75 cm as shown in Fig. 5.54. The crank CQ rotates in a clockwise direction with a constant speed of 300 r.p.m. Calculate the velocity and acceleration of the piston P at the instant when the angle θ = 30°.*

Sol. Given :

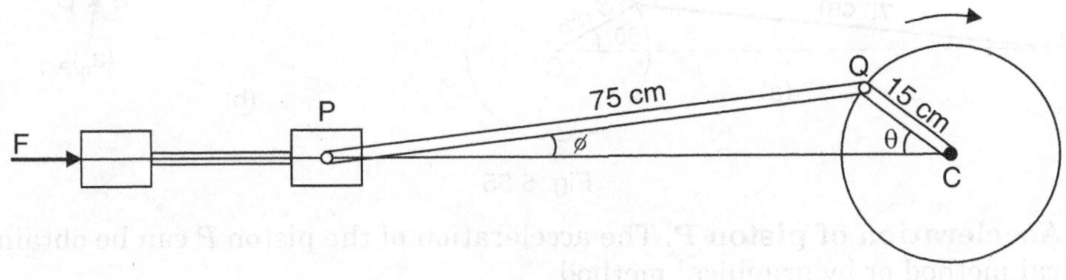

Fig. 5.54

Crank radius, $r = 15 \text{ cm} = 0.15 \text{ m}$

Length of connecting rod, $L = 75 \text{ cm} = 0.75 \text{ m}$

Speed of crank, $N = 300 \text{ r.p.m.}$

$\therefore$ Angular velocity of crank, $\omega = \dfrac{2\pi N}{60} = \dfrac{2\pi \times 300}{60} = 10\pi$ radian/s.

Angle, $\theta = 30°$.

(*i*) **Velocity of piston P.** The velocity of piston *P* can be obtained either by analytical method or by graphical method.

First find the value of angle ϕ. In triangle *PCQ*, using sine rule, we have

$$\frac{CQ}{\sin \phi} = \frac{PQ}{\sin \theta}$$

$$\therefore \quad \sin \phi = \frac{CQ}{PQ} \sin \theta = \frac{0.15}{0.75} \sin 30° = \frac{1}{5} \times \frac{1}{2} = \frac{1}{10} = 0.1$$

$$\therefore \quad \phi = \sin^{-1} 0.1 = 5.739°.$$

Graphical method. With *C* as centre, draw a circle of radius 0.15 m. Draw a line *CQ* through *C* making an angle of 30° with the horizontal as shown in Fig. 5.55. Take the distance *QP* = 0.75 m. Draw the normal to line *PC* passing through *P*. Produce line *CQ* to meet the normal through *P* at *O*. Then *O* is the instantaneous centre for the connecting rod *PQ* for the given position. Measure the distance *PO* and *QO*. They are :

PO = 0.495 m and *QO* = 0.84 m.

Then the velocity of piston *P* is given by equation (5.11).

$$\therefore \quad \text{Velocity of piston} = \frac{\omega \times r \times PO}{QO} \qquad [\because \; AO = PO \text{ and } BO = QO \text{ in eqn. (5.11)}]$$

$$= \frac{10\pi \times 0.15 \times 0.495}{0.84} = \mathbf{2.766\ m/s.} \quad \textbf{Ans.}$$

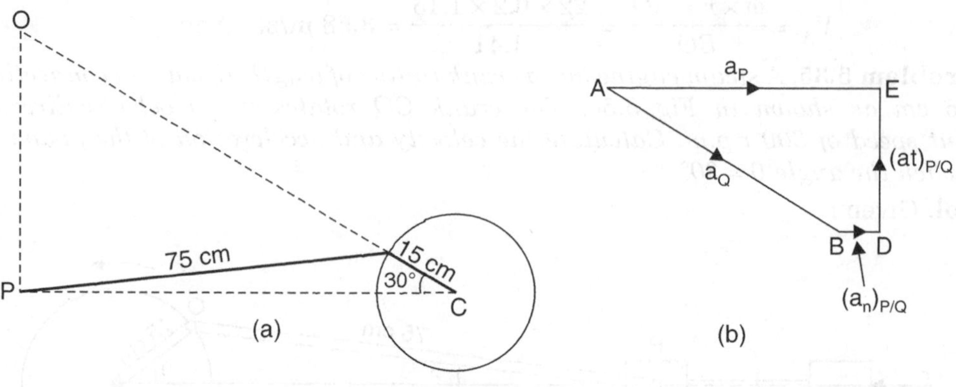

Fig. 5.55

(*ii*) **Acceleration of piston P.** The acceleration of the piston *P* can be obtained either by analytical method or by graphical method.

Graphical method. The crank rotates at uniform angular velocity, hence there will be no tangential linear acceleration. But the crank will have normal acceleration equal to $\dfrac{V^2}{r}$ or $\omega^2 r$ directed towards the centre *C*.

$\therefore$ Normal acceleration of Q or

$$a_Q = \omega^2 r = (10\pi)^2 \times 0.15 = 148.04 \text{ m/s}^2.$$

Tangential acceleration of P with respect to Q or $(a_t)_{P/Q}$

$$= L \times \frac{d\omega_{P/Q}}{dt} \text{ in the direction perpendicular to } PQ.$$

(Magnitude is unknown, only direction is known)

Normal acceleration of P with respect to Q in the direction towards Q from P or $(a_n)_{P/Q}$

$$= \omega_0^2 \times L$$

where ω_0 = Angular velocity of PQ for the given position of the PQ.

The value of ω_0 is obtained from equation (5.10) as

$$\omega_0 = \frac{\omega \times r}{BO} = \frac{10\pi \times 0.15}{QO} \qquad (\because \ \omega = 10\pi \text{ and } r = 0.15)$$

$$= \frac{10\pi \times 0.15}{0.84} = 5.60 \text{ rad/s. (Here point } Q \text{ corresponds to point B)}$$

(From Fig. 5.55 (a), QO = 0.84 m)

Substituting this value in equation (i), we get

$$(a_n)_{P/Q} = 5.60^2 \times 0.75 = 23.52 \text{ m/s}^2.$$

The direction of acceleration of P ($i.e.$, a_P) is in horizontal direction from P to C. Now the acceleration diagram is drawn as given below :

1. Take any point A. From A draw a line AB parallel to the line CQ. Cut AB = 148.04 m/s^2 to some scale.

2. From B, draw a line BD parallel to line PQ. Cut BD = 23.52 m/s^2.

3. Draw a normal to the line BD through point D. From point A, draw a line AE in the horizontal direction meeting the normal through D at point E.

4. Measure AE. This represents the acceleration of P.

Hence by measurement, acceleration of P

$$= \mathbf{143.50 \text{ m/s}^2}. \textbf{ Ans.}$$

Problem 5.36. *The length of connecting rod (BC) and crank (AB) in a reciprocating pump are 1125 mm and 250 mm respectively. The crank is rotating at 420 r.p.m. Find the velocity with which the piston will move, when the crank has turned through an angle of 40 from the inner dead centre. Also find the angular velocity of the connected rod.*

(U.P. Tech. University, 2002–2003)

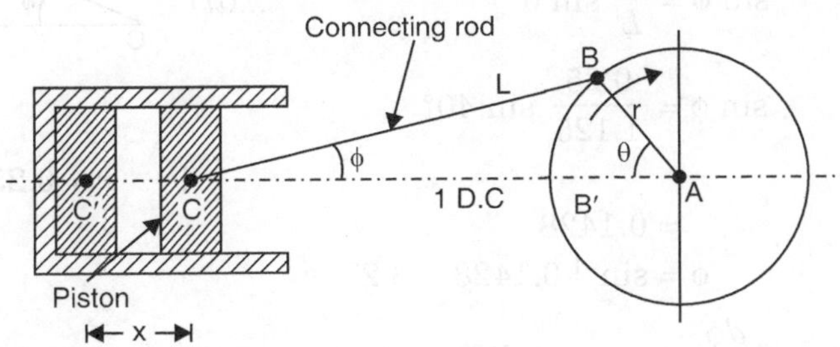

Fig. 5.56

Sol. Given :

Length of connecting rod, $L = BC = 1125 \text{ mm} = 1.125 \text{ m}$

Crank radius, $r = AB = 250 \text{ mm} = 0.25 \text{ m}$

Speed of crank, $N = 420$ r.p.m.

$\therefore$ Angular velocity of crank, $\omega = \dfrac{2\pi N}{60} = \dfrac{2\pi \times 420}{60} = 14\pi$ rad/s

Angle turned by crank, $\theta = 40°$

Find : (*i*) Velocity of piston and

(*ii*) Angular velocity of the connecting rod.

(*i*) *Velocity of piston*

The displacement of piston is given by

$$x = AC' - AC$$
$$= (C'B' + B'A) - AC$$
$$x = (L + r) - (L \cos \phi + r \cos \theta)$$

The velocity of piston is obtained by differentiating above equation w.r.t. time.

$$\therefore \quad V = \frac{dx}{dt} = \frac{d}{dt}[(L + r) - (L \cos \phi + r \cos \theta)]$$
$$= 0 - \left[L(- \sin \phi) \times \frac{d\phi}{dt} + r(- \sin \theta)\frac{d\theta}{dt}\right]$$

$$\left[\because \; L \text{ and } r \text{ are constant.} \;\therefore\; \frac{d}{dt}(L + r) = 0\right]$$

$$\therefore \quad V = L \sin \phi \times \frac{d\phi}{dt} + r \sin \theta \times \frac{d\phi}{dt} \qquad ...(i)$$

In the above equation, $\dfrac{d\phi}{dt}$ = Angular velocity of crank

$= \omega = 14\pi$ rad/s

$\dfrac{d\phi}{dt}$ = Angular velocity of connecting rod.

Let us find the value of ϕ in terms of θ.

From Fig. 5.56 (*a*), we have

$$BD = L \sin \phi = r \sin \theta$$

$$\therefore \quad \sin \phi = \frac{r}{L} \sin \theta \qquad ...(ii)$$

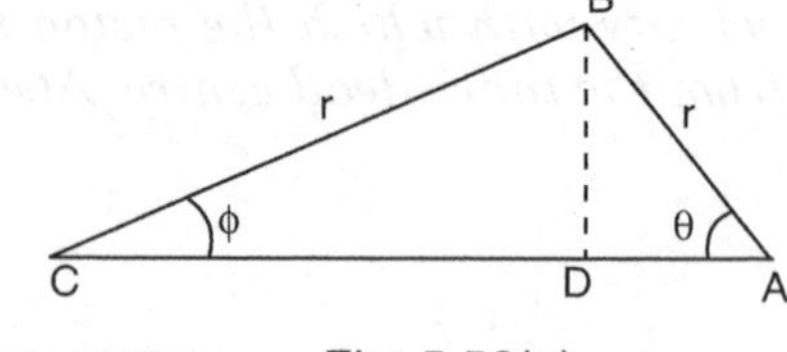

Fig. 5.56(a)

or $$\sin \phi = \frac{0.25}{1.125} \sin 40°$$

$$[\because \; r = 0.25, L = 1.125, \theta = 40°]$$

$$= 0.1428 \qquad ...(iii)$$

$$\therefore \quad \phi = \sin^{-1} 0.1428 = 8.2° \qquad ...(iv)$$

To find the value of $\dfrac{d\phi}{dt}$, we must differentiate equation (*ii*) w.r.t. time.

$$\therefore \qquad \frac{d(\sin\phi)}{dt} = \frac{d}{dt}\left[\frac{r}{L}\sin\theta\right]$$

or $$\cos\phi \times \frac{d\phi}{dt} = \frac{r}{L}\cos\theta \times \frac{d\theta}{dt}$$

$$\therefore \qquad \frac{d\phi}{dt} = \frac{r}{L}\frac{\cos\theta}{\cos\phi} \times \frac{d\theta}{dt}$$

In the above equation, $r = 0.25$, $L = 1.125$, $\theta = 40°$, $\phi = 8.2°$, $\frac{d\theta}{dt} = \omega = 14\pi$ rad/s.

$$\therefore \qquad \frac{d\phi}{dt} = \frac{0.25}{1.125} \times \frac{\cos 40°}{\cos(8.2°)} \times 14\pi$$

$$= \frac{0.25}{1.125} \times \frac{0.766}{0.9897} \times 14\pi = 7.56 \text{ rad/s} \qquad ...(v)$$

Substituting the values of $\frac{d\phi}{dt}$, $\sin\phi$, r, L, θ and $\frac{d\phi}{dt}$ in equation (*i*), we get

$$V = 1.125 \times 0.1428 \times 7.56 + 0.25 \times \sin 40° \times 14\pi$$
$$= 1.2145 + 7.06 = \mathbf{8.28 \text{ m/s.}} \quad \textbf{Ans.}$$

(*ii*) *Angular velocity of connecting rod*

It is equal to $\frac{d\phi}{dt}$. But $\frac{d\phi}{dt}$ from equation (*v*) is equal to 7.56 rad/s

$\therefore$ Angular velocity of connecting rod $= \frac{d\phi}{dt} =$ **7.56 rad/s. Ans.**

5.9. GENERAL PLANE MOTION

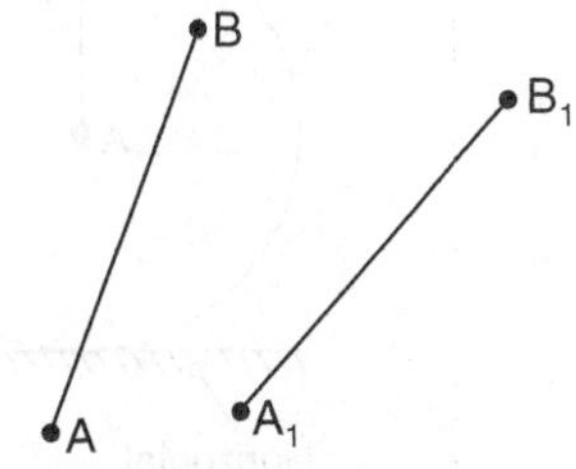

Fig. 5.57

General plane motion of a rigid body is the sum of the translation of the body and rotation of the body. The examples of general plane motion are rolling wheels, connecting rod of a reciprocating engine, links joining rotating elements of a machine. Hence

General plane motion = Motion of translation + Motion of rotation.

Consider a link AB, which moves from its initial position AB to A_1B_1 in a short interval of time as shown in Fig. 5.57. The link has neither entire motion of translation nor entire rotation, but a combination of the two. The motion of the link from the position AB to the position A_1B_1 may be regarded as to consist of :

1st Case

(*i*) A motion of entire translation from the position AB to the position A_1B^* so that A_1B^* is parallel to AB is as shown in Fig. 5.57 (*a*), and

(*ii*) A motion of entire rotation about A_1 from the position A_1B^* to the position A_1B_1.

2nd Case

(*i*) A motion of entire rotation about *A* from the position *AB* to the position AB^* as shown in Fig. 5.57 (*b*) and

(*ii*) A motion of entire translation from the position AB^* to the position A_1B_1.

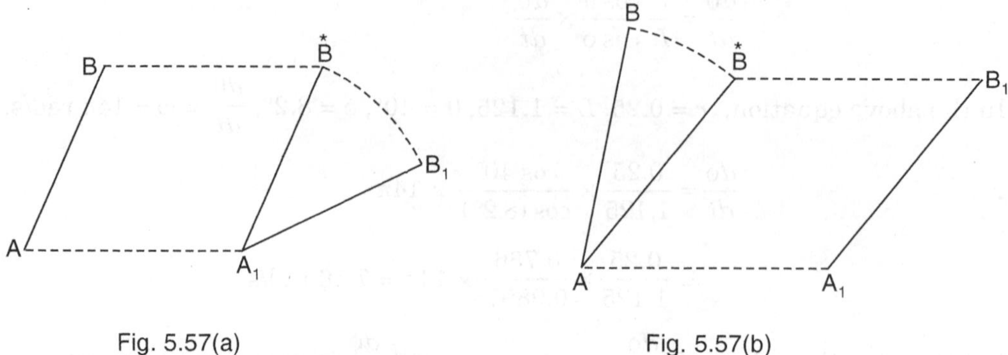

Fig. 5.57(a) Fig. 5.57(b)

3rd Case

The combined motion of translation and rotation of the link from its initial position *AB* to the position A_1B_1 is known as plane motion.

Problem 5.37. *Prove that the motion of a wheel rolling on a horizontal surface without slipping and the motion of a rod sliding against a wall at one end and the floor at the other end are the cases of plane motion.*

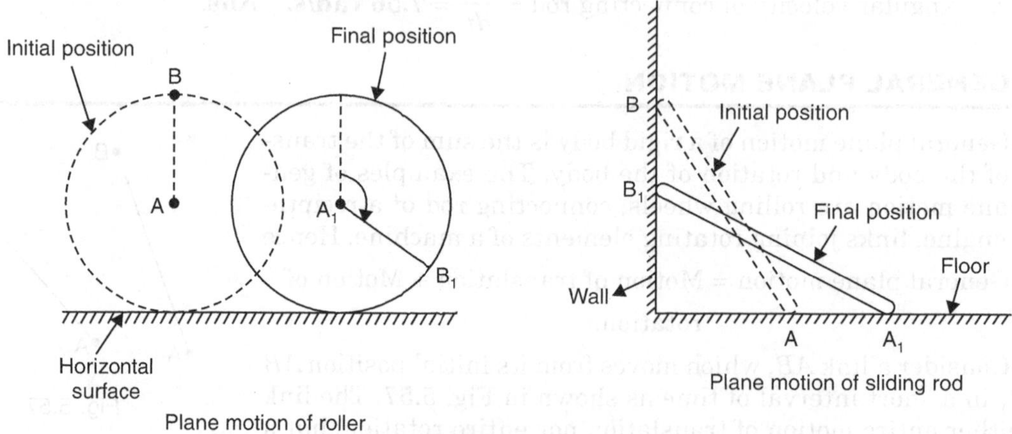

Fig. 5.58

Sol. Case I. *A wheel rolling on a horizontal surface without slipping.* [Refer to Fig. 5.59]

Plane motion means the motion of translation and motion of rotation simultaneously of a body at any particular instant.

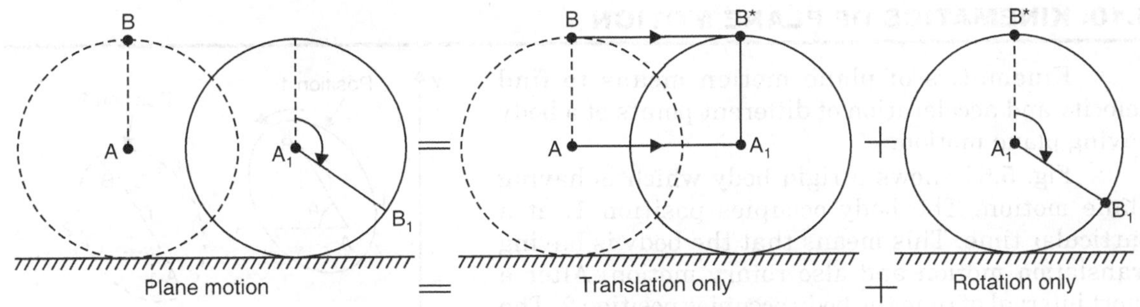

Fig. 5.59

The Fig. 5.59 shows a wheel rolling on a horizontal surface without slipping. The line AB moves from its initial position AB to $A_1 B_1$ in a short interval of time. The motion of line AB from the position AB to A_1B_1 may be regarded as to consist of :

(*i*) A motion of entire translation from the position AB to the position A_1B^* so that A_1B^* is parallel to AB.

(*ii*) A motion of entire rotation about A_1 from the position A_1B^* to the position A_1B_1 as shown in Fig. 5.59.

Case II. *A rod sliding against wall at one end and the floor at the other end.* [Refer to Fig. 5.60]

Fig. 5.60 shows a rod AB sliding against wall at one end and floor at other end. The rod AB moves from its initial position AB to final position A_1B_1 in a short interval of time. The motion of the rod from the position AB to A_1B_1 may be regarded as to consist of :

(*i*) A motion of entire translation from AB to the position A_1B^* so that A_1B^* is parallel to AB.

(*ii*) A motion of entire rotation about A_1 from the position A_1B^* to the position A_1B_1 as shown in Fig. 5.60.

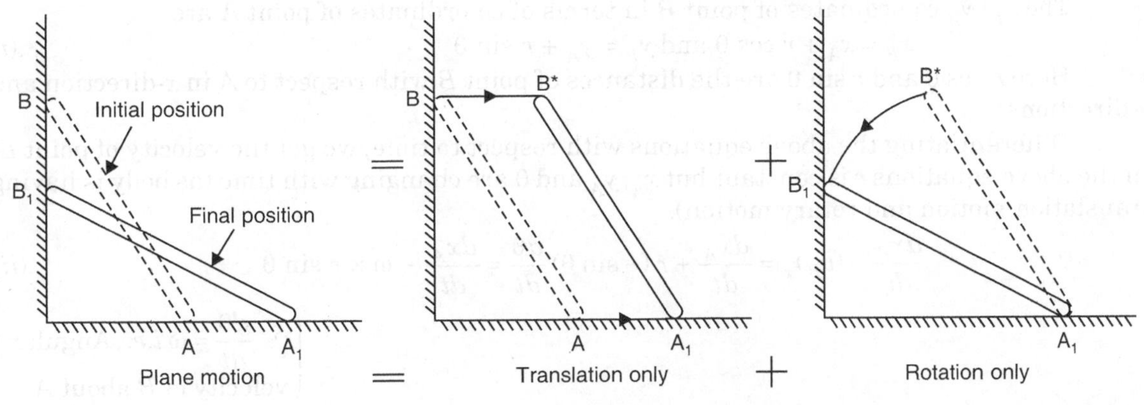

Fig. 5.60

5.10. KINEMATICS OF PLANE MOTION

Kinematics of plane motion means to find velocity and acceleration of different points of a body having plane motion.

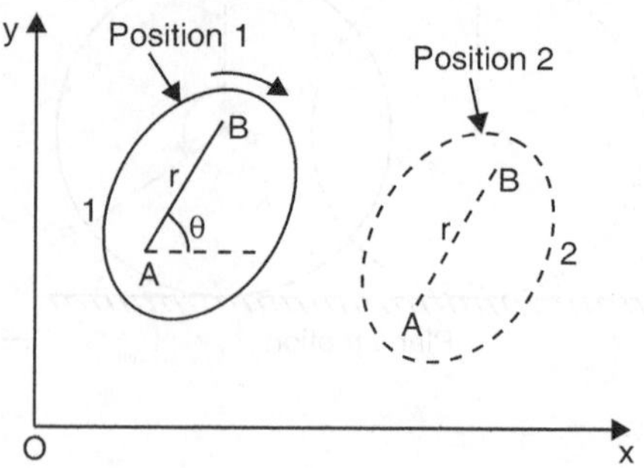

Fig. 5.61. Plane motion of a rigid body.

Fig. 5.61 shows a rigid body which is having plane motion. The body occupies position 1, at a particular time. This means that the body is having translation motion and also rotary motion. After a short interval of time the body occupies position 2. The distance between any two points is fixed. Let the distance between points A and B is r. This is constant in both positions. The inclination of line AB with x-axis is changing. This is not same for both positions. Thus the body is having translation motion and also rotational motion about point A. The point, about which the body is supposed to rotate, is known as pole or a reference point. Hence point A is a pole. Consider the body again in position 1, as shown again in Fig. 5.61(a) in a larger view.

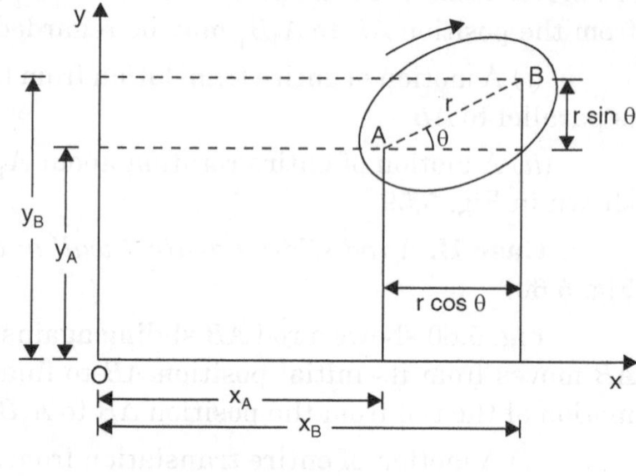

Fig. 5.61 (a)

Let x_A, y_A = Co-ordinates of point A

x_B, y_B = Co-ordinates of point B

r = Distance between AB, this is the distance of B with respect to A.

θ = Inclination of AB with x-axis

Let us find the velocity and acceleration of point B.

Velocity of point B

The x_B, y_B co-ordinates of point B in terms of co-ordinates of point A are.

$$x_B = x_A + r\cos\theta \text{ and } y_B = y_A + r\sin\theta \qquad ...(i)$$

Here $r\cos\theta$ and $r\sin\theta$ are the distances of point B with respect to A in x-direction and y-directions.

Differentiating the above equations with respect to time, we get the velocity of point B. In the above equations r is constant but x_A, y_A and θ are changing with time (as body is having translation motion and rotary motion).

$$\therefore \quad \frac{dx_B}{dt} = (v_B)_x = \frac{dx_A}{dt} + r(-\sin\theta)\frac{d\theta}{dt} = \frac{dx_A}{dt} - \omega \times r\sin\theta \qquad ...(ii)$$

$$\left(\because \frac{d\theta}{dt} = \omega \textit{ i.e.}, \text{ Angular velocity of } B \text{ about } A\right)$$

$$= (v_A)_x + (-\omega r\sin\theta) = (v_A)_x + (v_{B/A})_x \qquad ...(iii)$$

$$\frac{dy_B}{dt} = (v_B)_y = \frac{dy_A}{dt} + r\cos\theta \times \frac{d\theta}{dt} = \frac{dy_A}{dt} + \omega r\cos\theta \qquad ...(iv)$$

$$= (v_A)_y + (v_{B/A})_y \qquad ...(v)$$

Here ($-\omega r \sin\theta$) is the velocity of B with respect to A in x-direction and $\omega r \cos\theta$ is the velocity of B with respect to A in y-direction.

$\therefore \quad (-\omega r \sin\theta) = (v_{B/A})_x$ and $\omega r \cos\theta = (v_{B/A})_y$

Vector sum of $(v_B)_x$ and $(v_B)_y$ will give the velocity of B *i.e.*, v_B whereas the vector sum of $(v_A)_x$ and $(v_A)_y$ will be equal to v_A. Similarly vector sum of $(v_{B/A})_x$ and $(v_{B/A})_y$ will be equal to $v_{B/A}$.

$$\left[\because\ v_{B/A} = \sqrt{(v_{B/A})_x^2 + (v_{B/A})_y^2} = \sqrt{(-\omega r \sin\theta)^2 + (\omega r \cos\theta)^2} = \omega \times r\right]$$

$\therefore$ Vector sum of the above two equations (*iii*) and (*v*) gives

Velocity of B = Velocity of A $\mapsto$ Velocity of B with respect to A (Vector Sum)

or $$v_B = v_A \mapsto v_{B/A}\ (= \omega \times r)\ \text{(vector sum)} \quad ...(5.12)$$

where sign $\mapsto$ means that the vector v_A and $v_{B/A}$ are to be added geometrically. Also $v_{B/A}$ is equal to $\omega \times r$ where r is the distance between A and B and ω is the rotational velocity of B about point A.

ωr sin θ = (V_B/A)_x
ωr
ωr cos θ = (V_B/A)_y
V_B
V_B/A
φ
V_A
(V_A)_y
(V_A)_x

Fig. 5.61(b)

Acceleration of point B

The acceleration of point B is obtained by differentiating equations (*ii*) and (*iv*) with respect to time.

$$\therefore \quad \frac{dx_B^2}{dt^2} = (a_B)_x = \frac{d^2 x_A}{dt^2} - r\left(\omega \cos\theta \times \frac{d\theta}{dt} + \sin\theta \times \frac{d\omega}{dt}\right)$$

(r is constant but θ and ω are variable)

$$= (a_A)_x - r\,(\omega \times \cos\theta \times \omega + \sin\theta \times \alpha) \qquad \left[\because\ \frac{d\theta}{dt} = \omega \text{ and } \frac{d\omega}{dt} = \alpha\right]$$

$$= (a_A)_x - r\,(\omega^2 \cos\theta + \alpha \sin\theta)$$

$$= (a_A)_x - \omega^2 r \cos\theta - r \times \alpha \sin\theta \qquad ...(vi)$$

and $$\frac{d^2 y_B}{dt^2} = (a_B)_y = \frac{d^2 y_A}{dt^2} + r\left[\omega\,(-\sin\theta) \times \frac{d\theta}{dt} + \cos\theta \times \frac{d\omega}{dt}\right]$$

$$= (a_A)_y + r\,[\omega\,(-\sin\theta) \times \omega + \cos\theta \times \alpha] \qquad \left[\because\ \frac{d\theta}{dt} = \omega \text{ and } \frac{d\theta}{dt} = \alpha\right]$$

$$= (a_A)_y - \omega^2 r \sin\theta + r \times \alpha \cos\theta \qquad ...(vii)$$

In equations (*vi*) and (*vii*), the vector sum of $(a_A)_x$ and $(a_A)_y$ will give acceleration of A *i.e.*, a_A. Vector sum of $\omega^2 r \cos\theta$ and $\omega^2 r \sin\theta$ will give the normal acceleration of B with respect to A. Similarly vector sum of $r \times \alpha \cos\theta$ and $r \times \alpha \times \sin\theta$ will give the tangential acceleration of B with respect to A.

$\therefore$ Vector sum of the above two equations (*vi*) and (*vii*) gives

Acceleration of B = Acceleration of A $\mapsto$ Acceleration of B with respect to A

or $$a_B = a_A \mapsto a_{B/A}$$

where $a_{B/A}$ consists of normal acceleration $\omega^2 r$ and tangential acceleration $\alpha \times r$.

$$\therefore \quad a_B = a_A \mapsto a_{B/A}\ (= \omega^2 r + a \times r) \qquad ...(5.13)$$

Problem 5.38. *A link AB is moving in a vertical plane. At a certain instant, when the link is inclined at 30° to the horizontal, the point A is moving horizontally at 4 m/s, while B is moving vertically upwards. Find the velocity of B.*

Sol. Given :

Inclination of link AB with horizontal = 30°

Horizontal velocity of A, V_A = 4 m/s

Let V_B = Velocity of B in the vertically upward direction.

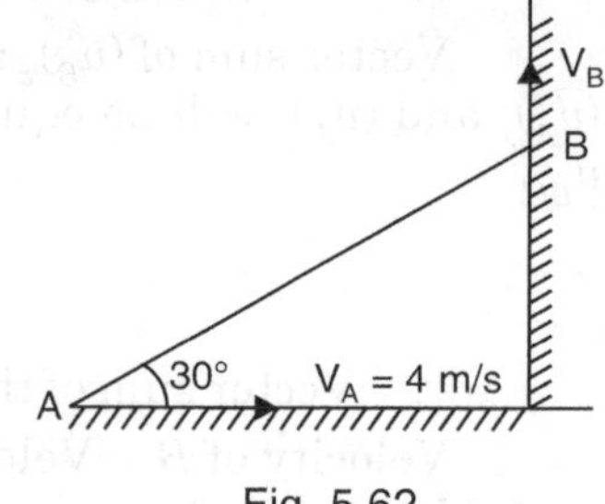

Fig. 5.62

The link AB is having plane motion.

The velocity v_A at point A is known. Hence this point can be chosen as the pole (or reference point). The velocity of point B with respect to A can be obtained by using equation (5.12).

$$\therefore \qquad v_B = v_A \mapsto v_{B/A} \; (= \omega \times r)$$

where ω = angular rotational velocity of B about point A and r = length AB.

The velocity v_B is vertically upwards, velocity v_A is in horizontal direction towards right of A and velocity $v_{B/A}$ is perpendicular to link AB at point B. The directions and magnitude of these velocities are shown below :

Fig. 5.62 (a)

Resolving the velocities in x and y directions, we get

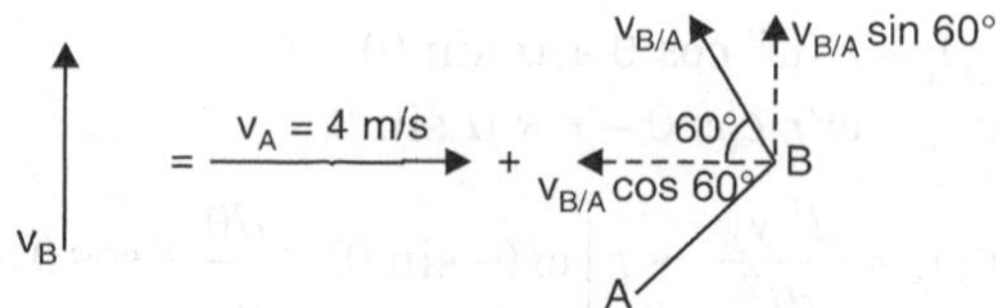

Fig. 5.62 (b)

Equating the velocity components in x and y direction, we get

For x-direction : $\quad 0 = v_A - v_{B/A} \cos 60°$

$$= 4 - v_{B/A} \times 0.5$$

$$\therefore \qquad v_{B/A} = \frac{4}{0.5} = 8 \text{ m/s}$$

For y-direction : $\quad v_B = 0 + v_{B/A} \sin 60°$

$$= 0 + 8 \times 0.866 = \mathbf{6.928 \text{ m/s.}} \quad \textbf{Ans.}$$

Problem 5.39. *A cylinder of radius 1 m rolls without slipping along a horizontal plane AB. Its centre has uniform velocity of 20 m/s. Find the velocity of point E and F on the circumference of the cylinder shown in Fig. 5.63.*

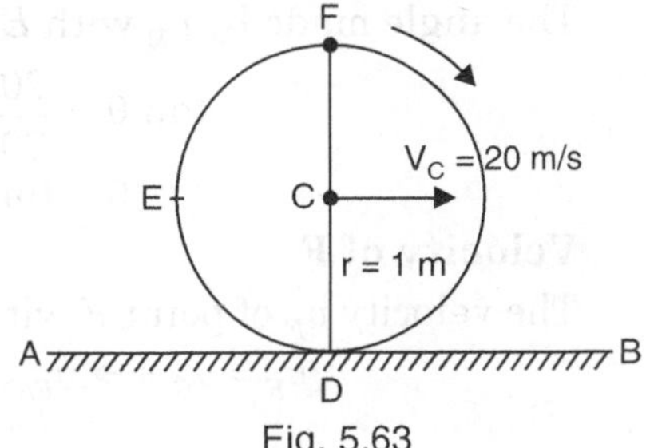

Fig. 5.63

Sol. Given : $r = 1$ m, $v_C = 20$ m/s

As the cylinder rolls without slipping, hence the point of contact D with the horizontal plane AB at any instant will have zero velocity.

The rolling cylinder without slipping is having plane motion. The velocity v_C at point C is known. Hence point C can be taken as the pole (or the reference point). The velocity of point D with respect to point C can be written as [Use equation (5.12)]

$$v_D = v_C \mapsto v_{D/C} =$$

($\omega \times r$, where ω = angular rotational velocity of DC about point C and $r = CD = 1$ m)

The velocity v_D at point D is zero. The velocity v_C at point C is 20 m/s and is in horizontal direction towards right of C and velocity $v_{D/C}$ is perpendicular to line CD at point D. The line CD is vertical and hence velocity $v_{D/C}$ will be in horizontal direction towards left of D. These velocities are shown in Fig. 5.63(a) below :

0 = v_C = 20 m/s → ⟼ ← $v_{D/C}$ (C, D)

Fig. 5.63 (a)

Resolving all velocities in x-direction, we get

$$0 = 20 + (-v_{D/C})$$

or $\quad v_{D/C} = 20 \quad$ But $\quad v_{D/C} = \omega \times r = \omega \times 1$

$\therefore \quad \omega \times r = 20 \quad$ or $\quad \omega = \dfrac{20}{r} = \dfrac{20}{1} = 20$ rad/s

Velocity of E

The velocity v_E of point E with respect to pole ($i.e.$, point C) can be obtained from equation (5.12) as

$$v_E = v_C \mapsto v_{E/C} =$$

($\omega \times r$, where ω = rotational angular velocity of E about point C = 20 rad/s and $r = EC = 1$ m)

= 20 m/s → + ↑ $v_{E/C}$ (E, C)

= 20 m/s → + ↑ $\omega \times r = 20 \times 1 = 20$ m/s (E)

Here $(v_E)_x = 20$ m/s (Equating velocities in x-direction)

$(v_E)_y = 20$ m/s

$$\therefore \quad v_E = \sqrt{(v_E)_x^2 + (v_E)_y^2}$$

$$= \sqrt{20^2 + 20^2} = 20 \times \sqrt{2} = \textbf{28.28 m/s}$$

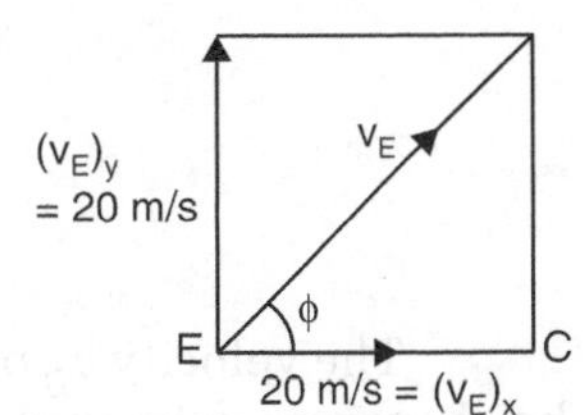

Fig. 5.63 (b)

The angle made by v_E with EC (or x-axis) is given by

$$\tan \theta = \frac{20}{20} = 1$$

$$\therefore \qquad \theta = \tan^{-1} 1 = \mathbf{45°.\ Ans.}$$

Velocity of F

The velocity v_F of point F with respect to pole (*i.e.*, point C) is written as

$v_F = v_C \leftrightarrow v_{F/C}$ (= $\omega \times r$, where ω = 20 rad/s and $r = CF$ = 1 m and $v_{F/C}$ = 20 × 1 = 20 m/s. This is ⊥ to line CF at point F).

= 20 m/s → ⟷ (F, 20 m/s →, C)

= 20 + 20 = **40 m/s. Ans.** (parallel to x-axis at point F)

Problem 5.40. *The crank of a reciprocating engine is rotating at 210 r.p.m. The lengths of the crank and connecting rod are 20 cm and 100 cm respectively. Find the velocity of the point A (i.e., velocity of piston), when crank has turned through an angle of 45° with the horizontal as shown in Fig. 5.64.*

Sol. Given :

Speed of crank, N = 210 r.p.m.

$$\therefore \text{ Angular velocity of crank, } \omega = \frac{2\pi N}{60} = 2 \times \frac{22}{7} \times \frac{210}{60} = 22 \text{ rad/s.}$$

Crank length or radius, r = 20 cm = 0.20 m

Length of connecting rod, L = 100 cm = 1.0 m

Angle turned by crank, θ = 45°.

Let us first calculate the angle ϕ.

Applying the sine rule for the triangle ABC shown in Fig. 5.64.

$$\frac{BC}{\sin \phi} = \frac{AB}{\sin \theta}$$

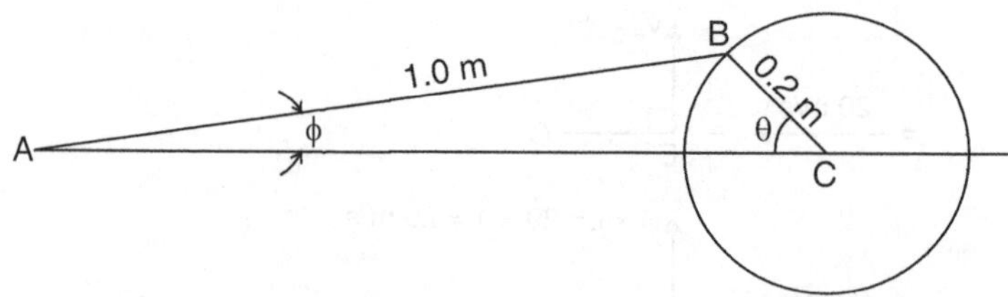

Fig. 5.64

$$\therefore \qquad \sin \phi = \frac{BC \sin \theta}{AB} = \frac{0.2}{1.0} \times \sin 45° = \frac{0.2}{1.0} \times 0.7071 = 0.1414$$

$$\therefore \qquad \phi = \sin^{-1} 0.1414 = 8.13°.$$

The velocity v_B of point B is known $v_B = \omega \times r$ = 22 × 0.2 = 4.4 m/s. Hence the point B can be chosen as reference point or pole. The connecting rod can be assumed to be rotating about point B.

The velocity v_A of point A with respect to point B (*i.e.*, with respect to pole) can be obtained from equation (5.12) as

$\therefore$ $$v_A = v_B \leftrightarrow v_{A/B} \text{ (vector sum)} \qquad \text{[Equation (5.12)]} \quad ...(i)$$

The velocity of point A (*i.e.*, piston) v_A is in horizontal direction towards right of A, when crank BC is moving clockwise and has turned an angle 45° as shown in figure. The velocity of point B, v_B is perpendicular to BC. The magnitude of v_B is 4.4 m/s. The velocity $v_{A/B}$ (Rotational velocity of connecting rod AB about point B is perpendicular to AB at point A and it is equal to $\omega_{AB} \times L$.

Now substituting in magnitude and direction of all velocities in equation (*i*), we get

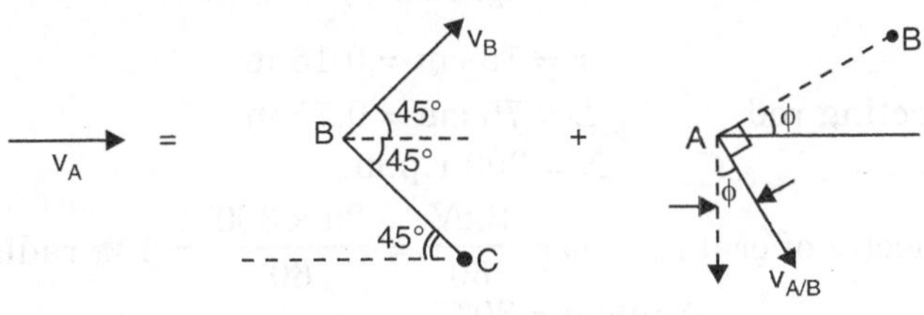

Fig. 5.64 (a)

$\therefore$ $$\cos \phi = 0.9899$$

Resolving all velocities in x and y direction, we get

Fig. 5.64 (b)

Equating the velocities in x and y directions,

For x-axis : $v_A = v_B \cos 45° + v_{A/B} \sin \phi$

$= 4.4 \cos 45° + v_{A/B} \times 0.1414$ $\quad (\because \; v_B = 4.4$ m/s ϕ and $\sin \phi = 0.1414)$

$= 3.11 + 0.1414\, v_{A/B}$ ...(*ii*)

For y-axis : $0 = v_B \sin 45° - v_{A/B} \cos \phi$

$= 4.4 \sin 45° - v_{A/B} \times 0.9899$

$= 3.11 - 0.9899\, v_{A/B}$

$\therefore$ $$v_{A/B} = \frac{3.11}{0.9899} = 3.143 \text{ m/s}$$

Substituting this value in equation (*ii*), we get

$$v_A = 3.11 + 0.1414 \times 3.143$$

$= \mathbf{3.55}$ **m/s. Ans.**

Problem 5.41. *A steam engine has a crank of radius 15 cm and connecting rod of length 75 cm as shown in Fig. 5.65. The crank CQ rotates in a clockwise direction with a constant speed of 300 r.p.m. Calculate the velocity and acceleration of the piston P at the instant when the angle θ = 30°.*

Sol. Given :

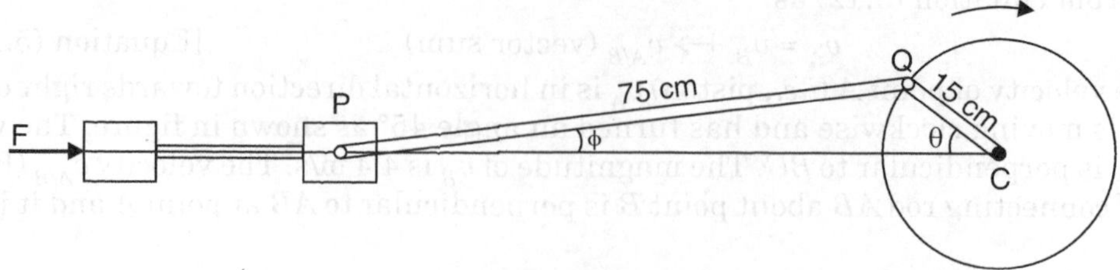

Fig. 5.65

Crank radius, $r = 15 \text{ cm} = 0.15 \text{ m}$

Length of connecting rod, $L = 75 \text{ cm} = 0.75 \text{ m}$

Speed of crank, $N = 300 \text{ r.p.m.}$

$\therefore$ Angular velocity of crank, $\omega = \dfrac{2\pi N}{60} = \dfrac{2\pi \times 300}{60} = 10\pi$ radian/s.

Angle $\theta = 30°$.

(*i*) **Velocity of Piston P.** The velocity of piston P can be obtained either by analytical method or by graphical method.

First find the value of angle ϕ. In triangle PCQ, using sine rule, we have

$$\frac{CQ}{\sin\phi} = \frac{PQ}{\sin\theta}$$

$$\therefore \quad \sin\phi = \frac{CQ}{PQ}\sin\theta = \frac{0.15}{0.75}\sin 30° = \frac{1}{5} \times \frac{1}{2} = \frac{1}{10} = 0.1$$

$$\therefore \quad \phi = \sin^{-1} 0.1 = 5.739°.$$

$$v_Q = \omega \times r = 10\pi \times 15 = 150\pi \text{ cm/s}$$

$$\sin\phi = 0.1, \ \phi = 5.739°, \ \cos\phi = 0.995$$

Using equation for velocity for plane motion and taking point Q as the pole, we get

$$v_P = V_Q \mapsto v_{P/Q} \qquad \text{...[Equation (5.12)]}$$

or

Fig. 5.65 (a)

or

Fig. 5.65 (b)

Equating the velocities in x and y directions,

Along x-axis : $v_P = v_Q \cos 60° + v_{P/Q} \sin \phi$

$= 150\pi \times 0.5 + v_{P/Q} \times 0.1$

$= 235.6 + 0.1\, v_{P/Q}$ ...(1)

$= 235.6 + 0.1 \times v_{P/Q}$

Along y-axis : $0 = v_Q \sin 60° - v_{P/Q} \cos \phi$

$= 150\,\pi \times 0.866 - v_{P/Q} \times 0.995$

$= 408.9 - v_{P/Q} \times 0.995$

or $v_{P/Q} = \dfrac{408.9}{0.995} = 410.95$ ...(2)

Substitute this value in (1), we get

$v_P = 235.6 + 0.1 \times 410.95 =$ **277.6 cm/s. Ans.**

Let us also find the angular velocity (ω_{PQ}) of connecting rod PQ. We know

$v_{P/Q} = \omega_{PQ} \times PQ$ But $v_{PQ} = 410.95$ m/s (calculated above)

$\therefore$ $410.95 = \omega_{PQ} \times 75$ $\therefore$ $\omega_{PQ} = \dfrac{410.95}{75} = 5.48$ rad/s

(*ii*) **Acceleration of Piston P.** Using equation of plane motion for acceleration and taking Q as the pole, we get

$a_P = a_Q \mapsto a_{P/Q}$ (which is $= r\omega^2 + r\alpha$ acting at P)

[Here $r = PQ$ and $\alpha = \omega \times PQ$. Radial acceleraton towards Q from $P = \omega_{PQ}^2 \times PQ = (5.48)^2 \times 75 = 2252$ cm/s^2 and Tangential acceleration at $P = PQ \times \alpha_{PQ} = 75\,\alpha_{PQ}$]

Vectorially the above equation becomes as

Fig. 5.65 (c)

But $a_Q = \omega^2 \times r = (10\pi)^2 \times 15 = 14804$ m/s^2 [here $r = 15$]

Resolving all accelerations in x and y-directions.

Along x-axis : $a_P = +\, a_Q \cos 30° + 2252 \cos \phi + 75\,\alpha_{PQ} \sin \phi$

$= 14804 \times 0.866 + 2252 \times 0.995 + 75 \times \alpha_{PQ} \times 0.1$

$= 12820 + 2240 + 75\,\alpha_{PQ}$ ...(3)

Along y-axis : $0 = -\, a_Q \sin 30° + 2252 \sin \phi - 75 \times \alpha_{PQ} \times \cos \phi$

$= -\,14804 \times 0.5 + 2252 \times 0.1 - 75 \times \alpha_{PQ} \times 0.995$

$= -\,7402 + 225.2 - 77.62\,\alpha_{PQ}$

$\therefore$ $\alpha_{PQ} = \dfrac{-\,7402 + 225.2}{77.62} = -\,92.46$

Substituting this value in (3), we get

$a_P = 12820 + 2240 + 7.5 \times (-\,92.46)$

$= 15060 - 693.45 =$ **14366.55 cm/s². Ans.**

Problem 5.42. *The lengths of connecting rod (BC) and crank (AB) in a reciprocating pump are 1125 mm and 250 mm respectively. The crank is rotating at 420 r.p.m. Find the velocity with which the piston will move, when the crank has turned through an angle of 40° from the inner dead centre. Also find the angular velocity of the connected rod.*

(U.P. Tech. University, 2002–2003)

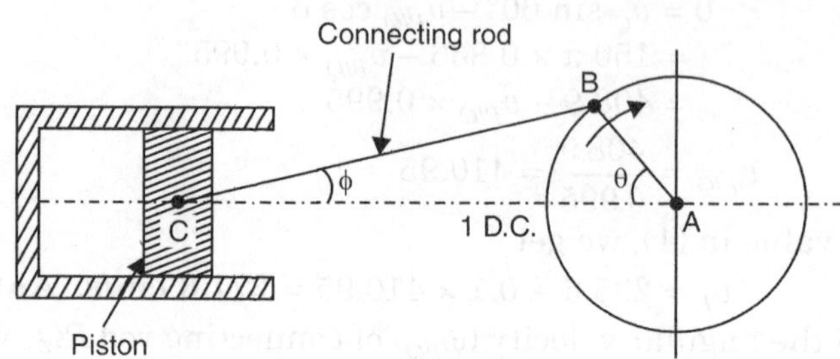

Fig. 5.66

Sol. Given :

Length of connecting rod, $L = BC = 1125 \text{ mm} = 1.125 \text{ m}$

Crank radius, $r = AB = 250 \text{ mm} = 0.25 \text{ m}$

Speed of crank, $N = 420$ r.p.m.

$\therefore$ Angular velocity of crank, $\omega = \dfrac{2\pi N}{60} = \dfrac{2\pi \times 420}{60} = 14\pi \text{ rad/s}$ ...(*i*)

Angle turned by crank, $\theta = 40°$

Find : (*i*) Velocity of piston and

(*ii*) Angular velocity of the connecting rod.

(*i*) *Velocity of piston*

Let us find the value of ϕ in terms of θ.

From Fig. 5.66(*a*), we have

$$BD = L \sin \phi = r \sin \theta$$

$\therefore$ $$\sin \phi = \frac{r}{L} \sin \theta \quad \text{...(ii)}$$

or $$\sin \phi = \frac{0.25}{1.125} \sin 40°$$

$[\because \; r = 0.25, L = 1.125, \theta = 40°]$

$= 0.1428$ ...(*iii*)

$\therefore$ $\phi = \sin^{-1} 0.1428 = 8.2°$ ...(*iv*)

and $\cos \phi = \cos 8.2° = 0.9897$

Fig. 5.66 (a)

$$v_B = \omega \times r = 14\,\pi \times 0.25 = 10.995 \simeq 11 \text{ m/s}$$

The link BC is having plane motion.

As velocity of point B is known. The link BC may be assumed as rotating about point B. Hence take point B as reference point. Then from the equation of plane motion for link BC, we get

$$v_C = v_B \leftrightarrow v_{C/B}$$

(where $v_{C/B}$ = rotational velocity of line BC about point $B = \omega \times r$, where $\omega = \omega_{BC}$ and $r = BC$)

The velocity v_C is along x-axis, velocity v_B is perpendicular to crank AB and $v_{C/B}$ is perpendicular to link CB acting at point C as shown in Fig. 5.66(b). The angle ϕ can be obtained by applying sine rule to triangle ABC as $\frac{AB}{\sin\phi} = \frac{BC}{\sin\theta}$ or $\sin\phi = \frac{AB}{BC} \times \sin\theta$

The above equation can be written in vector form (*i.e.*, the magnitude and direction of all velocities), we get

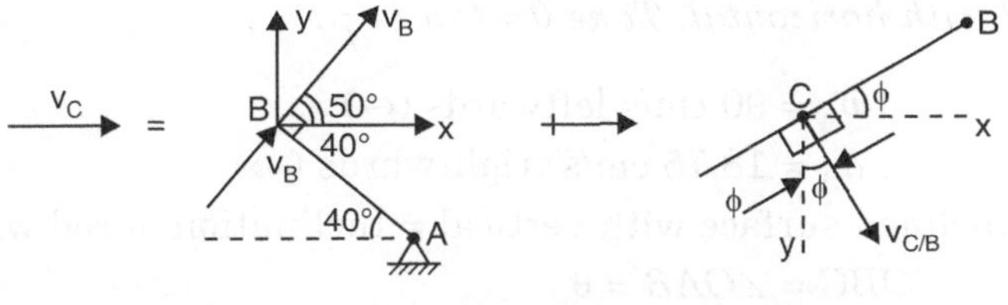

Fig. 5.66 (b)

Resolving all velocities in x and y-directions, we get

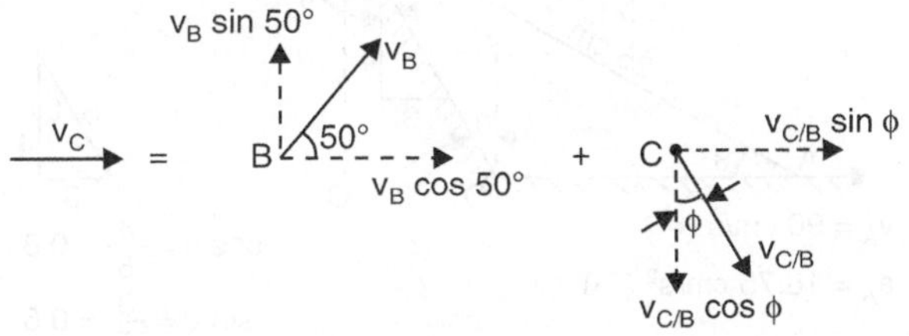

Fig. 5.66 (c)

Equating the velocities in x and y-directions,

Along x-axis : $v_C = v_B \cos 50° + v_{C/B} \sin\phi$

$= 11 \cos 50° + v_{C/B} \sin\phi$ [$\because$ $v_B = 11$]

$= 11 \cos 50° + v_{C/B} \times 0.1428$ [$\because$ $\sin\phi = 0.1428$]

$= 7.07 + v_{C/B} \times 0.1428$ ...(1)

Along y-axis : $0 = v_B \sin 50° - v_{C/B} \cos\phi$

$= 11 \sin 50° - v_{C/B} \times 0.9897$ [$\because$ $v_B = 11$ and $\cos\phi = 0.9897$]

$= 8.426 - 0.9897\, v_{C/B}$

or $v_{C/B} = \frac{8.426}{0.9897} = 8.514$ m/s ...(2)

Substituting this value of $v_{C/B}$ in equation (1), we get

$v_C = 7.07 + 8.514 \times 0.1428 = 8.285$ m/s ...(3)

$\therefore$ Velocity of piston, v_C = **8.285 m/s. Ans.**

(ii) Angular velocity of connecting rod (i.e., ω_{BC})

We know that $v_{C/B} = \omega_{BC} \times BC$

But $v_{C/B} = 8.514$ m/s from equation (2)

$\therefore \quad 8.514 = \omega_{BC} \times 1.125 \quad (\because \; BC = 1.125 \text{ m})$

$$\therefore \quad \omega_{BC} = \frac{8.514}{1.125} = \textbf{7.568 rad/s Ans.}$$

Problem 5.43. *The ends A and B of a slender rod of length 25 cm move in contact with a horizontal floor and an inclined surface as shown in Fig. 5.67. Determine the velocity and acceleration of end B if the end A has a velocity of 90 cm/s leftward and an acceleration of 18.75 cm/s² rightwards. The inclination of the inclined surface with vertical is equal to the inclination of the rod with horizontal. Take $\theta = \tan^{-1}\frac{3}{4}$.*

Sol. Given $v_A = 90$ cm/s leftwards (←)

$a_A = 18.75$ cm/s² rightwards (→)

Inclination of inclined surface with vertical = Inclination of rod with horizontal = θ

$\therefore \quad \angle OBC = \angle OAB = \theta$

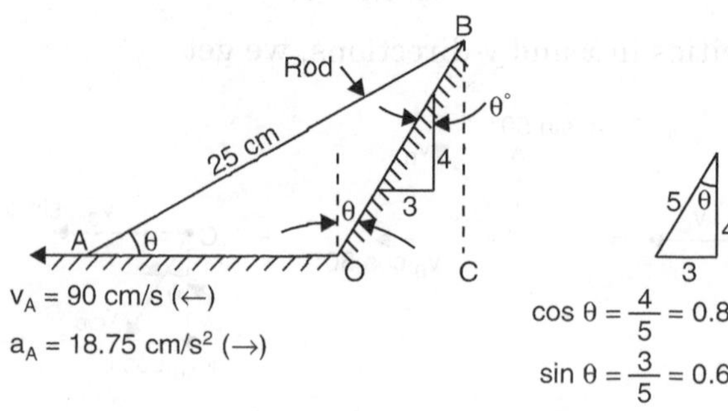

Fig. 5.67

Also $\tan\theta = \frac{3}{4}$ hence $\sin\theta = \frac{3}{5} = 0.6$ and $\cos\theta = \frac{4}{5} = 0.8$ (Refer to Fig. 5.67)

Let v_B = velocity of B and

a_B = Acceleration of B

Velocity of B(v_B)

As the velocity of A is given, hence this point can be considered as pole (or reference point). The rod AB is considered to be rotating about point A. Rod AB is having plane motion.

Now using the equation (5.12), we get

$$v_B = v_A \leftrightarrow v_{B/A} \quad \ldots(i)$$

where $v_{B/A}$ = relative rotational velocity of B with respect to A

$= \omega \times r$ where r = distance between A and B = 25 cm

$= \omega \times 25$ $\quad \omega$ = angular velocity of rod AB about pole A

$V_{B/A}$ is acting at B perpendicular to rod AB.

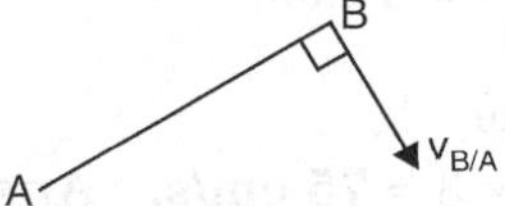

The velocity v_B acting along inclined surface downwards

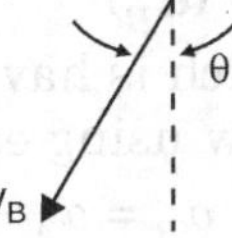

and $\quad v_A$ = 90 cm/s (given) (acting in horizontal direction towards left *i.e.*, $\overleftarrow{v_A}$)

The velocity of point B can now be determined graphically or analytically.

Analytical Method

$$v_B = v_A \mapsto v_{B/A} \text{ (vector sum)}$$

Substituting the values in magnitude and direction, we get

Fig. 5.67 (a)

Resolving the velocities in x and y-directions, we get

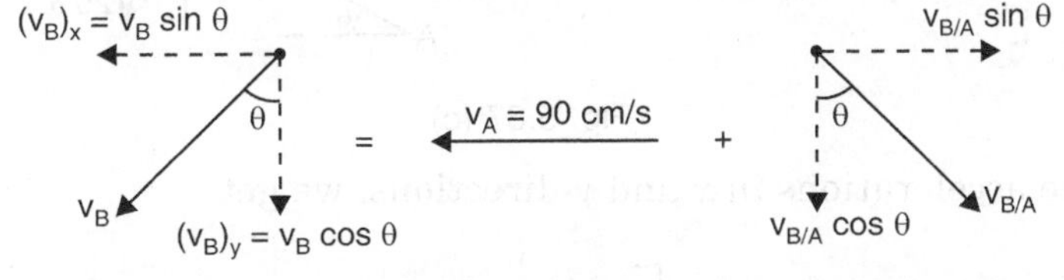

Fig. 5.67 (b)

Equating the horizontal and vertical components of velocity to both sides, we get

For horizontal direction,

$$-v_B \sin\theta = -90 + v_{B/A} \sin\theta$$

or $\quad v_B \times 0.6 = 90 - (25\omega) \times 0.6 \qquad [\because \sin\theta = 0.6 \text{ and } v_{B/A} = 25\,\omega]$

or $\quad v_B = \dfrac{90}{0.6} - 25\,\omega = 150 - 25\,\omega \qquad ...(1)$

For vertical direction

$$v_B \cos\theta = v_{B/A} \cos\theta$$

or $\quad v_B = v_{B/A} \qquad$ But $v_{B/A} = 25\,\omega$

$\therefore \quad v_B = 25\,\omega \qquad ...(2)$

Solving equations (1) and (2), we get

$150 - 25\,\omega = 25\,\omega \qquad$ [Equating the values of v_B in (1) and (2)]

or $\quad 150 = 50\,\omega$

$\therefore \qquad \omega = \dfrac{150}{50} = 3$ rad/s

From equation (2), $\quad v_B = 25\ \omega$

$= 25 \times 3 =$ **75 cm/s. Ans.**

Acceleration of B (a_B)

For the rod AB which is having plane motion, the rod is assumed to be rotating about the reference point A. Now using equation (5.30) for acceleration, we get

$$a_B = a_A \nrightarrow a_{B/A} \text{ (where } a_{B/A} = r\omega^2 + r\alpha\text{) (vector sum)} \qquad ...(ii)$$

where $\quad a_B$ = Acceleration of point B, which is acting along inclined surface downwards

a_A = Acceleration of point A, acting along horizontal surface towards right and = 18.75 cm/cm^2

$a_{B/A}$ = Relative rotational acceleration of B with respect to pole A. This consists of normal acceleration ($r\omega^2$ acting from B towards A) and tangential acceleration ($r \times \alpha$) acting perpendicular to AB. Here $r = AB = 25$ cm. These accelerations are acting at point B.

Now $\quad r\omega^2 = 25 \times 3^2 = 225$ cm/s^2

and $\quad r \times \alpha = 25\alpha$

Substituting the values in magnitude and direction of all acceleration in equation (*ii*), we get

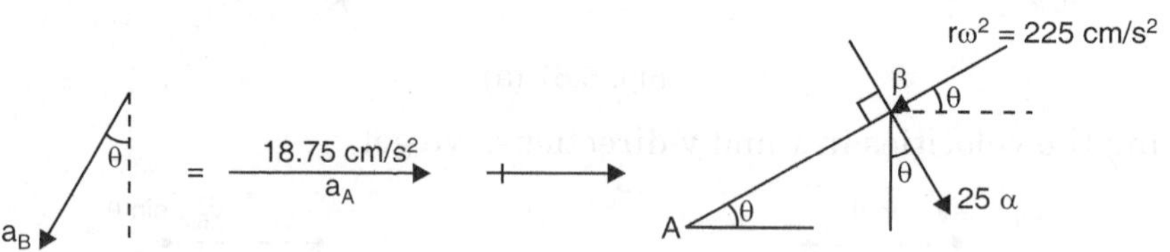

Fig. 5.67 (c)

Resolving the accelerations in x and y-directions, we get

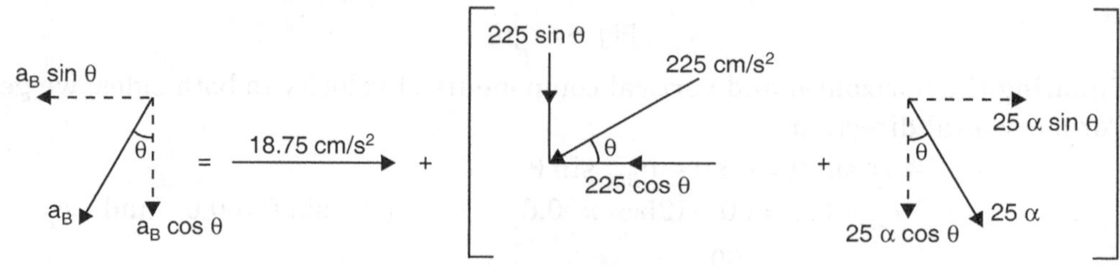

Fig. 5.67 (d)

Equating the horizontal and vertical components of all accelerations to both sides, we get

$$a_B \sin\theta = -\,18.75 + 225\cos\theta - 25\alpha\sin\theta \qquad \text{(For horizontal direction)}$$

and $$a_B \cos\theta = 0 + 225\sin\theta + 25\alpha\cos\theta \qquad \text{(For vertical direction)}$$

But $\sin\theta = 0.6$ and $\cos\theta = 0.8$. Hence above equations becomes

$$a_B \times 0.6 = -\,18.75 + 225 \times 0.8 - 25\alpha \times 0.6 \qquad ...(3)$$

and $$a_B \times 0.8 = 225 \times 0.6 + 25\alpha \times 0.8 \qquad ...(4)$$

From equation (3), $a_B = -\frac{18.75}{0.6} + \frac{225 \times 0.8}{0.6} - \frac{25\alpha \times 0.6}{0.6}$ [Dividing by 0.6]

$= -31.25 + 300 - 25\alpha$

$= 268.75 - 25\alpha$ ...(5)

From equation (4), $a_B = \frac{225 \times 0.6}{0.8} + \frac{25\alpha \times 0.8}{0.8}$ [Dividing by 0.8]

$= 168.75 + 25\alpha$ ...(6)

Equating equations (5) and (6), we get

$$268.75 - 25\alpha = 168.75 = 25\alpha$$

or $$268.75 - 168.75 = 25\alpha + 25\alpha$$

or $$100 = 50\alpha$$

$\therefore$ $$\alpha = \frac{100}{50} = 2 \text{ rad/s}^2$$

Substituting this value of α in equation (5), we get

$$a_B = 268.75 - 25 \times 2 = \mathbf{218.75 \text{ cm/s}^2}. \quad \textbf{Ans.}$$

Graphical Method (For velocity)

The following are the steps for drawing a velocity diagram.

(*i*) Take any point *o*. From *o* draw a horizontal line *oa* towards left from *o* to represent the velocity v_A = 90 cm/s to some suitable scale.

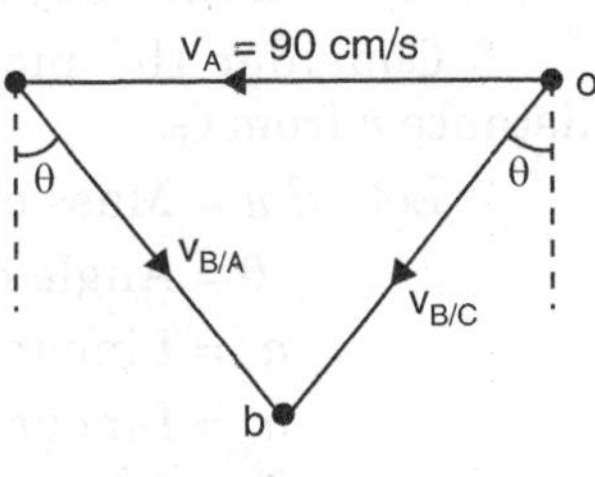

Fig. 5.67 (e)

(*ii*) From point *o*, draw a line *ob* parallel to the direction of velocity v_B (*i.e.*, parallel to inclined surface downwards) to represent the velocity of point *B*. The magnitude of v_B is not known.

(*iii*) From point a, draw a line *ab* perpendicular to line *AB* to represent the rotational velocity of *B* with respect to *A*. *i.e.*, velocity $v_{B/A}$. This line cuts the line *ob* at point *b*. Then *ob* represents the velocity of *b* *i.e.*, v_B. Measure line *ob*.

By measurement,

$$v_B = ob \times \text{scale} = \mathbf{75 \text{ cm/s. Ans.}}$$

$$v_{B/A} = \text{Length } ab \times \text{scale} = 75 \text{ cm/s} = \omega \times r$$

$$= \omega \times 25$$

$\therefore$ $$\omega = \frac{75}{25} = 3 \text{ rad/s.}$$

Graphical Method for Acceleration

(*i*) Take any point *o′*. From *o′*, draw a horizontal line *o′a′* towards right from *o′* to represent the acceleration of *A* *i.e.*, a_A = 18.75 cm/s².

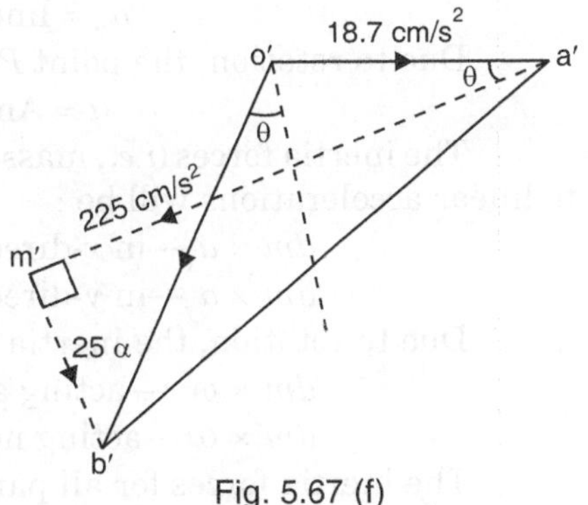

Fig. 5.67 (f)

(*ii*) From point *a′*, draw a line *a′m′* parallel to line *BA* to represent the normal component of acceleration of *B* with respect to *A*. Cut this line = 225 cm/s².

(*iii*) From point *m′*, draw *m′b′* perpendicular to *a′m′* to represent the tangential component. The magnitude is not known.

(*iv*) From point a', draw a line $a'b'$ parallel to inclined surface to represent the acceleration of B. This line cuts the line $m'b'$ at b'. Join $o'b'$. Then $o'b'$ represents the acceleration of B, Measure $o'b'$.

By measurement, $a_B = o'b' =$ **218.75 cm/s²**. **Ans.**

5.11. EQUATIONS OF MOTION FOR PLANE MOTION OF A RIGID BODY

Fig. 5.68 shows a rigid body having a plane motion subjected to several external forces

Let m = Mass of the body,

G = Centre of gravity of the body

F_x = Net force due to external forces in x-direction

F_y = Net force due to external forces in y-direction

M_G = Net moment about G due to external forces.

x_C and y_C = Co-ordinates of point G.

Let under the action of applied forces, the centre of gravity G of the body moves parallel to x-y plane.

Consider the motion of any particle P at a distance r from G.

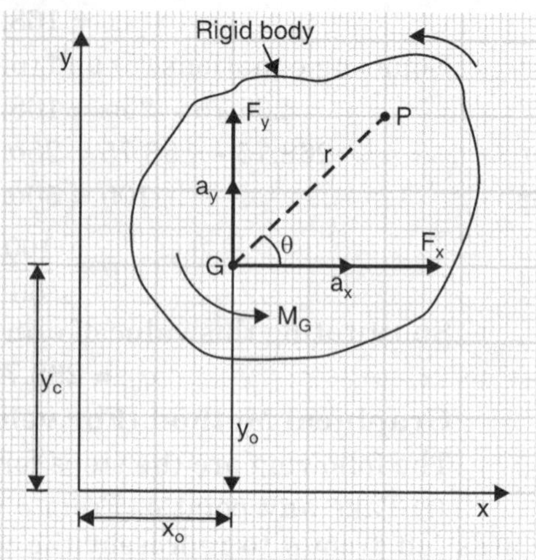

Fig. 5.68

Let dm = Mass of the particle P

θ = Angle made by line PG with x-axis

a_x = Linear acceleration of G in x-direction

a_y = Linear acceleration of G in y-direction

α = Angular acceleration of the body about point G.

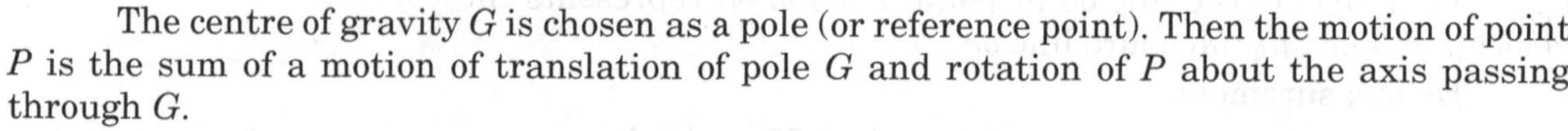

The centre of gravity G is chosen as a pole (or reference point). Then the motion of point P is the sum of a motion of translation of pole G and rotation of P about the axis passing through G.

$\therefore$ Motion of P = Translation of G + Rotation of P about G (vector sum)

Due to translation of the body the point P will also have ;

a_x = linear acceleration in x-direction,

a_y = linear acceleration in y-direction, and

Due to rotation, the point P will have,

α = Angular acceleration about G and ω = angular velocity

The inertia forces (*i.e.*, mass × acceleration = $dm \times a$) acting on the point P corresponding to linear accelerations will be :

$dm \times a_x$—in x-direction opposite to the direction of a_x

$dm \times a_y$—in y-direction opposite to the direction of a_y

Due to rotation, the inertia forces at point P are :

$dm \times \omega^2 r$—acting along line GP (normal component)

$dm \times \alpha r$—acting normal to GP as shown in Fig. 5.70.

The inertia forces for all particles in the body can be obtained by integrating the above inertia forces.

The equations of dynamic equilibrium of the body can be written as :

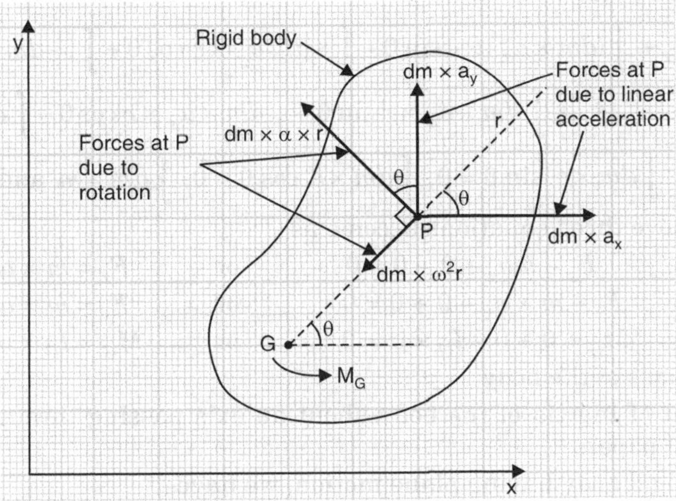

Fig. 5.69. Four forces are acting at *P*.

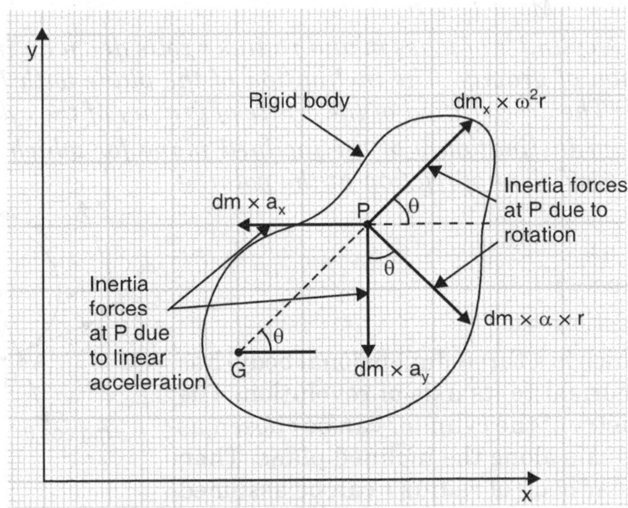

Fig. 5.70. Four inertia forces at *P*.

(*i*) Net external force in *x*-direction + Total Intertia force in *x*-direction = 0

$$\therefore \quad F_x + \left[-\int dm \times a_x + \int dm\, \omega^2 r \cos\theta + \int dm \times \alpha r \sin\theta\right] = 0$$

or

$$F_x - \int dm \times a_x + \int dm\, \omega^2 r \cos\theta + \int dm \times \alpha r \sin\theta = 0$$

For y-direction

$$F_y - \int dm \times a_y - \int dm \times \alpha r \cos\theta + \int dm \times \omega^2 r \sin\theta = 0$$

(*ii*) Net external torque due to external forces + Toal inertia torque about G due to inertia forces shoud be zero.

$$\therefore \quad M_G + \int (dm \times a_x) \times r \sin\theta + \int -dm\, a_y \times r \cos\theta + \int (-dm \times \alpha \times r) \times r = 0$$

or
$$M_G + \int dm \times a_x \times r \sin\theta - \int dm \times a_y \times r \cos\theta - \int dm \times \alpha \times r^2 = 0$$

But
$$\int dm \times r \sin\theta = 0, \int dm \times r \cos\theta = 0, \int dm = m \text{ and } \int dm \times r^2 = I_G$$

Hence the above three equations become as

$$F_x - ma_x + 0 + 0 = 0 \quad \text{or} \quad F_x = m \times a_x$$

$$F_y - m \times a_y - 0 + 0 = 0 \quad \text{or} \quad F_y = m \times a_y$$

and
$$M_G + 0 + 0 - I_G \times \alpha = 0 \quad \text{or} \quad M_G = I_G \times \alpha$$

where m = Total mass of the body

and I_G = Moment of Inertia of the body about an axis passing through G and normal to plane of paper.

Hence for a rigid body having plane motion, we have

$F_x = m \times a_x$, where F_x and F_y are the external forces applied and

$F_y = m \times a_y$ $\quad$ M_G is the external torque applied

and $M_G = I_G \times \alpha$.

Problem 5.44. *A cylinder of mass m and radius r rolls down an inclined plane without sliding due to the action of gravity. The inclination of the plane with horizontal is given by angle* θ. *Determine : (i) The acceleration of the centre of gravity of the cylinder down the plane*

(ii) The maximum inclination θ *of the inclined plane for which the cylinder will roll without sliding, if the coefficient of friction at the point of contact is* μ.

Sol. Given :

Mass = m ; radius = r ;

Coefficient of friction = μ

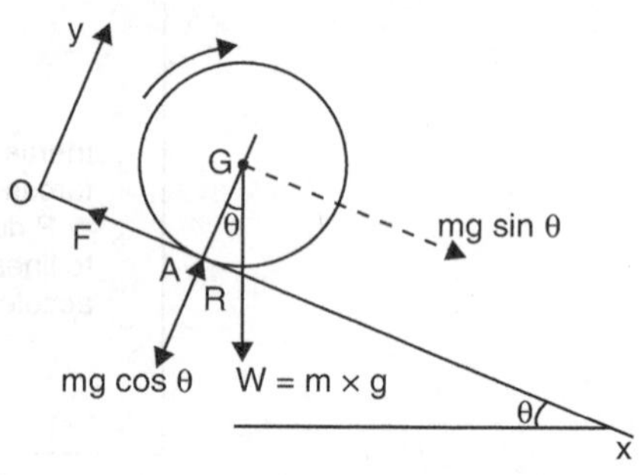

Fig. 5.71

The cylinder (which is a three-dimensional body) has a plane of symmetry normal to its axis of revolution. The vertical plane of symmetry of the cylinder can be chosen as the *x-y* plane, taking *x*-axis along the inclined plane. Then the motion of the circular body of radius r can be assumed in *x-y* plane. The forces acting on the circular body are :

(*i*) *Weight, W.* The components of the weight along *x*-axis (*i.e.*, along inclined plane) and *y*-axis are $W \sin\theta$ and $W \cos\theta$ respectively.

(*ii*) Force of friction, $F = \mu R$

(*iii*) Normal reaction, R

1st Part

Let a = Linear acceleration of C.G. of the body

α = Angular acceleration of the body

For a body, the equation of motion are (when the body is considered as a particle concentrated at its centre of gravity G to which all forces are applied parallel to their actual direction.)

(*i*) Net force + Inertia force = 0 Net force = Mass × Acceleration

(*ii*) Net torque about point G + Inertia torque = 0 Net torque = $I_G \times \alpha$

For x-direction : $W \sin\theta - F = m \times a$

or $mg \sin\theta - \mu R = m \times a$ ...(1) ($\because$ $W = mg$ and $F = \mu R$)

For y-direction : $W \cos\theta - R = 0$ (as acceleration normal to inclined plane is zero)

or $mg \cos\theta - R = 0$ ...(2)

For rotation of the cylinder about its geometric axis, we have

$$F \times r = I_G \times \alpha \quad ...(3)$$

The forces R and W pass through G. Hence net torque about point G is due to force F. This torque is equal to $F \times r$.

But I_G = Moment of Inertia about an axis passing through G and normal to x-y plane

$= \dfrac{mr^2}{2}$ for a cylinder.

and $F = \mu R$

$\therefore$ Equation (3) becomes as [substituting the value of F in equation (3)]

$$(\mu R) \times r = \frac{mr^2}{2} \times \alpha \quad \left(\text{But } a = \alpha \times r \;\therefore\; \alpha = \frac{a}{r}\right)$$

$$= \frac{mr^2}{2} \times \frac{a}{r}$$

$$\therefore \quad R = \frac{\left(\dfrac{mr \times a}{2}\right)}{\mu r} = \frac{m \times a}{2\mu}$$

Eliminate R from equation (1)

Substituting this value of R in equation (1), we get

$$mg \sin\theta - \mu \times \frac{ma}{2\mu} \times m \times a$$

or $$g \sin\theta - \frac{a}{2} = a$$

or $$g \sin\theta = \frac{a}{2} + a = \frac{3a}{2}$$

or $$a = \frac{\mathbf{2g \sin\theta}}{\mathbf{3}}.$$ **Ans.**

2nd Part

The actual force of friction can be obtained from equation (1)

From (1), $mg \sin\theta - F = m \times a$

or $$F = mg \sin\theta - m \times a$$

$$= m\,[g \sin\theta - a]$$

$$= m\left[g \sin\theta - \frac{2g}{3}\sin\theta\right]$$

$$= m \times \frac{g}{3}\sin\theta \quad ...(a)$$

For any inclination θ of the plane, the maximum possible frictional force is equal to μR.

$\therefore$ Maximum frictional force, $F_{max} = \mu R$

$$= \mu \times W \cos\theta \quad ...(b)$$

The cylinder will roll without sliding if actual force of friction (F) is less than the maximum possible force of friction (F_{max}). The frictional force given by (a) should be less than given by (b).

$\therefore$ Condition of rolling without sliding is ($F \le F_{max}$)

$$m \times \frac{g}{3} \sin\theta \le \mu W \cos\theta$$

$$\le \mu \times mg \cos\theta$$

or $$\sin\theta \le 3\mu \cos\theta$$

For $\mu = 0.192$, we have

$$\sin\theta \le 3 \times 0.192 \cos\theta$$

or $$\tan\theta \le 0.576$$

or $$\theta \le \tan^{-1} 0.576 \le 30°. \quad \textbf{Ans.}$$

HIGHLIGHTS

1. Force of friction always acts in the direction opposite to the direction of motion.
2. The maximum value of frictional force acting on a body, when the body is on the point of motion, is called limiting force of friction. It is denoted by F.
3. The force of friction, acting on a body when the body is moving, is called dynamic friction.
4. The ratio of the limiting force of friction (F) to the normal reaction (R) between two bodies is known as coefficient of friction. It is denoted by μ. Mathematically, $\mu = \frac{F}{R}$.
5. The angle made by the resultant of the normal reaction (R) and the limiting force of friction (F) with the normal reaction is known as angle of friction. It is denoted by ϕ.
6. The relation between angle of friction (ϕ) and coefficient of friction (μ) is expressed as $\tan\phi = \mu$.
7. The force of friction is always equal to μR, where R is normal reaction.
8. If a ladder is leaning against a smooth vertical wall, the force of friction between ladder and vertical wall will be zero.
9. The problems on ladder friction are solved by taking algebraic sum of horizontal and vertical components of the forces to be zero and also taking the moments of all forces about any point to be zero.
10. If a body is placed on a rough inclined plane and the angle of inclination of the plane is gradually increased till the body just starts sliding down the plane, the angle of the inclined plane, at which the body just begins to slide down the plane, is called angle of repose.
11. Angle of repose is equal to angle of friction.
12. If the inclination of the plane, with the horizontal is less than angle of friction, the body placed on the inclined plane will be always in equilibrium without any external force.
13. The minimum force required to drag a body of weight W placed on a rough horizontal plane, when the force is applied at an angle θ with the horizontal, is equal to $W \sin\theta$. And the angle θ will be equal to angle of friction for the least force.

14. The force required to drag a body of weight W, placed on a rough inclined plane of inclination α to the horizontal, when force is applied to the body in such a way that the force makes an angle θ with the inclined plane is given by

$$P = W \frac{\sin(\alpha + \phi)}{\cos(\theta - \phi)} \text{ ... when the body is on the point of moving up the plane.}$$

$$= W \frac{\sin(\alpha - \phi)}{\cos(\theta + \phi)} \text{ ... when the body is on the point of moving down the plane.}$$

where α = Inclination of the inclined plane with horizontal.
θ = Angle made by the force with the inclined plane.
ϕ = Angle of friction.

EXERCISE 5

(A) Theoretical Questions

1. Define the terms : Friction, limiting force of friction, coefficient of friction and angle of friction.

2. Explain the difference between coefficient of friction and angle of friction. *(AMIE Winter, 1981)*

3. (*a*) State the laws of static and dynamic friction. *[AMIE (S) 1982, (S) 1986]*
(*b*) State the laws of solid friction. *[AMIE Summer, 1983, (S) 1989]*

4. Prove that the angle of friction is equal to the angle of the inclined plane, when a solid body of weight W placed on the inclined plane, is about to slide down.

5. What do you mean by 'angle of repose' ? Prove that angle of repose is equal to the angle of friction.

6. A body of weight W is placed on an inclined rough plane. The inclination of the plane with the horizontal is less than the angle of friction. The body will
(*a*) be in equilibrium (*b*) move downwards
(*c*) move upwards (*d*) none of the above. **[Ans.** (*a*)**]**

7. If in the above question, the inclination of the plane with the horizontal is more than the angle of friction, the body will
(*a*) be in equilibrium (*b*) move downwards
(*c*) move upwards (*d*) none of the above. **[Ans.** (*b*)**]**

8. A body of weight W is placed on a rough inclined plane having inclination α to the horizontal. The force P is applied horizontally to drag the body. If the body is on the point of motion up the plane, prove that P is given by

$$P = W \tan(\alpha + \phi)$$

where ϕ = Angle of friction.

9. In the above question, if the body is on the point of motion down the plane, prove that the force P is given by

$$P = W \tan(\alpha - \phi).$$

10. A body of weight W is placed on a rough inclined plane having inclination α to the horizontal. A force P is applied to the body in such a way that it makes an angle θ to the inclined plane. Prove that the force required to drag the body when the body is on the point of motion up the plane is given by

$$P = \frac{W \sin(\alpha + \phi)}{\cos(\theta - \phi)}$$

where α = Inclination of the inclined plane with horizontal
θ = Inclination of the force with the inclined plane
ϕ = Angle of friction.

11. Prove that in question 10, (*i*) the force P will be minimum if the angle of inclination of the force with the inclined plane is equal to the angle of friction and (*ii*) least force is given by

$$P_{min} = W \sin(\alpha + \phi).$$

12. Define static and kinetic friction and state the laws of solid friction. (*AMIE Summer, 1983*)

13. Derive an expression for the least force required to drag a body on a rough horizontal plane.

(B) Numerical Problems

1. A body of weight 90 N is placed on a rough horizontal plane. Determine the coefficient of friction if a horizontal force of 63 N just causes the body to slide over the horizontal plane. [**Ans.** 0.7]

2. A body of weight 150 N is placed on a rough horizontal plane. If the coefficient of friction between the body and the horizontal plane is 0.4, determine the horizontal force required to just slide the body on the plane. [**Ans.** 60 N]

3. The force required to pull a body of weight 40 N on a rough horizontal plane is 15 N. Determine the coefficient of friction if the force is applied at an angle of 20° with the horizontal. [**Ans.** 0.404]

4. A body of weight 60 N is placed on a rough horizontal plane. To just move the body on the horizontal plane, a push of 18 N inclined at 20° to the horizontal plane is required. Find the coefficient of friction. [**Ans.** 0.255]

5. A pull of 60 N inclined at 25° to the horizontal plane, is required just to move a body placed on a rough horizontal plane. But the push required to move the body is 75 N. If the push is inclined at 25° to the horizontal, find the weight of the body and coefficient of friction. [**Ans.** 253.83 N, 0.238]

6. Find the least force required to pull a body of weight W placed on a rough horizontal plane, when the force is applied at an angle θ with the horizontal. [**Ans.** $W \sin \theta$]

7. A body of weight 450 N is pulled up an inclined plane, by a force of 300 N. The inclination of the plane is 30° to the horizontal and the force is applied parallel to the plane. Determine the coefficient of friction. [**Ans.** 0.192]

8. A body of weight 400 N is pulled up along an inclined plane having inclination 30° to the horizontal at a steady speed. If the coefficient of friction between the body and the plane is 0.3 and force is applied parallel to the inclined plane, find the force required. Find also the work done on the body if the distance travelled by the body is 10 m along the plane. [**Ans.** 303.92 N, 3039.2 N m]

9. An effort of 180 N is required just to move a certain body up an inclined plane of angle 15°, the force being parallel to the plane. If the angle of inclination of the plane is made 20°, the effort required, again applied parallel to the plane, is found to be 210 N. Find the weight of the body and coefficient of friction. [**Ans.** 386.68 N, 0.214]

10. A cord connects two bodies of weights 500 N and 1000 N. The two bodies are placed on an inclined plane and cord is parallel to inclined plane. The coefficients of friction for the weight of 500 N is 0.20 and that of 1000 N is 0.4. Determine the inclination of the plane to the horizontal and tension in the cord when the motion is about to take place, down the inclined plane. The body weight 500 N is below the body weighing 1000 N. [**Ans.** 18.417°, 63.086 N]

11. A uniform ladder of length 13 m and weighing 30 N is placed against a smooth vertical wall with its lower end 10 m from the wall. In this position the ladder is just to slip. Determine : (*i*) the coefficient of friction between the ladder and the floor, and (*ii*) frictional force acting on the ladder at the point of contact between ladder and floor. [**Ans.** (*i*) 0.602 (*ii*) 18.06 N]

12. A uniform ladder of length 10 m and weighing 20 N is placed against a smooth vertical wall with its lower end 6 m from the wall. The coefficient of friction between the ladder and floor is 0.45. Show that the ladder will remain in equilibrium in this position. What is the frictional force acting on the ladder at the point of contact between the ladder and floor ? [**Ans.** 7.5 N]

13. A uniform ladder of weight 800 N and of length 7 m rests on a horizontal ground and leans against a smooth vertical wall. The angle made by the ladder with the horizontal is 60°. When a man of weight 600 N stands on the ladder at a distance 4 m from the top of the ladder, the ladder is at the point of sliding. Determine the coefficient of friction between the ladder and the floor.
[**Ans.** 0.237]

14. A uniform ladder of weight 250 N and of length 5 m rests on a horizontal ground and leans against a rough vertical wall. The coefficient of friction between the ladder and floor is 0.3 and between the ladder and vertical wall is 0.2. When a weight of 900 N is placed on the ladder at a distance of 2 m from the top of the ladder ; the ladder is at the point of sliding.

Find : (*i*) The angle made by the ladder with horizontal.

(*ii*) Reaction at the foot of the ladder, and

(*iii*) Reaction at the top of the ladder. [**Ans.** (*i*) 61.52°, (*ii*) 1084.9 N, (*iii*) 325.47 N]

15. A uniform ladder of length 15 m rests against a vertical wall making an angle of 60° with the horizontal. Coefficient of friction between wall and ladder, and ground and ladder are 0.3 and 0.25. A man weighing 500 N ascends the ladder. How high will he be able to go before the ladder slips ? Find the weight necessary to be put at the bottom of the ladder so as to be just sufficient to permit the man to go to top. Assume weight of ladder as 850 N.
(*AMIE Winter, 1974, Old Scheme*) [**Ans.** 6.39 m, 607.6 N]

16. A uniform ladder 3 m long weighs 18 N. It is placed against a wall making an angle 60° with the floor as shown in Fig. 5.72. The coefficient of friction between the wall and ladder is 0.25 and that between the floor and ladder is 0.35. The ladder, in addition to its own weight, has to support a man weighing 90 N at its top at A. Calculate :

(*i*) The horizontal force F to be applied to the ladder at the floor level to prevent slipping.

(*ii*) If the force F is not applied what would be the minimum inclination of the ladder with the horizontal so that there is no slipping of it with the man at the top ?

(*AMIE Winter, 1977*) [**Ans.** (*i*) 16.6 N, (*ii*) 69°]

Fig. 5.72

17. A ladder 5 m long rests on a horizontal ground and leans against a smooth vertical wall at an angle of 70° with the horizontal. The weight of the ladder is 90 N and acts at its middle. The ladder is at the point of sliding, when a man weighing 75 N stands on a rung 3.5 m from the top of the ladder. Calculate the coefficient of the friction between the ladder and the floor.
[**Ans.** 0.15]

18. A uniform ladder of 7 m rests against a vertical wall with which it makes an angle of 45°. The coefficient of friction between the ladder and the wall is 1/3 and that between ladder and floor is 0.5. If a man whose weight is one-half of that of the ladder, ascends its, how high will it be when the ladder slips. [**Ans.** 5.25 m]

19. A body of weight 100 N is at rest on a horizontal plane. A horizontal force of 70 N just causes the body to slide.

Determine : (*i*) limiting force of friction and (*ii*) coefficient of friction. [**Ans.** (*i*) 70 N, (*ii*) 0.7]

20. The coefficient of friction between a body of weight 100 N and the rough horizontal plane on which the body rests is 0.3. Calculate the horizontal force required just to cause the body to slide over the horizontal plane.

If the body is loaded with an additional weight of 40 N, find the least horizontal force which will cause the body to slide. [**Ans.** (*i*) 30 N, (*ii*) 45 N]

QUESTION BANK

QUESTION BANK

UNIT 1

1. Determine the resultant of four forces tangent to the circle of radius 3 m shown in Fig. 1. What will be its location with respect to the centre of the circle.

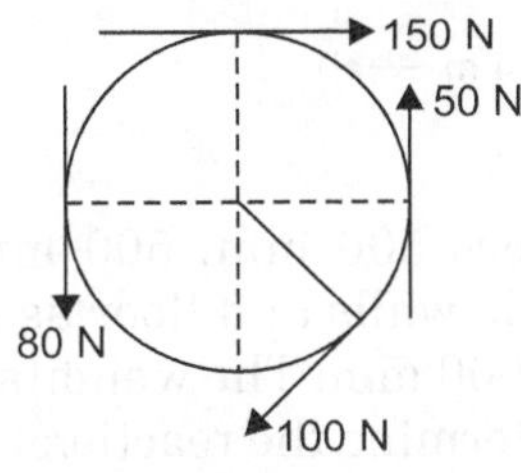

Fig. 1

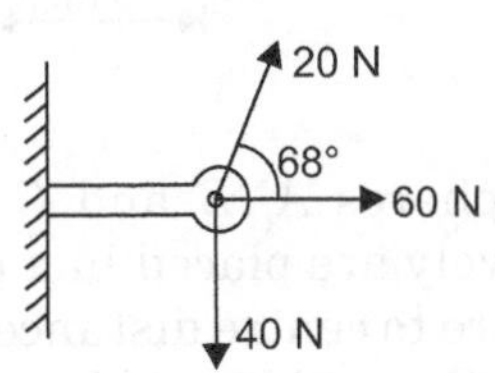

Fig. 2

2. Three wires exert the tensions indicated on the ring in Fig. 2. Assuming a concurrent system determine the force in a single wire which will replace three wires.
3. Determine algebraically the reactions on the beam loaded as shown in Fig. 3. Neglect the thickness and mass of the beam.

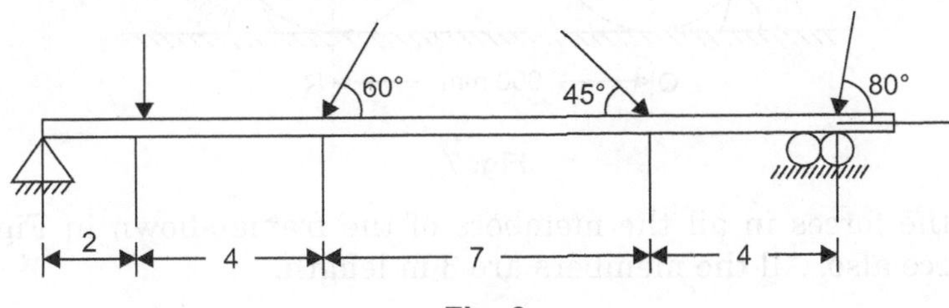

Fig. 3

4. A roller shown in Fig. 4 is of mass 150 kg. What force T is necessary to start the roller over the block A?

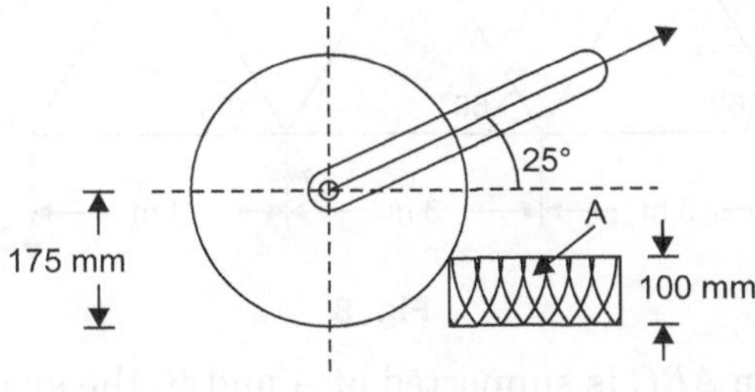

Fig. 4

5. Express in terms of θ, β and W the force T necessary to hold the weight in equilibrium (Fig. 5). Also derive an expression for the reaction of the plane on W. No friction is assumed between the weight and the plane.

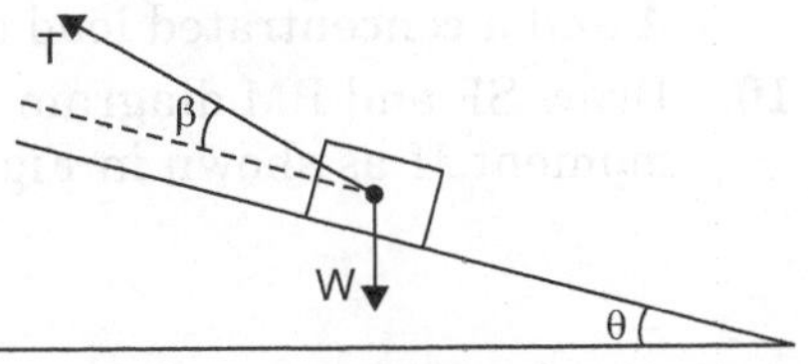

Fig. 5

6. A system of levers is shown supporting a load of 80 N (Fig. 6). Determine the reaction at A and B on the lever.

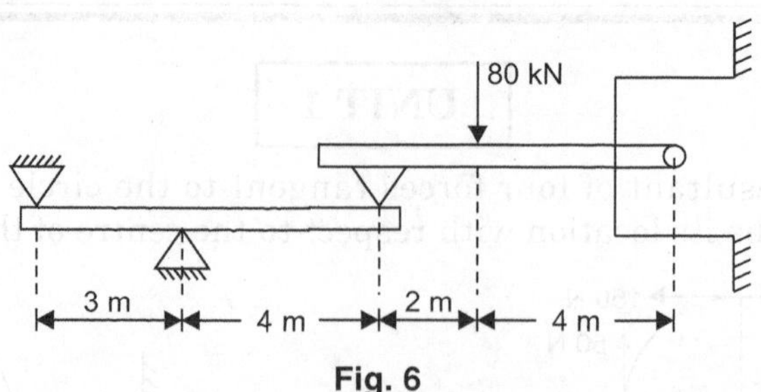

Fig. 6

7. Three spheres A, B and C having their diameters 500 mm, 500 mm and 800 mm respectively are placed in a trench with smooth side walls and floor as shown in Fig. 7. The centre to centre distance of spheres A and B is 600 mm. The weights of the cylinders A, B and C are 4 kN, 4 kN and 8 kN respectively. Determine the reactions at P, Q, R and S.

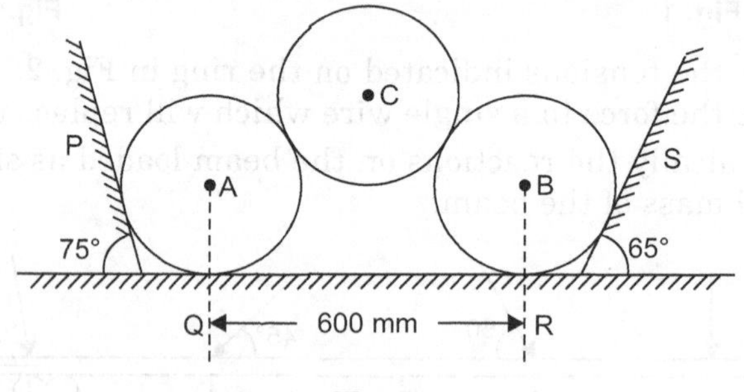

Fig. 7

8. Determine the forces in all the members of the frame shown in Fig. 8. Indicate the nature of force also. All the members are 3 m length.

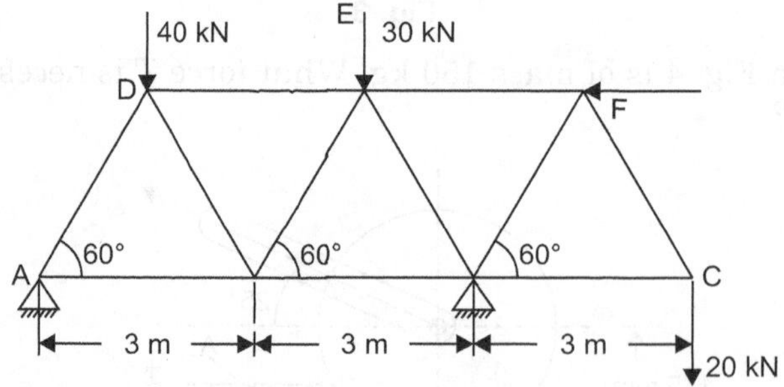

Fig. 8

9. An overhanging beam ABC is supported at A and B, the span AB is 6 m. The overhang BC is 2 m. It carries a uniformly distributed load of 30 kN/m over a length of 3 m from A and a concentrated load of 20 kN at free end. Draw SF and BM diagrams.

10. Draw SF and BM diagram for the beam with a central moment M as shown in Fig. 9.

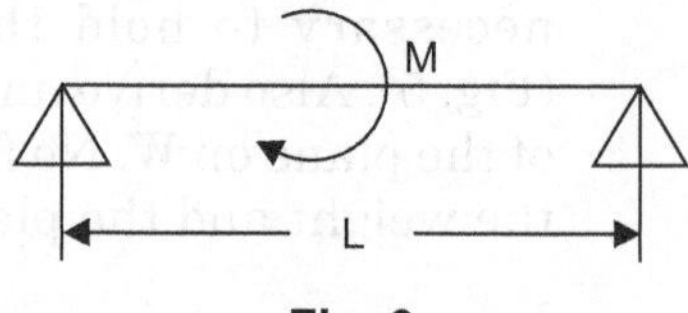

Fig. 9

11. State the principle of transmissibility. What will happen if the equivalent forces *F* and *F* acting on a rigid body are not in line ? Explain.

12. What is the 'moment of a couple' ? Find the magnitude of the couple for two forces F_1 and F_2 both equal in magnitude co-opposite in direction acting at d_1 and d_2 distance from a fixed point in the body.

13. What is a rigid body ? What will happen if (*i*) Force (*ii*) Couple is applied to (*a*) Rigid body (*b*) None Rigid body ?

14. State the conditions necessary for equilibrium of rigid body. What will happen of one of the conditions are not satisfied ?

15. "Any system of coplanar forces can be reduced to a force-couple system at an arbitrary point". Explain the above statement by assuming a suitable system.

16. A fixed crane of 1000 kg mass is used to lift 2400 kg crates. It is held in place by a pin at *A* and a rocker at *B*. The C.G. is located at *G*. Determine the components of reactions at *A* and *B* after drawing the free body diagram (Fig. 10).

[**Hint.** The reaction at pin will have a horizontal and vertical component while at rocker these will be horizontal reaction only.]

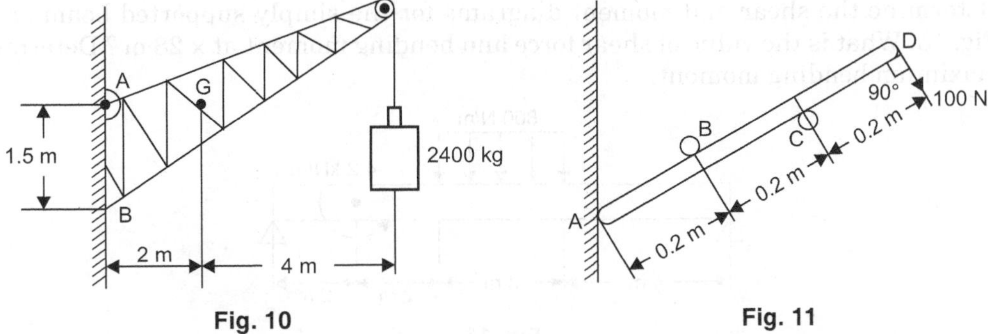

Fig. 10

Fig. 11

17. A light road *AD* is supported by frictionless pegs at *B* and *C* and rests against a frictionless wall at *A*. A force of 100 N is applied at end D. Determine the reactions at *AB* and *C* (Fig. 11).

18. A uniform rod *AB* of length 3*R* and weight *W* rests inside a hemispherical bowl of radius *R* as shown in Fig. 12. Neglecting friction determine the angle θ.

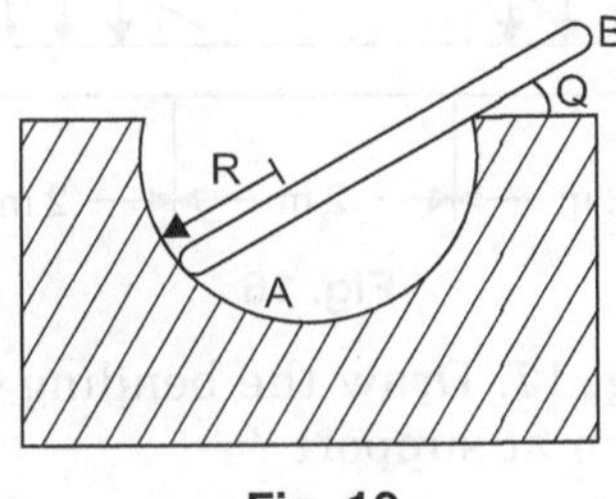

Fig. 12

19. State Varignon's theorem. How it can help on determination of moments. Can a single force balance a couple ? Explain.

20. Explain the following statements with example:

(*i*) A force has no moment about an axis parallel to its line of action.

(*ii*) A force has no moment about an axis that intersects its line of action.

21. Two parallel forces are acting at points A and B respectively are equivalent to a force of 100 N acting downwards at C and couple of 200 Nm. Find the magnitude and sense of forces F_1 and F_2 shown in Fig. 13.

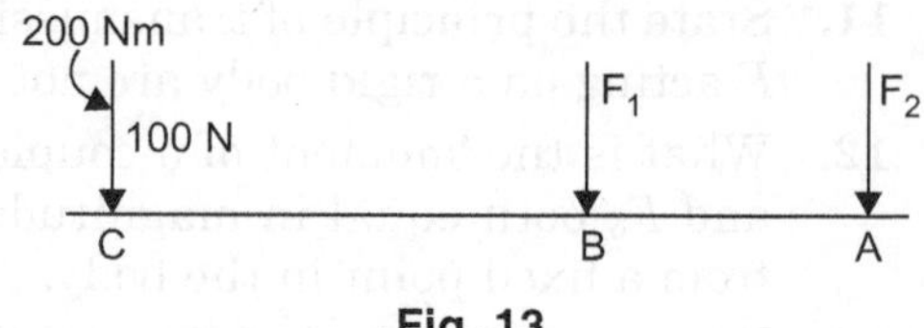

Fig. 13

22. Compute the reactions at A and B for the beam subjected to distributed and point loads (Fig. 14). State what type of beam it is ?

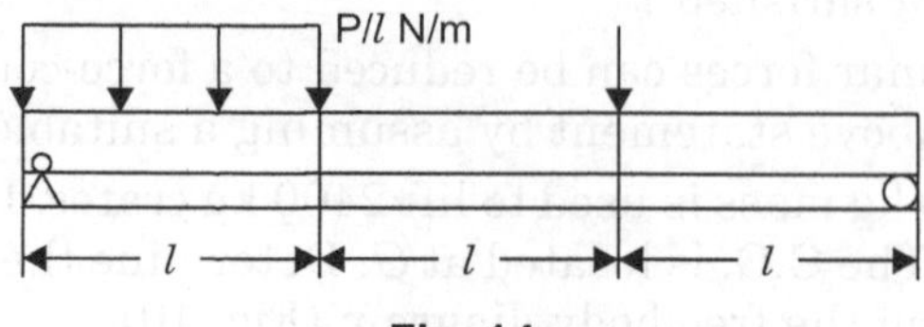

Fig. 14

23. What are statically determinate and indeterminate beams. Give examples of each along with figures.

24. Determine the shear and moment diagrams for the simply supported beam shown in Fig. 15. What is the value of shear force and bending moment at × 28 m ? Determine the maximum bending moment.

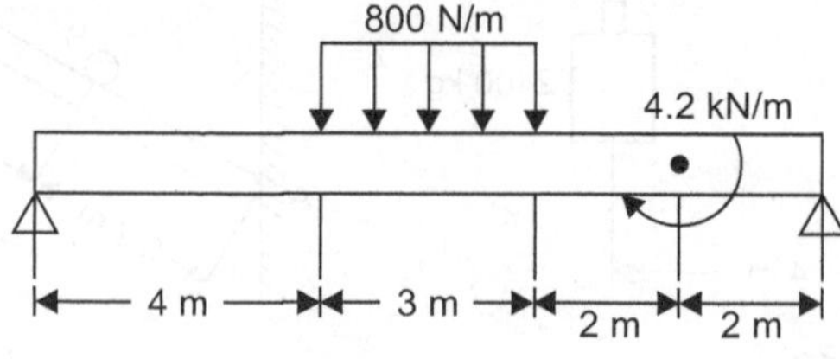

Fig. 15

25. Draw the shear force and bending moment diagram of the loaded simply supported beam (Fig. 16). Also find the maximum bending moment.

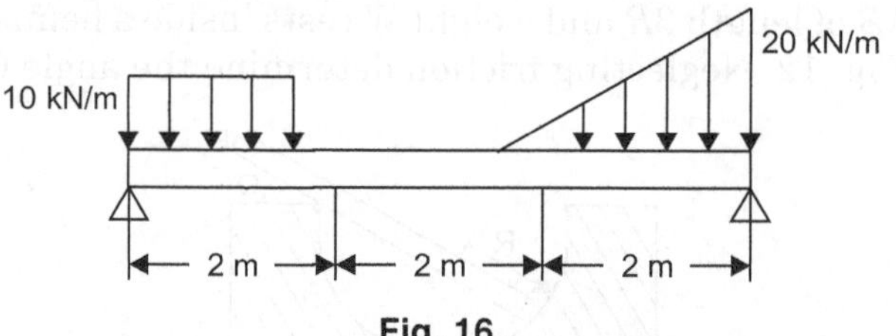

Fig. 16

26. A cantilever is shown in Fig. 17. Draw the bending moment diagram and shear force diagram. What is the reaction at support ?

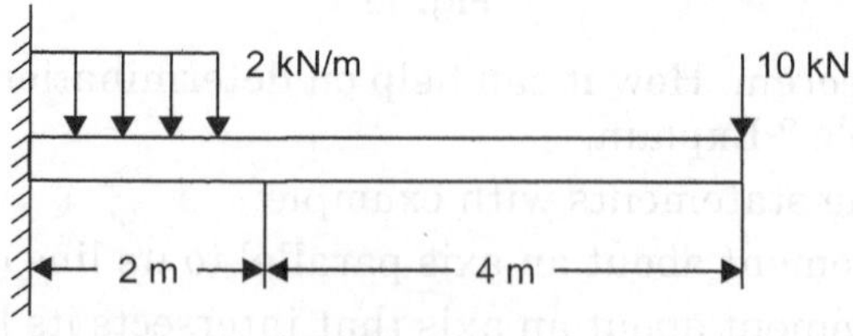

Fig. 17

27. A cantilever is shown in Fig. 18. Draw the shear force diagram and bending moment diagram.

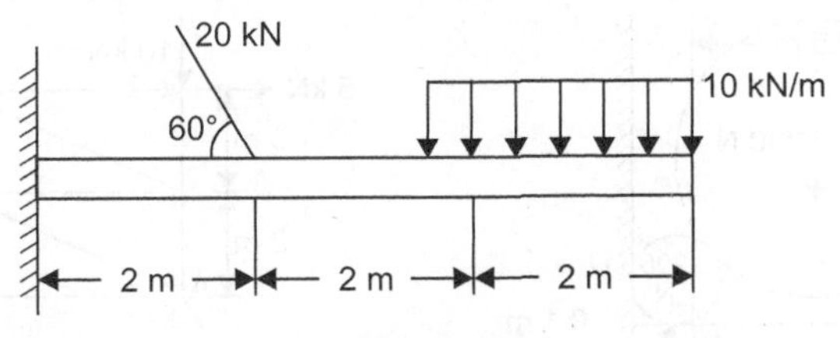

Fig. 18

28. Write the equations of equilibrium for a particle subjected to coplanar forces.

29. What do you mean by 1 Newton ?

30. What do you mean by equilibrium of a rigid body ?

31. How do you classify the force system ?

32. What do you mean by position time curve ?

33. A rigid bar *CD* is subjected to a system of parallel forces as shown in Fig. 19. Reduce the given system of force to an equivalent force couple system at *F*.

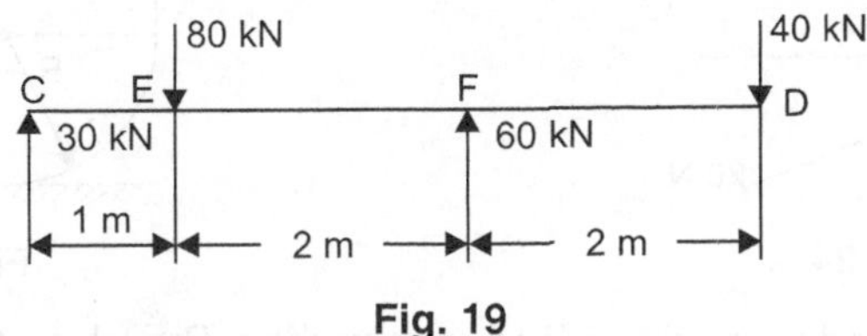

Fig. 19

34. Three spheres *A, B, C* are placed in a groove shown in Fig. 20. The diameter of each sphere is 100 mm. Sketch the free body diagram of *B*. Assume the weights of spheres *A, B, C* as 1 kN, 2 kN and 1 kN respectively.

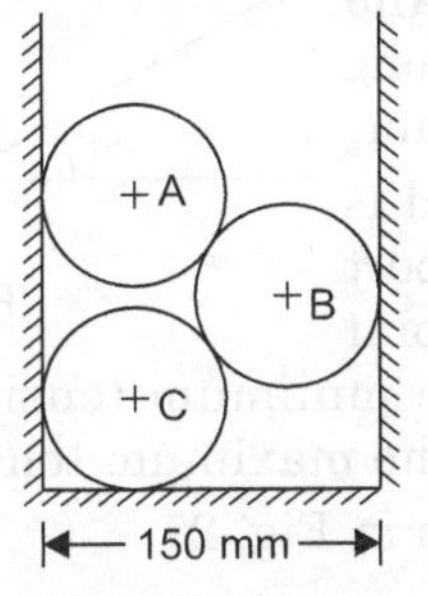

Fig. 20

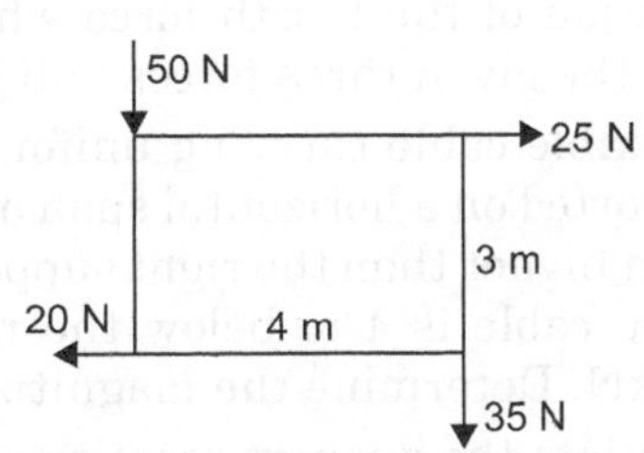

Fig. 21

35. A system of forces are acting at the corners of a rectangular block as shown in Fig. 21. Determine magnitude and direction of resultant.

36. Find the magnitude and direction cosines of the resultant of 3 space forces $4\hat{i} - 8\hat{j} + 10\hat{k}$; $-6\hat{k} - 5\hat{j} + 4\hat{k}$ and $3\hat{i} + 7\hat{j} - 2\hat{k}$.

37. For the cylinders shown in Fig. 22. Determine reactions at *C*.

38. Find the reactions at the supports (Fig. 23).

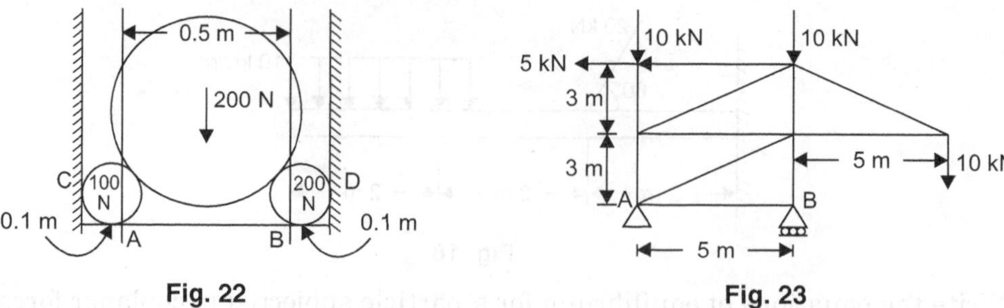

Fig. 22 Fig. 23

39. For the system shown in Fig. 24, find the additional single force required to maintain equilibrium.

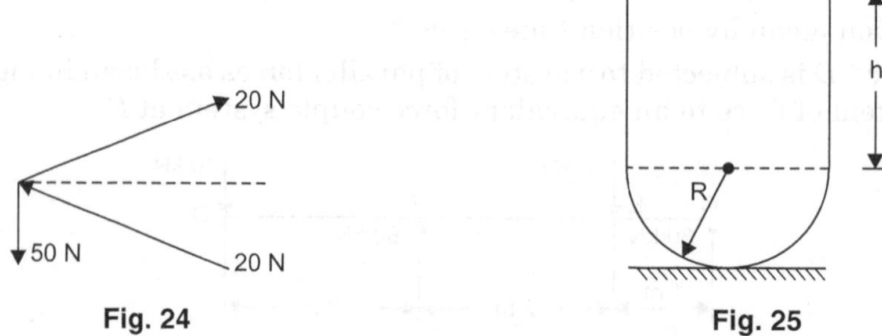

Fig. 24 Fig. 25

40. A uniform body is made up of a cylinder of radius R and rectangular prism of height h which has a common base (Fig. 25). What is the ratio of h to R of the body to be stable ?

41. Three forces act on a particle O as shown in Fig. 26. Determine the value of P such that the resultant of these three forces is horizontal. Find the magnitude and direction of the fourth force which when acting along with the given three forces will keep O in equilibrium.

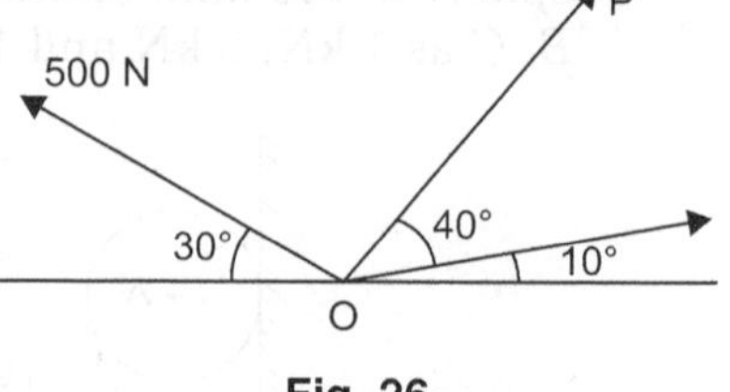

Fig. 26

42. A flexible cable carrying uniformly distributed load is supported on a horizontal span of 30 m. The left support is 5 m higher than the right support and the lowest point of the cable is 4 m below the right support. If the minimum tension in the cable is 180 kN. Determine the magnitude of udl and also the maximum tension in the cable.

43. Calculate the support reactions for the beam shown in Fig. 27.

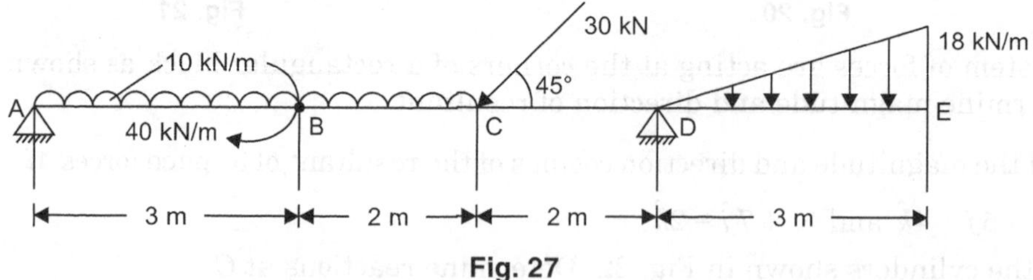

Fig. 27

44. Two concurrent forces of magnitudes 40 N and 30 N acting at origin O are directed respectively towards the points A (3, 0, – 3) and B (2, – 2, 4). Determine the magnitude of the resultant and its inclination with x, y-and z-axes.

45. Calculate the support reactions for the following. Refer Fig. 28 and Fig. 29.

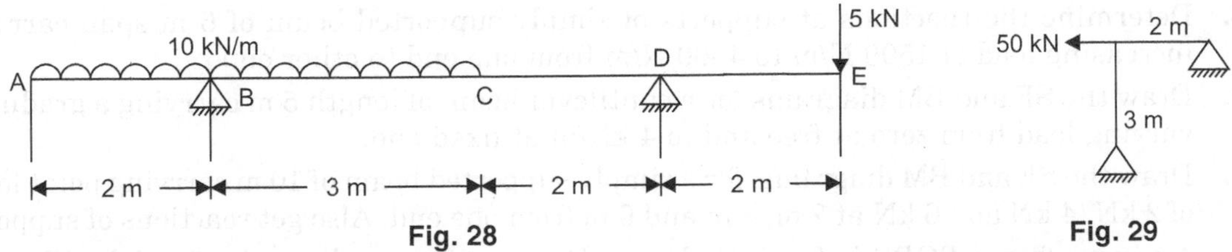

Fig. 28 Fig. 29

46. Two bodies of weights 25 N and 50 N are connected by a light inextensible string passing over a small smooth fixed pulley. Find the velocity of the system after the blocks move by 1 m, if they are initially at rest.

47. Fig. 30 shows a system of levers supporting a load of 500 N. Determine reactions at the supports *A* and *B*.

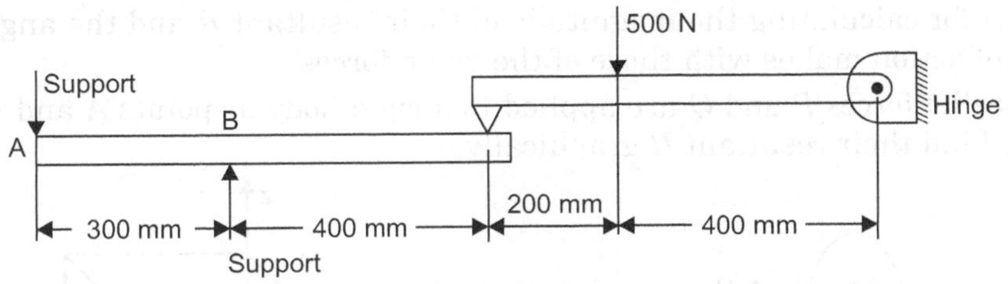

Fig. 30

48. (*i*) State the polygon law of forces of finding the resultant of several concurrent forces.

(*ii*) State Varignon's Theorem of Moments.

49. (*i*) State Lami's theorem.

(*ii*) A space force has components 20 N, 30 N and – 40 N along *x*, *y*-and *z*-axes respectively. Find the resultant and the unit vector along direction of the force.

50. State and explain the principle of transmissibility of forces.

51. Four forces of magnitude *P*, 2*P*, 5*P* and 4*P* are acting at a point. Angles made by these forces with *x*-axis are 0°, 75°, 150° and 225° respectively. Find the magnitude and direction of resultant force.

52. Four coplanar forces are acting at a point. Three forces have magnitudes of 20, 50 and 20 N at angles of 45°, 200° and 270° respectively. Fourth force is unknown. Resultant force has a magnitude of 50 N and acts among *X* axis at angle of 0°. Determine the unknown force and its direction from *x*-axis analytically and graphically.

53. Show that a given force *F* applied to a body at any point '*A*' can always be replaced by an equal force applied at another point '*B*' together with a couple.

54. Determine the resultant of parallel forces acting on a horizontal beam with magnitudes (– 10), (+ 40), (– 30), (+ 10), (+ 10) N at distances of 0, 2 m, 2.5 m, 3.5 m, and 4 m respectively.

55. Two cylindrical identical rollers *A* and *B*, each of weight *W* are supported by an inclined plane and vertical wall. Assuming all surfaces to be smooth, draw free body diagrams of (*i*) roller *A*, (*ii*) roller *B* and (*iii*) rollers *A* and *B* taken together.

56. A lamp of mass 1 kg is hung from the ceiling by a chain and is pulled aside by a horizontal chord until the chain makes an angle of 60° with ceiling. Find the tensions in chain and chord.

57. Explain 'action' and 'reaction' with the help of suitable examples.

58. Determine the reactions at supports of simply supported beam of 6 m span carrying increasing load of 1500 N/m to 4500 N/m from one end to other end.

59. Draw the SF and BM diagrams for a cantilever beam of length 5 m carrying a gradually varying load from zero at free and to 4 kN/m at fixed end.

60. Draw the SF and BM diagrams, for a simply supported beam of 10 m carrying point loads of 2 kN, 4 kN and 6 kN at 2 m, 4 m and 6 m from one end. Also get reactions of supports.

61. A square frame *PQRS* is fabricated out of four uniform rods, each of weight *W* and is stiffened by another rod *QS* of the same weight. Find the thrust in rod *QS* if the frame is suspended vertically from *P*.

62. A force of magnitude *F* is acting from a point (4, 1, 4) metres towards a point (– 3, 4, – 1) metres. Determine the moment of *F* about *x* and *y*-axis if it causes a moment of M_z = 1900 Nm.

63. If two given forces *P* and *Q*, acting under the angle θ, are applied to a body at *A*, find formulas for calculating the magnitude of their resultant *R* and the angles and which its line of action makes with those of the given forces.

64. Two parallel forces *P* and *Q* are applied to a rigid body at points *A* and *B* as shown in Fig. 31. Find their resultant *R* graphically.

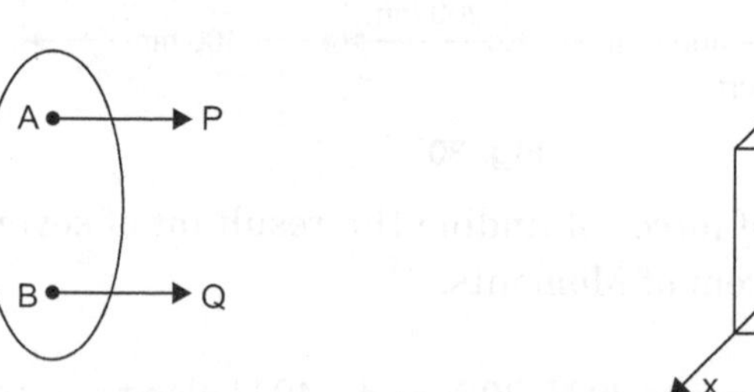

Fig. 31 **Fig. 32**

65. If the force *F* (Fig. 32) is 100 N and passes through a point whose coordinate is 4, 5 and 3. Determine the rectangular components of *F*. Also find the component of *F* in the *x*-*y* plane.

66. The 100 N force is applied to the bracket as shown in Fig. 33. Determine the components of *F* in

(*i*) the *x*- and *y*-directions

(*ii*) the *x*′- and *y*′-directions

(*iii*) the *x*- and *x*′-directions.

67. Determine the *x*- and *y*-components of the force exerted on the pin at *A* (Fig. 34).

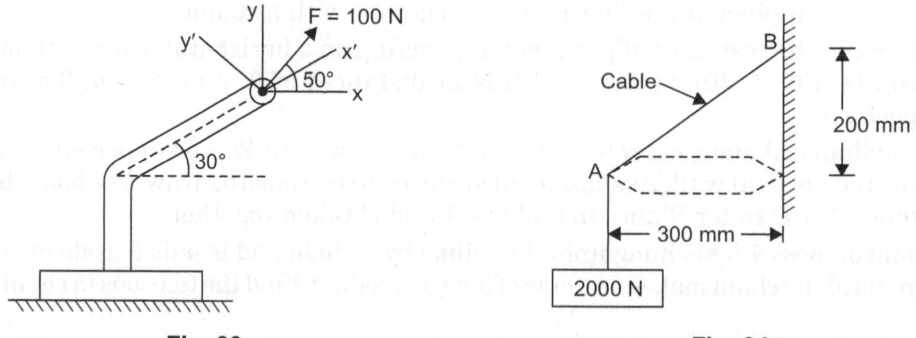

Fig. 33 **Fig. 34**

68. The two forces shown in Fig. 35, are to be replaced by an equivalent force R applied at the point P. Locate P by finding its distance x from AB and specify the magnitude of R and the angle O it makes with the horizontal.

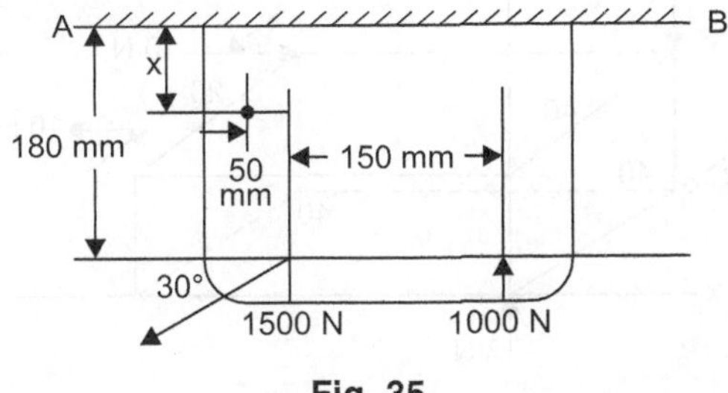

Fig. 35

69. A 100 A force which make an angle of 45 with the horizontal x-axis is to be replaced by two forces, a horizontal force F and a second force of 75 N magnitude. Find F.

70. Determine the direction cosine of the force F (Fig. 36). If the x component of the force F is 300 N. Find F.

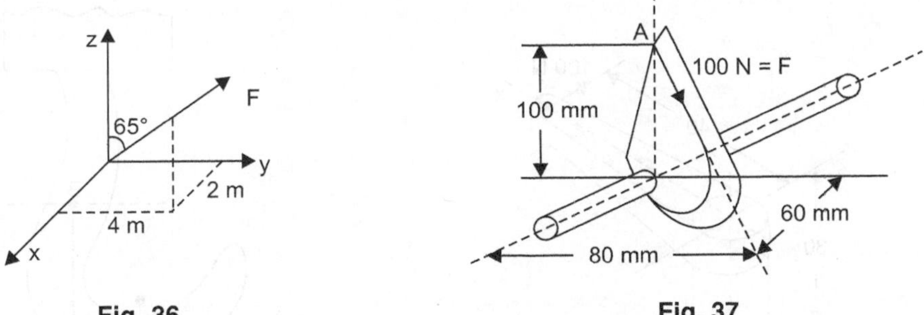

Fig. 36 Fig. 37

71. A force $F = 100$ N is applied to the arm shown in Fig. 37 which is attached to the rigid shaft. Determine the moment, M of F about the shaft axis.

72. A force F is applied at a point whose coordinates are x, y, z. Show that the moment of F about the x-axis is $M_x = F_{zy} - F_{yz}$ and write the expressions for the moments M_y and M_z about the y- and z-axes respectively.

73. Determine the moment of the 50 N force about the point A (Fig. 38).

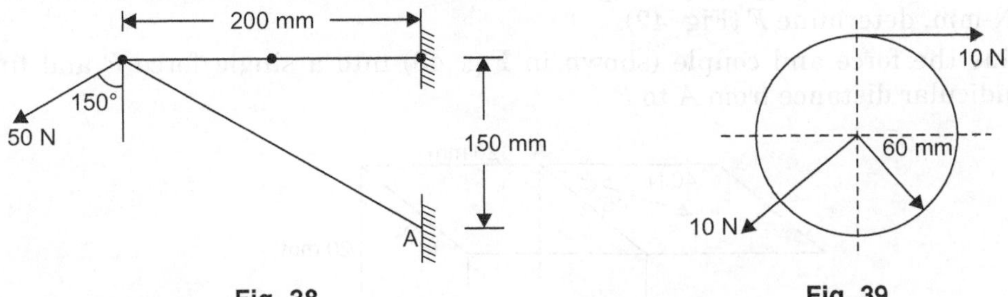

Fig. 38 Fig. 39

74. A wheel is subjected to two forces as shown in Fig. 39. Determine algebraically the coordinates of the point of the rim of the wheel about which the combined moment, M, of the two forces is a maximum. Find M for this point.

75. Determine the magnitude and direction of the vector *M* which represents the resultant of the two given couple (Fig. 40).

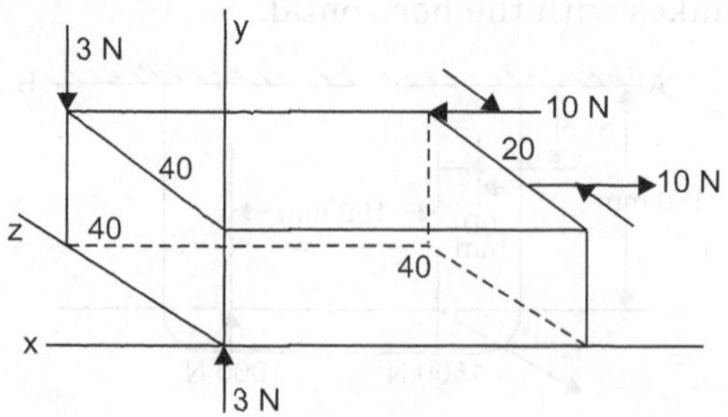

Fig. 40

76. A force of 100 N is applied to a lever which is attached to a fixed shaft as shown in Fig. 41. In determining the effect of the force on the shaft at a section such as *A* it is convenient to replace the force by a force at *A* and a couple *M*. Find the magnitude and direction of *M*.

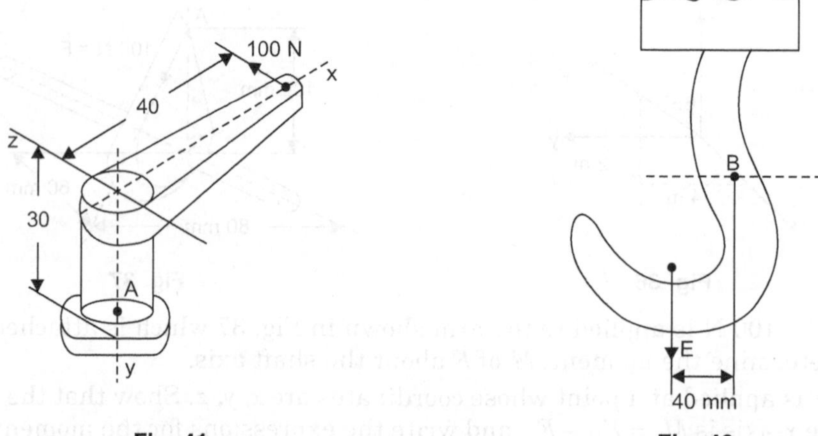

Fig. 41 Fig. 42

77. In designing the lifting hook, the forces acting on a horizontal section through *B* may be determined by replacing *F* by a equivalent force at *B* and a couple. If the couple is 3000 N-mm, determine *F* (Fig. 42).

78. Combine the force and couple (shown in Fig. 43) into a single force *F* and find the perpendicular distance from *A* to *F*.

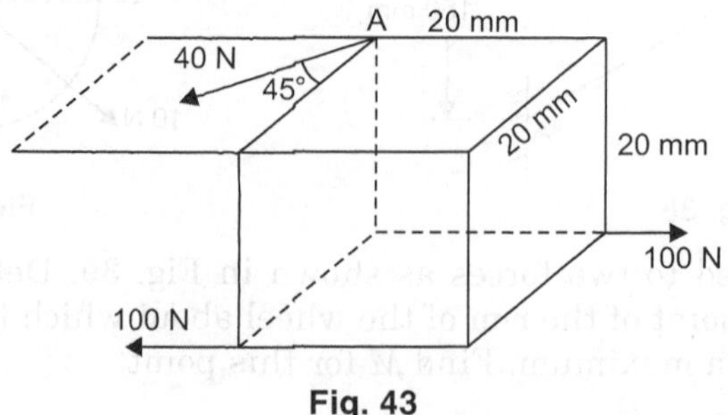

Fig. 43

79. Determine the x-, y- and z-components of the single couple vector M which is equivalent to the two couple shown in Fig. 44.

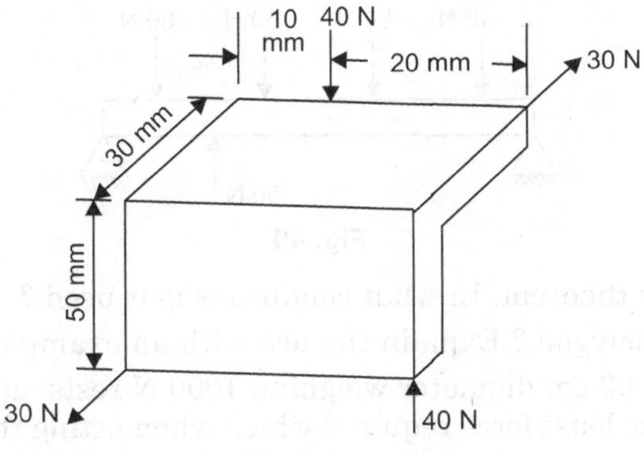

Fig. 44

80. Determine the resultant of the four forces acting on the plate shown in Fig. 45.

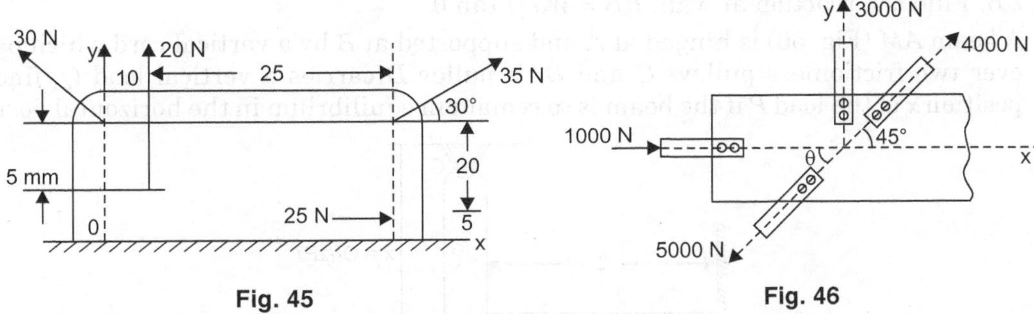

Fig. 45 **Fig. 46**

81. Determine the resultant R of the four forces transmitted to the gusset plate if $\theta = 45°$ (Fig. 46).

82. Replace the two parallel forces acting on the control level by a single equivalent force R (Fig. 47).

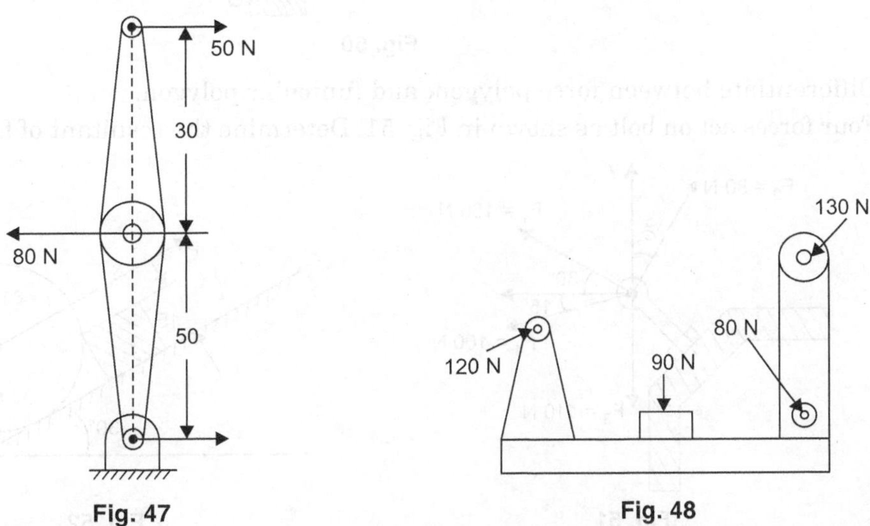

Fig. 47 **Fig. 48**

83. Determine by means of a force and string polygon the resultant of the four forces shown acting on the Fig. 48.

84. Determine the resultant R of the five vertical forces acting on the beam as shown in Fig. 49. By funicular polygon.

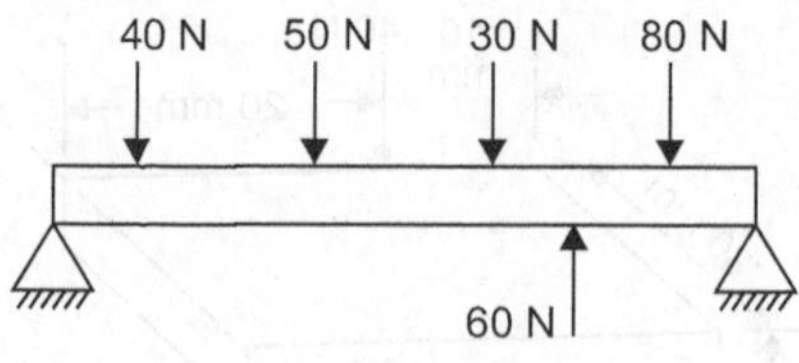

Fig. 49

85. Explain Varignon's theorem. In what conditions is it used ?

86. What is funicular polygon ? Explain the use with an example.

87. A uniform wheel of 60 cm diameter weighing 1000 N rests against rectangular obstacle 15 cm high. Find the least force required which when acting through centre of the wheel will just turn the wheel over the corner of the block. Find the angle of force with horizontal.

88. A prismatic bar AB of negligible weight is hinged at A, with end B resting against a vertical wall. A vertical load P is acting at a distance from A whereas l is the length of AB. Find the reaction at wall. $RB = Wa/l \tan \theta$.

89. A beam AM (Fig. 50) is hinged at A and supported at B by a vertical cord which passes over two frictionless pulleys C and D. If pulley D carries a vertical load Q, find the position x of the load P if the beam is to remain in equilibrium in the horizontal position.

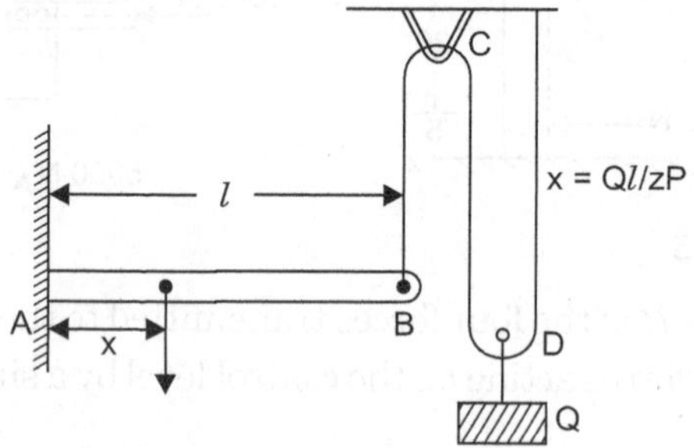

Fig. 50

90. Differentiate between force polygon, and funicular polygon.

91. Four forces act on bolt as shown in Fig. 51. Determine the resultant of forces on the bolt.

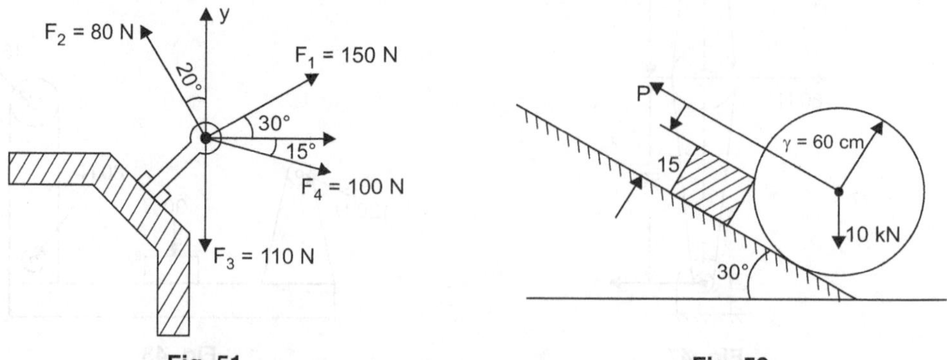

Fig. 51 **Fig. 52**

92. Determine the magnitude and direction of smallest force P required to start the wheel over the block (Fig. 52).

BEAM

93. Draw the shear force and bending moment diagram for the simply supported beam as shown in Fig. 53.

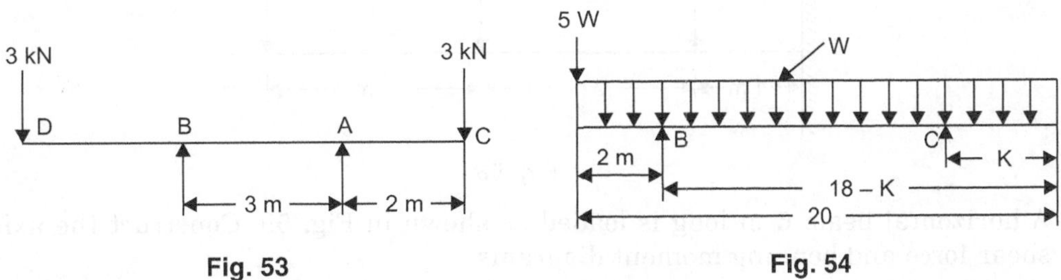

Fig. 53 Fig. 54

94. A beam *ABCD* 20 m long is loaded as shown in Fig. 54. The beam is supported at *B* and *C* and has a overhang of 2 m to the left of support *B* and an overhang of *K* metres to the right of support *C* which in the right hand half of the beam. Determine the value of *K* if the mid-point of beam is a point of inflexion and for this arrangement, plot BM and SF diagrams indicating the principal numerical values.

95. A horizontal beam, 30 m long, carries a uniformly distributed load of 10 kN/m over whole length and concentrated load of 30 kN at right end (Fig. 55). If the beam is freely supported at the left end, find the position of the second support so that the bending moment on the beam should be as small as possible. Draw the diagrams of shearing force and bending moment and insert the principal values. R_A, R_B = reactions at point *A* and *B*.

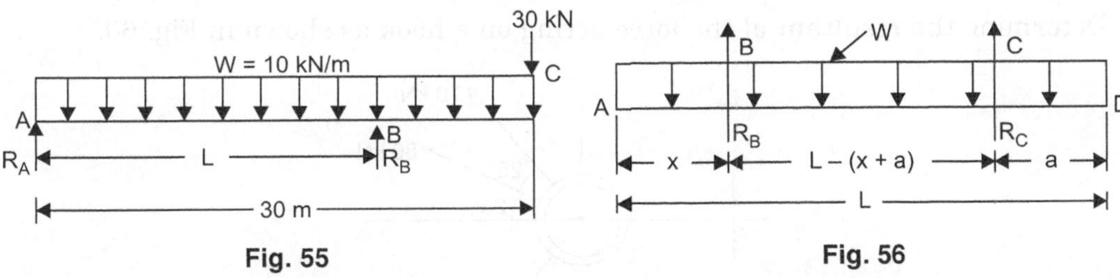

Fig. 55 Fig. 56

96. A girder of 10 m long carrying a uniformly distributed load of *w* per metre run is to be supported on two piers 6 m apart so that the greatest bending moment on girder shall be as shall as possible (Fig. 56). Find the distances of the piers from the end of the girder and maximum bending moment. Plot the BM and SF diagrams.

97. A simply supported beam of span *L* with *A*, *B* as the supporting points is loaded with a triangular load with intensity zero at one end to *w* per length at the other end (Fig. 57). Plot the bending moment and shear force diagram indicating the principle values.

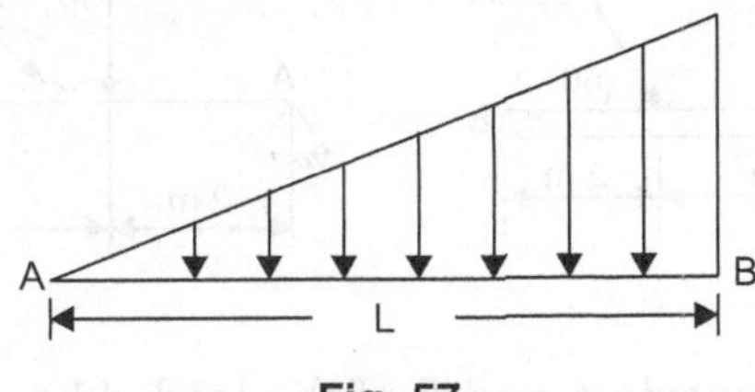

Fig. 57

98. Draw the SF and BM diagram for the beam as shown in Fig. 58. Also indicate the principal values on the diagram.

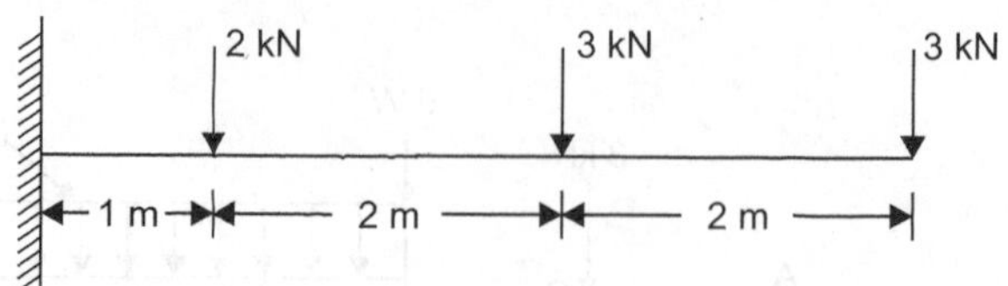

Fig. 58

99. A horizontal beam 6 m long is loaded as shown in Fig. 59. Construct the axial force, shear force and bending moment diagrams.

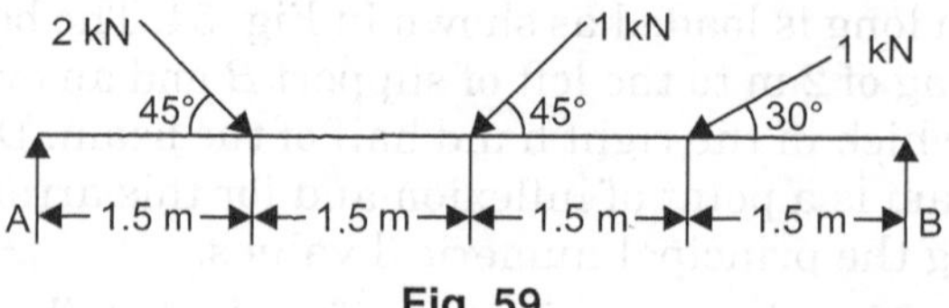

Fig. 59

100. Write the relation between the bending moment and shear force.

101. The angle between the two forces of magnitude 20 kN and is 60°, the 20 kN forces being horizontal. Determine the resultant in magnitude and direction if

(*i*) the force are pulls

(*ii*) the 15 kN force is rush and 20 kN force is a pull.

102. Determine the resultant of the force acting on a hook as shown in Fig. 60.

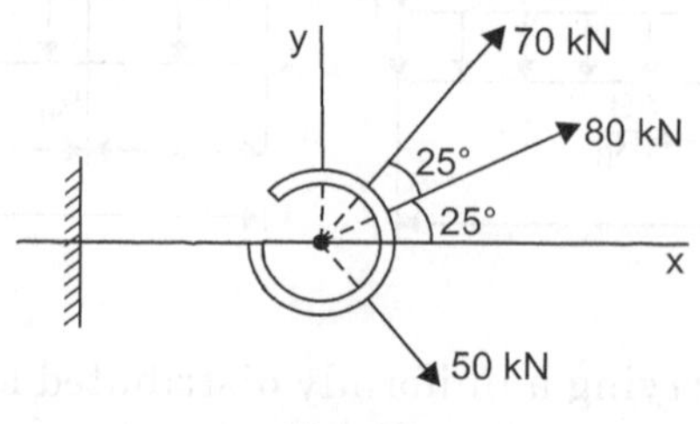

Fig. 60

103. Explain and prove the Varignon's theorem.

104. A system of loads acting on a beam is shown in Fig. 61. Determine the resultants of the loads.

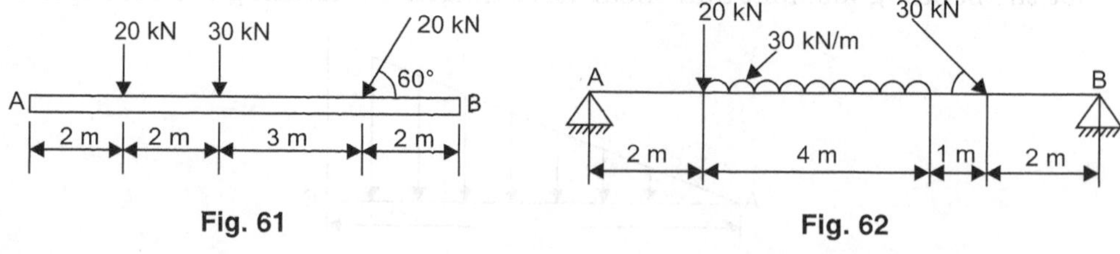

Fig. 61 **Fig. 62**

105. Find the reactions at supports *A* and *B* of the loaded beam shown in Fig. 62.

106. The cantilever is shown in Fig. 63. Determine the reaction when it is loaded as shown in Fig. 63.

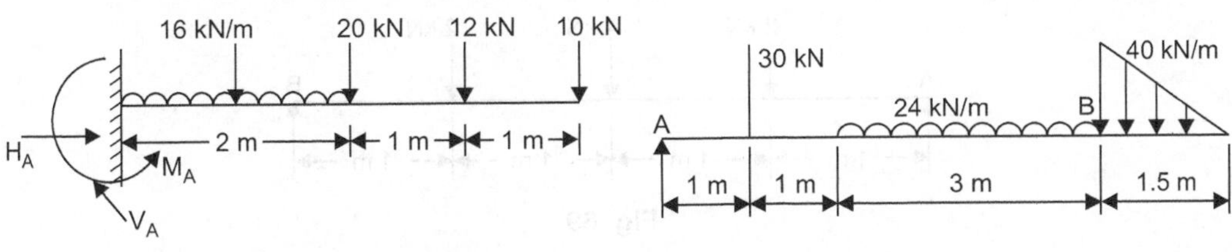

Fig. 63 **Fig. 64**

107. Determine the reactions at supports *A* and *B* of the overhanging beam shown in Fig. 64.

108. Determine the reactions at *A* and *B* of the overhanging beam shown in Fig. 65.

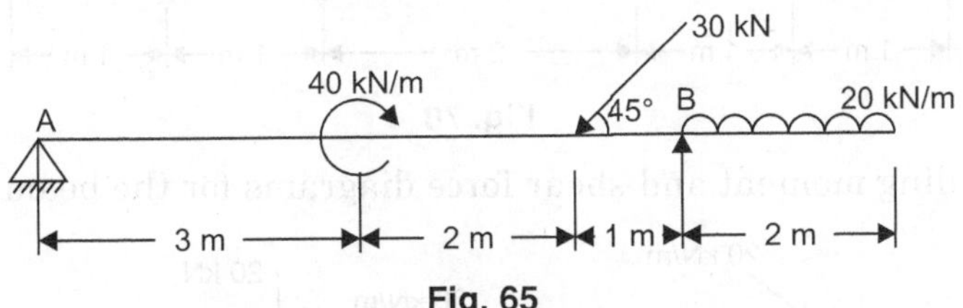

Fig. 65

109. Find the reactions at supports *A* and *B* of the loaded beam shown in Fig. 66.

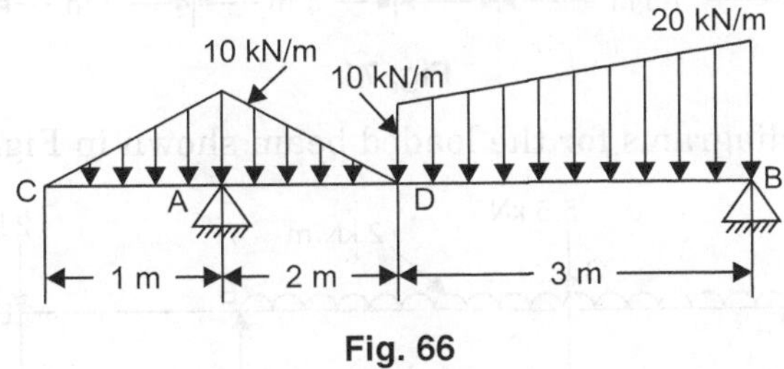

Fig. 66

110. Determine the reactions at *A* and *B* in Fig. 67.

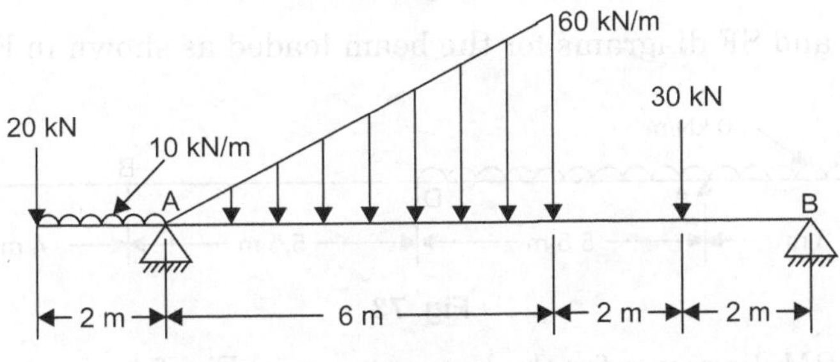

Fig. 67

111. Determine graphically the resultant of the parallel force system shown in Fig. 68.

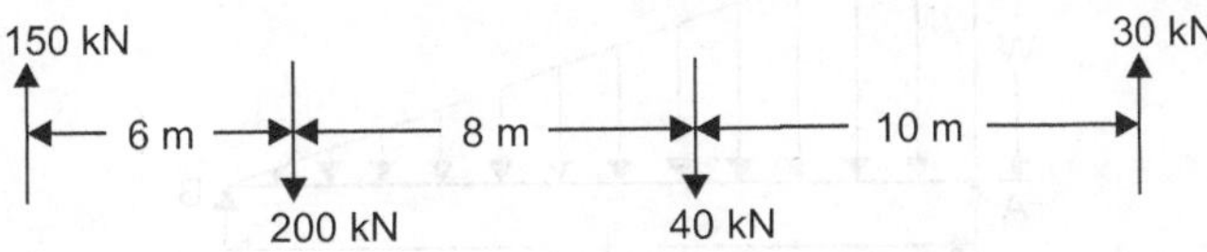

Fig. 68

112. Draw the SF and BM diagrams for simply supported beam loaded as shown in Fig. 69.

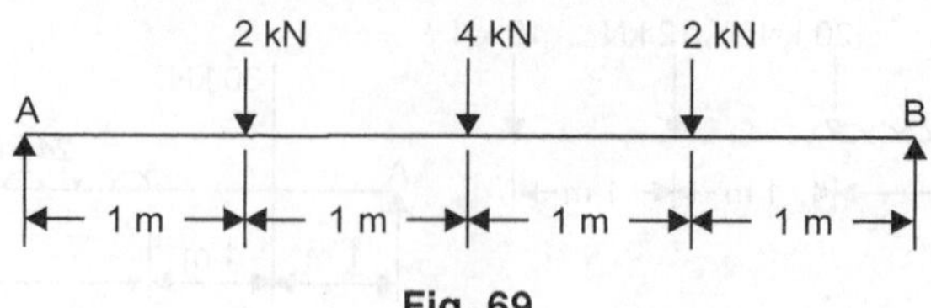

Fig. 69

113. Draw the SF and BM diagrams for the beam shown in Fig. 70.

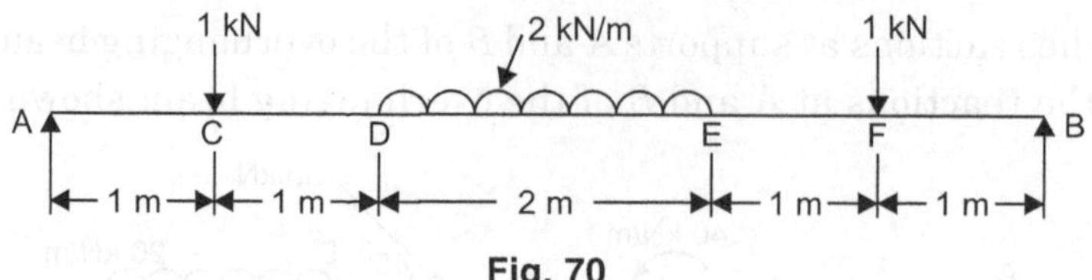

Fig. 70

114. Draw the bending moment and shear force diagrams for the beam shown in Fig. 71.

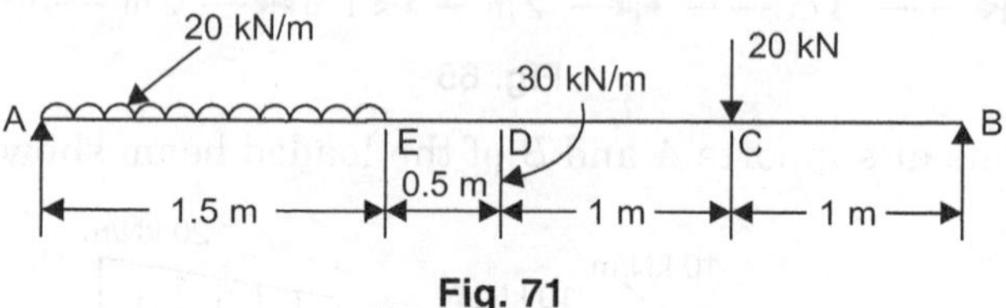

Fig. 71

115. Draw SF and BM diagrams for the loaded beam shown in Fig. 72.

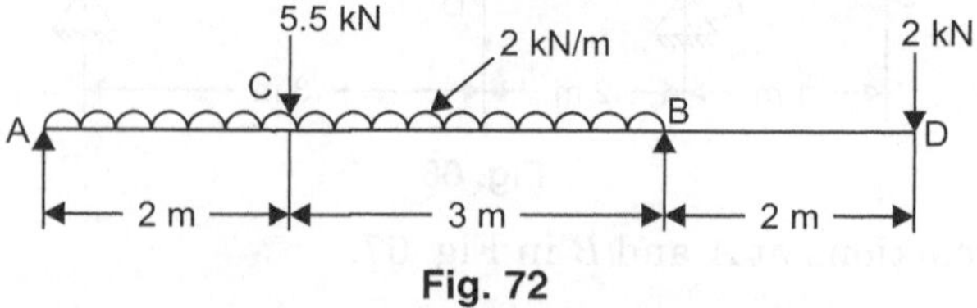

Fig. 72

116. Draw the BM and SF diagrams for the beam loaded as shown in Fig. 73.

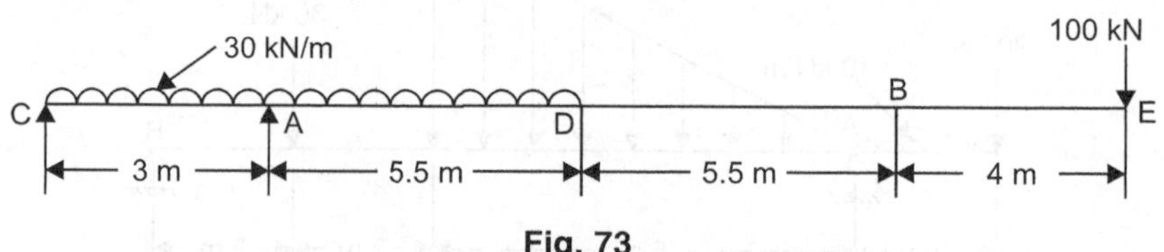

Fig. 73

117. Draw SF and BM diagrams for the beam shown in Fig. 74.

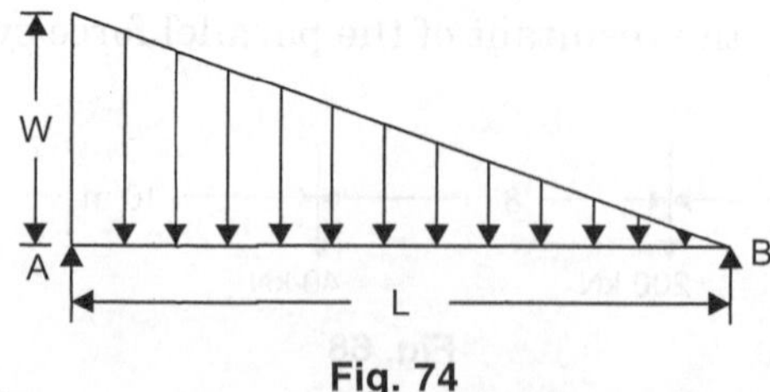

Fig. 74

UNIT 2

1. Determine the forces members *BD, CD* and *CE* of the fink truss shown in Fig. 75.

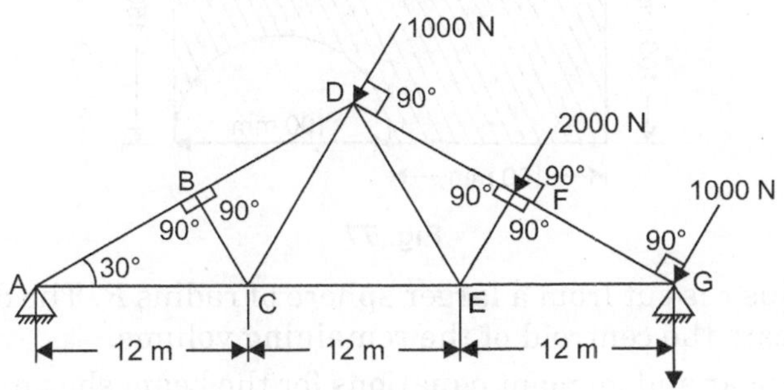

Fig. 75

2. (*a*) Define and explain the terms :
   Perfect frame,
   Imperfect frame,
   Deficient frame,
   Redundant frame.

   (*b*) What is the advantage of method of section over the method of joint ?
3. Prove that moment of inertia of a triangular section about the base of the section is given by

$$\frac{bh^2}{12}$$

where b = base of triangular section
h = height of triangular section.

4. Find the centre of gravity of the I-section shown in Fig. 76.

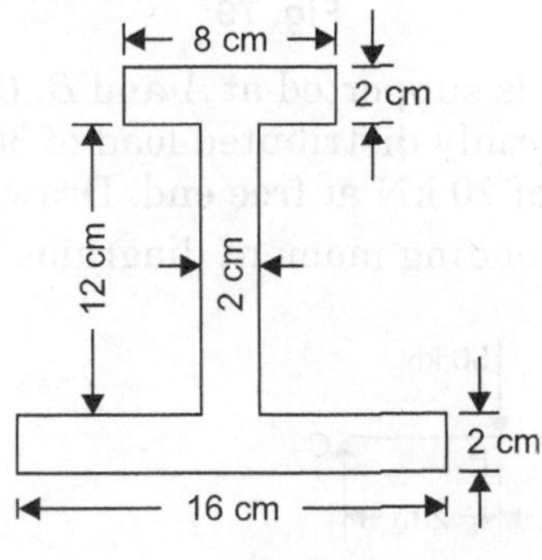

Fig. 76

5. A semi-circular area is removed from the trapezoid as shown in Fig. 77. Determine the centroid of the remaining area.

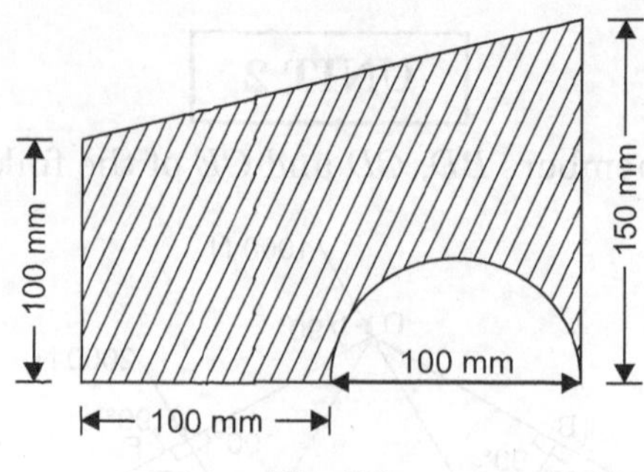

Fig. 77

6. A sphere of radius r is cut from a larger sphere of radius R. The distance between their centers is a. Locate the centroid of the remaining volume.
7. Determine the shear and moment equations for the beam shown in Fig. 78. Draw shear force and bending moment diagrams.

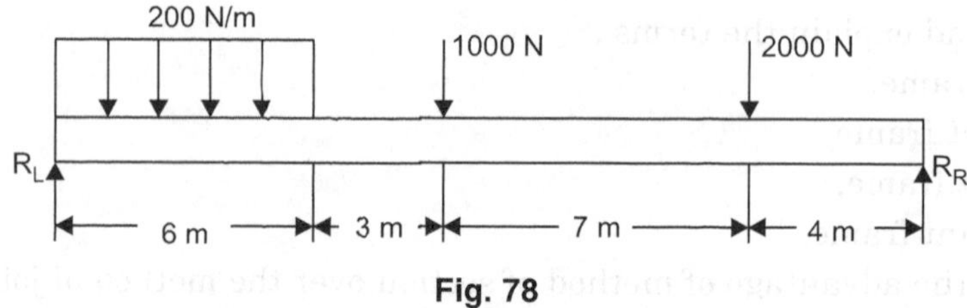

Fig. 78

8. Draw shear force and bending moment diagrams for the beam shown in Fig. 79. Find out the position and the magnitude for maximum moment.

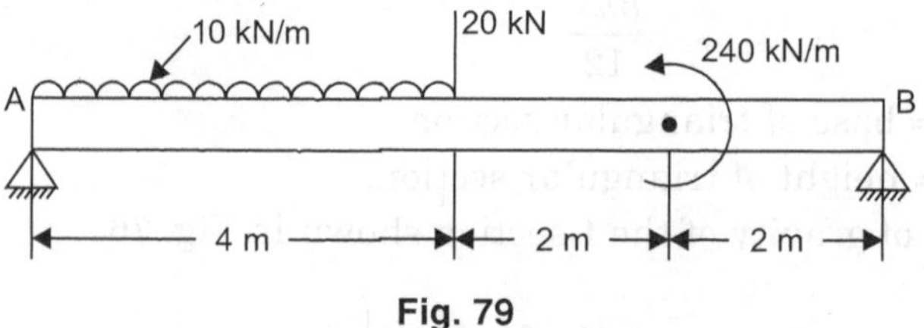

Fig. 79

9. An overhanging beam ABC is supported at A and B, the span AB is 6 m. The overhang BC is 2 m. It carries a uniformly distributed load of 30 kN/m over a length of 3 m from A and a concentrated load of 20 kN at free end. Draw SF and BM diagrams.
10. Draw the shear force and bending moment diagrams for the beam shown in Fig. 80.

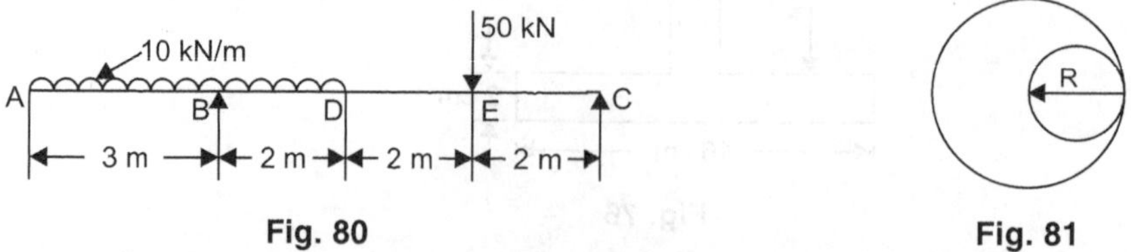

Fig. 80

Fig. 81

11. Find CC of Fig. 81 (smaller circle of diameter R is removed from bigger circle of radius R).
12. Prove that moment of inertia of a disc of radius R about z-axis perpendicular to its plane is $I_z = MR^2/2$. Thereby using theorem of I- and II-axis theorems, find moment of inertia of the disc about tangent axis in the plane.

13. Prove for belt and pulley system

$$\frac{T_1}{T_2} = e^{\mu\theta}$$

where T_1 and T_2 are the tension the belt and θ is contact angle and μ is coefficient friction.

14. What are trusses ? When can the trusses be rigid trusses. State the condition followed by simple trusses.

15. Where do you find trusses in use ? What are the various methods of analysis of trusses? What is basically found when analysis of a system is done ?

16. A truss is shown in Fig. 82. Find whether it is a simple truss. Use method of joints to determine the force in each member of truss. Also state which members are on tension and which it compression.

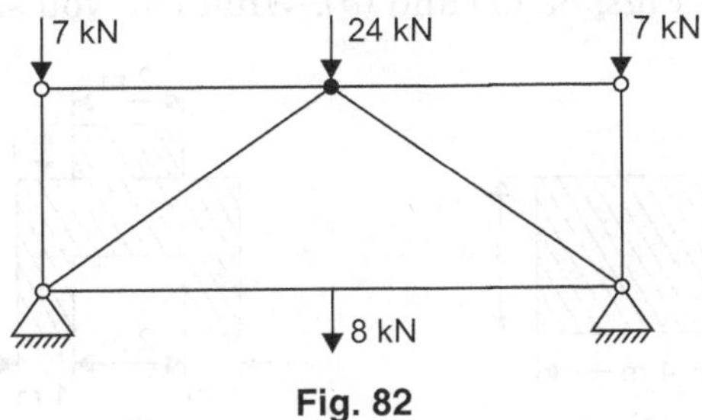

Fig. 82

17. Determine the force in members *EI, EG* and *BD* of the truss shown in Fig. 83.

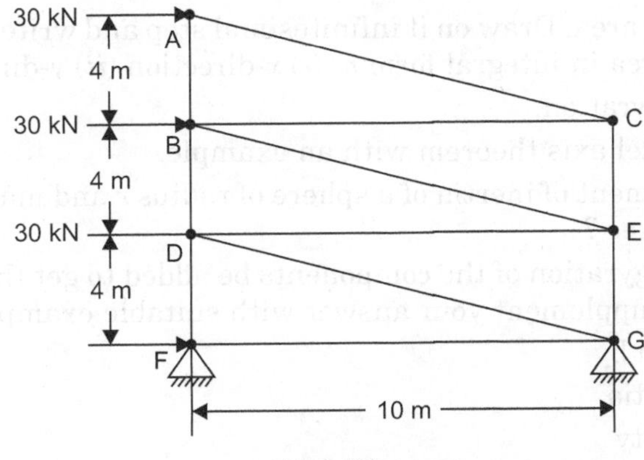

Fig. 83

18. Give examples of :

(*a*) An area with one line of symmetry.

(*b*) An area with two lines of symmetry.

(*c*) A body having one plane of symmetry.

(*d*) A body having two planes of symmetry.

(*e*) A body having three planes of symmetry.

19. Locate the centroid of the plane area shown in Fig. 84.

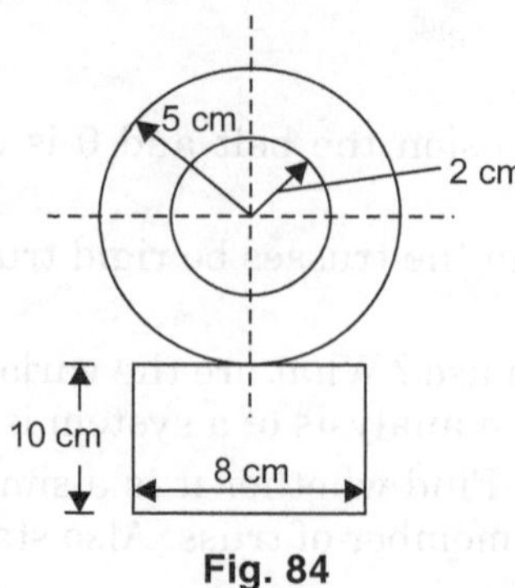

Fig. 84

20. Locate the centroid of Figs. 85 (*a*) and (*b*). What can you say about the change ?

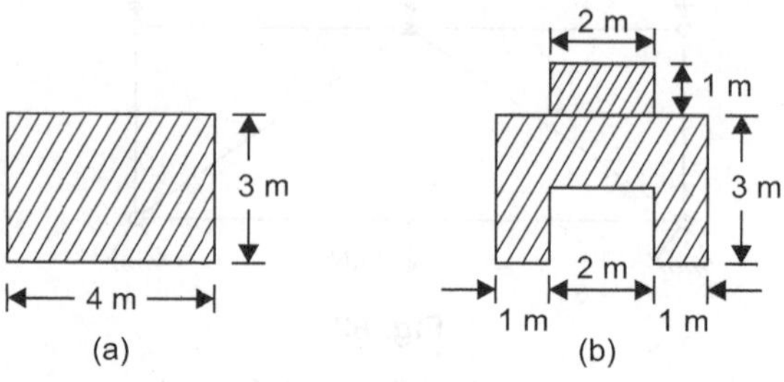

Fig. 85

21. Assume a suitable area. Draw on it infinitesimal step and write the formula for moment of inertia of the area in integral form *m* (*i*) *x*-direction (*ii*) *y*-direction.

22. Define radius of gyration.

23. Explain the parallel axis theorem with an example.

24. Find the mass moment of inertia of a sphere of radius r and mass m. Will it be different along x, y-and z-axes ?

25. Can the radius of gyration of the components be added to get the radius of gyration of a composite body. Supplement your answer with suitable example.

26. What do you mean by

(*a*) Product of inertia

(*b*) Center of gravity

(*c*) Centroid

(*d*) Composite bodies

(*e*) Principal moment of inertia ?

27. State assumption in truss analysis ?

28. How do you determine centroid of composite area ?

29. Locate the centroid of the lamina shown in Fig. 86 from base *AB*.

30. A thin circular plate is of thickness t and diameter d. If the density of the material of the plate is p. Calculate the radius of gyration of the mass of the plate about its diameter.

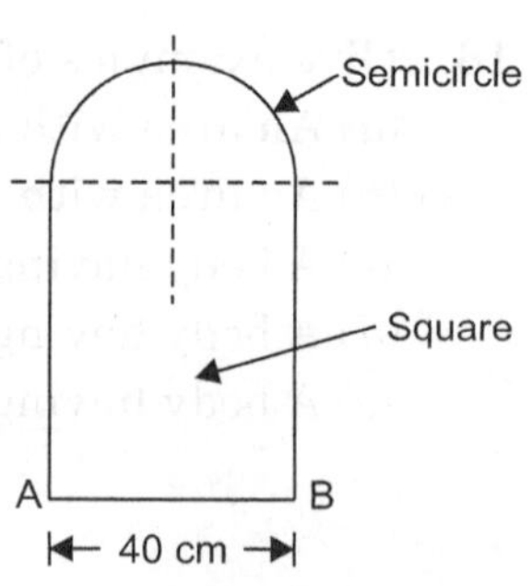

Fig. 86

31. The moment of inertia of a circular disc of radius r and mass m about any diameter is ($mr^2/4$). Calculate the moment of inertia about an axis passing through the center and perpendicular to its plane.

32. Two concurrent forces of magnitudes 40 N and 30 N acting at origin O are directed respectively towards the points A (3, 0, – 3) and B (2, – 2, 4). Determine the magnitude of the resultant and its inclination with x, y-and z-axes.

33. Locate the centroid for the plane section shown in Fig. 87, and determine the moment of inertia about axis AB.

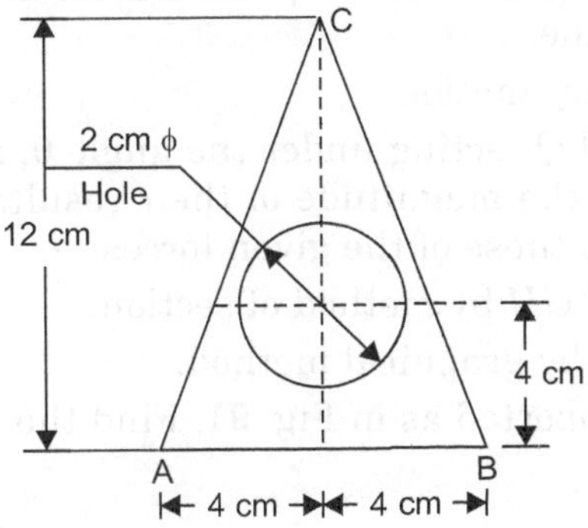

Fig. 87

34. (*i*) Distinguish between centroid and center of gravity.

(*ii*) For the system shown in Fig. 88, find the additional single force required to maintain equilibrium.

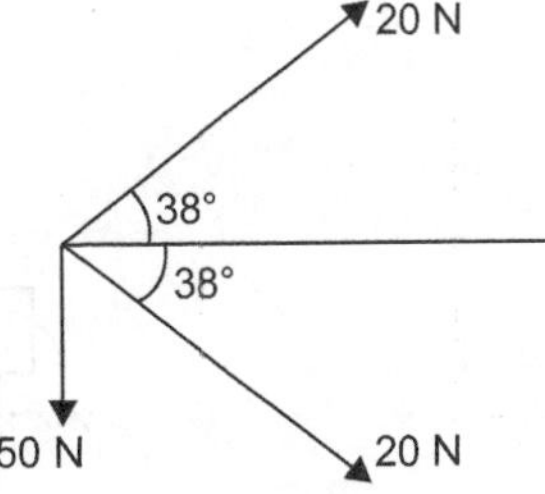

Fig. 88

35. Explain, why roller supports are used in case of steel trusses of bridge ?

36. Discuss the advantage of method of section over method of joints. Explain the use of method of reaction for getting forces in members of a trace.

37. Determine the forces in all members of a cantilever truss as shown in Fig. 89. Also determine the forces in members using graphical method and determine sections of cantilever.

38. Find center of gravity of L-section having limb of 120 mm and other 80 mm with their widths as 20 mm (Fig. 90).

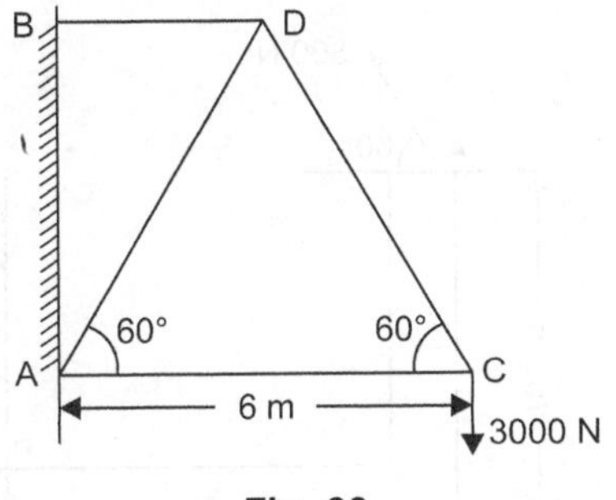

Fig. 89

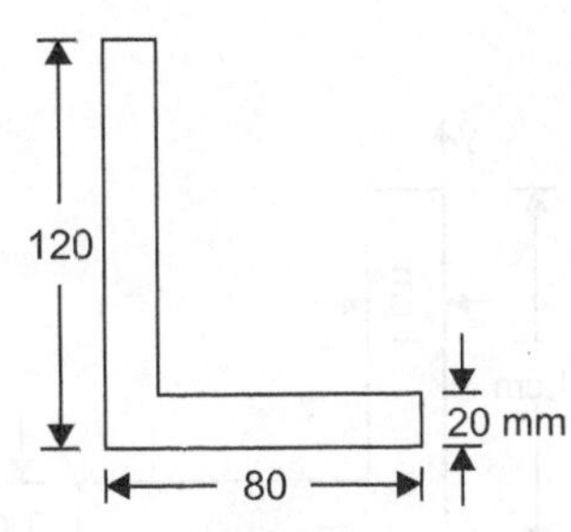

Fig. 90

39. State and prove the parallel axes theorem on moment of inertia for a plane area.

40. Show that moment of inertia of a rectangular section of $b \times d$ is $(bd^3/12)$ and $(bd^3/13)$ about horizontal axis passing through center of gravity and about horizontal axis passing through base respectively.

41. Obtain moment of inertia of a triangular section about an axis passing through centre of gravity of section and parallel to base.

42. Calculate the moment of inertia and radius of gyration of a solid sphere of diameter 50 cm and of mass 0.002 kg/cm^2, a circular lamina of diameter 50 cm and of mass 0.002 kg/m^2.

43. State the law of machine and obtain expressions for maximum mechanical advantage and efficiency of a machine.

44. Discuss the laws for rotary motion.

45. If two given forces P and Q, acting under the angle θ, are applied to a body at A, find formulas for calculating the magnitude of their resultant R and the angles which its line of action makes with those of the given forces.

46. Find the force in BG and GH by method of section.

47. Solve the above problem by graphical method.

48. A lever is loaded and supported as in Fig. 91. Find the support reactions R_1 and R_2.

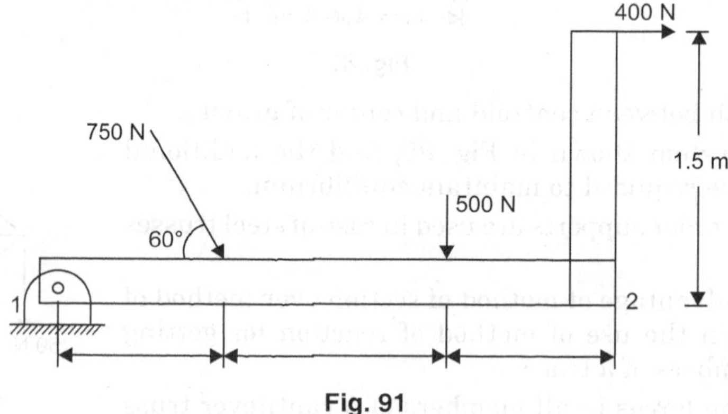

Fig. 91

49. Determine the moment of inertia about x-axis of an area under the sine curve the equation for which is $y = 6 \sin \frac{\pi r}{a}$.

50. Find the moment of inertia of L-section with respect to x- and y-axes (Fig. 92).

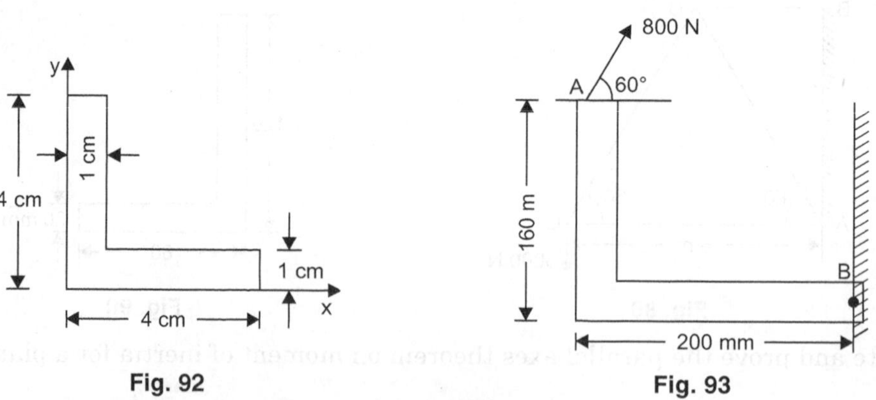

Fig. 92 **Fig. 93**

51. A force of 800 N acts on a bracket as shown in Fig. 93. Determine the moment of force about B.

52. Compute the moment about A of the force as shown in Fig. 94 :

(*a*) by the definition of moment of a force.

(*b*) by resolving the force into horizontal and vertical components.

(*c*) by resolving the force into components along AB and in direction perpendicular to AB.

(*d*) Determine the following from Fig. 95.

(*i*) moment about C of the force shown.

(*ii*) The perpendicular distance from C to the line of action of the force.

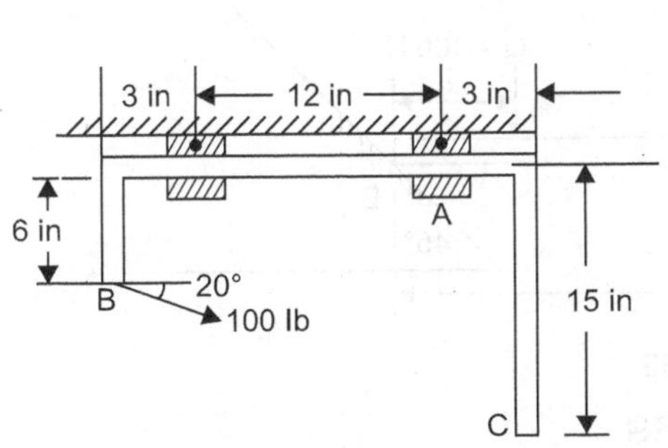

Fig. 94

Fig. 95

53. A rigid bar AB is supported in a vertical plane by a hinge at the end A and by a horizontal string attached to the bar as shown in Fig. 96. The end B of the bar carries a load of W Newton's. Find the tensile force in string. Neglect the weight of the bar.

54. Determine the forces exerted on the cylinder at A and B by a vertical spanner wrench due to a vertical force of 250 N applied to the handle (Fig. 97). Neglect friction at B.

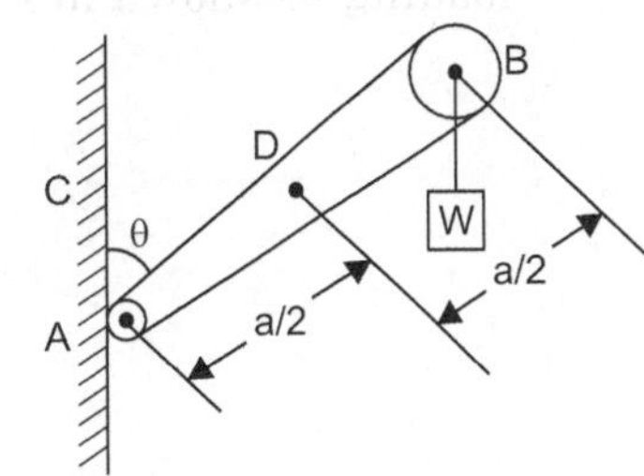

Fig. 96

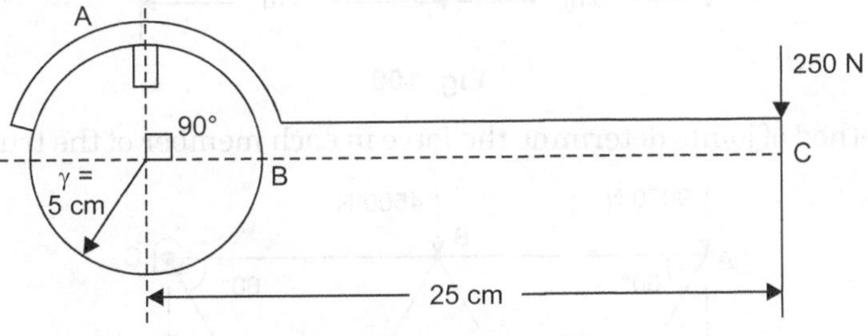

Fig. 97

55. A smooth cylinder of radius $r = 10.0$ cm resting on a horizontal surface supports a bar AB of length 30.0 cm which is hinged at point A (Fig. 98). The weight of the bar is 50 N. The cylinder is kept rolling away by a string AO of length 20.0 cm. Assuming all surfaces to be friction less, find the tensions in the string.

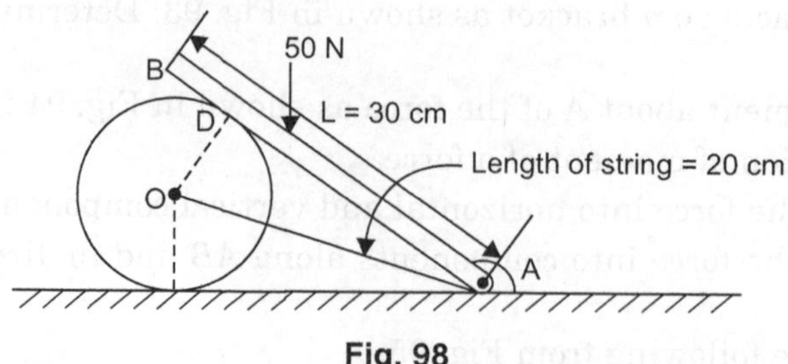

Fig. 98

56. A bar 2.0 m long and of negligible weight rests in horizontal position on two smooth inclined planes (Fig. 99). Determine the distance x at which the load Q = 100 N should be placed from point B to keep the bar horizontal.

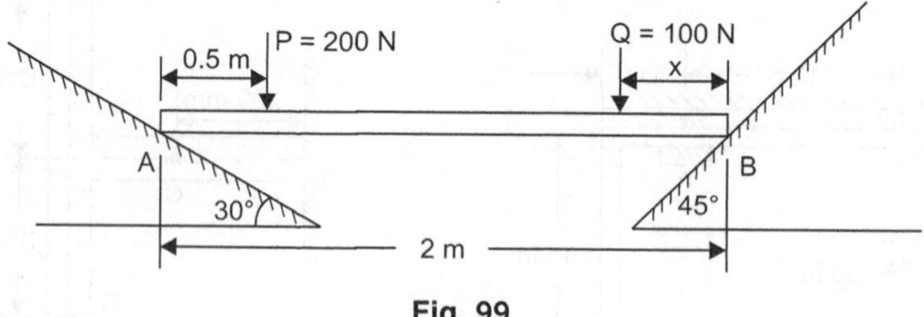

Fig. 99

TRUSS

57. Using the method of joint, find the axial forces in all the members of a truss with the loading as shown in Fig. 100.

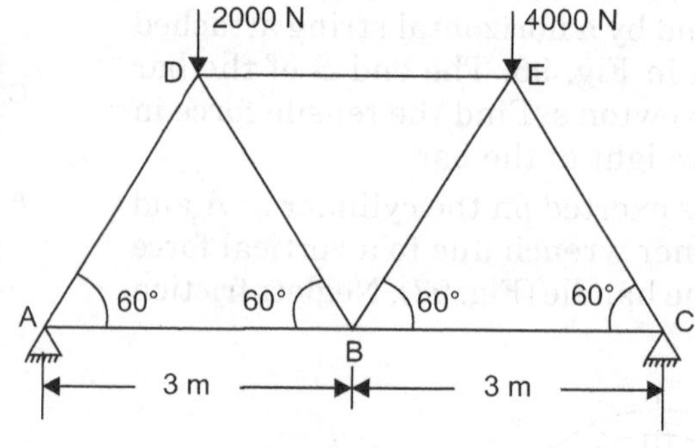

Fig. 100

58. Using the method of joints determine the force in each member of the truss shown in Fig. 101.

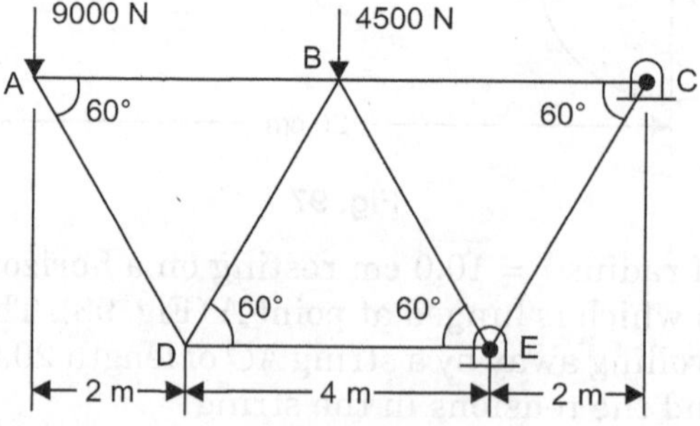

Fig. 101

59. Using the method of joints, determine the force in each member of truss shown in Fig. 102. State whether each member is in tension or compression.

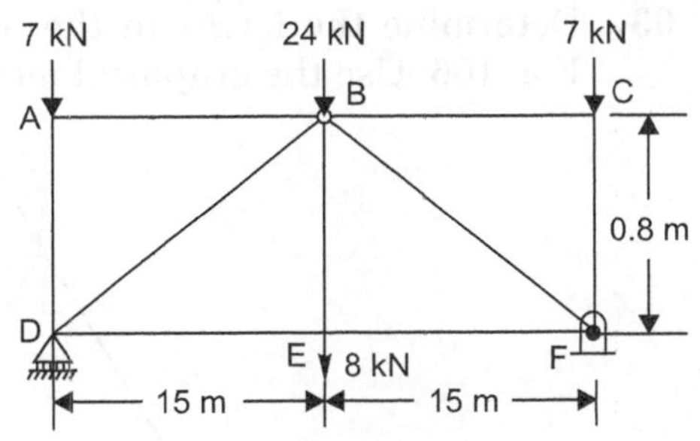

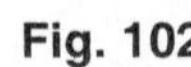

Fig. 102

60. A truss is located and supported as shown in Fig. 103. Determine the axial forces in member *CE*, *CG* and *FG*. Use the method of section.

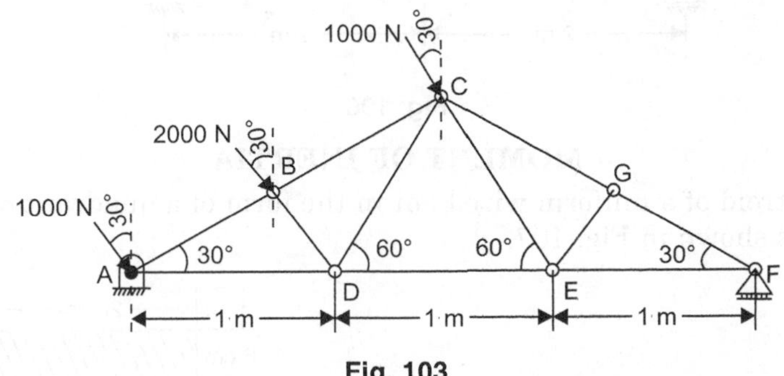

Fig. 103

61. A cantilever truss is loaded and supported as shown in Fig. 104. Find the value of *P* which would produce an axial force of magnitude 3 kN in the member *AC*.

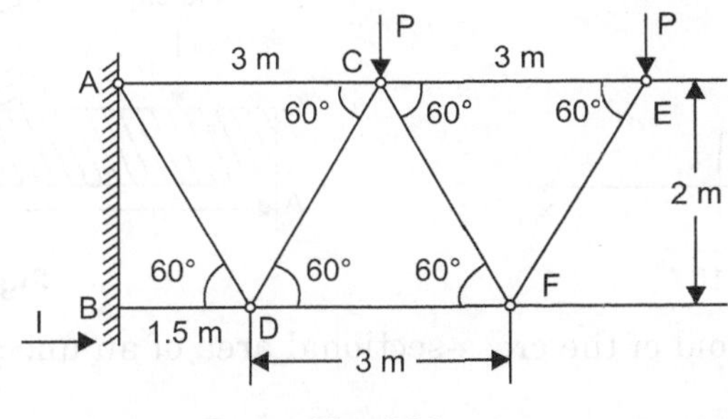

Fig. 104

62. Determine the force in *FH*, *GH* and *GI* of the roof truss shown in Fig. 105.

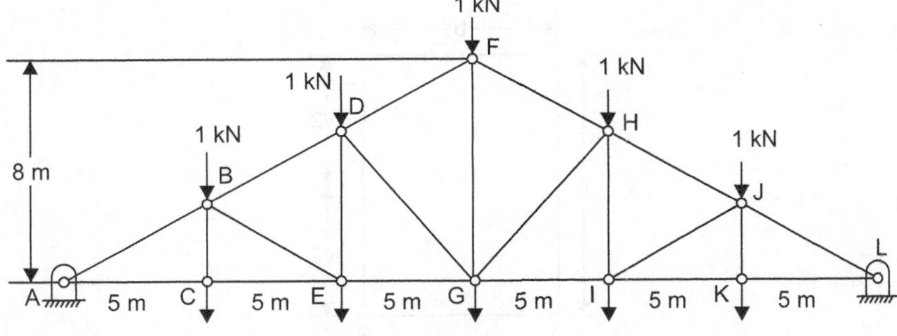

Fig. 105

63. Determine the forces in the members of a truss loaded and supported as shown in Fig. 106. Use the graphical method for calculation of the forces in members.

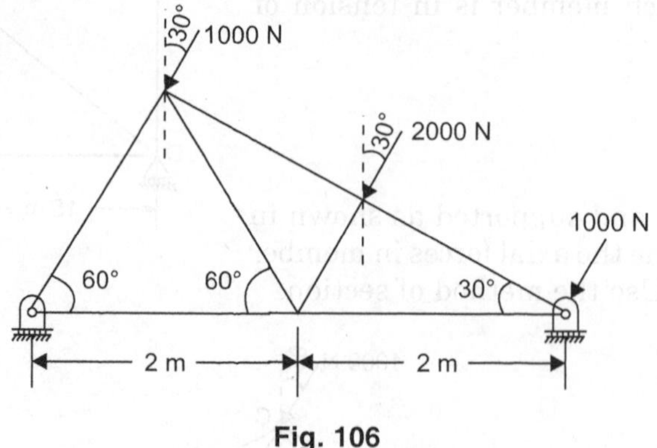

Fig. 106

MOMENT OF INERTIA

64. Find the centroid of a uniform wire bent in the form of a quadrant of the arc of a circle of radius *r* as shown in Fig. 107.

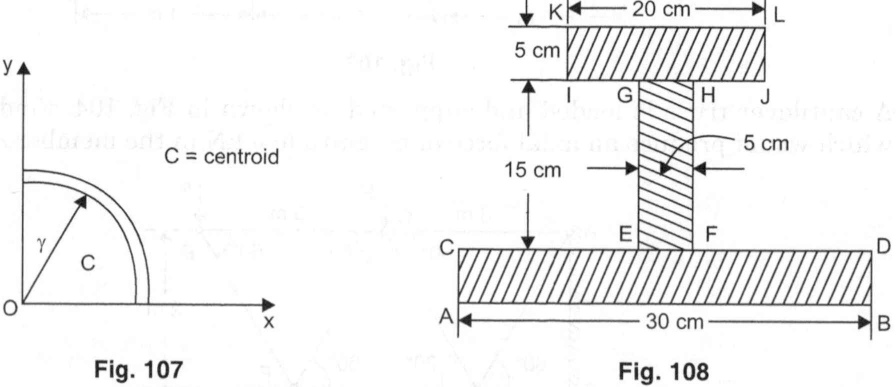

Fig. 107 Fig. 108

65. Determine the centroid of the cross-sectional area of an unequal I-section as shown in Fig. 108.

66. Write the parallel axis theorem. Find the moment of inertia of a rectangular cross-section about its centroidal axes as shown in Fig. 109. Also find its moment of inertia about the base *AB*.

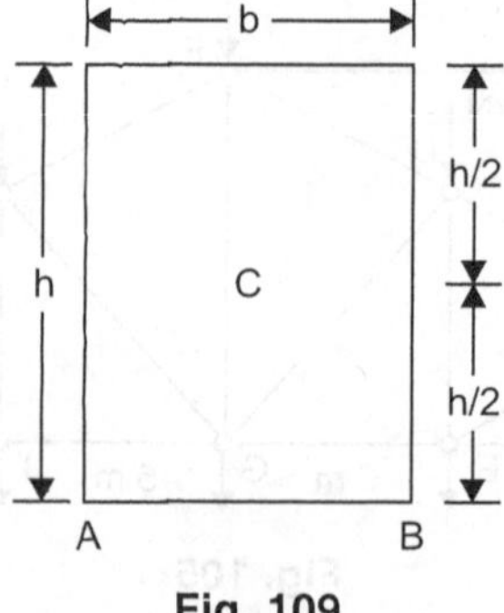

Fig. 109

67. Determine the moments of inertia of a circular area about the centroidal axes $I_x = \pi D^4/64$ where D = diameter of the circle.

68. Find the moment of inertia of a plate with a circular hold about its centroidal x-axis as shown in Fig. 110.

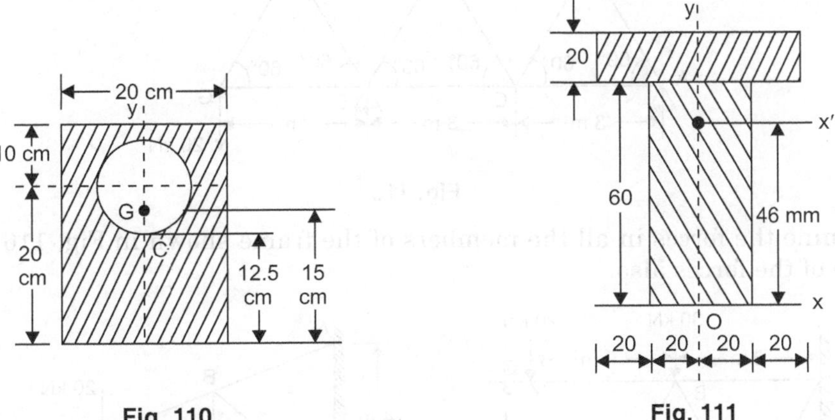

Fig. 110 Fig. 111

69. Determine the moment of inertia of the area of T-section as shown in Fig. 111, with respect to the centroidal x-axis.

70. Determine the product of inertia of a rectangular area about the x and y-axes as shown in Fig. 112.

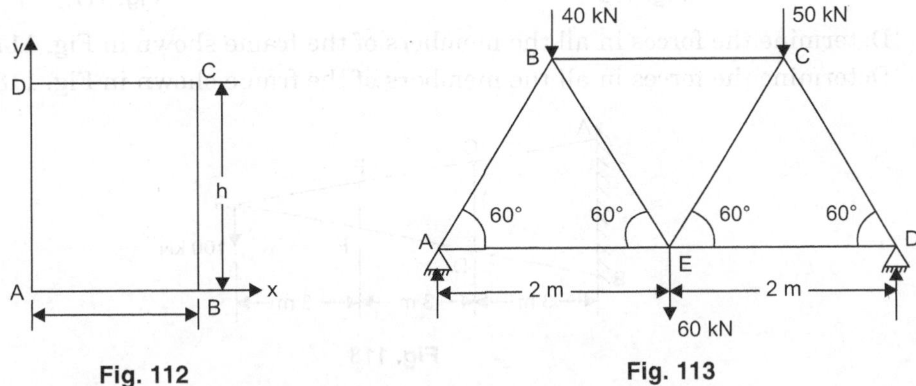

Fig. 112 Fig. 113

71. Determine the forces in all the members of the truss shown in Fig. 113. All inclined members are at 60° to horizontal and length of each member is 2 m.

72. Find the forces in all the members of the truss shown in Fig. 114.

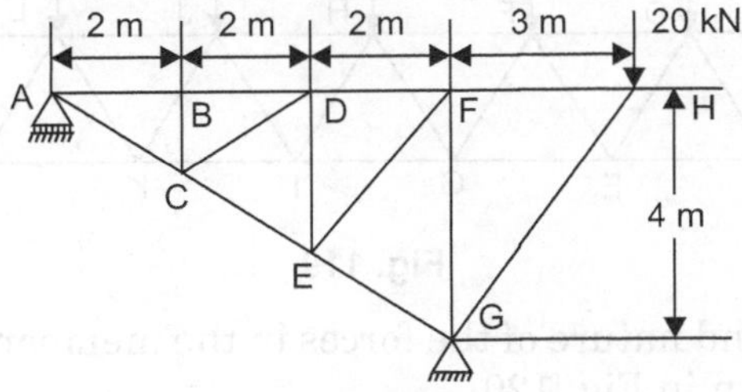

Fig. 114

73. Analyses the truss shown in Fig. 115. All the members are of 3 m length.

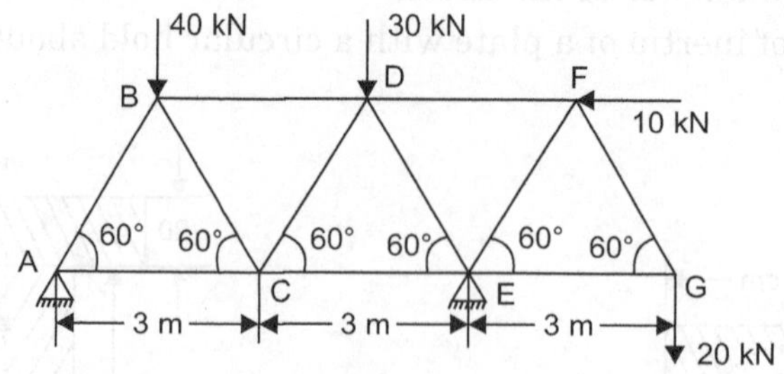

Fig. 115

74. Determine the forces in all the members of the frame shown in Fig. 116. Indicate the nature of the forces also.

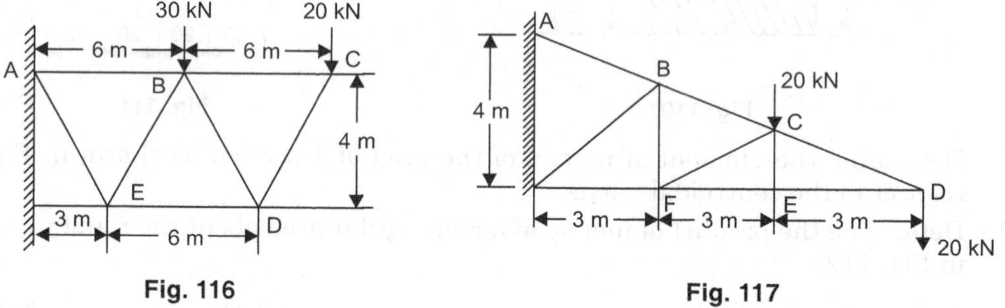

Fig. 116

Fig. 117

75. Determine the forces in all the members of the frame shown in Fig. 117.

76. Determine the forces in all the members of the frame shown in Fig. 118.

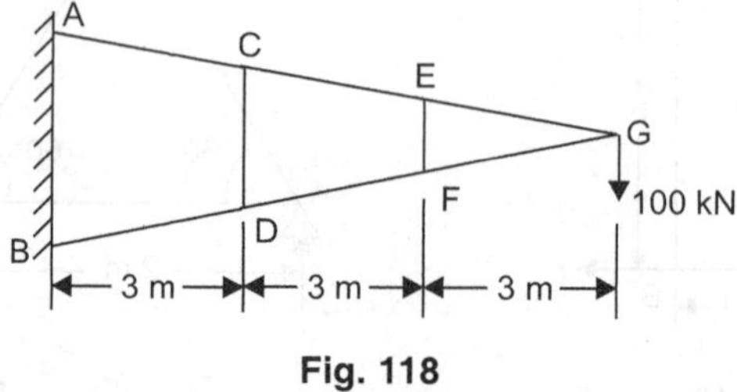

Fig. 118

77. Determine the forces in the members *FH, HG* and *GI* in the truss shown in Fig. 119. Each load is 10 kN and all triangles are equilateral triangles.

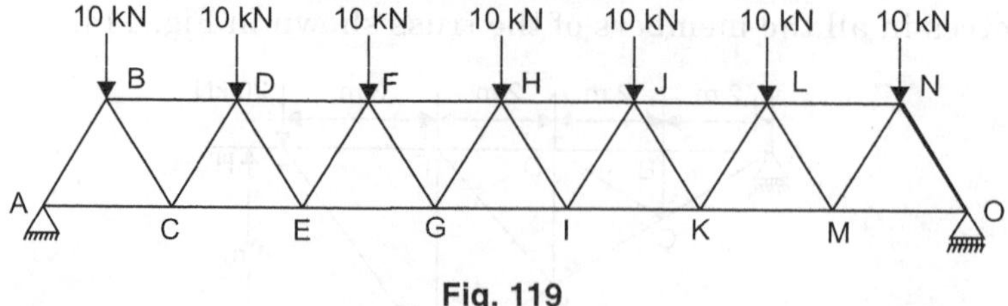

Fig. 119

78. Find the magnitude and nature of the forces in the members U_3, U_4, L_3, L_3 and U_4, L_3 of the loaded truss shown in Fig. 120.

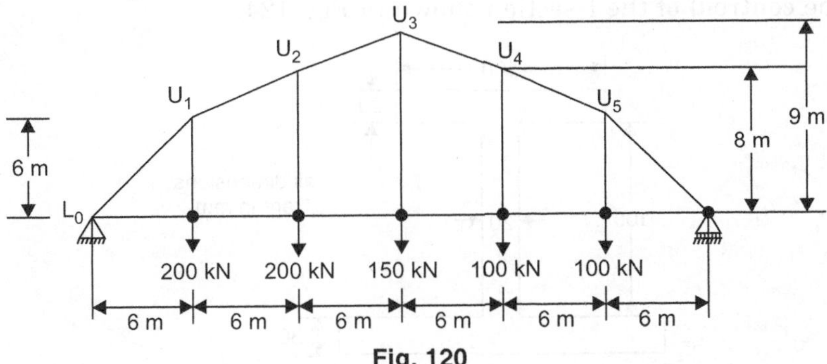

Fig. 120

79. Find the forces in the member *FG* of the triangular truss shown in Fig. 121.

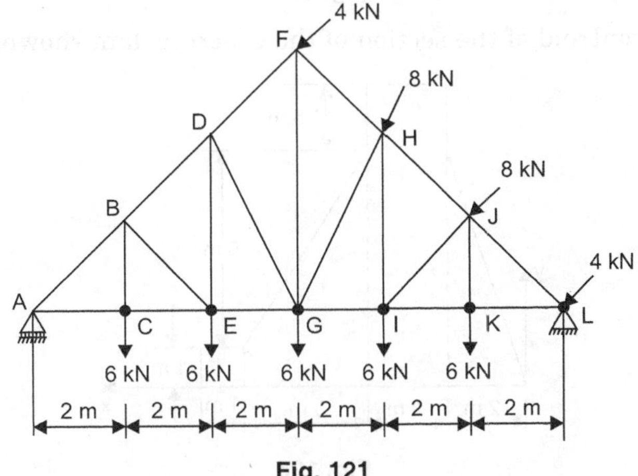

Fig. 121

80. Find the force in the member *CE, CD* and *BD* of the truss shown in Fig. 122.

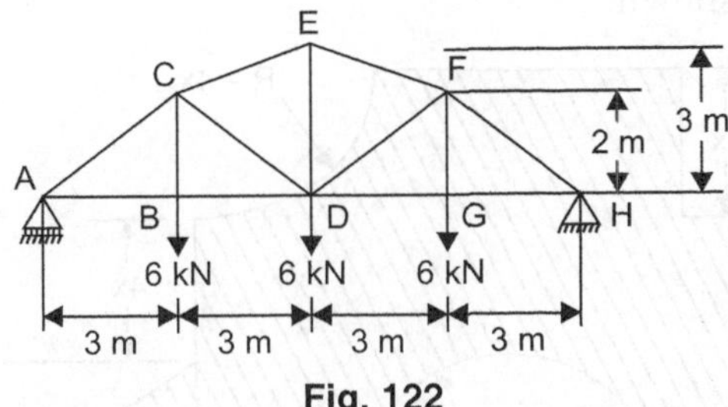

Fig. 122

81. Determine the centroid for the following cases by the first principal :

(*i*) Centroid of a triangle

(*ii*) Centroid of a semicircle.

82. Find the centroid of the unequal angle 200 × 150 × 12 mm shown in Fig. 123.

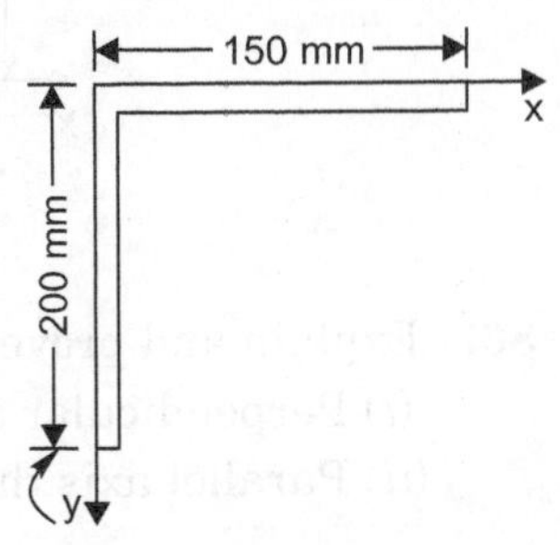

Fig. 123

83. Locate the centroid of the I-section shown in Fig. 124.

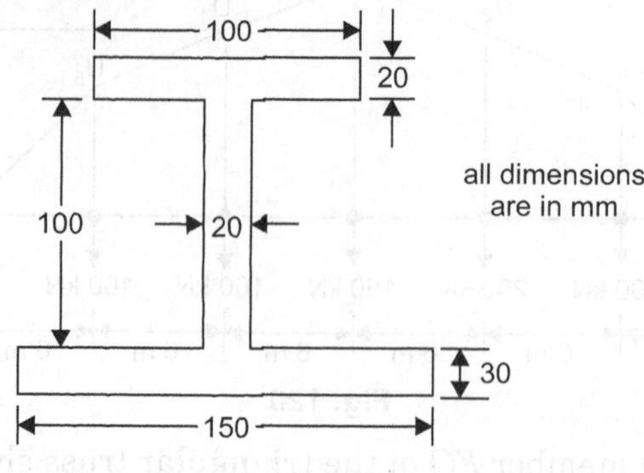

Fig. 124

84. Determine the centroid of the section of the concrete dam shown in Fig. 125.

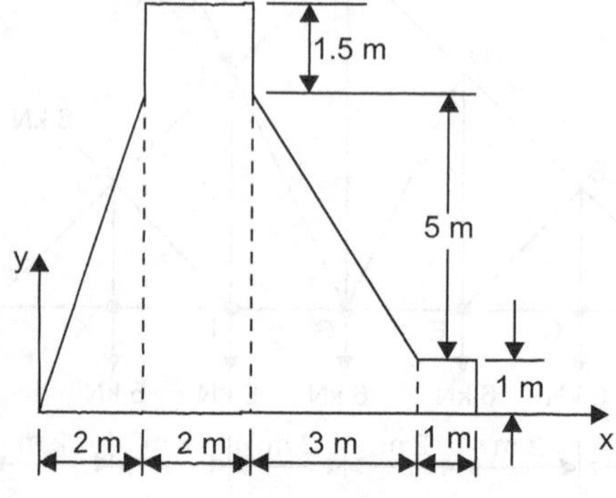

Fig. 125

85. Determine the coordinates of the centroid of the plane area shown in Fig. 126. With reference to the axis shown.

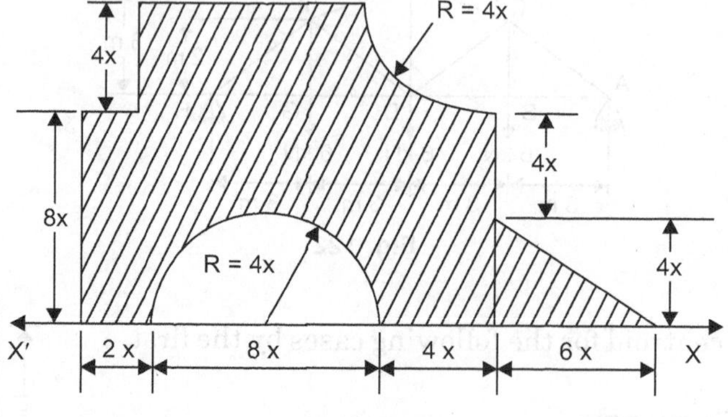

Fig. 126

86. Explain and prove the following theorems :

(*i*) Perpendicular axis theorem

(*ii*) Parallel axis theorem.

87. Determine the moment of inertia of the L-section show in Fig. 127. About its centroidal axis parallel to the legs. Also find the polar moment of inertia.

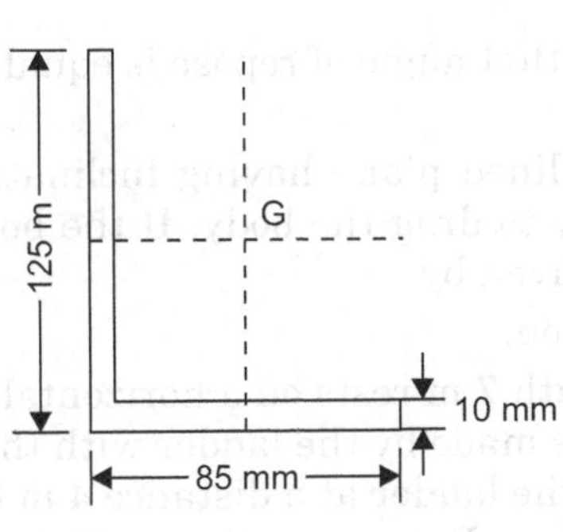

Fig. 127

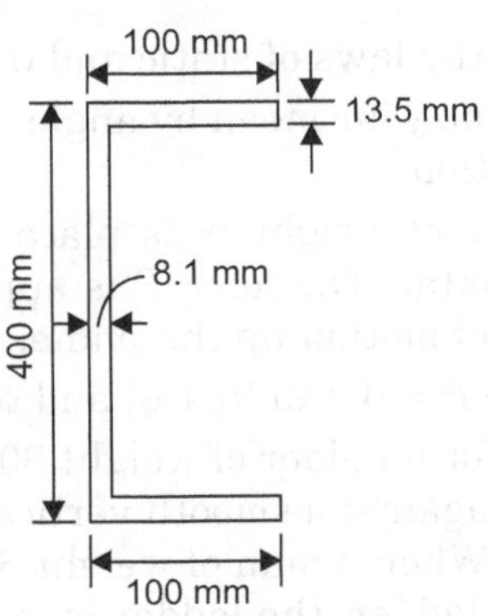

Fig. 128

88. Compute the moment of inertia of the channel section shown in Fig. 128. About centroidal axis *x-x* and *y-y*.

89. Determine the moment of inertia of the I-section about centroidal axis *x-y* and *y-y* (see Fig. 129).

90. Determine the mass-moment of inertia of a uniform rod of length L about its

(*a*) centroidal axis normal to rod

(*b*) axis at the end of the rod and normal to it.

91. Determine the moment of inertia of a solid space of radius of about its diametrical axis.

92. Find the product of inertia for the plane attached area about the axis *x-x* and *y-y* shown in Fig. 130. All dimensions are in mm.

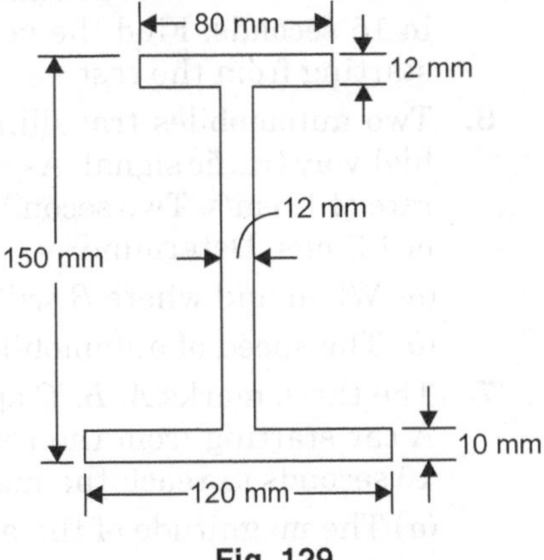

Fig. 129

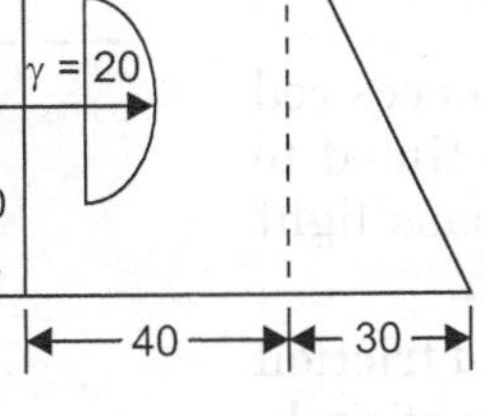

Fig. 130

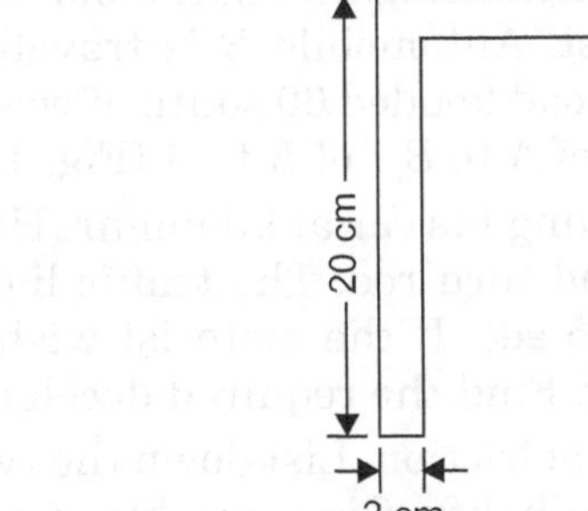

Fig. 131

93. Find the principal moments of inertia and directions of principal axis for the angle section shown in Fig. 131.

UNIT 3

1. State the laws of static and dynamic friction.
2. What do your mean by angle of repose ? Prove that angle of repose is equal to the angle of friction.
3. A body of weight W is placed on a rough inclined plane having inclination α to the horizontal. The force P is applied horizontally to drag the body. If the body is on the point of motion up the plane, prove that P is given by
   where $P = W \tan(\alpha + \phi)$ and ϕ = Angle of friction.
4. A uniform ladder of weight 800 N and the length 7 m rests on a horizontal ground and leans against a smooth vertical wall. The angle made by the ladder with the horizontal is 60. When a man of weight 600 N stands on the ladder at a distance 4 m from the top of the ladder, the ladder is at the point of sliding. Determine the coefficient of friction between the ladder and the floor.
5. A flywheel 460 mm in diameter is brought uniformly from rest up to a speed of 280 rpm in 15 seconds. Find the velocity and accelerations of a point on the run 1 second after starting from the rest.
6. Two automobiles travelling in the same direction in adjacent lanes are stopped at a highway traffic signal. As the signal turns green, automobile. A accelerates at a constant rate of 1.0 m/s. Two seconds later automobile B starts and accelerates at a constant rate of 1.3 m/s. Determine
   (*a*) When and where B will overtake A.
   (*b*) The speed of automobiles at that time.
7. The three marks A, B, C spaced at a distance of 100 m are made along a straight load. A car starting from the rest and accelerating uniformly passes the mark A and takes 10 seconds to reach the maker B and further 8 seconds to reach mark C. Calculate :
   (*a*) The magnitude of the acceleration of the car
   (*b*) The velocity of the car at A
   (*c*) The velocity of the car at B
   (*d*) The distance of the mark A from the starting point.
8. Automobile A is travelling 32 km/h along a straight load headed northwest. Automobile B is travelling 112 km/h along a straight load headed 60 south of west. What is the relative velocity of A to B ? of B to A (Fig. 132) ?

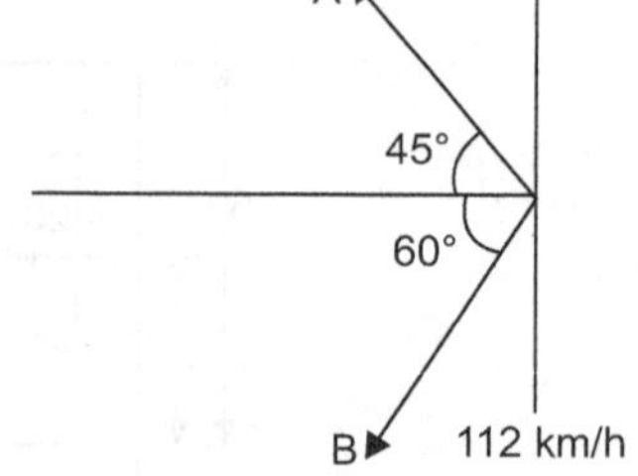

Fig. 132

9. A motorist is driving his car at 90 km/hr. He observes red light 240 m ahead turn red. The traffic light is timed to remain red for 15 sec. If the motorist wish to pass light without stopping. Find the required deceleration.
10. Explain the laws of friction. List down the types of friction encountered in daily life. Give suitable examples of each.
11. What are the different types of friction ? What do the friction forces develop ? How much can friction be minimized ?
12. “Friction is both desirable and undesirable” Explain with examples.
13. Explain the law of dry friction. On which factors does the coefficient of friction depend and what efforts can be done to minimize it ?

14. A body on contact with a surface is being pulled along it with force increasing from zero. How does the state of motion of a body change with force ? Draw a graph and explain.

15. The figure shows a body in contact with a surface and coefficient of static friction being *P* is being applied. Draw the free body diagram of it. Replace the normal force and frictional force by a resultant force and derive the expansion for angle of state friction *Q*.

16. A block weighing 200 N is in contact with an inclined plane. Will the block move under its own weight (Fig. 133) ? Determine the minimum force applied (*i*) parallel (*ii*) perpendicular to the plane to prevent the motion down the plane. What force *P* will be required to just cause the motion up the plane $\mu = 0.25$?

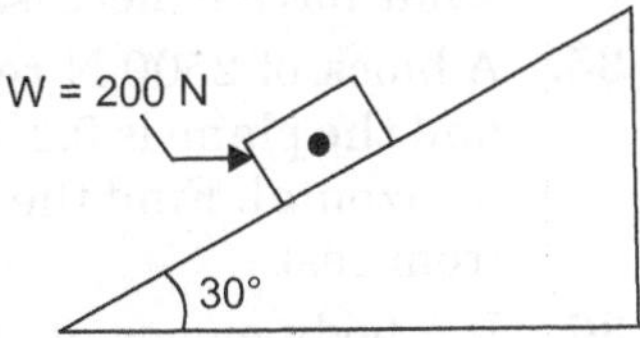

Fig. 133

17. A metal box weighing 10 kN is pulled along a level surface at uniform speed, by applying a horizontal force of 3500 N. If another box of 6 kN is put on top of this box, determine the force required.

18. What is cone of friction ? Explain the significance of cone of friction.

19. A flat belt is used to transmit a torque from pulley *A* to pulley *B* (Fig. 134). The radius of each pulley is 50 mm and the coefficient of friction is 0.3. Determine the largest torque that can be transmitted if the allowable belt tension is 3 kN.

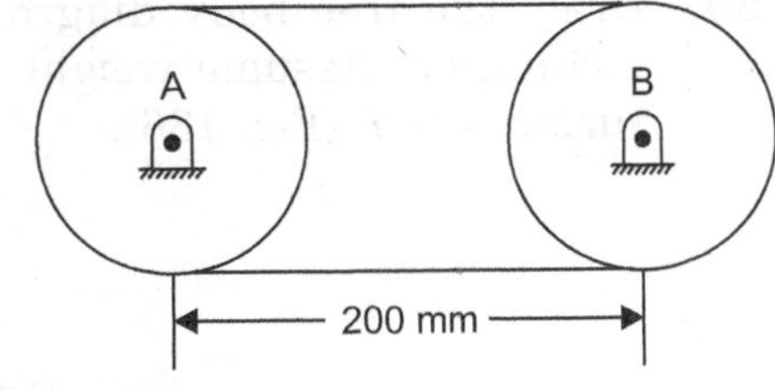

Fig. 134

20. What do you mean by self-looking screw ?

21. Write the equation for free falling motion ?

22. Define average velocity.

23. What do you understand by uniformly accelerated rotation ?

24. Define periodic time of satellite.

25. Write the equation of motion for a plane motion.

26. What do you mean by "body cent rode" and "space cent rode" ?

27. Write the equation in rectilinear motion, if $a = f(t)$.

28. Give the ratio of tension in tight side (T_2) to slack side (T_1).

29. A belt drives a pulley of 200 mm diameter such that the ratio of tensions in the tight side and slack side is 1.2. If the maximum tension in the belt is not to exceed 240 kN, find the safe power transmitted by the pulley at a speed of 60 rpm.

30. A body of weight 50 kN rests in limiting equilibrium on a rough plane, whose slope is 30°. The plane is raised to a slope of 45°. What force, applied to the body parallel to the inclined plane, will support the body on the plane ?

31. A body weighing 500 N is just moved along a horizontal plane by a pull of 100 N making 45° with horizontal. Find the coefficient of friction.

32. A vehicle starts from rest with a linear motion and comes to rest after 20 sec. The velocity time graph is shown in Fig. 135. Calculate the total distance travelled by the vehicle.

33. A train starts from rest and is accelerated with an acceleration of 0.163 m/sec^2. It then moves with a retardation of 0.652 m/sec^2 and comes to rest after

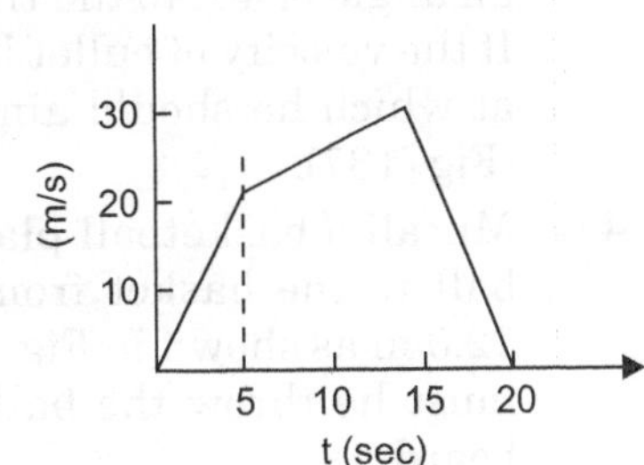

Fig. 135

travelling a distance of 1000 m. Calculate (*a*) distance between the starting point and the point where the train started retarding (*b*) velocity of the train when it commences retardation.

34. A screw jack has a pitch of 12 mm with a mean radius of thread equal to 22.5 mm. A lever 600 mm long is used to raise a load of 1800 kg. If the coefficient of friction is 0.10, what force is necessary when applied normal to the lever at its free end ?

35. A block of 2500 N rests on a horizontal plane. The coefficient of friction between block and the plane is 0.3. The block is pulled by a force of 1000 N acting at an angle 30° to the horizontal. Find the velocity of the block after it moves over a distance of 30 m, starting from rest.

36. If a body moves at the rate of 45 kmph for 6 minutes, 90 kmph for 10 min and finally 8 kmph for 3 minutes. What is the average velocity in the total interval ?

37. A car travels along a straight track between two points at a mean speed of 60 m/sec. It returns at a mean speed of 40 m/sec. What is the mean speed for the round trip ?

38. Distinguish between coefficient of friction and angle of friction.

39. Draw the free body diagram of block *B* shown below indicating forces and their inclination. Assume weight of block *B* as *W* and coefficient of friction on all contact surfaces as *P* (Fig. 136).

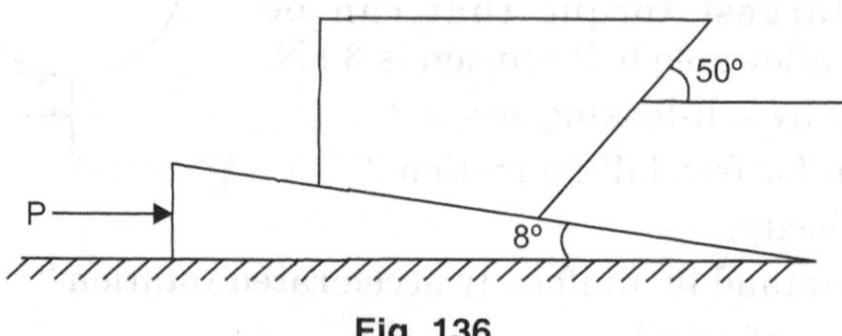

Fig. 136

40. The acceleration of a particle is defined by the relation $a = 1\,t - 4$. Knowing that $v =$ 4 m/s., when $t = 2$ s and $v = 1$ m/s when $t = 1$ s, determine the constant "write the equation of motion when $x = 0$ at $t = 3$ s."

41. A body moving with uniform acceleration is observed to travel 33 m in eighth second and 53 m in the thirteenth second of its travel. Calculate the velocity at start and uniform acceleration.

42. A man travelling in a train moving with a speed of 72 kmph. First at an object moving away from the train at right angles to it with a speed of 54 kmph. The line connecting the object and the man makes an angle of 40° to the train at the time of shooting. If the velocity of bullet is 680 kmph. Find the angle at which he should aim in order to hit the object (Fig. 137).

Fig. 137

43. Murali a basketball player 1.80 m tall throws the ball to the basket from a horizontal distance of 12.0 m as shown in Fig. 138. If he shoots at 35° angle horizontally, at what initial speed must he throw the ball so that the ball goes through the Hoop without striking the board.

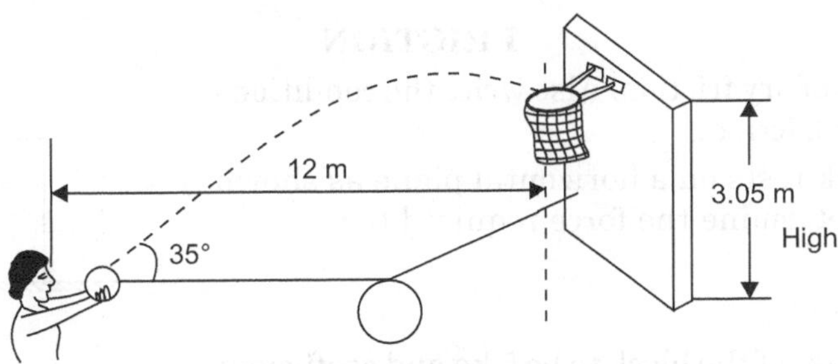

Fig. 138

44. Force of 200 N is required just to move a certain body up an inclined plane of angle 15°, the force being parallel to plane. If angle of indication is made 20° the effort again required parallel to plane is found 250 N. Determine the weight of body and coefficient of friction.

45. Explain how a wedge is used for raising heavy loads ? Also give principal.

46. Obtain the expression for minimum force required to drag a body on rough horizontal plane.

47. Force required to pull a body of 50 N on a rough horizontal plane is 20 N. Determine the coefficient of friction if force is applied at an angle of 25° with x-axis.

48. Show that efficiency of a screw jack for raising a load W is given by :

$$\eta = \tan \alpha \tan (\alpha + \phi)$$

where α = angle of screw, ϕ = angle of friction.

49. A crew jack can lift a load of 5 kN. Screw jack has screw with square threads of two threads to 1.2 cm. If coefficient of friction between nut and screw is 0.09 and outer diameter of screw is 6 cm, find the force required at the end of handle of length 70 cm to lift the load.

50. Determine the effort needed for lifting a load of 1550 N using a screw jack of efficiency 55%. The effort is applied at the end of a handle of length 60 cm and pitch of screw thread is 1 cm.

51. Show that the condition for a self locking machine is that its efficiency is less than 50%.

52. Show that the maximum efficiency of a screw jack is given by $\eta_{max} = 1 - \sin \phi/1 + \sin \phi$ at $\alpha = 45 - \phi/2$ where α is helix angle and ϕ is the angle of friction.

53. For a flywheel knotting at 200 rpm. Initially and at 150 rpm after 10 seconds determine the number of revolutions made by flywheel and time taken before it comes to rest. Take retardation as uniform.

54. Establish the relationship between the linear velocity of a point on a body rotating about fixed axis and its angular velocity. Also describe the two.

55. Find the speed of shaft driven with the belt by an engine running at 300 rpm. The thickness of belt is 1 cm, diameter of engine pulley is 50 cm and that of shaft is 30 cm.

56. A truck of mass 4000 kg is travelling at 40 km/hr on a plane road. It is brough to rest in 20 m, determine the average force of resistance acting on truck.

57. Discuss the laws for rotary motion.

58. A particle moves along a straight line with an acceleration given by $a = (4t^2 - 2)$, where a is in m/s^2 and t is in s.

When $t = 0$, particle is located 2 m to the left of origin and when $t = 2$ seconds, particle is 20 m to left or origin. Determine the position of particle when $t = 6$ seconds.

FRICTION

59. Write the laws of dry friction. Also write the conditions of the limiting friction.

60. A wooden block rests on a horizontal plane as shown in Fig. 139. Determine the force required to :

(*i*) Pull it

(*ii*) Push it

Assume the mass of the block to be 5 kg and coefficient of friction M = 0.4.

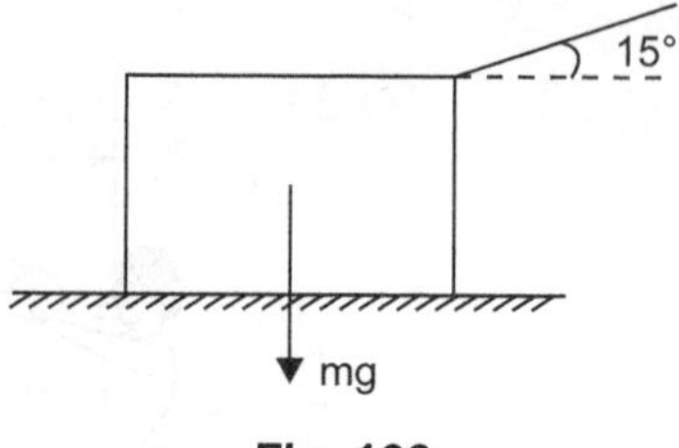

Fig. 139

61. A 7.0 m long ladder rests against a vertical wall making an angle of 45° as shown in Fig. 140. If a man, whose weight is one half of that of the ladder, climbs it, at what distance along the ladder is about to slip ?

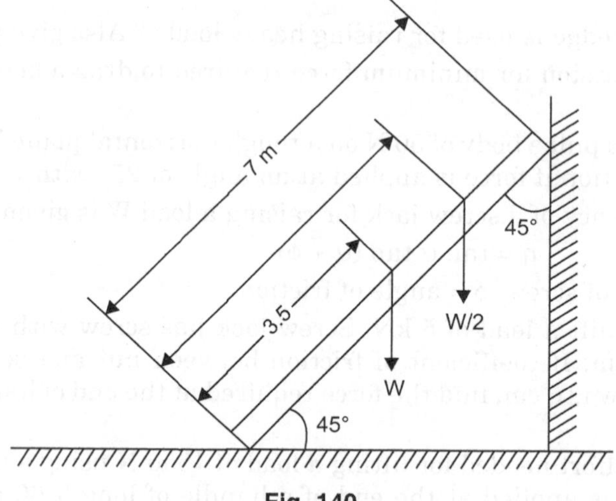

Fig. 140

62. The coefficient of friction between the ladder and wall is 1/3 and that between ladder and floor is 1/2.

63. A four wheel drive can, as shown in Fig. 141, has a mass of 2000 kg with passengers. The roadway is inclined at angle θ with the horizontal. If the coefficient of friction between the tyres and the road is 0.3, what is the maximum inclination θ that can climb ?

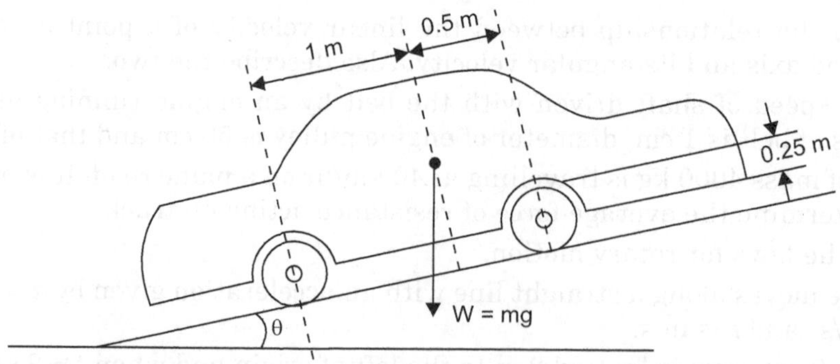

Fig. 141

64. The two ends of a circular rod AB of mass M are supported by a circular ring in a vertical plane as shown in Fig. 142. The length of rod is such so as to subtend an angle of 90° at the centre of the ring. The coefficients of friction at the points of contacts A and B are μ each. What is the greatest angle of inclination θ that the rod can make with the horizontal in the condition of equilibrium ?

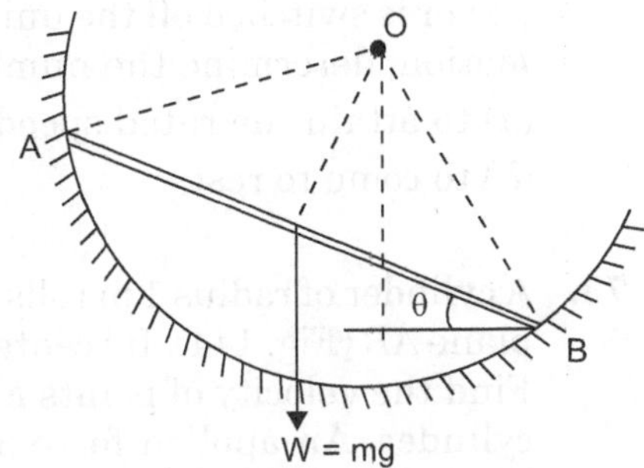

Fig. 142

65. Homogeneous cylinder of weight W rests on a horizontal floor in contact with a wall (Fig. 143). If the coefficient of friction for all contact surfaces be μ, determine the couple M acting on the cylinder which will start counter clockwise rotation.

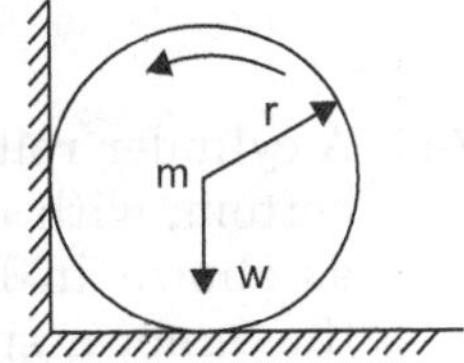

Fig. 143

66. (*a*) Find the length of the belt in case of :
(*i*) Open belt drive
(*ii*) Crossed belt drive
(*b*) Also establish the following relation for open belt drive $\frac{l_1}{l_2} = e^{\mu\theta}$

μ = Coefficient of friction
θ = Angle of lip
T_1 = Tension in belt on tight side
T_2 = Tension in belt on slack side.

67. What do you mean by centrifugal tension in belt ? Derive the formula or condition for the maximum power transmitted by belt considering the centrifugal tension.

68. A belt 100 mm wide and 8.0 mm thick transmitting power at a belt speed of 160 m/minute. The angle of lap for smaller pulley is 165° and coefficient of friction is 0.3. The maximum permissible stress in belt is 2 MN/m^2 and mass of the belt is 0.9 kg/m. Find the power transmitted and the initial tension in the belt.

69. Find the power transmitted by belt (cross type) drive connecting two pulleys of 45.0 cm and 20.0 cm diameters which are 1.95 m apart. The maximum permissible tension in the belt is 1 kN, coefficient of friction is 0.20 and speed of larger pulley is 100 rpm.

70. A screw jack has square threads of mean diameter of 10.0 cm and a pitch of 1.25 cm. Determine the force that must be applied to the end of 50 cm lever (*i*) to raise (*ii*) to lower a weight of 50 kN. Find the efficiency of the jack. It is self locking.
Assume U = 0.20.

71. Power is to be transmitted between two shafts by a *v*-belt whose mass is 0.9 kg/m length. The angle of lap is 170° and groove angle is 45°. Assuming the coefficient of friction between the belt and pulley to be 0.17. Find (*a*) Velocity of the belt for maximum power and (*b*) Power transmitted at this velocity.

KINEMATICS OF RIGID BODIES

72. A grinding wheel is attached to the shaft of an electric motor of rated speed of 1800 rpm. When power is switched on the unit attains the rated speed of 1800 rpm when the

power is switched off the unit comes to rest in 90 second. Assuming uniformly accelerated motion, determine the number of revolution the unit tums.

(*a*) to attain the rated speed

(*b*) to come to rest.

73. A cylinder of radius 1 m rolls without slipping along a horizontal plane *AB* (Fig. 144). Its center has a uniform velocity of 20 mile. Find the velocity of points *E* and *F* on the circumference of the cylinder. An applied force 300 N acting 11 to inclined plane. The coefficient of friction is = 0.2. The body moves from the rest. Determine.

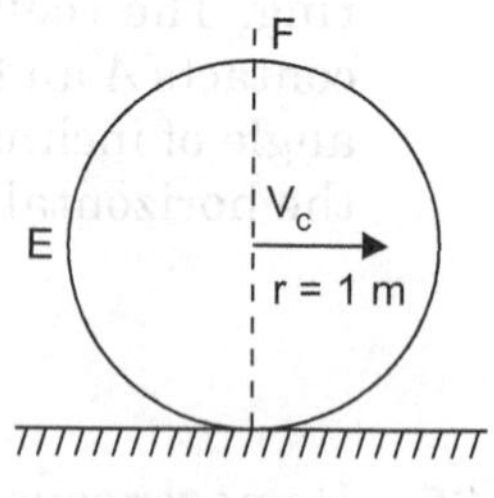

Fig. 144

74. A cylinder roller is in contact as its top and bottom, with two conveyer belt *PQ* and *RS* as shown in Fig. 145. If the belts run, at the uniform speed of V_1 = 3 m/sec and V_2 = 2 m/sec, find the linear velocity and the angular velocity of the roller. The diameter of roller may be assumed to be 40 cm.

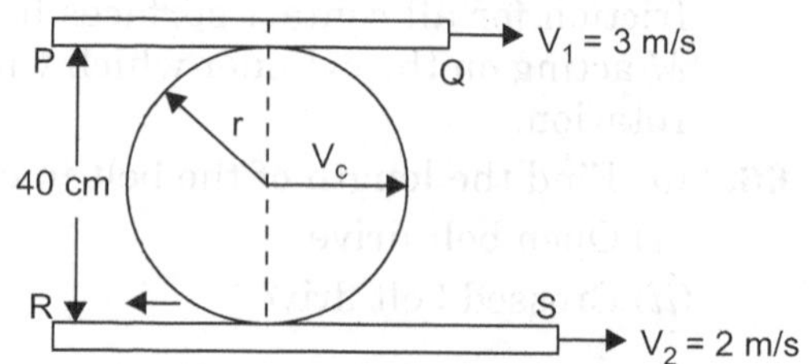

Fig. 145

75. A bar *AB* of length 1 m has its ends *A* and *B* constrained to move horizontally and vertically as shown in Fig. 146. The end *A* moves with constant velocity of 3 m/s horizontally. Find

(*a*) the angular velocity of bar

(*b*) the velocity of the end *B*

(*c*) the velocity of mid-point *C* of the bar at the instant when the axis of the bar makes an angle of 30° with the horizontal axis.

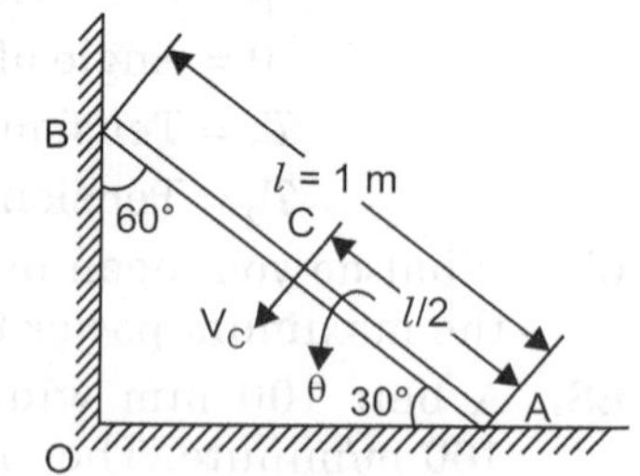

Fig. 146

76. A reciprocating engine mechanism is as shown in Fig. 147. The crank *OA* has a constant angular velocity of *N* = 300 rpm. The crank *OA* is of length *r* = 12.0 cm and the connecting rod *AB* is of length *l* = 60 cm. Find

(*a*) the angular velocity of connecting rod

(*b*) the velocity of the piston when the crank makes an angle of θ = 30° horizontally. Will the velocity of piston be maximum when θ = 90° ?

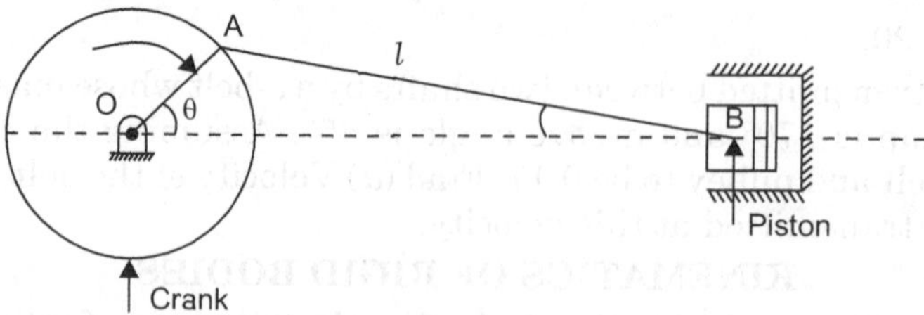

Fig. 147

77. What do you mean by I-center ? How is I-center located for a body having the motion in plane ? Describe with a suitable example.

78. A compound wheel rotates without shipping as shown in Fig. 148. The velocity of center C is 1 m/sec. Find the velocities of points A, B and D.

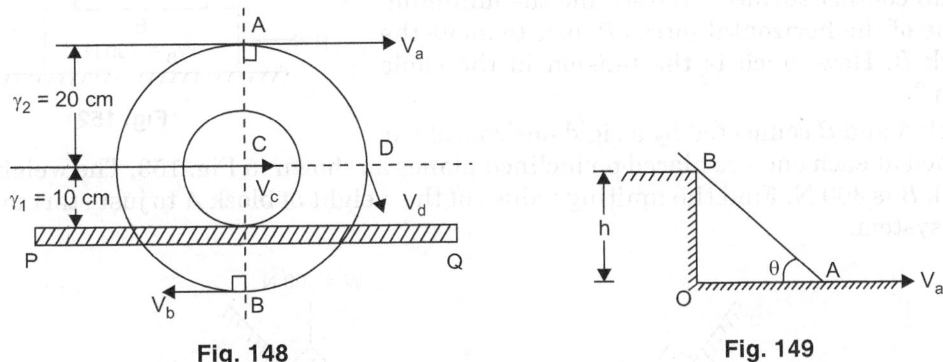

Fig. 148

Fig. 149

79. A bar AB rests at the edge of a wall at some point C with its end A resting on a horizontal floor as shown in Fig. 149. If the end A moves with a constant velocity V_a, find angular velocity of the bar.

80. In a three bar mechanism $ABCD$ (Fig. 150), the bar AB rotates clockwise with an angular velocity of 5 rad/sec. Find the angular velocities of the bar BC and CD, when the bar AB makes an angle of 30° with the horizontal, bar CD makes an angle of 60° and bar BC is horizontal.

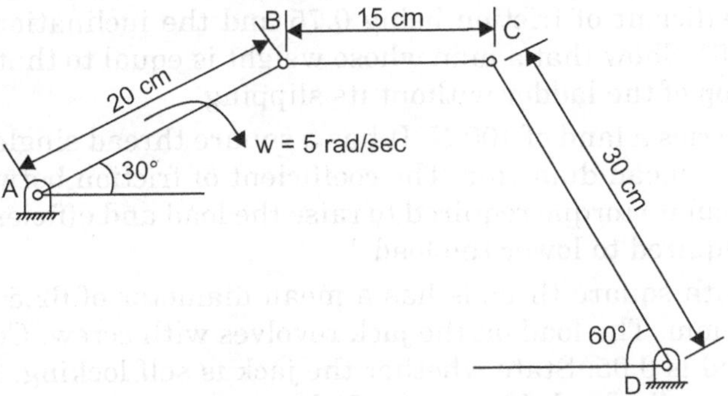

Fig. 150

81. Block A weighing 1000 N rests over block B which weighs 2000 N as shown in Fig. 151. Block A is tied to wall with a horizontal string. If the coefficient of friction between A and B is 1/4 and between B and floor is 1/3, what should be the value of P to move the block B. If

(*a*) P is horizontal

(*b*) P is an angle of 30° with the horizontal.

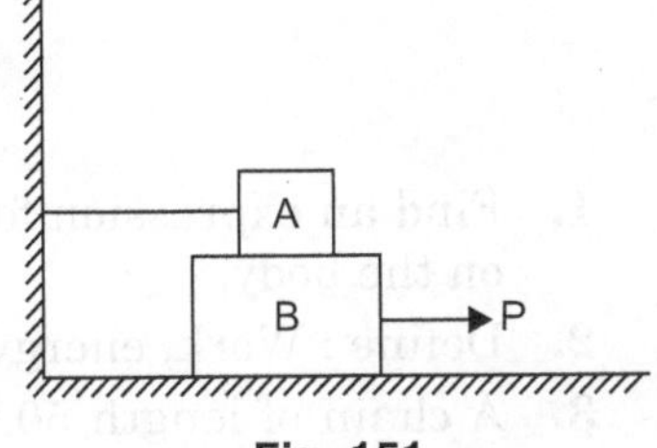

Fig. 151

82. A weight 500 N just starts moving down a rough inclined plane supported by a force 200 N acting parallel to the plane and it is at the point of moving up the plane when pulled by a force of 300 N paralleled to the plane. Find the inclination of the plane and the coffee of friction between the inclined plane and the weight.

83. Block A of weight 520 N rests on the horizontal top of block B having weight 700 N as shown in Fig. 152. Block A is tied to a support c by a cable at 30° horizontally. Coefficient of friction is 0.4 for all contact surfaces. Determine the minimum value of the horizontal forces P, just to move the block B. How much is the tension in the cable then ?

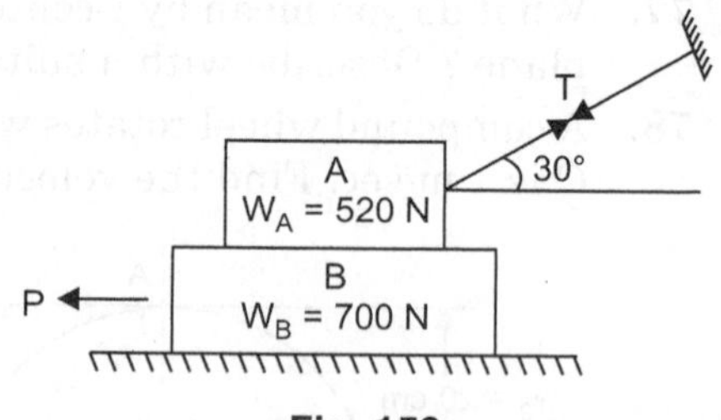

Fig. 152

84. Block A and B connected by a rigid horizontal bar planed at each end are placed on inclined planes as shown in Fig. 153. The weight of the block B is 300 N. Find the limiting values of the weight of block A to just start motion of the system.

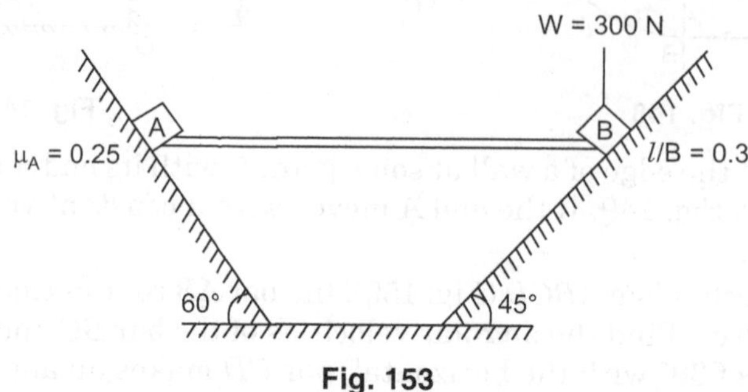

Fig. 153

85. A uniform ladder rests with one end against a smooth vertical wall and the other on the ground, the coefficient of friction being 0.75 and the inclination of the ladder on the ground being 45°. Show that a man whose weight is equal to that of the ladder can just ascend to the top of the ladder without its slipping.

86. A screw jack carries a land of 400 N. It has a square thread single start screw of 20 mm pitch and 50 mm mean diameter. The coefficient of friction between the screw and its nut is 0.27. Calculate torque required to raise the load and efficiency of the screw. What is the torque required to lower the load ?

87. A screw jack with square threads has a mean diameter of 62.5 mm. The pitch of the threads is 12.5 mm. The load on the jack revolves with screw. Coefficient of friction at the screw thread is 0.05. State whether the jack is self locking. If yes, find the torque necessary to lower the load. If it is not, find the torque, which must be applied to keep the load from descending ?

UNIT 4

1. Find an expression for the work done and of a rotating body when a torque T is acting on the body.

2. Define : Work, energy, impulse, moment of momentum.

3. A chain of length 50.5 m is placed on a pulley whose radius is 16 cm. One end of the chain hangs down 40 m below the point it leaves the pulley. A man holds the other end of the chain and pulls it down until the length of the chain on both sides of the pulley is equal. Find the work done by the main. Given, the weight of the chain is 10 N/m length and pulley is frictionless.

4. A body is weighing 196.2 N slides up a 30° inclined plane under the action of an applied force 300 N acting parallel to the inclined plane. The coefficient of friction is equal to 0.2. The body moves from the rest. Determine :
   (*a*) Acceleration of the body.
   (*b*) Distance travelled by body in 4 seconds.
   (*c*) Velocity of body after 4 seconds.
   (*d*) Kinetic energy of the body after 4 seconds.
   (*e*) Work done on the body in 4 seconds.
   (*f*) Momentum on the body after 4 seconds.
   (*g*) Impulse applied in 4 seconds.
5. Two bodies weighing 300 N and 450 N are hung to the two ends of a rope passing over an ideal pulley as shown in Fig. 154. With what acceleration will the heavier body come down ? What is the tension in the string ?

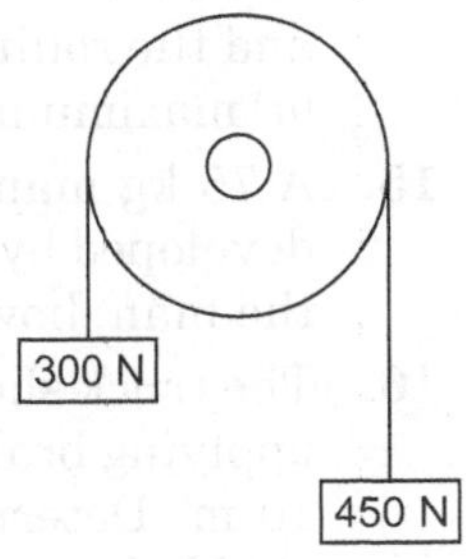

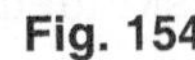

Fig. 154

6. Find the acceleration of bodies and tension in the string joining *A* and *B* shown in Fig. 155.

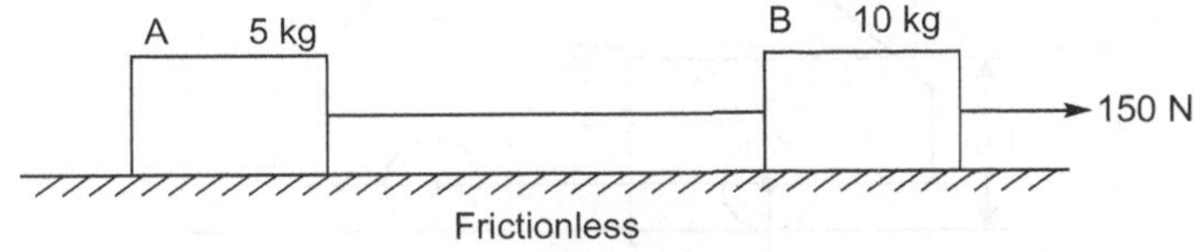

Fig. 155

7. A particle moves along a straight line. Its motion is represented by
$$S = 16t + 4t^2 - 3t^3$$
where *S* is displacement in metres and *t* is time in seconds, determine
   (*a*) displacement, velocity and acceleration 2 seconds after start
   (*b*) displacement and acceleration when velocity is zero.
8. A filled drum of mass *M* and radius *R* and height *H* is placed on the back of a truck. The truck is gradually increasing its acceleration (Fig. 156). Using D' Alembert's principle find the acceleration when the drum will topple down. Assume friction to be zero.

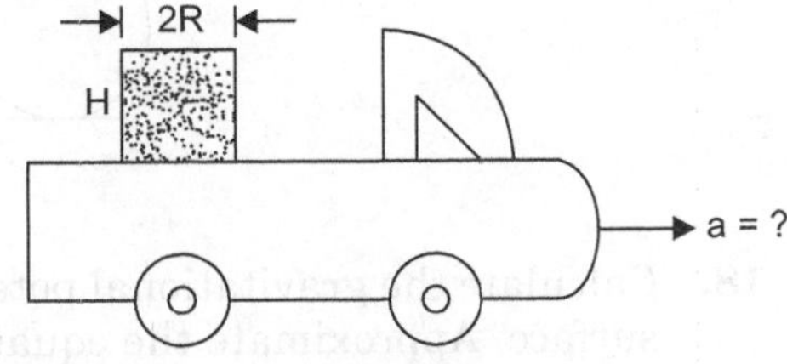

Fig. 156

9. A gun weighing 300 kN fires a 5 kN projectile with a velocity of 300 m/s. With what velocity will the gun recoil ? If the recoil is overcome by an average force of 600 kN, how far will the gun go in the reverse direction ? How long will it take ?
10. A 8 kg body moving to the right at a speed of 4 m/s strikes a 1 kg body that is moving to the left at a speed of 10 m/s. If the coefficient of restitution is 0.2, find the velocities of the bodies after impact.
11. A ball is dropped from a height *h* on ground with coefficient of restitution *e*. Find the expression for total distance travelled.

12. A rotor of an electric motor uniformly accelerates to a speed of 1800 rpm from rest for 5 seconds and then immediately power is switched off and the rotor decelerates uniformly. If the total time elapsed from start to stop is 12.3 second, determine the number of resolutions made while

(*i*) acceleration (*ii*) deceleration.

13. Write short notes on

(*a*) D' Alembert's principle (*b*) Virtual work.

14. A 1500 kg automobile travels 200 m while being accelerated at a uniform rate from 50 to 75 km/hr. During the entire motion, automobile is travelling on a horizontal road, and the rolling resistance is equal to 2% of the weight of the automobile. Determine the (*a*) maximum power required (*b*) power required to maintain a constant speed of 75 km/hr.

15. A 75 kg man runs up a 6 m high flight of starts in 85 (*a*) What is the average power developed by the man ? (*b*) If a 60 kg woman can develop 80% of the power developed by the man, how long will it take for her to run up a 4 m high flight of stairs.

16. The truck shown in Fig. 157 was moving at a speed of 40 m/sec when brakes applied. On applying brakes the wheels of the car stopped rotating and the car skidded to halt after 10 m. Determine the normal reaction of the friction force at each wheels as the car skidded to rest.

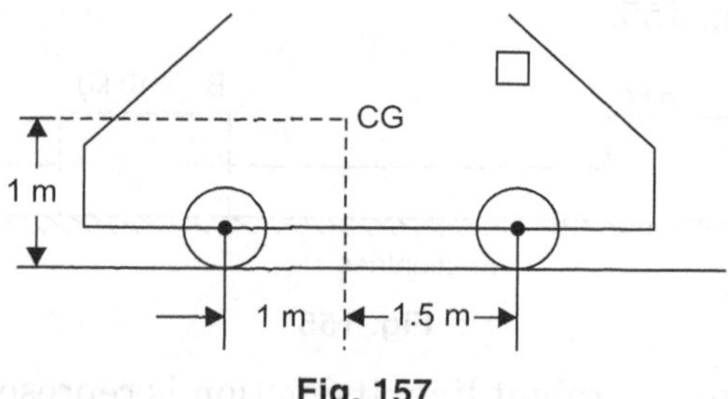

Fig. 157

17. The device shown in Fig. 158 is used to throw ball. The spring is compressed by 5 cm on loading. Find the velocity at the time of leaving assuming no friction.

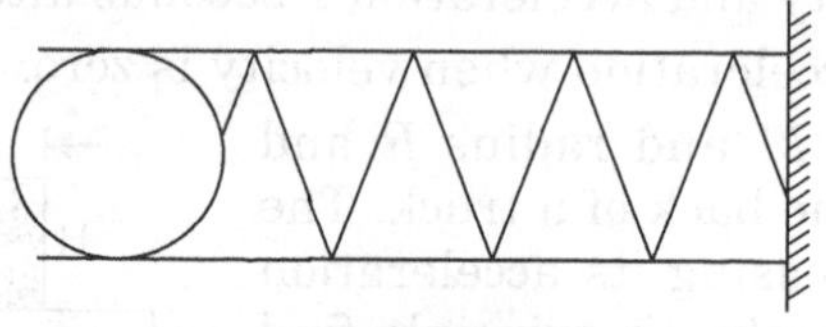

Fig. 158

18. Calculate the gravitational potential energy of body situated at height h from the earth surface. Approximate the equation for $h << R$.

19. A 4 kg slider is released from rest at A and slides with negligible friction down the circular path in vertical plane (Fig. 159). Determine (*a*) the velocity V of the slider as it reaches the bottom B and (*b*) the maximum deformation in the spring.

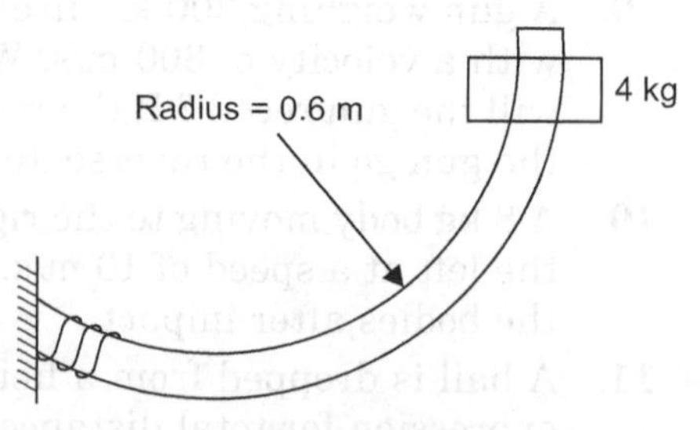

Fig. 159

20. (*a*) State end explain D' Alembert's principle.

(*b*) A cord is coiled around a homogeneous disk of radius $r = 0.5$ m and mass $m = 15$ kg. If the cord is pulled upwards with a force $T = 180$ N.

Determine the acceleration at the centre of the disk.

21. (*a*) Using equation $K - = 1/2\ mv$ derive the expression for $K - 0$ in rotation.

(*b*) The rotor of an electric motor has an angular velocity of 5600 rpm. When the load and power is cut-off. The 50 kg rotor which has a radius of gyration of 180 mm. Knowing that the kinetic friction has a moment of 3.5 Nm determine the number of revolution that the rotor will complete before coming to rest.

22. The flywheel of a punching machine has a man of 300 kg and a radius of gyration of 600 mm. Each punching operation requires 2500 J of work (*i*) If the speed of flywheel just before punching is 300 rpm, determine the speed just after punching (*ii*) If a combat 25 Nm is applied to the staff of flywheel determine the number of rest executed before the speed again becomes 300 rpm (*iii*) Number of punching operation that can be done of further torque is supplied to the flywheel shaft.

23. State the principal of virtual work. Explain it giving a suitable example.

24. State the principal of conservation of momentum for two distinct cases.

25. Use Newton's second law to derive the principle of impulse and momentum.

26. Two swimmers *A* and *B* of mass 75 kg and 50 kg respectively dive off the end of a 250 kg boat. Each swimmer drives so that his relative horizontal velocity work to the boat is *e* m/sec. If the boat is initially at rest determine the final velocity if (*i*) Two swimmers dive simultaneously (*ii*) Swimmer *A* dives first (*iii*) Swimmer *B* dives first.

27. When is a collision called an impact ? Evolve the equation of principle of conservation of momentum for two bodies indirect central impact.

28. Define central impact.

29. Name some forces which do not work.

30. Write the equation of real work.

31. Define simple harmonic motion and give an example.

32. State law of conservation of angular momentum.

33. Define kinetics.

34. A particle of mass 1 kg moves in a straight line under the influence of a force which increases linearly with time at the rate of 60 N per second. Initial force at $t = 0$ is 40 N. Determine the acceleration and velocity of the particle after a lapse of 5 seconds of its starting from rest at origin.

35. A cannon of 1 kg is fired with a velocity of 200 m/s from a gun mounted carriage weighing 2000 kg. Find the velocity of the gun carriage.

36. Two spherical balls *A* and *B* impending on it with a velocity of 20 m/s. If $e = 0.75$, determine the velocities of the balls after impact.

37. A shell of mass 14 kg moving with a velocity of 15 m/s explodes and breaks into two pieces of 10 kg and 4 kg. If the former comes to rest after explosion. Determine the velocity of the latter.

38. A shot of mass 300 kg is fired with a velocity of 350 m/s from a gun of mass 35000 kg. Find the velocity with which the gun recoils.

39. A tennis ball moving with a velocity of 25 m/s strikes the court at an angle of 70° to the vertical. If $e = 0.75$, find the velocity and direction of motion after impact.

40. If the velocity-time diagram is as given in Fig. 160, draw the acceleration-time diagram.

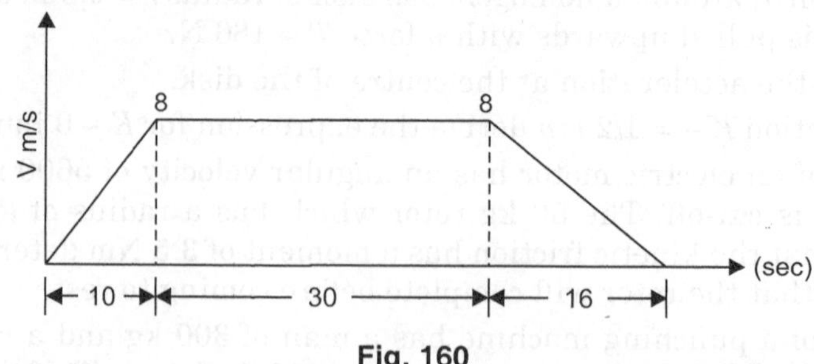

Fig. 160

41. A ball of mass 18 kg moving with a velocity of 5 m/s strikes another ball of mass 10 kg moving in the opposite direction with a velocity of 10 m/s. If the coefficient of restitution is 0.75. Calculate the velocity of each ball after impact and also loss in kinetic energy due to impact.

42. (*a*) State the principle of impulse and momentum.

(*b*) State parallel axis Theorem for moment of a plane area.

43. (*a*) Distinguish between the space diagram and free body diagram.

(*b*) D' Alembert's principle.

44. Determine the acceleration of a railway wagon moving on a railway track if fraction force exerted by wagon weighing 50 kN is 2000 N and the frictional resistance is 5 N per kN of wagon's weight.

45. Show that the tension in cables supporting a lift when lift is moving up is given by $T = w\ (1 + f/g)$ where w = weight of lift f = uniform acceleration, and T = tension in cables.

46. Explain the D' Alembert's principle for a rigid body in plane motion. Also define the principle.

47. Show that angular momentum of a rotating body is equal to the product of moment of inertia and angular velocity.

48. Estimate the total kinetic energy of a uniform homogeneous cylinder rolling without slip on a horizontal level surface with translation velocity of 40 cm/sec. Take its weight as 0.3 N and radius as 30 cm.

49. Determine tensions in strings and acceleration of blocks P and Q weighing 300 N and 60 N connected through string and passing over frictionless weightless pulleys as shown in Fig. 161.

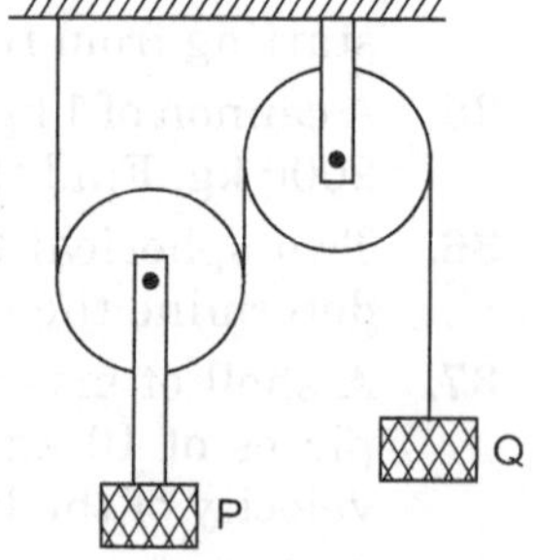

Fig. 161

50. Explain the principle of virtual work as applied to a system in equilibrium.

51. A uniform ladder of weight 180 N rests with its upper end against a smooth vertical wall and its foot on a rough horizontal ground making an angle of 60° with ground. Find force of friction of ground using virtual work method.

52. What do you understand by slip of a belt ? Show how does it affect the speed of driven shaft.

53. A bullet of mass 5 gm is fired into a body of mass 0.75 kg suspended by a 1 m long string. Bullet gets embedded in the body and body swings through an angle of 10° due to impact. Determine the velocity of bullet.

54. Obtain expressions for work done and power of rotating body upon which a torque T is acting.

KINETICS OF RIGID BODIES

55. A cylinder of mass m and radius r resting on an inclined plane is related from rest and rolls down the inclined plane without slipping (Fig. 162). Determine :

(*a*) The acceleration of its center of mass

(*b*) The maximum angle θ of the inclined plane for the body will roll without slipping.

(*c*) Maximum velocity of the center of the cylinder after it has rolled a distance of 1 m. Assume the coefficient of static friction $\mu = 0.192$.

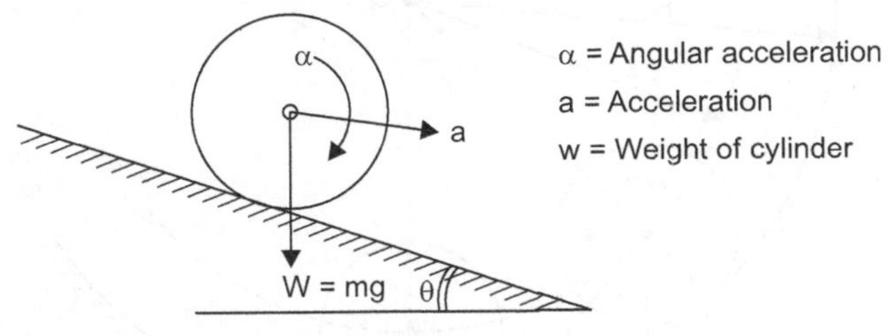

Fig. 162

56. What would be the acceleration of the center of the cylinder if it were to move down the inclined plane under frictionless conditions ?

57. A uniform spherical ball of mass m and radius r is projected along a rough horizontal plane with an initial linear velocity of V^0 and zero angular acceleration (Fig. 163). The coefficient of friction between the ball and the plane is μ. Determine the time after which the ball will start rolling without sliding. Also find the linear and angular velocities of the ball at the time.

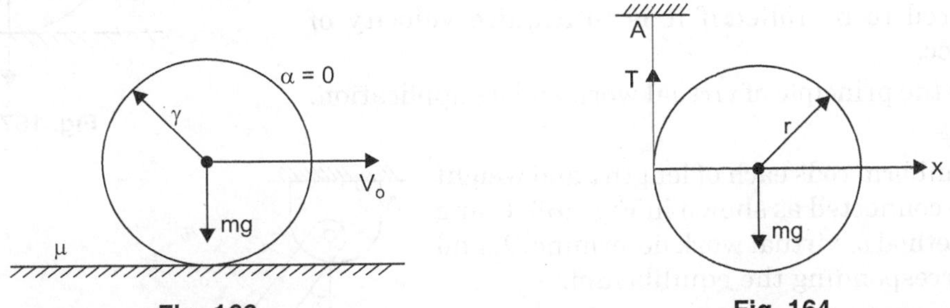

Fig. 163 **Fig. 164**

58. A right circular cylinder of mass m and radius r is suspended from a cord that is wound around its circumference (Fig. 164). If the cylinder is allowed to fall freely, find the acceleration of its mass center G and tension in the cord.

59. A thin uniform bar of mass m and length L is suspended from two vertical inextensable strings (Fig. 165). If the right hand string BD is cut, find the tension in the left string AC and the angular acceleration of the bar.

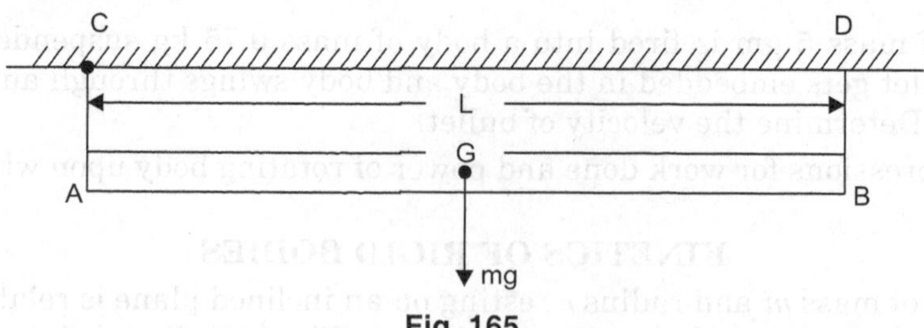

Fig. 165

60. Write the D' Alembert's principle in plane motion that is in dynamic equilibrium.

61. A circular cylinder of mass m and radius r rolls without slipping on a circular path of radius R as shown in Fig. 166. Find the period of oscillations of a cylinder when it is displaced slightly from the equilibrium position.

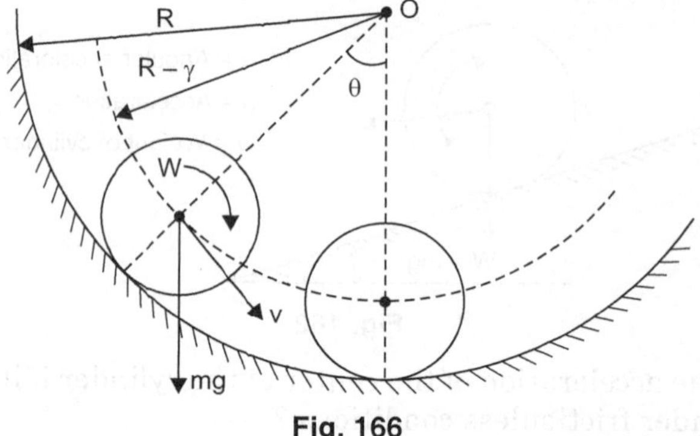

Fig. 166

62. A roller of mass $m = 600$ kg and radius $r = 0.25$ m is pushed with a constant force $p = 850$ N on a rough horizontal plane as shown in Fig. 167. If roller starts from rest and rolls without slipping, find the distance required to be rolled if it is to acquire velocity of 3 m/sec.

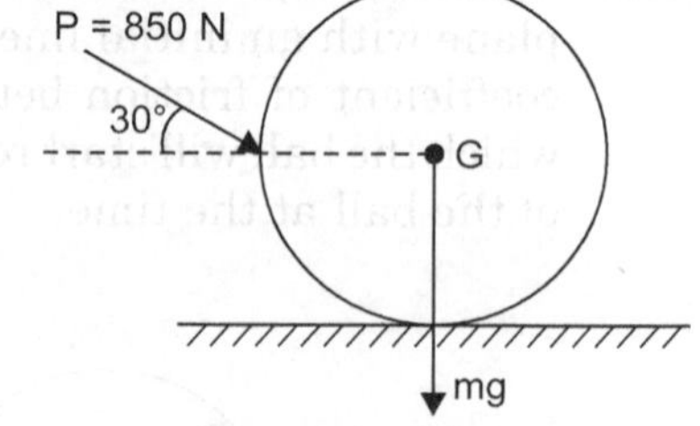

Fig. 167

63. Write the principle of virtual work and its application.

64. Two uniform rods each of length l and weight W are connected as shown in Fig. 168. Using the method of virtual work determine D_1 and D_2 corresponding the equilibrium.

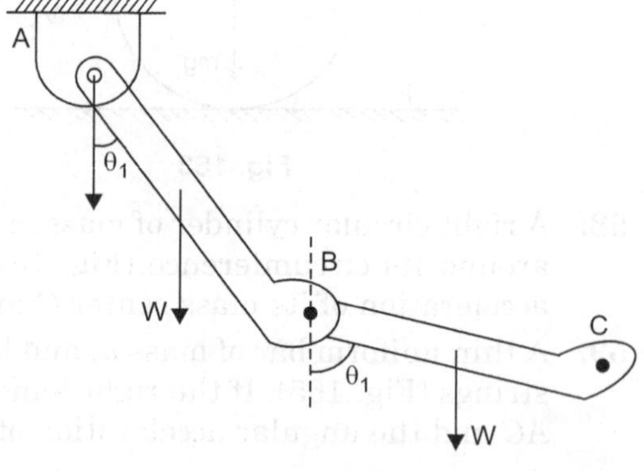

Fig. 168

65. A hexagonal frame is made up of sir bars of identical length and cross-section (Fig. 169). Each bar has a weight of W Newton. The rod AB is fixed in horizontal position. A string joins the mid-points of the bars AB and DE. Using the principle of virtual work find the tensions in the string.

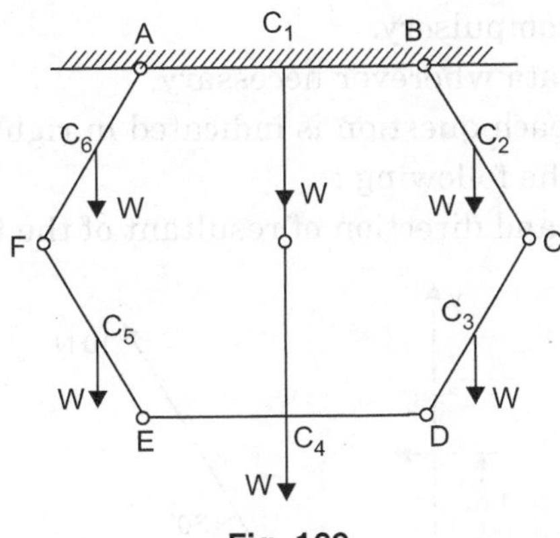

Fig. 169

MODEL QUESTION PAPER

Time : 2 Hrs. *Max. Marks : 50*

Notes : 1. All questions are compulsory.

2. Assume suitable data wherever necessary.

3. Marks allotted to each question is indicated in right hand column.

1. Attempt any FOUR of the following : [3.5 × 4]

(*a*) Find the magnitude and direction of resultant of the forces shown in Fig. 1.

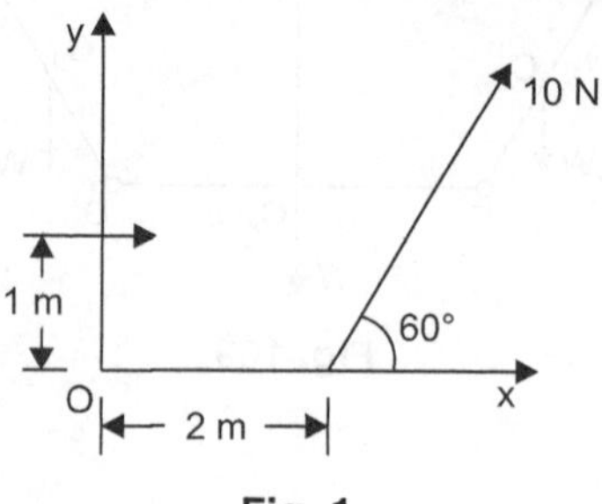

Fig. 1

(*b*) State and Prove Varignon's theorem.

(*c*) Draw the free body diagram of the beam shown in Fig. 2 and find the reactions at the supports.

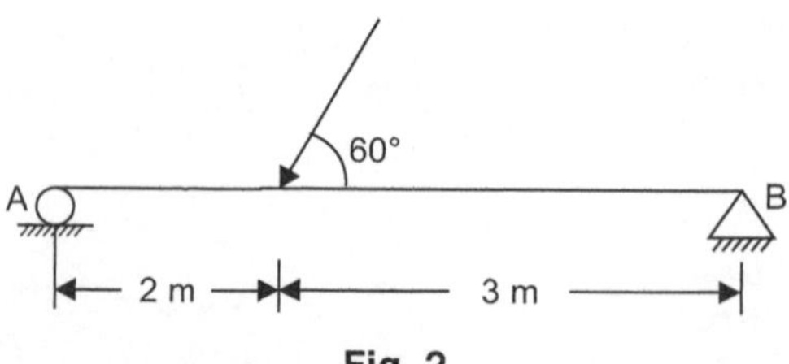

Fig. 2

(*d*) Draw the shear force diagram (SFD) of the beam shown in Fig. 3. Calculate the maximum shear force in the beam.

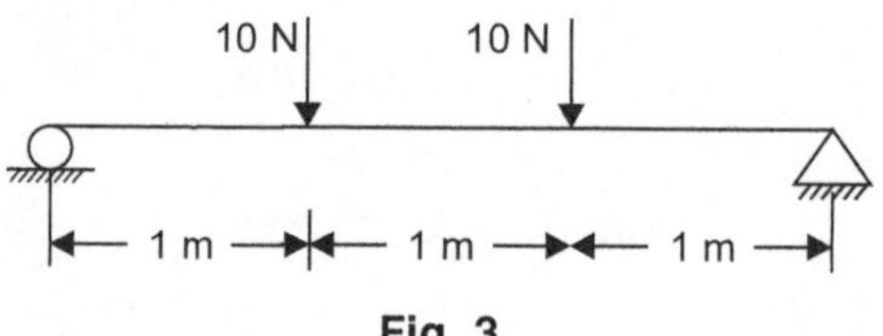

Fig. 3

(*e*) Draw the Bending Moment Diagram (BMD) of the beam shown in Fig. 4. Calculate the maximum bending moment in the beam.

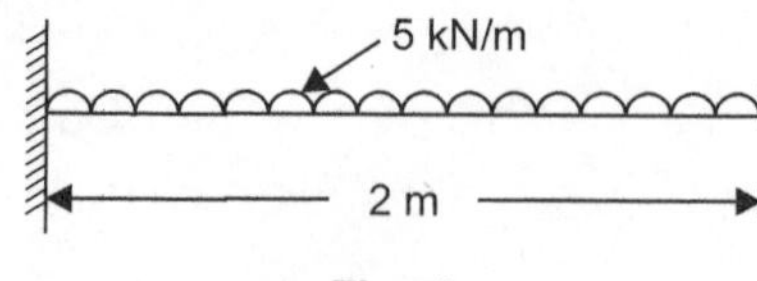

Fig. 4

(*f*) Determine the point of application and the direction of a 20 N force, that lies in the plane of the square plate shown in Fig. 5, so that this force creates the greatest counter clockwise movement about point *A*. What is the value of this moment ?

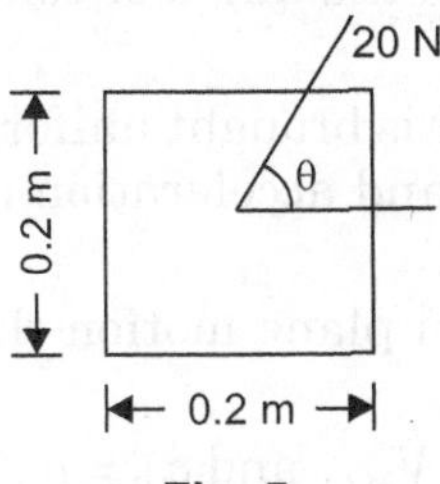

Fig. 5

2. Attempt any TWO of the following : [6 × 2]

(*a*) Find the centroid $\bar{x}$ and $\bar{y}$ of the area shown in Fig. 6.

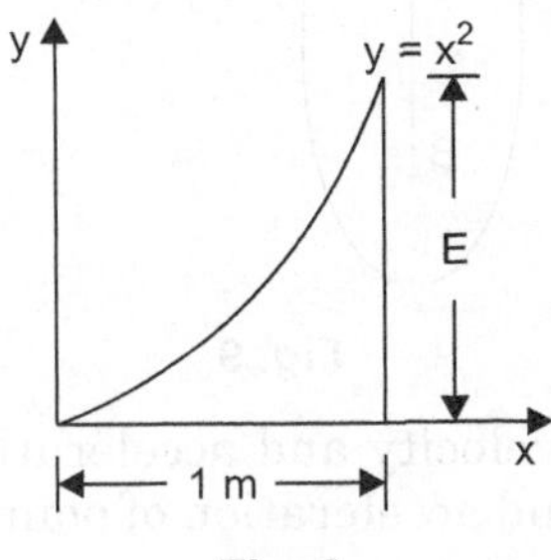

Fig. 6

(*b*) Find the moment of inertia of the section shown in Fig. 7 about *x*- and *y*-axes.

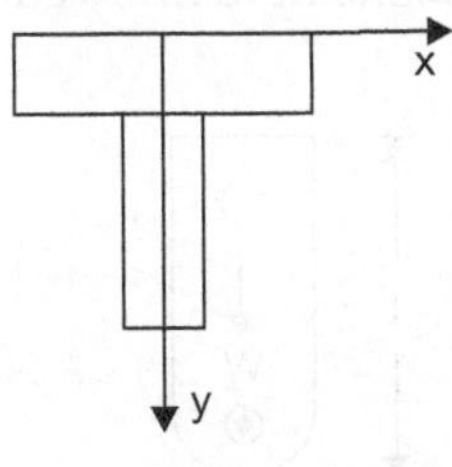

Fig. 7

(*c*) Find the forces in all the members of the truss shown in Fig. 8.

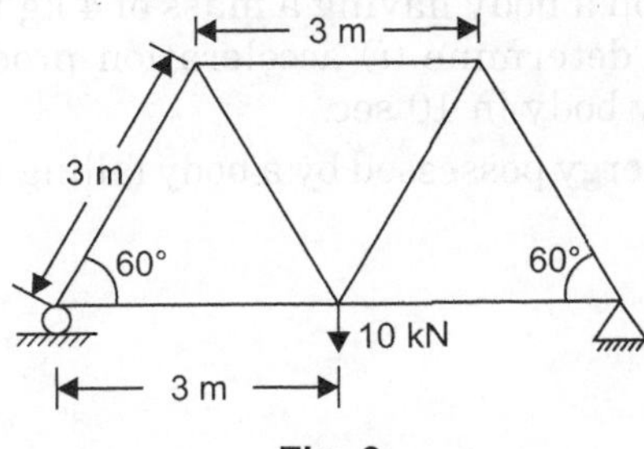

Fig. 8

3. Attempt any TWO of the following : [6 × 2]

(*a*) Show that the ratio of Belt Tension in a Belt drive is given by $T_1/T_2 = e^{\mu\theta}$, where T_1 and T_2 are tensions on tight and slack sides respectively, μ is coefficient of friction between belt and pulley, θ is the angle of contact between belt and pulley. Neglect weight of the belt.

(*b*) A flywheel 50 cm in diameter is brought uniformly from rest up to a speed of 280 rpm in 15 sec. Find the velocity and acceleration of a point on the rim one second after starting from rest.

(*c*) Show that for a rigid body in plane motion the velocity and acceleration of a point are given by

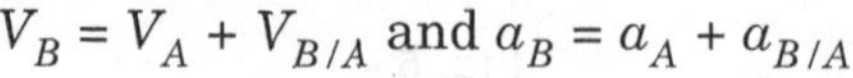

$$V_B = V_A + V_{B/A} \text{ and } a_B = a_A + a_{B/A}$$

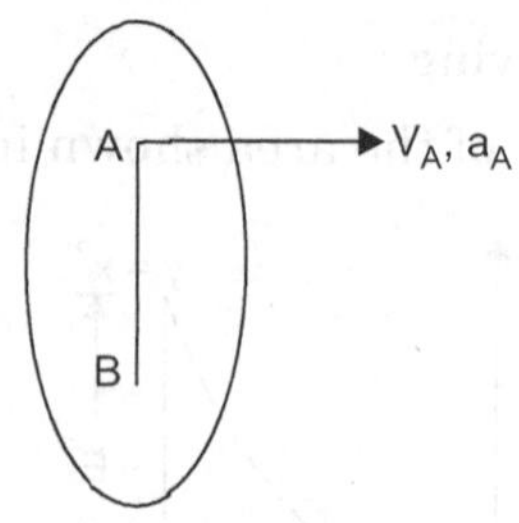

Fig. 9

where V_A and a_A are the velocity and acceleration of point A in the body and $V_{B/A}$ and $a_{B/A}$ are the velocity and acceleration of point B with respect to A at that instant.

4. Attempt any TWO parts of the following : [6 × 2]

(*a*) A slender rod of weight W and length L is pinned at one end to a horizontal plane. The rod initially in vertical position is allowed to fall. What will be its angular speed when it strikes the floor ?

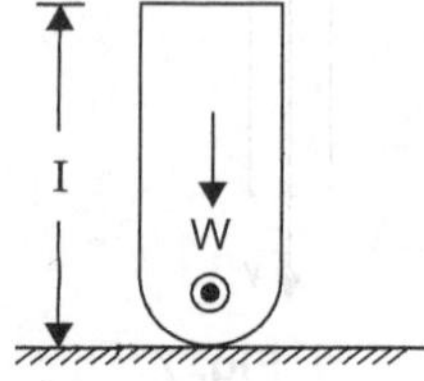

Fig. 10

(*b*) A force of 100 N acts on a body having a mass of 4 kg for 10 sec. If the initial velocity of the body is 5 m/s, determine (*i*) acceleration produced in the direction of force (*ii*) distance moved by body in 10 sec.

(*c*) Show that the total energy possessed by a body falling under gravity remains constant at all positions.